PHYSICAL BIOCHEMISTRY

PHYSICAL BIOCHEMISTRY: PRINCIPLES AND APPLICATIONS

Second Edition

David Sheehan

Department of Biochemistry
University College Cork
Ireland

A John Wiley & Sons, Ltd, Publication

Wiley-Blackwell is an imprint of John Wiley & Sons, formed by the merger of Wiley's global Scientific, Technical and Medical business with Blackwell Publishing.

Registered office: John Wiley & Sons Ltd, The Atrium, Southern Gate, Chichester, West Sussex, PO19 8SQ, UK

Other Editorial Offices:
9600 Garsington Road, Oxford, OX4 2DQ, UK

111 River Street, Hoboken, NJ 07030-5774, USA

For details of our global editorial offices, for customer services and for information about how to apply for permission to reuse the copyright material in this book please see our website at www.wiley.com/wiley-blackwell

Reprinted with corrections August 2009

Library of Congress Cataloging-in-Publication Data

Sheehan, David, 1958–
Physical biochemistry : principles and applications / David Sheehan. – 2nd ed.
p. ; cm.
Includes bibliographical references and index.
ISBN 978-0-470-85602-4 (hb) – ISBN 978-0-470-85603-1 (pb) 1. Physical biochemistry. I. Title.
[DNLM: 1. Biophysics. 2. Biochemistry. 3. Chemistry, Phisical. QT 34 S541p 2008]

QD476.2.S42 2009
572′.43–dc22 2008046672

ISBNs: 9780470856024 (HB)
9780470856031 (PB)

A catalogue record for this book is available from the British Library.

Set in 9/11pt Times by Aptara Inc., New Delhi, India

First Impression 2009

To the memory of my father, Patrick Sheehan (1917–2003)

Contents

Preface

The first edition of *Physical Biochemistry: Principles and Applications* set out to describe the physical basis and some examples of applications of key physically-based techniques used in Biochemistry and other areas of molecular life science research. In the last decade there has been a noticeable renaissance in some traditional techniques such as X-ray diffraction, ultracentrifugation and electrophoresis in a variety of formats. In the same time-frame 'hyphenated' techniques (e.g. LC-tandem MS) have become much more mainstream and some instrumentation has become available in desktop formats making quite sophisticated analysis possible even in the nonspecialist lab. The emphasis was on a largely nonmathematical treatment at a level appropriate to students in the penultimate year of a Biochemistry course with a view to making these techniques comprehensible and accessible to a level intermediate between a general Biochemistry textbook and a specialist text. The feedback I have had from many readers is that this goal was largely achieved.

My task with this second edition was to retain as much as possible of the description of physical principles whilst updating and integrating new material into a reasonably compact volume. The first edition was strongly influenced by the effects of the then-recent completion of the Human and other large-scale genome sequencing projects. At the time Proteomics and other '-omics' technologies were still relatively new paradigms and bioinformatics approaches largely the province of the specialist. It is remarkable how quickly these technologies have now become embedded in many areas of biochemical research so that they are now perceived as part of the mainstream. In the second edition I have dedicated new chapters to proteomics and bioinformatics, respectively, to reflect this changed situation and to emphasize how interconnected physical and computational techniques have now become.

The second edition required choices to be made and these are inevitably influenced by one's perception of what is needed and useful for students to know at the start of their scientific journey into molecular life science. In making these choices I have continued to be guided by what I perceive to be the most generally-used and helpful techniques but I accept that specialists in one or more technique may disagree.

In preparing this second edition I had help from many of the colleagues listed in the preface to the First Edition. In addition, I must thank Dr Rebecca Green, School of Pharmaceutical Sciences, University of Nottingham, UK, for her invaluable comments on surface plasmon resonance (Chapter 3). I am also indebted to the excellent staff at Wiley's especially my editors, Celia Carden and Fiona Woods for their unfailing encouragement and understanding. However, any errors in the text are my own. I earnestly hope that the reader will find something interesting and thought-provoking in this volume and be encouraged to explore these very powerful approaches in their work.

Prof David Sheehan
University College Cork
June 2008

Chapter 1
Introduction

Objectives

After completing this chapter you should be familiar with:

- *The* **special chemical conditions** *often required by biomolecules.*
- *Importance of the use of* **buffers** *in the study of physical phenomena in biochemistry.*
- **Quantification** *of physical phenomena.*
- **Objectives** *of this volume.*

This volume describes a range of physical techniques which are now widely used in the study both of biomolecules and of processes in which they are involved. There will be a strong emphasis throughout on *biomacromolecules* such as proteins and nucleic acids as well as on *macromolecular complexes* of which they are components (e.g. biological membranes, ribosomes, chromosomes). This is because such chemical entities are particularly crucial to the correct functioning of living cells and present specific analytical problems compared to simpler biomolecules such as monosaccharides or dipeptides. Biophysical techniques, give detailed information offering insights into the structure, dynamics and interactions of biomacromolecules.

Life scientists in general and biochemists in particular have devoted much effort during the last century to elucidation of the relationship between structure and function and to understanding how biological processes happen and are controlled. Major progress has been made using *chemical* and *biological* techniques which, for example, have contributed to the development of the science of *molecular biology*. However, in the last decade *physical* techniques which complement these other approaches have seen major development and these now promise even greater insight into the molecules and processes which allow the living cell to survive. For example, a major focus of life science research currently is the *proteome* as distinct from the *genome*. This has emphasized the need to be able to study the highly-individual structures of biomacromolecules such as proteins to understand more fully their particular contribution to the biology of the cell. For the foreseeable future, these techniques are likely to impact to a greater or lesser extent on the activities of most life scientists. This text attempts to survey the main physical techniques and to describe how they can contribute to our knowledge of biological systems and processes. We will set the scene for this by first looking at the particular analytical problems posed by biomolecules.

1.1 SPECIAL CHEMICAL REQUIREMENTS OF BIOMOLECULES

The tens of thousands of biomolecules encountered in living cells may be classified into two general groups. Biomacromolecules (e.g. proteins; nucleic acids) are characterized by high molecular mass (denoted throughout this text as *relative molecular mass*, M_r) and are generally unstable under extreme chemical conditions where they may lose structure or break down into their chemical building blocks. Low molecular weight molecules are smaller and more chemically robust (e.g. amino acids; nucleotides; fatty acids). Within each group there is displayed a wide range of water-solubility, chemical composition and reactivity which is determined by complex interactions between physicochemical attributes of the biomolecule and solvent. These attributes are the main focus for the techniques described in this volume and reflect the highly individual function which each molecule performs in the cell (Tables 1.1 and 1.2).

Notwithstanding the great range of form and structure, we can nonetheless recognize certain attributes as common to all biomolecules. The first and most obvious is that all of these molecules are produced in living cells *under mild chemical conditions* of temperature, pressure and pH. Biomacromolecules are built up from simpler building block molecules by covalent bonds formed usually with the elimination of water. Moreover, *biomolecules are continuously synthesized and degraded* in cells in a highly regulated manner. It follows from this that many biomolecules are especially sensitive to extremes of temperature and pH which may present a problem in their handling prior to and during

Table 1.1. Some important physical attributes of biomolecules amenable to study by biophysical techniques

Physical attribute	Technique
Mass	MS (Chapters 4 and 10) Electrophoresis (Chapters 5 and 10) Gel filtration (Chapter 2)
Volume/density	Gel filtration (Chapter 2) Centrifugation (Chapter 7) Pulsed field gel electrophoresis (Chapter 5)
Charge	Ion exchange chromatography (Chapter 2) Electrophoresis/chromatofocusing (Chapters 5 and 10) MS (Chapters 4 and 10)
Shape	Chromatography (Chapter 2) Electrophoresis (Chapters 5 and 10) Crystallization (Chapter 6) Centrifugation (Chapter 7)
Energy	Spectroscopy (Chapter 3) Biocalorimetry (Chapter 8)

Table 1.2. Some important chemical attributes of biomolecules which may be used in study by biophysical techniques

Chemical attribute	Technique
Composition	MS (Chapters 4 and 10) Spectroscopy (Chapter 3)
Molecular structure	Spectroscopy (Chapter 3) Crystallization (Chapter 6)
Covalent bonds	MS (Chapters 4 and 10) Bioinformatics (Chapter 9) Electrophoresis (Chapters 5 and 10)
Noncovalent bonds	Chromatography (Chapter 2) Electrophoresis (Chapters 5 and 10) Spectroscopy (Chapter 3)
Native/denatured structure	Chromatography (Chapter 2) Electrophoresis (Chapter 5) Proteomics (Chapter 10)
Solubility	Chromatography (Chapter 2) Crystallography (Chapter 6) Precipitation (Chapter 7)
Complex formation	Electrophoresis (Chapter 5) MS (Chapters 4 and 10) Spectroscopy (Chapter 3) Crystallography (Chapter 6)

any biophysical analysis. Since biomolecules result from a long process of biological evolution during which they have been selected to perform highly specific functions, a very close relationship has arisen between chemical structure and function. This means that, even at pH and temperature values under which the molecule may not be destroyed, it may function suboptimally or not at all.

These facts impose limitations on the chemical conditions to which biomolecules may be exposed during extraction, purification or analysis. In most of the techniques described in this volume, sample analytes are exposed to a specific set of chemical conditions by being dissolved in a solution of defined composition. Whilst other components in the solution may also be important in individual cases as will be discussed below (Section 1.3.2), three main variables govern the makeup of this solution which are discussed in more detail in the following section.

1.2 FACTORS AFFECTING ANALYTE STRUCTURE AND STABILITY

In practice, most of the biophysical procedures described in this volume use conditions which have been optimized over many years for thousands of different samples. These robust conditions will normally maintain the sample in a defined structural form facilitating its separation and/or analysis. However, some procedures (e.g. chromatography, capillary electrophoresis, crystallization) may require case-by-case optimization of conditions. Before embarking on a detailed analysis of a biomolecule using biophysical techniques it is often useful to know something about the stability of the sample to chemical variables, especially pH, temperature and solvent polarity. This knowledge can help us to design a suitable solvent or set of chemical conditions which will maximize the stability of the analyte for the duration of the experiment and may also help us to explain unexpected results. For example, we sometimes find loss of enzyme activity during column chromatography which may be partly explained by the chemical conditions experienced by the protein during the experiment. Moreover, many of the techniques described in this volume are actually designed to be *suboptimal* and to take advantage of *disruption* of the normal functional structure of the biomolecule to facilitate separation or analysis (e.g. electrophoresis, HPLC, MS).

A good indication of the most stabilizing conditions may often be obtained from knowledge of the biological origin of the biomolecule. It is also wise to assess the structural and functional stability of the analyte over the range of experimental conditions encountered in the experiment during its likely time-span.

We can distinguish two main types of effects as a result of variation in the chemical conditions to which biomolecules are exposed. *Structural effects* reflect often irreversible structural change in the molecule (e.g. protein/nucleic acid denaturation; hydrolysis of covalent bonds between building blocks of which biopolymers are composed). *Functional effects* are frequently more subtle and may be reversible (e.g. deprotonation of chemical groups in the biomolecule resulting in ionization; partial unfolding of proteins). A detailed treatment of these effects on the main classes of biomolecules is outside the scope of the present volume but a working knowledge of the likely effects of these conditions can be very useful in deciding conditions for separation or analytical manipulation.

1.2.1 pH Effects

pH is defined as the negative log of the proton concentration:

$$\mathrm{pH} = -\log[\mathrm{H}^+] \tag{1.1}$$

Because both the H^+ and OH^- concentrations of pure water are 10^{-7} M, this scale runs from a *maximum* of 14 (strongly alkaline) to a minimum of 0 (strongly acidic). As it is a log scale, one unit reflects a *10-fold* change in proton concentration. Most biomacromolecules are *labile* to alkaline or acid-catalyzed hydrolysis at extremes of the pH scale but are generally *stable* in the range 3–10. It is usual to analyse such biopolymers at pH values where they are structurally stable and this may differ slightly for individual biopolymers. For example, proteins normally expressed in lysosomes (pH 4) are quite acid-stable while those from cytosol (pH 7) may be unstable near pH 3. Aqueous solutions in which sample molecules are dissolved usually comprise a *buffer* to prevent changes in pH during the experiment. These are described in more detail in Section 1.3 below.

Many biomolecules are *amphoteric* in aqueous solution that is they can accept or donate protons. Some chemical groups such as inorganic phosphate or acidic amino acid side-chains (e.g. aspartate) can act as *Brønsted acids* and donate protons:

$$\mathrm{AH} \underset{k_{-1}}{\overset{k_1}{\rightleftarrows}} \mathrm{A}^- + \mathrm{H}^+ \tag{1.2}$$

Other groups such as the imidazole ring of histidine or amino groups can act as *Brønsted bases* and accept protons:

$$\mathrm{B}^- + \mathrm{H}^+ \underset{k_{-1}}{\overset{k_1}{\rightleftarrows}} \mathrm{BH} \tag{1.3}$$

The position of equilibrium in these protonation/deprotonation events may be described by an *equilibrium constant*, K_a:

$$K_a = \frac{k_1}{k_{-1}} = \frac{[\mathrm{A}^-]\cdot[\mathrm{H}^+]}{[\mathrm{AH}]} \tag{1.4}$$

pK_a ($-\log K_a$) is the pH value at which 50% of the acid is protonated and 50% is deprotonated. The *Henderson–Hasselbach equation* describes variation of concentrations of A^- and AH as a function of pH:

$$\mathrm{pH} = \mathrm{p}K_a + \frac{\log[\mathrm{A}^-]}{[\mathrm{AH}]} \tag{1.5}$$

Functional groups present as structural components of biomolecules (e.g. amino acid side-chains of proteins; phosphate groups of nucleotides) will have distinct K_a values which may differ slightly from the value found in other chemical circumstances (e.g. the K_a values of amino acid side-chains in polypeptides differ from those in the free amino acid). Some biomolecules can contain *both* acidic and basic groups within their structure (e.g. proteins) while particular chemical structures found in biomolecules may be *polyprotic*, that is capable of *multiple* ionizations (e.g. phosphate). Such biomolecules may undergo a complex pattern of ionization resulting in varying net charge on the molecule. *pH titration curves* for biomolecules allow us to identify pK_a values (Figure 1.1).

Since protonation-deprotonation effects are responsible for the charges on biomacromolecules which maintain their solubility in water, their solubility is often *lowest* at their *isoelectric point*, pI, the pH value at which the molecule has no net charge. These can also be determined by titration using methods described in Chapter 5 (Section 5.5.3; Figure 5.24).

While the pH scale reflects the situation in aqueous solution, many microenvironments encountered in living cells are quite nonpolar (*see* below). Good examples include biological membranes and water-excluding regions of proteins (e.g. some enzyme active sites). In these environments, protonation/deprotonation properties of chemical groups may deviate widely from those observed in aqueous solution. For example, catalytic residues of many enzymes frequently display pK_a values which are perturbed far from those normal for that residue in water-exposed regions of proteins.

1.2.2 Temperature Effects

Three main effects of temperature on biomolecules are important for the biophysical techniques described in this volume. These are effects on structure, chemical reactivity and solubility. Heat can disrupt noncovalent bonds such as

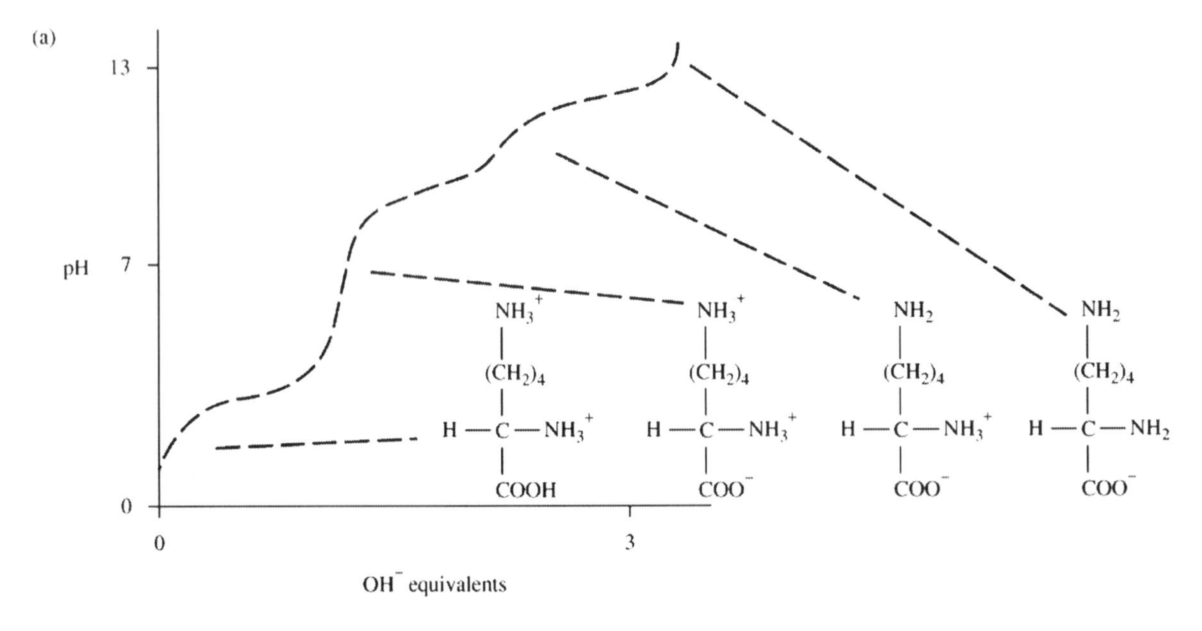

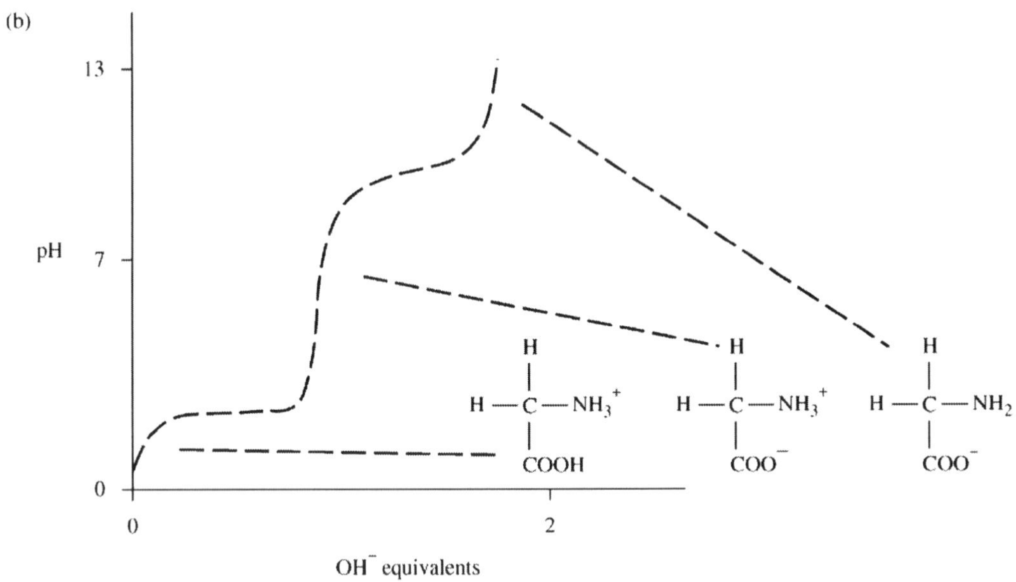

Figure 1.1. pH titration curves, (a) Lysine. Four protonation states are possible for lysine as shown. Three pK_a values are evident from this at pH values of 2.18, 8.95 and 10.53. The pI of lysine is 9.74. (b) Glycine. Note that only three protonation states exist for glycine compared to four for lysine. pK_a values are at pH values of 2.34 and 9.6 which differ slightly from the corresponding ionisations in (a). Glycine has a pI of 5.97

hydrogen bonds which are especially important in the structure of biomacromolecules. This can lead to denaturation of proteins and DNA or to disruption of multimolecular complexes in which they may be involved. Moreover, since covalent bonds linking building block molecules (e.g. peptide bonds; glycosidic bonds; 3′, 5′-phosphodiester bonds) have generally lower bond energies than bonds within such building blocks, extensive heating can result in disintegration of the covalent structure of biomacromolecules. Thus proteins can break down into component peptides or nucleic acids into smaller polynucleotide fragments as a result of exposure to heat.

Secondly, all chemical reaction processes obey the Arrhenius relationship:

$$k = A\,e - E_a/RT \quad (1.6)$$

where k is the rate constant for the process, A is a constant, E_a is the activation energy of the reaction, R is the Universal gas constant and T is absolute temperature. This relationship arises from large changes in the number of activated molecules available for reaction as a result of change in temperature. The exponential dependence on temperature means that small changes in T can result in large effects on the rate constant, k.

Thirdly, temperature usually increases the solubility of molecules in a solvent as well as the rate of diffusion through the solvent. This is because heat increases the average kinetic energy of solvent molecules. In the case of water, this is accompanied by extensive breakdown of water-water hydrogen bonds which increases the solute capacity of a given volume of water. Thus, for example 8 M urea is soluble at 30 °C while the limit of solubility is closer to 5 M at 4 °C. Kinetic energy effects are also important in situations involving biological membranes because the phospholipid bilayer of which they are composed becomes increasingly fluid at higher temperature.

Temperature is therefore usually tightly controlled during biophysical experiments. In dealing with biomacromolecules in particular, it is generally not possible to use temperatures higher than 80 °C and, in most cases, much lower temperatures are used. Moreover, samples such as proteins or nucleic acids are normally stored under refrigerated conditions to maximize their stability. This is achieved with the aid of liquid nitrogen (−196 °C) or with refrigerators set at −80 or −20 °C. Particular care must be taken in handling crude biological extracts since hydrolases such as proteases and nucleases present in these will be active in the range 18–37 °C and result in extensive degradation of proteins and nucleic acids. This can be avoided by maintaining low temperatures near 4 °C during manipulation of sample and by cooling buffer solutions before dissolving biological samples.

Most biomolecules are optimally active at temperatures similar to those experienced in the biological source from which they were obtained. For example, proteins from thermophilic bacteria are especially heat stable compared to corresponding proteins from mesophilic bacteria while mammalian proteins are optimally active around 37 °C.

1.2.3 Effects of Solvent Polarity

Polarity arises from unequal affinity of atoms bonded together for shared electrons called electronegativity. Apart from fluorine (with an electronegativity value of 4), oxygen is the most electronegative element in the periodic table (electronegativity value of 3.5) leading, for example, to oxygens of –OH groups tending to be partially negatively charged (δ^-) while hydrogens tend to be partially positively charged (δ^+). Water (which is itself polar) interacts *ionically* with polar functional groups such as –OH by hydrogen bonding (Figure 1.2). Organic molecules composed mainly of carbon and hydrogen, however, tend to be nonpolar as these atoms have similar electronegativities (2.5 and 2.1, respectively). In general, polar biomolecules dissolve readily in polar solvents such as water while those which are nonpolar dissolve in nonpolar solvents (e.g. trichloromethane).

Biomolecules lacking strongly electronegative elements such as oxygen and nitrogen and consisting mainly of carbon and hydrogen tend to be principally nonpolar (e.g. fatty acids; sterols; integral membrane proteins). Conversely, those containing oxygen, sulfur and nitrogen tend to be mainly polar (e.g. monosaccharides; nucleotides). Biomacromolecules often contain distinct structural regions some of which may be polar while others may be nonpolar.

Since water is the main biological solvent, most biomolecules (or parts of biomolecules) have been selected by evolution to interact with it in particular ways either by attraction or repulsion. Polar regions strongly *attracted* to water are called *hydrophilic* while nonpolar regions which are *repulsed* by water are called *hydrophobic*. In the living cell, biomolecules adopt a structure determined to a large degree by the extent to which they are hydrophobic/hydrophilic. For example, biological membranes are made up of phospholipid bilayers which spontaneously form when phospholipid molecules are dissolved in water. The polar heads of phospholipids are on the exterior in contact with water while the nonpolar fatty acid components are on the interior of the bilayer protected from water. Cytosolic proteins express hydrophilic groups on their surface whilst folding in such a manner that hydrophobic groups are protected from exposure to water in the interior of the protein (Chapter 6). Membrane-bound proteins such as hormone receptors expose hydrophobic groups to the interior of biological membranes and hydrophilic groups to the exterior.

In extracting, analyzing and purifying biomolecules these intricate structural interactions are often lost which can result in aggregation, precipitation or loss of structure and, hence, of biological activity. If it is desired to retain biological activity we use aqueous solutions to handle largely hydrophilic biomolecules, nonpolar solvents to dissolve mainly hydrophobic samples and detergent solutions for molecules which possess both classes of groups. Many of the individual techniques described in this volume use specific solvent systems of distinct polarity/nonpolarity but it may occasionally be necessary to design individual solvent systems to take account of the requirements of particular biomolecules. Examples include column chromatography

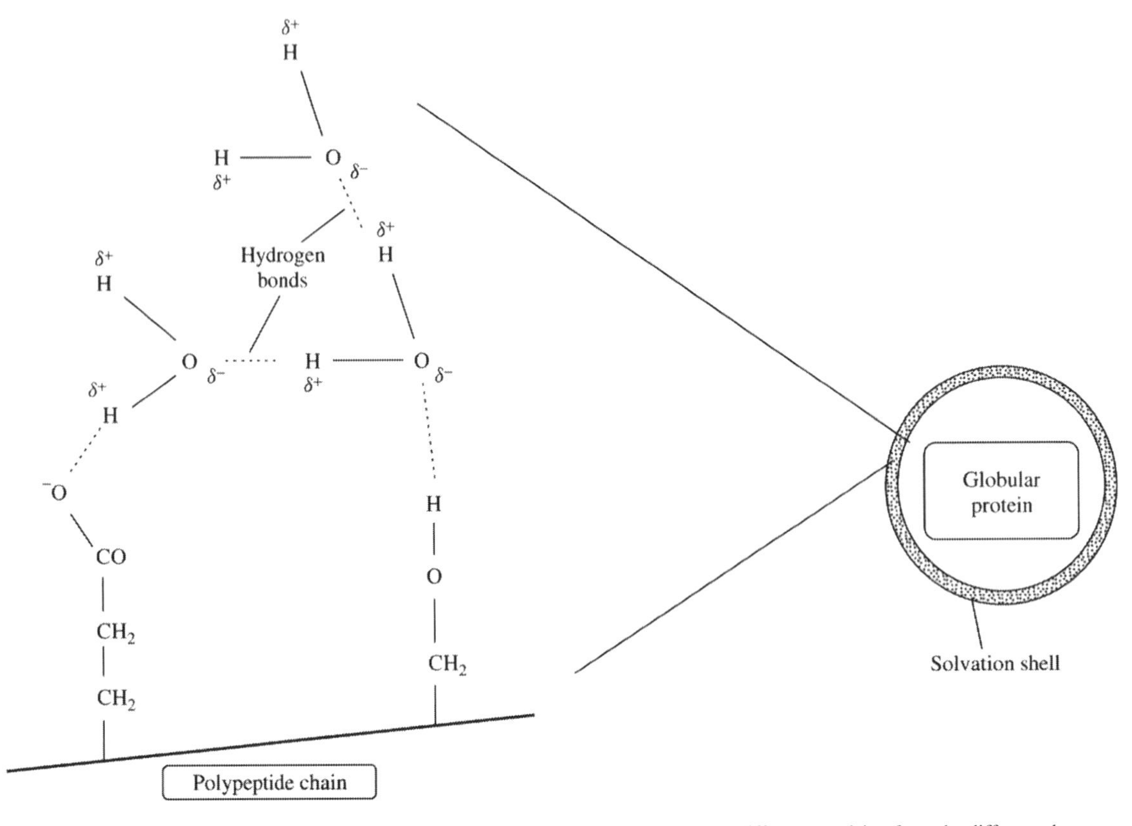

Figure 1.2. Hydrogen bonding. Water molecules hydrogen bond because of partial charge-differences arising from the different electronegativities of oxygen and hydrogen. Hydrogen bonds (dashed lines) are shown both in bulk water and between water and amino acid side-chains of glutamic acid and serine residues of a polypeptide. Such ionic interactions maintain a solvation shell of water around the surface of globular proteins and other hydrated biomacromolecules

(Chapter 2), spectroscopy (Chapter 3) and capillary electrophoresis (Chapter 5).

1.3 BUFFERING SYSTEMS USED IN BIOCHEMISTRY

A buffer is an aqueous solvent system designed to maintain a given pH. In the context of biochemical work, the main function of buffers is to resist any tendency for pH to rise or drop during the experiment. This can happen during any process which might release or absorb protons from solution such as, for example, during an enzyme-catalyzed reaction or as a result of electrochemical processes such as electrophoresis. A secondary but often crucial role for a buffer is to maximize the stability of biomolecules in solution. Frequently, additional molecules are dissolved in the buffer to help it to do this and these are discussed in more detail below.

1.3.1 How Does a Buffer Work?

Any aqueous solution containing both A^- and AH (Section 1.2 above) is, in principle, capable of resisting change in pH. This is because, if protons are generated in the solution, they can be neutralized by A^-:

$$A^- + H^+ \rightarrow AH \qquad (1.7)$$

Conversely, if alkali is generated in the solution (which would tend to *remove* protons), it can be neutralized by AH:

$$AH + OH^- \rightarrow A^- + H_2O \qquad (1.8)$$

In practice, most buffers consist of mixtures either of a weak acid and its salt or of a weak base and its salt.

Of course, the ability of a buffer to resist change in pH is finite, especially if the number of protons involved is especially large. This limit is represented by the *buffering capacity* of the buffer, β. This is defined as the number of moles of $[H^+]$ which must be added to a liter of the buffer to decrease the pH by one unit. It can be mathematically calculated from the following equation:

$$\beta = \frac{2.3 \cdot K_a \cdot [H^+] \cdot [C]}{(K_a + [H^+])^2} \qquad (1.9)$$

where [C] is the *sum* of the concentrations of A^- and AH. This relationship means that *buffering capacity increases with buffer concentration* so that, for example 100 mM acetate buffer has 50-fold greater buffering capacity than 2 mM. It can be demonstrated experimentally that β reaches a maximum at pH values equal to pK_a which is when the concentrations of A^- and AH are approximately equal. This means that buffers work best at pH values around their pK_a. In practice, most buffers are effective one pH unit above and one below their pK_a so, for example acetate buffers ($pK_a = 4.8$) are useful in the pH range 3.8 to 5.8, although most effective around pH 4.8.

1.3.2 Some Common Buffers

A selection of buffers commonly used in biochemistry is given in Table 1.3. Some of these buffer components are of biological origin (e.g. glycine; histidine; acetate). *Good's buffers* were developed by N.E. Good to facilitate buffering in the pH range 6–10.5. Because of their complicated chemical names, these buffers are more usually known by abbreviations (e.g. Pipes; Hepes; Mops). Inspection of Table 1.3 shows that buffers are available which span the range of interest for physical studies of biomolecules (i.e. pH 3–10). When preparing buffers, it is essential that *both* the concentration and pH are correct since these are the two variables critical to buffering capacity (Equation (1.9)).

A number of problems can arise with particular buffers which can limit their use in specific cases. For example, several buffers interact with divalent metals (e.g. phosphate binds Ca^{2+}; Tris reacts with Cu^{2+} and Ca^{2+}) and should be avoided in cases where this is important to interpretation of the experiment. Tris buffers are especially sensitive to temperature which can result in the same buffer giving a slightly different pH at different temperatures. Phosphate buffers are particularly susceptible to bacterial contamination if stored for long periods of time (although this can be avoided by including a low concentration of sodium azide as a preservative). Some buffer components (e.g. EDTA) may give high absorbance readings which can affect detection during processes such as chromatography. Volatile buffer components (e.g. formic acid; bicarbonate; triethanolamine) can be lost from the buffer over time leading to a gradual change in pH and buffering capacity.

1.3.3 Additional Components Often Used in Buffers

In addition to buffer components such as weak acids/bases and their salts, buffers frequently contain a range of other components of which a selection is shown in Table 1.4. These may be necessary to maintain stability of the biomolecule, to control levels of metal ions, to ensure reducing/oxidizing conditions or to keep the biomolecule dissolved and/or denatured. We will observe that some of the agents tabulated in Table 1.4 are used in more than one of the techniques described in this book and therefore represent generally useful tools for the manipulation of chemical conditions to which biomolecules are exposed.

Table 1.3. Some common buffers used in biochemistry

Buffer	pK_a
Phosphate[a]	0.85, 1.82, 6.68
Histidine[a]	1.82, 5.98, 9.17
Phosphoric acid[a]	1.96, 6.7, 12.3
Formic acid	3.75
Barbituric acid	3.98
Acetic acid	4.8
Pyridine	5.23
Bis TRIS [Bis-(2-hydroxy-ethyl)imino-tris-(hydroxy-methyl)-methane]	6.46
PIPES [1,4-piperazinebis-(ethanesulphonic acid)]	6.8
Imidazole	7.0
BES [*N*, *N*-*Bis*(2-hydroxy-ethyl)-2-amino-ethane-sulphonic acid]	7.15
MOPS [2-(*N*-morpholino)propane-sulphonic acid]	7.2
HEPES [*N*-2-hydroxyethyl-piperazine-*N'*-2-ethane-sulphonic acid]	7.55
TRIS [hydroxymethyl)amino-methane]	8.1
TAPS [W-tris(hydroxymethyl)methyl-2-aminopropane sulphonic acid]	8.4
Boric acid	9.39
Ethanolamine	9.44
CAPS [3-(cyclohexylamino)-l-propane-sulphonic acid]	10.4
Methylamine	10.64
Dimelhylamine	10.75
Diethylamine	10.98

[a]There are polyprotic with several pK_a values.

1.4 QUANTITATION, UNITS AND DATA HANDLING

1.4.1 Units Used in the Text

Physical measurements usually result in *quantification* of some property of a molecule or system such as those tabulated in Table 1.1. Various systems of internationally-agreed units have been used historically to record these measurements but, throughout this book, the *Systeme Internationale* (SI) system, the most currently-agreed scientific

Table 1.4. Additional reagents sometimes added to buffers

Chemical	Function
2-Mercaptoethanol	Reducing agent
Dithiothreitol	Reducing agent
Sodium borohydride	Reducing agent
Divalent metals plus O_2	Oxidising agents
Performic acid	Oxidising agent
Leupeptin	Protease inhibitor
Phenylmethyl sulphonyl fluoride (PMSF)	Serine protease inhibitor
Ethylene diamine *NNN′N′* tetra-acetic acid (EDTA)	Metal chelator/ metalloprotease inhibitor
Ethylene glycol-bis(β-aminoethyl ether) *NNN′N′* tetraacetic acid (EGTA)	Calcium chelator
Urea	Denaturing agent
Guanidinium hydrochloride	Denaturing agent
Sodium dodecyl sulphate (SDS)	Anionic detergent
Cetyltrimethyl ammonium chloride	Cationic detergent
3-(3- cholamidopropyl)dimethy lammonio)-1-propane sulphonate (CHAPS)	Zwitterionic detergent
Triton X-100	Nonionic detergent
Digitonin	Nonionic detergent
Protamine K	Binds DNA

quantification system will be used. The main units are tabulated in Appendix 1. This system has the advantage of great *internal consistency* which removes any need for conversion factors. For example, units of distance (meters) readily relate to units of velocity (meters/second).

In addition to SI units, some measurements used are *operational measurements* commonly employed in the life science literature. These are units which have gained wide acceptance in the international science community but which are not strictly part of the SI system. Relative molecular mass (M_r) is expressed as Daltons (Da) or multiples thereof (e.g. kDa) with 12 Da being equivalent to twelve atomic mass units (i.e. the mass of ^{12}C). In the case of nucleic acids, base pairs (bp) or multiples thereof (e.g. kbp, Mbp) are used as units of mass. Interatomic distances such as bond-lengths are generally given as angstroms (Å) with 1 Å corresponding to 10^{-10} m (i.e. 0.1 nm).

Concentrations are given mainly in molarity (1 M solution being Avogadro's Number of molecules dissolved in 1 l of solvent) although occasionally they are expressed as percentages of weight/weight (% w/w) or weight/volume (% w/v). Thus, 10% (w/v) would represent a solution of 10 g per 100 ml while 10% (v/v) would represent a solution of 10 ml per 100 ml.

The most commonly used temperature scale in the text is the Celsius scale although absolute temperatures (in units of Kelvin, K) are specifically referred to by T (e.g. Equation (1.6) above).

1.4.2 Quantification of Protein and Biological Activity

Most of the techniques described in this text are used to separate or analyse biomolecules or mixtures containing them. In carrying out this kind of experimentation it is crucial to know exactly *how much* sample is being applied since most of the systems described are highly loading-sensitive. In the case of pure samples this may not be a problem but many samples encountered in biochemistry may be quite crude and heterogeneous. A common strategy for quantifying such samples is to estimate their *protein content* (e.g. by the *Bradford* method, Figure 3.18, or by one of the other methods mentioned in the bibliography at the end of this chapter) and to load a standard amount of protein. Since the ratio of protein to the other components of the mixture is fixed, this normally ensures uniform loading. Similarly, when quantifying the biological activity of a sample (e.g. enzyme activity, antibody content, antiviral activity) it is often useful to express this as *specific activity* that is units/mg protein. This is a measure which is *independent* both of sample volume and sample concentration.

The majority of the approaches described measure *relative* properties of biomolecules rather than *absolute* properties. Examples of this would include M_r estimation by mass spectrometry, gel filtration and electrophoresis, pI estimation by isoelectric focusing, secondary structure estimation of proteins by circular dichroism and determination of chemical shifts in NMR spectroscopy. For this reason, a common strategy found in many of the techniques is to *compare* the sample being analysed to a series of well-characterized standard molecules using well-established procedures which have been optimized for that particular method. It is important to understand that measurements obtained in this way are therefore highly dependent on standard measurements being of good quality and that this may vary somewhat from method to method.

1.5 THE WORLDWIDE WEB AS A RESOURCE IN PHYSICAL BIOCHEMISTRY

1.5.1 The Worldwide Web

The worldwide web was originally devised as a *distributed computer network* for the military capable of withstanding nuclear attack! In the last decade, it has grown to include millions of individual entries called *web pages* containing information on almost every conceivable subject. These web pages exist on a computer somewhere in the world but can in principle be accessed by other computers through the web. At the time of writing, (March, 2008) it is estimated that 20% (1320 million) of the Earth's 6606 million population have access to the web (http://www.internetworldstats.com/stats). We can connect to web pages on the web with an appropriate *browser* such as *Internet Explorer*. However, due to the sheer mass of material being constantly added to and changed on the web, we normally use a *search engine* to find pages on specific named topics. *Google* is a good example of a general purpose search engine. It should be remembered that no search engine gives 100% coverage so results from a search could represent as few as 25% of the total possible pages on a given topic.

The 'address' of a particular web page is given by a *uniform resource locator* (URL). Examples of URLs include http://www.google.com for google and http://alta-vista.com for altavista. The prefix http:// is to tell the receiving computer that it can expect a communication in *hypertext transfer protocol* – the most common format allowing one computer to communicate with another. The rest of the URL defines a location, that is a computer containing the relevant file. The ending .html which often occurs in URLs signifies *hypertext markup language*, the language in which web pages are written.

1.5.2 Web-Based Resources for Physical Biochemistry

The web provides several resources of use in Physical Biochemistry. Individual web pages are available which describe various experimental techniques thus complementing published work such as review articles and textbooks. There are also *databases* which are archives of one particular category of information. Examples would include sequence databases, databases of NMR spectra, the three dimensional structure database and databases of two-dimensional electrophoresis patterns. The best databases are *curated* (i.e. they are looked after and regularly updated by some reputable body) and they are *annotated* (which means each entry contains extra information such as literature citations, references to other related entries, etc.). These features make databases part of the daily life of modern molecular life scientists. Even though many resources on the web are not peer-reviewed in the way that say research articles are, most authoritative databases achieve the same result by maintaining a close link with the peer-reviewed literature. Conversely, it is becoming increasingly common for research articles to be submitted to journals in electronic format and for peer-reviewed articles to appear on the web long before the paper version. A third set of very useful resources on the web is made possible by the availability, often as *freeware* (i.e. for free!) of computer programs which help us to analyse or represent our data differently. Examples would include graphics programs which allow us to view the three dimensional structures of biomacromolecules encoded in *protein databank* (PDB) files (Chapters 6 and 9) or hydropathy plots which allow us to identify hydrophobic regions of amino acid sequences (Chapter 9).

The ever-closer links between molecular life sciences and *information technology* (IT) is represented in the relatively new discipline of *bioinformatics* which is introduced in Chapter 9. In this book relevant URLs for web-based resources are given at the end of each chapter.

In addition to text and programs, the wordwide web can be searched for images or videos using the standard search engines. A word of caution about using this type of searching in an academic context. The fact that material is on the web does not absolve us as scientists from respecting copyright law so permission should always be obtained to reproduce images, text or videos obtained from the web just as we would in using such content from a published source. Secondly, we should always take care to refer back to the primary literature as this is the bedrock of modern science and is likely to remain so as long as rigorous peer-review prevails.

1.6 OBJECTIVES OF THIS VOLUME

All of the techniques mentioned in this book merit one or more volumes to describe fully their potential for the future of life science research. In the bibliography at the end of each chapter the reader will find a list of such specialist texts and it is hoped that the present book will act as a general introduction to specialist biophysical techniques. In addition, recent review articles are cited which will bring the reader more up-to-date on specific applications of individual techniques. It is not the intention of the text to supplant such specialist literature but rather to guide students towards a greater understanding of the potential of biophysical approaches to biochemistry.

A chapter is devoted to each technique or group of techniques which describes the physical basis, advantages, limitations and opportunities it offers. This is presented in a generally nonmathematical way to maximize its accessibility (more detailed treatments may be found in the specialist texts). Moreover, the relationship between techniques is strongly emphasized because several combinations of individual techniques often offer advantages over single experimental approaches. In particular, recent advances have seen the combination of techniques such as mass spectrometry, chromatography, electrophoresis and spectroscopy as *hyphenated* or *multi-dimensional* analytical techniques. Care has also been taken to emphasize how biophysical approaches often complement biological and chemical experimentation to give a fuller understanding of biochemical systems.

Specific examples of applications of the approaches described are given in boxes throughout the text. These are meant to give a flavour of their versatility and power for the solving of many different types of problems in biochemistry. The bibliography contains many more examples such as, for example, applications in clinical laboratories and in industry. Articles and books (e.g. laboratory manuals) containing practical hints to novices contemplating using these techniques are also cited.

Finally, it is hoped that this book will furnish the student with sufficient understanding to allow them to understand and grasp as-yet undeveloped biophysical approaches which may appear in the next decade or so by noticing the common factors underlying the methods described as well as their diversity.

REFERENCES

Buffers and pH

Voit, E.O. and Ferreira, A.E.N. (1998) Buffering in models of integrated biochemical systems. *Journal of Theoretical Biology*, **191**, 429–37. A description of the effects of including buffers in modelling of biochemical systems.

Good, N.E. and Izawa, S. (1972) Hydrogen ion buffers for photosynthesis research. *Methods in Enzymology*, **XXIV**, 53–68. A description of Good's buffers.

Grady, J.K., Chasteen, N.D. and Harris, D.C. (1988) Radicals from Good's buffers. *Analytical Biochem.*, **173**, 111–5. Further reading on Good's buffers.

Units and quantities

Kotyk, A. (ed.) (1999) *Quantities, Symbols, Units and Abbreviations in the Life Sciences*. Humana Press, Totowa, NJ, USA. A guide to standard usage of units and quantities in the life sciences.

Methods for protein estimation

Bradford, M. (1976) A rapid and sensitive method for the quantitation of microgram quantities of protein utilizing the principle of protein-dye binding. *Analytical. Biochem.*, **72**, 248–54. Original description of the Bradford method.

Hartree, E.F. (1972) Determination of protein: A modification of the Lowry method that gives a alinear photometric response. *Analytical Biochem.*, **48**, 422–7. A later modification of the Lowry method.

Lowry, O., Rosebrough, A., Farr, A. and Randall, R. (1951) Protein measurement with the Folin phenol reagent. *J. Biol. Chem.*, **193**, 265–75. Original description of the Lowry method.

Williams, G.A., Macevilly, U., Ryan, R. and Harrington, M.G. (1995) Semiautomated protein assay using microtitre plates – some practical considerations. *British Journal of Biomedical Science*, **52**, 230–1. Description of problems encountered in miniaturising protein assays with microtitre plates.

Bioinformatics

Roberts, E., Eargle, J., Wright, D. and Luthey-Schulten, Z. (2006) MultiSeq: Unifying sequence and structure data for evolutionary analysis. *BMC Bioinformatics*, **7**, Art. No 382. This on-line journal article introduces MultiSeq, a tool for combining sequence and structure data for proteins including an excellent general discussion of current bioinformatic issues.

Lesk, A. (2005) *Introduction to Bioinformatics*, 2nd edn, Oxford University Press, Oxford, UK. An excellent, comprehensive and clear description of modern bioinformatics.

Some useful web sites

The SI system at National Institute of Standards and Technology (USA): http://www.physics.nist.gov/cuu/Units/.

An illustrated site in "chemguide" on ionization and acid-base chemistry by Jim Clark: http://www.chemguide.co.uk/physical/acideqiamenu.html.

Aquasol solubility database: http://www.pharm.arizona.edu/aquasol/index.html.

Useful bioinformatics sites: National Library of Medicine (USA): http://www.ncbi.nlm.nih.gov/.

European Bioinformatics Institute (UK): www.ebi.ac.uk.

Uniprot site (US/Europe): http://www.ebi.uniprot.org/index.shtml.

Chapter 2
Chromatography

Objectives

After completing this chapter you should be familiar with:

- *The* **physical basis** *of chromatography.*
- *The* **chemical basis** *of the principal chromatography methods used in biochemistry.*
- **Performance criteria** *which can be used to compare chromatography systems.*
- *The range of different* **chromatography formats** *used in biochemistry.*
- *How one might approach* **design of a purification protocol**, *for example to purify a specific protein of interest.*

Living cells contain hundreds of thousands of distinct chemical species. These include large molecules like proteins, nucleic acids, lipids as well as lower molecular weight molecules which act as building blocks for biopolymers or as components of complex metabolic pathways. Some of these molecules are present in only trace amounts (e.g. intermediates in enzyme mechanisms) whilst others are present in abundance (e.g. structural proteins). Moreover, some components are present only at certain stages of the cell-cycle, whilst others are present at approximately constant levels. Study of individual chemical components of cells can, therefore, give us an insight into many fundamental cellular processes and help us to understand the dynamics of cell composition and function.

One approach to the study of individual chemical species is to separate them from each other by analytical or preparative chromatography. Originally, this technique was used by Tswett (1903) in the separation of plant pigments (*chromatography* comes from the Greek, *chroma*, meaning *colour*) but we now know that it is applicable to all chemical species, whether coloured or not. Because of the large range of size, shape and hydrophobicity found in biomolecules, it is to be expected that no one chromatography technique will suffice for all separations. In this chapter, the basic principles of chromatography will be described to explain why different molecules are separable. Some examples of the main chromatographic techniques used in Biochemistry are then given to illustrate how biomolecules are separated in practice.

2.1 PRINCIPLES OF CHROMATOGRAPHY

2.1.1 The Partition Coefficient

When applied to any two-phase system (e.g. liquid–liquid, liquid–solid), a molecule may partition between the phases (Figure 2.1). The precise ratio of concentration achieved is ultimately determined by inherent thermodynamic properties of the molecule (in turn, a function of its chemical structure) and of the phases. In the case of a liquid–liquid system, the relative solubility of the molecule in each liquid will be very important in determining partitioning. In a liquid–solid system, different sample molecules may adsorb to varying degrees on the solid phase. Both partition and adsorption phenomena are possible in a column system and this is called column chromatography. In column chromatography, one phase is maintained stationary (the stationary phase) while the other (the mobile phase) may flow freely over it. We can express the concentration ratio in such a system as the partition coefficient, K:

$$K = \frac{C_s}{C_m} \tag{2.1}$$

where C_s and C_m are the sample concentrations in the stationary and mobile phases, respectively. When a mixture made up of several components is applied to such a two-phase system, each component will have its own individual

Physical Biochemistry: Principles and Applications, Second Edition David Sheehan

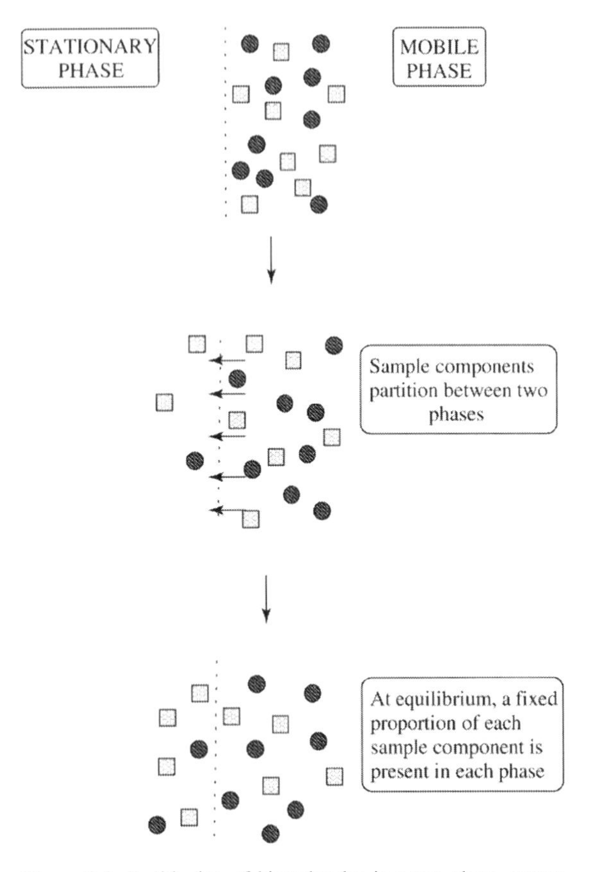

Figure 2.1. Partitioning of biomolecules in a two-phase system. Two components are represented by circles and squares, respectively. The two phases could be an aqueous buffer and a solid stationary phase. The two samples have very different partition coefficients in this experimental system.

partition coefficient. As a result, each will interact slightly differently with the stationary phase and, because of different partitioning between phases, will migrate through the column at different rates. Since K will be directly affected by the precise experimental conditions (e.g. temperature, solvent polarity) the chromatographer may vary these to optimize separation. In column chromatography, we therefore exploit what are often tiny differences in the partitioning behaviour of sample molecules to achieve their efficient separation.

2.1.2 Phase Systems Used in Biochemistry

In chromatographic systems used in biochemistry the stationary phase is made up of solid particles or of solid particles coated with liquid. In the former case, chemical groups are often covalently attached to the particles and this is called bonded phase liquid chromatography. In the latter case, a liquid phase may coat the particle and be attached by noncovalent, physical attraction. This type of system is called liquid–liquid phase chromatography. A good example of liquid–liquid phase chromatography is silica coated with a nonpolar hydrocarbon (e.g. C-18 reverse phase chromatography; Section 2.4.3). Commonly, the particles are composed of hydrated polymers such as cellulose or agarose. Such particles may be immobilized in a column (Section 2.3.1) and washed with mobile phase. They offer good flow characteristics and possess sufficient mechanical strength and chemical inertness for the chromatography of biomolecules. Because biomolecules have evolved to function in an aqueous environment, it is usually necessary to use aqueous buffers as the mobile phase if we require the molecule to retain its native structure (e.g. in the purification of active enzymes). If the native structure is not required, however, then it is possible to use more 'nonbiological' conditions such as organic solvents (e.g. in purification of peptides by reverse phase chromatography; Section 2.4.3).

Liquid–solid or liquid–liquid phases are the most common phase systems used in biochemistry. However, in specialized situations other phases may be used. For example, gas-solid and gas-liquid phases are used in gas chromatography (GC; Section 2.1.4). Regardless of the precise phase composition, chromatographic separation is a direct result of the different K values of each sample component.

2.1.3 Liquid Chromatography

To minimize loss of biological activity, separations are often carried out in aqueous buffers below room temperature. Low temperatures are especially important in the chromatography of cell extracts during, for example, protein purification. This reduces protease activity which might otherwise destroy the protein of interest. Chromatography with liquid mobile phases is called *liquid chromatography* (LC).

LC uses an experimental system outlined in Figure 2.2. Separation takes place in a column which contains the stationary phase. The volume and shape of the column will depend on the amount of sample to be separated and on the mode of chromatography to be used. Buffer is stored in a reservoir and is pumped through tubing onto the column. Appropriate valves allow the convenient injection of sample into this flow or the formation of gradients with a second buffer if required. The stationary phase is packed in the column and, as the sample passes through the bed of stationary phase, separation occurs. In *partition chromatography* modes, the sample separates into individual components as it passes through the stationary phase (e.g. *gel filtration*; Section 2.4.2). In *adsorption chromatography* modes, however, it is necessary first to load the entire sample and later to fractionate it. A good example of adsorption chromatography is *ion exchange chromatography* where the sample

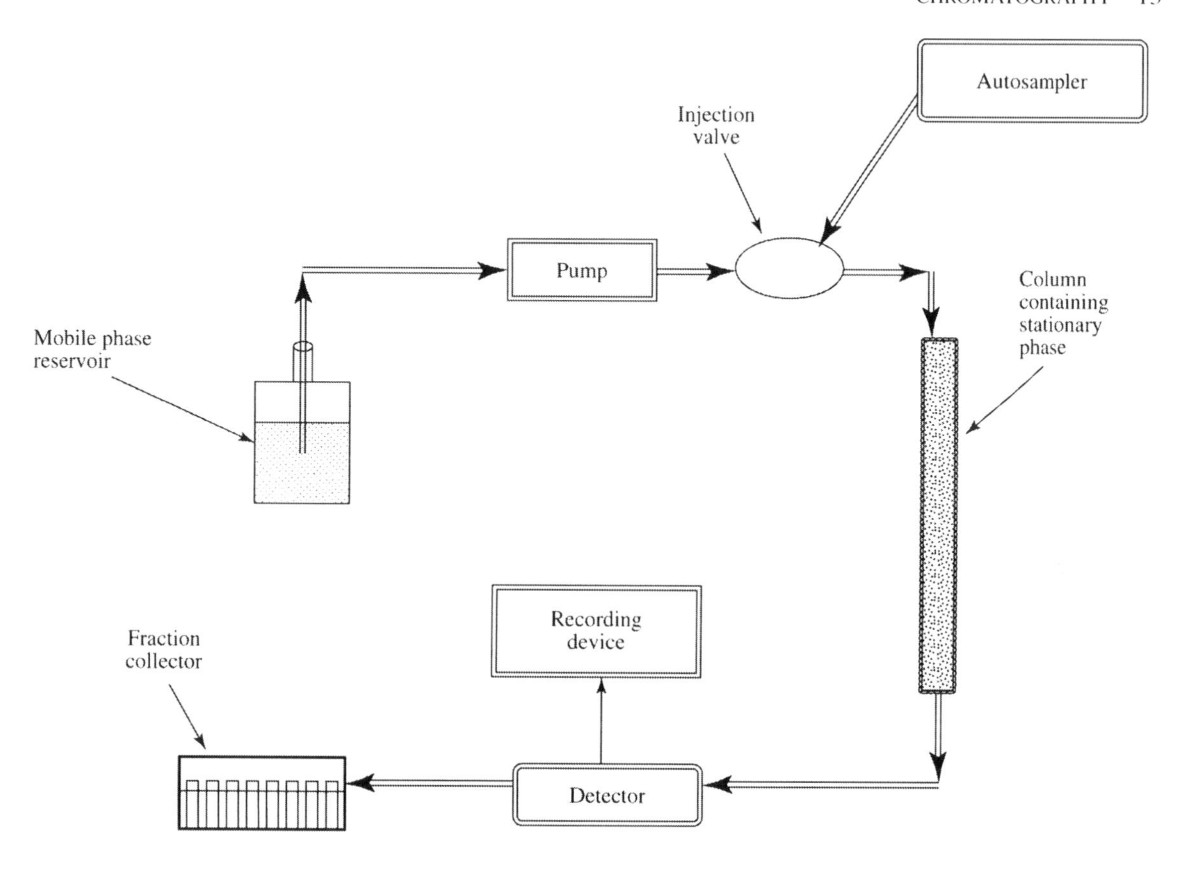

Figure 2.2. A typical liquid chromatography system. The direction of flow is shown by arrows. Sample is loaded *via* injection through a valve. If a large number of samples are required an *autosampler* may be used to reload the column repetitively after each chromatography.

components are eluted by means of a gradient of competing salt counterions after all the sample has been loaded onto and washed completely through the column (Sections 2.3.4 and 2.4.1). Separated sample components elute from the column and are detected. A typical LC separation is shown in Figure 2.3.

2.1.4 Gas Chromatography

Many biomolecules are sensitive to high temperatures that can lead to destruction of structure and function (*see* Section 1.2.2). However, some molecules of importance in biochemistry may be converted to derivatives that are structurally stable, though volatile, in the temperature range 200–250 °C. Good examples are trimethylsilated sterols and carbohydrates (esterified at their hydroxyl groups), and methylated esters of fatty acids. A second category of molecules (e.g. ethanol) is stable and volatile at somewhat lower temperatures without derivatization. Both of these groups of molecules may be analysed by *gas chromatography* (GC, also sometimes called *gas-liquid chromatography*; GLC). This is a form of partition chromatography in which a volatile sample is carried in an inert gas mobile phase (the *carrier gas*) such as nitrogen, helium or argon and applied to a narrow column (0.1 to 0.5 cm diameter) ranging in length from one to thirty meters that contains a liquid stationary phase. Sample is introduced to the system by injection from a syringe through a rubber septum into an injection port. This port is held approx. 10 °C above the column temperature to ensure efficient volatilization. The column is located in an oven that maintains a high temperature. Control of the temperature in this oven is crucial to successful chromatography and gradients of temperature may be used to achieve good separation. Because the sample components partition differently between the phases and because the column is so long, they separate efficiently from each other. The column may either be packed with solid particles coated with a liquid phase or it may be a *capillary column* in which the

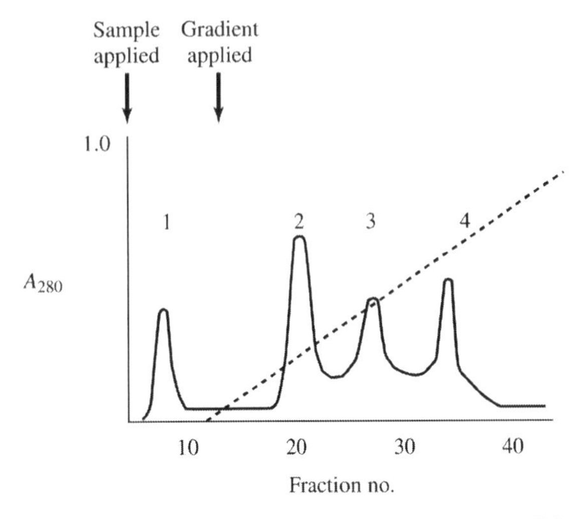

Figure 2.3. A typical liquid chromatography experiment. This shows separation of a group of isoenzymes by ion exchange chromatography. A 0–100 mM NaCl gradient (dashed line) is used with detection at 280 nm. This wavelength is often used to detect proteins. Note that peak 1 contains unbound material while proteins in peaks 2 to 4 are bound with progressively greater affinity.

inner wall of the column itself is coated with liquid. The latter have much smaller inner diameters (e.g. 25 μm) and are generally longer (10–100 m) than packed columns.

Sample components may be detected using flame ionization, flame photometry or thermal conductivity detectors. In flame photometry detectors, the sample is combusted to produce electrons and other ions that produce an electric current while in flame photometry detectors the change in conductivity of a wire is altered by the sample. As with the injection port, detectors are held approx. 10 °C above the temperature of the column. If the separated sample components are to be collected for further analysis (or for an interfaced method such as *mass spectrometry*; *see* Section 4.4.3), the carrier gas flow may be split before it enters the flame ionization detector.

Varying column temperature, gas flow and type of column may optimize GC separations.

2.2 PERFORMANCE PARAMETERS USED IN CHROMATOGRAPHY

Varied types (e.g. GC, LC) and modes (e.g. gel filtration, ion exchange) of chromatography are used in biochemistry. It is therefore very useful to be able to compare and quantify separations so that we can optimize them for a particular purpose. For example, in an analytical HPLC separation, we might wish to replace a particular column with one from a different manufacturer or with a slightly different stationary phase. The question would then arise, how effective is the new column compared to the old one? The yardsticks that allow us to make such comparisons we will call *performance parameters*.

2.2.1 Retention

When chromatography is carried out as described in Figures 2.2 and 2.3, the various sample components separate on the column because they are *retained* to different extents. We can describe this retention in terms of time (taking zero as the time of sample injection) or volume (calculated from volume at sample injection). These are called *retention time* (t_R) and *retention volume*, (V_R) respectively (*see* Figure 2.4). We can interconvert these terms using Equation (2.2):

$$V_R = f \cdot t_R \tag{2.2}$$

where t_R is the retention time (min), V_R the retention volume (ml) and f is the flow rate (ml/min). Under a single set of experimental conditions (temperature, column composition and dimensions, etc.), each component of the sample will have a characteristic retention (i.e. t_R or V_R) that can often be used to identify it in different samples. In general, retention is proportional to the length of the column used and is inversely proportional to the flow rate. Moreover, because the response of the detector usually relates to the concentration of the sample components, we can quantify each component by measuring their peak areas.

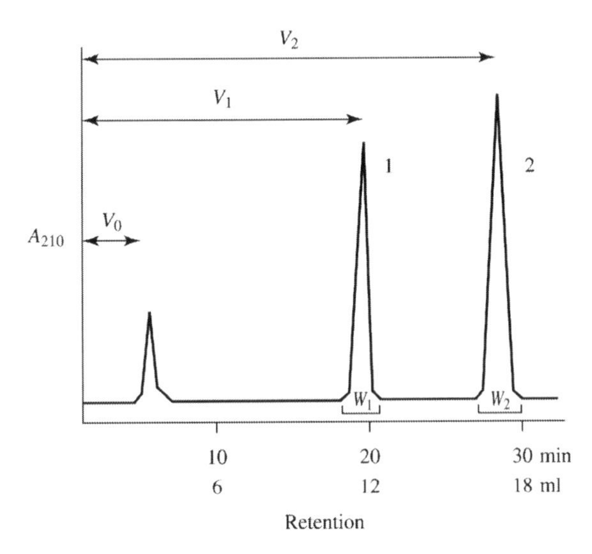

Figure 2.4. Retention in column chromatography. A typical chromatography trace showing the separation of two components; 1 and 2. The retention volumes of the components are shown by V_1 and V_2, respectively, while the base peak widths are denoted by W_1 and W_2. The void volume is denoted by V_0.

Sample components move through the column more slowly than does mobile phase. We can therefore define a *retardation factor* (R.F.) for each component as follows:

$$\text{R.F.} = \frac{V_{\text{sample}}}{V_{\text{mobile phase}}} \tag{2.3}$$

where V denotes rate of movement.

Even if the sample did not adsorb onto or partition into the stationary phase, it would take a finite amount of time before it eluted from the column. The volume required for the injected solvent merely to pass, unretained, through the column and to be eluted is called the *void* or *excluded volume* (V_0, *see* Figure 2.4). This is the contained volume of the column, plus that of injection valve and tubing, *minus* the volume of the polymeric component of the column packing. Retention volumes greater than V_0 are said to occur in the *included volume* and it is here that the sample may be fractionated.

2.2.2 Resolution

A sample chromatographic separation of two components of a mixture, 1 and 2, is shown in Figure 2.4. As the two peaks are 'resolved' from each other, we use the term *resolution*, R, to describe their separation. This important performance parameter in chromatography allows us to compare how effective a given column is in separating a mixture. R has a definite mathematical meaning (Equation (2.4)) that may be calculated from the trace shown in Figure 2.4:

$$R = \frac{V_2 - V_1}{(W_1 + W_2)/2} \tag{2.4}$$

W is the base width of the peak of each sample component of retention volume, V. It is clear from this equation that the best resolution chromatographies are those in which the peaks are *narrow* (i.e. small W) and *well separated* (e.g. large $V_2 - V_1$).

2.2.3 Physical Basis of Peak Broadening

If a sample of a single component is applied to a chromatography system in a small volume, it might be expected to elute in a very sharp peak of similar volume. This is not usually observed, however, due to a phenomenon called *band broadening*, that is the sample elutes in a much larger volume than that in which it was applied. The considerations summarized in Equation (2.4) indicate that this is a major factor in determining the R-value of a given chromatography column. To understand why some chromatography experiments achieve separation while others do not, it is necessary to understand the physical basis of band broadening. The applied sample may interact with the stationary phase, resulting in *diffusion* of the sample or it may become involved in *mass transfer phenomena* within the mobile or stationary phase. These are represented diagrammatically in Figure 2.5.

Eddy diffusion arises from a tendency of the mobile phase to experience *eddies* or local regions of circular flow similar to that observed in whirlpools. Mobile phase passes through channels in the stationary phase (usually composed of small particles) that will vary somewhat in size, some being wide and some narrow. Circular flow in narrow channels is generally slower than that in wide channels. Since sample will diffuse through a variety of channel sizes, it will be gradually distributed between fast-flowing, wide channels and slow-flowing, narrow channels. As a band of sample moves through the stationary phase it is broadened as a result of this diffusion.

Mass transfer phenomena arise from distribution of sample molecules within either the mobile or stationary phases. *Mobile phase mass transfer* occurs because mobile phase adjacent to the particle moves more slowly than mobile phase in the middle of the channel. This is due to frictional resistance to flow of the mobile phase by the particle.

Stagnant mobile phase mass transfer arises because different sample components enter channels that are not orientated in the same direction as the main flow through the column. These molecules are retained for varying amounts of time in the *stagnant* mobile phase in these channels. Therefore, the sample will be retained to varying extents. It is also possible for mass transfer to occur within the stationary phase in *stationary phase mass transfer* (i.e. for sample to enter the surface of the stationary phase).

2.2.4 Plate Height Equation

One way to envisage the chromatographic separation of a sample is mentally to divide the column into very thin, horizontal sections that we call *theoretical plates*. If we take a simple example of a two-component mixture in which the components have partition coefficients of 1 and 0.1, respectively, and follow how it behaves as it moves through a small number of succeeding plates we can easily see how separation depends on the difference in component partition coefficients (Example 2.1). The *theoretical plate number*, N, is an important performance indicator of a chromatography column and is directly related to the surface area of the particles of which the stationary phase is composed. It may be calculated from the relationship between a chromatography peak's retention time, t_R and width at its base, W, using Equation (2.5):

$$N = 16.\left(\frac{t_R}{W}\right)^2 \tag{2.5}$$

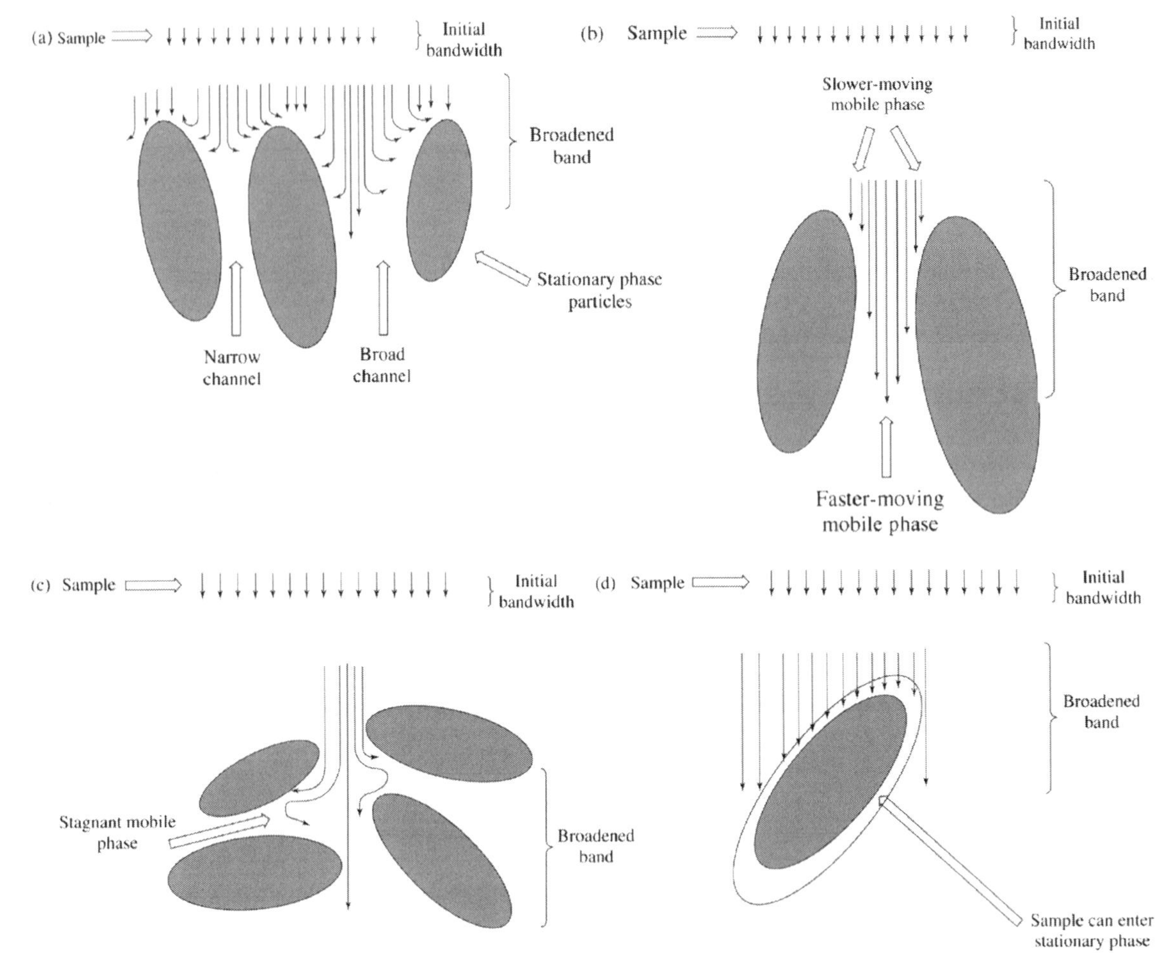

Figure 2.5. Physical causes of band broadening. (a) Eddy diffusion, (b) Mobile phase mass transfer, (c) Stagnant mobile phase mass transfer, (d) Stationary phase mass transfer. All of these may simultaneously contribute to broadening of the comparatively narrow initial bandwidth of applied sample.

The higher the value of N, the better the chromatographic performance we can expect to obtain from a column. This number may be compared between different columns and may even be used to compare bioseparations achieved in other types of experimental systems (e.g. electrophoretic separations; *see* Section 5.9.1). In practice, it may be difficult to measure W accurately so an alternative relationship may be used in which the *peak width at half the peak height*, W_h, is used (Equation (2.6)):

$$N = 5.54\left(\frac{t_R}{W_h}\right)^2 \tag{2.6}$$

Two chromatography columns of different length (e.g. 15 and 30 cm) might contain the same number of theoretical plates (e.g. a typical 15×0.46 cm C-8 reverse phase column contains approx. 13 000 theoretical plates). Clearly, the shorter column would be regarded as giving better chromatographic performance. We can express the number of theoretical plates as a function of column length with the *plate height number*, H, also sometimes known as the *height equivalent to a theoretical plate*, HETP (Equation (2.7)):

$$H = \frac{L}{N} \tag{2.7}$$

where L is column length. The better the chromatographic performance of a column, the smaller this number should be. In our example of columns of 15 and 30 cm length, the H value of the former (1.15 μm) will be half that of the latter (2.3 μm).

Example 2.1 *Theoretical plates* in chromatography

Theoretical plate is a term borrowed from the theory of distillation in which fractionation of samples between different phases occurs. In column chromatography, a theoretical plate may be imagined as a very thin section within which molecules can partition between mobile and stationary phases. The *thinner* this section, the better the chromatographic performance of the system. The concept of theoretical plates is useful in understanding how the partitioning behaviour of sample components determines their chromatographic separation.

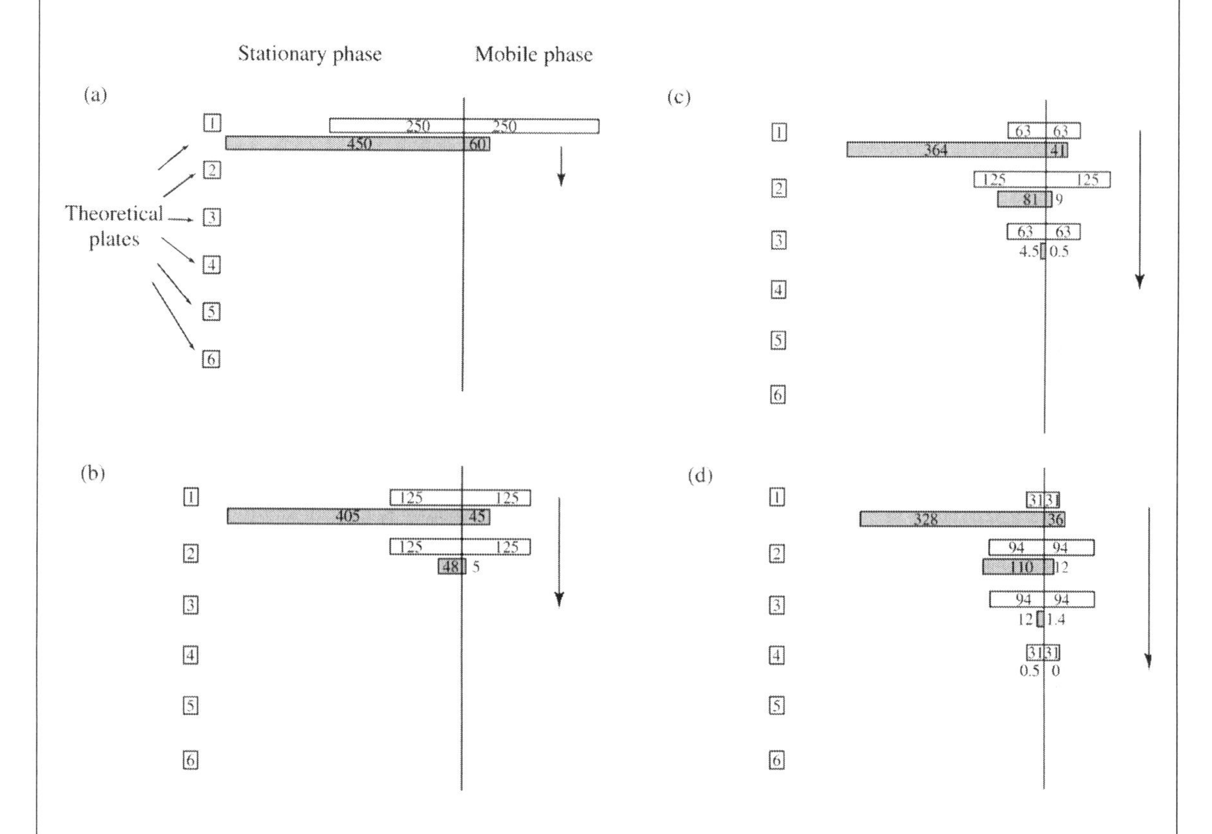

A sample made up of 1000 molecules of two types of components (500 of each) is applied to a chromatography system consisting of a mobile and stationary phase. These differ from each other in the value of their partition *coefficient,* k, which is the ratio of the concentration of the molecule in the stationary phase to that in the mobile phase. In this example, the values for k are 1.0 (white bars) and 0.1 (grey bars), respectively. This indicates that the molecule represented by the white bars partitions 1 : 1 between the phases while that represented by the grey bars partitions 9 : 1. Panels a) to f) show expected partitioning of the two components between the stationary and mobile phases as they progress through the first six theoretical plates.

(a) In the first plate the molecules with $k = 1.0$ partition evenly between the two phases (i.e. 250 in each) while the molecules with $k = 0.1$ partition very differently (i.e. 450 in stationary phase and 50 in mobile phase). The molecules in the mobile phase pass freely to the next theoretical plate (arrow).

(b) In this plate, they again partition 1 : 1 and 9 : 1 according to their values of k. Molecules left behind in the stationary phase of the first plate also partition in these ratios (i.e. 250 → 125 : 125, etc.). As molecules move through succeeding plates, those in the mobile phase constantly redistribute themselves according to their partition coefficients as do those left behind in stationary phase of earlier plates.

(g) Shows the profile of molecules in mobile phase as they might elute from the system (ignoring band broadening effects described in Section 2.2.3). The molecules represented by the grey bars are strongly retained by the stationary phase while those represented by the white bars are much less strongly retained. However, the components have clearly begun

(Continues)

(Continued)

to separate as a result of passage through only six theoretical plates. It should be noted that a real chromatography system would contain many thousands of theoretical plates. This example shows how differences in k underlie chromatographic separation.

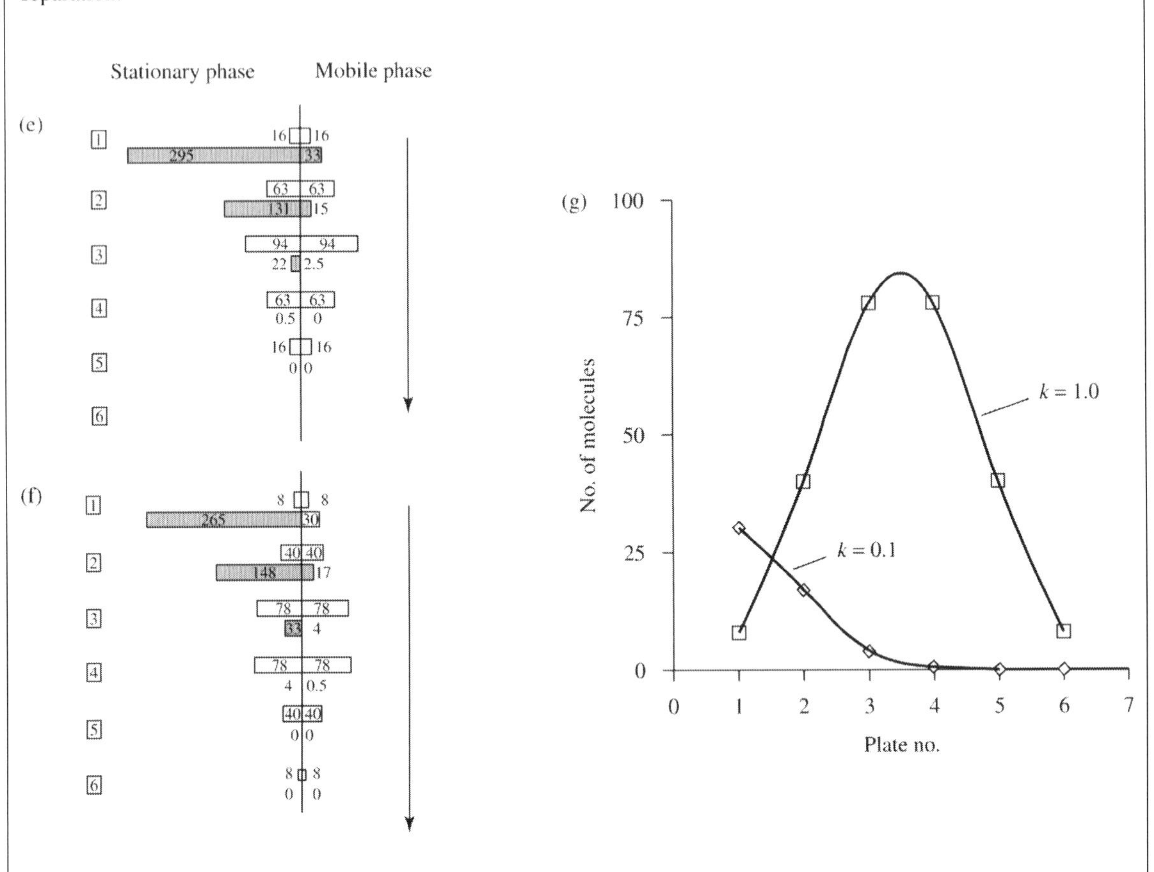

Since W is the *denominator* in Equation (2.5), there is a close relationship between the number of theoretical plates in a chromatography column and the extent of band broadening experienced by sample as it passes through the column. A column containing a large number of theoretical plates would be expected to cause little band broadening to sample while the opposite is true for columns containing a small number of theoretical plates.

We can write an *overall plate height equation* for a chromatographic system as follows (Equation (2.8)):

$$H = H_1 + H_2 + H_3 + H_4 \tag{2.8}$$

where H_1–H_4 are individual plate height values for the main causes of band broadening described in the previous section (H_1, eddy diffusion; H_2 mobile phase mass transfer, etc.). Since in liquid chromatography H_4 (the contribution of stationary phase mass transfer) is small, we can generally ignore it. For such a system the overall plate height is, therefore, given by Equation (2.9):

$$H = C_e \cdot d_p + \frac{C_m \cdot d_p^2 \cdot v}{D_m} + \frac{C_{sm} \cdot d_p^2 \cdot v}{D_m} \tag{2.9}$$

where C_e, C_m, C_{sm} and C_s are *coefficients* for the particular class of particle used as stationary phase, d_p is particle diameter, d_f is the thickness of stationary phase film, v is the flow rate of mobile phase, D_m and D_s represent the diffusion coefficients of sample in the mobile and stationary phases, respectively.

We can simplify this relationship as follows (Equation (2.10)):

$$H = d_p\left(\frac{C_c + C_m \cdot d_p \cdot v}{D_m} + \frac{C_{sm} \cdot d_p \cdot v}{D_m}\right) \quad (2.10)$$

The principal factors leading to extensive peak broadening, and hence high values for H, are therefore; the class of stationary phase selected; the particle diameter d_p; the flow rate v; the molecular size of the sample molecules (which will affect D_m). The chromatographer can achieve lower values of H by judicious choice of stationary phase particle and, especially, by using particles with small diameters. This is the physical basis for high resolution in systems such as HPLC (Section 2.6) and FPLC (Section 2.7). It should also be noted that separations can also be potentially improved by using slower flow rates and elevated temperature (which tends to increase D_m).

A plot of H versus the flow rate of the mobile phase, v, is called *the van Deemter curve* (Figure 2.6). This may be used to fit experimental data to allow comparison of different stationary phases and to determine the optimum flow rate of the mobile phase giving lowest values of H in a particular chromatography system. It is based on an *empirical* relationship between H and three factors (Equation (2.11)):

$$H = A + \frac{B}{v} + C \cdot v \quad (2.11)$$

where A represents eddy diffusion, B is *longitudinal diffusion* (i.e. diffusion along the direction of flow) and C describes mass transfer effects. Slower flow rates give better chromatography and this is reflected in the partial dependence of H on the product of C and v (i.e. a *decrease* in $C \cdot v$ gives a smaller H). At extremely slow flow rates, H *increases*, as a result of the partial dependence on the ratio B/v. A very small v is the denominator of the B/v term that would therefore tend to be very *large* in this region of the plot, giving a large value for H. The point where these two tendencies balance represents the optimum value for v. In practice, this value is usually found to occur at prohibitively low flow rates that would result in extremely long analysis times. For this reason, flow rates higher than the optimum value are used allowing shorter analysis times, albeit at the expense of loss of separation efficiency.

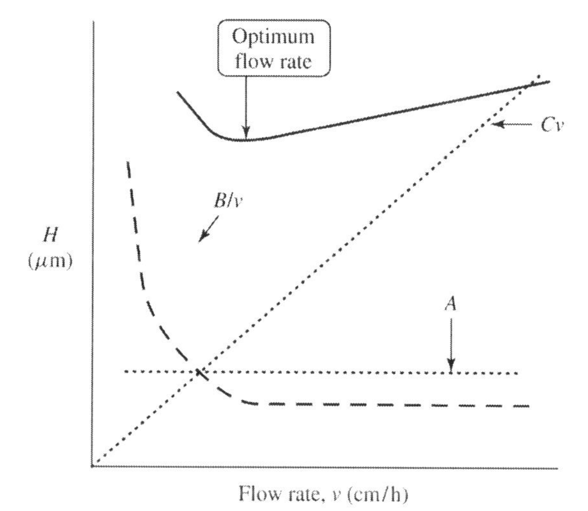

Figure 2.6. The van Deemter curve. A plot of H versus flow rate, v, is shown (solid line). Variation of the three components of Equation (2.16) with v is shown individually.

2.2.5 Capacity Factor

When there are several components in a sample separated by chromatography as described in Figure 2.4, we can write, for each component, an equation for retention volume (Equation (2.12)):

$$V_1 = V_0\,(1 + k_1) \quad (2.12)$$

where V_0 is the void volume defined in Section 2.2.1 and k_1 is the *capacity factor* for component 1. This capacity factor is partly dependent on partition coefficient and can easily be calculated from chromatograms using Equation (2.12). Since separation of component 1 from component 2 is determined largely by their different partition coefficients as shown in Example 2.1 we can further define a *separation factor*, α, for this separation (Equation (2.13)):

$$\alpha = \frac{k_1}{k_2} \quad (2.13)$$

This is also a performance indicator since high separation factor values denote good chromatography while low separation factors are associated with poor separation.

2.2.6 Peak Symmetry

Many chromatographic separations achieve *baseline* separation; that is no overlap between individual peaks in the chromatogram. Moreover, it is often assumed that the peaks will possess a *Gaussian* shape. This is the ideal situation and may allow us to calculate N directly from the peak (Equations (2.5) and (2.6)). In practice, *peak symmetry* may be significantly distorted from Gaussian. This arises from a variety of factors such as the use of nonlinear flow-rates, incomplete separation of peaks (e.g. the presence of 'shoulders' on peaks, an important clue to peak contamination) and the use of gradient elution techniques all of which are frequently encountered in practical chromatography. Peak symmetry can, therefore, give us valuable clues as to the performance of chromatography systems.

2.2.7 Significance of Performance Criteria in Chromatography

Performance criteria may be used for the comparison of otherwise very different chromatography systems. They are independent of the chemical nature of the adsorption of sample to the stationary phase (Section 2.4) because they depend primarily on physical interactions between sample components and the stationary phase. It is possible, therefore, to use these criteria meaningfully to compare, for example, reverse phase HPLC separations to open column ion-exchange chromatography with DEAE-cellulose. This comparison is also independent of the nature of the sample components, being equally applicable to peptides and sterol derivatives, for example. For 'good' chromatography we would require high resolution (R), a large number of theoretical plates (N) in the column, a low plate height number (H) and high values of the separation factor (α). Poor chromatography may be readily recognized by these criteria and improved purification quantified by their use.

2.3 CHROMATOGRAPHY EQUIPMENT

2.3.1 Outline of Standard System Used

A wide variety of experimental systems may be used to achieve chromatographic separation. The degree of complexity of the system depends on the degree of automation used. However, the general arrangement is as shown in Figure 2.2. The simplest, manual chromatography set-up for open column chromatography (Section 2.5) may be assembled in the laboratory using plastic bottles as *reservoirs* and glass columns which may be blown onto sintered glass funnels. Detection in such a system may be achieved by making manual absorbance measurements after separation. In such a system, flow may be driven simply by gravity or else by use of a peristaltic pump. Inexpensive glass and plastic columns are widely available from commercial suppliers.

The principal advantage of automation is the possibility of achieving more reproducible separation. A modest degree of automation is possible by the use of peristaltic pumps, on-line detectors and fraction collectors. However, the introduction of specialized microprocessor-controlled chromatography workstations has made possible the sequential separation of multiple samples, fine-tuning of chromatography programs to optimize resolution, automatic peak collection and calculation of peak area and symmetry as well as overlaying of chromatograms for comparison.

2.3.2 Components of Chromatography System

Buffers and other solutions required for preparation of mobile phase should be filtered carefully before use to remove particulate matter and other contaminants. For high-resolution chromatography, they should also be degassed although automatic *sparging* with oxygen-free nitrogen or helium gas is available in some systems. Mobile phase is pumped from a reservoir onto the column. Where more than one reservoir contributes to the mobile phase (e.g. in *binary*, *ternary* and *quaternary* gradients), flow from these reservoirs is mixed together in a *mixer* before arrival at the column. In open column chromatography (*see* Section 2.5), flow may be achieved by gravity or the use of peristaltic pumps. In FPLC (*see* Section 2.7) and HPLC (Section 2.6) high-precision pumps drive mobile phase through the system. Separation takes place in a column that contains the stationary phase. This may be constructed from glass, plastic or steel depending on the pressure of the system. In HPLC, a further column called a *guard column* may precede the chromatography column in the flow. The function of this is to protect the latter column from clogging and to prolong its useful life.

On-line detection may be by absorbance, fluorescence, refractive index, electrochemical or some other physicochemical measurement. Absorbance measurements at 280 nm will detect most proteins although, to be sure not to miss proteins or peptides poor in aromatic residues, detection in the range 205–220 nm (which detects peptide bonds which have a λ_{max} at 214 nm) is often preferable. Fluorescence and refractive index detection allow greater sensitivity and are especially suitable for analytical chromatography. A plot of detector signal versus the retention of sample components is called an *elution profile*. Separated peaks may be collected by means of a *fraction collector*.

Before chromatography, it is advisable to filter or centrifuge sample to remove particulates that could potentially clog the column. Sample introduction is by direct flow onto the column (in open column chromatography) or by injection through a sample injection port into the mobile phase stream in high-resolution systems. For multiple samples an *autosampler* is used and this is especially useful in analytical separations. Since open column separations are generally quite slow, it is often desirable to carry them out at lower temperatures such as in a cold-room or a refrigerated cabinet to minimize loss of biological activity. Higher resolution chromatography may usually be performed at room temperature.

2.3.3 Stationary Phases Used

As mentioned in Section 2.1.2, hydrated polymers are most commonly used in chromatography of biochemical samples. A wide range of these and other *packings* are now commercially available and a selection of these, together

Table 2.1. Some stationary phases used in chromatography

Packing	Composition	Application
DEAE/CM-cellulose	Polysaccharide (cellulose)	Ion exchange chromatography, especially early in purification scheme
Glutathione-agarose	Polysaccharide (agarose)	Affinity chromatography and purification of GST fusion proteins
IDA-agarose	Polysaccharide (agarose)	IMAC (see Section 2.4.6)
Sephacryl S-300	Polysaccharide (dextran/ bis acrylamide)	Gel filtration of proteins in Mass range 10–1500 kDa
Sephadex G-25	Polysaccharide (dextran)	Desalting of protein extracts by gel filtration
Superose 12	Polysaccharide (agarose)	Gel filtration FPLC in the Mass range 1–300 kDa
C-18 Silica	Silica	Reversed-phase HPLC of tryptic peptides
POROS	Poly(styrene-divinylbenzene)	Perfusion chromatography
DEAE/CM-MemSep	Polysaccharide (cellulose)	Membrane-based ion exchange chromatography of proteins

with their applications, is shown in Table 2.1. The stationary phase is required to be mechanically stable (especially to high pressures and liquid shear forces) and chemically inert under the chromatography conditions used. Often, it is also desirable that it is capable of acting as a support for chemical groups possessing desirable properties for chromatography such as hydrophobicity, ion exchange or affinity ligands (Section 2.4).

2.3.4 Elution

Elution from or *development of* chromatography columns may be achieved in three possible ways, depending on the specific nature of chemical interaction between sample components and stationary phase. These are illustrated in Figure 2.7. *Continuous flow elution* or *isocratic elution* is where mobile phase flowing through the column is continuous both in flow rate and in composition. This is especially

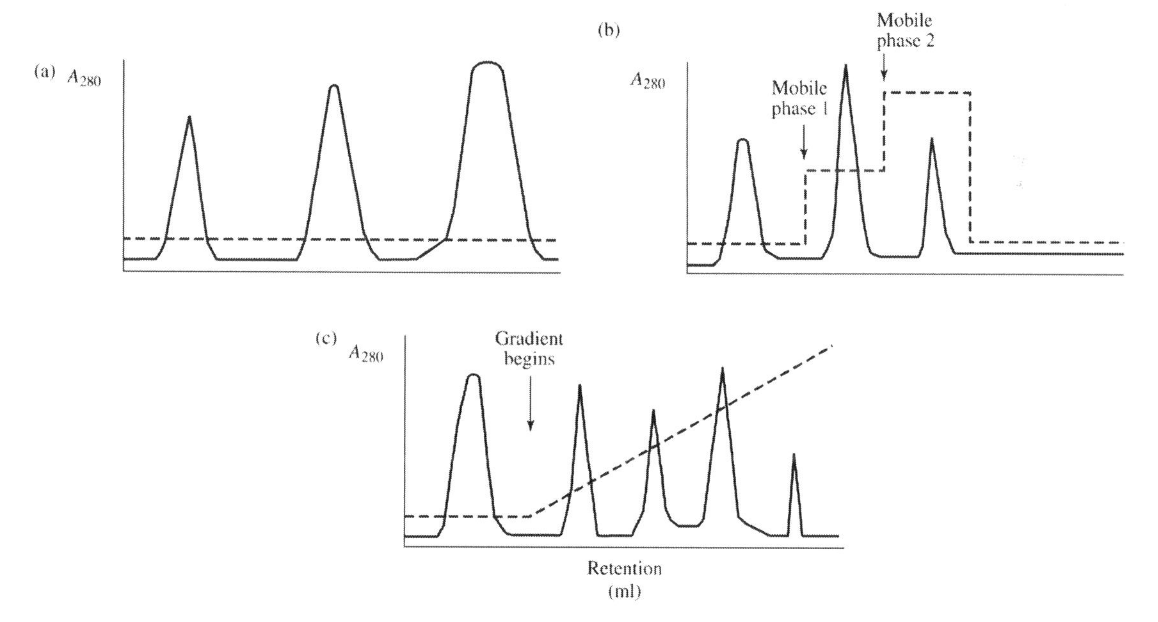

Figure 2.7. Elution from stationary phase. Mobile phase is shown as dashed lines. This may vary in composition to give different pH, ion strength or hydrophobicity as described in text. (a) Continuous flow elution. Sample components separate in a situation where they possess different inherent affinities for the stationary phase and where the mobile phase composition and flow-rate remain constant. (b) Batch flow elution. Adsorbed sample components may be selectively eluted by washing column sequentially with a range of mobile phases. (c) Gradient elution. Adsorbed material is separated by the application of a gradient composed of two or more buffers in which mobile phase composition is continuously varied. More complex gradients (e.g. hyperbolic gradients) may be generated by continuously varying percent buffer B in the mobile phase.

applicable in partition chromatography systems. It is usually found that *leading* (i.e. early eluting) peaks are sharp but poorly resolved with this type of elution while *trailing* (i.e. late eluting) peaks, by contrast, are well separated but tend to have undergone extensive peak broadening. This is known as the *general elution problem*.

Where sample components have *adsorbed* to the column packing, they often have quite different, individual, affinities for the stationary phase. A description of the major types of adsorption chromatography used in biochemistry is given in Section 2.4 and it will be clear from this that these different affinities arise from inherent structural characteristics of the sample components. The chromatographer may use this to separate them from each other. *Batch flow elution* can be used in such a situation and involves the *stepwise* introduction of *different* mobile phases in which the pH, polarity, salt concentration or some other component are varied. Conditions may be selected where one sample component remains bound to the stationary phase while another elutes, usually in a small volume.

A variation on this is *gradient elution*, where pH, polarity, salt concentration are varied *continuously* rather than stepwise. This type of elution overcomes the general elution problem mentioned earlier, ensuring high-resolution separations throughout the chromatogram within a relatively short range of elution. A *gradient maker* is required for this procedure. Automation allows the creation of complex gradients with a variety of shapes (Figure 2.7). Gradients are prepared by continuously mixing two different buffers (although *ternary gradients*, prepared from three different buffers are also possible) usually referred to as buffers 'A' and 'B', respectively. The resulting mixture acts as the mobile phase. Gradient elution would need to be performed to determine elution conditions for batch flow elution. This kind of practice is often used when *scaling up* chromatography for preparative separation such as in *downstream processing* of proteins (Section 2.5.2). In some cases, combining various gradient steps with isocratic elution may optimize separation of poorly resolved peaks.

Elution with ionic compounds such as NaCl and KCl in ion exchange chromatography (Section 2.4.1) may require high salt concentrations. In downstream processing of proteins this salt will need to be removed after chromatography which can be expensive and time consuming. An alternative approach is stepwise elution with *displacers*, that is molecules with a higher affinity for the stationary phase than the target protein of interest. These molecules may be of low or of high molecular weight (e.g. heparin and starch derivatives, respectively) and offer the advantage that they induce elution of proteins in highly concentrated peaks of pure material at relatively low displacer molar concentrations (e.g. 40 mM).

Because of the sensitivity of biomolecules such as proteins to their surrounding environment, it is possible to alter the charge, hydrophobicity and overall structure of their surface by varying pH, ion strength, and so on of the mobile phase. As a consequence of this, the elution profile obtained at a particular pH will vary if the pH of the mobile phase is altered. Since proteins are highly individual in this sensitivity to their environment, they will each be affected differently by changes in pH. Thus, by performing chromatography with a variety of stationary phases under a range of conditions of pH and ion strength, we can optimize conditions for separation.

2.4 MODES OF CHROMATOGRAPHY

So far, we have described chromatography systems mainly in terms of their resolution and performance criteria. This is useful because it helps us to understand why the *physical* characteristics (especially particle diameter) of the particles of which the stationary phase is composed are so important (Section 2.2). However, the *chemical* basis underlying interaction between sample components and the stationary phase provides a complementary description of the principal chromatography systems used in biochemistry. We may define this interaction as the *mode* of chromatography. In the following sections we will look at relatively low resolution open column chromatography and the more high resolution FPLC, HPLC and other systems (Sections 2.5–2.9). All of these systems, regardless of the very different physical characteristics of their stationary phases, are available in a variety of chromatography *modes*. We can illustrate this situation by the matrix shown in Figure 2.8. Any chromatography system may be located on this matrix in terms of only two parameters; chromatography mode and format (i.e. stationary phase differing in *physical* characteristics such as particle diameter). This matrix is useful in that it may be easily extended to accommodate new formats and modes of chromatography. In this section, the main modes of chromatography will be described.

2.4.1 Ion Exchange

Proteins have a net charge on their surface (Section 1.2.1) arising mainly from amino acid side chains some of which possess either a positive or negative charge at a given pH. Around pH 7.0 *acidic* amino acids (glutamic and aspartic acids) contribute negative charges while *basic* residues (lysine and arginine) contribute positive charges. At acidic pH, most proteins possess net positive charge while at alkaline pH they possess net negative charge. Positive and negative charges can cancel each other out so that, at a particular pH value, the protein may possess net zero charge even though it contains some positive and some negative charges. The pH at which a given protein possesses zero net charge is called the *isoelectric point*, *pI* (Figure 2.9). Since the

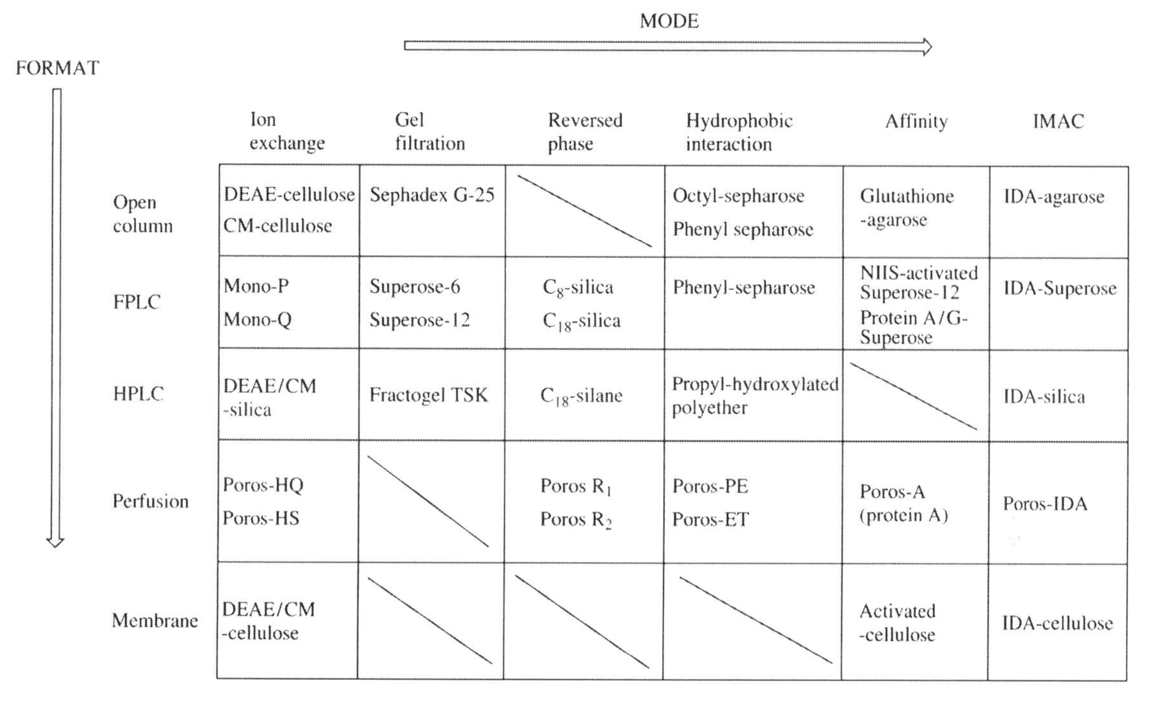

MODE →

FORMAT ↓

	Ion exchange	Gel filtration	Reversed phase	Hydrophobic interaction	Affinity	IMAC
Open column	DEAE-cellulose CM-cellulose	Sephadex G-25		Octyl-sepharose Phenyl sepharose	Glutathione -agarose	IDA-agarose
FPLC	Mono-P Mono-Q	Superose-6 Superose-12	C_8-silica C_{18}-silica	Phenyl-sepharose	NHS-activated Superose-12 Protein A/G-Superose	IDA-Superose
HPLC	DEAE/CM -silica	Fractogel TSK	C_{18}-silane	Propyl-hydroxylated polyether		IDA-silica
Perfusion	Poros-HQ Poros-HS		Poros R_1 Poros R_2	Poros-PE Poros-ET	Poros-A (protein A)	Poros-IDA
Membrane	DEAE/CM -cellulose				Activated -cellulose	IDA-cellulose

Figure 2.8. Matrix illustrating the relationship between different chromatography systems in terms of chromatography mode and format. Selected commonly-used stationary phases are shown.

surface charge, structure and pI are dependent on the amino acid sequence of the protein, they may be regarded as characteristic physical properties that may be used to separate proteins from each other.

Ion exchange chromatography is a form of adsorption chromatography in which charged proteins or other biomolecules are exchanged for small ions of like charge originating in salts (e.g. Na^+, Cl^-). These ions are attached

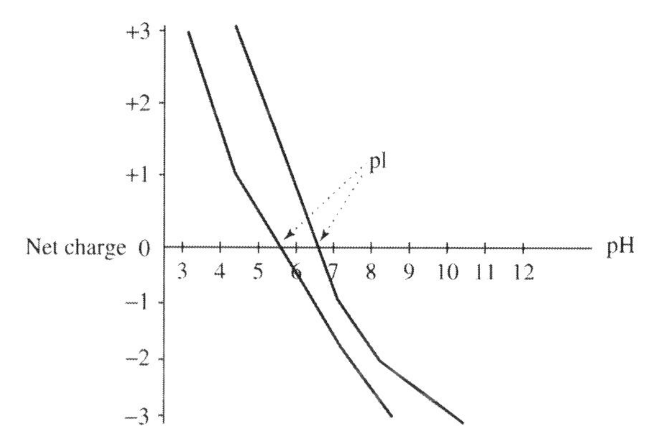

Figure 2.9. Surface charge of proteins. The net charge on the surface of two proteins (pI values of 5.5 and 7.5, respectively) are shown at a range of pH values. Note how this varies with pH.

to chemical structures on the surface of the stationary phase called *ion exchange* groups or *ion exchangers*. Two types of ion exchanger are common in biochemistry; *Anion exchangers* (which exchange negatively charged ions or anions) and *cation exchangers* (which exchange positively charged ions or cations; Figure 2.10). It is important to note that this exchange process is *independent* of mass, shape or other physical characteristics of the molecule to be separated. Proteins that are otherwise very different might, coincidentally, display the same behaviour on ion exchange chromatography: Conversely, isoenzymes (which may be very similar in terms of primary structure and biological activity) might behave very differently.

Ion exchangers are attached to a wide variety of solid *supports* such as cellulose, agarose and vinylbenzene. A selection of these is shown in Table 2.2. The physical properties of the stationary phase as a whole (i.e. ion exchanger plus support) will determine whether high or low resolution is achievable with a given experimental system as previously described in Section 2.2. Some ion exchangers are regarded as *weak*, that is functioning best over a comparatively narrow pH range, while others are *strong*, that is functioning over a wider pH range. Exchangers with quaternary amino or sulfonic acid groups behave as strong anion and cation exchangers, respectively, while aromatic/aliphatic amino and carboxylic acid groups are weak anion and cation

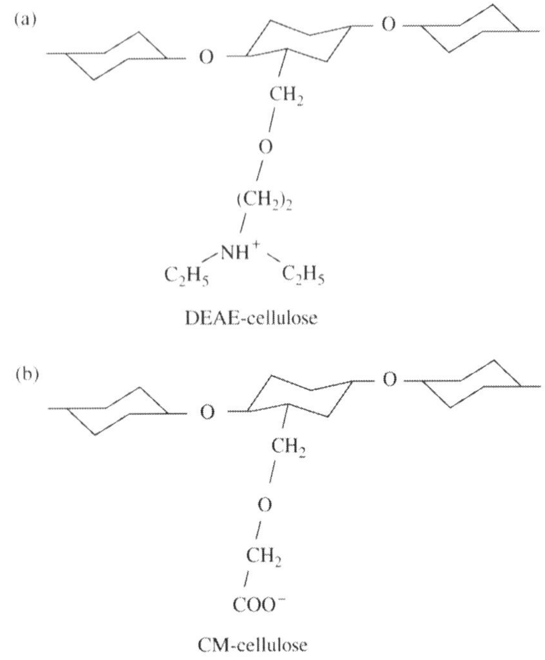

(a)

DEAE-cellulose

(b)

CM-cellulose

Figure 2.10. Ion exchangers. The structure of (a) diethylaminoethyl (DEAE)-cellulose, an anion exchanger and (b) carboxymethyl (CM)-cellulose, a cation exchanger.

exchangers, respectively. Choice of ion exchange stationary phase is heavily influenced by knowledge of the pI of the protein of interest.

In an ion exchange chromatography experiment (Figure 2.11), sample is applied to a stationary phase which has been *charged* with the ion to be exchanged; the *counterion* (e.g. Cl^-). The protein (and any other species of like charge in the sample) may exchange with this counterion, binding to the ion exchanger. It is important that the sample is free of components of like charge since these are usually present in large molar excess relative to the concentration of protein. The ion exchanger will not distinguish between proteins possessing a single positive charge and, for example, a Na^+ ion also present in the sample. *Desalting* is carried out before ion exchange either by *gel filtration chromatography* (Section 2.4.2), by *dialysis* (Section 7.6.1) or by *centrifugal filtration* (Section 7.6.2). Elution of bound proteins is achieved by reversing the process of binding and, again, exchanging a counterion for protein. This is usually carried out by applying a large excess of a salt (e.g. NaCl) containing the counterion in the mobile phase. Because proteins have different net charge, they may bind to an ion exchanger at a given pH with a variety of strengths, that is some proteins may bind strongly whilst others bind weakly or not at all. We can take advantage of this to separate proteins by applying salt in a continuous gradient. Weakly bound proteins will elute first from such a system while strongly bound proteins elute later.

Proteins possess some charge at most pH values (their net charge is zero only at the pH corresponding to their pI). By deliberately varying pH we may selectively alter net charge on their surface. Even though two proteins may behave identically at a given pH, they are unlikely to do so at all pH values. By carrying out separations across a range of pH, therefore, we can resolve proteins that might cochromatograph at a single pH. In exploring chromatographic behaviour across a range of pH values it should be borne in mind that many proteins are structurally unstable at extremes

Table 2.2. Commonly-used ion exchange groups

Group	pH range	Chemical structure
	Cation exchangers	
Sulphopropyl (SP)	2–12	$-(CH_2)_2-CH_2-SO_3^-$
Methyl sulphonate (S)	2–12	$-CH_2-SO_3^-$
Carboxymethyl (CM)	6–11	$-O-CH_2-COO^-$
	Anion exchangers	
Quaternary ammonium (Q)	2–12	$-CH_2-N^+(CH_3)_3$
Diethylaminoethyl (DEAE)	2–9	$-O-CH_2-CH_2-NH^+(C_2H_5)_2$
Quaternary aminoethyl (QAE)	2–12	$-O-(CH_2)_2-N^+(C_2H_5)_2-CH_2-CHOH-CH_3$

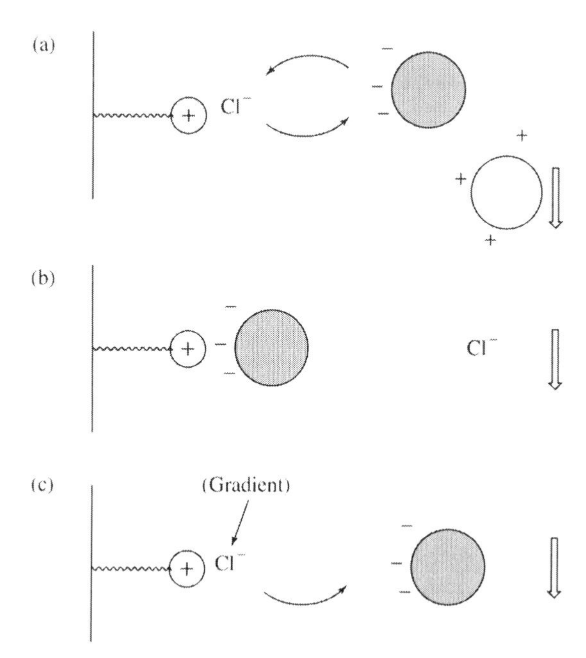

Figure 2.11. A typical ion exchange experiment. (a) Cl^- is attached to positively-charged ion exchanger. (b) Negatively-charged protein exchanges with Cl^- and binds to the ion exchanger. (c) A competing Cl^- (from a NaCl gradient) elutes the protein. Note that positively-charged proteins pass straight through the stationary phase and that proteins with different net charge are separated by the gradient of NaCl (Figure 2.7).

of pH. This may cause fouling of the stationary phase bed and also result in loss of biological activity of the protein to be purified. Preliminary experiments to determine pH stability and solubility of the protein to be purified are, therefore, best carried out before ion exchange chromatography.

As well as achieving separation of fully active proteins under native conditions with ion exchange chromatography, it is also possible to carry out the experiment under *denaturing* conditions. For example, peptides from a chemical or enzymatic digest of a protein may be separated in the presence of 4–7 M urea. Urea is a *chaotropic agent* that denatures proteins by decreasing hydrophobic interactions within polypeptides and by disrupting ionic interactions (especially hydrogen bonding) within and between protein subunits or individual peptides in a digest. When ion exchange is performed in the presence of urea, subunits and peptides will no longer be bound together by interactions such as salt bridges and hydrogen bonds. Although a polar molecule, urea possesses zero net charge so it will not itself interact with the ion exchanger.

2.4.2 Gel Filtration

Proteins and other biomolecules often differ greatly from each other in their mass and shape. *Gel filtration chromatography* (also known as *molecular sieve* or *size exclusion chromatography*) takes advantage of this difference by retaining biomolecules of a given size range and fractionating them in a manner related to their mass in a form of partition chromatography. This is possible because the stationary phase employed in this method is made up of a gel consisting of beads containing pores of a defined and narrow size range. These will allow only biomolecules *below* a particular mass to diffuse into the pore. This is referred to as the cutoff value or *exclusion limit* (Figure 2.12) of the stationary phase. A variety of different stationary phases, each with its own particular cutoff value is commercially available and a selection of these is summarized in Table 2.3. Once inside the bead, the biomolecule will be retained momentarily in

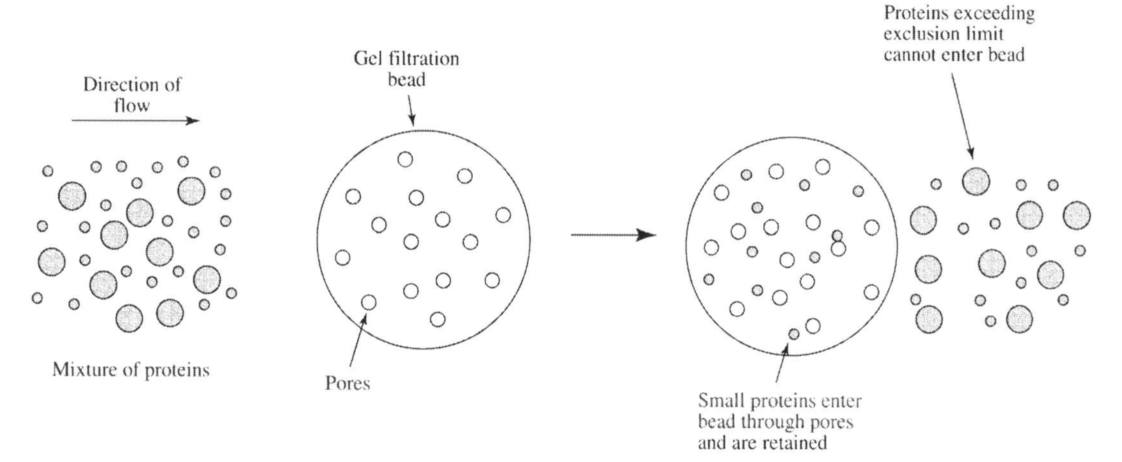

Figure 2.12. Gel filtration chromatography of proteins. Note that only proteins smaller than the bead exclusion limit may enter the bead. Larger proteins are excluded. Therefore, proteins elute in inverse order of native mass.

Table 2.3. Gel filtration media. A selection of commonly-used gel filtration media with their individual fractionation range and composition

Resin	Fractionation range kDa (globular proteins)	Application
	SEPHADEX-(dextran)	
G-10	<0.7	Desalting
G-25	1–5	Desalting
G-50	1.5–30	Peptide separation
G-75	3–80	Protein fractionation
G-100	4–150	Protein fractionation
G-150	5–300	Protein fractionation
G-200	5–600	Protein fractionation
	SEPHAROSE-(agarose)	
6B	10–4000	Protein fractionation
4B	60–20 000	Protein fractionation/ligand immobilisation
2B	70–40 000	Fractionation of nucleic acids, particles and viruses
	SEPHACRYL (dextran crosslinked with *N*,*N*′-methylene bisacrylamide)	
S-100 HR	1–100	Peptide/protein fractionation
S-200 HR	5–250	Protein fractionation
S-300 HR	10–1500	Protein fractionation
S-400 HR	20–8000	Protein fractionation of molecules with extended structures
S-500 HR	80–80 000	Large molecules and small particles
	BIOGEL (polyacrylamide)	
P-2	0.1–1.8	Desalting
P-4	0.8–4	Desalting
P-6	1–6	Peptide separation
P-10	1.5–20	Fractionation of small proteins
P-30	2.5–40	Protein fractionation
P-60	3–60	Protein fractionation
P-100	5–100	Protein fractionation
P-150	15–150	Protein fractionation
P-200	30–200	Protein fractionation
P-300	60–400	Fractionation of extended structures
	SUPERDEX-(dextran covalently bonded to highly crosslinked agarose)	
75	3–70	FPLC gel filtration
200	10–600	FPLC gel filtration
SUPEROSE-(cross-linked agarose)		
12	1–300	FPLC gel filtration
6	5–5000	FPLC gel filtration
	FRACTOGEL-(polyvinyl chloride)	
TSK HW-40	0.1–10	HPLC gel filtration (peptides)
TSK HW-55	1–700	HPLC gel filtration (proteins)
TSK G2000SW	5–100	HPLC gel filtration (proteins)
TSK G3000SW	10–500	HPLC gel filtration (proteins)
TSK HW-65	50–5000	HPLC gel filtration (proteins)
TSK HW-75	500–50 000	HPLC gel filtration (complexes)

Example 2.2 Determination of native mass of a protein by gel filtration

Gel filtration beads covering a particular molecular mass range are selected and packed into a column (Table 2.3). Unlike other chromatography formats column geometry is important in this experiment. Gel filtration columns are usually long relative to their column diameter. However, the low tolerance of gel filtration resins to hydrostatic pressure places an upper limit on the volume of resin which may be used (i.e. the column cannot be too long). Conversely, the loading volume of sample is also limited (usually 1–2% of the total column volume) which means that the column volume cannot be too small either. Once packed, the void volume of the column is determined by passing a solution of blue dextran through it and measuring its elution volume. Standard proteins of known native M_r are then applied to the column to calibrate it. These may be applied singly or in pairs (provided the two proteins are well-separated on the column and can clearly be distinguished by some other technique such as spectroscopy or electrophoresis). In selecting standards it is important that they are comparable to the sample protein under investigation (i.e. that they have the same general shape characteristics).

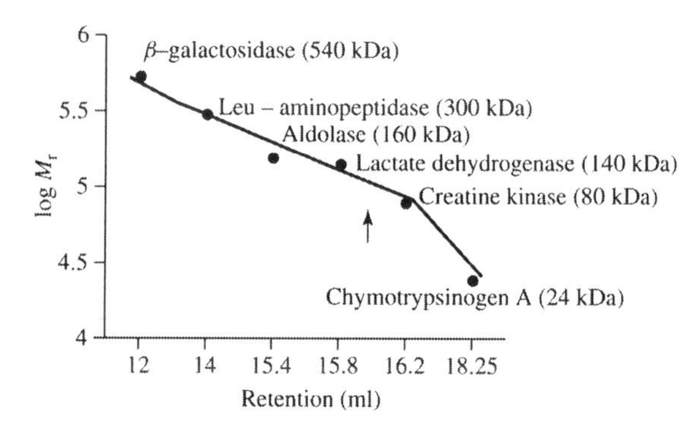

The elution volumes (retention) of sample and standard proteins were determined on Superose 12. There is a linear relationship between log molecular mass and elution volume and this relationship allows creation of a standard curve for the column. From this plot, native mass of the sample protein may be estimated as 100 kDa. Note that the plot is slightly curved at the ends, suggesting a deviation from linearity. Comparisons should only be made in the linear part of the plot and, if necessary, the experiment should be repeated using a resin with a slightly different fractionation range. Electrophoretic analysis (Section 5.3.1) indicated that this protein had a subunit mass half that of its native mass (i.e. 50 kDa) suggesting a dimeric quaternary structure.

channels passing through it. As the sample progresses through a bed of such stationary phase, smaller molecules will be retained more readily than larger ones. Molecules of greater mass than the exclusion limit will pass freely through the column, eluting in a single, unresolved, peak at the void volume. Molecules within the *fractionation range* of the gel, however, will be retained in a manner generally inversely proportional to their mass, that is smaller molecules will be retained longest. The fractionation range and resolution achievable with such a chromatography system will depend greatly on the composition and uniformity of the population of beads in the stationary phase.

Gel filtration chromatography has three main applications in biochemistry. It may be used in a routine way to remove low molecular weight components from samples. For example, as pointed out in the previous section, it is possible to *desalt* protein solutions by passage through a column of Sephadex G-25 (fractionation range 1–5 kDa). This application takes advantage of the *cutoff value* of these beads 5 kDa) rather than of their fractionation range. Such chromatography is also useful for rapid exchange into a buffer of different composition with minimum dilution of protein sample.

A second application depends on the fact that the elution volume of proteins from gel filtration columns depends on their mass. Thus, if we determine the elution volumes of proteins of known mass on a given gel filtration column, we can easily determine the mass of a protein of unknown mass by plotting log M_r versus elution volume (Example 2.2). This mass is the *native* mass of the protein. It is important to note that this experiment makes a number of important assumptions. Firstly it assumes that the protein has similar shape to the standards used (i.e. proteins of known mass). This is acceptable if both the protein and standards are globular, for example. However, molecules with large *axial ratios* (i.e. rod-like in shape rather than globular; e.g. collagen) behave as if they have a much greater mass on gel filtration. Where little is known about the overall shape of the protein, care should be taken in interpreting gel filtration data. A further point to note is that some proteins may experience secondary ionic or ion exchange interactions with the stationary phase beads. This can result in the molecule being retained on gel filtration in a way that may not depend on mass alone. One means of avoiding such effects is to carry out the experiment in the presence of 100–300 mM NaCl or a similar salt.

The mass of individual polypeptides or *subunits* of which *oligomeric* proteins are composed may be determined under denaturing conditions by a number of techniques such as SDS polyacrylamide gel electrophoresis (Chapter 5) or mass spectrometry (Chapter 4). Determination of native mass may help us to determine the quaternary structure of proteins by comparison with results from denaturing experiments.

A third application of gel filtration chromatography is in *protein purification*. Since the method depends on molecular mass, it may be used to separate proteins based on differences in their mass. An important limitation to this approach is the *loading volume* that can be achieved. Generally, this is a maximum of approximately 5% of the bed volume of the column. This technique is therefore often used near the end of a purification strategy as a *polishing* step. Neither gel filtration nor ion exchange chromatography alone is capable of achieving complete purification of proteins. However, since it is unlikely that two proteins will have the same net charge and mass, combining the two approaches can often achieve impressive purification (Section 2.10.1).

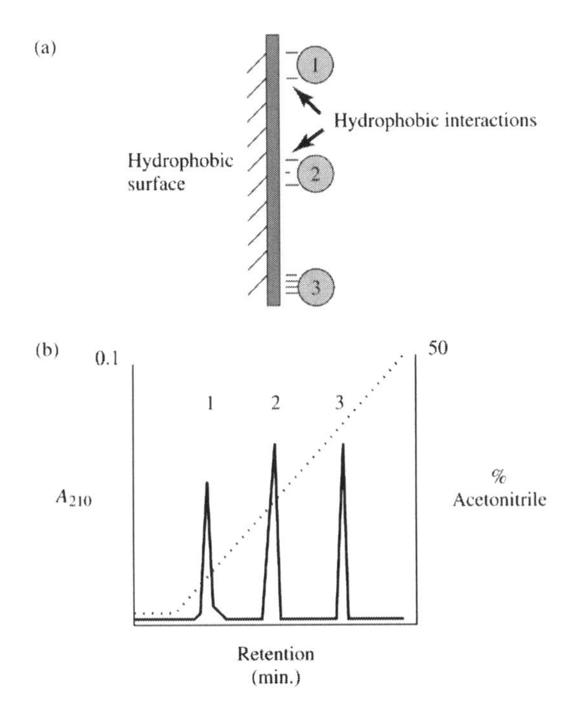

Figure 2.13. Reversed phase chromatography. (a) Samples interact with nonpolar stationary phase by hydrophobic interaction and are eluted by increase in hydrophobicity of mobile phase. The three sample components shown (1 to 3) have increasing hydrophobicity. (b) Elution profile obtained with sample components 1 to 3%. Acetonitrile is shown by dashed lines.

2.4.3 Reversed Phase

As well as possessing characteristic mass, shape and surface charge, biomolecules may also differ from each other in terms of their net *hydrophobicity*. Even soluble proteins will often possess regions of their structure that are rather hydrophobic. These regions will depend on the primary structure of the protein and may be used to separate otherwise very similar proteins or peptides. If we immobilize hydrophobic groups such as hydrocarbon chains on a stationary phase, sample components may bind to the chains as a result of these hydrophobic parts of their structure. This is the basis of *reversed phase chromatography*. In this type of partition chromatography, we designate the stationary phase by the length of hydrocarbon chain used (e.g. C-4, C-8, C-18). In general, shorter length chains are used for chromatography of proteins while longer ones are more appropriate for peptides because of their greater relative hydrophobicity. The conditions required for chromatography of proteins, in particular, frequently leads to their denaturation. Accordingly, reversed phase chromatography has many applications in analysis of proteins but is not very useful for the purification of biologically active proteins.

In this type of chromatography (Figure 2.13) sample is applied in a polar solvent such as water, methanol, acetonitrile or tetrahydrofuran or mixtures of these. Sample components attach to the hydrophobic chains by *hydrophobic* interactions. Sometimes, the sample components separate isocratically. More often, however, the column is developed with a gradient of *increasing* solvent (e.g. acetonitrile) and *decreasing* water (thus giving a gradual increase in mobile phase hydrophobicity). When the mobile phase becomes sufficiently hydrophobic, individual sample components elute from the stationary phase. *Polar* sample components elute first from this system (i.e. they show *weakest* hydrophobic interactions with hydrocarbon chains) while *nonpolar* components elute later (i.e. they show *strongest* hydrophobic interactions with hydrocarbon chains). Chromatography is frequently carried out in the presence of *ion pair reagents*. These contain a *counterion* (i.e. an ion of opposite charge to the sample component to be separated). In

C-18 reversed phase chromatography of peptides, for example, 0.1% trifluoroacetic acid (TFA) is included in the mobile phase. The two oppositely charged species (i.e. peptide plus TFA) form an *ion pair* that has sufficient hydrophobic character to be retained by the stationary phase. Inclusion of ion pair reagents strongly affects the chromatography of ionic species in the sample but has no effect on the behaviour of nonionic sample components.

To a greater extent than in most other chromatography systems, retention of sample components in reversed phase chromatography is highly sensitive to the composition of the mobile phase. Comparatively small changes in polarity, temperature, pH and the presence of ion pair reagents in the mobile phase can exert profound effects on the chromatography achieved. This allows extremely high-resolution separations of otherwise quite similar sample components (Example 2.3).

An interesting approach to reversed phase chromatography involves the use of commercially available cartridges filled with stationary phase. These are relatively cheap and disposable, may be used with a syringe thus obviating the need for a column, pump, and so on and have a large capacity. Moreover, many of these cartridges may be placed in series, thus allowing loadings to be increased indefinitely. Since the reversed phase in these cartridges is identical to that used in reversed phase HPLC (Sections 2.6.2 and 2.10.2), it is possible to reproduce closely separations that have been performed on an analytical column. A good example of this type of experiment is the quantitative purification of particular peptides from proteolytic digests for further structural or functional analysis.

A much less widely used, though related technique to reversed phase is *normal phase chromatography* (Figure 2.14). In this system, the stationary phase is polar (e.g. alkylamine bonded to silica) while the mobile phase is nonpolar (e.g. heptane) and molecules elute in inverse order compared to reversed phase chromatography in a gradient of increasing polarity (e.g. methanol). This technique is especially useful for samples with low water-solubility. Historically, this type of system (nonpolar mobile and polar stationary phases) was introduced first but was later largely superseded by a system *reversing* these phases (i.e. polar mobile and nonpolar stationary phases), hence the term 'reversed phase' chromatography.

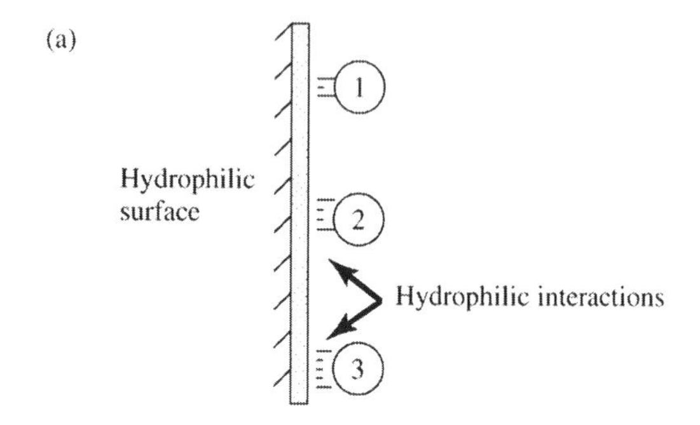

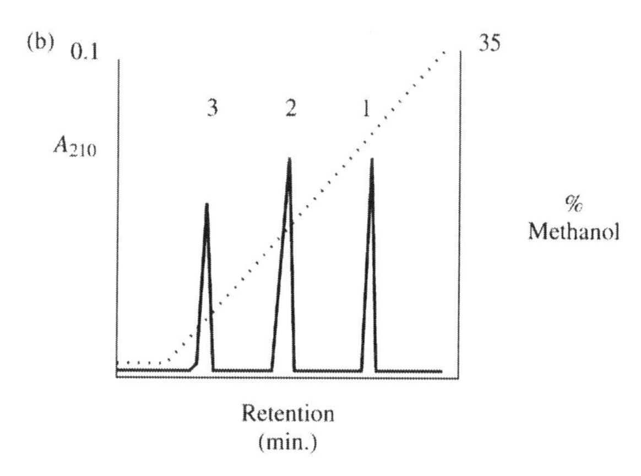

Figure 2.14. Normal phase chromatography. (a) Samples interact with a polar stationary phase in the presence of *N*-heptane and are eluted by increasingly polar mobile phase. (b) Elution profile obtained with sample components 1 to 3 identical to those in Figure 2.13. Note that these elute in inverse order compared to reversed phase chromatography percent. Methanol is shown by dashed lines.

2.4.4 Hydrophobic Interaction

Soluble proteins require a *solvation shell* of water on their surface to maintain solubility in aqueous solution. This water masks hydrophobic groups that also exist on the protein surface. In the presence of reagents that are capable of binding water from the solvation shell (e.g. $(NH_4)_2SO_4$), it is possible to disrupt solvation and expose these hydrophobic groups. Using a hydrophobic stationary phase such as provided by bonded octyl (strongly hydrophobic) or phenyl (more weakly hydrophobic) groups, it is possible to promote hydrophobic interactions between protein and such a bonded phase. This is the basis of the adsorption chromatography technique called *hydrophobic interaction chromatography*. Protein samples are applied to columns containing octyl or phenyl groups in the presence of a large concentration (approx. 2 M) of a highly polar solvent such as $(NH_4)_2SO_4$. In such solvents, hydrophobic interactions are strongly promoted between proteins and the stationary phase. Applying a *decreasing* gradient of solvent polarity, (e.g. 2-0 M $(NH_4)_2SO_4$) gradually disrupts hydrophobic interactions, thus separating proteins (with different net hydrophobicity) from each other. Alternatively, elution may be achieved by the use of a hydrophobic *displacer* molecule such as a nonionic detergent (e.g. Triton X-100), an aliphatic amine (e.g. butylamine) or an aliphatic alcohol (e.g. butanol). Since

Example 2.3 Tryptic peptide mapping of glutathione transferase isoenzymes by reversed phase HPLC

Many proteins such as globins and immunoglobulins are expressed in living systems as complex *protein families*. Within an individual family there is close similarity in primary structure and we are often interested in identifying the extent of this similarity. Some proteolytic enzymes cleave substrate proteins specifically at bonds involving particular amino acids, thus generating a population of peptides which is directly dependent on the amino acid sequence of the substrate protein. Experiments in which these peptides are separated from each other are called *peptide mapping* experiments. Even closely-related proteins within a family would be expected to produce *peptide maps* containing some common peptides and some peptides specific to each protein.

An example of a complex protein family is the glutathione transferases (GSTs), a group of detoxification enzymes which catalyze conjugation reactions between foreign (usually electrophilic) chemicals (xenobiotics) and the intracellular tripeptide glutathione. Two isoenzymes were identified which showed close similarity in substrate specificity, tissue expression and electrophoretic behaviour. One of these was a homodimer of the 3-subunit (rGST M3-3) while the other was composed solely of the 4-subunit (rGST M4-4). Notwithstanding their similarities, these isoenzymes could be separated by ion exchange chromatography. Samples of each were purified, digested with trypsin (which cleaves specifically at peptide bonds containing lysine or arginine) and peptide maps resolved by reversed phase C-18 HPLC.

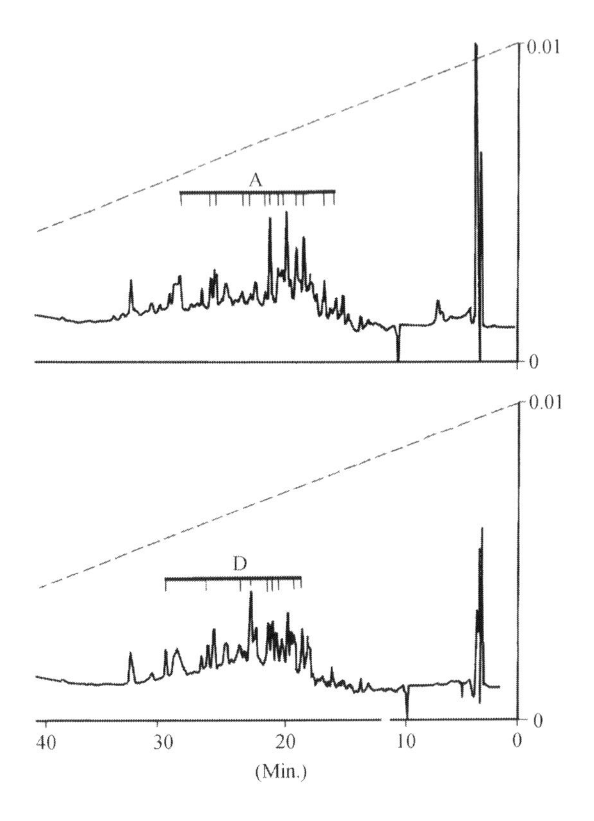

The maps for rGST M3-3 (a) and rGST M4-4 (b) from rat liver could easily be compared. The column dimensions were 0.39 × 30 cm and the particle size was 10 μm. Retention is represented from right to left (flow-rate; 1 ml/min) and detection was at 254 nm. Peptides unique to the 3-subunit are labeled 'A', those unique to the 4-subunit are labeled 'D' and unlabelled peptides were common to both proteins. The mobile phase contained 0.05% TFA and comprised a 0–60% acetonitrile gradient (diagonal solid line). Commencement of chromatography is marked with arrows. This type of comparison is also possible with other techniques such as mass spectrometry (Chapter 4) and electrophoresis (Chapter 5). Reversed phase HPLC is especially useful for analysis of tryptic peptide maps since this protease generates a large number of small peptides differing widely from each other in hydrophobicity. Other cleavage treatments (e.g. CNBr which cleaves at methionine) can generate fewer and larger fragments.

$(NH_4)_2SO_4$ precipitation is frequently an early stage in protein purification, this type of chromatography may be conveniently used subsequent to this step in a protein purification strategy. Important experimental variations in this technique include; using stationary phases of differing hydrophobicities (e.g. phenyl or octyl), different buffer concentrations and composition, varying pH (lower pH favours binding) and temperature (higher temperature favours binding).

There is considerable similarity between this technique and reversed phase chromatography. Both depend essentially on hydrophobic interactions between sample components and a bonded stationary phase and on solvent polarity. They differ from each other, however, in the precise experimental conditions used to bring these interactions about and this has consequences for their applications in biochemistry. In reversed phase chromatography, elution arises as a result of the nature of the mobile phase, that is generally quite hydrophobic. A consequence of this is that proteins are *denatured* in this type of system. This means that reversed phase chromatography depends on the *total hydrophobicity of the protein* and is especially useful in *analytical* applications since this is an inherent property of the protein determined by its primary structure. This technique is, however, not appropriate where it is desired to retain a protein's biological activity. In hydrophobic interaction chromatography, interactions depend mainly on gentle disruption of the solvation shell, that is hydrophobic interactions only occur *between groups on the surface of the protein* and the stationary phase. These are determined largely by the tertiary structure of the protein and, again, will be different for each protein. This technique, therefore, *retains native protein structure* and function. It is consequently widely used in protein purification as a *preparative* technique (Example 2.4).

2.4.5 Affinity

Recognition of different chemical shapes and structures is a fundamental and widespread property of biomolecules. For example, an enzyme is capable of recognizing its substrate and distinguishing it from other molecules that may be chemically similar. This type of biospecific recognition is known as *affinity* and advantage may be taken of this in the purification of enzymes and other biomolecules in the adsorption chromatography technique called *affinity chromatography*. The fundamental principle of this technique is immobilization of a small molecule or *affinity ligand* on a stationary phase and application of sample containing the biomolecule to be purified to this phase. Because of the highly specific nature of biological recognition phenomena, only the molecule to be purified will bind to the stationary phase and other molecules will pass freely through the chromatography system. Examples of ligand-molecule pairs applicable to this kind of experiment are given in Table 2.4. As a result of its high degree of specificity, this type of chromatography is regarded as one of the most effective means of purifying proteins, often in a single step.

Table 2.4. Examples of group-specific ligands used in affinity chromatography

Ligand	Molecule purified
Substrates	Enzymes
Inhibitors	Enzymes
Cofactors	Enzymes
Avidin	Biotin-containing enzymes
Antibodies	Antigen
Antigen	Immunoglobulins
Proteins A and G	Immunoglobulins
Hormone	Receptor/binding protein
Glutathione	GST fusion proteins
Soyabean lectins	Glycoproteins
Oligo dT	PolyA mRNA
Oligo A	PolyU RNAs
Lysine	rRNA

Unlike many of the other chromatography techniques so far described it is necessary to design a *specific stationary phase* for each protein to be purified which will bind this protein and none other. This requires considerable thought and experimentation but, in view of the high *purification factors* (the ratio of specific activity of protein to be purified at each step of the purification to that of the crude extract) achievable with this technique, it is usually worthwhile. Because of the gentle conditions employed, the technique is especially appropriate where it is desired to purify a molecule retaining full biological activity. For example, enzyme molecules that may have become inactivated during preparation of cell extract will not normally bind to a resin containing an immobilized substrate or inhibitor. Only fully active enzyme will bind, thus ensuring specific selection of fully active protein from the extract. This will maximize specific activity at this step of the purification, in turn, maximizing the purification factor achieved.

An important aspect of the technique is *immobilization of the ligand*. Immobilization techniques take advantage of functional groups on the ligand such as $-NH_2$, $-SH$, $-COOH$ and $-OH$. $-OH$ groups on the stationary phase (usually agarose) require *activation* to render them chemically reactive. The affinity ligand is then immobilized on the activated resin. The main techniques used to achieve activation are shown in Figure 2.15 while methods for immobilization are shown in Figure 2.16. Epoxy- and CNBr-activated resins are commercially available.

Example 2.4 Hydrophobic interaction chromatography of snake venom Phospholipase A_2

Phospholipases are responsible for a wide range of pharmacological effects including haemolysis of red blood cells, hypertension and prevention of blood coagulation. They are important components in snake venom and may contribute to hemorrhage in bitten patients. In order to study these enzymes in detail, for example to screen possible antidotes or other treatments, it is desirable to obtain them in purified form.

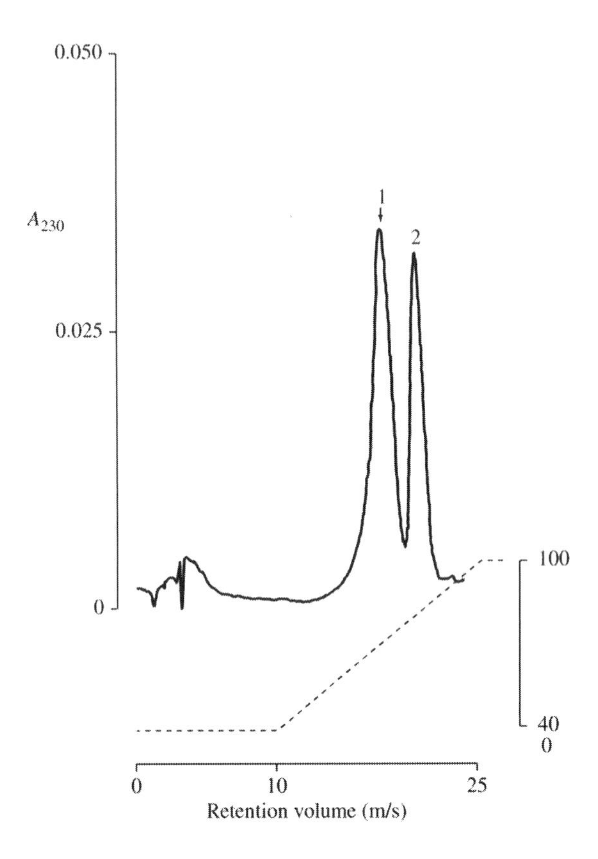

Venom obtained from the snake *Pseudechis papuanus* was fractionated into several protein peaks by Mono-S cation exchange FPLC chromatography. One of these peaks displayed phospholipase A2 activity and this was further fractionated into two peaks by hydrophobic interaction FPLC chromatography on phenyl-superose. The column had a volume of one ml and was equilibrated at pH 7.2 in 2 M $(NH_4)_2SO_4$. Sample was applied to the column in 2 M $(NH_4)_2SO_4$. Phenyl-superose was developed with a gradient of 2–0 M $(NH_4)_2SO_4$ as shown by dashed line (buffer B was 20 mM Tris-Cl, pH 7.2 containing no $(NH_4)_2SO_4$). Peak 1 contained phospholipase A2 activity and was pure by the criterion of electrophoresis. Reproduced from Laing *et al.* (1995) Characterization of a purified phospholipase A_2 from the venom of the papuan black snake (*Pseudechis papuanus*). *Biochimica et Biophysica Acta,* **1250**, 137–143, by permission of Elsevier.

Sometimes, it is necessary to introduce some distance between the ligand to be immobilized and the stationary phase particle itself. This is to avoid *steric hindrance* by the solid phase particle. A good example of this would be where a protein to be purified cannot bind to a ligand attached directly to the stationary phase because of the geometry of its ligand binding site. Some commercially available stationary phases are provided in a *derivatized* form (e.g. AH- and CH-agarose; *see* Figure 2.16). In these, a *spacer arm* 6-carbon atoms long is provided by hexane. At the end of the spacer is either an $-NH_2$ (AH-agarose) or –COOH (CH-agarose) group. Ligand may be attached at these groups by formation

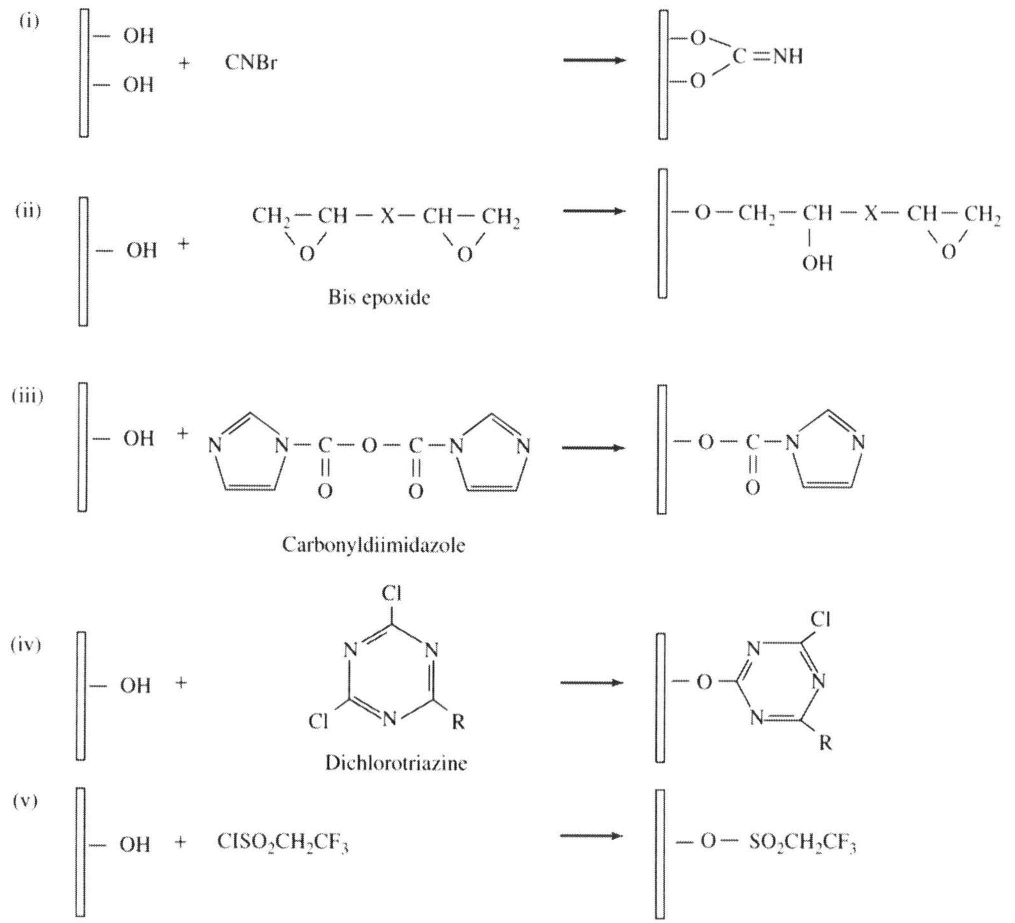

Figure 2.15. Activation of agarose for affinity chromatography. Important methods for activation of agarose are; (i) CNBr, (ii) epoxy, (iii) carbonyldiimidazole, (iv) dichlorotriazine, (v) trecyl.

of an amide linkage with ligand –COOH or $–NH_2$ groups, respectively, in the presence of dicyclohexylcarbodiimide. By use of diamines of varying lengths, it is also possible in the laboratory routinely to create a range of spacer arms of differing lengths using resins activated as shown in Figure 2.15 and to couple –COOH groups of ligands at one of their terminal $–NH_2$ groups with dicyclohexylcarbodiimide (Figure 2.16). For immobilized macromolecules (e.g. antibodies, antigens) use of spacer arms is usually not necessary.

Since contaminants may sometimes bind to the stationary phase in a non specific and low affinity manner, it is best to wash the column with a *wash buffer* containing 100–200 mM NaCl after the sample has passed through the column. Elution of specifically bound molecules can sometimes present difficulties due to too great an affinity for the immobilized ligand. One approach to eluting such molecules is to change the pH or ion strength by applying a stationary phase with a different buffer composition. For very strongly bound samples, the use of *chaotropic* reagents such as urea may have to be considered. However, the use of harsh conditions may lead to loss of biological activity of the protein to be purified. The likely strength of binding and specificity of the molecule to be purified for the immobilized ligand are important criteria in the design of the stationary phase to be used in this technique. A gentler approach is to elute the bound molecule with an excess of the affinity ligand in the mobile phase. Soluble and immobilized versions of the ligand *compete* with each other for the binding site on the protein, thus gently eluting it from the stationary phase. A variation of this approach which is useful in situations where closely related proteins (e.g. isoenzymes) have bound to a single immobilized ligand is to create an *affinity gradient*,

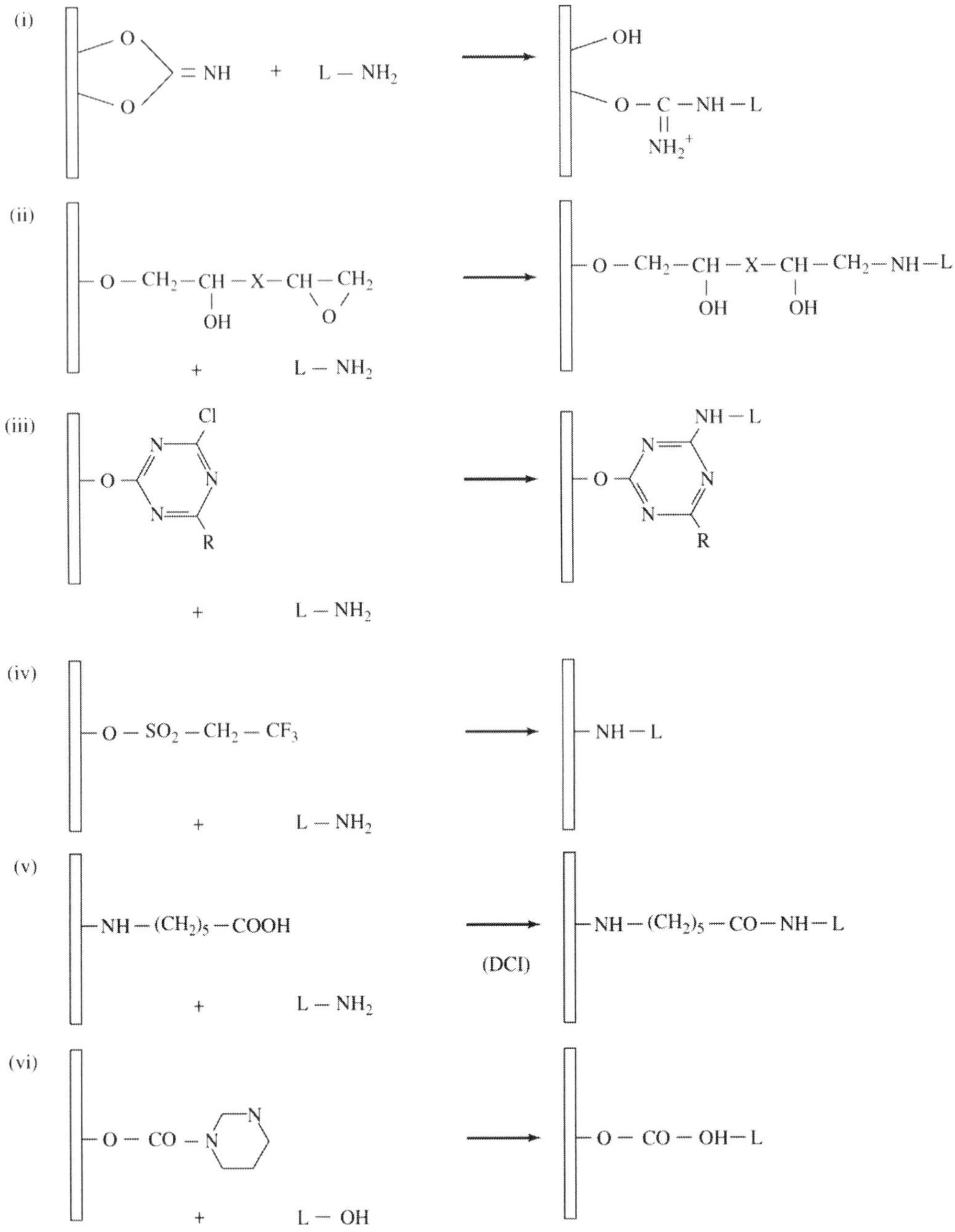

(i)

O, O =NH + L — NH_2 ⟶ OH, O — C(=NH_2^+) — NH — L

(ii)

O — CH_2 — CH(OH) — X — CH — CH_2 (O) + L — NH_2 ⟶ O — CH_2 — CH(OH) — X — CH(OH) — CH_2 — NH — L

(iii)

O — triazine(Cl, R) + L — NH_2 ⟶ O — triazine(NH — L, R)

(iv)

O — SO_2 — CH_2 — CF_3 + L — NH_2 ⟶ NH — L

(v)

NH — $(CH_2)_5$ — COOH + L — NH_2 ⟶ (DCI) NH — $(CH_2)_5$ — CO — NH — L

(vi)

O — CO — N(imidazole) + L — OH ⟶ O — CO — OH — L

Figure 2.16. Coupling of affinity ligands. Attachment of $-NH_2$ groups to(i) CNBr-activated resin, (ii) epoxy-activated resin, (iii) dichlorotriazine-activated resin, (iv) tresyl-activated resin, (v) 6-aminohexanoic-acid-agarose (CH-agarose). Attachment of –OH groups to (vi) carbonyldiimidazole-activated resin. Attachment of –SH groups to (vii) thiopropyl-agarose. Attachment of –COOH groups to (viii) 1,6-diaminohexane-agarose (AH-agarose). Couplings (v) and (viii) are carried out with dicyclohexylcarbodiimide (DCI) which promotes amide bond formation between –COOH and $-NH_2$ groups. (*Continued*)

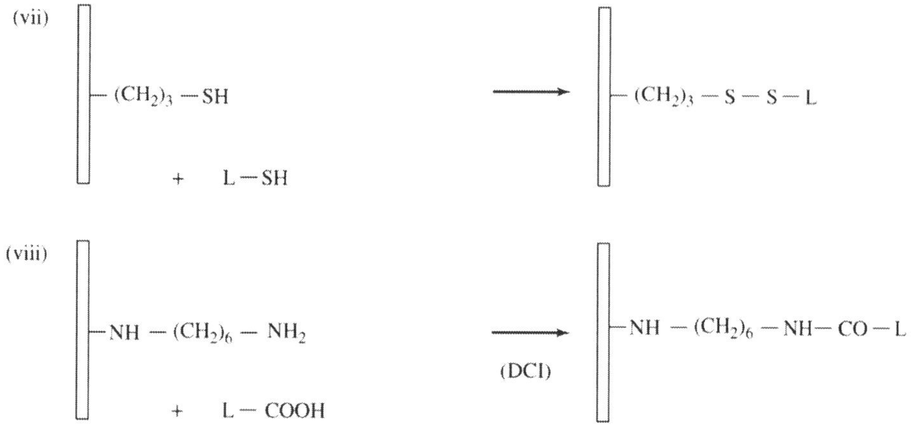

(vii) $-(CH_2)_3-SH$ + $L-SH$ ⟶ $-(CH_2)_3-S-S-L$

(viii) $-NH-(CH_2)_6-NH_2$ + $L-COOH$ ⟶ (DCI) $-NH-(CH_2)_6-NH-CO-L$

Figure 2.16. (*Continued*)

where the concentration of competing ligand is gradually increased in the mobile phase. Since individual isoenzymes may differ from each other in affinity for the ligand (e.g. isoenzymes may have differing Km's for their substrates or Ki's for their inhibitors), they may be competed off the stationary phase at different points in the gradient.

A popular application of affinity chromatography is in the recovery of proteins expressed in an *E. coli* expression system as *fusion proteins* with *glutathione transferases* (GSTs; Example 2.5). The GST component of the fusion binds to a glutathione-agarose column and the fusion protein may thus be purified from other *E. coli* proteins. Subsequently, it is collected by elution with the affinity ligand, glutathione. The expressed protein may be released by proteolytic cleavage with thrombin, and the GST is removed by a second passage through the affinity column.

2.4.6 Immobilized Metal Affinity Chromatography

A variation on conventional affinity chromatography and an alternative approach to the use of GST fusion expression systems is *immobilized metal affinity chromatography* (*IMAC*; also sometimes referred to as *metal chelate chromatography*). In this technique, metals are immobilized on a stationary phase and metal-binding regions of proteins bind to this ligand. Metals such as Cu, Zn, Ca, Ni and Fe may be conveniently immobilized by chelation to a group such as that provided by *iminodiacetic acid* (IDA) bonded to a stationary phase such as agarose or silica. A nitrogen and two oxygens in IDA coordinate with the metal leaving three sites available for binding to proteins (Figure 2.17). Amino acid side chains most likely to coordinate to metals are those of histidine, cysteine and tryptophan. A limitation of this approach is the fact that the metal may be gradually leached from the stationary phase due to the weakness of interaction with IDA. Other groups are now commercially available which coordinate at *four* sites on the metal thus facilitating more stable binding to the stationary phase. Metals may be regarded as *group-specific ligands*, that is a variety of proteins can bind to a single metal. Specificity is achieved by varying the elution procedure, especially the mobile phase characteristics such as pH, and in the choice of gradient. Proteins bind best to IMAC stationary phases at neutral to alkaline pH values and in the presence of 0.5–1 M NaCl (which avoids nonspecific electrostatic interactions). If desired, a given metal may be stripped from the stationary phase by treatment with EDTA and replaced with a different metal. Elution is usually achieved by either a decreasing pH gradient or by displacement with an electron donor such as imidazole. In selecting mobile phase buffer components for IMAC, molecules containing primary or secondary amines should be avoided, especially around neutrality.

An application of IMAC to purification of proteins expressed in bacteria (Example 2.6) is the use of the *His-tag system*, that is the creation of a fusion protein between the protein to be purified and a chain of six histidine residues (*His-tag*) that provides a binding site for an IMAC stationary phase with four metal coordination sites (*see* above). Combination of the use of six histidines and a more stably bound metal gives 1000-fold greater specificity of binding than that observed with IDA-agarose. As with the GST fusion protein, the His-tag may be removed by protease treatment at a cleavage site included in design of the fusion (e.g. enterokinase at the N-terminus or carboxypeptidase at the C-terminus with an arginine included to terminate digestion). This may be followed by a second passage through the IMAC stationary phase to remove the His-tag. Often, the tag does not greatly affect the properties of the protein to which it is fused, so it is not necessary to remove it.

Example 2.5 GST fusion protein expression system; an application of affinity chromatography

To purify a protein of interest, it is usual to identify a tissue or cell-type which is rich in this protein and to use this as the starting material for a purification protocol. However, some proteins are normally expressed in cells at very low concentration or they may be transiently expressed for short periods of the cell cycle. It may be difficult to purify enough of such a protein for structural or functional analysis using the natural site of expression as a starting point. An alternative approach to this problem is to clone the gene for the protein of interest and to express the protein in a suitable expression system. This experiment also lends itself to the expression of *derivatives* of the protein such as mutants and *chimeras* (fusion proteins). A major drawback to this approach is the possibility that the cells in which the protein is expressed may not be capable of correct post-translational processing. A further problem is the fact that the protein still requires to be laboriously purified by column chromatography methods.

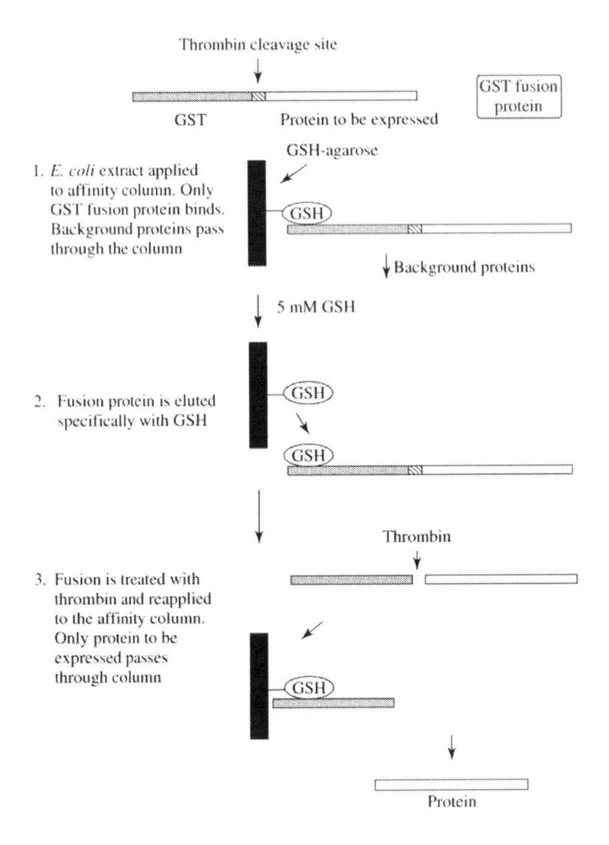

Purification of a cloned protein from an expression system may be facilitated by constructing a fusion between the gene of interest and the gene for a protein known to bind to a specific affinity ligand. A popular choice for this experiment is GST since this protein binds specifically and quantitatively to glutathione (GSH)-agarose.

Protein is expressed in *E. coli* as a fusion with GST in an appropriate expression vector. In the illustration, this protein is represented in white, the GST moiety in grey and the thrombin cleavage site is hatched.

(1) This fusion selectively binds to the affinity resin GSH-agarose facilitating the removal of background *E. coli* proteins which are simply the proteins normally expressed by the *E. coli* cell in the absence of the fusion. (2) The fusion protein is then eluted by 5 mM GSH which competes with the GSH affinity ligand for the GSH binding site of the GST moiety. (3) The purified fusion protein is collected and treated with thrombin which releases the GST moiety from the fusion. A second passage through the affinity column removes the GST, resulting in pure protein. In principle, this approach can be taken to improve the expression/purification of any protein capable of being cloned.

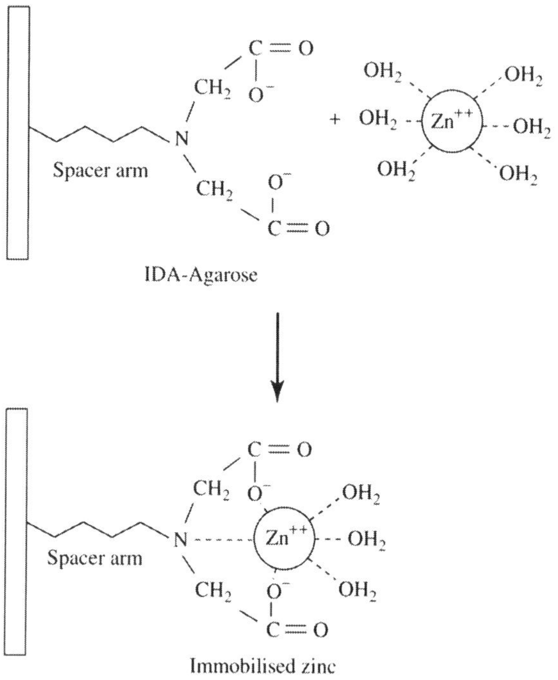

Spacer arm

IDA-Agarose

Spacer arm

Immobilised zinc

Figure 2.17. Immobilization of metal on IDA-agarose.

2.4.7 Hydroxyapatite

Hydroxyapatite is a crystalline form of calcium phosphate with the chemical formula $Ca_{10}(PO_4)_6(OH)_2$. Usually, this material is equilibrated with dilute (e.g. 0.1 mM) phosphate buffers and significant numbers of phosphate ions become immobilized on the resin as shown in Figure 2.18. It is possible for positively charged proteins to bind to the phosphate ions and these may easily be eluted by altering the phosphate, salt or divalent metal (e.g. Ca^{2+} or Mg^{2+}) concentrations of the eluting buffer. The latter metals neutralize the negative charges of the phosphate ions. Negatively charged proteins may also bind to hydroxyapatite by interaction of carboxylate side chains with Ca^{2+} in the resin. However, these proteins are also simultaneously electrostatically repulsed by bound phosphate ions. Elution of these latter proteins is by application of a gradient of an ion that binds more strongly to Ca^{2+} such as phosphate or fluoride. In practice, hydroxyapatite is first equilibrated with dilute phosphate buffer and then washed sequentially with 5 mM $MgCl_2$ (possibly as a 0–5 mM gradient) to remove basic proteins, 1 M NaCl (or a 0–1 M gradient) to remove neutral proteins and lastly a gradient of 0.1–0.3 mM phosphate to elute acidic proteins.

Because of its difficult handling and flow properties and limited capacity, hydroxyapatite is not as popular as some of the other modes of chromatography described in this section. However, since it adsorbs and desorbs proteins in a unique manner, the technique often allows purification of proteins that may have proved difficult to separate by the other chromatography modes and provides a valuable complement to them.

2.5 OPEN COLUMN CHROMATOGRAPHY

2.5.1 Equipment Used

The modes of chromatography described in Section 2.4 may be performed in a chromatography system operating at atmospheric pressure, using stationary phase particles with relatively large (e.g. 50–150 μM) diameters. Such a system is called *open column chromatography* or *low performance liquid chromatography*. This gives relatively low resolution and is slow compared to the more high-resolution systems described in the following sections (Sections 2.6–2.9). However, there are also a number of compelling advantages to the use of open column chromatography. Stationary phases and other equipment used are cheap making this approach especially convenient for large-scale preparative chromatography. This lends itself to steps used *early* in protein purification procedures (Section 2.10.1), where a priority is the decrease of total protein in the extract to facilitate subsequent application to high-resolution systems.

Columns and tubing used at low pressure may be constructed out of cheap and readily available materials. Generally, plastic reservoirs and tubing are used for mobile phase. Stationary phases are retained in glass or plastic columns either by a high porosity glass sinter or else by placing plastic wool at the bottom of the column.

A degree of standardization of flow rate is possible with a peristaltic pump although flow rates achieved will be limited by the hydrodynamic properties of the stationary phase. Since this is generally composed of polysaccharide-based gels that are quite *soft*, they are limited in the hydrostatic pressures which they can accommodate without compression. Most systems flow adequately under gravity although the flow rate may vary somewhat during chromatography as mobile phase level in the reservoir drops (Figure 2.19). When using gravity flow, a *safety loop* is usually included in the system design. The lowest point of the loop carrying flow to the column is deliberately placed *lower* than the outflow from the column. If the mobile phase reservoir becomes unexpectedly exhausted during the experiment, flow will cease as soon as mobile phase reaches the lowest point of the loop. This protects the gel bed from accidentally drying out.

An on-line ultraviolet detector gives a continuous trace of protein elution and use of a fraction collector allows peak collection. It is also possible to measure ultraviolet

Example 2.6 IMAC chromatography of a metal-binding protein using His-TAG

An alternative, though similar approach to that described in Example 2.5 is the use of immobilized metal affinity chromatography (IMAC) which takes advantage of metals immobilized on an affinity matrix. A fusion of the gene coding for the protein to be purified (white) with a sequence of six histidine residues (grey) called a His-tag is constructed which contains an enterokinase cleavage site (hatched) as shown in the illustration. This fusion is expressed in an *E. coli* expression system.

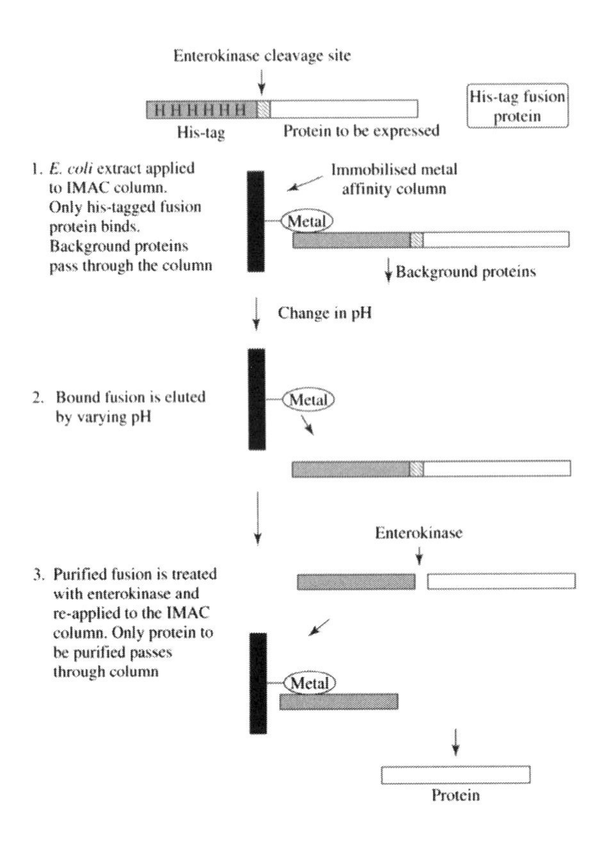

(1) When proteins from such an expression system are applied to the IMAC column, the *background* (i.e. *E. coli*) proteins simply pass straight through the column. The N-terminal $(histidine)_6$ tag recognizes metal immobilized on the IMAC stationary phase and binds allowing convenient, one-step removal of untagged contaminants. (2) The purified fusion is collected by varying the pH of the elution buffer (usually making it more acidic). Note that stability to a range of pH is therefore essential for proteins to be purified by this method. (3) The His-tag may be removed, if necessary, by treatment with enterokinase followed by a second passage through the IMAC column. Frequently, the presence of a His-tag has no structural or functional effect on the protein so it is often left in place.

The protease cleavage sites used in this type of experiment are deliberately chosen because they occur rarely in proteins thus avoiding inadvertent proteolytic cleavage of the protein to be purified.

absorbance after separation on a bench spectrophotometer and to plot the elution profile manually. Since this chromatography can be quite slow (hours to days), it is usually necessary to carry it out either in a cold-room or a refrigerated cabinet (4 °C).

Depending on the mode of chromatography to be employed, some attention should be given to column design. For gel filtration chromatography, long, narrow columns are necessary to maximize resolution whereas for most types of adsorption chromatography, shorter and wider columns may

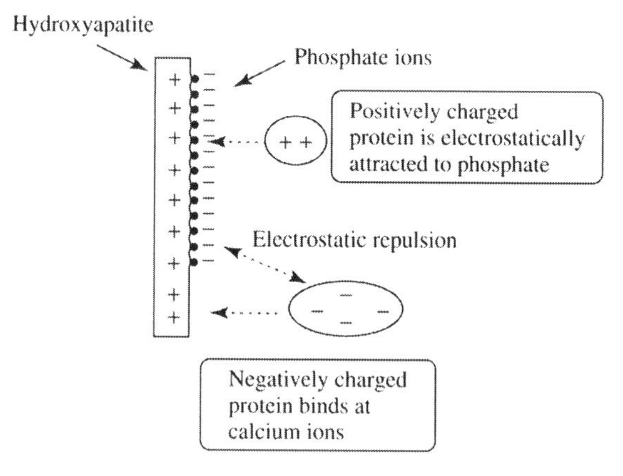

Figure 2.18. Hydroxyapatite chromatography. Ca^{2+} ions of hydroxyapatite attract phosphate ions from buffer (small circles). These, in turn, electrostatically attract positively-charged proteins. Negatively-charged proteins must compete with phosphate to bind at Ca^{2+}, overcoming electrostatic repulsion. Development of the column is as described in text.

be used. There is, however, a limit to the length of column it is possible to use in gel filtration chromatography since, as mentioned above, soft gels may be subject to compression if exposed to large hydrostatic pressures which will in turn lead to decreased flow rates.

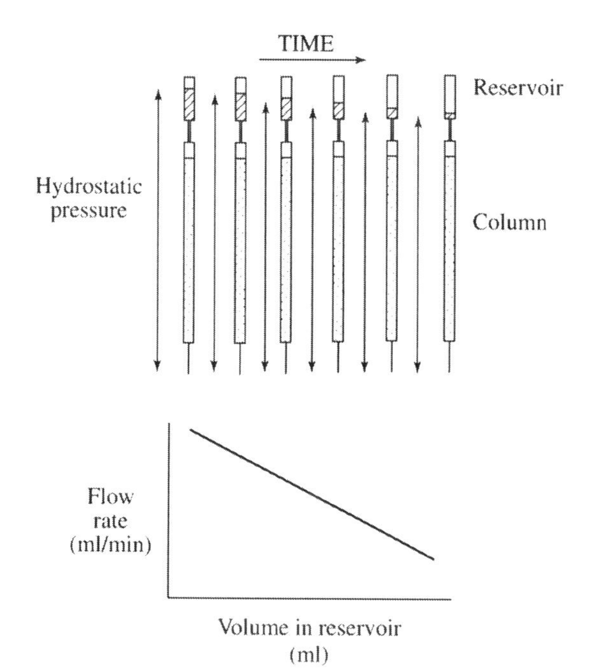

Figure 2.19. Decrease in flow-rate as a result of drop in mobile phase level in reservoir. Hydrostatic pressure across the column during chromatography is represented by arrows.

2.5.2 Industrial Scale Chromatography of Proteins

The cheapness and large capacity of open column chromatography systems make them especially appropriate to industrial scale protein purification. Modern production of proteins for pharmaceutical use is extensively regulated by agencies such as the US *Food and Drug Administration* (FDA). Protein drugs are amongst the most expensive to produce industrially and there has been a remarkable increase in FDA approvals for protein drugs in recent years (21 products approved in 2005 alone and 92 in the period 2000–2005). Other important applications of proteins include diagnostics, use as laboratory reagents and as additives in food and brewing. These latter proteins may have significantly *less* demanding purity requirements which make them relatively cheaper to produce. Expression of proteins of interest in a wide range of *heterologous* systems (e.g. expression of human genes in microbial expression systems) is also now widespread.

As a result of these developments, production of proteins has become of major importance to the pharmaceutical and chemical industries in recent years. This has led to the development of a new area of activity called *downstream processing of proteins*. Most proteins currently produced commercially, arise either from microbial fermentations or from animal cell culture. Major limitations in this activity are the handling of large volumes, stability of the protein to pH and heat, incorrect folding and other post-translational processing of the protein (as a result of expression in a heterologous system). Inert, stainless steel containers are used instead of glass vessels for the storage and handling of the much larger volumes of extracts and buffers required. These may be sterilized and are robust enough for process work. Extracts and buffers used in purification are normally *pumped* through stainless steel or plastic piping rather than manually transferred as in the laboratory scale purification. These pipes also are amenable to sterilization and cleaning.

In general, the same overall techniques described in Section 2.4 are used but the *scale* of the column is much larger. Commercially available columns are constructed mainly from reinforced plastic or steel and can accommodate many liters of stationary phase. Usually, industrial scale columns retain similar length to those used in the laboratory scale but the column diameter is increased greatly. By *stacking* columns one on top of the other, it is possible to put in place hundreds of liters of stationary phase without extensive bed compression. An alternative column design is offered by *radial flow chromatography*, where sample flows across rather than down the length of the column. This design facilitates higher flow rates than conventional columns of the same volume. Sample enters the column from the periphery, flows across the stationary phase and then flows into a channel running through the centre of the column.

Automated systems for control of downstream processing such as BioPilot are now commercially available. In developing an industrial scale separation, it is usual to begin with the laboratory scale and then to deal with effects of scale-up with a *pilot scale purification* (an intermediate scale between laboratory scale and industrial scale). Only once these stages have been completed does industrial scale purification become a realistic proposition. As each stage of purification is developed and scale increases, the financial cost of purifying each unit of protein decreases significantly due to economies of scale. A major determinant in this cost is the *concentration of protein* in the original extract. Considerable thought should therefore be given to choice of expression system used for industrial scale purification.

2.6 HIGH PERFORMANCE LIQUID CHROMATOGRAPHY (HPLC)

2.6.1 Equipment Used

A major contributor to the overall plate height equation is *stagnant mobile phase mass transfer*, H_3 (*see* Section 2.2.4 and Equation (2.8)). This represents diffusion of sample components into and out of the *stagnant mobile phase* of the stationary phase. This term is directly proportional to the square of particle diameter (d_p^2). It is to be expected, therefore, that smaller values of d_p would lead to large reductions in H and much improved chromatographic performance. One physical reason for this is that the *surface area* available for adsorption/desorption in such particles is proportionately much greater than in larger particles. This is why the theoretical plate number, N, is directly related to the surface area of the stationary phase particle. Adsorption/desorption kinetics and consequent chromatographic performance are, therefore, expected to be significantly improved in stationary phases with small values of d_p.

Smaller diameter particles also result in greatly *reduced* flow rates, however, since they give greater resistance to mobile phase flow. This results in *back pressure* on the stationary phase which, in turn, could cause serious band broadening and long retention times due to distortion of the stationary phase bed. Back pressure increases in inverse proportion to d_p^2. In open column chromatography, the stationary phase particles used (Section 2.5) are made up mainly of polysaccharide gels which are mechanically *weak*. Therefore, even if such particles were produced with small diameters, they would not be sufficiently strong to withstand the high pressures required to achieve high resolution chromatography. Silica-based particles (diameter 5–10 μm), by contrast, allow the use of high flow rates achieved by actively pumping mobile phase through the stationary phase *under high pressure* (up to 55 MPa). This is called *high performance liquid chromatography* (HPLC). A combination of high pressure with small particle diameter achieves high resolution in this type of chromatography format. The high pressure required by the system dominates design and construction of equipment used (Figure 2.20). In practice, below d_p values of 3–5 μm the pressure required for flow in HPLC systems becomes impractical.

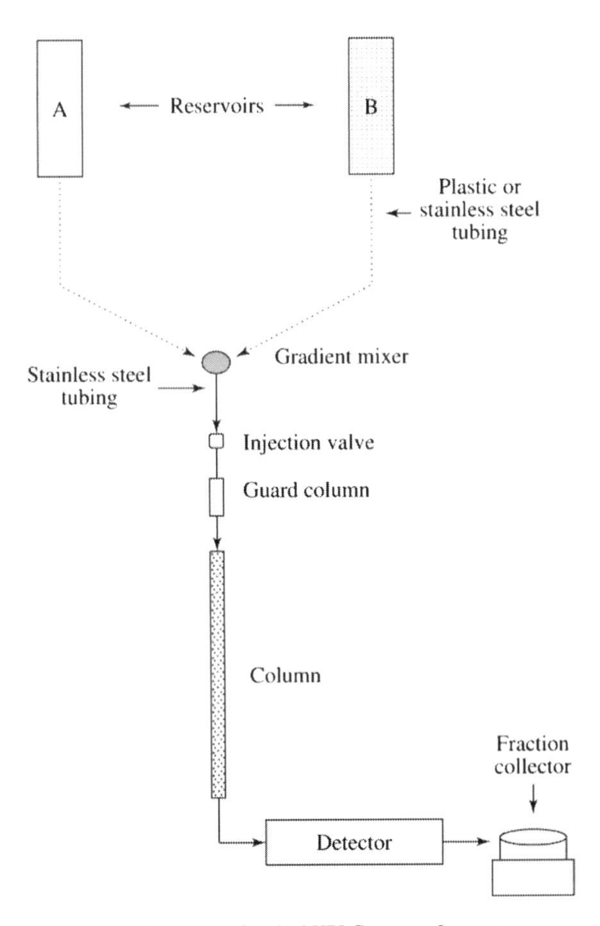

Figure 2.20. Overview of typical HPLC system. Low-pressure region of the system where tubing may be plastic is denoted by dashed line. Solid lines denote stainless steel tubing required due to high pressure.

Sample is injected through a neoprene/Teflon *septum* from a syringe which may be operated *manually* or *automatically* from an *autosampler*. The latter system is especially appropriate for analytical applications where multiple, repetitive analyses may be required.

In order to achieve a stable baseline in elution profiles, it is important that flow of mobile phase be very stable (i.e. *pulse*-free). In HPLC, either *constant volume* (also known as *constant displacement*) or *constant pressure* pumps are used to achieve pulse-free flow of mobile phase. The latter maintain a constant pressure regardless of changes in stationary phase resistance to flow during chromatography. If this resistance increases, however, then the flow rate decreases automatically to maintain constant pressure. Constant volume pumps are more popular in HPLC, because they maintain a constant flow rate through the stationary phase during chromatography, increasing pressure if necessary, to overcome increased stationary phase resistance. The most common type of constant volume pump used is the *reciprocating pump* which uses a piston to deliver a fixed volume of solvent onto the column in repeated cycles of filling and emptying. These pumps can introduce pulses into the flow but *pulse dampeners* are built into them to minimize this.

In most HPLC chromatography, the mobile phase consists of a *mixture* of two or more components (A and B; e.g. acetonitrile and water) applied to the column to achieve rapid separation. The HPLC instrument therefore contains a *mixer* for efficient mixing of A and B. This may be of two general types. *Dynamic mixers* stir A and B together but have the disadvantage of a large dead volume. *Static mixers*, by contrast, mix A and B by an eddy diffusion process and normally have small dead volumes.

Mobile phase is pumped around the system through stainless steel tubing which is required as a result of the high pressure used. The *column* in which stationary phase is packed is also of stainless steel construction with internal diameters of 2–5 mm for analytical separations and lengths of 5 to 30 cm. However, industrial-scale *preparative* HPLC columns may have dimensions of up to 15 × 100 cm. Porous plugs of Teflon or stainless steel retain the stationary phase in the column.

A variety of forms of detection are used in HPLC. Variable wavelength UV detection is especially popular in analysis of biochemical samples. For proteins, detection may be carried out at 280 nm (which is specific for aromatic amino acids) or 220 nm (specific for peptide bonds) while 260 nm is useful for detection of nucleic acids. Other types of detectors are fluorescence and refractive index detectors. An example of a specialized approach to detection is *postcolumn derivatization* which may be used, for example, in amino acid analysis. Amino acids are separated chromatographically and then are derivatized with *o*-phthalaldehyde (OPA) *postcolumn*. The OPA conjugates formed are strongly fluorescent and may be detected with high sensitivity.

The dimensions of HPLC columns can vary widely. A traditional column with an inner diameter of 2–5 mm might have a flow-rate range of approximately 1–10 ml/min. However, increasingly finer columns are being used in high-resolution separations with corresponding reduction in achievable flow-rates: *Capillary* columns have inner diameters in the range 100 μm–1 mm (flow-rate range; 0.4–200 μl/min); *Nanobore* columns have inner diameters in the range 25–100 μM (flow rate range; 25–4000 nl min). Nanobore HPLC achieves very stable submicroliter flow-rates which has facilitated interfacing with other analytical methods such as mass spectrometry. These are especially important and useful in situations where only minute quantities of protein are available for analysis such as in proteomics (Chapter 10).

2.6.2 Stationary Phases in HPLC

The particles of which the stationary phase is composed have a rigid, solid structure rather than that of a soft gel such as is used in open column chromatography. Three main types of particle are used (Figure 2.21). *Microporous* particles (5–10 μM) contain microscopic pores which run right through the particle, giving a large available surface area, and are often composed of silica or aluminium. *Pellicular* (also known as *superficially porous*) particles (40 μm) are also porous but these are coated on an inert core composed of glass or some other hard material. Porous stationary phases especially useful for proteins and peptides may be composed of polymethacrylates or poly (styrene-divinylbenzene). These display excellent pH-stability

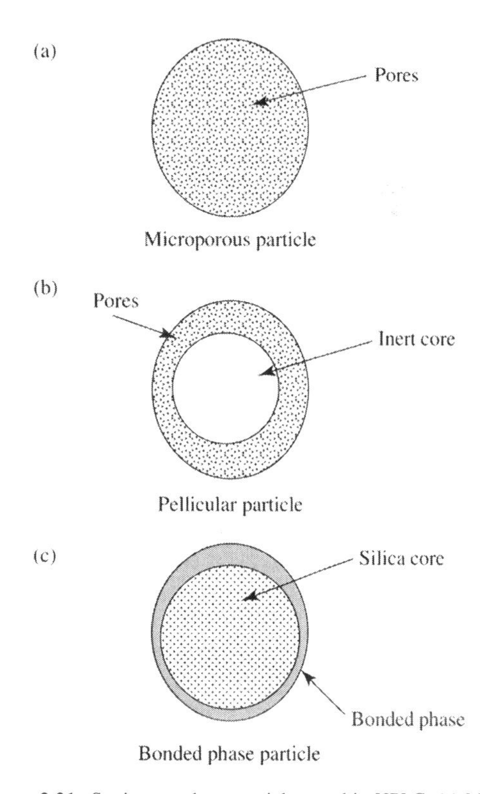

Figure 2.21. Stationary phase particles used in HPLC: (a) Microporous, (b) Pellicular and (c) Bonded phases.

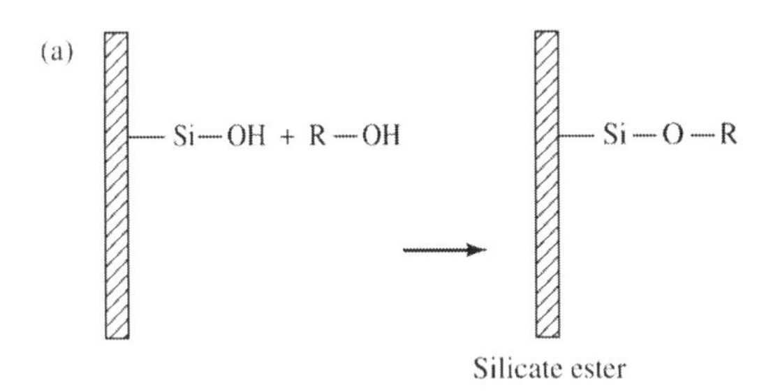

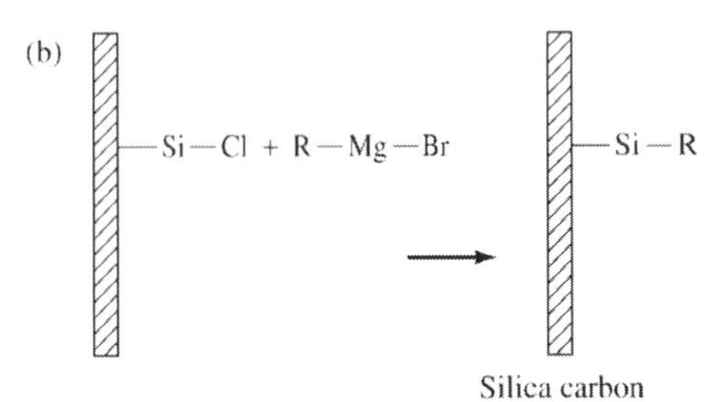

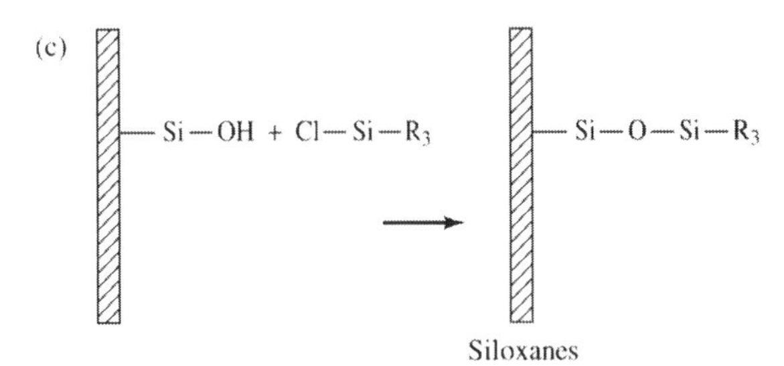

Figure 2.22. HPLC bonded phases. (a) Silicate esters, (b) Silica-carbon and (c) Siloxanes.

properties (from pH 2–12) and are mechanically very stable. The third type of particles are *bonded phases* in which the stationary phase is *chemically bonded* to an inert support such as silica (Figure 2.22). A disadvantage of silica-based stationary phases is that they can sometimes give cation exchange effects and that they have limited stability above pH 9 and below pH 2. A selection of some of the more popular HPLC stationary phases is given in Table 2.5.

Stationary phase particles are packed with a *high pressure slurrying procedure* which involves forming a *slurry* of stationary phase in a solvent of equal density and pumping it into the column under high pressure. This produces a dense, continuous bed of stationary phase which, ideally, is free of cracks and other imperfections.

2.6.3 Liquid Phases in HPLC

It is essential that the mobile phase used in HPLC be of high purity and be chemically unreactive. Under elevated pressure, air or gas in solvents can form bubbles which could damage stationary phase packing and interfere with chromatographic analysis. It is usually necessary to *degas* the solvents before chromatography, either by *purging* (also sometimes called *sparging*) with gas or by exposing the solvent to a vacuum. To avoid *fouling* or blocking of columns by particulate matter suspended in sample and solvents used, it is also usual to filter these through a 5 μm filter and to introduce a *guard column* into the system itself *before* the chromatography column (Figure 2.20).

A variety of mobile phase solvents may be used in HPLC. Choice of these is often a matter of experiment but they can be organized into a list reflecting their ability to displace adsorbed sample components called an *elutropic series* (Table 2.6). The value of eluent strength increases, generally, with increase in polarity. A further important factor in choosing a mobile phase solvent is that it should not interfere with detection of sample components. Some solvents shown in Table 2.6 display strong UV absorbance and allowance should be made for this if using this type of detection.

2.6.4 Two Dimensional HPLC

It is often not possible to achieve a complete separation of a biochemical extract in a single chromatogram. This is because there may be close similarity between components in terms of polarity, molecular mass, charge,

Table 2.5. Selected HPLC stationary phases

Stationary phase	Description	Application
Silica C-18[a]	Bonded phase	Reversed-phase chromatography of peptides
Silica C-8	Bonded phase	Reversed phase chromatography of proteins
Fractogel TSK HW-55[b]	Porous poly vinyl chloride	Gel filtration of proteins
CM/DEAE-silica[c]	Bonded phase	Ion exchange chromatography
Fractogel TSK HW-65-ligand[d]	Porous polyvinyl chloride with immobilised ligand	Affinity chromatography

[a]For details of chemical bonding see Figure 2.22.
[b]See also Table 2.3.
[c]For chemical structure of ion exchange groups see Table 2.2.
[d]See Figure 2.16 for details of ligand immobilisation chemistry.

Table 2.6. Elutropic series of HPLC solvents

Solvent	Eluent strength
n-Pentane	0
n-Hexane	0.01
Cyclohexane	0.04
Carbon tetrachloride	0.18
2-Chloropropane	0.29
Toluene	0.29
Ethyl ether	0.38
Chloroform	0.4
Tetrahydrofuran	0.45
Acetone	0.56
Ethyl acetate	0.58
Dimethylsulphoxide	0.62
Acetonitrile	0.65
2-Propanol	0.82
Ethanol	0.88
Methanol	0.95
Ethylene glycol	1.11

pI, and so on. However, if separation based on one physicochemical parameter (e.g. charge characteristics) is followed by separation based on a second parameter (e.g. molecular mass), extremely high resolution can be achieved. This is the concept underlying two dimensional HPLC (2-D HPLC). Such multidimensional separations have a very large *peak capacity* (total number of peaks theoretically distinguishable) as follows:

$$P_{MD} = P_1 \times P_2 \times P_3 \ldots \quad (2.14)$$

Where P_{MD}, P_1, $P_2 \ldots$ are the peak capacities of multidimensional, first dimension, second dimension, and so on separations. Common pairings of dimensions include ion exchange – reversed phase and gel filtration-capillary electrophoresis. As a fraction elutes from dimension 1, it is automatically passed to dimension 2. In this way extremely high resolution separations of complex mixtures of peptides or proteins are possible. As well as coupled columns, single columns containing two types of packing are possible. An example is an ion exchanger mixed with a reversed phase stationary phase. 2-D HPLC has potential in the field of proteomics where it complements two dimensional electrophoresis (Chapter 10).

2.7 FAST PROTEIN LIQUID CHROMATOGRAPHY

2.7.1 Equipment Used

The high resolution and impressive versatility of HPLC have made it a widespread technique in biochemistry, analytical and organic chemistry and many other areas of scientific investigation. It is especially useful in separation and analysis of *low molecular weight biomolecules* such as metabolic intermediates, building blocks of biopolymers and in the analysis of peptide mixtures. Although principally used in analytical applications, the technique may also be applied to the preparation of pure sample components, up to and including industrial scale, provided columns with suitable capacities are available. It is fair to say, however, that the technique does not usually lend itself readily to the purification of complex biopolymers such as proteins in their active form. There are a number of reasons for this. Firstly, some modes of chromatography commonly performed in a HPLC format (e.g. reversed-phase chromatography) require the use of organic solvents which will denature proteins. Secondly, although large capacity HPLC columns are available, they are very expensive and small capacity analytical columns are much more common. These latter columns place severe limitations on loadings achievable in HPLC (up to a maximum of approx. 200 μg, depending on the column).

Pharmacia Biotech (currently GE Healthcare) pioneered *Fast Protein Liquid Chromatography* (FPLC) to address the particular needs of active biopolymer purification. Although this chromatography format (Figure 2.23) was originally developed for protein purification, it has since been used for the purification of a wide variety of other biomolecules including DNA (plasmids, restriction fragments and oligonucleotides), nucleotide derivatives (e.g. NAD, ATP), carbohydrates as well as peptides and other low molecular weight biomolecules.

The system uses piston-driven, reciprocating pumps. Each contains a pair of pistons operating in opposite directions (one filling as the other empties) which avoids pulse formation. Because the pressure of the system is significantly *lower* than that of HPLC (approx. 1–2 MPa), components are constructed from glass and plastic rather than of steel as in HPLC. Sample is loaded by syringe into a *loading loop* (25–500 μl) or by a peristaltic pump into a *superloop* (10 ml). The contents of these loops are then pumped onto the column. Stationary phase is provided in *prepacked* high capacity (e.g. up to 20 mg per ml stationary phase) columns including superose and superdex (crosslinked agarose; average d_p, 13 μm) for gel filtration, affinity and IMAC, silica with bonded phases for reversed phase (d_p, 5 μm) and mono beads (a hydrophilic polyether) with bonded ion exchange and chromatofocusing groups (d_p, 10 μm). These phases are highly compatible with aqueous buffers and salts commonly used in bioseparation. Automated valves and a modular design allow convenient assembly of the various components of the system. A program controller which may be further interfaced with a personal computer and a range of dedicated software options allow a very high degree of automation which facilitates rapid and reproducible chromatography.

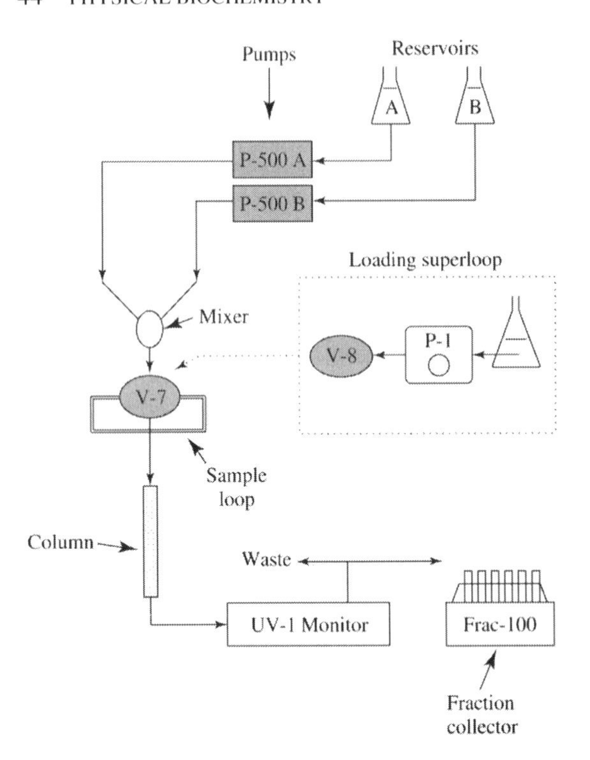

Figure 2.23. Outline of FPLC format for high resolution chromatography (adapted from Doonan, 1996). Small sample volumes (25–500 μl) may be loaded from a fixed volume sample-loop while larger volumes (several ml) may be loaded from a superloop *via* the arrangement shown.

2.7.2 Comparison with HPLC

This chromatography format has a number of *similarities* to HPLC; small particle diameter stationary phases are used (5–13 μm); it is available in the various chromatography modes described in Section 2.4; mobile phase flow takes place in a *pressurized* system; sample and buffer components require filtration before chromatography. Important *differences* with HPLC, though, arise from the fact that this chromatography format was designed *specifically with the needs of biopolymers in mind*. Mainly aqueous solvents are used and the stationary phases have a relatively large loading capacity. Moreover, the system operates at much lower pressures than HPLC which simplifies the assembly and rearrangement of components for particular purposes as leaks in the system are less common than in HPLC.

2.8 PERFUSION CHROMATOGRAPHY

2.8.1 Theory of Perfusion Chromatography

Mass transfer effects within stationary phase particles are major contributors to band broadening in liquid chromatography. This is decreased by *reduction* of particle size but, as pointed out (Section 2.6), a lower limit of 3–5 μm is imposed by practical limitations of pressure on the system. *Perfusion chromatography* provides an alternative means of addressing this problem and combines unusually high flow rates with a novel type of stationary phase particle which allows extremely rapid separation. In order to understand how this comes about, we need to review some aspects of the basic theory of chromatography described in Section 2.2.

Mass transfer effects were given as H_2 and H_3 in Equations (2.8) and (2.9) and these equations may be combined into a simplified overall plate height equation as follows:

$$H = \frac{C' \cdot d_p^2 \cdot v}{D_m} \tag{2.15}$$

where C' represents the *sum* of C_{sm} and C_m. It is clear from this that H *increases* with flow rate, thus giving increased band broadening at high flow rates. It is expected from the Van Deemter equation (Equation (2.11); Figure 2.6) that, *at very high flow rates* (in the range 1000–10 000 cm h^{-1}), the mass transfer effects term would dominate such that:

$$H \sim C \cdot v \tag{2.16}$$

This is because, at large v values, the product of C and v would also be large while the contribution of eddy diffusion is generally modest in liquid chromatography and that for longitudinal diffusion (B/v) would *decline* with increasing flow rate.

The diffusion coefficient in the mobile phase, D_m, can be divided into *diffusive* and *convective* components such that:

$$D_m = D + \frac{v_{pore} \cdot d_p}{2} \tag{2.17}$$

where D is the diffusion coefficient and v_{pore} the flow rate in the pores of the stationary phase particle. High resolution chromatography formats such as HPLC and FPLC normally use $v \sim 400$ cm h^{-1} and, in such systems, diffusive transport dominates. However, at high values of v, the convective term may become much more important such that:

$$D_m \sim \frac{v_{pore} \cdot d_p}{2} \tag{2.18}$$

There is a proportional relationship between v_{pore} and v:

$$v_{pore} = \text{const.} \cdot v \tag{2.19}$$

We can, therefore, insert Equation (2.19) into ((2.15):

$$H = \frac{2 \cdot d_p^2 \cdot v}{\text{const.}\, v \cdot d_p} \quad (2.20)$$

$$= \frac{2 \cdot d_p}{\text{const.}} \quad (2.21)$$

At very high values of v, H becomes largely independent of v and *intraparticle diffusion becomes unimportant relative to intraparticle convection* for protein samples. If it were possible in practice to achieve such high flow rates of mobile phase through stationary phase particles with small values of d_p, then it would be possible to achieve high resolution separations (low values of H) *in extremely short times* (as a consequence of high flow rates) which would be independent of flow rate.

Perfusion chromatography allows us to realize this objective in practice using *perfusion*. This involves a stationary phase particle which permits the use of very high flow rates with little back-pressure. *Perfusive particles* contain large (~1 μm) *channels* through which rapid diffusion of mobile phase is facilitated (Figure 2.24). At the flow rates used in perfusion chromatography, convective flow is some ten times greater than diffusion. These channels are also sometimes called *perfusive pores* or *through-pores*. Separations obtained with this stationary phase agree well with the theoretical considerations described above.

2.8.2 Practice of Perfusion Chromatography

Particles used in perfusion chromatography are called POROS and are composed of poly (styrene-divinylbenzene). They have small d_p, (10, 20 and 50 μm) and the stationary phase (which can contain ion exchange, reversed phase, hydrophobic interaction, IMAC and other affinity groups; Table 2.7) is applied as a thin film to this support. Two types of pores transect this particle. Very large *through pores* (6000–8000 Å) are responsible for rapid flow through the particle. They contain a very small amount of the overall surface area of the particle, however. *Diffusive pores* are much smaller (500–1000 Å) and connect the through-pores. In these latter pores, which are very short (<1 μm), diffusion is more important than convective flow and their surface represents the bulk of the particle surface-area. This porous network with its large surface area contributes to the very high capacity of POROS particles.

Perfusion chromatography may be performed in columns packed with POROS particles in open column, HPLC or FPLC chromatography systems. Separation of proteins in as little as 15–60 s is possible because of the high flow rates achievable with this stationary phase. An interesting adaptation of the procedure is *immunodetection*, where the steps of an ELISA assay are performed extremely rapidly in a flow-through column format using immobilized antibodies to a molecule of interest.

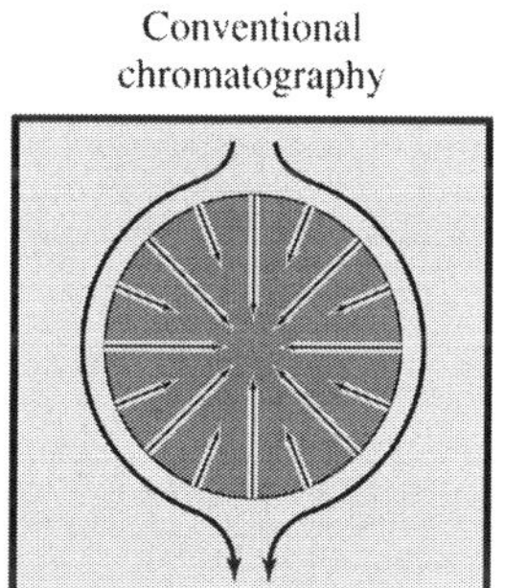

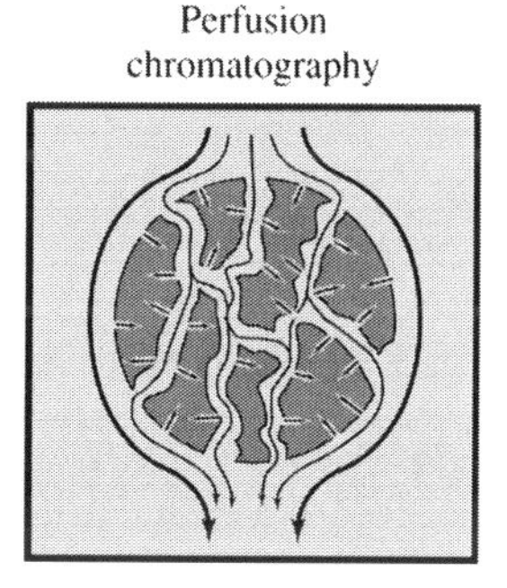

Figure 2.24. A Perfusion chromatography particle. Mobile phase flow is denoted by arrows in a conventional and perfusion chromatography particle. Note improvement in mobile phase flow due to perfusive channels.

2.9 MEMBRANE-BASED CHROMATOGRAPHY SYSTEMS

2.9.1 Theoretical Basis

So far, we have focused on chromatography systems which use stationary phases made up of *particles* with diameter, d_p. This parameter is crucial to achieving high resolution separations but small d_p values lead to a requirement for high pressure chromatography. Moreover, a lower limit of 3–5 μm is placed on d_p as a result of impracticably high back pressures below this value. An alternative approach is the use of *membrane-based* rather than *particle-based* stationary phases. In membrane-based chromatography, the stationary phase is composed of thin, porous membranes which are layered on top of each other. These may be coated with ion exchange, affinity or other groups, providing a range of chromatography modes. The membranes are contained in a cartridge which may be used in place of a column in open column, FPLC and HPLC formats.

When a sample is applied to such a system, it is found that diffusion does not impose a limit on mass transfer to the stationary phase. This process seems to occur largely by *convection* resulting in much faster adsorption kinetics (e.g. 200–300-times faster than in agarose-based systems). Moreover, by layering many membranes on top of each other (to a thickness of ~10 mm) and by increasing membrane diameter, it is possible to achieve high capacities in the cartridge (15–30 mg protein). It is also possible to use very high flow rates (up to 10 ml/min in routine use) without generating high back pressures. The reason for these unusual

Table 2.7. Principal types of Perfusion Chromatography stationary phases

Resin	Chemical group	Use
HQ	Quaternised polyethyleneimine	Strong anion exchanger useful for preparative applications
QE	Quaternised polyethyleneimine	Strong anion exchanger as alternative lo POROS HQ
DEAE	Diethylaminoethyl	Weak anion exchanger (pH 3–9)
PI	Polyethyleneimine	Weak anion exchangere (pH 3–9)
HS	Sulphopropyl	Strong cation exchanger useful for preparative applications
SP	Sulphopropyl	Strong cation exchanger as alternative to POROS HS
S	Sulphoethyl	Strong cation exchanger useful for separation of very hydrophobic and/or basic proteins/peptides
CM	Carboxymethyl	Weak cation exchanger (pH range 3–8)
HP2	High-density phenyl	Hydrophobic interaction resin useful for weakly hydrophobic proteins
PE	Phenyl ether	Hydrophobic interaction (moderately hydrophobic proteins
ET	Ethyl ether	Hydrophobic interaction resin useful for strongly hydrophobic proteins
OH	Hydroxyl	Can be activated for affinity chromatography
AL	Aldehyde	Preactivated for coupling of primary amines
EP	Epoxide	Preactivated for coupling primary/secondary amines –SH, or –OH groups
NH	Primary amine	Pre-activated for coupling proteins sugars, and ligands with poor stability under reducing conditions
HY	Hydrazide	Preactivated for site-specific immobilization of IgG
A	Recombinant protein A	Isolation of IgG especially human and humanised antibodies
G	Recombinant protein A	Isolation of mouse IgG
HE	Heparin	Purification of coagulation proteins DNA-binding proteins and lipoproteins
MC	Imido-diacetate	Binding of metals (IMAC; his-tagging)
Rl	Base poly(styrene divinyl benzene)	Reversed-phase chromatography of very hydrophobic proteins or peptides
R2	Base poly(styrene divinyl benzene)	Reversed-phase chromatography of hydrophobic proteins or peptides
Poroszyme	Trypsin, endoproleinase glu-C, pepsin and pa pain	Rapid digestion of proteins to produce peptides
Immunodetection Cartridges:	In addition to EP. AL A, G and HE chemistries (see above).	
BA	Streptavidin	Noncovalent binding of biotinylated ligands
XL	Protein G	Adsorption of anitbody followed by covalent crosslinking

observations seems to be that there is a broad pore size distribution in the membrane which is responsible for high flow rates while large *transport pores* contribute to a high surface area available for adsorption within the membrane. In these respects, the system resembles (particle-based) perfusion chromatography (Section 2.8).

2.9.2 Applications of Membrane-Based Separations

A number of cartridges suitable for membrane-based chromatography are commercially available (Figure 2.25). These cover the principal modes of chromatography described in Section 2.4 with the exception of gel filtration. Affinity applications are especially popular in membrane chromatography with antibodies, metals and other ligands being immobilized to the membrane essentially as described in Sections 2.4.5 and 2.4.6. A feature of membrane-based affinity chromatography is that it appears to give significantly higher efficiency of *ligand utilization* than agarose-based systems. This is due primarily to the greater degree of convection allowing more access to immobilized ligands. A further feature of membrane-based systems is that they are not as limited in their geometry as are column-based systems. Filtration systems are available in a wide variety of designs, unlike chromatography columns. It is possible to imagine downstream processing, for example, in a membrane-based

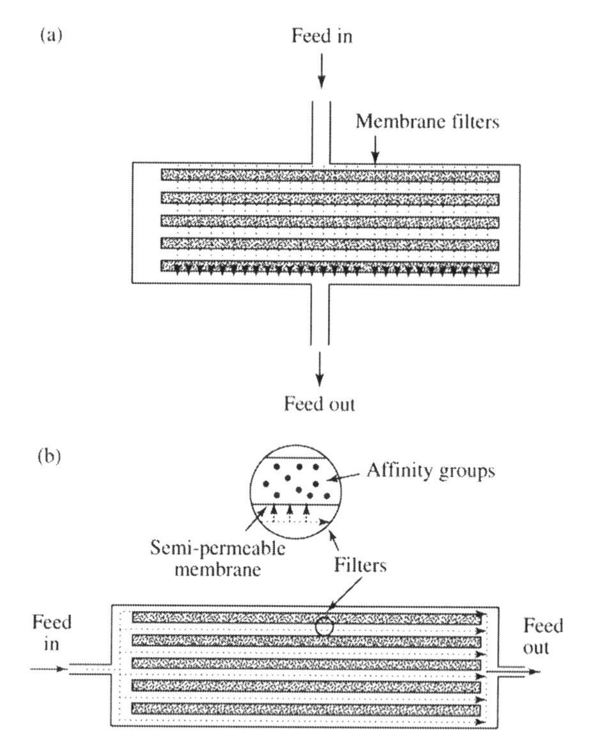

Figure 2.25. Membrane filtration chromatography. (a) Conventional membrane design, (b) Cross-flow hollow fiber filtration unit. Note that, in the cross-flow filtration module, proteins must first pass through a semipermeable membrane to reach affinity groups which may be attached to gel beads thus ensuring filtration of sample to be purified.

system which would not be possible in a column due to limitations imposed by column dimensions, particle properties and their effects on flow rate, back-pressure and resolution. An example of the type of unusual design formats available would include *cross-flow filtration* (Figure 2.25).

2.10 CHROMATOGRAPHY OF A SAMPLE PROTEIN

2.10.1 Designing a Purification Protocol

The combination of steps used in purification of a particular protein is called a *purification protocol*. Protein purification is often a matter of balancing variables which may be inter-related in a very complex way. The most important of these are:

- *Stability properties of the target protein.* Proteins differ considerably from each other in terms of their stability to temperature, pH and oxidation. Either when removed from the cell or else when expressed in the alien environment of a heterologous expression system, the protein may denature, losing its structural integrity and associated biological activity. This may not matter if only a small sample of protein is required for structural studies such as protein sequencing. Usually, though, structurally intact, active protein is required. This sensitivity of proteins can be taken advantage of in the purification protocol. Where proteins display unusual stability to heat and pH, it is possible to use heat/pH treatments selectively to denature contaminant proteins without affecting the protein of interest. Inadvertent release of proteases from subcellular organelles during preparation of cell extracts is a complication associated with cell disruption (e.g. release of cathepsins from lysosomes) resulting in proteolytic degradation of the protein of interest. This may be prevented by the inclusion of *protease inhibitors* in homogenization and chromatography buffers. An example is the inclusion of the metal chelator ethylenediaminetetraacetic acid (EDTA) to minimize metalloproteinase activity.
- *Concentration.* The intracellular protein concentration is of the order of 200 mg/ml. When extracting proteins from cells, we *dilute* the protein concentration (often to less than 1 mg/ml) and this can affect the structural stability of proteins. Since protein purification is a process of retaining the activity of interest whilst decreasing protein concentration, this can mean that biological activity may *decrease* during purification. We can detect this as a decrease of specific activity (activity/mg protein) from step to step. This may be minimized by addition of a *stabilizer* to the extract such as bovine serum albumin which can keep the protein concentration high enough to maintain activity.
- *Expression system.* When using heterologous expression systems expression levels in the range 5–20% of total cell protein can be achieved. This may prove toxic to the cell and the cell may lack enzymes or other proteins necessary for correct post-translational processing. As a result, the protein may be expressed as insoluble *plaques* or *inclusion bodies* (also sometimes called *refractile bodies*). Since these have an unusually high density, they may be separated from the rest of the cell proteins by centrifugation (Section 7.3). Subsequently, plaques can be dissolved with the addition of urea and the protein refolded correctly. In this way, inappropriate expression of protein in plaques can be *included* in the purification protocol. A further option offered by heterologous expression systems is the possibility of expressing the protein as a *fusion* including polypeptides capable of recognizing affinity-ligands such as glutathione or immobilized metals (Examples 2.5 and 2.6). Inclusion of a protease cleavage site allows the subsequent removal of the affinity site from the purified fusion.
- *Inadvertent removal of cofactors or proteins essential for activity.* If there is a reduction in specific activity during purification, consideration should be given to the possibility of the inadvertent removal of low M_r

cofactors such as metals or coenzymes. This can be checked by adding candidate cofactors to highly purified preparations. It is also possible that a protein may require the presence of other proteins in a multi-protein *complex* for full activity. These other proteins also may be inadvertently removed during purification, leading to an apparent decrease in specific activity. This finding may provide the first direct evidence for the existence of a protein complex.

- *Intrinsic chemical and biological properties.* A wide variety of chromatography modes are available to us which take advantage of different chemical and biological properties of proteins (Section 2.4). The strategy adopted in the design of a purification protocol, is to take advantage of small differences in such properties to maximize the yield of the protein of interest while minimizing the levels of other proteins at each step.
- *Protocol design.* Most chromatography modes are available in a variety of chromatography formats which, as we have seen, may be of two general types: high-capacity and low resolution such as in open column chromatography or low-capacity and high resolution as in high performance chromatography methods (Sections 2.5–2.10). In designing a purification protocol, we take advantage of the matrix of chromatography modes and formats available to us to *minimize the number of steps* required for purification, to *maximize the purity of the final product* and to *retain the protein in its native state*, thus preserving its biological properties. A key factor in achieving this is the *order* in which the steps are carried out.

Early steps of the protocol often involve decreasing the protein concentration of the extract to allow the use, subsequently, of high resolution methods. Open column chromatography methods are appropriate for these early steps. Once total protein has been reduced to milligram amounts, more high performance methods may be used where advantage is taken of high resolution separation. It is wise to perfect each step by assessing a variety of conditions to maximize the purification factor achieved. For example, in ion exchange chromatography, it is wise to carry out trial separations with a range of gradients and at a variety of pH values. In affinity chromatography, a variety of ligands and elution techniques might be assessed. It may also be advantageous to study the effect of altering the precise order of the chromatography steps to maximize biological activity and purity of the protein of interest.

Many of the methods described in this book can be used to assess the purity of the protein at each step of the purification protocol. Of particular importance in this regard are analytical chromatography and electrophoretic methods (Chapter 5). The number and nature (e.g. M_r, hydrophobicity) of contaminants at each step of the protocol is assessed. These findings facilitate taking advantage of differences in biophysical properties between the protein of interest and contaminant proteins (e.g. large differences in M_r or hydrophobicity).

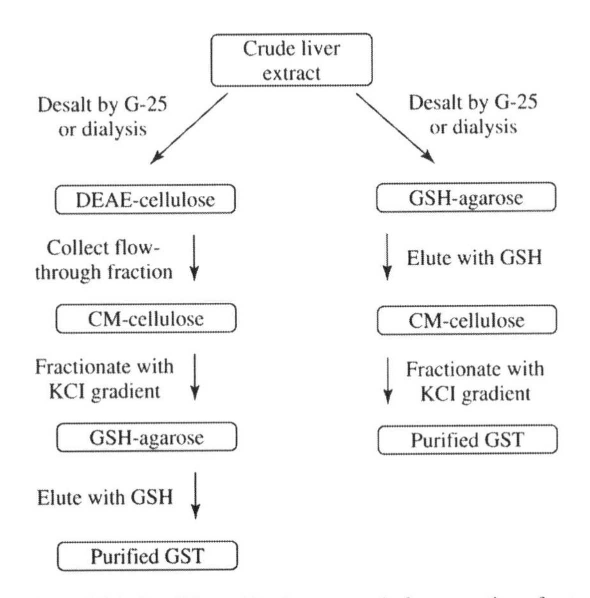

Figure 2.26. Possible purification protocols for separation of rat liver cytosolic GSTs. Note how the order in which steps are carried out may be varied.

2.10.2 Ion Exchange Chromatography of a Sample Protein: Glutathione Transferases

Glutathione transferases (GSTs) are mainly cytosolic enzymes which serve to protect cells from foreign chemicals which have the potential to be genotoxic. They accomplish this in two main ways: by catalyzing a conjugation reaction with the intracellular antioxidant tripeptide, glutathione (GSH) and by noncatalytic binding of a wide range of endogenous and foreign chemicals. This latter property may explain why the GSTs are the most abundant cytosolic proteins in mammalian liver comprising 2–5% of total protein. In common with other drug detoxification enzymes, GSTs exist in multiple isoenzymes which need to be purified in order to study their individual properties. This presents a challenge since isoenzymes may display considerable structural similarities to each other. This is the kind of purification problem which frequently presents itself in biochemistry. An overview of possible purification protocols for rat liver GSTs is given in Figure 2.26.

One of the principal methods for fractionation of GST isoenzymes is ion exchange chromatography (Example 2.7). Before application to a cation exchange column, however, it is advantageous to pass the extract through an anion exchange column (DEAE-cellulose) since most cytosolic proteins have anionic pIs. Prior to application of extract to either ion exchange resin, it is necessary to desalt by passage

Example 2.7 Ion exchange chromatography of glutathione transferase isoenzymes

At all pH values apart from their *isoelectric point* (pI) proteins possess either positive or negative net surface charge (Figure 2.9). Advantage can be taken of difference in surface charge to achieve separation of otherwise very similar proteins by ion exchange chromatography. Depending on their net charge, proteins can exchange with either positive ions such as Na^+ or negative ions such as Cl^- on *cation* and *anion* exchange resins, respectively (Figure 2.11). These ions are electrostatically attracted to *oppositely* charged groups on the resin such as CM- and DEAE-, respectively (Figure 2.10; Table 2.2). This experiment is highly dependent on pH since this affects the charge of amino acid side-chains on the protein surface.

Ion exchange chromatography is particularly useful in achieving separation of structurally-related proteins such as those forming complex protein families as described in Example 2.3. The GSTs are a good example of such a family since they are present in most mammalian tissues in the form of multiple isoenzymes. The bulk of the cytosolic GSTs may be separated from other non-GST proteins by passage of an extract through a glutathione (GSH)-agarose affinity column as described for GST fusion proteins in Example 2.5. Individual isoenzymes can then be routinely separated by ion exchange chromatography.

The affinity extract was applied to CM-Cellulose at pH 6.7 and the column was developed with a 0–75 mM KCl gradient. Open circles denote activity with 1-chloro-2,4-dinitrobenzene (CDNB) while filled circles are activity with an alternative substrate, 1,2-dichloro-4-nitrobenzene (DCNB; GSTs differ in their preference for these and other substrates). Elution positions of isoenzymes were determined by electrophoretic analysis across peaks of activity and also by the presence of activity either with one or both substrates: AA, rGST A1-1; A, rGST M3-3; B, rGST A1-2; C, rGST M3-4. KCl gradient was monitored by measurement of conductivity (dashed line).

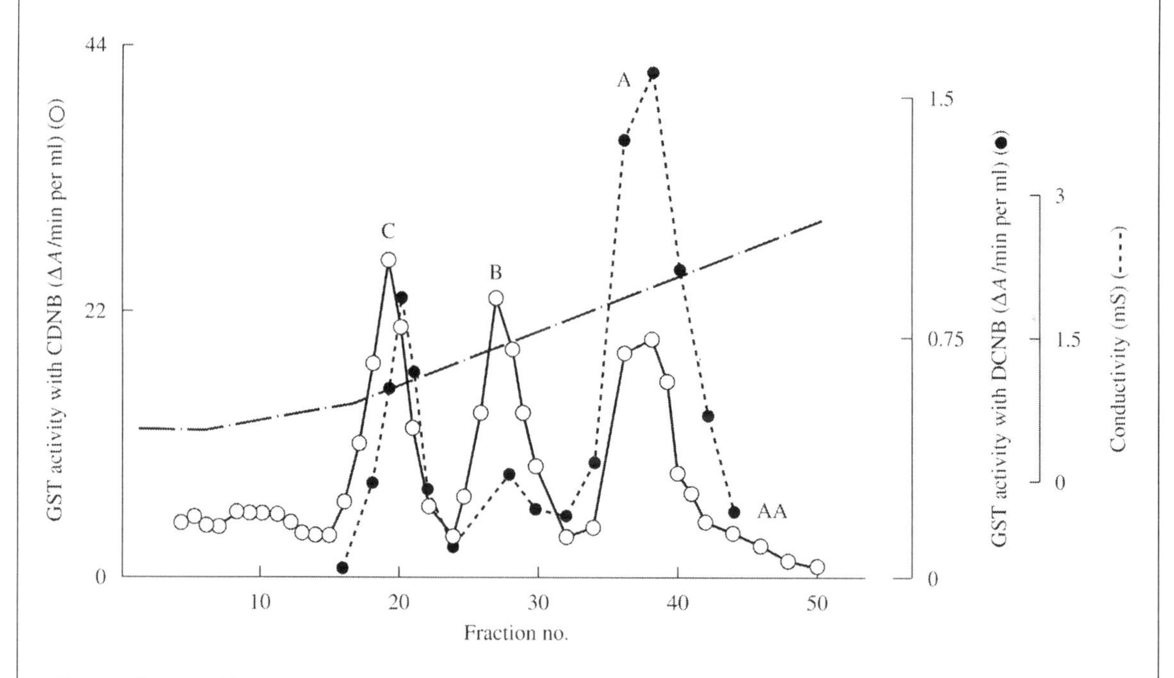

By carrying out this experiment at other pH values, a slightly different elution profile would be obtained. Moreover, by varying the salt gradient (e.g. using 0–100 mM KCl) the separation of individual peaks may be optimized. This example illustrates an important principle in protein purification. Namely, that proteins which behave identically in one chromatographic system such as binding to an affinity resin, may behave very differently in another system (e.g. cation exchange). Protein purification protocols exploit these often very small differences in chromatographic behaviour.

Reproduced from Sheehan, D. and Mantle, T.J. (1984), Evidence for two forms of ligandin (YaYa dimers of glutathione S-transferase) in rat liver and kidney. *Biochemical Journal* **218**, 893–897, by permission of Portland Press.

through Sephadex G-25 (i.e. by gel filtration). CM-cellulose ion exchange chromatography of cytosolic GST activity is shown in Example 2.7. These enzymes have generally basic pIs, since they bind to CM groups at pH 6.7. They also bind to the ion exchange group with different affinities and we can fractionate them with a 0–75 mM KCl gradient. If the proteins were more strongly bound we might have needed to use a higher KCl concentration (perhaps 500 mM). This chromatography is not enough completely to purify individual isoenzymes since they still contain significant amounts of non-GST protein. These may be conveniently removed by passage through a GSH-agarose column which retains only GSTs in the mixture (i.e. by affinity chromatography). Passage of 5 mM GSH through the affinity column elutes purified GST. This purification may also be analysed by SDS PAGE and conveniently managed by use of a *purification table*, which summarizes the relative effects of each step in the protocol on total protein and enzyme activity.

It should be noted that this is not the only protocol possible for GST purification. Possible points of variation include: rearranging the order of steps; varying pH of ion exchange steps to alter the precise pattern obtained; introducing high performance to the protocol (FPLC); taking advantage of differences in K_m^{GSH} between them by applying GSH as a gradient (0–5 mM) rather than as a 5 mM wash (*affinity gradient elution*; Section 2.4.5).

2.10.3 HPLC of Peptides From Glutathione Transferases

Individual isoenzymes of GST are often closely-related structurally with up to 99% identity in amino acid sequence between pairs of isoenzymes. It is often required to compare newly-discovered isoenzymes by *peptide mapping* with trypsin. Where a large degree of sequence identity exists, this results in a number of common peptides on C-18 reversed phase HPLC. This approach allows comparison within complex protein families (e.g. haemoglobins, immunoglobulins). A comparison of the peptide maps generated by tryptic digestion of GST subunits 3 and 4 is shown in Example 2.3. These subunits possess approximately 80% amino acid sequence identity and are capable of dimerizing to generate up to three isoenzymes (two homodimers and a heterodimer). In humans, a structurally related GST is missing from a significant number of the population (the *null phenotype*) and this may predispose some smokers to more rapid development of lung cancer. Reversed phase HPLC analysis allowed the visualization of peptides common to both subunits and also the identification of subunit-specific peptides. Sequencing of these could facilitate further studies such as the design of oligonucleotides for cloning of genes coding for the different proteins. More detailed comparisons are possible by use of other proteolytic agents or by carrying out high resolution analysis of peptide maps using techniques such as mass spectrometry (Chapter 4) and capillary electrophoresis (Section 5.10). It is now possible to interface both of these groups of procedures with HPLC peptide separation.

REFERENCES

General reading

Dorsey, J.G., Cooper, W.T., Siles, B.A. *et al.* (1996) Liquid chromatography: Theory and methodology. *Analytical Chemistry*, **68**, R515–68. A comprehensive review of liquid chromatography.

Mondal, K. and Gupta, M.N. (2006) The affinity concept in bioseparation: Evolving paradigms and expanding range of applications. *Biomolecular Engineering*, **23**, 59–76. A thought-provoking review on the affinity concept in bioseparation including precipitation and chromatography.

Sheehan, D., Meade, G., Foley, V. and Dowd, C. (2001) Structure, function and evolution of glutathione S-transferases: Implications for the classification of non-mammalian members of an ancient enzyme superfamily. *Biochemical Journal*, **360**, 1–16. A review of the GST superfamily.

Shi, Y., Xiang, R., Horváth, C. and Wilkins, J.A. (2004) The role of liquid chromatography in proteomics. *Journal of Chromatography A*, **1053**, 27–36. A review on aspects of LC relevant to proteomics.

Purification of proteins

Doonan, S. (ed.) (1996) *Protein Purification Protocols*. Methods in Molecular Biology, Vol. **59**, Humana Press, Totowa, NJ, USA. An introduction to protein chromatography and purification.

Gallant, S.R., Vunnum, S. and Cramer, S.M. (1996) Optimization of preparative ion-exchange chromatography of proteins: Linear gradient separations. *Journal of Chromatography A*, **725**, 295–314. A review of prediction of chromatographic performance during protein chromatography.

Jungbauer, A. (2005) Chromatographic media for bioseparation. *Journal of Chromatography A*, **1065**, 3–12. A review of chromatography packings.

Modes of chromatography

Afeyan, N.B., Fulton, S.P. and Regnier, F.E. (1991) Perfusion chromatography packing materials for proteins and peptides. *Journal of Chromatography*, **544**, 267–79. A review of perfusive packings in various chromatography modes.

Aizawa, T., Koganesawa, N., Kamakura, A. *et al.* (1998) Adsorption of human lysozyme onto hydroxyapatite – identification of its adsorbing site using site-directed mutagenesis. *FEBS Letters*, **422**, 175–8. Excellent example of application of hydroxyapatite chromatography.

Chaga, S. (2001) Twenty five years of immobilized metal ion affinity chromatography: Past, present and future. *Journal of Biochemical and Biophysical Methods*, **49**, 313–34. A review of IMAC chromatography.

Cramer, S.M. (1991) Displacement chromatography. *Nature*, **351**, 251–2. An introduction to use of displacers in ion-exchange chromatography.

Hengen, P.N. (1995) Methods and reagents – purification of His-tag fusion proteins from *Escherichia coli*. *Trends in Biochemical Sciences*, **20**, 285–6. Description of use of His-tag affinity purification.

Kawasaki, T. (1991) Hydroxyapatite as a liquid chromatographic packing. *Journal of Chromatography A*, **544**, 147–84. Review of hydroxyapatite in chromatography focusing on mode of interaction with sample components.

Ladiwala A., Rege, K., Breneman, C.M. and Cramer, S.M. (2005) *A priori* prediction of adsorption isotherm parameters and chromatographic behavior in ion exchange systems. *Proceeding of the National Academy of Sciences, USA*, **102**, 11710–15. Consideration of factors critical to protein behaviour in ion exchange chromatography.

Laing, G.D., Kamiguti, A., Wilkinson, M.C. *et al.* (1995) Characterization of a purified phospholipase A2 from the venom of the papuan black snake (*Pseudechis papuanus*). *Biochimica et Biophysica Acta*, **1250**, 137–43. An excellent example of application of hydrophobic interaction chromatography.

Lienqueo, M.E. and Mahn, A. (2005). Predicting protein retention time in hydrophobic interaction chromatography. *Chemical Engineering and Technology*, **28**, 1326–34. A review on hydrophobic interaction chromatography.

Regnier, F.E. (1991) Perfusion chromatography. *Nature*, **350**, 634–5. An overview of perfusion chromatography.

Sheehan, D. (2008) Ion exchange chromatography, in *Molecular Biomethods Handbook* (eds J. Walker and R. Rapley), 2nd edn, Humana Press, Totowa, NJ, USA. Description of protein ion exchange chromatography.

Winzor, D.J. (2003) Exclusion chromatography. *Journal of Biochemical and Biophysical Methods*, **56**, 15–52. A review of the origin and uses of size exclusion (gel filtration) chromatography.

HPLC

Bellot, J.C. and Consoret, J.S. (1991) Liquid chromatography modelling: a review. *Process Biochemistry*, **26**, 363–76. An overview of the main models of chromatographic separation.

Claessens, H.A. and van Straten, M.A. (2004) Review on the chemical and thermal stability of stationary phases for reversed phase liquid chromatography. *Journal of Chromatography A*, **1060**, 23–41. A good review on recent developments in reversed phase HPLC packings.

Gritti, F. and Guiochon, G. (2005) Critical contribution of nonlinear chromatography to the understanding of retention mechanism in reversed-phase liquid chromatography. *Journal of Chromatograpy A*, **1099**, 1–42. Review of retention mechanism in reversed-phase HPLC.

Hamada, J.S. (1997) Large-scale high-performance liquid chromatography of enzymes for food applications. *Journal of Chromatography A*, **760**, 81–7. A review of industrial-scale HPLC especially of enzymes relevant to the food industry.

Lundell, N. and Schreitmuller, T. (1999) Sample preparation for peptide mapping – a pharmaceutical quality-control perspective. *Analytical Biochemistry*, **266**, 31–47. A review of issues in peptide mapping as a means of identifying structural changes in proteins.

McNay, J.L. and Fernandez, E. (1999) How does a protein unfold on a reversed-phase liquid chromatography surface? *Journal of Chromatography A*, **849**, 135–48. A study of the mechanism of interaction between proteins and reversed phase HPLC packings.

Wang, H. and Hanash, S. (2003) Multi-dimensional liquid phase based separations in proteomics. *Journal of Chromatography B*, **787**, 11–18. A review of multidimensional HPLC.

Membrane chromatography

Barut, M. Podgornik, A., Brne, P. and Strancar, A. (2005). Convective interaction media short monolithic columns: Enabling chromatographic supports for the separation and purification of large biomolecules. *Journal of Separation Science*, **28**, 1876–92. Membrane chromatography of high Mr molecules

Charcosset, C. (2006). Membrane processes in biotechnology: An overview. *Biotechnology Advances*, **24**, 482–92. A review of membrane-based processes in protein biotechnology.

Podgornik, A., Barut, M., Jankar, J. *et al.* (1999). High-performance membrane chromatography of small molecules. *Analytical Chemistry*, **71**, 2986–91. Membrane chromatography of low Mr molecules.

Tejeda, A., Ortega, J., Magana, I. and Guzman, R. (1999) Optimal design of affinity membrane chromatographic columns. *Journal of Chromatography A*, **830**, 293–300. An overview of column design issues in membrane chromatography.

Tennikova, T.B. and Svec, F., (1993) High-performance membrane chromatography – highly efficient separation method for proteins in ion-exchange, hydrophobic interaction and reversed-phase modes. *Journal of Chromatography A*, **646**, 279–88. A review of membrane chromatography of proteins.

Downstream processing

Chapman, T. (2005) Protein purification: Pure but not simple. *Nature*, **434**, 795–8. A product review on industrial scale protein purification.

Cramer, S.M. and Jayaram, G. (1993). Preparative chromatography in biotechnology. *Current Opinion in Biotechnology*, **4**, 217–25. Review of industrial-scale chromatography.

Hamada, J.S. (1997) Large-scale high-performance liquid chromatography of enzymes for food applications. *Journal of Chromatography A*, **760**, 81–7. A review of industrial scale HPLC.

Hjerten, S., Li, Y.M., Liao, J.L. *et al.* (1992). Continuous beds – High-resolving, cost-effective chromatographic matrices. *Nature*, **356**, 810–11. An overview of efficient matrices for industrial-scale chromatography.

Prince, K. and Smith M. (2004) Purification process scale-up, in *Protein Purification Protocols* (ed. P. Cutler), Humana Press, Totowa, NJ, USA, pp. 463–80. A discussion of issues arising in scale-up.

Roque, A.C.A., Lowe, C.R. and Taipa, M.A. (2004) Antibodies and genetically engineered related molecules: Production and purification. *Biotechnology Progress*, **20**, 639–54. A critical review of downstream processing of Biopharma protein products.

A useful web site

Separationsnow.com: http://www.separationsnow.com/.

Chapter 3
Spectroscopic Techniques

Objectives

After completing this chapter you should be familiar with:

- *The* **Electromagnetic spectrum,**
- *The concept of* **transition** *between energy levels,*
- *The* **variety** *of spectroscopic techniques available,*
- **Applications** *of spectroscopic techniques in biochemistry.*

Electromagnetic radiation spans a wide range of energy levels in a continuum called the *electromagnetic spectrum*. Each part of this spectrum may be characterized by a specific range of *wavelengths*. Living organisms are constantly exposed to a wide range of electromagnetic radiation from the Sun or other stars. Man-made sources of radiation include heated filaments of the type found in ordinary light bulbs. We can select for different parts of the electromagnetic spectrum using a prism or similar device. Depending on its energy content, radiation from different parts of the electromagnetic spectrum may interact with biomolecules in a wide variety of ways. Electromagnetic radiation therefore provides a useful means to probe the chemical structure of biomolecules. In this chapter we will look at the electromagnetic spectrum and describe some of the experimental information it is possible to derive from studying interactions between electromagnetic radiation and biomolecules.

3.1 THE NATURE OF LIGHT

3.1.1 A Brief History of the Theories of Light

Light is composed of electric and magnetic fields, which are mutually perpendicular and which *radiate* out from a source in all directions (Figure 3.1). It is therefore a form of *electromagnetic radiation*. A *wave* description of electromagnetic radiation resulted from the work of Maxwell, Hertz and others in the nineteenth century. In this description, electric and magnetic fields are propagated through space as *wave functions* which may be characterized by *wavelength*, λ (the distance from one part of the wave to the corresponding position on the next wave) and *frequency*, ν (the number of times a wave passes through a fixed point in space every second). These parameters are related to the *energy* content of the wave, E, by Equations (3.1) and (3.2):

$$E = \frac{h \cdot c}{\lambda} \tag{3.1}$$

$$E = h \cdot \nu \tag{3.2}$$

where h is Planck's constant and c is the speed of light. These equations demonstrate that there is a *fixed* relationship between the energy of a particular type of electromagnetic radiation on the one hand and its wavelength/frequency on the other. High energy radiation is characterized by *short* wavelengths and *high* frequency while lower energy radiation is characterized by *long* wavelengths and *low* frequencies. As shown in Figure 3.1, waves also have characteristic *amplitude* which is the maximum value the electric or magnetic vector can have. When two waves are superimposed, they produce a resultant wave which has amplitude equal to the *sum* of that of the individual waves. This is the *principle of superposition* and it is characteristic of all wave functions.

Electromagnetic radiation from the sun covers a wide range of wavelengths in a continuum called the *electromagnetic spectrum* (*see* below Section 3.2.1). Light visible to humans as the colour-range red to violet ($\lambda = 720$–400 nm) represents only a tiny part of this spectrum called the *visible* range. This is because protein receptors in the eye are sensitive only to a comparatively small set of wavelength values. When white light from the sun impacts on a material such as a painting or a flower, some visible wavelengths are *absorbed*. What we experience as *colour* is simply wavelengths in the visible part of the electromagnetic spectrum which happen to be *reflected* from materials of a particular chemical composition. To understand how light interacts with matter and especially with biomolecules, we need to review the physical nature of light. The development of our

Physical Biochemistry: Principles and Applications, Second Edition David Sheehan

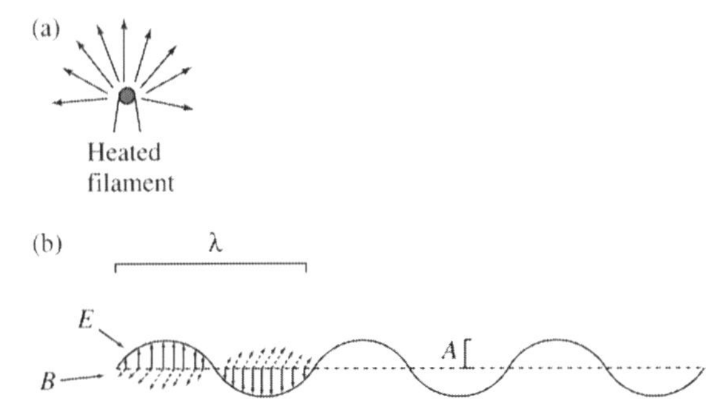

Figure 3.1. Light is a form of electromagnetic radiation. (a) When a filament is heated, light is emitted or *radiated* in all directions. (b) Light consists of mutually perpendicular electric (E, solid arrows) and magnetic (B, dashed arrows) vectors which are wave functions. The wave may be characterized by wavelength (λ) and amplitude (A).

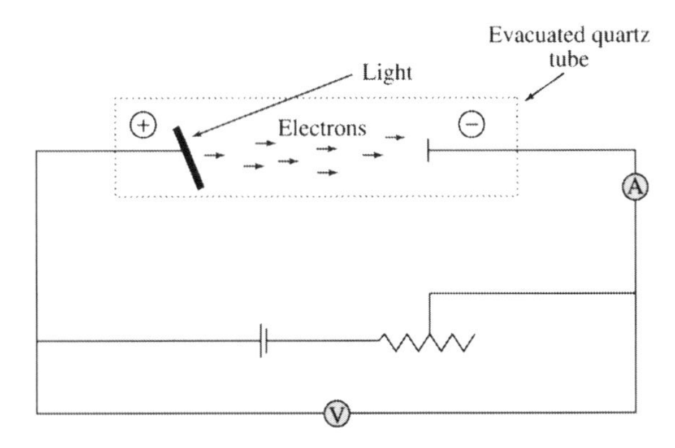

Figure 3.3. Apparatus for the photoelectric effect experiment. When light of a particular intensity (I) and frequency (ν) is shone on a metal plate, electrons are released. These cause a current which may be detected by an ammeter (A). A retarding (or stopping) voltage (V) may be provided which stops the flow of electrons to the cathode. The value of this stopping voltage can be determined with light of different I and ν values (Figure 3.4).

understanding of light may be traced by describing three classic experiments on the interaction of light with matter.

In the seventeenth century, Sir Isaac Newton demonstrated that it was possible to separate light of different colours from the Sun's white light (Figure 3.2). He made a tiny perforation in the window-blind of his study in Cambridge. This selected for a thin beam of white light. When he passed this beam through a triangular piece of glass called a *prism*, he could separate white light into a range of colours. This phenomenon is now called *diffraction* and the prism in this experiment serves as a *monochromator* (i.e. a device to select for light of single wavelength). Newton's conclusion that white light is really a mixture of light of different colours was quite controversial in its time since, from antiquity, it had always been assumed that white light was 'pure' and colours were 'impure'. In 1801, Thomas Young

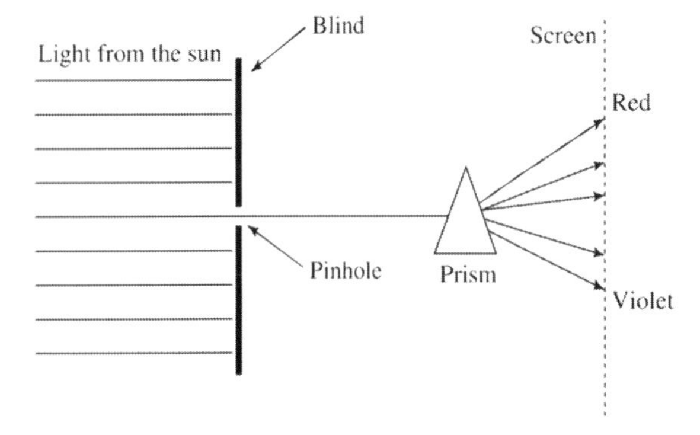

Figure 3.2. Diffraction of white light. Newton selected for a thin beam of white light by making a tiny hole in the blind of his study. After passage through a prism, this light was separated into light of different colours denoted by arrows.

demonstrated that diffraction is a consequence of *interference* and is characteristic of wave functions. It was therefore concluded that *light is a wave phenomenon.*

However, in the early twentieth century two experiments were performed which appeared to contradict a wave description of light. The first of these was called the *photoelectric effect experiment*. This involved light of a single wavelength being shone on a metal plate which formed part of an electric circuit (Figure 3.3). The frequency and *intensity* (i.e. amplitude) of this light could be varied. Electrons ejected from the metal plate could be detected as a current which could be measured in the electrical circuit (hence *photoelectric effect*). It had been expected that, when the frequency (i.e. energy) of light was continuously increased, release of electrons would occur in a continuous fashion. However, this was not observed (Figure 3.4). Instead, a *threshold frequency* needed to be reached before electrons were ejected. Below this threshold, *regardless of the intensity of the light*, no electrons were detected. Above the threshold frequency, it was expected that the current measured would increase with the intensity of the radiation. Instead, it was found that, irrespective of this intensity, the *kinetic energy* of each electron released appeared to be constant for a given metal. Increasing intensity of light indeed released more electrons but these had the same average kinetic energy. The threshold frequency has a characteristic value for each metal, *regardless of the intensity of the light*. A plausible explanation for these findings was that light energy was taken up in minute 'packets' or *quanta*. This led to the *quantum theory of light* proposed by Max Planck and Albert Einstein.

The third experiment, which is generally similar to that of Newton involved passing an electric current through a tube

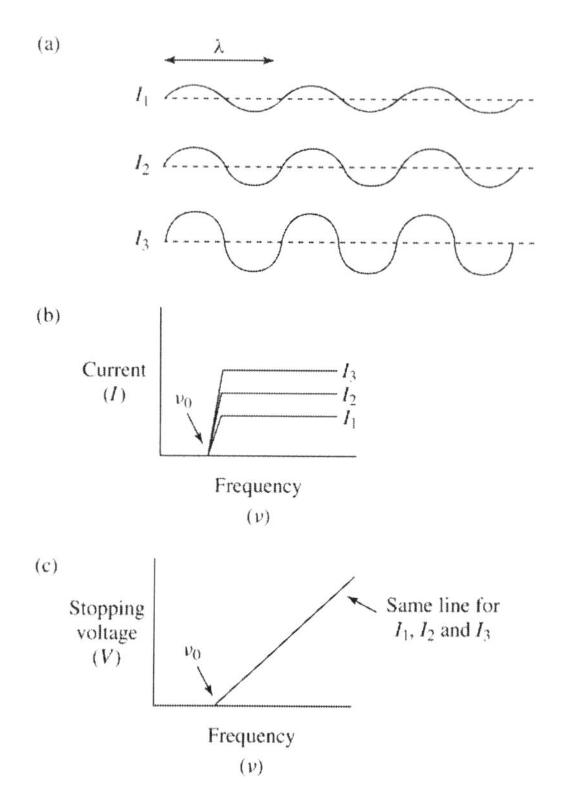

Figure 3.4. The photoelectric effect experiment. (a) Light of different intensity (I_1 to I_3) has the same frequency (ν) and wavelength (λ). (b) Current measured in apparatus shown in Figure 3.3 at a variety of ν values. Regardless of the light intensity, current is only detected at values above the threshold frequency ν_0. (c) Stopping voltage increases linearly with ν. Different metals give a line of identical slope but differing ν_0 values. This plot shows that V is proportional to ν at values above ν_0.

filled with hydrogen gas (Figure 3.5). This generated the release of electromagnetic radiation of a much more *limited* range of wavelength values than that of the Sun. The *emitted* radiation could be detected on a photographic plate. This was called the *hydrogen emission experiment* and the spectrum of radiation emitted was called the *hydrogen emission spectrum*. A wave description of light would predict that a *continuum* of wavelengths would be found in this emission spectrum reflecting a range of possible energy values for the hydrogen atom. The emission spectrum actually observed, however, consisted of a series of *discrete* lines suggesting that hydrogen had a number of discrete energy-levels. Neils Bohr (1913) rationalized this by postulating that the electron of the hydrogen atom had *quantized* energy levels which could adopt values of:

$$\frac{n \cdot h}{2\pi} \tag{3.3}$$

where n is the *electronic quantum number*. Bohr explained the hydrogen emission experiment as follows (Figure 3.6). Electrical energy stimulates promotion of electrons from a *ground state* ($n = 1$) to an *excited state* (e.g. $n = 2, 3, 4$, etc.). These electronic energy levels are *quantized* (i.e. of only certain allowed energy values). When the electrical stimulus is removed, the excited electron can return to the ground state with the emission of light energy. The wavelengths of the hydrogen emission spectrum (i.e. the individual lines on the photographic plate) correspond to *transitions* between the various electronic energy levels ($N = 2 \rightarrow N = 1$; $N = 3 \rightarrow N = 1$, etc.) called *electronic transitions*.

3.1.2 Wave-Particle Duality Theory of Light

The quantum theory of Planck and Einstein led to the suggestion that light had a *particulate nature* and could be regarded as consisting of tiny particles called *photons*. This particulate nature, however, appeared to contradict the diffraction of light and the fact that light passes freely through a vacuum which are consistent with a wave description of light. The competing wave and particulate theories were unified in 1923 by Louis de Broglie in his *wave-particle duality theory*. This postulated that light had *both* wave and particle natures. Moreover, de Broglie postulated that particulate matter also had some characteristics of a wave nature. This is summarized in *de Broglie's relationship*:

$$\lambda = \frac{h}{p} \tag{3.4}$$

where p is the momentum (mass × velocity) of a moving particle such as an electron. We can use this relationship to compare the relative magnitude of the wave and particle natures of a tiny object such as an electron with a macroscopic object such as a golf ball (Example 3.1). This comparison shows that *both* objects have *both* wave and particle natures but that the particulate nature dominates the golf ball while the wave nature dominates the electron.

3.2 THE ELECTROMAGNETIC SPECTRUM

3.2.1 The Electromagnetic Spectrum

We have observed that the human eye may only detect a comparatively small part of the electromagnetic spectrum (Figure 3.7). In fact, some insects see parts of the spectrum which are invisible to us because the protein receptors in their eyes can detect a slightly different range of wavelengths and this is important in recognition of suitable

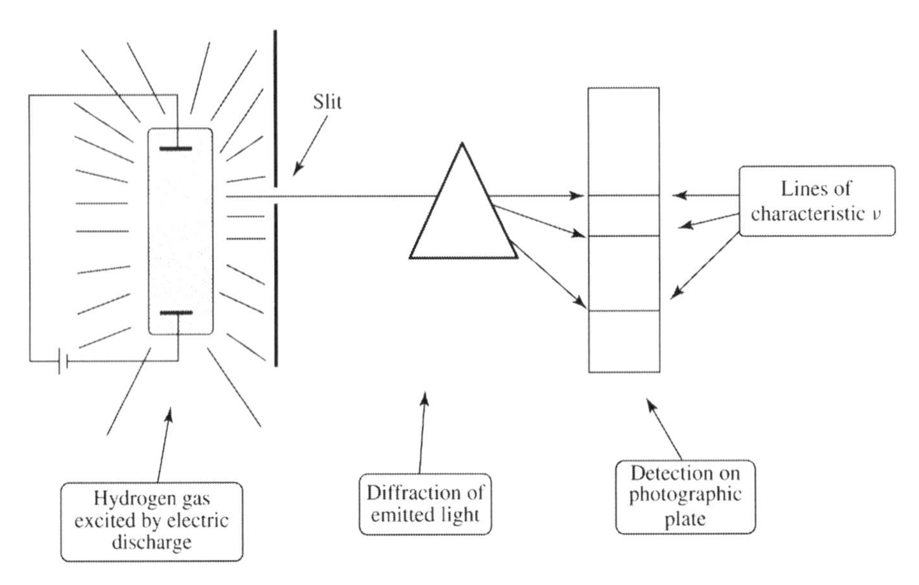

Figure 3.5. The hydrogen emission experiment. When excited by an electric discharge, hydrogen gas emits light. This may be diffracted through a prism and detected on a photographic plate. A series of lines of particular ν and λ are observed rather than a continuum. A completely different spectrum is obtained for helium and other elements.

flowers by honey bees, for example. Nonvisible parts of the electromagnetic spectrum are of relevance to biochemistry because they interact with matter, especially biomolecules, in particular ways. Important types of nonvisible radiation include *infrared* (i.e. *below* red), *ultraviolet* (i.e. *above* violet), *microwave*, *radiowave* and *X-rays*. The precise nature of the interaction of these different types of electromagnetic radiation with matter differs and hence we need specific equipment or experimental approaches for each. However, they are all part of the same electromagnetic spectrum, differing from each other in terms of their wavelength, frequency and energy (Equations (3.1) and (3.2)).

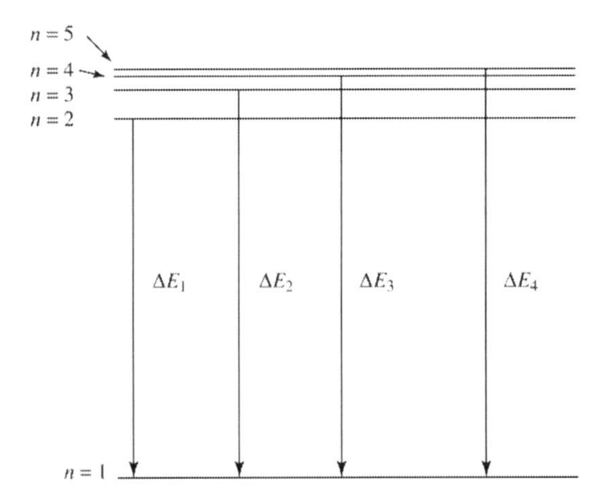

Figure 3.6. Electronic transitions. The hydrogen emission experiment may be explained by postulating that electrons are promoted from their ground state ($n = 1$) to an excited state (e.g. $n = 2$–5). Return of an electron to the ground state involves an electronic transition between quantized energy levels which releases energy (ΔE) corresponding to a light frequency, ν, by the relationship $\Delta E = h\nu$. These beams of light are detected as lines on the photographic plate used in the hydrogen emission experiment.

3.2.2 Transitions in Spectroscopy

The foregoing allows us to understand that biomolecules may be expected to absorb light energy to promote electrons from their ground to excited states. In order to return to the ground state, it is necessary for the electron to *lose* the extra energy it has acquired. This extra energy has a value which may be represented as ΔE (i.e. the energy value of the excited state *minus* that of the ground state; Figure 3.6). Because only certain values are allowed for the various energy levels, the Hydrogen emission experiment results in a number of distinct frequencies of radiation as a consequence of Equation (3.2):

$$\begin{aligned} \Delta E_1 &= h \cdot \nu_1, \\ \Delta E_2 &= h \cdot \nu_2, \\ \Delta E_3 &= h \cdot \nu_3, \end{aligned} \tag{3.5}$$

where ν_1, ν_2, and so on represent the frequencies resolved on the photographic plate. These frequencies occupy the visible and ultraviolet parts of the electromagnetic spectrum.

Example 3.1 De Broglie's relationship and the wave-particle duality

De Broglie's relationship:

$$\lambda = \frac{h}{p}$$

Where h is Planck's constant (6.626×10^{-27} erg s) implies that *all* matter has *both* a wave and a solid (i.e. particle) nature. We can illustrate this by considering an electron and a golf-ball traveling at the same velocity; 100 m/s. The electron has a mass of 9.11×10^{-28} g while the golf-ball is much larger (10 g). The momentum of each may be calculated by multiplying mass by velocity as follows and de Broglie's relationship can be used to calculate the wavelength associated with the wave nature of each:

Property	Electron	Golf ball
Mass (g)	9.11×10^{-28}	10
Velocity (m/s)	100	100
Momentum (g m/s)	9.11×10^{-26}	1000
Wavelength (cm)	7.27	6.626×10^{-22}

From this table, it is clear that the wave properties of the electron are more significant than its particle properties (e.g. momentum). Conversely, the particle properties of the golf ball are quantitatively more significant than its wave properties. The de Broglie relation means therefore that wave properties are more pronounced at the atomic level while particle properties are more significant at the macroscopic level.

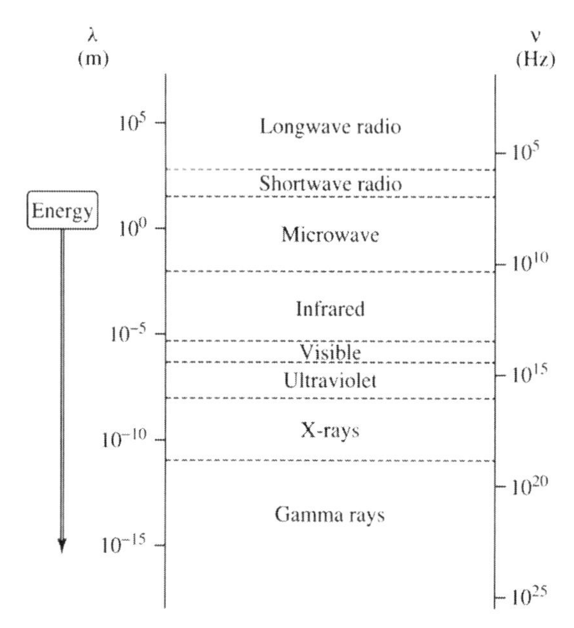

Figure 3.7. The electromagnetic spectrum. The electromagnetic spectrum forms a continuum of wavelengths (λ) and frequencies (ν). The various categories overlap with each other with the exception of visible light. Note that visible light represents only a small part of the spectrum. Because the energy of each category of electromagnetic radiation differs, they interact with matter in distinct ways.

This phenomenon of transition between energy levels is not confined to electrons, however. For example, chemical bonds can occupy a variety of *vibrational energy levels* and atoms connected by single covalent bonds may *rotate* relative to each other through a number of *rotational energy levels* (Figure 3.8). These energy levels are also *quantized*, having only a set number of allowed values. Transitions can occur between them corresponding to particular frequencies as described in Equation (3.5). Because energy increments between vibrational energy levels are much *smaller* than those for electronic energy levels, the frequencies of radiation corresponding to these transitions are in the infrared part of the spectrum (i.e. *longer* λ and *lower* ν than the visible and ultraviolet). Similarly, energy increments between rotational energy levels are *smaller* even than those between vibrational energy levels and correspond to frequencies in the microwave part of the spectrum (i.e. *longer* λ and *lower* ν than the infrared).

We will see in this chapter that transition between energy levels is a common feature of the interaction between biomolecules and electromagnetic radiation. However, due to instrumental and experimental limitations, only some of the transitions are of use in the study of biomolecules. The energy increments (ΔE) and frequencies associated with transitions in the range useful in biochemistry are summarized in Table 3.1.

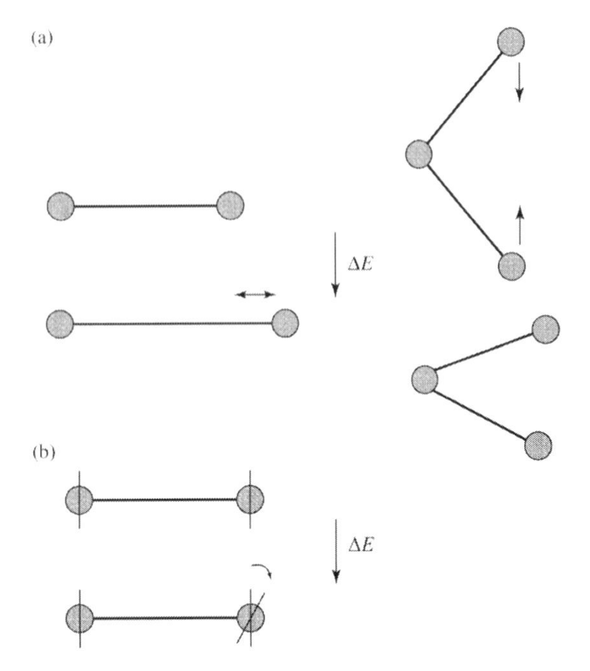

Figure 3.8. Non-electronic transitions. (a) Vibrational transitions involve alternating between a number of allowed bond lengths and bond angles, of differing energy. The energy increment between vibrational energy levels is denoted by ΔE. (b) Rotational transitions. Atoms bonded together can rotate relative to each other. The different rotation states have distinct energies and the energy increment between them is denoted by ΔE.

3.3 ULTRAVIOLET/VISIBLE ABSORPTION SPECTROSCOPY

3.3.1 Physical Basis

We have seen that light from the Sun is white but, when it impacts on a coloured object, some of this light is *absorbed* and it is the *reflected* or nonabsorbed light which gives us the sensation of colour. If light in the utraviolet/visible part of the electromagnetic spectrum is passed through a sample in solution, some light energy may also be absorbed

Table 3.1. Transitions in spectroscopy

Type of transition	Part of spectrum	ΔE (J/mole)	v (Hz)
Electronic	X-ray	4×10^6	10^{18}
Electronic	Ultraviolet/visible	400×10^3	10^{15}
Vibrational	Infrared	40×10^3	10^{13}
Rotational	Microwave	40	10^{11}
Nuclear spin	Radiowave	4×10^{-4}	10^8

(Figure 3.9). Molecules (or parts of molecules) capable of absorbing light are called *chromophores*. This is known as an *absorption spectroscopy experiment* and it is of use because the particular frequencies at which light is absorbed are affected by both the *structure* and the *environment* of the chromophore. Light energy is used to promote electrons from the ground state to various excited states as shown in the *Morse diagram* in Figure 3.10. Each chemical structure absorbs slightly different frequencies of light since each has a characteristic electronic structure (i.e. pattern of electron distribution). The ground and excited electronic energy levels each contain many vibrational energy levels differing from each other by smaller energy increments than those between the electronic energy levels (ΔE). Excited electrons can return to the ground state by *vibrational transitions* through these smaller energy increments (Figure 3.10). This excess energy is lost in collision with solvent molecules. The light energy absorbed therefore appears ultimately as heat in the solution.

The absorption phenomenon may be *quantified* by the *Beer–Lambert law* (Equation (3.6)):

$$\log \frac{I_0}{I} = \varepsilon \cdot c \cdot l \qquad (3.6)$$

where I_0 is the intensity of *incident* light, I is the intensity of *transmitted* light, c is the molar concentration and l is the length of the light path (usually 1 cm). ε is the *molar extinction coefficient*. The term [log I_0/I] is the *absorbance* (A_λ) at a particular wavelength or frequency. A plot of A_λ or ε versus wavelength or frequency is known as an *absorption spectrum* (Figure 3.11). Wavelengths corresponding to maxima in such spectra are denoted by λ_{max}. Some useful λ_{max} and ε values are tabulated in Table 3.2. Because each electronic energy level consists of several vibrational energy levels, a range of wavelengths is absorbed rather than one fixed wavelength. Under standard conditions, this spectrum is a fixed property of a pure chromophore and may therefore be used in *identification* of previously-characterized molecules. Moreover, since absorbance is directly dependent on molar concentration, it may be used as a measure of *concentration* provided a standard curve for that chromophore is also available. This and other uses of the Beer–Lambert Law are described in Section 3.3.3.

Absorbance is an especially appropriate scale for measuring interactions between light and molecules in solution in that, when no light is absorbed, this term has a value of 0. When light is absorbed, the absorbance value *increases*. In theory, the parameter is linear and can reach a maximum of infinity. But, since absorbances of 1, 2 and 3 correspond to 10, 1 and 0.1% of light transmitted, respectively, in practice it is usually not possible to measure absorbance accurately above 3.0. Moreover the accuracy of quantification

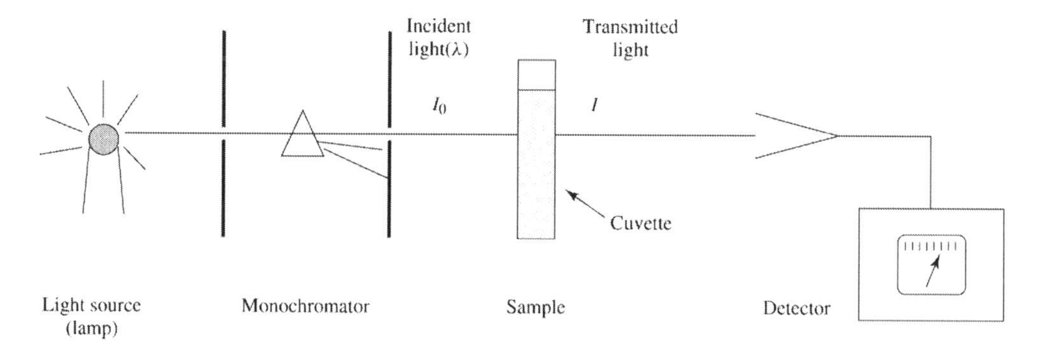

Figure 3.9. Ultraviolet/visible absorption spectroscopy experiment. Light of a single wavelength (λ) and intensity (I_0) is passed through a sample held in a cuvette. Some of this light may be absorbed by the sample. This is detected as a decreased intensity of transmitted light (I) compared to incident light. Log I_0/I is measured as absorbance.

decreases at high absorbance and the most linearly accurate measurements of absorbance are therefore obtained in the range 0–1.0.

Some wavelengths are particularly useful in the study of biomolecules (Figure 3.11; Table 3.2). Amino acids have a strong absorbance around 210 nm and this is frequently used to detect peptides. The aromatic amino acids, tyrosine and tryptophan have relatively strong absorbance at 280 nm while nucleic acids absorb strongly at 260 nm. These wavelengths are therefore widely used in studies of proteins and nucleic acids, respectively. Reduction of NAD^+ to NADH + H^+ causes a major increase in absorbance at 340 nm which is taken advantage of in assay of oxidoreductase enzymes.

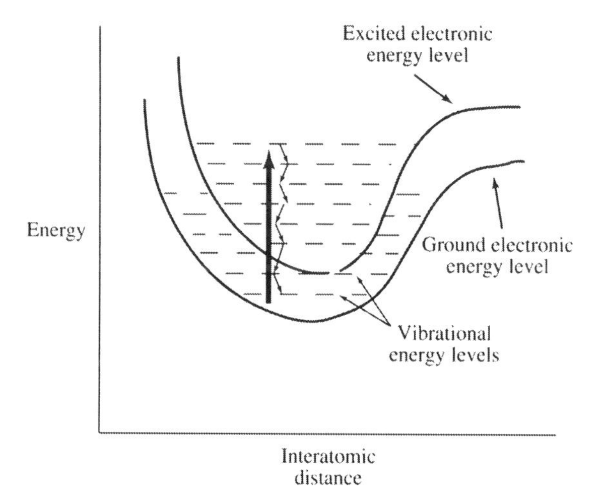

Figure 3.10. Physical basis of absorbance. The electrons of a chromophore may be promoted to a higher energy level as a result of absorption of light energy (large arrow). The electron returns to the ground state by passing through various energy levels (small arrows). Vibrational energy levels of the ground and excited electronic energy levels overlap in flexible molecules. The small increments of energy released as the electron undergoes vibrational transition are lost in collisions with solvent molecules and appear as heat.

The absorbance spectrum for a chromophore under standard conditions is only partly determined by its chemical structure. The *environment* of the chromophore also affects the precise spectrum obtained. The most important environmental factors affecting absorption spectra are *pH*, *solvent polarity* and *orientation* effects. These effects are especially important in studies of biopolymers such as proteins and nucleic acids where chromophores may act as *reporter molecules* which can give information about their immediate environment. These effects also mean that it is essential to determine absorption spectra under defined conditions of pH and solvent composition.

Protonation/deprotonation effects resulting from pH changes or oxidation/reduction affect electron distribution in chromophores. This often results in dramatic differences between the absorption spectra of the protonated and deprotonated forms of the chromophore (Figure 3.12). Comparison of the spectra for protonated and deprotonated tyrosine reveal wavelengths called *isosbestic points* where absorbance of both forms of the chromophore are identical. This can be useful if it is desired to quantify the total chromophore in solutions of different pH. It is often possible to find a wavelength where only one form of the chromophore (prototonated or deprotonated, reduced or oxidized) has a strong absorbance while the other does not. Measurements of absorbance at this wavelength across a range of pH may be used to measure the pKa of the relevant group.

Solvent polarity also affects the absorption spectrum determined for chromophores (Figure 3.13). Alternative solvents to water include aqueous solutions of dimethylsulphoxide, dioxane, ethylene glycol, glycerol and sucrose. Chromophores frequently give a slightly different spectrum in such solvents compared to water and this experiment is

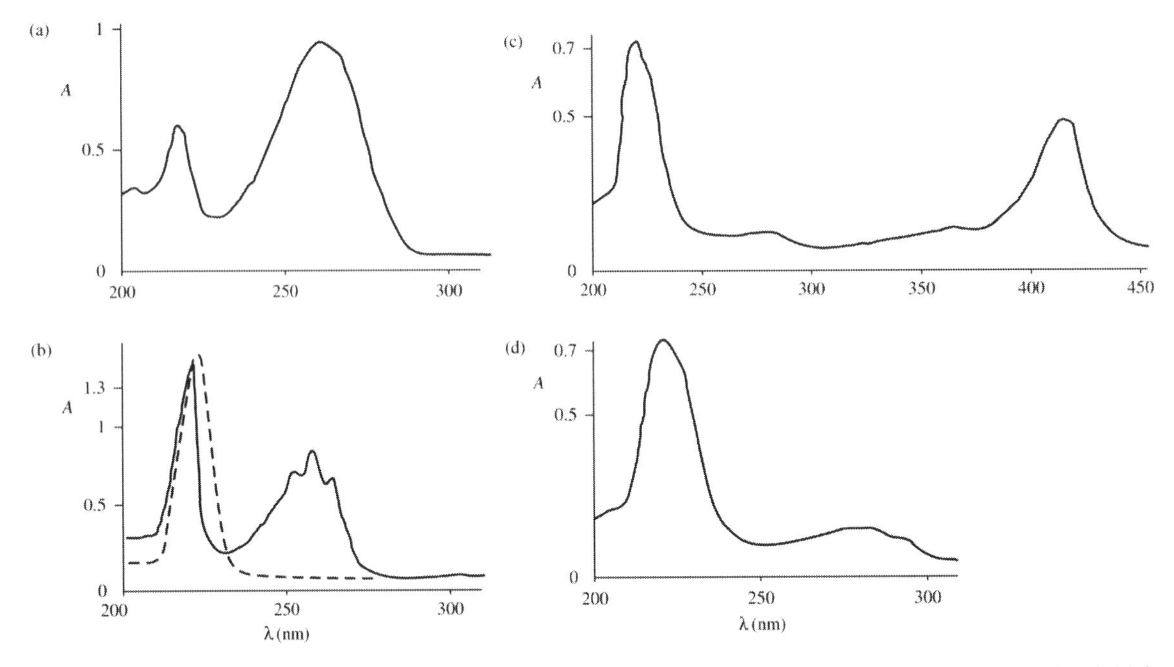

Figure 3.11. Absorption spectra of some biomolecules. Absorbance (A) was measured at pH 5.0 and is plotted against wavelength (λ) for some representative biomolecules. (a) 66 μM adenosine monophosphate (AMP). Note the strong absorbance at 260 nm which is characteristic of nucleotides. (b) 5 mM phenylalanine (solid line) and 0.7 mM histidine (dashed line). Note the λ_{max} at 257 nm which is characteristic for phenylalanine. (c) 70 μg/ml cytochrome C. Note that cytochrome C has a red colour due to the presence of a haem group and has a λ_{max} at 410 nm. (d) 65 μg/ml lysozyme.

known as *solvent perturbation*. They are particularly useful in determining whether or not a chromophore is in contact with the solvent (e.g. on the surface of a protein or membrane or buried in the interior of the protein/membrane).

Orientation effects are a spectroscopic consequence of the relative *geometry* of neighbouring chromophore molecules. A good example is the *hypochromicity* of nucleic acids (Figure 3.14). A solution of free nucleotides has a higher A_{260} (i.e. absorbance at a wavelength of 260 nm) than an identical concentration assembled into a single-stranded polynucleotide. Double stranded nucleic acids, in turn, have a lower absorbance at this wavelength than single stranded polynucleotides. For this reason, absorbance measurements are useful in monitoring assembly or denaturation of nucleotides *in vitro*.

Table 3.2. Some λ_{max} values useful in biochemistry

Chromophore	λ_{max} (nm)	ε ($mM^{-1} \cdot cm^{-1}$)
Tryptophan	280	5.6
	219	47.0
Tyrosine	274	1.4
Phenylalanine	257	0.2
Adenosine	260	14.9
DNA[a]	260	6.6
RNA[a]	260	7.4
Bovine serum albumin	280	40.9

[a]Calculated per mM of 330 Da repeating units.

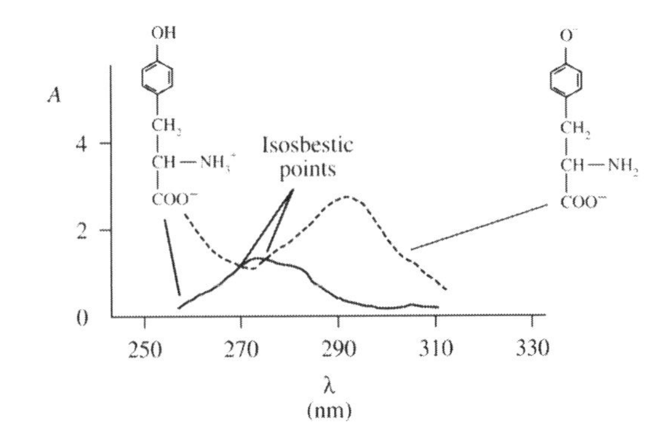

Figure 3.12. Absorbance spectra of tyrosine at pH 6 and 13. Absorbance spectra of 1 mM tyrosine at pH 6 (solid line) and 13 (dashed line) are shown. The structure of the major form of the chromophore at each pH is indicated. Note the λ_{max} at 295 nm for the deprotonated tyrosine anion which is missing from the spectrum of tyrosine. Isosbestic points are wavelengths where the absorbance of both protonated and deprotonated forms of the chromophore are identical.

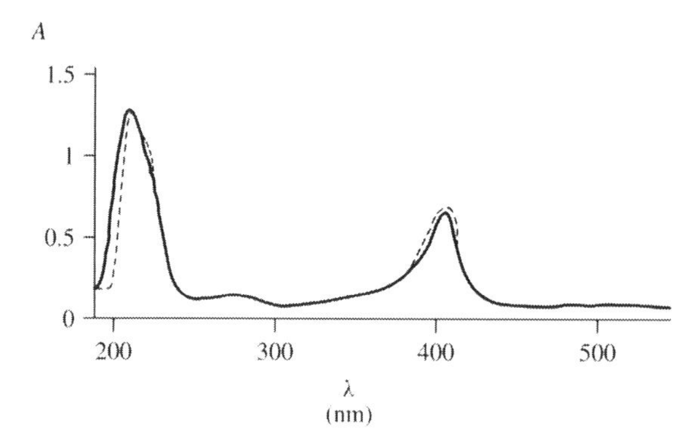

Figure 3.13. Solvent perturbation of haemoglobin. Absorption spectra are shown for 0.1 mg/ml haemoglobin in the presence (dashed line) and absence (solid line) of 20% sucrose. The solvent of reduced polarity (20% sucrose) enhances absorbance near 410 nm while decreasing absorbance near 220 nm. Note the similarity of these spectra to those of cytochrome C due to the presence of a haem group in both proteins.

In most cases changes in spectra as a result of environmental effects are obvious on inspection. It is also possible to plot a *difference spectrum* which may be obtained by *subtracting* one spectrum (A) from the other (B), that is $\varepsilon_A - \varepsilon_B$ or $A_{\lambda A} - A_{\lambda B}$ versus wavelength or frequency (Figure 3.15). This facilitates identification of wavelengths or frequencies where differences between the two spectra are greatest (Section 3.6.4; Example 3.5).

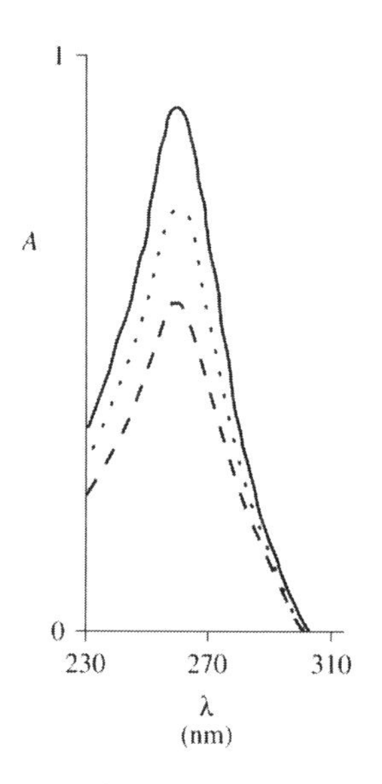

Figure 3.14. Hypochromicity of nucleic acids. This is an example of an orientation effect in spectroscopy. Spectra in the region of 260 nm are shown for identical concentrations of free nucleotides (solid line), single-stranded DNA (dotted line) and double-stranded DNA (dashed line). As the nucleotides are assembled into progressively more ordered structures, A_{260} decreases.

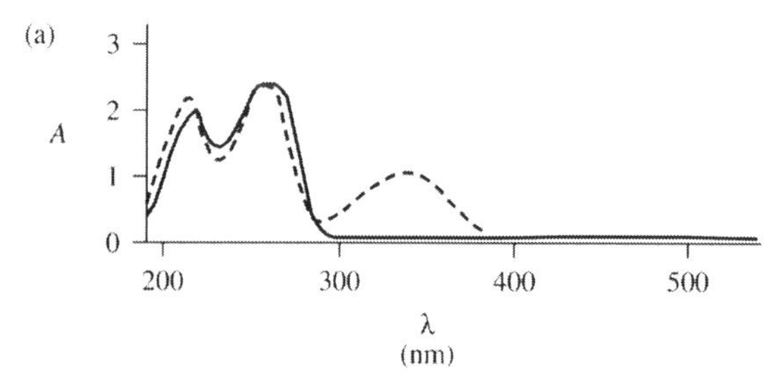

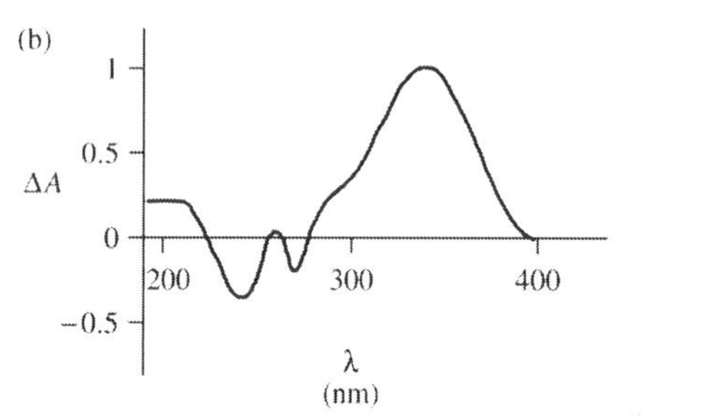

Figure 3.15. Difference spectroscopy. (a) Absorption spectra are shown for 0.2 mM NAD^+ (solid line) and NADH (dashed line). Note the λ_{max} at 340 nm specific for the reduced NADH. (b) Difference spectrum, that is plot of $\Delta A(A_{NADH} - A_{NAD+})$ versus λ.

Another way in which absorption spectra may be presented is as *derivative* spectra. This involves plotting the rate of change of absorbance dA_λ as a function of λ. It is particularly useful for high-resolution comparison of spectra.

3.3.2 Equipment Used in Absorption Spectroscopy

Absorption spectra are measured in a *spectrophotometer*, the basic outline of which is shown in Figure 3.9. Electromagnetic radiation is generated by a *lamp* which contains a metal filament through which an electric current flows. For wavelengths in the visible range a *tungsten-halogen* lamp (290–900 nm) is used while *deuterium* lamps provide both ultraviolet and visible radiation (210–370 nm). The light emitted from these sources will consist of a wide variety of wavelengths. It is necessary to select a single wavelength by passing the light through a monochromator of some sort. In modern instruments, this is usually a diffraction grating which achieves the same effect as the prism used by Newton. This monochromatic light is then used as the incident light in the absorption experiment.

Transmitted light is detected by a photodetector of which the most popular are photomultiplier tubes and photodiode array detectors. Photomultiplier tubes convert light

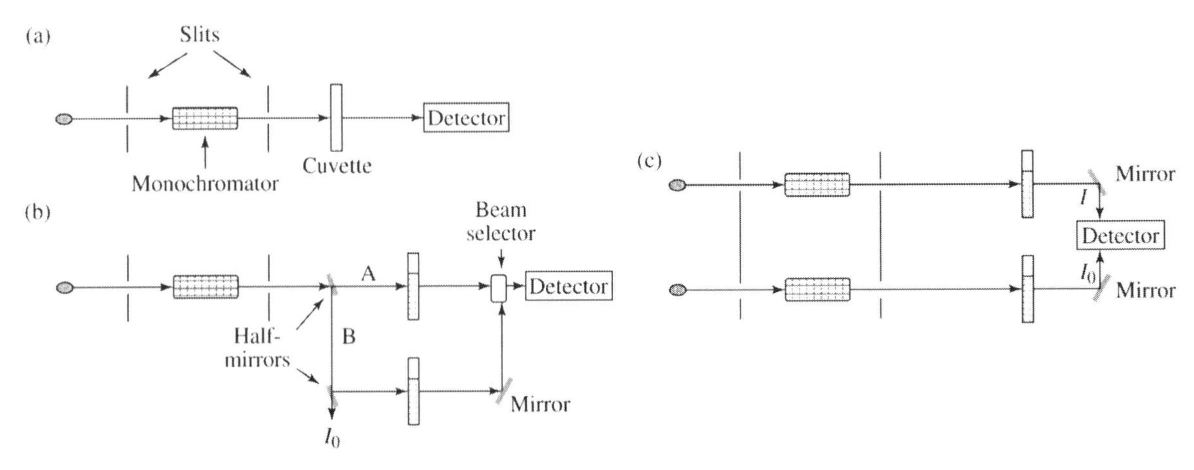

Figure 3.16. Some formats for spectrophotometers. (a) Single-beamed. (b) Split-beamed. Half-mirrors allow some incident light to pass through while reflecting the rest. Two beams of light (A and B) are therefore generated from a single source. One of the cuvettes provides a blank measurement which is automatically subtracted from that of the cuvette containing sample. Note that, at any given moment in time, the detector only measures light from beam A or B as determined by the beam selector. (c) Dual-beamed. Two sources generate individual beams.

intensity into an electric signal by amplifying the signal with a cascade of electrons. Photodiode arrays are cheaper alternatives to photomultiplier tubes and are composed of silicon crystals arranged in a linear array which are sensitive to light in the λ range 190–820 nm. Absorption of light by a photodiode produces a current proportional to the number of photons absorbed. Photodiodes allow extremely fast detection of light intensity. This type of detector is a feature of the diode array spectrophotometer in which the detector consists of an array of several hundred individual photodiodes. Each individual photodiode is assigned a particular wavelength across the range of the instrument. In the case of instruments with 1 nm resolution, each photodiode is assigned a wavelength 1 nm apart. In an instrument with 2 nm resolution each photodiode is assigned wavelength values 2 nm apart. This detector offers much faster scanning of spectra and greater sensitivity than is achievable with photomultiplier detectors.

A variety of design formats are available for spectrophotometers including single beam, split (double) beam and dual beam instruments (Figure 3.16). Use of a single, fixed wavelength may be appropriate in some applications. This is achieved by replacing the monochromator with filters which allow a single wavelength of incident light pass through to the sample. Examples of this include microtitre plate readers (which are used to take readings from 96-well microtitre plates) and variable wavelength detectors of the type used in chromatography (Chapter 2).

Sample is contained in tubes called cuvettes which may be constructed from a number of materials. Glass or plastic cuvettes are useful in the visible part of the spectrum but have the disadvantage that they may absorb ultraviolet radiation. Quartz cuvettes do not absorb in the ultraviolet but are fragile and expensive. Disposable plastic cuvettes which do not absorb at wavelengths greater than 280 nm are now commercially available. Plastic cuvettes may be damaged by exposure to organic solvents which can limit their usefulness in certain types of experiments. Because the cuvette (and/or solvent) may have its own absorption spectrum, it is essential to determine a blank spectrum of cuvette plus solvent and to subtract this from that of solvent containing the chromophore to obtain a true spectrum for the chromophore. Frequently this is achieved in a dual beam instrument with one beam passing through the blank cuvette and the other through the analytical sample.

Modern spectrophotometers are usually connected to a computer which facilitates storage, presentation and analysis of spectroscopic data.

3.3.3 Applications of Absorption Spectroscopy

Since many biomolecules possess distinct absorption spectra, absorption spectroscopy has a wide range of applications in biochemistry. Advantage is frequently taken of the Beer–Lambert law (Equation (3.6)) to carry out quantitative measurements. If the value of ε is known, it is possible to calculate concentrations directly from absorbance readings at specific wavelengths. In enzyme assays (Figure 3.17) the concentration of either substrate or product may be measured to allow calculation of rates of

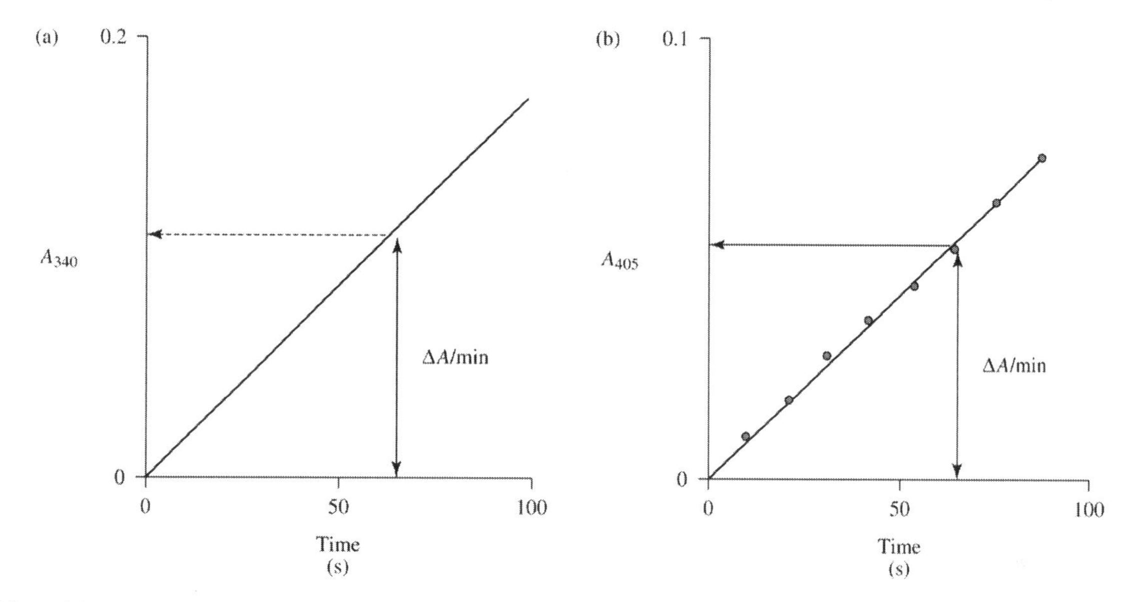

Figure 3.17. Use of spectroscopy in enzyme assays. (a) Continuous assay of a dehydrogenase producing NADH + H^+. The rate (μMoles/min) may be calculated from ε_{NADH} and the ΔA_{340} per minute (dashed arrow). (b) Stopped assay of alkaline phosphatase. This enzyme produces 4-nitrophenol from 4-nitrophenyl phosphate. A sample is taken from the assay at various times (points on graph) and NaOH is added. This halts the enzymatic reaction by rapidly increasing the pH and, simultaneously, converts 4-nitrophenol into the 4-nitrophenolate anion which has a strong absorbance at 405 nm. The rate is calculated as in (a).

enzyme-catalyzed reactions. These measurements may be *continuous* (i.e. the reaction is monitored continuously in the light-beam of a spectrophotometer) or *stopped* (i.e. the reaction is halted at various time-points and spectroscopic measurements are performed separately from the reaction producing product or consuming substrate). For example, reduction of NAD to NADH + H^+ by oxidoreductase enzymes is conveniently followed by continuous measurement of A_{340}. NADH has a strong λ_{max} at this wavelength while NAD does not (Figure 3.15). Other quantitative applications of absorbance measurements include measurement of concentrations of biomolecules with the aid of a *standard curve* (Figure 3.18).

A second type of application involves structural studies of biopolymers such as proteins and DNA. These are complex biomacromolecules and their assembly and unfolding are of considerable research interest. Chromophores, which are part of these molecules, are very sensitive to their immediate environment. We can often follow assembly/denaturation processes in such molecules, therefore, simply by monitoring absorbance changes at particular wavelengths, since these processes will alter the precise environment of the chromophore (Figure 3.19).

Under standard conditions, absorption spectra are characteristic for specific biomolecules. Such spectra may therefore provide a useful chemical 'fingerprint' for comparison of biomolecules (especially of low molecular mass) to each other. This makes possible confirmation of identity between molecules perhaps purified from different sources. Conversely, differences between spectra are good evidence for structural difference.

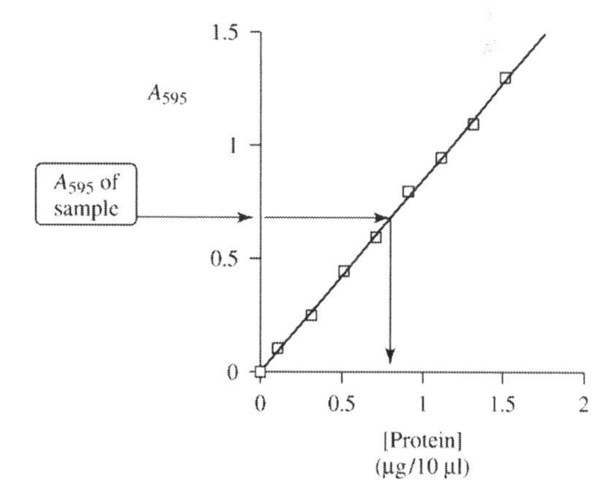

Figure 3.18. Determination of protein concentration using a standard curve – the Bradford assay. Coomassie brilliant blue has a λ_{max} at 596 nm when it complexes with protein. A standard curve is constructed with various known concentrations of bovine serum albumin. The concentration of an unknown sample may be estimated by comparison with this curve (arrows). The use of a standard curve for the relationship between A_λ and concentration depends on the Beer–Lambert law. It is commonly used for estimations in biochemistry.

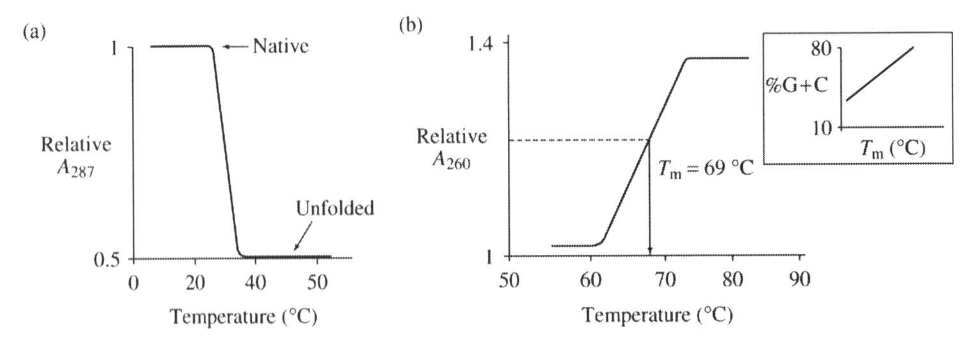

Figure 3.19. Spectroscopic determination of denaturation/refolding in biopolymers. (a) Thermal denaturation of ribonuclease A as determined by a decrease in A_{287}. Since unfolded protein has a lower A_{287} than folded, this gives a measure of denaturation. (b) Thermal denaturation of *Pseudomonas aeruginosa* DNA. Denaturation leads to an increase in A_{260} due to the hypochromicity of DNA. The melting temperature (T_m) may be calculated as the half-way point between completely folded and completely unfolded DNA (dashed line). There is a direct correlation between T_m and the % G + C content of DNA (inset).

Chromophores such as aromatic amino acids and nucleotide bases form part of the structure of biomacromolecules such as proteins and DNA. For this reason, they are referred to as *intrinsic* chromophores. It is also possible chemically to attach artificial groups with strong absorption spectra to proteins. Such groups are *extrinsic* chromophores and may be used as reporter groups because their spectra will be affected by their specific environment. Ideally, reporter groups should have a single site of attachment to the target macromolecule and should not affect its normal structure and function.

Absorption measurements are frequently the basis of detection methodologies in biochemistry. Such measurements may be made *directly* or *indirectly*. The variable wavelength detector used in chromatography systems mentioned above is a good example of direct spectroscopic detection (Section 3.3.2). Measurements at 210, 280 and 260 nm allow convenient on-line detection of peptides, proteins and DNA, respectively. Most biomolecules have relatively low intrinsic absorbance values, however, which limit their sensitivity in detection systems. Moreover, not every molecule in a complex mixture will have the same value for ε at a given wavelength. For these reasons, indirect measurements of a particular extrinsic chromophore with a strong absorption are widely used in detection systems. Examples include *chromogenic* substrates used in the detection of enzymes such as horseradish-peroxidase and β-galactosidase (Table 3.3). These are nonbiological substrates with weak absorption spectra which are enzymatically converted into chromophores with strong spectra. Conjugation of such enzymes to specific antibodies provides the basis for highly sensitive *immunodetection* in commercial diagnostic kits. High sensitivity is conferred by *signal amplification*, that is the production of many molecules of product by a single molecule of enzyme.

3.4 FLUORESCENCE SPECTROSCOPY

3.4.1 Physical Basis of Fluorescence and Related Phenomena

The *Morse diagram* shown in Figure 3.10 explains the apparent disappearance of light energy in absorption spectroscopy in terms of appearance of small increments of heat energy which are lost in kinetic collisions in solution. Because vibrational energy levels of the ground and excited states of most chromophores overlap, they can return to the ground state by a series of *nonradiative transitions*. Some chromophores are quite rigid and inflexible molecules however, and may have a limited range of vibrational energy levels. In such molecules, the vibrational energy levels of the excited state often do not overlap with those of the ground state (Figure 3.20). When chromophores of this type absorb light, it is not possible for them to return to the ground state by simply losing their excess energy as heat. Instead, they undergo a *radiative transition* in which a portion of the absorbed energy is re-emitted as light. This phenomenon is called *fluorescence* and such chromophores are called *fluors* or *fluorophores*. Since at least some of the light energy initially absorbed is lost in transitions between vibrational energy levels, the light energy emitted is always of longer wavelength (i.e. lower energy) than that absorbed. This means that fluors have a characteristic *fluorescence* or *emission spectrum* as well as a characteristic absorbance spectrum (Figure 3.21).

The *excited state lifetime* of a fluor (i.e. the period of time spent in the excited state) is usually very short (ranging from 0.5 to 8 ns). However, in certain cases the period of time spent in the excited state may be significantly longer ranging up to 2 s. The latter situation can arise as a consequence of a phenomenon associated with electrons called

Table 3.3. Some chromogenic enzyme substrateas

Substrate		Product
p-Nitrophenyl phosphate	PHOSPHATASE → (P_i)	λ_{max} = 405 nm
2-Hexadecanoylamino-4-nitrophenylphosphorylcholine	SPHINGOMYELINASE → (Phospherylcholine)	λ_{max} = 410 nm
5-Bromo-4-chloro-3-indolyl-*β*-D-galactopyranoside (X-Gal)	*β*-GALACTOSIDASE → (Galactose)	Product is blue against a white background (activity stain)
3, 5, 3,′ 5′-Tetramethylbenzidine (TMBZ)	PEROXIDASE → (H_2O_2, 2 H_2O)	λ_{max} = 450 nm

Note: These substrates have negligible absorbance at λ_{max} but, when converted into product by the enzymes, a strong absorbance is detected at this wavelength.

magnetic spin which will be described in more detail later (Section 3.7). In most atoms, pairs of electrons in a single orbital have spins oriented in *opposite* directions as a result of the *Pauli exclusion principle*. Such atoms are said to exist in a *singlet state* and the electrons are said to be *paired*. If an electron is promoted to a higher energy orbital however, then its spin can be oriented either in the same or opposite orientation as the electron left in the original orbital (i.e. *parallel* or *antiparallel*). When placed in a magnetic field, this atom can have three distinct energy levels which may be designated −1, 0 and +1, that is it now exists in a *triplet state*. Some heavy atoms possessing even numbers of electrons in their singlet ground state can have two unpaired electrons stably occupying different orbitals in a triplet state of slightly higher energy than the ground state (e.g. molecular oxygen). In such atoms, it is possible for a transition to

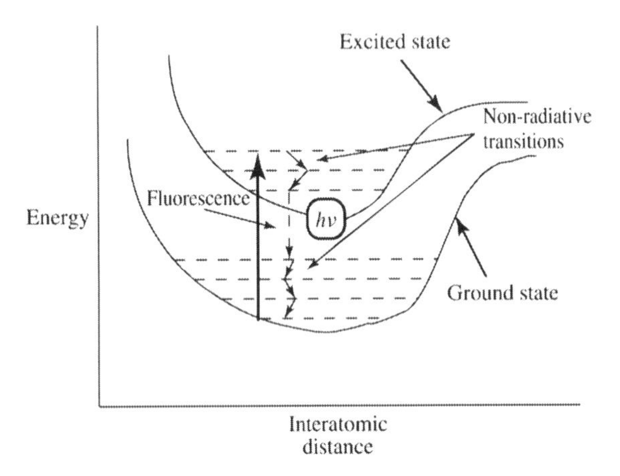

Figure 3.20. Physical basis of fluorescence. If a chromophore has a comparatively rigid structure with few vibrational energy levels such that the electronic energy levels of the ground and excited states do not overlap (*cf.* Figure 3.10), a radiative transition may occur (dashed arrow) which is called fluorescence. This results in emission of radiation of frequency ν, and wavelength λ. Such molecules are called fluors. Phenomena related to fluorescence include phosphorescence and chemiluminescence (*see* text).

occur between the singlet ground state (where the electrons are antiparallel and paired) to an excited state (where the electrons are parallel and unpaired). When this molecule returns to the ground state a radiative triplet-singlet transition occurs which is a much *slower* process than fluorescence called *phosphorescence*. Radiation emitted as a result of this transition is at an even longer wavelength than radiation emitted during fluorescence because the lowest-energy

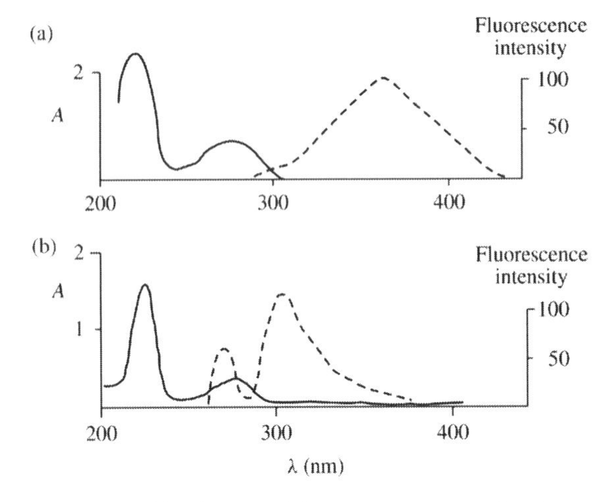

Figure 3.21. Absorbance and fluorescence spectra. Absorbance (solid line) and fluorescence (dashed line) spectra of (a) tryptophan and (b) tyrosine.

triplet state typically has only slightly higher energy than the lowest-energy singlet state.

A third phenomenon resulting in radiative transitions called *chemiluminescence* occurs in molecules which can be promoted to an excited state as a result of a chemical reaction and which then return to the ground state with the emission of light. Molecules such as *luciferin* and *luminol* (Table 3.4) show this property and, when used as substrates by the enzymes *firefly luciferase* and peroxidase, respectively, emit light which may be measured or detected (Section 3.4.2). These reactions are highly efficient and may be readily incorporated into sensitive assay systems or in staining of protein or nucleic acid blots (Figure 3.22). For example the level of ATP (or of any biomolecule which may be coupled to ATP production) may be quantified as follows:

$$\text{Luciferin} + \text{ATP} + O_2$$
$$\downarrow \quad \text{(Firefly Luciferase)}$$
$$\text{Oxyluciferin} + \text{AMP} + \text{pyrophosphate} + CO_2 + \text{light}$$

This assay is extremely sensitive allowing detection at the pM/fM level. Since peroxidase-conjugated antibodies are extensively used in immunodetection systems, luminol in particular has wide applications in this general area (e.g. Section 5.10.2). Both phosphorescence and chemiluminescence are related to fluorescence in the sense that all three phenomena result in emission of light.

Fluors and molecules capable of chemiluminescence are often found to possess complex ring structures which are largely responsible for their rigidity (Table 3.4). Most components of biopolymers are nonfluorescent and this means that fluorescence may give information about processes occurring in specific parts of the biopolymer. For example, the most important *intrinsic* fluors in proteins are tryptophan, tyrosine and phenylalanine residues (in practice, tryptophan and tyrosine give stronger spectra than phenylalanine). Fluorescence of tyrosine is frequently quenched as a result of proton transfer in the excited state (*see* below). These residues have average occurrences of only 3.5, 1.1 and 3.5% in proteins, respectively. Accordingly, many proteins with extensive absorbance spectra may have only one or two fluorescent residues and advantage can be taken of this in the study of biochemical processes involving the protein. The bases of DNA nucleotides and of some enzyme cofactors (e.g. NAD) are also intrinsic fluors although they display generally weak fluorescence spectra. In such cases, *extrinsic fluors* may be used to carry out structural and functional studies. These are nonbiological molecules which possess fluorescence properties (Table 3.4).

Table 3.4. Chemical structures of some common fluors and chemiluminescent compounds

Compound	Structure	Use
Tyrosine	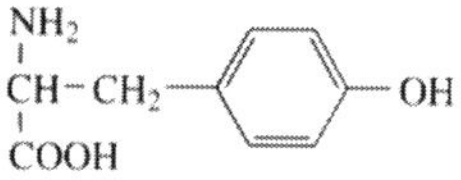	Intrinsic fluor (proteins)
Tryptophan	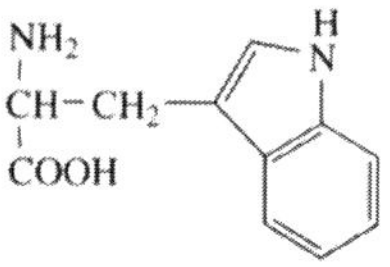	Intrinsic fluor (proteins)
1-Anilino-8-naphthalene sulphonate (ANS)	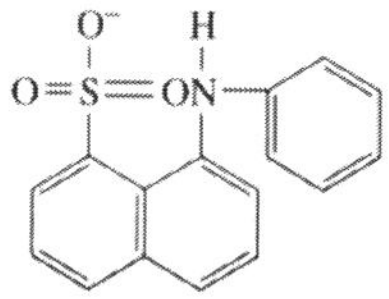	Extrinsic fluor (proteins)
Fluorescein	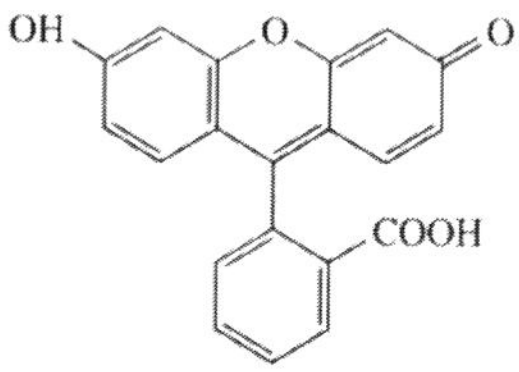	Extrinsic fluor (proteins)
Ethidium bromide	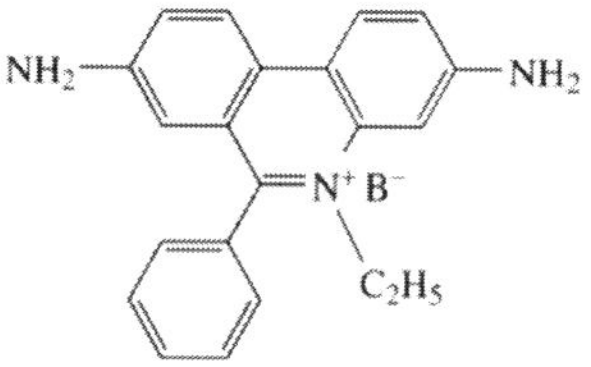	Extrinsic fluor (DNA)
Acridine orange	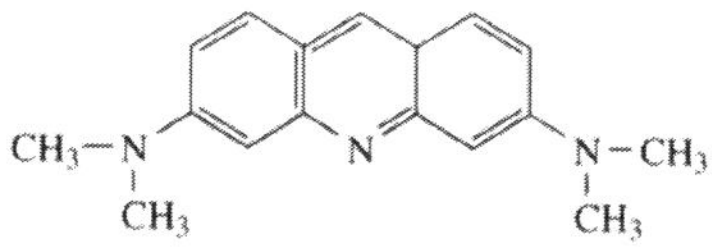	Extrinsic fluor (DNA)
Luminol (3-Aminophthalhydrazine)	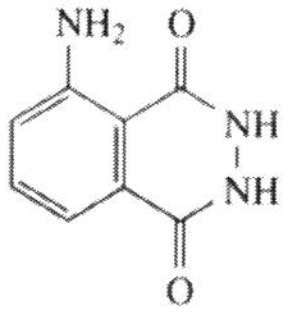	Chemiluminescent substrate (peroxidase)
Luciferin	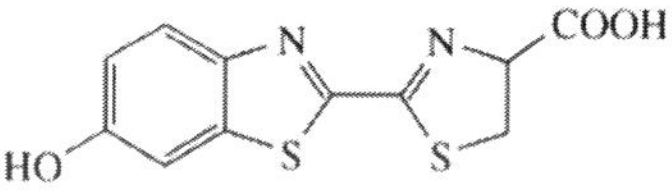	Chemiluminescent substrate (firefly luciferase)

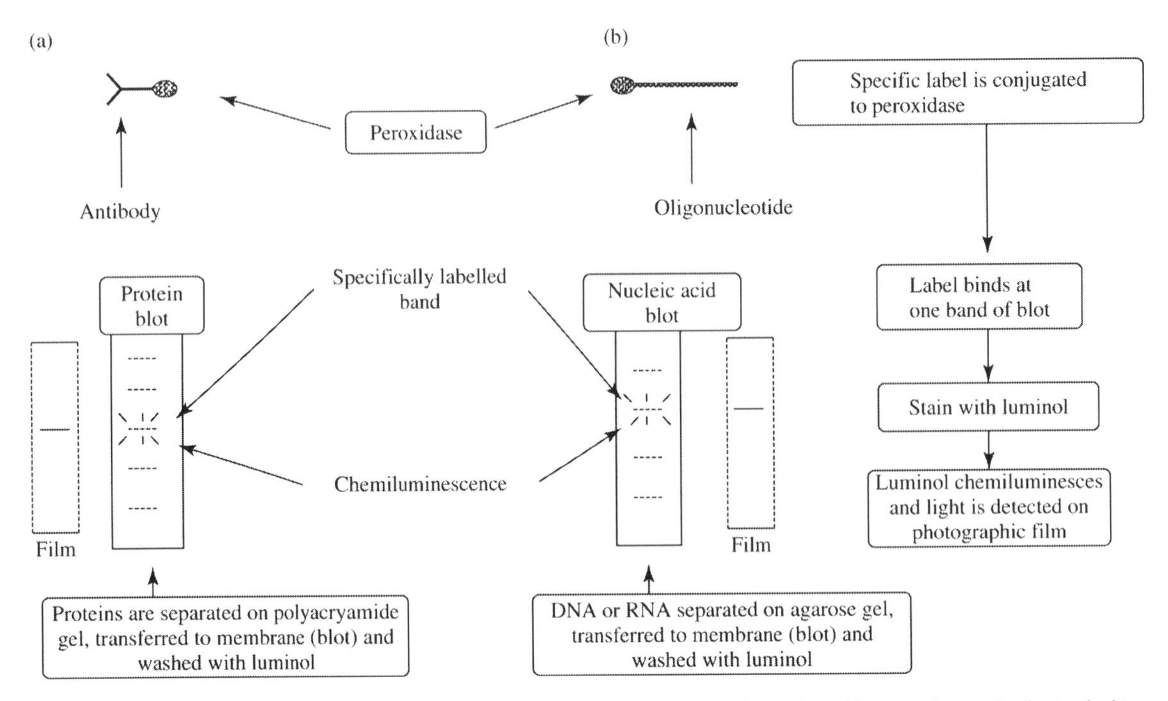

Figure 3.22. Use of chemiluminescence in specific staining of protein and nucleic acid blots. Peroxidase may be covalently attached to an antibody or oligonucleotide as shown. After transfer of (a) protein or (b) nucleic acids to a suitable membrane (Chapter 5), specific proteins or nucleic acids may be visualized by staining with luminol. Light emitted by chemiluminescence may be detected on photographic film.

3.4.2 Measurement of Fluorescence and Chemiluminescence

Fluorescence measurements are performed in a *spectrofluorimeter*, an outline diagram of which is shown in Figure 3.23. An *incident beam* of radiation of a given wavelength is passed through a sample cuvette containing the fluor. *Emitted radiation* (of longer wavelength than that of the incident beam) is detected by a photomultiplier tube. This is generally similar to the design of the spectrophotometer previously described (Figure 3.9). The main differences are that emitted radiation is detected at 90° to the direction of the incident light beam and that a second monochromator is required to select for the different wavelength of the emitted light. Since fluorescence is emitted in all directions from the fluor, this design excludes inadvertent detection of the incident beam. A consequence of this arrangement is that cuvettes used in fluorescence measurements must have four clear sides whereas those used in absorbance measurements may often have *frosted* or *blackened* side-walls.

Chemiluminescence is measured in a *luminometer* (Figure 3.24). The reaction leading to light emission is carried out in a cuvette and the wavelength and intensity of emitted light is detected. When part of an *in situ* detection system (e.g. western blotting), the emitted light can be detected on photographic film.

Only a proportion of the light energy originally absorbed is emitted as radiation, since some energy may be lost in vibrational transitions as previously described (Figure 3.20). Two further processes can diminish or *quench* the amount of light energy emitted from the sample. *Internal quenching* is due to some intrinsic structural feature of the excited molecule involving, for example, structural rearrangement. *External quenching* arises either from interaction of the excited molecule with another molecule present in the sample or else absorption of exciting or emitted light by another chromophore present in the sample. All forms of quenching result in nonradiative loss of energy. The amount of energy lost due to both vibrational transitions and internal quenching will be a property of the particular molecule since these processes arise as a result of chemical structure. External quenching may be due to contaminants present in preparations or, alternatively, may be deliberately introduced into the experiment. Acrylamide, iodide and ascorbic acid are examples of molecules capable of acting as external quenchers.

We can quantify fluorescence by the *quantum yield*, Q:

$$Q = \frac{\text{Number of photons emitted}}{\text{Number of photons absorbed}} \tag{3.7}$$

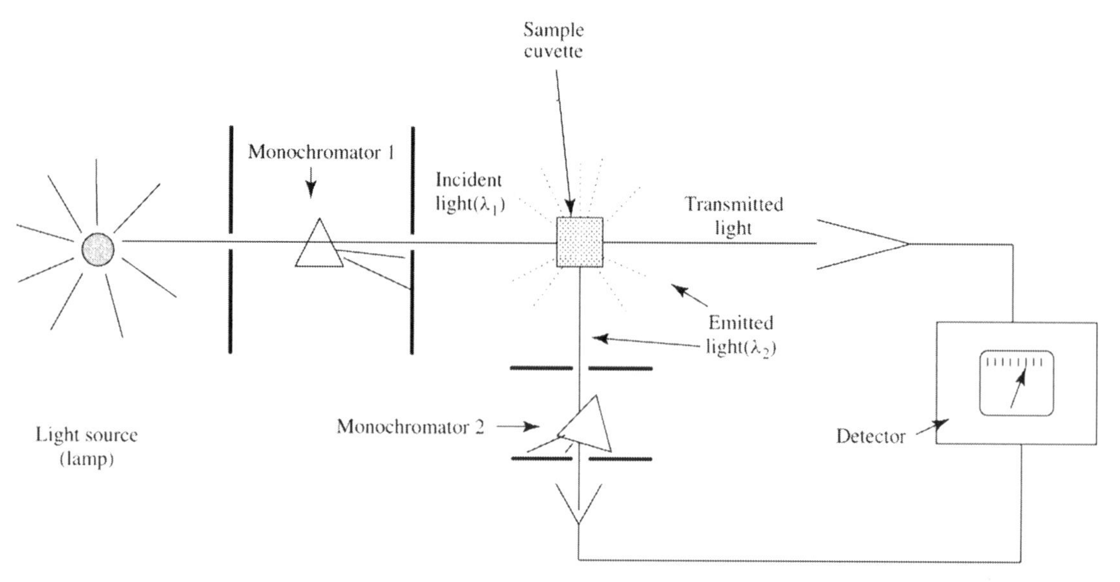

Figure 3.23. Ultraviolet/visible spectrofluorimeter. Light of a single wavelength (λ_1) is passed through and absorbed by a sample held in a cuvette as described in Figure 3.9. At certain values of λ, light is absorbed and the molecule fluoresces. Fluorescent light is emitted in all directions (dashed lines) and is measured at 90° to the incident beam. Emitted light, which is of longer wavelength (λ_2) than absorbed light, is passed through a second monochromator and detected.

Under a given set of conditions, Q will usually have a fixed value for a particular fluor, with a maximum possible value of 1. However, because it is experimentally difficult to measure Q accurately, we often use *relative* measurements of fluorescence in practice. The possibility of external quenching has the important practical consequence that great care must be taken in carrying out fluorescence measurements to ensure the absence of quenchers from the sample and all solutions used. These may therefore require prior purification and filtration.

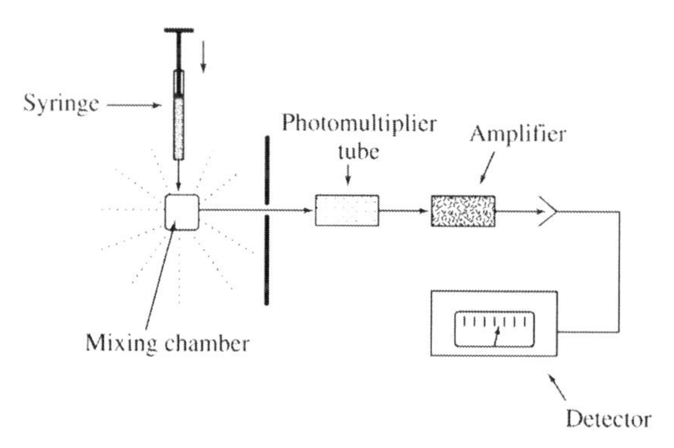

Figure 3.24. Outline design of a luminometer. Reactants (e.g. firefly luciferase and luciferin) are introduced into a mixing chamber by a syringe or similar device. When chemiluminescence occurs, light is emitted in all directions (dashed lines). A beam of such light is selected by a slit and collected by a photomultiplier tube. The electrical signal from this is amplified and passed to a detector. Note that the reaction chamber is protected from background light and is temperature-controlled.

Fluorescence is usually much more sensitive than absorbance spectroscopy since the lower limit of detection of a spectrophotometer is determined by the ability of the instrument to distinguish between I_0 and I (Equation (3.6)). By contrast, the lower limit of detection of a *spectrofluorimeter* is set by the ability of the instrument to detect light (i.e. to distinguish I from 0). Use of a more powerful light-source and a more sensitive detector increases the sensitivity of spectrofluorimeters while giving little improvement in sensitivity of spectrophotometers. Moreover, most chromophores lack fluorescence properties altogether and even large macromolecules such as proteins may contain a small number of intrinsic fluors. A further advantage of fluorescence over absorption spectroscopy is the fact that fluorescence spectra are even more sensitive to environmental effects than absorption spectra. For these reasons fluorescence is generally more sensitive and specific than absorbance.

3.4.3 External Quenching of Fluorescence

External quenching may involve direct contact between the excited molecule and external quencher. This is possible by two main mechanisms. *Dynamic quenching* involves

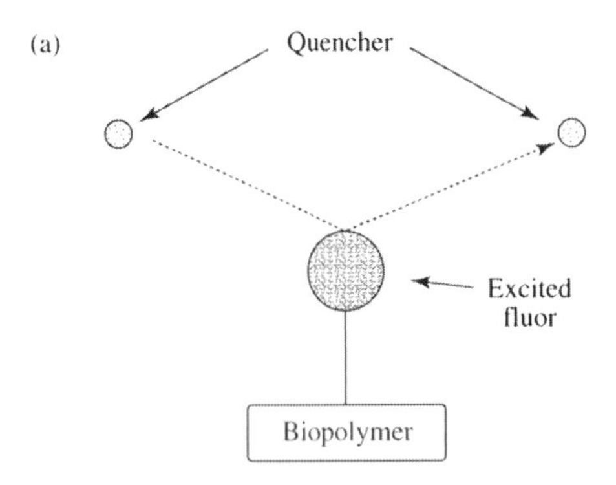

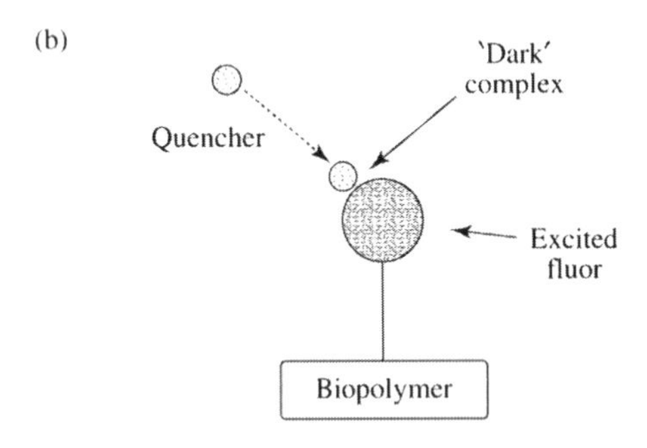

Figure 3.25. Mechanisms of fluorescence quenching. (a) Collisional or dynamic quenching. The quencher collides with the excited fluor leading to loss of some energy from the excited state as kinetic energy. (b) Static quenching. The quencher and excited fluor form a stable complex called a 'dark' complex. Some energy from the excited state is lost during this process.

collision between the two molecules with the fluor losing energy as kinetic energy. *Static quenching* involves a more long-lasting formation of a complex between the fluor and quencher, which is sometimes referred to as a *dark* complex (Figure 3.25). It is often difficult to distinguish these two processes experimentally, but the following discussion will show how this may be achieved.

An excited fluor (X^*) can lose energy by three general processes described by the following relationships:

1. The fluor can fluoresce (Fl):

$$X* \xrightarrow{k_1} X + \text{light} \quad (3.8)$$

2. The fluor can experience internal quenching (IQ):

$$X* \xrightarrow{k_2} X \quad (3.9)$$

3. The fluor can experience external quenching (EQ) in the presence of a specific concentration of external quencher $[q]$:

$$X^* + q \xrightarrow{k_3} X_q \quad (3.10)$$

These processes are governed by rate constants $k_1 - k_3$, respectively. The overall fluorescence measured, F, is given by the ratio of fluorescence to the sum of these three possibilities:

$$F = \frac{\text{Fl}}{\text{Fl} + \text{IQ} + \text{EQ}} \quad (3.11)$$

If little internal or external quenching occurs, the quantum yield may approximate closely to 1 (although, depending on the fluor, it is usually less than 1 due to nonradiative vibrational transitions as discussed above). In the presence of an external quencher, the value of F will be *smaller* than in the absence of quencher. The rate of each of the three processes can be calculated from simple reaction kinetics and substituted into Equation (3.11):

$$F = \frac{k_1 \cdot [X^*]}{k_1 \cdot [X^*] + k_2 \cdot [X^*] + k_3 \cdot [X^*] \cdot [q]} \quad (3.12)$$

where square brackets denote concentration. This relationship can be simplified further by canceling out the common term $[X^*]$:

$$F = \frac{k_1}{k_1 + k_2 + k_3 \cdot [q]} \quad (3.13)$$

The fluorescence in the *absence* of q (i.e. $[q] = 0$), F_0, is given by:

$$F_0 = \frac{k_1}{k_1 + k_2} \quad (3.14)$$

The ratio between F_0 and F is therefore:

$$\frac{F_0}{F} = \frac{k_1 + k_2 + k_3 \cdot [q]}{k_1 + k_2} \quad (3.15)$$

$$= 1 + \frac{k_3}{k_1 + k_2} \cdot [q] \quad (3.16)$$

The value for $(1/k_1 + k_2)$ is the excited state lifetime of the fluor and would be expected to be constant for a particular molecule. It can be designated as T. Therefore:

$$\frac{F_0}{F} = 1 + k_3 \cdot [q] \cdot T \quad (3.17)$$

This is known as the *Stern–Volmer relationship* and it is used widely in analysis of fluorescence quenching data. The product of k_3 and T is defined as the *dynamic Stern–Volmer constant*, K_{SV}:

$$K_{SV} = k_3 \cdot T \qquad (3.18)$$

A plot of the ratio of fluorescence intensity in the absence of external quencher (F_0) to that in the presence of external quencher (F) as a function of concentration of external quencher ([q]) is expected to be *linear* with a slope equal to K_{SV} and an intercept of 1. Such a graph is known as a *Stern-Volmer plot* and an example is shown in Figure 3.26. In practice such plots often show deviations from linearity (i.e. curving) because of a variety of factors such as heterogeneity in fluor excited state, a mixture of quenching mechanisms and deviation from ideal behaviour. Such situations are beyond the scope of the present discussion but, in analyzing such processes, modified forms of the Stern–Volmer equation are used.

Where quenching arises as a result of a collisional process, k_3 is the *dynamic quenching rate constant*. However, when a stable bimolecular complex is formed leading to quenching:

$$F + q \rightarrow Fq \qquad (3.19)$$

The ratio between F_0 and F is given by;

$$\frac{F_0}{F} = 1 + K_{SV} \cdot [q] \qquad (3.20)$$

where K_{SV} is the *association constant* governing the complex formed in 3.19. Dynamic and static quenching therefore give Stern–Volmer plots of the same mathematical form and the two mechanisms cannot be distinguished by Stern–Volmer plots alone. In practice, it is necessary to measure F_0/F under a range of conditions of temperature and viscosity.

A major difference between dynamic and static quenching is the fact that temperature affects the two processes in opposite ways as described in more detail below. Dynamic quenching depends on collisions between the excited fluor and the quencher and results in a *decrease* in the excited lifetime, T. It is a *diffusion-controlled* process which *increases* with temperature. Static quenching, on the other hand, results from formation of a complex between a portion of the fluor and the quencher which leaves the bulk of the fluor unquenched. This process does not affect T and, moreover, occurs *less efficiently* at higher temperatures since the fluor-quencher complex is likely to be less stable under these conditions. It is to be expected, therefore, that static quenching decreases while dynamic quenching increases with temperature and that T is unaffected in static quenching but decreases in dynamic quenching.

Solvent viscosity strongly affects dynamic quenching and can be used experimentally to distinguish it from static quenching. The dynamic quenching rate constant, k_3, is related to the bimolecular rate constant for collision, k_0, by the *Smoluchowski equation*:

$$k_3 = \gamma \cdot k_0 \qquad (3.21)$$

where γ is the *quenching efficiency* which has values in the range 0–1. k_0 can be related to physical attributes of the fluor/quencher pair by the following relationship:

$$k_0 = \frac{4 \cdot \pi \cdot N \cdot X \cdot (D_f + D_q)}{1000} \qquad (3.22)$$

where N is Avogadro's number, X is the distance of closest approach between the fluor and quencher (i.e. the sum of their molecular radii) and D_f and D_q are the diffusion coefficients of the fluor and quencher, respectively. For dilute

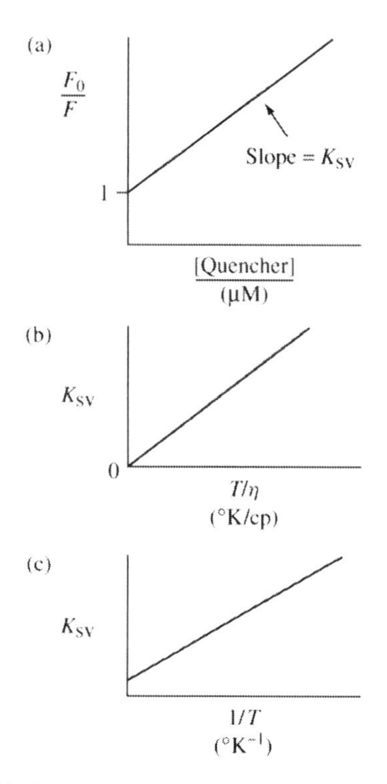

Figure 3.26. Experimental distinction of mechanism of quenching. (a) Stern–Volmer plot for both static and dynamic quenching is often linear. (b) A linear dependence of K_{SV} on T/η is diagnostic of dynamic quenching. (c) A linear dependence of log K_{SV} on $1/T$ is diagnostic of static quenching.

solutions, the diffusion coefficient, D, can be approximated by the *Stokes–Einstein equation* as follows:

$$D = \frac{k \cdot T}{6 \cdot \pi \cdot \eta \cdot r} \tag{3.23}$$

where η is the viscosity of the solution, r is the radius of the molecule, k is Boltzmann's constant and T is absolute temperature. Thus, for an efficient quencher obeying only a dynamic quenching mechanism, it is expected that k_3 *should vary linearly with the ratio* T/η.

By contrast, a quencher obeying only a static quenching mechanism gives a slope determined by the association constant, K_{SV}, which *would be expected to obey the van 'tHoff equation* since it is an *equilibrium constant*:

$$\frac{d \log K_{SV}}{d(1/T)} = \frac{\Delta H}{2.3 \cdot R} \tag{3.24}$$

where ΔH is the enthalpy change associated with complex formation and R is the Universal Gas Constant.

This means that measurements of fluorescence quenching may be made under varying conditions of temperature and viscosity. In the case of solely dynamic quenching the slope of Stern-Volmer plots is expected to vary linearly with T/η while in solely static quenching, its log varies linearly with $1/T$ (Figure 3.26). In practice, a particular fluor/quencher pair may interact by a combination of these two mechanisms. For details of analysis of such complex situations literature in the bibliography at the end of this chapter should be consulted.

Fluorescence spectra are strongly influenced by exposure to solvent and/or quenchers present in the solvent. This fact underlies much of the practical usefulness of fluorescence spectroscopy in biochemistry. For example, slight changes in the spatial position or solvent exposure of intrinsic protein fluors such as tryptophan can result in either an increase or a decrease of quantum yield as well as qualitative changes in the spectrum such as a shift to longer wavelengths (a *red shift*). Fluorescence measurements may be performed at single emission wavelengths or, more informatively, may be acquired across the whole spectrum. Moreover, fluorescence measurements may be made under *steady-state* conditions (i.e. at a single time) or may be *time-resolved* (i.e. measurements are taken at various times along a time-course). In order to give an indication of the versatility of this technique, we will now describe some of the main applications of fluorescence in biochemistry.

3.4.4 Uses of Fluorescence in Binding Studies

Binding of ligands to proteins frequently causes changes to their three-dimensional structure. Examples of this include the binding of substrates, inhibitors, cofactors or allosteric modulators to enzymes or of hormones to receptors. If this structural change has an effect on the *environment* of an intrinsic or extrinsic fluor in the protein, this can result in measurable changes in the fluorescence spectrum. Provided that the fluor has a unique location in the protein, such changes of fluorescence at a particular wavelength (ΔF) can be used to determine the dissociation constant (K_D) of the protein for the ligand where K_D is a measure of the affinity of the protein for the ligand (Figure 3.27):

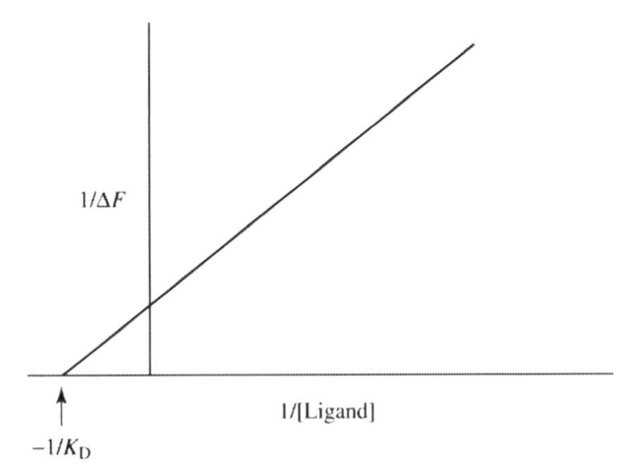

Figure 3.27. Use of intrinsic fluorescence measurements for determination of ligand K_D in binding proteins. Changes in intrinsic fluorescence (ΔF) as a result of ligand binding to a protein are determined. Plots of $1/\Delta F$ versus 1/[Ligand] are linear. The K_D may be determined from the intercept of this line on the abscissa (Example 3.2).

$$P + L \rightleftarrows PL$$

$$K_D = \frac{[P] \cdot [L]}{[PL]} \tag{3.25}$$

Some extrinsic fluors such as 1-anilino 8-naphthalene sulphonate (ANS; Table 3.4) are only strongly fluorescent when bound to proteins. ANS is attracted to and will bind at slightly hydrophobic pockets on the protein surface. Competition experiments can be carried out in which this fluor *competes* with other ligands (Figure 3.28). Fluorescence measurements of both intrinsic and extrinsic fluors can therefore be used to estimate K_D values for ligands in proteins (Example 3.2). This technique complements other methods of K_D determination such as *equilibrium dialysis* (Chapter 7; Section 7.2.1).

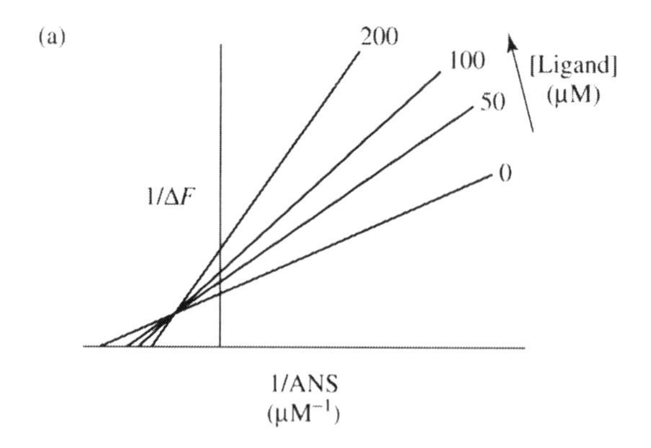

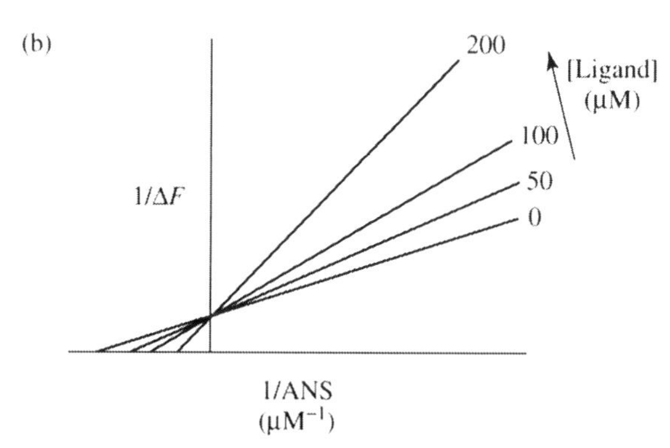

Figure 3.28. Competition between ANS and ligand for binding site on protein. Change in fluorescence (ΔF) is measured in the presence of various concentrations of ANS and ligand. (a) ANS and ligand bind at different sites. (b) ANS and ligand compete for the same site (lines intersect on ordinate). In practice, other patterns of intersecting lines may be observed but the pattern observed in (b) is diagnostic for competitive binding.

3.4.5 Protein Folding Studies

Fluorescence is one of the main spectroscopic techniques used in the study of protein folding (for a more detailed discussion of this topic *see* Chapter 6). Folded proteins are very compact structures in which individual residue side-chains adopt highly-defined spatial locations. When a protein unfolds from the native (N) to the unfolded (U) state, these side-chains become free to rotate around C_α and C_β-bonds and move freely in solution. Tryptophan (the main intrinsic fluor of proteins) is normally buried in the interior hydrophobic core of the protein. When the protein unfolds, tryptophan comes into contact with solvent water and may also be able to interact with quenching agents.

The effects of unfolding on intrinsic fluorescence are individual for each protein but four main experimental parameters are usually observed; (1) The fluorescence intensity may *increase or decrease* as a result of unfolding, thus causing an increase or a decrease in Q. (2) The fluorescence lifetime, T, may decrease as a result of dynamic quenching (e.g. encounters with disulfides or amines) or be unchanged as a result of static quenching. (3) The radiation emitted from an unfolded protein is usually of *lower energy and therefore of longer wavelength* (i.e. the spectrum is *shifted* to the red end of the spectrum). (4) The fluor is less constrained in terms of rotational movement around C_α and C_β-bonds in the unfolded state than in the folded state. This means that interactions between the fluor and *plane-polarized light* (Section 3.5) are different in the folded and unfolded versions of the protein. A parameter called *polarization* is a good measure of this conformational mobility of tryptophan and is *lower* in unfolded proteins than in folded proteins.

A further measure of solvent-exposure of tryptophan is quenching of fluorescence by an external quencher such as potassium iodide. This agent, which may be dissolved in solvent water, has a strong negative charge which precludes it from diffusing into the hydrophobic interior of folded or partially-folded proteins. Quenching of fluorescence by KI is therefore good evidence for exposure of tryptophan to solvent during protein unfolding.

Fluorescence is not the only experimental measure of protein folding since absorbance spectroscopy and circular dichroism (CD; Section 3.5.3) are also widely used. If protein folding approximates closely to a two-state model (i.e. N → U), these independent methods should give coincident estimates of unfolding. However, differences in estimates between the different techniques may be important experimental evidence for deviation from a two-state model and for the existence of other species such as folding intermediates (Chapter 6).

3.4.6 Resonance Energy Transfer

Since a fluor has characteristic λ_{max} values in both its absorbance and emission spectra, it is possible to establish an experiment in which the emission λ_{max} of one fluor (A) overlaps with the absorbance λ_{max} of a second fluor (B). If these separate fluors (extrinsic or intrinsic) have unique locations in a protein or macromolecular complex, it is possible for emission light energy from fluor A to be absorbed by fluor B and to be emitted as part of B's emission spectrum (Figure 3.29). This phenomenon is called *resonance energy transfer* and, since it is strongly dependent on the *distance*, R, between the fluors, it may be used to measure distances in proteins, membranes and macromolecular assemblies especially in the range 10–80 Å. The *efficiency* of the transfer process, E, may be expressed in a number of ways as follows.

$$E = \frac{k_T}{k_T + k_f + k'} \tag{3.26}$$

Example 3.2 Use of fluorescence in binding studies

The glutathione transferases (GSTs) have extensive binding properties which may help to explain why they are the most abundant cytosolic proteins in rat and human liver, comprising some 3–5% of total protein. They are able to bind foreign compounds such as drugs and dyes as well as a range of naturally-occurring, generally hydrophobic ligands such as cholic acid and bile salts. The most abundant GSTs possess a small number (2–4) of tryptophan residues which act as intrinsic fluors. Binding of ligands can be conveniently followed by measuring fluorescence quenching. This is because ligand binding causes a slight change in the location of tryptophan. If the change in fluorescence intensity (ΔF) is plotted against the ligand concentration as a double-reciprocal plot, the dissociation constant (K_D) of the protein for the ligand can be conveniently determined. The following graph shows a typical plot for binding of bilirubin to GST 1–2.

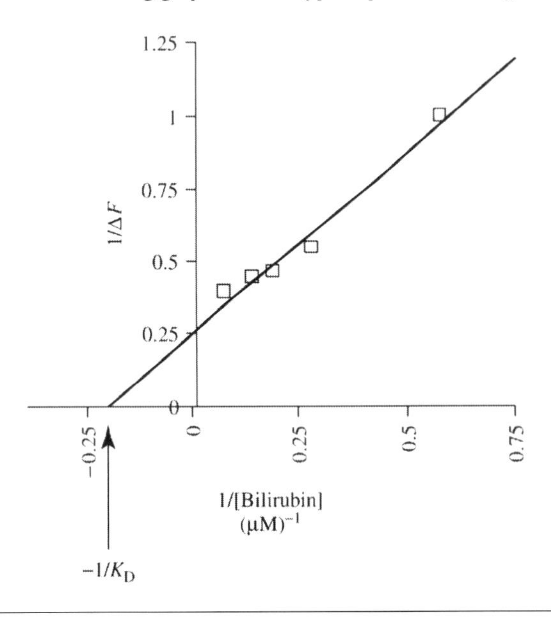

where k_T, k_f and k' are, respectively, the *rates* of transfer of excitation energy, of fluorescence and the sum of all other de-excitation processes such as vibrational transitions.

$$E = \frac{R_0^6}{R_0^6 + R^6} \tag{3.27}$$

where R is the distance between the donor and acceptor molecules and R_0 is a constant related to the donor-acceptor pair which can be calculated from their absorption and emission spectra. In practice, E can be determined either from the fluorescence intensity (F) or the excited state lifetime (T) of the donor determined in the presence (da) and absence (d) of the acceptor as follows:

$$E = 1 - \frac{F_{da}}{F_d} \tag{3.28}$$

$$E = \frac{1 - T_{da}}{T_d} \tag{3.29}$$

Once the value of E has been determined, R can be calculated from Equation (3.27) once R_0 is known. Measurements of distances determined by this method agree well with those from independent structural determinations made using techniques such as X-ray diffraction in crystals and NMR (Chapter 6).

An advantage of resonance energy transfer is that the technique can be used to probe the topology of membrane-bound proteins which are not amenable to study by crystallographic methods and which are difficult to study by NMR. An example is the P-glycoprotein which is responsible for the multiple drug resistance (MDR) phenotype, a major problem in cancer chemotherapy. The technique is also useful in high-sensitivity detection of *changes* in structure of biomacromolecules and macromolecular complexes since such changes result in altered values for R. Recent examples of such experiments include studies of conformational change in contractile proteins (e.g. actin and myosin) on ligand binding, mapping of the active site of enzymes (e.g. serine proteases), monitoring duplex and tetraplex

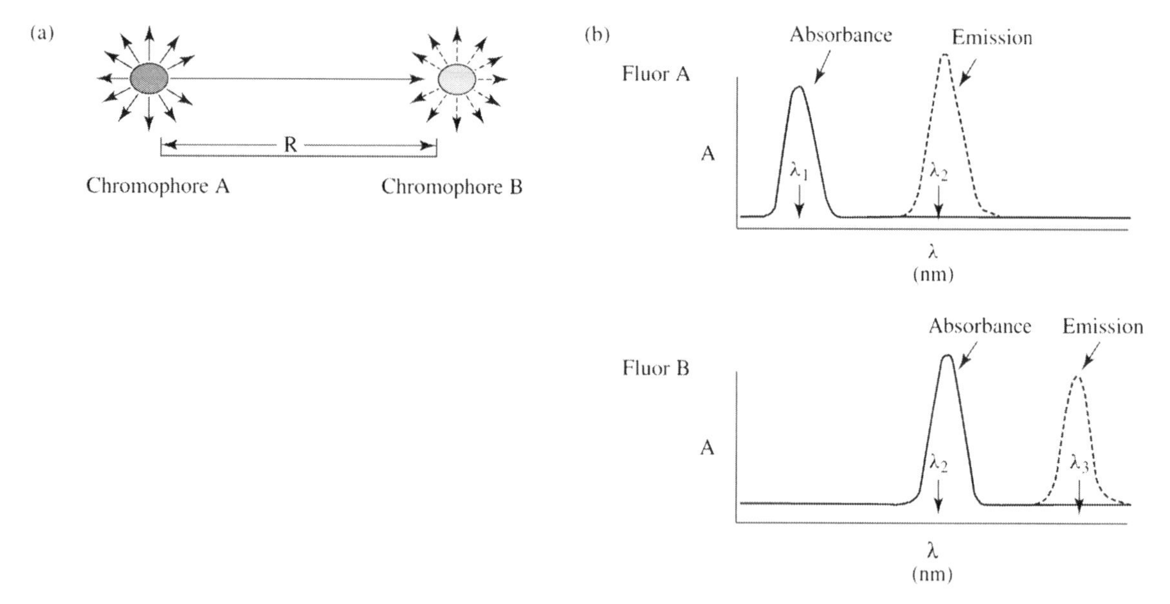

Figure 3.29. Resonance energy transfer in fluorescence. (a) Resonance energy transfer involves the transfer of light energy from one chromophore to another. The efficiency of this process is proportional to $1/R^6$, where R is the distance between chromophores. Light from chromophore A (solid arrows) is absorbed by chromophore B. (b) If the emission λ_{max} of fluor A overlaps with the absorbance λ_{max} of fluor B (e.g. at λ_2), this energy can be absorbed ad re-emitted by the latter (dashed arrows in (a)) at λ_3 allowing R to be calculated from Equation (3.27).

formation in DNA and measurement of conformational flexibility of chromosomes.

3.4.7 Applications of Fluorescence in Cell Biology

Because of the high sensitivity of fluorescence, certain fluors have found a wide range of applications in *microscopy*. Their use frequently results in a particularly low level of background staining, that is a low signal : noise ratio. When fluors are used in this way, this experiment is called *fluorescence microscopy*. It allows the detection and visualization of specific cell components such as chromosomes, organelles, cytoskeleton and membrane-bound proteins expressed on the surface of individual cell types (e.g. receptors). This information is particularly useful in identifying the location in or on the cell of particular proteins or nucleic acids or macromolecular complexes containing them.

Acridine orange is a fluor which has similar molecular dimensions to a DNA base-pair. It therefore intercalates readily into double-stranded DNA and is now known to be a potent mutagen. At low ratios of acridine orange:polynucleotide, intercalation into double-stranded DNA results in a large increase in fluorescence with no effect on emission λ_{max}. In the presence of single-stranded polynucleotides such as RNA, however, there is significantly less fluorescence and a shift in λ_{max} towards the red end of the spectrum. If cells are stained with a high concentration of acridine orange they therefore emit green light where there is an abundance of double-stranded DNA (e.g. the nucleus) while they emit red light where there is an abundance of single-stranded RNA (e.g. the cytoplasm). It is possible to distinguish the nucleus of any cell using this simple experiment.

More specific visualization of cell structure is possible with *immunostaining*. In this technique, cells are prepared for microscopy (often by freezing or fixing) followed by *staining* with an antibody coupled to a fluor. This is called *direct labeling*. More commonly, an antibody which is not fluorescently labeled (the *primary* antibody) is used which is then detected by a *secondary* antibody which is fluorescently labelled (Figure 3.30). The secondary antibody is specific for the primary antibody. This is called *indirect labelling* and it has the advantage that a single secondary antibody (e.g. antigoat IgG) may be used to detect a wide range of primary antibodies (i.e. *all* goat IgGs). These secondary antibodies are widely available commercially. Moreover, because several secondary antibody molecules bind to a single primary antibody, indirect labelling is more sensitive than direct labeling. In essence this approach is similar to

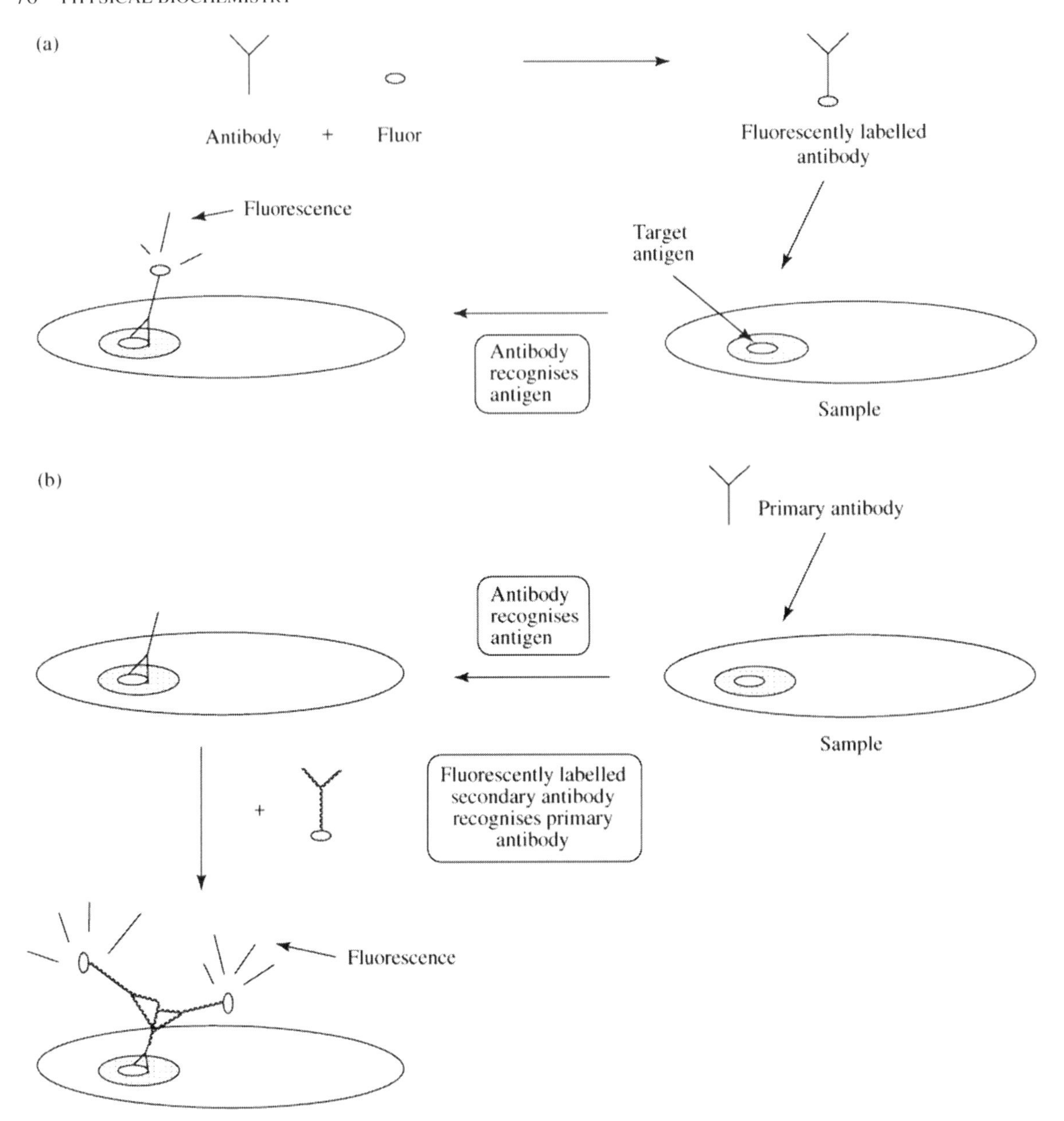

Figure 3.30. Immunofluorescence microscopy. (a) Direct immunostaining. (b) Indirect immunostaining. Secondary antibodies are raised to constant parts of immunoglobulins from a particular class and species. Note the amplification of signal possible with indirect immunostaining.

the ELISA assay and to dot blotting techniques described elsewhere (Chapter 5).

An example of this procedure is the use of fluorescein isothiocyanate (FITC) covalently conjugated to an antibody raised to a specific antigen (e.g. a structural component of the cytoskeleton). When a cell is stained with FITC-labelled antibody, only the target protein will be labelled. Light emitted from the stained sample passes through a *filter* which only selects for light of the emission λ_{max}. In this way we can 'see' the cytoskeleton in the cell. A second antibody with a different specificity can be fluorescently labelled in the same way and used to stain the same preparation. The two antibodies are usually labelled with distinct fluors possessing different emission λ_{max} values. By changing the filter

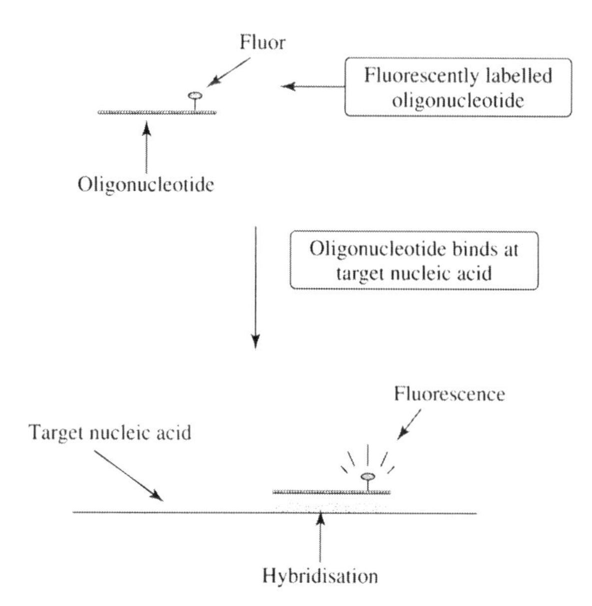

Figure 3.31. *In situ* hybridization with fluorescently labelled oligonucleotide. A fluor such as fluorescein is incorporated into an oligonucleotide of defined sequence. The labelled oligonucleotide recognizes a complementary sequence in the target nucleic acid. (DNA or RNA) and hybridizes to it. Fluorescence allows visualization of this hybrid in a microscope. This technique allows us to determine the location of specific nucleic acids in cell and tissue samples.

appropriately, it is possible to take photographs of a single sample which will reveal the subcellular location and distribution of the two target molecules. This allows us to identify proteins or other structures which form part of a single complex in or on the cell.

An alternative strategy for specific labelling of sites in the cell is offered by the technique of *in situ hybridization.* This is sometimes referred to as the *FISH* technique (*fluorescence* in situ *hybridization*). Short DNA sequences, *oligonucleotides,* can be used to stain cell preparations (Figure 3.31). An oligonucleotide can be designed and synthesized to hybridize specifically to a particular piece of DNA or RNA (e.g. part of a gene, a ribosome or a virus) according to the rules of Watson–Crick base pairing. When coupled with the use of *in situ* polymerase chain reaction (which allows amplification of a target sequence), this technique gives a sensitive and specific means of locating DNA or RNA in cells. If pyrimidine nucleotides labelled with fluorescein are incorporated into the oligonucleotide, double-stranded hybrids with the target sequence will be visible. Non-fluorescent oligonucleotide labels such as *digoxigenin* and *biotin* can be used in the same way but these must be detected with the aid of fluorescently-labelled antibodies as described above. This technique has been used in a wide range of studies in biochemistry including elucidation of chromatin structure, the detection of viruses and in *positional cloning* (i.e. the location of defective genes in chromosomes by phenotype mapping without prior knowledge of the functional basis of disease). *In situ* hybridization can also be used in combination with immunostaining to demonstrate, for example, colocalization of mRNA and translated protein or to identify the location of viruses in viral infections.

3.5 SPECTROSCOPIC TECHNIQUES USING PLANE-POLARIZED LIGHT

3.5.1 Polarized Light

Electromagnetic radiation consists of mutually-perpendicular electric and magnetic vectors (Figure 3.1). In a beam of radiation originating from the sun or any other light-source, the electric vectors are randomly orientated around the beam axis as shown in Figure 3.32. If, however, the electric vectors in one plane could be selected a beam of *plane polarized* or *linearly polarized* light would result. It is possible to achieve this by passing unpolarized light through a filter called a *polaroid.* Plane polarized light may be thought of as being composed of

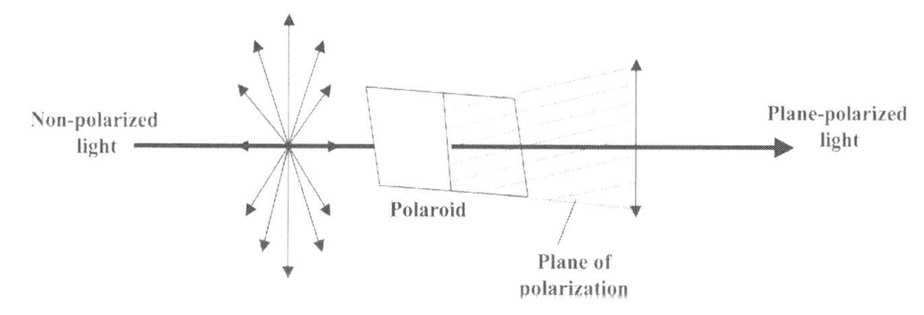

Figure 3.32. Plane polarized light. The electric vectors of a beam of light are shown as arrows. In nonpolarized light, these vectors are orientated randomly through 360°. A polaroid filter selects for a single orientation of electric vectors resulting in a plane of polarization. This is plane-polarized light which is also known as linearly polarized light.

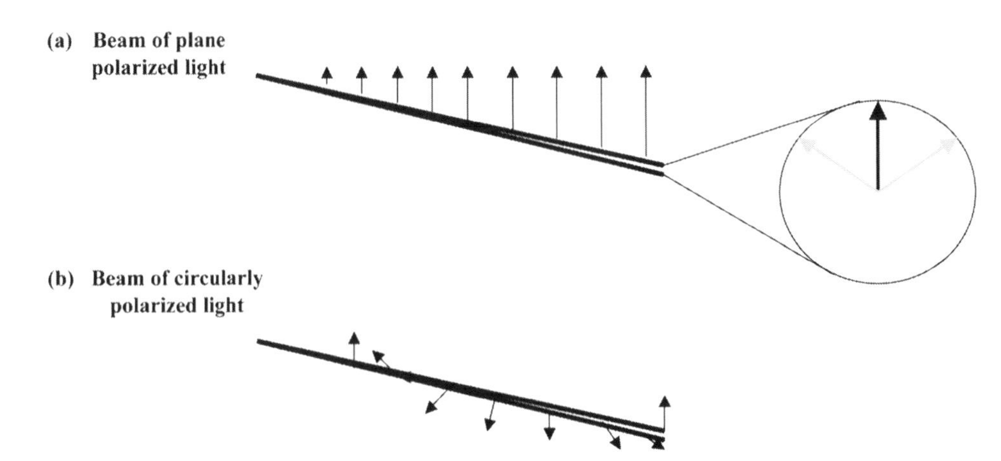

Figure 3.33. Circularly polarized light. (a) The electric vector of plane polarized light (solid arrows) is composed of two equal and opposite circularly polarized components (dashed arrows). (b) The electric vector of circularly polarized light describes an elliptical path along the direction of propogation.

two circularly polarized vectors of opposite direction but of equal strength (Figure 3.33). These are called, respectively, *left* and *right circularly polarized* components of plane polarized light. As described below (Section 3.5.4), we can select for left or right circularly polarized light. If we traced the path of the electric vector along the axis of a beam of circularly polarized light, it would retain equal magnitude but describe a helical path (Figure 3.33).

3.5.2 Chirality in Biomolecules

Carbon has four valencies and, when bound to four distinct chemical groups, has two possible isomeric forms which may be referred to as D and L (Figure 3.34). Pairs of isomers which are *nonsuperimposable mirror images* of each other are called *enantiomers* or *optical isomers*. The $\alpha-C$ of the amino acid alanine is called an *optical centre* or a *centre of asymmetry* to denote the fact that the two isomers arise from different arrangements around this atom. In some molecules there may be more than one optical centre resulting in four, six or more possible stereoisomers (Figure 3.34). Some of these structural variants are obviously stereoisomers but are not related to each other as enantiomers (i.e. nonsuperimposable mirror images of each other). These are referred to as *diastereoisomers*.

The existence of enantiomeric pairs is known as *chirality* and molecules capable of existing as enantiomers are called *chiral molecules*. 'Chiral' derives from the Greek word for hand (*cheir*) because hands are good examples of nonsuperimposable mirror image objects. L- and D-enantiomers are conventionally assigned by comparison with the reference compound L-glyceraldehyde. An alternative notation called the RS convention gives the *absolute* configuration of enantiomers and more information on this is available in the bibliography at the end of this chapter.

The phenomenon of chirality is especially important in biochemistry because many biomolecules are chiral. In biological systems there is usually a selection for one of a pair of enantiomers. For example most amino acids in living systems are L-enantiomers with D-enantiomers occurring only very rarely (e.g. in antibiotic peptides). Similarly, most monosaccharides are D-enantiomers with L-enantiomers occurring rarely. By contrast, most chemical reactions carried out in solution result in a 50 : 50 mixture of D- and L-enantiomers.

Chirality has several important consequences for the structure, shape and functional properties of biomolecules. For example, L-amino acids result in the right-handed helices observed in proteins. D-amino acids result in left-handed helices which are themselves mirror images of right-handed helices while heteropolymers of a mixture of D- and L-amino acids cannot form helices at all. Proteases are capable of hydrolyzing peptide bonds in polypeptides composed of all-L-amino acid substrates but cannot hydrolyze those of all-D composition. Conversely, synthetic proteases which can be made chemically with all-D-amino acids (e.g. the HIV protease) can cleave all-D polypeptide substrates but not those of an all-L amino acid composition. Indeed, the synthetic HIV protease is itself a mirror-image of the natural all-L enzyme showing that chirality is maintained throughout the hierarchy of protein structure.

It has also been discovered that a pair of enantiomers may display distinct toxicity to humans with one enantiomer being toxic while the other is nontoxic. A well-known example of this is the fertility drug *thalidomide* which was widely

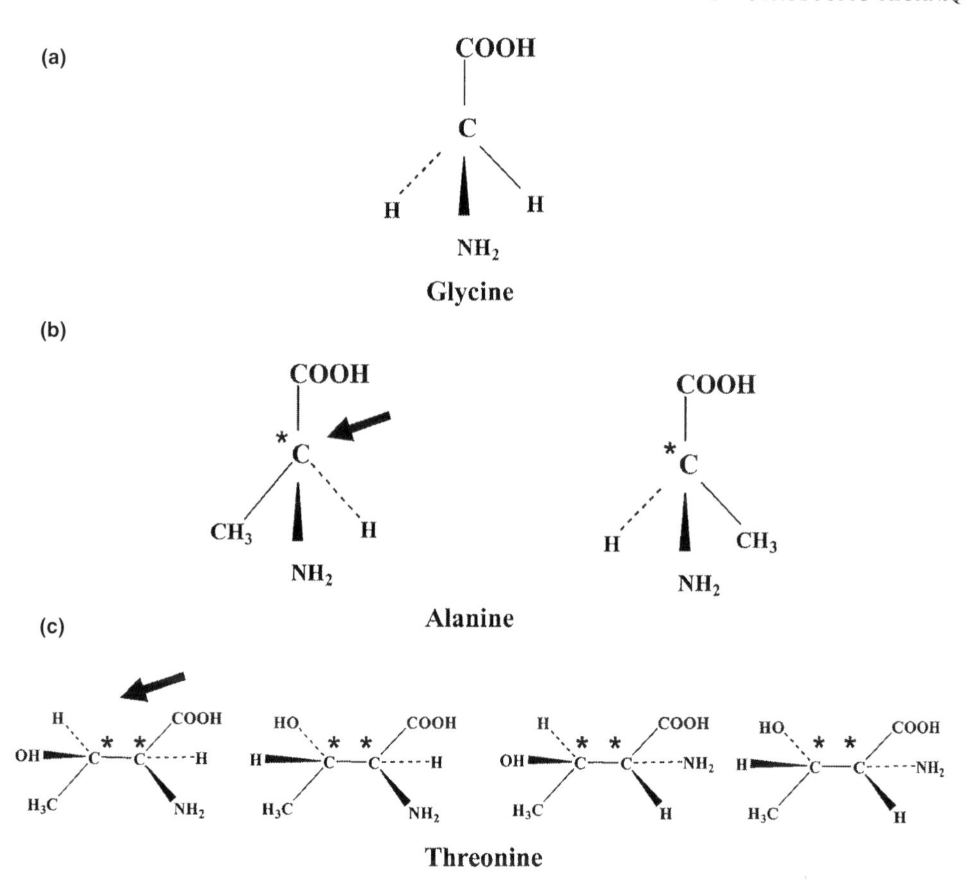

Figure 3.34. Chirality in amino acids. (a) Glycine is an achiral molecule because the four substituents to the α-C are not all chemically different. All the 19 other amino acids commonly found in proteins are chiral. (b) Alanine is a chiral molecule because it has a centre of asymmetry (*). In nature, only the L-enantiomer is commonly found. (c) Threonine has two centres of asymmetry and accordingly has four possible enantiomers, only one of which (arrow) is commonly found in nature.

prescribed in the late 1950s. One enantiomer of this compound was nontoxic while the other was *teratogenic* leading to deformity or lack of limbs in babies born to patients treated with the drug. For this reason there is considerable interest in the pharmaceutical industry in the possible use of enzymes for *enantioselective synthesis* of new drugs.

3.5.3 Circular Dichroism (CD)

In the early nineteenth century the French physicist Jean Baptiste Biot observed that solutions of some organic molecules appeared to *rotate* the plane of polarization of plane polarized light, a phenomenon referred to as *optical rotatory dispersion* (ORD; Figure 3.35). We now know that this is a consequence of the fact that each enantiomer of an optically active molecule *interacts differently with left and right circularly polarized light*. By convention, laevorotation is designated as (−) while dextrorotation is denoted by (+). A 50:50 mixture of equal amounts of D- and L-enantiomers does not rotate the plane of polarization nor does a solution of an achiral molecule such as glycine.

Light passing through a chromophore solution may interact with the sample in two main ways. The light may be *refracted* or delayed on passage through the solution or it may be *absorbed*. Refraction is quantified by the *refractive index*, n, of the solution while absorption is quantified by the *molar extinction coefficient*, ε. If the light is plane polarized and the sample is optically active, each enantiomer may interact *differently* with the left and right circularly polarized components of the light beam.

ORD arises from the fact that there is a specific refractive index for left (n_L) and right (n_R) circularly polarized light:

$$n_L \neq n_R \tag{3.30}$$

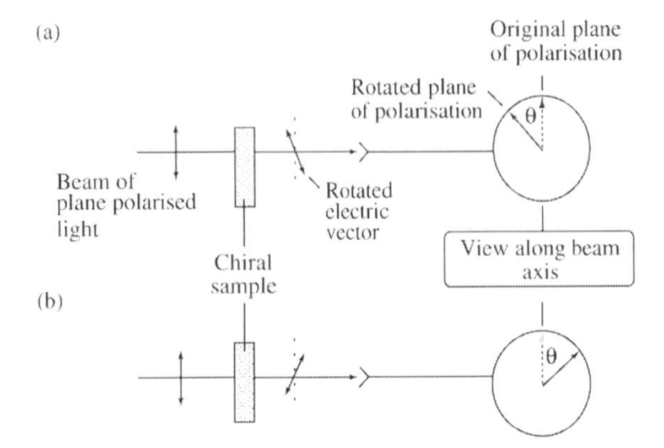

Figure 3.35. Optical Rotatory Dispersion (ORD). When a beam of plane polarized light is passed through a sample of a single enantiomer (i.e. a chiral sample), the plane of polarization is rotated through an angle, θ. (a) L-enantiomers rotate the plane to the left. (b) D-enantiomers rotate the plane to the right. Equal concentrations of an enantiomer rotate the plane of polarization through equal (though opposite) angles, θ, so that a 50 : 50 mixture does not rotate the plane at all.

The difference in refractive index at any wavelength may be expressed as Δn. An ORD spectrum is a plot of Δn against wavelength (λ).

Similarly, optically active samples have distinct molar extinction coefficients for left (ε_L) and right (ε_R) circularly polarized light. This is called *circular dichroism* (CD):

$$\varepsilon_L \neq \varepsilon_R \tag{3.31}$$

The difference between ε_L and ε_R may be expressed as $\Delta\varepsilon$. Combining Equations (3.31) and (3.6) (the Beer–Lambert law) means that there is a difference in the absorbance of left and right circularly polarized light (ΔA):

$$\Delta A = \Delta\varepsilon \cdot c \cdot l \tag{3.32}$$

If $\Delta\varepsilon$ or ΔA or *ellipticity* (*see* below) is plotted against wavelength (λ), a CD spectrum is obtained. The CD spectrum of one enantiomer is a mirror image of that of the other and is related to the corresponding ORD spectrum (and *vice versa*) by a mathematical transformation called the *general Kronig–Kramers transformation*. Both ORD and CD spectra are evidence for optical activity in the sample and both reflect structure of molecules in the sample, especially of chiral biopolymers such as proteins and nucleic acids (Figure 3.36). In practice, ORD has now largely been superseded by CD spectroscopy.

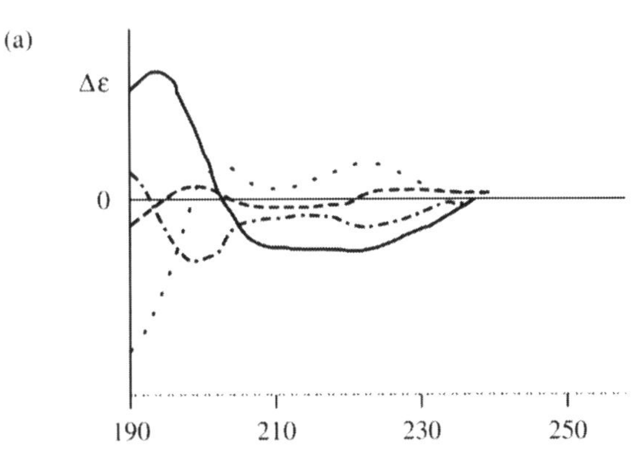

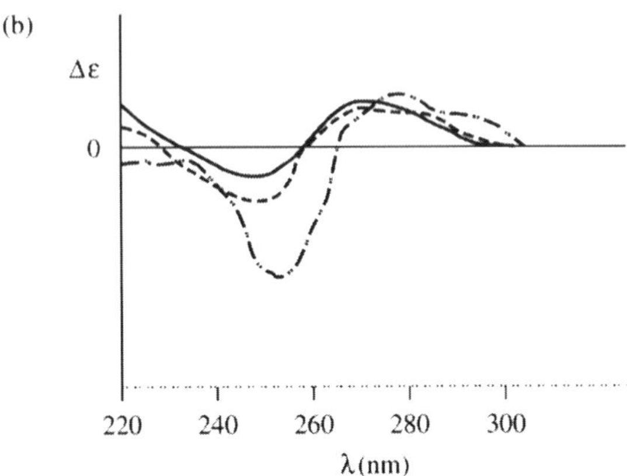

Figure 3.36. Circular dichroism (CD) spectra of biopolymers. Secondary structure strongly affects CD spectra of biopolymers. (a) Standard spectra of α-helix (–), β-sheet (—), random coil (- - -) and β-turns (– - -) of polypeptides. (b) Spectra of double-stranded DNA of varying G + C content. 26% (—), 72% (—) and 100% (-···). Modification of Figures 2.5 and 2.9 in Roger and Norden (1997) *Circular Dichroism and Linear Dichroism*, by permission of Oxford University Press.

3.5.4 Equipment Used in CD

CD spectra are measured in a special type of spectrophotometer called a *CD spectropolarimeter* of which an outline design is shown in Figure 3.37. Since CD depends on differential absorbance, a means of selectively exposing sample to left and right circularly polarized light is necessary. This is achieved by passing a beam of plane polarized light through a *photoelastic modulator* which is normally a quartz crystal subjected to an oscillating electric field. The effect of this is to vary the circular polarization of the beam passing through the modulator alternately from left to right with a frequency of some 50 kHz whilst maintaining a constant light intensity.

Differential absorption of left and right circularly polarized light is detected at a photomultiplier and converted into *ellipticity*, θ, which has units of millidegrees. This term arises from the fact that selective absorption of one of the

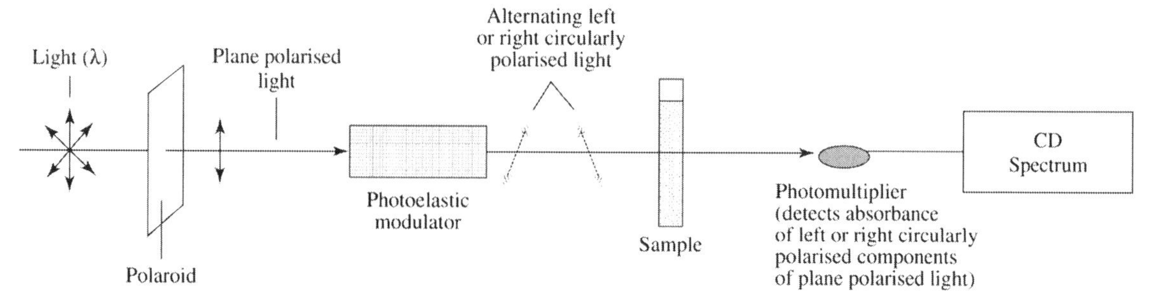

Figure 3.37. CD spectropolarimeter. A photoelastic modulator selects at any time for either left or right circularly polarized components of plane polarized light. It alternates between these at a frequency of 50 Hz. Selective absorption of either left or right circularly polarized light is detected at the photomultiplier and gives a CD spectrum for the sample. Samples which are achiral or composed of 50 : 50 mixtures of enantiomers would give no detectable spectrum in this instrument (i.e. $\Delta A = 0$ at all λ).

circularly polarized components of plane polarized light has the consequence that the resultant electric vector traces an elliptical path around the axis of the beam and is said to be *elliptically polarized* (Figure 3.38). In a CD spectropolarimeter, the two light beams are not in fact recombined but a photomultiplier detector converts incident light intensity into an electric current composed partly of alternating current (AC) and partly of direct current (DC) components. The DC component is related to total light absorption by the sample while the AC component is a direct measure of CD. This arrangement facilitates separate absorption measurements of the right and left circularly polarized components of plane polarized light.

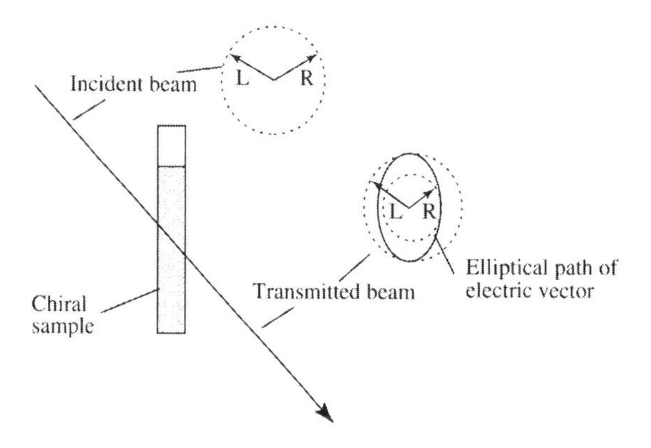

Figure 3.38. Ellipticity. A beam of plane polarized light consists of electric vectors of equal amplitude. When recombined, the resultant vector from the left (L) or right (R) circularly polarized components of plane polarized light would describe a circle (incident beam). If this light is passed through a chiral sample, one circularly polarized component is selectively absorbed. This lowers the amplitude associated with that component (R in the sample above). When recombined, the resultant electric vector now traces an elliptical path. The degree of ellipticity is a measure of circular dichroism.

Ellipticity may be converted to units of absorbance using the following equation:

$$\Delta A = \frac{\theta}{32\,982} \tag{3.33}$$

3.5.5 CD of Biopolymers

The synthetic homopolymer poly-L-lysine adopts a random coil structure at neutral and acid pH values. However, at high pH values, a mainly α-helical conformation is adopted which may be converted to predominantly antiparallel β-sheet by gentle heating. Each of these three forms of poly-L-lysine gives a characteristic CD spectrum in the range 190–240 nm and the fact that similar spectra are obtained for homopolymers of other amino acids suggests that they arise predominantly from asymmetry of the polypeptide backbone. Measurement of CD spectra in proteins and peptides of unknown secondary structure is often used (by analogy with such homopolypeptides) to estimate empirically their percentages of common secondary structure features. Because of the fact that they depend on comparison with simple homopolymers and exclude effects due to amino acid side-chains, such estimates are not as directly meaningful as those arising from more direct structural methods such as X-ray diffraction and two-dimensional NMR (Chapter 6). However, CD spectra are obtainable with a wide range of proteins and peptides at relatively low sample concentrations in aqueous solution and thus avoid some of the limitations of these methods. Moreover, methodologies for analyzing spectra in terms of the main secondary structure conformations are now very reliable.

In practice, CD spectra provide a generally useful index of structure in proteins (Example 3.3). Changes in CD

Example 3.3 Circular Dichroism studies of structural stability of tetrameric p53

p53 is a tumour suppressor protein which is the most commonly mutated protein in human tumours. It is therefore of considerable interest to an understanding of why some cells develop into cancer cells leading to the formation of tumours. These mutations make this protein dysfunctional in over 50% of human tumour samples from which it has been sequenced. A functionally important part of the structure of p53 is the tetramerization domain which includes residues 325–355. In intact p53, this domain is responsible for forming a tetramer between four individual polypeptides. Even when expressed as a short peptide, this tetramerization activity is conserved resulting in the smallest known protein tetramer.

In this example, CD was used together with fluorescence (Section 3.4) and differential scanning calorimetry (DSC; Chapter 8) in a study of the structural stability of this tetramerization domain. The following plot shows how ellipticity of a small concentration (2.5 μM) of the protein varies with temperature. The intersection of all the traces at a single point called the *isodichroic* point is characteristic for a two-state system such as a protein only existing in either a folded or unfolded state (Section 3.4.5). As the temperature is increased from 21 to 81 °C, the ellipticity measured in the region of 220 nm decreases closer to 0.

Such measurements allow calculation of the fraction of the protein unfolded. This allowed determination of unfolding curves at various concentrations of protein in the range 0.5–20 mM as shown in the following plot.

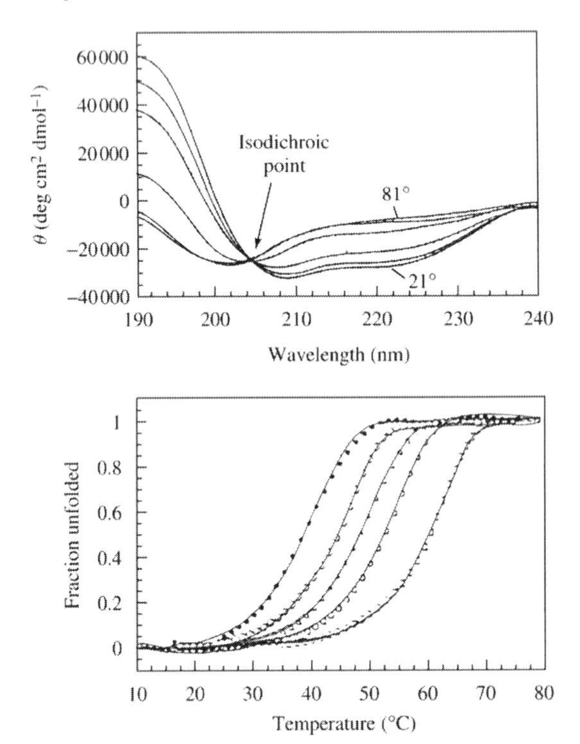

These data were combined with DSC measurements to confirm that this model system unfolds according to a two-state model from native (N) to unfolded (U) states;

$$N \rightleftarrows U$$

This is consistent with a structural model derived from two-dimensional NMR (Chapter 6) which indicates that hydrophobic interactions along the intersubunit interface is responsible for the bulk of stabilization of the tetrameric structure in this important protein.

Source: (From Johnson *et al.*, 1995) Reprinted with permission from Johnson, C.R. *et al.*, *Biochemistry* **34**, 5306–5316 © 1995 American Chemical Society.

spectra may therefore be taken to reflect perturbation of structure. For example, binding of a ligand to a protein or protein folding/unfolding can be conveniently followed by CD spectroscopy. A particular advantage of CD in this regard is the short time-scale of the CD measurement compared to that of other spectroscopic measurements. Since absorption of a UV photon takes some 10^{-15} s (compared to 10^{-9} s for the radiowave radiation used in NMR; Section 3.7), CD measurements may be expected to detect ligand-protein effects of shorter duration than those detectable by NMR. Even though many ligands may be intrinsically achiral, their unique orientation on binding to a protein coupled with their specific interaction with the protein can cause them to acquire CD properties for the duration of the interaction. This phenomenon is known as *induced* CD. Examples of this phenomenon include CD spectra induced on binding of bilirubin to bovine serum albumin and of dicumarol derivatives to α-1 acid glycoprotein.

The second major class of biopolymers which has been studied by CD is the nucleic acids. Of the nucleotide's structural components, only the pentose sugar is chiral. Mainly as a result of the presence of this sugar, nucleotides are intrinsically chiral structures and give measurable, albeit weak, CD spectra. As the level of structural order increases, however (e.g. polymerization of nucleotides into polynucleotides followed by assembly into duplex structures such as double-stranded DNA and tRNA), the asymmetry of the system and hence the strength of the CD spectrum obtained also increases. CD is therefore a useful measure of structure in nucleic acids. Differences are observed in CD spectra due to variation of nucleotide sequence, GC composition, stacking of bases in different forms of DNA (i.e. A-, B- and Z-DNA), formation of macromolecular assemblies such as ribosomes and nucleosomes and ligand binding to DNA. As with proteins, induced CD is also possible with DNA such as when anthracene-9-carbonyl-N^1-spermine binds to oligonucleotides of defined sequence.

Many chromophores associated with biomacromolecules are themselves chiral and CD measurements of their spectra can give information on their interactions with proteins, DNA or macromolecular complexes.

3.5.6 Linear Dichroism (LD)

Electronic transitions underlying absorption of ultraviolet and visible light occur in individual molecules in a particular direction or *axis*. Molecules dissolved in an aqueous solvent are *randomly* arranged so that the sample will give the same absorption spectrum regardless of the direction of a beam of incident light (Figure 3.39). If the molecules were all orientated in a single direction, however, different absorption spectra would be obtained depending on the direction of the light beam. In *linear dichroism* this fact is

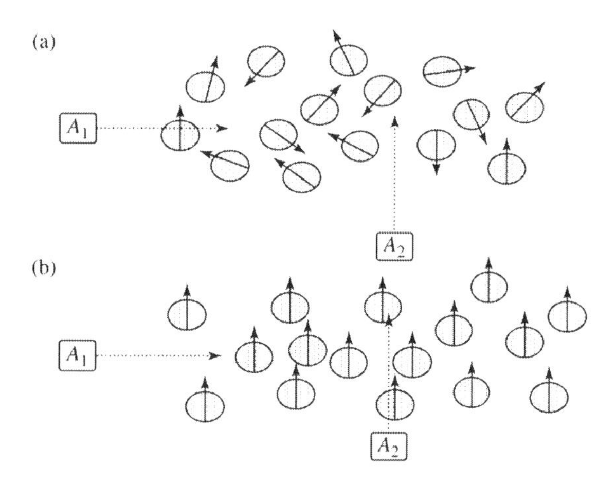

Figure 3.39. Linear dichroism (LD) in orientated samples. The axis of a single spectroscopic transition in each sample molecule is denoted by a solid arrow. A_1 and A_2 refer to absorbances corresponding to this transition in directions shown by dashed arrow. (a) Randomly orientated sample. Axes of transitions are randomly arranged. $A_1 = A_2$. (b) Orientated sample. If all sample molecules are orientated in a single direction, $A_1 \neq A_2$.

exploited by exposing *orientated* samples to plane polarized light. Extreme cases are presented where the incident beam is *parallel* to the axis of a particular transition and where the beam is *perpendicular* to the transition axis. Linear dichroism (LD) is the differential absorbance of these incident beams as follows:

$$\mathrm{LD} = A_{\parallel} - A_{\perp} \tag{3.34}$$

where the incident beam is perfectly parallel to the axis of the transition, LD > 0. If the polarization is perfectly perpendicular to this axis, then LD < 0.

In LD spectroscopy, samples are orientated in a single axis relative to the incident linearly polarized light and it is possible to vary the relative orientation of sample and light such that the incident light is parallel or perpendicular to this axis. A wide variety of physico-chemical means of achieving this orientation have been described (Figure 3.40). In the *stretched polymer* technique, orientation is achieved by absorbing sample onto a polymer (e.g. polyvinylalcohol) either before or after stretching it in a particular direction. The long axis of the sample molecule aligns in the stretch direction. Other popular techniques include *flow orientation*, *electric field orientation* and *squeezed gel orientation*.

3.5.7 LD of Biomolecules

LD spectroscopy is well-suited to the study of certain sample molecules such as fibrous proteins and DNA since these

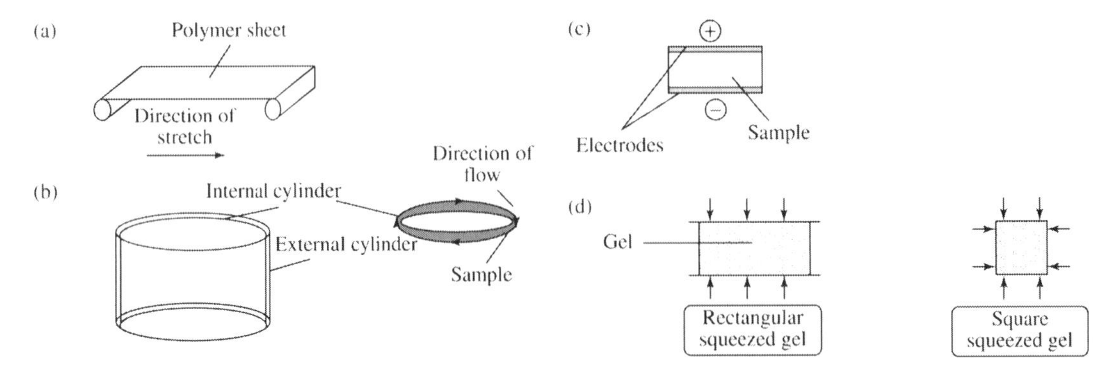

Figure 3.40. Some orientation methods used in linear dichroism. (a) Stretched polymer orientation. Sample is absorbed to stretched polymer either before or after stretching in a particular direction. The long axis of sample molecules aligns in this direction. (b) Flow orientation. A cylindrical coquette flow cell consists of an internal and external cylinder. Sample is placed in the gap between these cylinders. The inner cylinder is rotated causing the sample to flow by viscous drag as shown by arrows. The long axis of sample molecules aligns with direction of flow. (c) Electric field orientation. An electric field is created around the sample by parallel electrodes, Excessive heating of sample can be avoided by use of pulsed fields as described in Chapter 5. (d) Squeezed gel orientation. Protein is embedded in a gel which is mechanically squeezed either bidirectionally or unidirectionally. This results in a rectangular or square squeezed gel.

possess a high axial ratio (Figure 5.36) and are readily orientated in a particular direction. Information has been obtained from such studies about the relative orientation of DNA bases, intercalation of agents such as ethidium bromide into DNA, binding of drugs at the major and minor DNA grooves and flexibility of DNA (Example 3.4). 'Naturally orientated' biopolymers such as membrane proteins have also been extensively studied by LD and this has yielded information about the orientation of chromophores in photosynthetic reaction centres during excitation. As with CD, the short timescale of LD measurements makes this technique especially appropriate to such investigations. LD has also been used in studies of chromatin structure and of binding of proteins to DNA.

A limitation of LD in the study of biomolecules is the necessity for orientated samples. Globular proteins are particularly difficult to orientate and LD has not been as widely used in investigations of such samples.

3.6 INFRARED SPECTROSCOPY

Table 3.1 shows that energy increments (ΔEs) of *lower* value than those involved in electronic transitions correspond to frequencies in the infrared part of the spectrum (i.e. $\lambda = 10^3$–10^5 nm; *see* Section 3.2.2). *Vibrational* energy levels are quantized in the same way as electronic energy levels and, if infrared light of a frequency corresponding exactly to the ΔE between two vibrational energy levels is absorbed by a molecule; it may undergo a *vibrational transition*. As shown earlier (Figure 3.8), the various vibrational energy levels are mainly due to differences in bond length and angle which are possible due to stretching and bending of bonds.

3.6.1 Physical Basis of Infrared Spectroscopy

When atoms come together to form a covalent bond, they undergo an electronic rearrangement which involves two competing sets of forces. The positively-charged nuclei of the two atoms tend to *repel* each other (*electrostatic repulsion*) while the nucleus of each atom is *attracted* to the negatively-charged electrons of the other (*electrostatic attraction*). The mean distance settled on between the atoms (i.e. the bond length) is a reflection of a point of balance between these competing attractive and repulsive forces and, in general, will be characteristic for a particular chemical bond. For example, in polypeptides C^{α}–C bonds are 1.52 Å while N–C^{α} bonds are 1.45 Å. It is possible to plot the energy of a molecule as a function of the interatom distance, r, in a *Morse diagram* (Figure 3.41). Most molecules in a population at rest are at the *minimum* of this diagram, but individual bond lengths can vary by ± 0.5 Å and bond angles by $\pm 5°$ at room temperature. We can think of each chemical bond, therefore as existing in a particular *vibrational energy level* characterized by a particular bond length, bond angle and electron density.

However, if radiation of an appropriate frequency is passed through the sample, it is possible for the molecule to undergo a transition to a higher vibrational energy level by absorption of radiation. As described in Section 3.2.2, vibrational energy levels are *quantized* in the same way as electronic energy levels although the ΔE values associated with vibrational transitions are much smaller than those

Example 3.4 Linear Dichroism study of ligand binding to DNA

4′,6′-Diamidino-2-phenylindole (DAPI) binds to the minor groove of AT-rich double-stranded DNA to form highly fluorescent complexes. This type of interaction with DNA is of major interest since it is known that chemical compounds can interact in a variety of ways with DNA such as the formation of adducts or intercalation between base-pairs. These interactions can result in important biological effects such as mutagenesis and/or carcinogenesis. DAPI is a particularly interesting model compound in this regard as it has found widespread use in fluorescent labelling of cells and chromosomes. In this example, LD was used to probe the interaction of DAPI with heteropolymers of the form $(A\text{-}T)_n$. LD spectra were obtained for a range of ratios of DAPI:DNA as follows.

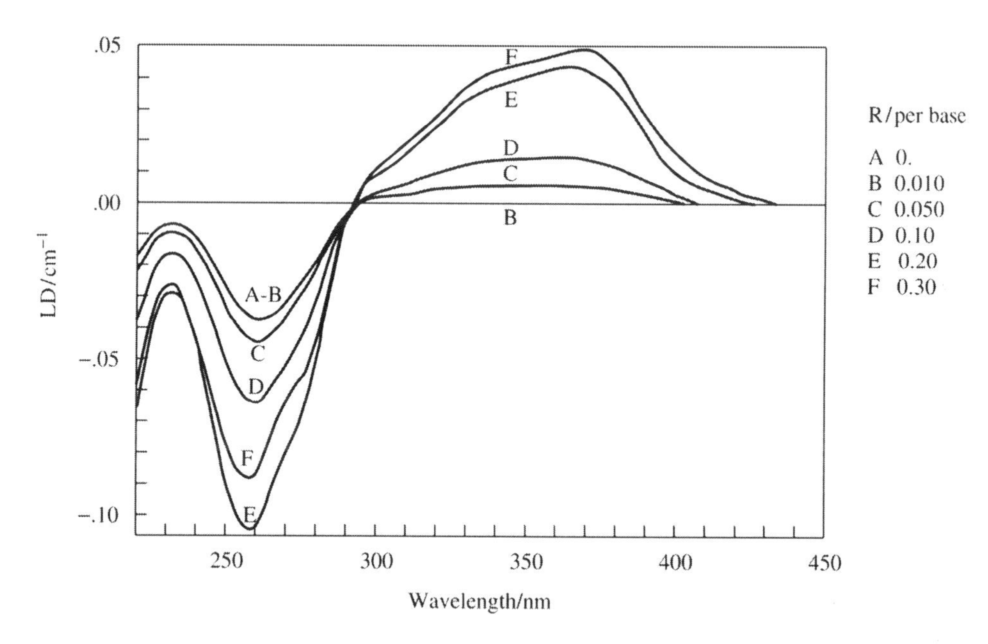

These spectra show a strong negative value in the region around 260 nm which is consistent with the planes of the bases being perpendicular to the helix axis. The strong positive spectra in the region 300–400 nm are due to the DAPI itself. These data are consistent with DAPI binding to the DNA at an angle of 43–46° at low DAPI:DNA ratios. At higher ratios, there is some 'stiffening' of the DNA and much higher affinity binding of DAPI suggesting that two different modes of binding operate depending on the ligand concentration and that there is a change in structure in DNA as a result of ligand binding; that is allosterism. This illustrates how LD can provide detailed information on ligand binding to DNA.

Source: (From Eriksson *et al.*, 1993) Reprinted with permission from Eriksson, S. *et al.*, *Biochemistry* **32**, 2987–2998 © 1993 American Chemical Society.

associated with electronic transitions. A vibrational transition from the ground state to the *first* excited state due to absorption of infrared light is called a *fundamental* absorption and the frequency, ν, associated with this is called the *fundamental frequency*. Whilst other transitions are also possible (e.g. from the ground to the second or third excited state), they occur much less frequently and represent weak absorbance.

An *infrared spectrum* consists of a plot of absorbance *versus* frequency or *wavenumber* ($1/\lambda$). In comparison with absorbance spectra in the ultraviolet/visible range, infrared spectra of small molecules consist of narrow lines rather than broad peaks (Figure 3.42). However, because of the large number and variety of bonds in macromolecules, infrared spectra of proteins and DNA consist of a small number of broad peaks.

Water gives a strong infrared spectrum because of its high absorption coefficient in the infrared range and high concentration (55 M). For this reason, infrared spectroscopy may be carried out either on nonaqueous samples prepared as dry films or on samples dissolved in alternative solvents such as D_2O or chloroform. It is also possible to

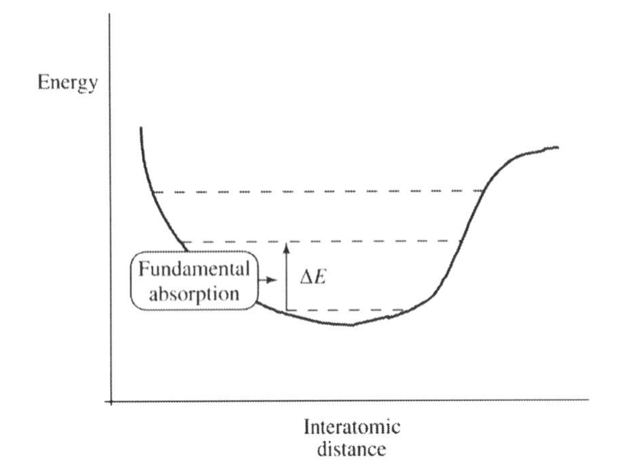

Figure 3.41. Physical basis of infrared spectroscopy. When infrared light of an appropriate frequency is absorbed, the molecule may be promoted through an energy increment ΔE to the first vibrational energy level which will have a different value for either bond length or bond angle. This is called the fundamental absorption. Different bond types have individual vibrational energy levels which has the consequence that the infrared absorption spectrum is characteristic for a specific chemical structure.

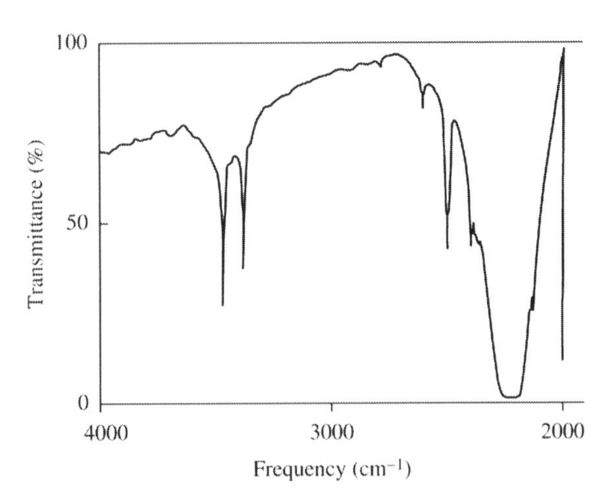

Figure 3.42. Infrared spectrum of KCN. The spectrum of potassium cyanide in the frequency range 2000–4000 cm^{-1} is shown. One hundred percent transmittance is equivalent to an absorbance of 0. Note the sharpness of the lines corresponding to transitions. An infrared spectrum of a molecule such as a protein usually consists of much broader bands. Gans (1971) *Vibrating Muscles: An Introduction to the Interpretation of Infrared and Raman Spectra*, Chapman & Hall, with kind permission from Kluwer Academic Publishers.

subtract the spectrum due to the water solvent by *difference spectroscopy* (Figure 3.15) but this requires high concentrations of protein or DNA (5–20%). These conditions are generally unsuitable for the study of biomacromolecules in their native state and this limits the practical application of simple infrared spectroscopy to their study. However, the technique has become standard for the study of low molecular mass biomolecules and yields structural information complementary to that available from other methods such as mass spectrometry (Chapter 4), CD (Section 3.5) and NMR (Section 3.7). Study of biomacromolecules by infrared spectroscopy requires more elaborate techniques which are described below (Sections 3.6.4 and 3.6.5, respectively).

3.6.2 Equipment Used in Infrared Spectroscopy

The overall design of an infrared spectrophotometer is very similar to that of the ultraviolet/visible spectrophotometer shown in Figure 3.9. The main differences are in the technology used for diffracting light and for detection of transmitted light. The monochromator used in an infrared spectrophotometer is usually a *diffraction grating*. It will be recalled that either a prism *or* a diffraction grating may be used in the ultraviolet-visible spectrophotometer. The detector component of an infrared spectrophotometer is a *thermocouple* rather than a photocell normally used in a ultraviolet-visible spectrophotometer. This basic infrared spectrophotometer suitable for collecting spectra from small molecules is called a *dispersive* or *grating* infrared spectrophotometer.

3.6.3 Uses of Infrared Spectroscopy in Structure Determination

In principle, any chemical bond of a given type (e.g. C–H) might be expected to have a fundamental absorption identical to that of any other bond of the same type. In practice, the chemical *environment* of the bond has an effect on the precise frequency of absorption, since this may alter the bond's electron density. Studies of infrared absorption from a large number of molecules of known chemical structure suggest that absorbance of infrared light near particular frequencies are characteristic for specific chemical groups (Table 3.5). These *group frequencies* allow us to determine aspects of the molecular structure of small molecules from their infrared absorbance pattern alone. It should be noted that many of the chemical groups detectable by infrared spectroscopy give little or no absorbance in the ultraviolet or visible parts of the spectrum. A further advantage of infrared spectroscopy is the variety of sample forms which can be analysed. It is possible to obtain infrared spectra of solutions, films or powders, and of crystalline samples.

While dispersive infrared spectroscopy has limited applicability to the study of biopolymers generally, it has found some applications in structural studies on DNA. The formation of hydrogen bonds and *tautomerization* between C=O (*keto*) and C–OH (*enol*) forms of nucleotide bases have detectable effects on infrared spectra.

Table 3.5. Group frequencies in the infrared region

Chemical group	Frequency (cm^{-1})
$-CH_3$	1460
$-CH_2-$	2930
	2860
	1470
C–H	3300
–C–C–	1165
–C=O	1730
–C–H (in CH_3)	2960
	2870
–C–H (in CHO)	2870
	2720
H (on benzene ring)	3060
–CN	
–O–O–	1200–1100
–OH	3600
$-NH_2$	3400
$=CH_2$	3030
–SH	2580
–C=N–	1600
C–Cl	725
C=S	1100

3.6.4 Fourier Transform Infrared Spectroscopy

The dispersive infrared spectrometer described in Section 3.6.2 allows scanning of comparatively short parts of the infrared spectrum in each measurement. The complete spectrum needs to be scanned segment by segment which is a comparatively slow process. Moreover, such measurements have serious limitations for the study of DNA and proteins as mentioned above. An alternative approach is the incorporation of a *Michelson interferometer* in a *Fourier Transform infrared (FTIR)* spectrometer which has now superseded dispersive instruments in biochemistry. The basic design of this is shown in Figure 3.43. An incident IR beam may be split by passage through a *half-mirror* or *beam splitter* made up of a material (e.g. crystalline KBr or a thin mylar film) which will *transmit* half of the light and *reflect* the other half. This beam splitter is positioned at 45° to the incident beam which is parallel to a *stationary* mirror. The beam transmitted by the beam splitter shines on a *moving* mirror arranged at 90° to the incident beam.

The two beams are *reflected* by the fixed and moving mirrors, respectively and recombine to form a *resultant* beam. If the two beams are *in phase*, the resultant beam is the *sum* of the two beams (in accordance with the *principle of superposition* mentioned in Section 3.1.1). This is a process called *constructive interference* (Figure 3.44). If the two beams are *out of phase* then the resultant beam is *weaker* than the sum of the two beams which is called *destructive interference*. Whether constructive or destructive interference occurs depends on the relative positions of the two mirrors at any moment in time. In practice, a number of scans are performed in the absence of sample and then repeated with sample present. These scans are very rapid and cover a wider part of the infrared spectrum than is possible in a dispersive spectrometer. The interference patterns obtained are added together and analysed by a mathematical tool called the *Fourier transform*. It should be noted that this is also used to analyse the diffraction pattern of X-rays in protein crystallography and in the analysis of multi-dimensional NMR spectra (Chapter 6; Appendix 2).

FTIR spectroscopy is essentially an interference-based technique rather than one based on absorption. Each molecule gives a characteristic FTIR spectrum which reflects its chemical structure. Moreover, proteins and DNA are amenable to FTIR analysis since the effects of solvent water can be largely excluded. The technique is applicable to most classes of biomolecules including lipids, glycolipids and oligosaccharides.

FTIR has proved particularly useful in the study of the structure and dynamic properties of proteins and peptides (Example 3.5). Hydrogen bonding between peptide bonds underlies secondary structure in proteins. Studies of the infrared spectra of synthetic homo-polypeptides led to the empirical identification of a number of peaks or *amide bands* associated with such hydrogen bonds. For example, an absorption in the region 1597–1672 cm^{-1} (the amide I band) is predominantly associated with stretching vibrations of the C=O bond. Other amide bands arise from a combination of effects. For example, the amide II band (1480–1575 cm^{-1}) is due partly to N–H bending (50%) and partly to C–N stretching (40%). Since hydrogen bonding strongly affects the electron density of atoms involved in the peptide bond, it is to be expected that amide bands could be used as a measure of secondary structure in synthetic polypeptides. In proteins, FTIR measurements have also proved useful for the identification of helices, turns and β-sheet secondary structure in both fibrous and globular proteins although interpretation of such measurements is more complicated than for peptides. These complications are introduced by overlap between different absorption maxima, slight deviations from ideal secondary structure in many proteins and limitations in data analysis. Notwithstanding these limitations, FTIR provides a sensitive means of probing secondary structure in proteins. In particular, structural changes due to alterations in pH, ion strength, pressure or temperature, ligand binding, adsorption to solid surfaces, aggregation and folding may readily be detected and analysed by FTIR. The technique is complementary to X-ray crystallography, CD/LD and fluorescence measurements and is suitable for a wide range of samples including membrane proteins.

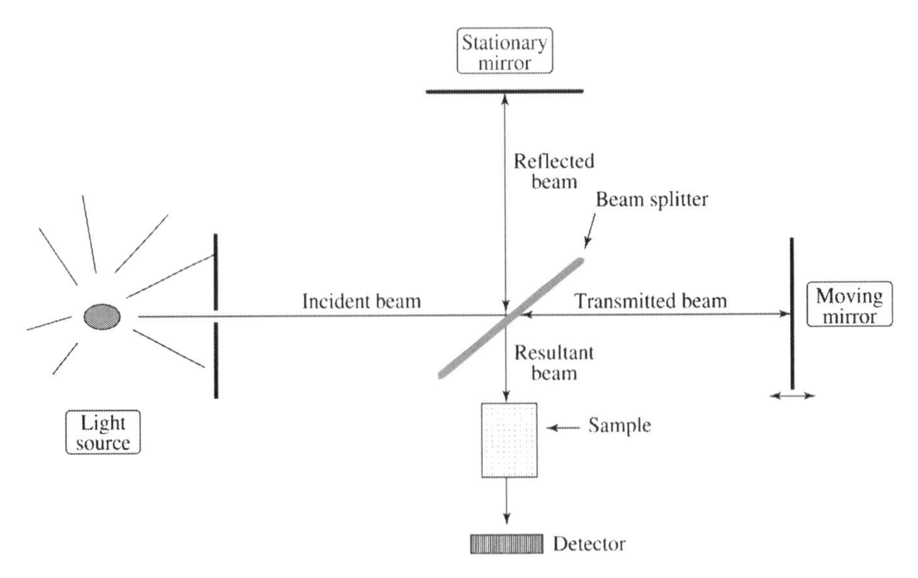

Figure 3.43. The Michelson interferometer as used in FTIR spectroscopy. The beam splitter divides an incident beam into a transmitted and reflected beam, respectively. These are reflected by a stationary mirror and a moving mirror, respectively. The reflected beams are recombined at the beam splitter and the resultant beam passes through the sample. The reflected and transmitted beams can interfere with each other (Figure 3.44) depending on the relative positions of the stationary and moving mirrors. The resultant beams detected are analysed by the Fourier Transform (Appendix 2) to generate an FTIR spectrum which is characteristic of the sample.

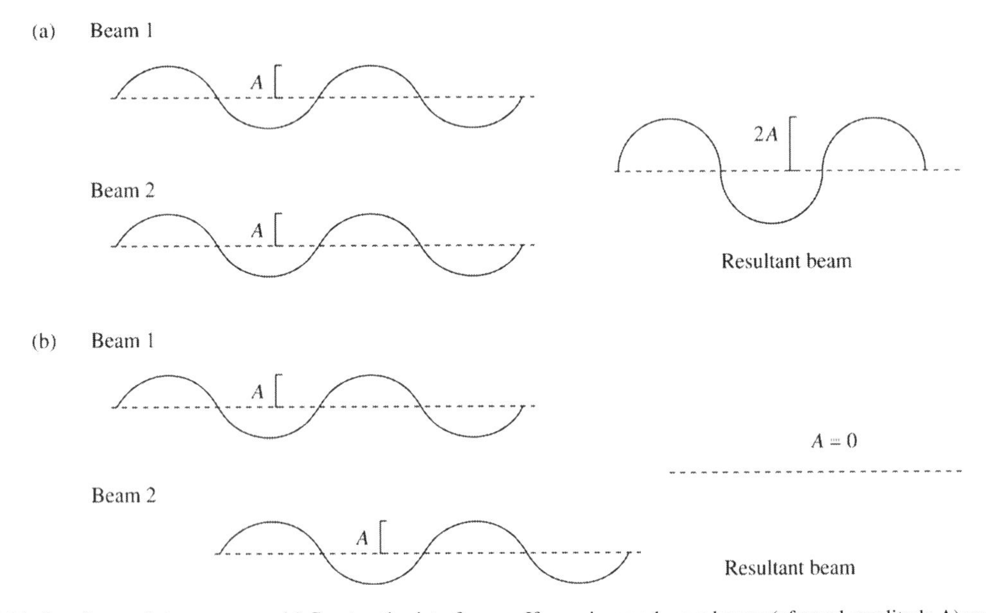

Figure 3.44. Interference between waves. (a) Constructive interference. If superimposed wave beams (of equal amplitude A) are in phase, they reinforce each other such that the resultant beam has amplitude equal to the sum of that of the superimposed waves (i.e. 2A). (b) Destructive interference. If superimposed beams are 180° out of phase, they weaken each other such that the resultant beam has amplitude of 0. Where the beams are less than 180° out of phase, the resultant beam would have amplitude between 0 and 2A.

Example 3.5 Characterization of photoconversion of the green fluorescent protein of the jelly fish *Aequorea victoria* by FTIR

The jelly fish, *Aequorea victoria*, possesses a green fluorescent protein (GFP) which functions as the secondary emitter of chemiluminescence from this organism (Section 3.4.1). The chromophore involved in this process is formed by cyclization of the tripeptide Ser-Tyr-Gly which makes up residues 65–67 followed by dehydrogenation of Tyr-66. This chromophore can exist in a neutral form ($\lambda_{max} = 395$ nm; GFP_{395}) or an anionic form ($\lambda_{max} = 480$ nm; GFP_{480}). Exposure to ultraviolet light converts most of the chromophore from the neutral to the anionic form. Once formed, the anionic form requires several hours to reconvert back to the neutral form. The following drawing shows the absorption spectra of the two forms of the protein.

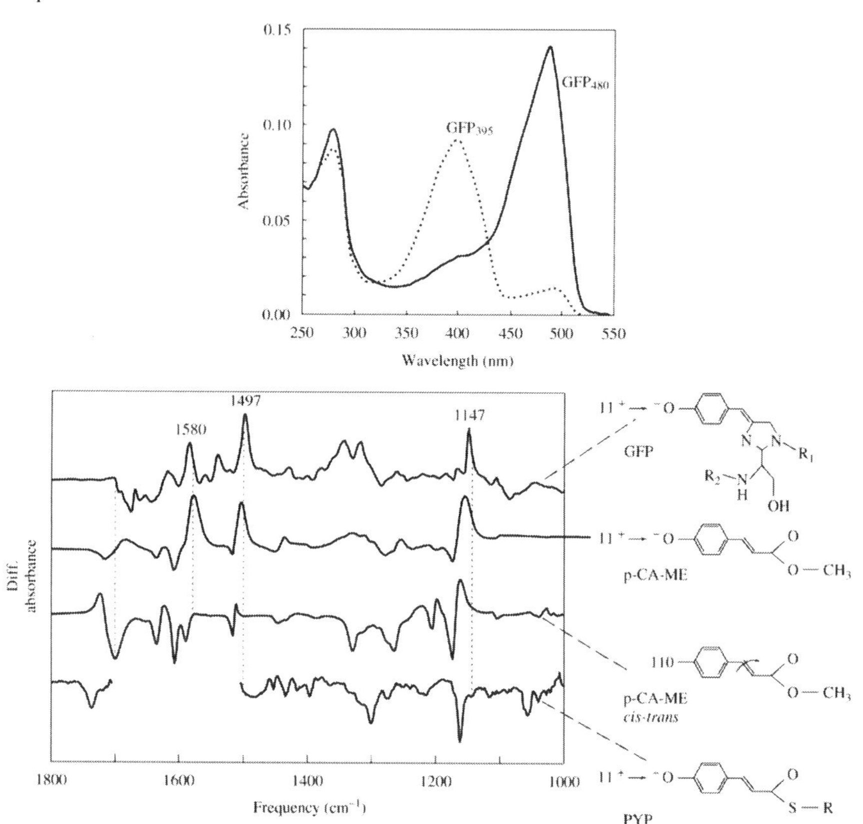

Structures determined by X-ray diffraction (Chapter 6) of wild type and mutant forms of this protein corresponding to GFP_{395} and GFP_{480} suggested that Glu-222 might function as a proton acceptor.

FTIR spectroscopy is highly sensitive to alterations in protonation state between protein conformations. In particular, difference FTIR spectroscopy where the FTIR spectrum of one protein conformation is subtracted from the other is informative about events such as protonation/deprotonation. In the present example, an FTIR spectrum was recorded for GFP_{395} and, after photoconversion, this was subtracted from the spectrum for GFP_{480}. The difference FTIR spectrum for GFP is compared in the following illustration with those of photoactive yellow protein (PYP, another photoactive protein) and p-coumaric acid (pCA), a model chromophore. This allowed allocation of peaks at 1580, 1497 and 1147 cm^{-1} to the phenolic ring common to all three chromophores.

If the $-COO^-$ group of a Glu residue were protonated during photoconversion, this would be expected to give rise to a new peak near 1730 cm^{-1} which would appear in the difference spectrum as a minimum. In PYP a Glu residue is known to be protonated during photoconversion and a minimum due to this is evident in the difference spectrum for this protein (arrow). The fact that there is no corresponding minimum in the spectrum for GFP allowed the conclusion that less than 0.1 Glu residues of the GFP protein changes protonation state on photoconversion. Thus, using FTIR difference spectroscopy, it was possible to exclude the suggestion that Glu-222 acts as a proton acceptor during this process.

Source: (From van Thor *et al.*, 1998) Reprinted with permission from Thor *et al.*, *Biochemistry* **37**, 16915–16921 (c) 1998 American Chemical Society.

Hydrogen bonding also underlies secondary structure in DNA and RNA and in DNA-RNA complexes. FTIR spectroscopy has found widespread use in the study of many biological processes involving nucleic acids. These studies range from *in vitro* experiments with synthetic oligonucleotides or their chemical analogues to investigations of tumour samples from cancer patients. The effect of nucleotide composition and sequence on base-pairing, the process of DNA-RNA hybridization and potential of DNA/RNA ratio in 'grading' the severity of cancer in human patients have all recently been investigated.

3.6.5 Raman Infrared Spectroscopy

In addition to absorbance of light, sample molecules can also *scatter* light and this can be detected at 90° to the direction of the incident beam. Scattering of monochromatic light is of two general types (Figure 3.45). *Rayleigh scattering* occurs when the frequency of the scattered light is the same as that of the incident light. Most of the light scattered by a sample results from this process. Much less commonly, the frequency of the scattered light may be *greater* or *less* than that of the incident light. This is a phenomenon known as *Raman scattering* and it forms the basis of *Raman spectroscopy*. Both of these phenomena occur with light of all frequencies and if the incident beam is composed of visible light then the scattered light will also be in the visible range. Because Raman scattering occurs with such a low probability, it is usual to use a high-energy laser emitting light in the high-frequency part of the visible spectrum as the incident light (Section 3.9).

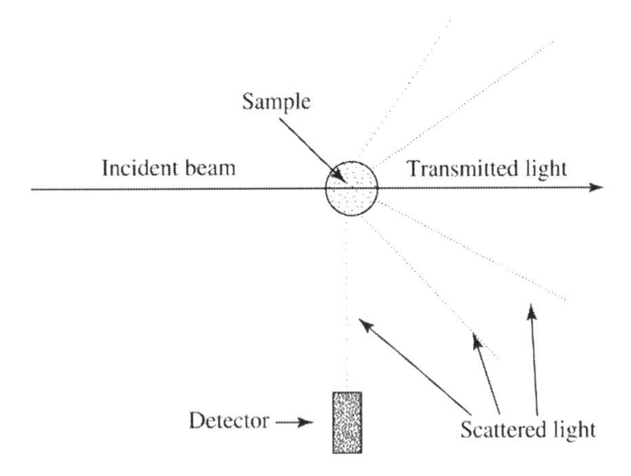

Figure 3.45. Light scattering. Incident light passes through sample solution (solid arrow). A proportion of this light is also scattered in all directions (dashed lines). This scattered light is mostly of the same frequency as the incident beam (Rayleigh scattering). However, occasionally the scattered light has a different frequency to the incident beam (Raman scattering). Measurement of these latter scattered beams is the basis of Raman spectroscopy.

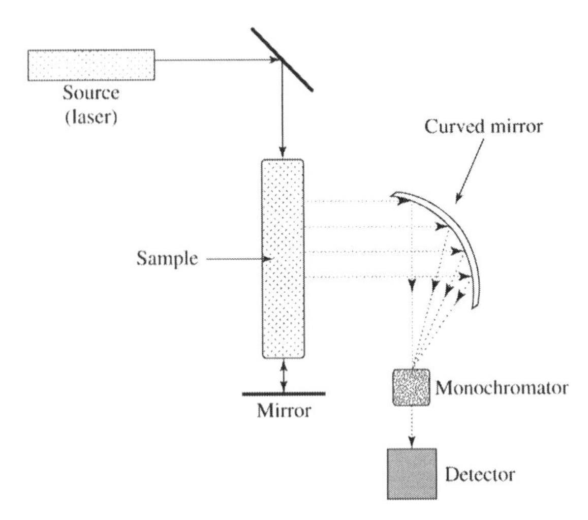

Figure 3.46. Raman spectrometer. Monochromatic light is passed through a concentrated sample. Scattered light (dashed lines) is collected with the aid of a curved mirror. This is then diffracted through a monochromator which allows the identification of frequencies due to both Rayleigh and Raman scattering. Note that the lower mirror reflects transmitted light back through the sample to double the intensity of scattered light.

The relevance of Raman spectroscopy to infrared spectroscopy lies in the fact that the energy-differences between incident and Raman scattered light corresponds to vibrational transitions. Raman spectroscopy is therefore a means of studying vibrational transitions, even though the light used is usually in the visible part of the spectrum. Moreover, since the phenomenon is based on scattering rather than absorption, it is possible to measure Raman spectra in aqueous solutions without the problems encountered due to water absorption in dispersive infrared spectroscopy.

The experimental apparatus used to measure Raman spectra is shown in Figure 3.46. Since the effects measured occur with such low probability, it is necessary to use highly-concentrated samples (i.e. in the range 20 mg/ml) to detect scattered light. Moreover, it is essential that the sample does not *aggregate* under these conditions as this may change the properties of the spectrum obtained.

In Raman scattering, a small amount of energy is transferred from the incident light (frequency, ν_i) to promote the sample molecule from the ground state to an excited vibrational level. In other words light energy *causes a vibrational transition*. Since $E = h\mu$ (Equation (3.2)), this *loss* of energy causes a small *decrease* in the frequency of the scattered light to a value ν. This decrease can be expressed

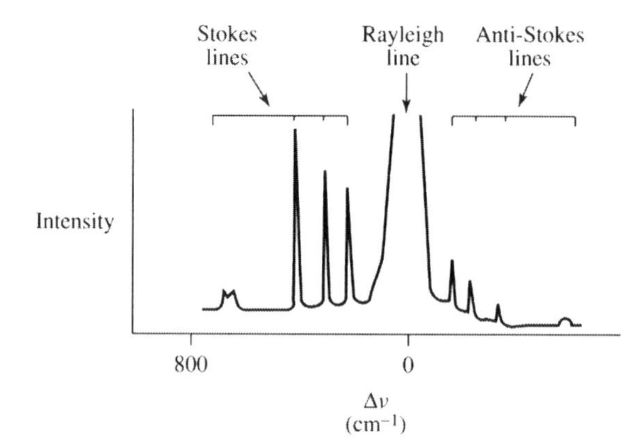

Figure 3.47. Raman spectrum of CCl_4. Light scattered by Rayleigh scattering has a $\Delta\nu$ of 0. Lines corresponding to $\nu - \nu_i$ are called Stokes lines and these have a characteristic values for each chemical structure. Corresponding, though much weaker, lines are observed at $\nu + \nu_i$ which are called anti-Stokes lines.

as $\Delta\nu$:

$$\Delta\nu = \nu - \nu_i \tag{3.35}$$

where ν is the frequency of Raman scattered light. Passage through a monochromator allows diffraction of scattered light into its various frequencies and a Raman spectrum is frequently a plot of light intensity *versus* $\Delta\nu$ (Figure 3.47). Most of the scattered light has the *same* frequency as the incident light, ν_i, that is $\Delta\nu = 0$. This is called the *Rayleigh line* of the spectrum. However, other lines in the spectrum have measurable $\Delta\nu$ values which depend on the structure of the sample molecule. These are called *Stokes lines*.

A second way in which the sample molecule and incident light can interact is where some molecules existing in the first excited vibrational energy level *impart* energy to the incident light thus *decreasing* the frequency of Raman scattered light. This gives rise to lines in the spectrum of opposite sign but identical $\Delta\nu$ to the Stokes lines which are called *anti-Stokes lines*. Because most molecules in a population are in the ground state, vibrational transitions leading to a decrease in frequency are statistically much more likely than those leading to an increase in frequency. Therefore, anti-Stokes lines are usually much *weaker* than Stokes lines. It should be clear from this that a sample molecule will give a Raman spectrum characteristic for its chemical structure. Since this spectrum is due to vibrational transitions, the information contained in such a spectrum is no different to that contained in a dispersive infrared spectrum. However, Raman spectroscopy has the important practical advantage that it is not affected by solvent water.

In addition to studies on samples prepared in solution, it is also possible to measure Raman spectra on samples prepared as dry fibres or as single crystals. Thus it is possible to compare Raman spectra collected in the presence and absence of solvent water. Structure elucidation of biomacromolecules involves collection and analysis of X-ray diffraction data from samples also prepared as fibres or crystals (Chapter 6). Raman spectroscopy therefore provides a convenient means of confirming structural data obtained from X-ray crystallography.

3.7 NUCLEAR MAGNETIC RESONANCE (NMR) SPECTROSCOPY

3.7.1 Physical Basis of NMR Spectroscopy

Atoms and molecules have a variety of *quantized* energy levels (e.g. electronic, vibrational and rotational). Many spectroscopic techniques take advantage of transitions between these energy levels with different ΔE values being related to particular frequency-ranges of the electromagnetic spectrum by Equation (3.2).

A further property of atoms which is quantized is *magnetic spin* and *nuclear magnetic resonance* (NMR) spectroscopy is a spectroscopic technique which takes advantage of this fact. To explain the physical basis of NMR we can use the hydrogen nucleus, ^1H, as a good example (this type of NMR is called *proton NMR*). However, as will become clear (Section 3.7.2), magnetic spin is a property of many different types of atoms. The ^1H nucleus may be regarded as a spinning positive charge (Figure 3.48). This generates a magnetic field which will have a *magnetic spin moment*, μ. If an external magnetic field (field-strength, B_0) is applied to such a nucleus, it can orientate itself either with (*parallel*) or against (*antiparallel*) this field in much the same way

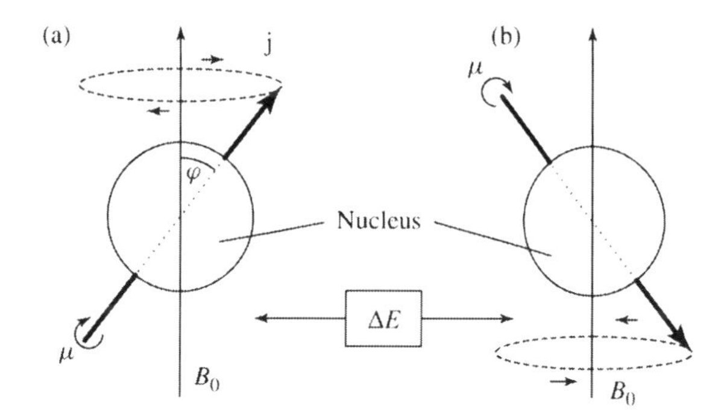

Figure 3.48. Physical basis of NMR. A spinning nucleus generates a magnetic field with a spin moment, μ, which generates an angular momentum, *j*. When placed in an external magnetic field (B_0), the nucleus can align (a) with or (b) against the field. The difference in energy between these orientations is ΔE which corresponds to frequencies in the radiowave part of the spectrum.

as a bar-magnet does in the Earth's magnetic field on the macroscopic scale. These two orientations are referred to as *spin states* and are distinguishable by their different *spin quantum numbers*, m_I, which are, respectively, $-1/2$ and $+1/2$. The magnetic spin moment 'wobbles' or *precesses* around the axis of the external magnetic field by an angle, θ, and rotates around this axis with a particular frequency, ω, which is called the *Larmor frequency*.

The *potential energies* of the two spin states are given by Equations (3.36) and (3.37):

1. (Low energy spin state; $m_I = -1/2$)

$$E = -\mu \cdot B_0 \cdot \sin\theta \tag{3.36}$$

2. (High energy spin state; $m_I = +1/2$)

$$E = \mu \cdot B_0 \cdot \sin\theta \tag{3.37}$$

The energy *difference*, ΔE, between them is therefore given by;

$$\Delta E = 2\mu \cdot B_0 \cdot \sin\theta \tag{3.38}$$

In fact, at room temperature the energy difference between the two spin states is very small and the low energy orientation is only marginally favoured over the high energy one (Figure 3.49). However, Equation (3.38) shows that ΔE varies in a manner which is *directly proportional* to the applied magnetic field. Early work on biomolecules used magnetic field strengths of only approximately 40 MHz. Modern NMR *spectrometers* (Section 3.7.5) use much larger field strengths (in the range 500–800 MHz) which give rise to larger ΔE values and yield NMR spectra of much higher resolution.

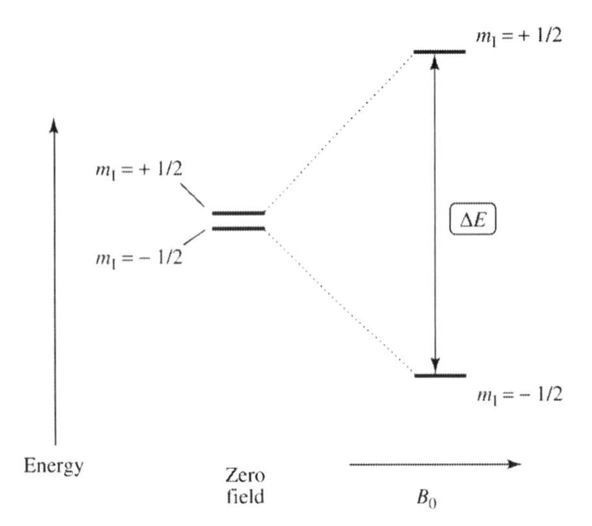

Figure 3.49. Effect of applied magnetic field on spin states. In the absence of an applied magnetic field (i.e. zero field), the energy difference (ΔE) between the two spin states is very small. The stronger the field strength (B_0) of an applied magnetic field becomes the larger ΔE (Equation (3.38)).

Combining Equations (3.38) and (3.2) shows that the electromagnetic frequency, ν, corresponding to a transition between these energy levels is:

$$\nu = \frac{2\mu \cdot B_0 \cdot \sin\theta}{h} \tag{3.39}$$

where h is Planck's constant.

NMR spectroscopy depends on absorption of electromagnetic radiation from the radiowave part of the spectrum (Figure 3.7) causing the nucleus to undergo a transition from a low to a high energy spin state. The precise value of ν required for this transition depends on both the *identity* of the nucleus (Section 3.7.2) and on its precise chemical *environment* (Section 3.7.3). Because of this, NMR spectra can yield precise information on the structure/composition of biomolecules and on processes in which they are involved (e.g. chemical reactions).

In order to undergo an NMR transition at a particular value of ν, a specific set of circumstances called the *resonance condition* needs to exist. As previously shown (Figure 3.1), electromagnetic radiation of frequency ν has an associated oscillating magnetic field. This field (of field strength B_1) may be regarded as *rotating around* the spinning nucleus in a plane at right angles to the applied external field, B_0 (Figure 3.50). However, it is much weaker than the applied external

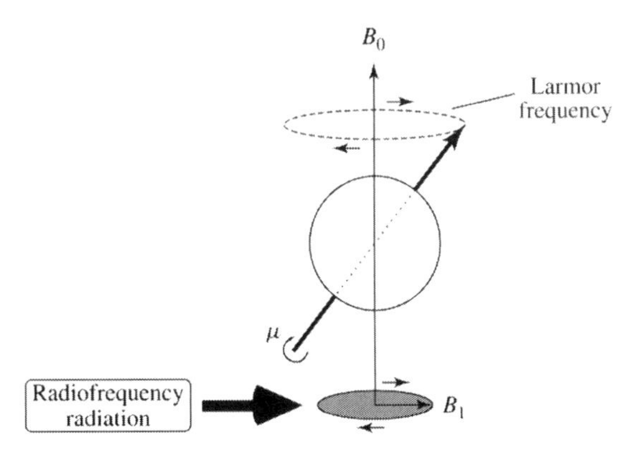

Figure 3.50. The resonance condition. Exposure of the nucleus to radiofrequency radiation sets up a magnetic field (of field strength, B_1, shown in grey) which has a frequency of oscillation. The resonance condition occurs when this frequency equals the Larmor frequency of the spin magnetic moment. Transition between spin states only occurs at the resonance condition.

field (i.e. $B_1 << B_0$). The resonance condition is achieved when the *frequency of rotation* of the field represented by B_1 equals the Larmor frequency. The relationship between the Larmor frequency and the radiowave frequency is given by:

$$\omega = 2\pi \cdot \nu \quad (3.40)$$

This requirement for the resonance condition explains the inclusion of the word 'resonance' in 'nuclear magnetic *resonance*'. Originally, this was achieved experimentally by holding the Larmor frequency at a fixed value (i.e. by exposing the sample to a continuous wave of electromagnetic radiation of frequency ν) and varying the strength of the applied magnetic field (B_0). However, modern NMR spectrometers monitor many radiowave frequencies simultaneously by exposing the sample to a *pulse* composed of several radiowave frequencies.

3.7.2 Effect of Atomic Identity on NMR

So far we have considered only the nucleus of ^{1}H. Because of the abundance of ^{1}H in biological systems, proton NMR is especially informative about the structure and functioning of biomolecules and is probably the most popular single form of biological NMR. However, in contrast to most other spectroscopic techniques, a major strength of NMR is that other atoms may be studied in the same way as protons allowing us to obtain 'atom-specific' information. Nuclei which are amenable to study by NMR have a characteristic magnetic spin, I (Table 3.6). This is a consequence of *pairing* of individual spins due to neutrons and protons which each have m_I values of 1/2. Nuclei possessing *even atomic mass and even atomic number* have no net spin and therefore $I = 0$ (e.g. ^{12}C, ^{16}O). Several nuclei occurring in Biology have $I = 1/2$ (e.g. ^{1}H, ^{13}C, ^{31}P), while others have $I = 1$ (e.g. ^{2}H and ^{14}N) or >1 (e.g. ^{35}Cl; $I = 3/2$). This means that, in addition to being atom-specific; NMR information is also *isotope-specific*, allowing us to obtain distinct NMR spectra for ^{1}H and ^{2}H, for example.

Table 3.6. Magnetic properties of some nuclei important in biochemistry

Nucleus	I	Natural abundance (%)	γ rad·s^{-1} T^{-1}	NMR ν at T = 2.3488 (MHz)
^{1}H	1/2	99.98	26.752	100
^{2}H	1/2	0.015	4.107	15.35
^{12}C	0	98.9	—	—
^{13}C	1/2	1.10	6.7283	25.144
^{14}N	1	99.63	1.9338	7.224
^{16}O	0	99.76	—	—
^{32}S	0	95.02	—	—
^{31}P	1/2	100	10.8394	40.481
^{35}Cl	3/2	75.77	2.642	9.798
^{15}N	1/2	0.37	−2.7126	10.133

Each nucleus possessing spin (i.e. $I = 0$), undergoes a characteristic transition from a lower to a higher spin state at the resonance condition:

$$\Delta E = \frac{\gamma}{2\pi} \cdot B_0 \quad (3.41)$$

where γ is a constant property of the nucleus called the *magnetogyric ratio* (sometimes also called the *gyromagnetic ratio*). In practice, this means that we can study individual nuclei by varying the field strength thus generating proton NMR spectra, ^{13}C NMR spectra, and so on.

3.7.3 The Chemical Shift

If our discussion of NMR was to finish at this point, we would have a picture of a specific resonance condition for each nucleus possessing magnetic spin determined by the value of the magnetogyric ratio, γ, for that nucleus. What makes NMR especially informative, however, is the fact that the precise radiation frequency, ν, corresponding to the resonance condition for each type of nucleus at a given applied magnetic field strength can be affected by its immediate chemical environment. For example, a proton forming part of a $-CH_2$ group will achieve the resonance condition at a different radiation frequency to a proton forming part of a $-CH_3$ group or an $-NH_2$ group. This is due to the effect of the magnetic fields of nearby nuclei on that of the nucleus undergoing the transition and is known as *chemical shielding*. The effect is to alter the magnetic field experienced by the nucleus (B_{eff}) such that:

$$B_{eff} = B_0(1 - \sigma) \quad (3.42)$$

where σ is a *shielding constant*.

This phenomenon operates over short distances (approx. 3–8 Å) and allows us to identify characteristic frequencies corresponding to particular chemical groups. For example, a $-CH_2$ group in one chemical structure (e.g. the amino acid serine) will achieve the resonance condition at a similar ν to the $-CH_2$ group of a monosaccharide. *Intramolecular* shielding effects like these can be used to determine the chemical structure of small molecules from their NMR spectra alone (*see* the example of ethanol below). 'Though-space' shielding effects are also possible, however, between adjacent chemical groups. For example, if a metal (Table 3.7) happened to be located near the nucleus under investigation or if two amino acid side-chains were brought into proximity in a folded protein, this would cause chemical shielding.

Table 3.7. Proton NMR chemical shift values for some common chemical groups encountered in biomolecules (nucleus under investigation is denoted by arrow)

Chemical group	Chemical shift (ppm)
TMS	0
↓ CH_3—Metal	−0.5–0
↓ CH_3—CH_2—	0.8–1
CH_3—↓CH_2—CH_2—	1.2–2
↓ CH_3—C(=O)—	1.8–2.2
↓ H—C≡C—	2.2–3
↓ CH_3—O—	3.2–3.8
↓ CH_3—O—	3.5–4.5
>C=↓CH_2	4.7–5.5
–CH=CH–	4.5–7
Benzene	7

We shall see later on in this text (Chapter 6) that through-space shielding effects such as these make it possible for us to determine three-dimensional structures of proteins and other biomacromolecules from NMR measurements.

Largely for instrumental reasons, it is difficult to measure ν values accurately. To standardize measurements between different NMR spectrometers and different experimental conditions, it is usual to include a *reference* compound (normally tetramethylsilane, TMS) with the sample to be analysed. The frequency corresponding to the resonance condition for each transition in the sample is then expressed as the *chemical shift*, δ, in *parts per million* (ppm) as follows:

$$\delta = \frac{\nu_S - \nu_{Ref}}{\nu_{Ref}} \times 10^6 \quad (3.43)$$

where ν_S and ν_{Ref} are the frequencies of radiowave radiation corresponding to the resonance condition of the sample and reference nucleus, respectively. TMS has a chemical shift of 0 and each chemical group has a particular value range

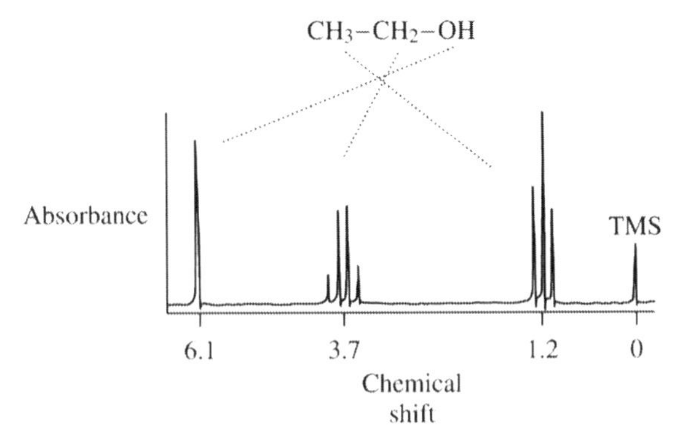

Figure 3.51. Proton NMR spectrum of ethanol. An NMR spectrum consists of a plot of absorbance intensity versus chemical shift (d). Peaks arising from each of the three types of proton in the structure are labelled. In the case of $-CH_3$ and CH_2 groups, multiple peaks arise as a result of spin coupling. Butler and Harrod (1989), *Inorganic Chemistry: Principles and Applications*, Addison Wesley Longman, Reproduced with permission.

as illustrated in Table 3.7. Generally speaking, *electron-withdrawing* groups bonded to the group under investigation tend to *increase* the chemical shift associated with that group while electropositive centres (e.g. metals) tend to *decrease* it. Note that the chemical shift may have negative *or* positive values.

An NMR spectrum consists of a plot of the intensity of absorbance of radiowave radiation as a function of chemical shift (ppm). A proton NMR spectrum for ethanol is shown in Figure 3.51 and it is possible to identify chemical shifts for each type of proton (i.e. $-CH_3$, $-CH_2$, –OH) in this spectrum. Since there is only one type of bonding structure which would explain these chemical shifts, this NMR spectrum would allow is to write down the structure of this molecule if it were not previously known. More definitive identification of structure in such spectra is made possible by a phenomenon known as *peak splitting*. In Figure 3.51, this is clear in the parts of the spectrum due to the $-CH_3$ and $-CH_2$ groups where three and four peaks are visible, respectively, rather than the single peak we might have expected.

3.7.4 Spin Coupling in NMR

Peak splitting in NMR spectra arises from a phenomenon known as *spin coupling* or *J coupling*. We can understand this if we look at the splitting of the $-CH_2$ 'peak' into four peaks. This is due to coupling of the spins of the adjacent $-CH_3$ group. The three protons of the $-CH_3$ group can have a spin of either +1/2 or −1/2 as previously described (Figure 3.49). Accordingly, we can write the possible spin arrangements of these protons as *up* (↑) or *down* (↓) as shown in Table 3.8. Two of these arrangements occur singly while

Table 3.8. Spin coupling between $-CH_2$ and $-CH3$ groups in ethanol

Arrangement	Intensity
CH_2 resonances coupled with $-CH_3$	
All up ↑↑↑	1
Two up ↑↑↓ ↑↓↑ ↓↑↑	3
One up ↓↑↓ ↑↓↓ ↓↓↑	3
All down ↓↓↓	1
CH_3 resonances coupled with $-CH_2$	
All up ↑↑	1
One up ↓↑ ↑↓	2
All down ↓↓	1

Note: Resonances associated with the $-CH_2$ group are affected by the spins of protons on the nearby $-CH_3$ group and *vice versa*. This is called spin-spin coupling. The number and type of possible spin arrangements on the coupled group is shown with arrows. The NMR peak due to the $-CH_2$ group splits in the ratio 1 : 3 :3: 1 while that due to the $-CH_3$ group splits 1 : 2 : 1. This ratio depends on the number of possible spin arrangements in the coupled nuclei.

two occur in equivalent arrangements three times. The *net spin* experienced in the resonance condition for the $-CH_2$ group has therefore four possibilities which occur in a ratio 1 : 3:3 : 1. This explains why the $-CH_2$ group gives rise to a characteristic pattern of four peaks in a ratio 1 : 3:3 : 1.

Generalizing from this example, the number of peaks to be expected from a nucleus due to an adjacent equivalent nucleus (or set of *equivalent* nuclei with the same I value) is given by the following equation:

$$\text{Number of peaks} = (2 \times n \times I) + 1 \tag{3.44}$$

where n is the number of equivalent nuclei in the chemical group. For the $-CH_2$ group resonance condition the value of peaks expected as a result of coupling to the nearby $-CH_3$ protons is $(2 \times 3 \times 1/2) + 1 = 4$ peaks. For the $-CH_3$ group we would predict $(2 \times 2 \times 1/2) + 1 = 3$ peaks due to coupling with the protons of $-CH_2$. The actual spacing (in Hertz) between each peak in an NMR spectrum is called the *coupling constant* or *J constant*. The *closer* together the coupled nuclei are in the sample molecule, the *stronger* the coupling and hence the *greater* the J constant. The number of bonds across which the coupling occurs is conventionally denoted as $^3J\ ^1H - ^1H$ (in our example of ethanol) to denote the fact that the coupled nuclei (^1H) are separated by three bonds.

The *relative intensity* of each peak in such multiplets (i.e. the area of each peak) may be found as the *coefficients* of a binomial expansion of the form $(1 + x)^n$ where n is the number of nuclei coupled to the resonance condition being measured. For the $-CH_2$ and $-CH_3$ groups, these are, respectively:

$$CH_2;\quad (1 + x)^3 = 1 + 3x + 3x^2 + 1x^3 \tag{3.45}$$

$$CH_3;\quad (1 + x)^2 = 1 + 2x + 1x^2 \tag{3.46}$$

where the coefficients are 1 : 3:3 : 1 ($-CH_2$) and 1 : 2:1 ($-CH_3$) as we have just seen (Table 3.8 and Figure 3.51). The coefficients arising from peak splitting due to spin coupling with n nuclei of spin 1/2 can be calculated from the above binomial or, more conveniently, may be obtained from *Pascal's triangle* (Table 3.9).

What about a situation where two nuclei (A and B) with $I \neq 0$ form part of a single chemical group? We can calculate the number of peaks to be expected from such a group as follows:

$$\text{Number of peaks} = (2 \cdot n \cdot I + 1)_A \times (2 \cdot m \cdot I + 1)_B \tag{3.47}$$

where n and m refer to the number of each nucleus and I to their magnetic spin. For example, the chemical shift due to an NH_2 group ($n_N = 1$; $m_H = 2$; $I_N = 1$; $I_H = 1/2$) would be expected to result in $(2 \times 1 \times 1 + 1) \times (2 \times 2 \times 1/2 + 1) = 9$ peaks.

It should be obvious from this that a unique pattern of peak splitting of chemical shifts would be expected for most small molecules such as amino acids, monosaccharides and small peptides which allows determination of their chemical structure by NMR spectroscopy alone.

3.7.5 Measurement of NMR Spectra

NMR spectra are determined experimentally in a specialized type of spectrophotometer called an NMR spectrometer (*see* Figure 3.52 for an outline design). The main unique feature of this instrument compared to spectrophotometric devices already described in this chapter is the powerful magnet which is responsible for inducing a magnetic field, B_0 in the x-axis. Radiowave frequency radiation is generated by a *transmitter* and may either be of a single frequency or in

Table 3.9. Pascal's triangle

n	Relative intensities
0	1
1	1 : 1
2	1 : 2 : 1
3	1 : 3 : 3 : 1
4	1 : 4 : 6 : 4 : 1
5	1 : 5 : 10 : 10 : 5 : 1
6	1 : 6 : 15 : 20 : 15 : 6 : 1

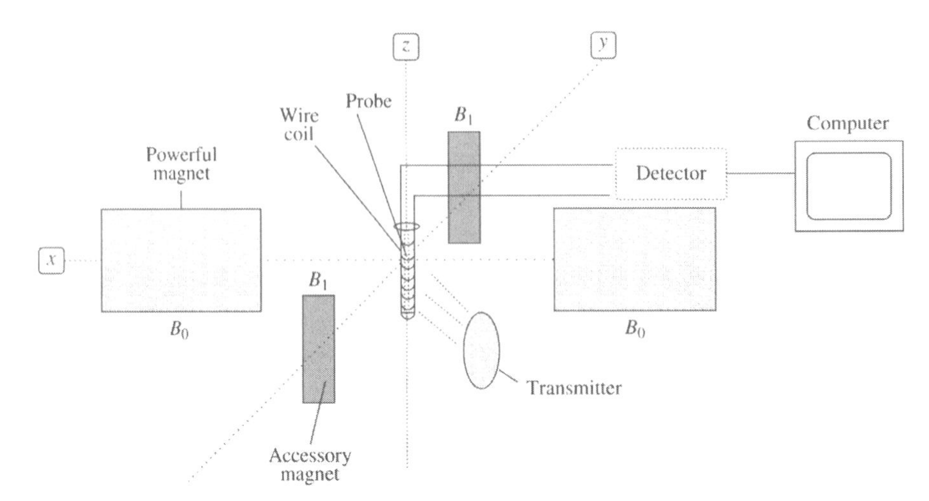

Figure 3.52. Outline design of an NMR spectrometer. The sample is placed in a tube called the probe. This is surrounded by wire coil. A powerful magnetic field (of field strength B_0) is generated in the x-axis. Radiofrequency radiation is generated by the transmitter which generates a magnetic field orientated along the y-axis. When the frequency of this field is the same as the Larmor frequency, the resonance condition is met. Absorption of radiation induces a current in the wire coil along the z-axis which is electronically detected. This current is proportional to the intensity of absorbance.

the form of a pulse of frequencies (in the case of *Fourier Transform* NMR spectrometers). A magnetic field (B_1) of a frequency *near* the Larmor frequency is generated either by this radiowave radiation alone or with the aid of a small accessory magnet. This field is orientated at right angles to the field generated by B_0, that is along the y-axis. The sample is placed in a tube called the *probe* which is surrounded by a coil of wire. When the resonance condition is achieved in the sample, a current is induced in the coil of wire which is at right angles to both magnetic fields, that is along the z-axis. This current is analysed by computer and converted into an intensity of absorbance value. More sophisticated NMR spectrometers are used to generate *multidimensional NMR* such as *two-dimensional NMR*. This is discussed in more detail in Chapter 6.

NMR is a particularly versatile analytical technique suitable for both low and high molecular weight molecules. Structural and dynamic information can be obtained therefore on samples ranging in size from amino acids and peptides to proteins of masses up to 30 kDa (using multidimensional NMR; Chapter 6). While strongest spectra are obtained with liquid samples, it is also possible to analyse solid samples such as crystals. Indeed, a technique closely related to NMR called *magnetic resonance imaging* (MRI) facilitates analysis of the whole human body (or component parts such as head, legs and torso) as an aid to clinical diagnosis.

Chemical shifts obtained from nuclei forming component parts of biomolecules are affected by their chemical environment. NMR is therefore especially sensitive to the involvement of nuclei in hydrogen bonding or their proximity to aromatic and carbonyl groups. The technique has been widely used for this reason in studies of protein folding (Chapter 6), ligand binding, molecular mobility within proteins, position and role of metals in metalloproteins and in studies of pH effects on proteins. Another set of applications, ^{31}P NMR focuses on measurements of phosphorylated biomolecules. This makes possible the study of effects on biological membranes (which are rich in phospholipids), signal transduction systems and processes such as glycolysis and oxidative phosphorylation. Example 3.6 describes a study of this kind on creatine phosphokinase activity in intact mitochondria.

3.8 ELECTRON SPIN RESONANCE (ESR) SPECTROSCOPY

Electron spin resonance (ESR) spectroscopy is based on an identical physical principle to NMR. For this reason, several aspects of ESR spectroscopy are similar to NMR but the following description also highlights some important points of difference.

3.8.1 Physical Basis of ESR Spectroscopy

ESR arises from the fact that an *unpaired electron* also possesses a spin magnetic moment when placed in a powerful magnetic field. Samples containing unpaired electrons are

Example 3.6 Determination of creatine phosphokinase activity in intact mitochondria by 31P NMR

In heart mitochondria it is known that ADP produced by creatine phosphokinase in the following reaction has preferential access to oxidative phosphorylation.

$$\text{Creatine} + \text{ATP} \xrightarrow{\text{creatine phosphokinase}} \text{phosphocreatine} + \text{ADP}$$

A convenient method for monitoring phosphorylated molecules such as adenine nucleotides and phosphocreatine is offered by ^{31}P NMR. Because this is an 'atom-specific' technique of high sensitivity, it is possible to monitor changes in levels of phosphorylated molecules in crude extracts, intact organelles (e.g. mitochondria, intact cells and even intact organs such as perfused heart.

A good illustration of the power of this technique is offered by a study of the rate of creatine phosphokinase activity in intact, functional mitochondria from skeletal muscle. The effect of external ADP and ATP concentrations on this activity was measured by incubating mitochondria in (a) 0.5 mM ATP or (b) 0.4 mM ADP. NMR peaks could be assigned to; (1) inorganic phosphate, (2) phosphocreatine, (3–5) γ, α and β - phosphate of ATP, (6) and (7) β and α-phosphate of ADP, respectively in a time-course from 3.84 to 30.71 min. as shown in the following spectra.

These data allowed quantification of the rate of production of phosphocreatine (peak 2 in (a) and (b) at a range of ATP and ADP concentrations. This data which is summarized in plot c) below demonstrated that the K_m of creatine phosphokinase for each nucleotide differed (K_m^{ADP} = 63 μM and K_m^{ATP} = 28 μM) while V_{max} for each nucleotide remained the same.

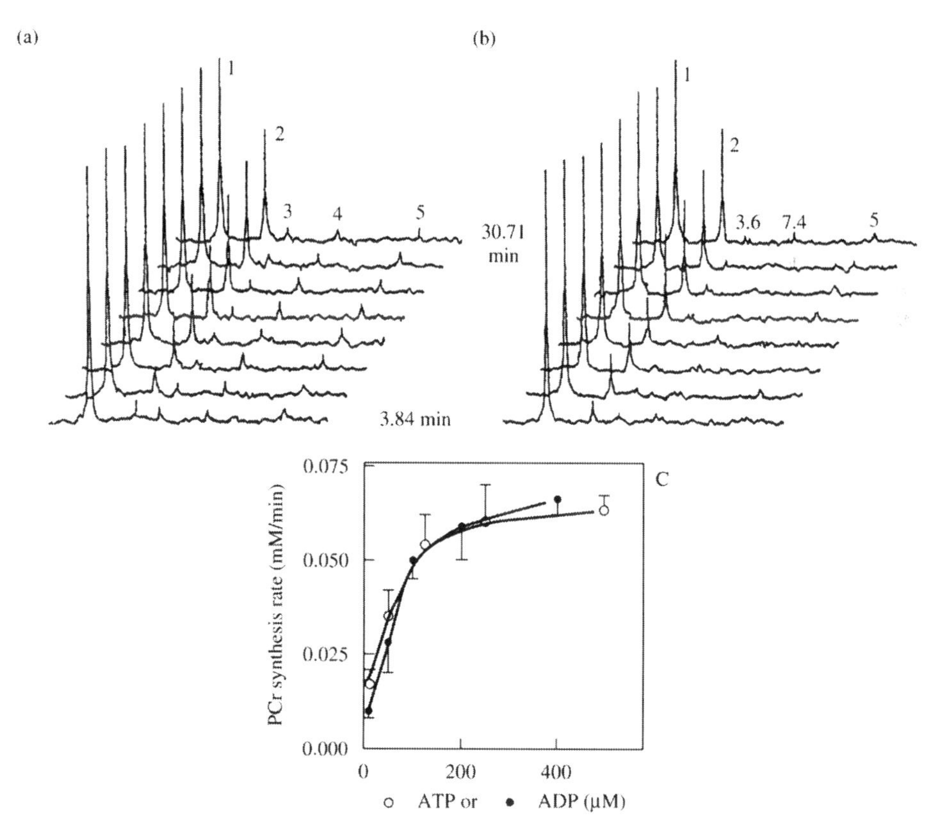

This example demonstrates how NMR allows detailed information to be gathered from a complex biochemical system by taking advantage of the sensitive and atom-specific nature of the technique.

Source: (From Kernec *et al.*, 1996) Kernec, F. *et al.*, (1996) Biochemical and Biophysical Research Communications, *Biochemical and Biophysical Research Communications* **225**, 819–825. Reproduced with permission of Academic Press, Inc.

said to be *paramagnetic* and ESR is also sometimes referred to as *electron paramagnetic resonance* (EPR). Like many nuclei, this spin magnetic moment is quantized with allowed spins of only $\pm 1/2$. However, the spin magnetic moment associated with an unpaired electron is approximately 1000-fold greater than that associated with ^{1}H. The energy difference (ΔE) between the two spin states of the electron therefore corresponds to the *microwave* part of the electromagnetic spectrum rather than the radiowave associated with NMR (Figure 3.7). This has the important consequence that we can measure ESR spectra *independently* of NMR spectra and *vice versa* since each technique avails of a distinct part of the electromagnetic spectrum.

The ΔE value relates to the magnetic field strength, B_0, as follows:

$$\Delta E = h \cdot \nu = g \cdot \beta \cdot B_0 \qquad (3.48)$$

where β is a constant called the *Bohr magneton* and g is *the Lande splitting factor*. For a free electron *in vacuo*, this splitting factor has a value of 2.0023. However, it can have different values for various paramagnetic samples. Moreover, in solid samples (e.g. crystals); g can have different values depending on the *orientation* of the paramagnetic species.

3.8.2 Measurement of ESR Spectra

Because of their common physical basis, ESR and NMR share several points of similarity. Both techniques depend on a frequency of electromagnetic radiation corresponding to a specific resonance condition. In both techniques, this frequency is determined partly by the *identity* of the chemical group containing the magnetic spin moment. In the case of ESR, individual chemical groups cause *shielding* of the unpaired electron, resulting in an altered value for g. Thus, experimentally-determined values of g may be used to identify the paramagnetic chemical group. *Peak splitting* can also arise in ESR as a result of spin coupling with nearby nuclei, especially the nucleus actually carrying the unpaired electron. The *number* of peaks we would expect is given by:

$$\text{Number of peaks} = 2I + 1 \qquad (3.49)$$

where I is the spin quantum number of the nucleus (Table 3.6). The relative intensities of each peak also follow a binomial expansion as with NMR and may be determined from Pascal's triangle (Table 3.9). By analogy with Equation (3.47), if the unpaired electron was associated with two nuclei (A and B), the number of peaks resulting would be a product of the splitting induced by each:

$$\text{Number of peaks} = (2 \cdot n \cdot I + 1)_A \times (2 \cdot m \cdot I + 1)_B \qquad (3.50)$$

where n and m refer to the number of each nucleus and I_A and I_B are their respective spins (Table 3.6). Thus, if an unpaired electron was associated with an N-H group, six peaks would be expected [*that is* $(2 \times 1 \times 1/2 + 1) \times (2 \times 1 \times 1 + 1) = 6$]. In practice, these six peaks would be organized as three doublets. As with NMR, the strength of coupling is related to the spacing between peaks which is called the *isotropic hyperfine coupling constant*, a. The larger the value of a, the greater the coupling and the larger the spacing between split peaks. Because of the limited number of combinations of nuclei found in biological ESR, peak splitting also helps unambiguously to identify the nature of the chemical group containing the unpaired electron if this is not previously known.

ESR spectra are measured in an ESR spectrometer which is generally similar in design to an NMR spectrometer (Figure 3.52). The main differences are that ESR spectrometers expose samples to microwave rather than radiowave radiation and produce *derivative* rather than absorbance spectra (*see* below). These differences mean that distinct spectrometers are required for each magnetic resonance technique.

Despite the many similarities, ESR also shows several points of difference to NMR spectroscopy. Largely for instrumental reasons, ESR spectra are conventionally measured as the *first derivative* (dA) of the absorbance (A) as a function of frequency (ν). The relationship between A and dA is shown diagrammatically in Figure 3.53. 'Peaks' in ESR spectra are therefore referred to as *lines* consisting of a peak and trough. ESR spectra are also considerably more *simple* and *sensitive* than NMR spectra. Species

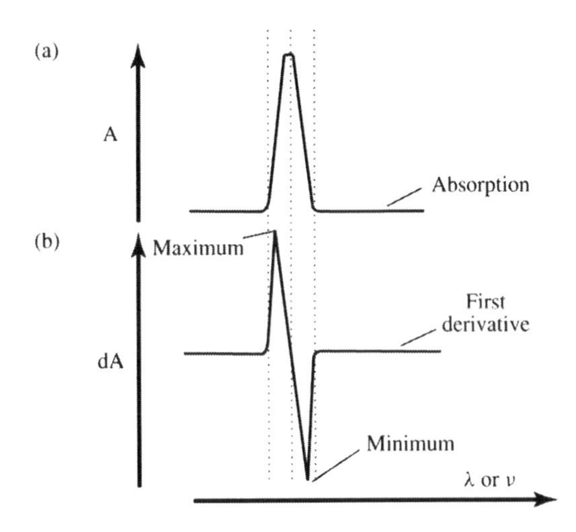

Figure 3.53. ESR spectra are recorded as first derivatives of absorption. (a) An absorbance spectrum is a plot of A versus λ or ν. (b) A first derivative plot of (a). This is a plot of rate of change of A. Note that the maximum and minimum points of this plot correspond to the half-way points of (a).

containing unpaired electrons are quite rare in biology which means that much fewer peaks are found in an ESR spectrum of a biological sample than in an NMR spectrum of the same sample. Good examples of paramagnetic species include molecular oxygen (O_2), free radical intermediates formed during enzyme catalysis and metals forming part of metalloproteins. ESR spectra are measurable at micromolar concentrations while NMR spectra are usually obtained in the millimolar range. In these respects, the relationship of ESR to NMR is reminiscent of that between absorbance and fluorescence spectroscopy in that fluors are rarer than chromophores in biomolecules and are detectable at much lower concentrations.

3.8.3 Uses of ESR Spectroscopy in Biochemistry

ESR is a particularly useful technique in biological situations involving unpaired electrons. Any process either involving or producing such species may be conveniently followed by ESR. For this reason ESR has been widely used in studies of the production of free radicals and in biological processes involving transition metals. Good examples of these investigations include studies of the electron transport chain in mitochondria, kinetic mechanisms of metalloenzymes and the process of photosynthesis in plant cells.

As well as studying naturally-occurring paramagnetic samples, it is possible to introduce nonbiological paramagnetic species into biomolecules. These *spin probes* are stable molecules which naturally contain unpaired electrons (Figure 3.54). If they can be attached at a unique site (e.g. in a protein or a membrane), they can act as *reporter molecules*. Unpaired electrons are especially sensitive to their freedom of movement. The more *restricted* the movement of a paramagnetic centre, the *broader* the ESR lines corresponding to this centre (Figure 3.55). Experimentally, movement can be restricted by crystallization (Chapter 6), lowering temperature, increased viscosity (e.g. by adding glycerol; Chapter 7) and by inclusion of the sample in a phospholipid bilayer.

One of the most useful applications of ESR is therefore in identifying and quantifying *dynamic mobility* of molecules such as phospholipids, peptides, proteins and drugs in biological systems, especially in membranes. Changes in mobility, for example due to binding of a protein at the cell surface, can be detected as a result of effects on ESR spectra. ESR also provides an elegant means of identifying transmembrane domains in membrane proteins by *site-directed spin labelling* (Example 3.7). Cysteine residues may be used to replace individual amino acids in any protein by site-directed mutagenesis. Cysteine-specific spin probes (Figure 3.54) can be attached at this cysteine in each mutant and the mutant proteins can then be incorporated in a phospholipid bilayer. A clear band-broadening effect is seen in ESR spectra when the mutated residue becomes part of the transmembrane domain. Such experiments can provide important confirmation of predictions obtained from computer algorithms.

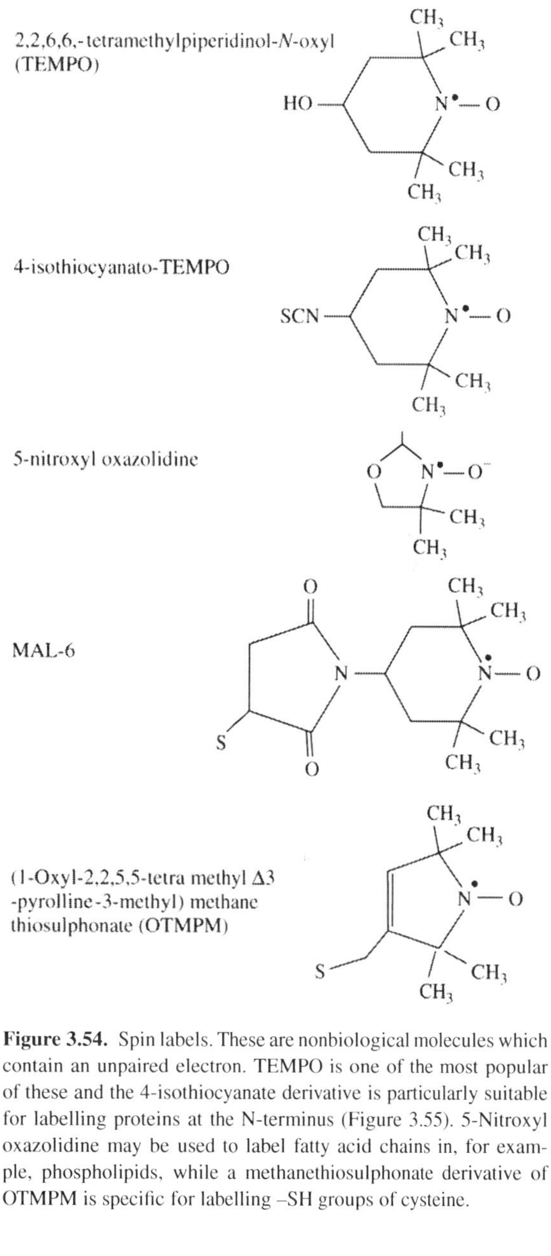

2,2,6,6,-tetramethylpiperidinol-*N*-oxyl (TEMPO)

4-isothiocyanato-TEMPO

5-nitroxyl oxazolidine

MAL-6

(1-Oxyl-2,2,5,5-tetra methyl Δ3 -pyrolline-3-methyl) methane thiosulphonate (OTMPM)

Figure 3.54. Spin labels. These are nonbiological molecules which contain an unpaired electron. TEMPO is one of the most popular of these and the 4-isothiocyanate derivative is particularly suitable for labelling proteins at the N-terminus (Figure 3.55). 5-Nitroxyl oxazolidine may be used to label fatty acid chains in, for example, phospholipids, while a methanethiosulphonate derivative of OTMPM is specific for labelling –SH groups of cysteine.

3.9 LASERS

In order to maximize signal : noise ratios and thus instrumental sensitivity, many manufacturers now use beams of narrow

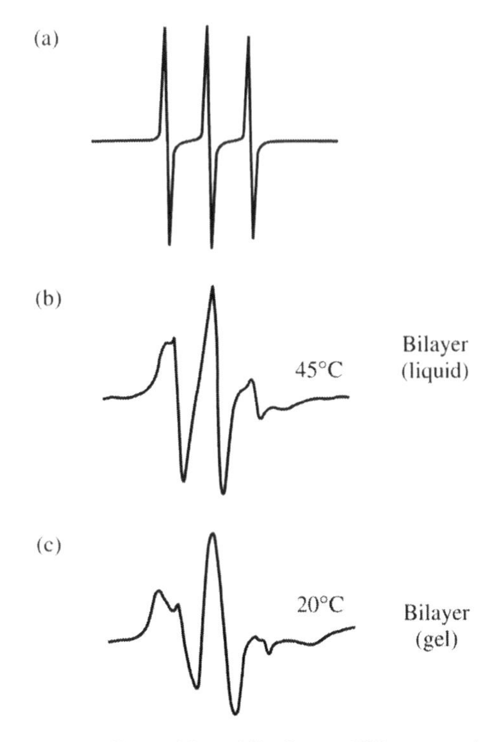

Figure 3.55. Effects of immobilization on ESR spectra. A protein was spin-labelled with 4-isothiocyanato-TEMPO (Figure 3.54). ESR spectra shown are for (a) free spin label, (b) spin-labelled protein in a phospholipid bilayer at 45 °C and (c) spin-labelled protein in a phospholipid bilayer at 20 °C. The phospholipids in the bilayer undergo a gel-liquid transition at 41 °C. Note how much broader the lines are as degree of immobilization increases.

and intense light called *laser beams* in certain items of laboratory equipment. Good examples of this include automated raman spectrometers (Section 3.6.5), surface plasmon resonance (Section 3.10), MALDI mass spectrometers (Section 4.1.3), DNA sequencers and image analyzers (Chapter 5). Laser means *light amplification by stimulated emission of radiation*. There follows a brief description of the origin and uses of laser beams.

3.9.1 Origin of Laser Beams

We have already seen that when atoms undergo an electronic transition from a higher to a lower energy level they emit light (Section 3.2.2). In a light-source such as a star or a light-bulb, light is emitted *randomly* from many thousands of individual atoms and has no particular *direction*. Moreover, there is no direct *phase* relationship between the light from any one atom and that from another (Figure 3.44). This light is said to be *noncoherent*. *Coherent* light, on the other hand, is light where the beam from every atom is *both in phase and in the same direction*. Lasers depend on the production of coherent light from a material arising from a process known as *stimulated emission*.

Under normal conditions, only a small number of a population of atoms exists in an excited state. By a process known as *pumping*, it is possible to achieve an *inversion* of this situation, called a *population inversion*, where a majority of the atoms in the population exist in a partially stable excited state. Depending on the characteristics of the *active material*, a number of pumping mechanisms are possible, but for crystalline and liquid materials *optical pumping* is one of the most important (Figure 3.56). If a photon of a given frequency enters this active material, an atom may return to the ground state by emitting a second photon of the same frequency, phase and direction as the incident photon (Figure 3.57). This pair of photons can stimulate emission from two further excited atoms resulting in four photons which can, in turn, stimulate emission of eight photons, and so on. In a form of chain reaction, each atom returning to the ground state by a process of constructive interference (Figure 3.44) *adds to* the amplitude of a unidirectional beam. This amplitude is proportional to the number of atoms in the population. Since there might be 10^{16} atoms in even a small amount of active material, an intensely strong and fine beam of coherent light results from this process (Figure 3.58). Reflection of the beam back and forth through the active material between two mirrors increases further the output of power. One of the mirrors is half-silvered thus reflecting half of the photons back into the active material to impact on any atoms missed during the first passage.

Of course, not every material is capable of sustaining a population inversion. The first laser was developed by Ted Maiman in 1960 using a single crystal of ruby which consists of sapphire (Al_2O_3) grown with a small amount of chromium. Since then, a wide variety of gases, liquids and solids have been used as active materials producing laser beams covering a wide range of energies. Because the active materials in these laser sources have several energy levels, photons of a variety of frequencies can result from the stimulated emission process. Incorporation of a prism into the design facilitates selection of single wavelengths and lasers of this type are called *tunable lasers* (Figure 3.59).

3.9.2 Some Uses of Laser Beams

Laser beams are useful in any situation where an intense and narrow beam of light of defined energy is required. They have found many applications in electronics (e.g. in fibre optics) and in medicine (e.g. in eye surgery) and some of the items of modern equipment found in life science laboratories which incorporate lasers described elsewhere in this text have been mentioned. Lasers make possible a number of novel high-resolution approaches to the study of

Example 3.7 Site-directed spin labeling of bacteriorhodopsin

Bacteriorhodopsin is a light-driven proton pump consisting of seven transmembrane α-helices. Although the gross structure of this protein is known, the detailed orientation of the helices and the exact location of interhelical loops are unclear. One possible way to study a problem of this type would be to classify individual amino acid side-chains as being (1) buried in the interior of the protein, (2) exposed on the protein surface but in the interior of the membrane or (3) exposed on the protein surface in contact with the extracellular solution.

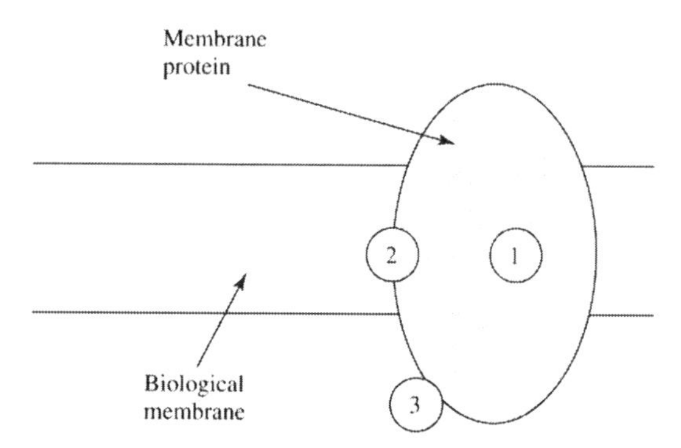

Site-directed spin labelling involves replacing each residue in the region of interest, in turn, with cysteine by site-directed mutagenesis thus generating a family of mutant proteins. Each variant of the protein is then individually treated with a cysteine-specific spin label such as OTMPM (Figure 3.54). This facilitates measurement of an ESR signal for each amino acid side-chain in turn. The ESR signal can be affected by spin relaxing agents, which enhance the relaxation rate of the spin label and this type of experiment is known as spin-label relaximetry.

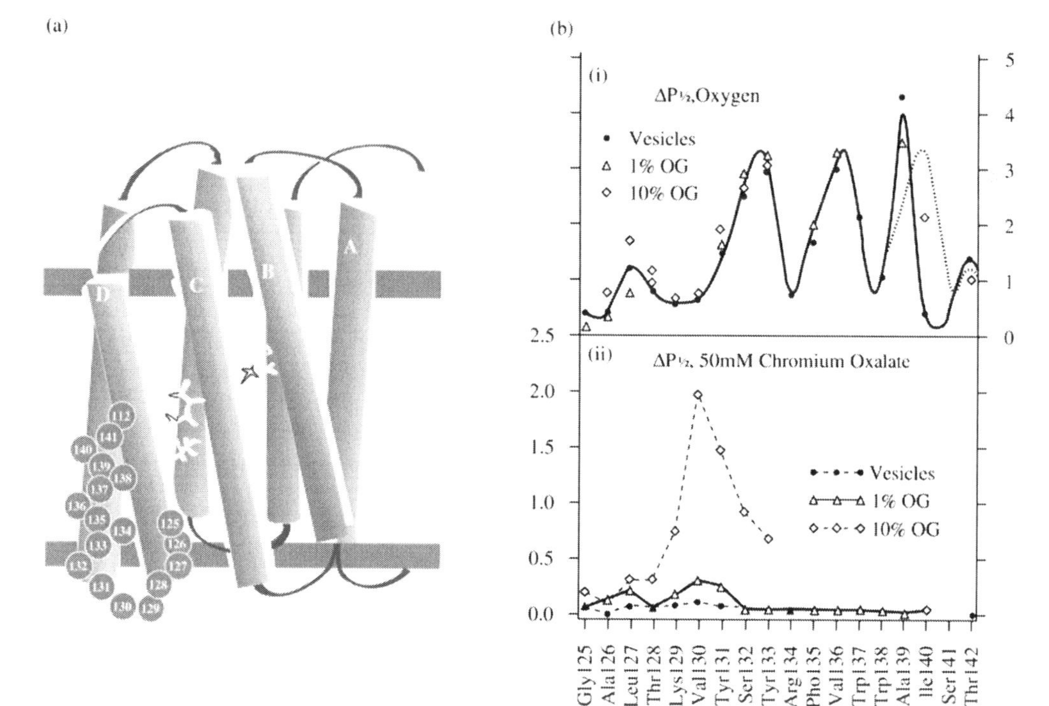

A variety of spin relaxing agents are available with individual solubility in aqueous solution and in the lipid components of biological membranes. Exposure of a specifically spin labelled mutant to spin relaxing agents can be quantified by

(Continues)

(Continued)

an accessibility parameter, $\Delta P_{1/2}$. This value is high in situations where the spin-label is accessible and low where it is not. In this example, molecular oxygen (O_2) is an effective spin relaxing agent capable of dissolving both in aqueous solution and in biological membranes. Chromium oxalate, on the other hand, is only soluble in aqueous solution. A spin-labelled mutant which is relaxed by both of these relaxing agents is likely to be on the surface of the protein in touch with the extracellular surface (i.e. 3 above) while a mutant only affected by O_2 is likely to be on the surface of the protein but surrounded by the biological membrane (i.e. 2 above). Spin-labelled mutants not affected by either relaxing agent represent residues likely to be buried in the interior of the protein (i.e. 1 above).

A series of X → Cys mutants in the sequence region 125–142 were generated and spin labelled with OTMPM. Independent methods were used to confirm that these mutant proteins had essentially identical structure to the wild-type. The mutants were then individually exposed to molecular Oxygen and to chromium oxalate and the accessibility parameter determined. The results are as follows.

This experiment allowed identification of residues 129–131 as being an interhelical region exposed to the cell surface and thus sensitive to chromium oxalate. Treatment with O_2 revealed a regular pattern in the region131–138 of exposure to lipid followed by exposure to protein with a periodicity of 3.6 residues (i.e. an α-helix). This is an approach which is generally applicable to membrane-bound proteins including receptors and ion-channels which can also be used to follow three-dimensional structural perturbations.

(From Millhauser, 1992) Fig 2 from Milhauser, G.L., selective placement of electron resonance spin labels: New structural methods for peptides and proteins. *Trends in Biochemical Sciences* **17**, 448–452.

biomolecules which are either not feasible or else not as sensitive using noncoherent light sources.

One of the most important applications of lasers is in *confocal microscopy* which makes possible the visualization of macromolecular complexes and immuno-labeled proteins in cells. A laser beam is focused on a particular point in a cell or tissue sample with very high precision. An image of a plane through the sample can be constructed by scanning the sample in a process called *laser scanning*. The ability to focus a fine laser beam on individual cells also makes it possible to excite either natural fluors or extrinsic fluors attached to sites in or on the cell. This technique is known as *laser-induced fluorescence* and it can be used, for example, to detect tumours, dermal lesions or atherosclerotic plaques in human tissue samples.

Recent developments in the application of lasers to biochemistry have even made possible the illumination and study of single molecules. Microscopic plastic beads can be trapped in tightly-focused laser beams called *optical traps*. If these beads are attached to a single biomolecule (e.g. a DNA molecule or protein), the optical trap can be used as an 'optical tweezers' to manipulate the single molecule. This makes possible the study of the mechanics, fluorescence and other properties of single molecules under conditions of very low signal : noise ratio since the information obtained is not averaged over a whole population of molecules. This experimental approach has enormous potential for the study of motion and kinetics of biomolecules crucial to cell processes. Special interest has focused on DNA-protein interactions and molecular 'motors' such as myosin, kinesin and F_1-ATPase. Insights which have been obtained from this type of study include the finding that RNA polymerase can 'pull' DNA during transcription with a force which is

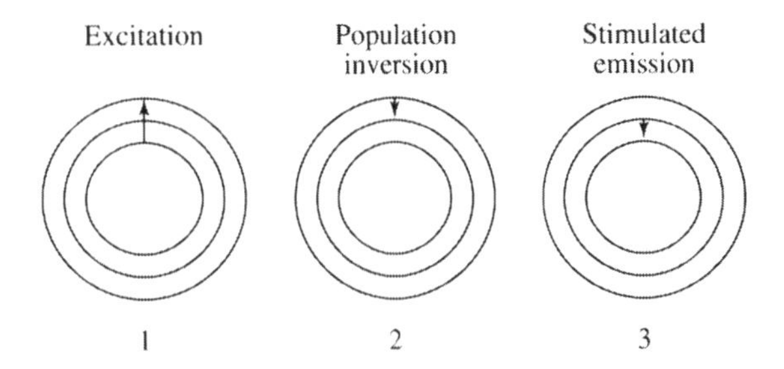

Figure 3.56. Population inversion by optical pumping. The atoms of the active material have two excited electronic energy levels. (1) The electron is promoted to the higher energy level by a flash of light. (2) It can spontaneously drop to an intermediate energy level. In this energy level the atom is stable for some time allowing build-up of a large number of atoms at this energy, i.e. population inversion. (3) Impact of a photon causes stimulated emission with a return to the ground state (Figure 3.57).

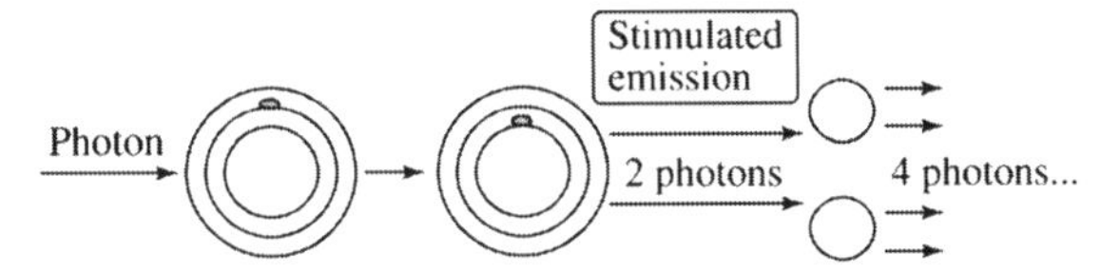

Figure 3.57. Principle of the laser. A photon of light returns the electron to the ground state by stimulated emission. This releases two photons of the same direction and phase. These impact on two atoms resulting in stimulated emission of four photons, and so on. Ultimately, this results in a fine beam of coherent light.

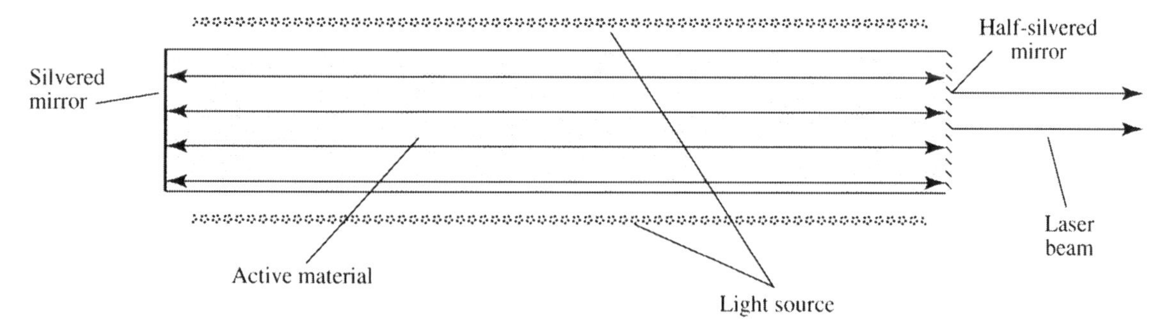

Figure 3.58. Outline design of a laser device. Active material is surrounded by a light source (e.g. a flashlamp). This emits an intense flash of light to initiate the optical pumping process. Light emitted by stimulated emission is passed back and forth through the cylinder of active material by reflection between a silvered and half-silvered mirror (arrows). Half of these photons pass through the half-silvered mirror and are emitted from the laser source as an intense laser beam.

four times as strong as that exerted by myosin during muscle contraction. Combination of this approach with fluorescence resonance energy transfer (Section 3.4.5) is likely to make possible the measurement of distances in the range 5–9 nm.

Another laser-based method allowing visualization of atomic-level resolution in single biomolecules and at membrane surfaces is *atomic force microscopy* (AFM) which is also sometimes called *scanning probe microscopy*. This technique (Figure 3.60; Example 3.8) brings a cantilever tip physically close to the surface to be imaged. The apex of this tip is very tiny (of the order of a few nm to tens of nm in radius) giving the technique its nanoscale resolution. An ionic repulsive force is exerted upwards on the tip which bends the cantilever upwards. The amount of bending is measured by reflection of a laser spot on to a detector. This technique allows exquisitely sensitive imaging of nanosurfaces including biological membranes. By attaching ligands or proteins to the tip, protein-ligand and protein-protein interactions can be sensitively determined under physiological conditions.

3.10 SURFACE PLASMON RESONANCE

Surface plasmon resonance (SPR) is an optical technique which depends on changes in refractive index near metal surfaces. When two surfaces, one a metal and the other a dielectric material are exposed to a beam of plane-polarized light of wavelength, λ, a longitudinal charge density wave (a *surface plasmon*) is propagated along the interface between them (Figure 3.61). This only happens when one of the surfaces is a metal and works best with silver, gold, copper and aluminium. This is because metals contain free oscillating

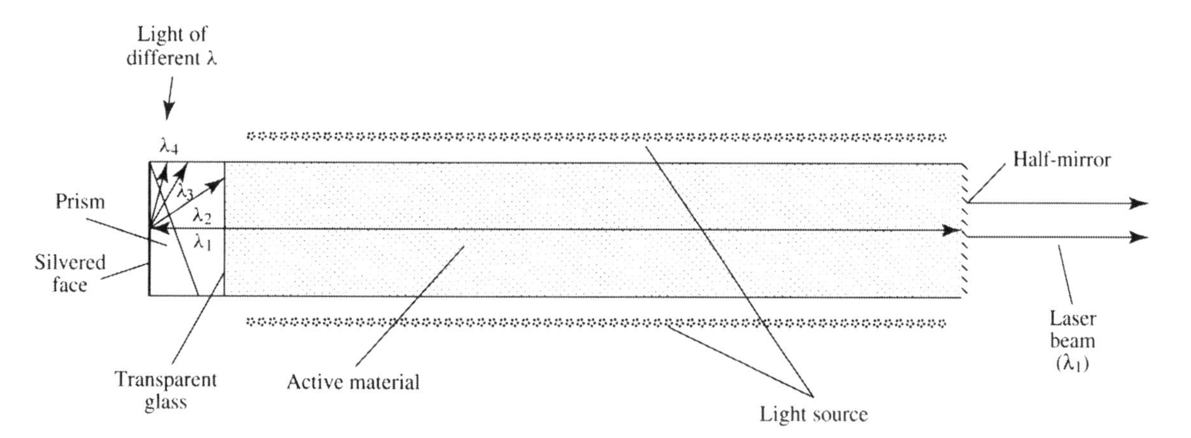

Figure 3.59. Outline design of a tunable laser. The mirror at one end of the tube is transparent allowing light to impact on a silvered prism. This diffracts the light into individual wavelengths (e.g. $\lambda_1 - \lambda_4$). By varying the angle of the prism, any value for λ may be selected for reflection back into the active material (in this example it is λ_1). The laser beam resulting from this will consist only of light of the selected wavelength

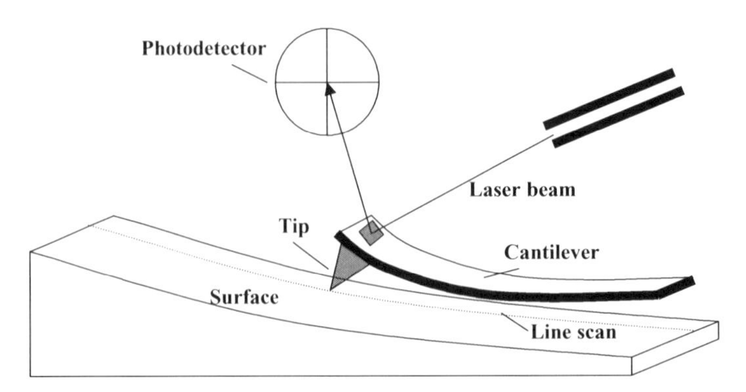

Figure 3.60. Atomic force microscopy. A very fine (nm scale) tip on a flexible cantilever is brought near to the object to be imaged. An upwards repulsive force bends the cantilever which is detected as movement of a reflected laser spot on a photodetectior. This movement allows measurement of the force experienced at the tip. Combining such scans allow construction of three-dimensional images.

Example 3.8 Transfection of plasmid DNA into cells detected by atomic force microscopy

The cantilever tip of an AFM probe is very tiny but can be functionalized by addition of ligands or proteins such as antibodies and streptavidin. In this experiment, the tips (10 nm diameter) were decorated with plasmid DNA encoding an enhanced green fluorescent protein (EGFP) (see also Example 3.5). After rinsing, the tips were immediately brought near to cultured human embryonic kidney cells. The following images show A. the tip approaching the cell, B. the tip penetrating the cell and C. the undamaged tip being removed from the cell. The point of penetration is detected by a sharp inflection (*) in the force versus tip-cell distance graph.

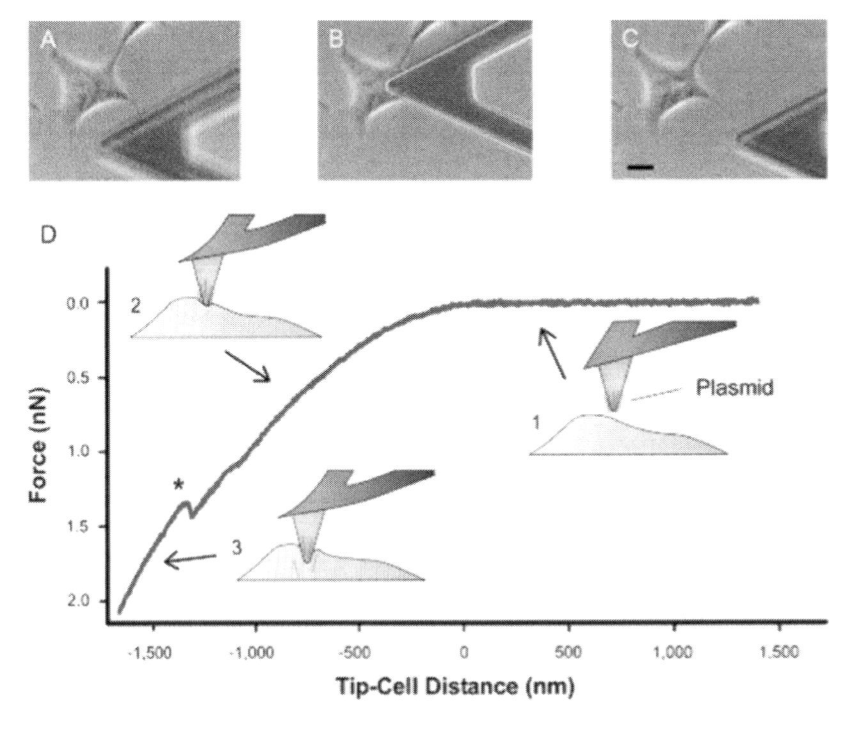

Success of the transfection was confirmed by expression of the EGFP.

(Reproduced with permission from Ceurrier *et al.*, 2007, *Biochemical and Biophysical Research Communications* **355**, 819–825).

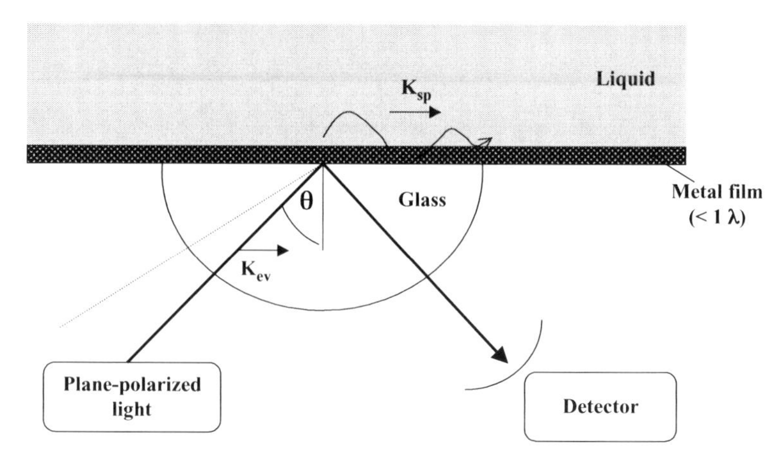

Figure 3.61. Surface Plasmon resonance. Plane-polarized light (wavelength λ) arriving at a thin layer (thickness <λ) of metal between a more (glass) and less (liquid or air) optically dense material is reflected by total internal reflection. An evanescent wave enters the metal interface to a depth <λ. At a particular angle of incidence, θ, the evanescent wave (vector K_{ev}) couples with free oscillating electrons (plasmons; vector K_{sp}) within the metal. Energy absorbed in this process is detected as a sharp reduction in intensity of the reflected light at a specific value for θ. K_{sp} depends strongly on the refractive index of the liquid or air immediately above the metal layer to a depth <300 nm.

electrons called *plasmons*. When light traveling through an optically dense medium such as glass arrives at an interface with a lower optical density (e.g. liquid), it is reflected back into the more optically dense medium, a phenomenon called *total internal reflectance*. However, a component of the incident light, the *evanescent wave*, penetrates into the less dense medium to a distance smaller than λ. In SPR, two optically dense surfaces are used (e.g. glass and liquid) one of which is coated with a layer of metal of thickness less than λ. The value for the wave vector of the evanescent field, K_{ev} is given by:

$$K_{ev} = \frac{\mu_0}{c} \cdot n_g \cdot \sin\theta \qquad (3.51)$$

where μ_0 is the frequency of incident light, n_g is the refractive index of the glass, θ is the angle of incidence (see Figure 3.61) and c is the speed of light *in vacuo*. The wave vector of a surface plasmon, K_{sp}, approximates to:

$$K_{sp}^2 = \frac{\omega_0}{c} \cdot \frac{\delta_m \cdot n_s^2}{\delta_m + n_s^2} \qquad (3.52)$$

where δ_m is the dielectric constant of the metal film and n_s is the refractive index of the dielectric medium, in this case liquid. The evanescent wave can couple by resonance (hence surface plasmon *resonance*) with plasmons in the metal when $K_{ev} = K_{sp}$. This happens only at specific values of θ. Some energy is lost from the incident light as a result which causes a reduction in the intensity of the reflected light which can be detected.

Inspection of Equation (3.52) shows that K_{sp} is highly dependent on the refractive index of the medium above the metal film. This holds up to a distance of approximately 300 nm above the metal. Any process altering n_s can be sensitively detected by SPR so the technique has found applications in the study of kinetics and thermodynamics of binding processes (e.g. protein-ligand, protein-protein). Because SPR only reports on events <300 nm above the metal layer, it gives meaningful measurements of binding processes in real time, over a wide dynamic range with direct detection and no requirement for labels or special optical characteristics in the samples studied.

3.10.1 Equipment Used in SPR

Traditional SPR instruments use plane-polarized light of a single wavelength, λ, and measure the sharp decrease in intensity of the reflected beam (Figure 3.62). This happens at a particular value of θ, and can vary also with λ. This design also allows variation in flow rates and temperature. The decrease in intensity of the reflected beam is measured as decreased *reflectivity* by a two-dimensional detector array. The surface of the sensor may be coated with a protein, a lipid bilayer or a spin-coated polymer. A typical binding experiment is shown in Figure 3.63. The sensor surface is first coated with a suitable buffer which allows the SPR instrument to establish a baseline. Protein is then added to

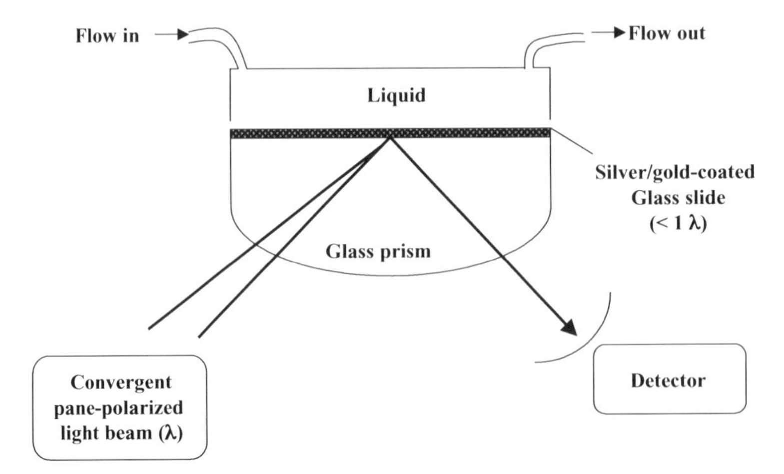

Figure 3.62. A typical SPR instrument. A convergent beam of plane-polarized light (wavelength, λ) is focused through a glass prism onto a silver/gold-coated glass slide and reflected to a detector. Solutions can pass through a flow-cell at defined values of flow-rate and temperature. Protein flowing through the flow-cell can alter the surface immediately above the metal layer.

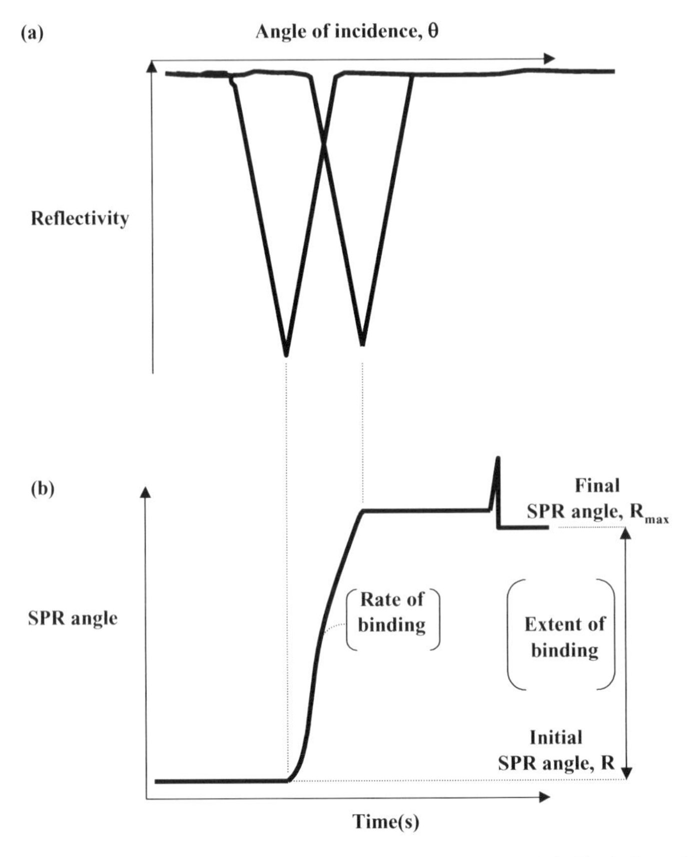

Figure 3.63. Schematic of SPR protein binding experiment. The sensor is equilibrated with buffer before addition of a protein solution. This changes n_s immediately above the sensor surface giving a sharp decrease in reflectivity. Protein binding is followed as a change in SPR angle (θ) over time. When the surface becomes saturated, the SPR angle reaches a maximum. Loosely bound protein is removed by washing with buffer. Extent of adsorption is given by the difference between initial and final SPR angles while the rate of binding may be measured from the steepest part of the positive slope.

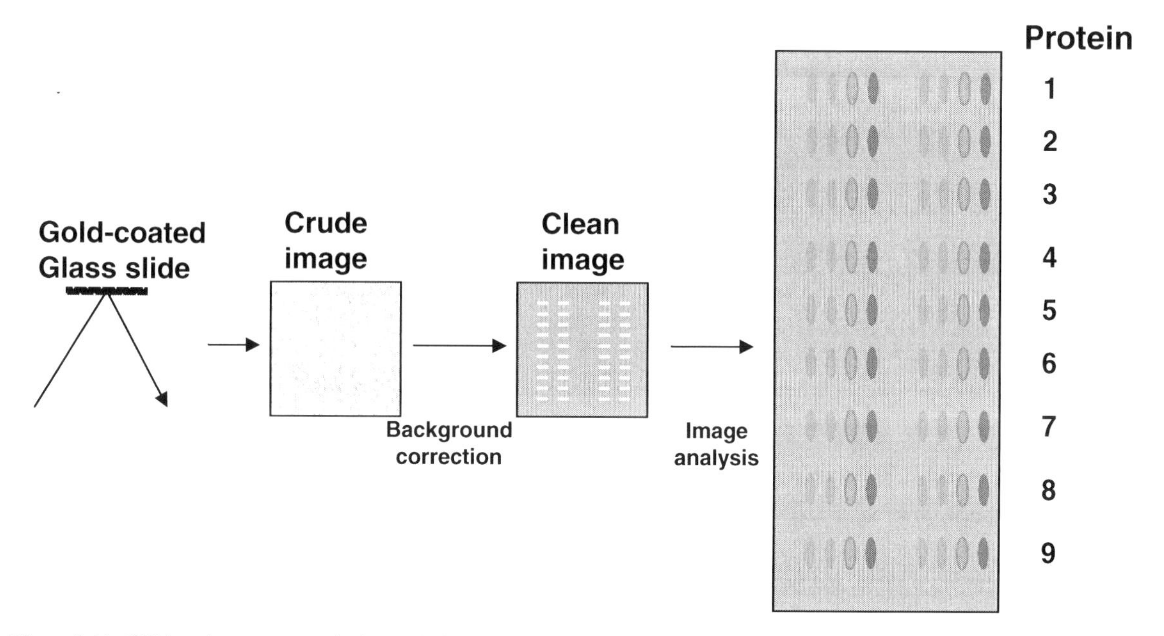

Figure 3.64. SPR imaging apparatus. A plane polarized laser light source which provides monochromatic coherent light to a metal-coated glass slide. Reflected light is collected on a charge-coupled device camera and recorded. A correction is made for background to yield a "clean image".

the sensor surface causing a change in n_s and a sharp decrease in reflectivity. When protein saturates the surface, n_s again changes with a second decrease in reflectivity. Buffer is then added to the flow cell which removes loosely-bound protein. The instrument converts reflectivity measurements into a real-time adsorption profile describing binding of the protein to the surface. The extent of adsorption is given by the difference between initial and final SPR angles while the rate of adsorption is measured from the steepest positive gradient of the adsorption curve.

This experimental set-up has the disadvantage of limited sample throughput. For this reason, there is now much interest in combining SPR with array and imaging technologies to allow simultaneous measurement of thousands of biomolecular interactions. An *SPR imaging apparatus* consists of a plane-polarized laser light source which provides monochromatic coherent light to a metal-coated glass slide (Figure 3.64). Reflected light is collected on a charge-coupled device camera and recorded. As the intensity of reflected light is linearly proportional to that of the incident light, a high-intensity laser is best as the light source. However, it can be difficult in a protein array to achieve uniform illumination with a laser source which results in signal variation requiring background correction. Therefore, the crude image initially obtained must undergo a background correction with image analysis software to produce a clean image of the array. SPR imaging instruments can achieve a spatial resolution down to 10 μm, giving the capability to resolve 10 000 spots in a 1.4 cm^2 array (Figure 3.64).

3.10.2 Use of SPR in Measurement of Adsorption Kinetics

The experimental setup shown in Figure 3.62 allows introduction of components and measurement of their binding to the sensor surface (Example 3.9). The rate of adsorption of an analyte is determined from the steepest part of the adsorption profile (Figure 3.63). This is dominated by two factors; the intrinsic kinetics of binding and mass transport through a boundary layer above the sensor. The process of diffusion to the sensor surface is generally much slower than the intrinsic rate of adsorption and thus is rate-limiting. However, this rate varies with flow rate while the intrinsic kinetics of binding do not. The rate of diffusion, J_d, is given by:

$$J_d = \frac{D}{x_d}(A_b - A_s) \tag{3.53}$$

where D is the diffusion coefficient, A_b the bulk concentration of analyte, A_s the surface concentration of analyte and

Example 3.9 Surface plasmon resonance identifies a plasmin binding site in blood factor VIII

A plasma protein called factor VIII is essential for clotting of blood and is defective or deficient in people suffering from the genetic disease haemophilia A. The protein is synthesized as a large precursor consisting of six domains (A1-A2-B-A3-C1-C2). This is cleaved by a protease at the B-A3 junction generating heavy (A1-A2) and light (B-A3-C1-C2) chains which circulate in the blood as a complex associated with another protein called the von Willebrand factor. Factor VIII is converted into its active form, factor VIIIa, by limited proteolytic cleavage by either of the proteases thrombin or Xa at Arg-372 and Arg-740 of the heavy chain to produce A1 and A2 subunits. Cleavage at Arg-1689 of the light chain produces an A3-C1-C2 subunit. Active factor VIIIa results from cleavage at Arg-372 and Arg-1689. Plasmin and other proteases can inactivate factor VIII by cleavage at Arg-372.

This study used SPR to identify the sites within domain A2 where plasmin interacts. Plasmin (7 ng/mm^2) was immobilized on the sensor chip and various factor VIII components were added for 4 min. These included intact heavy chain, intact light chain, A1 and A2 domains. After this, running buffer was changed over 2 min. The SPR data are shown below:

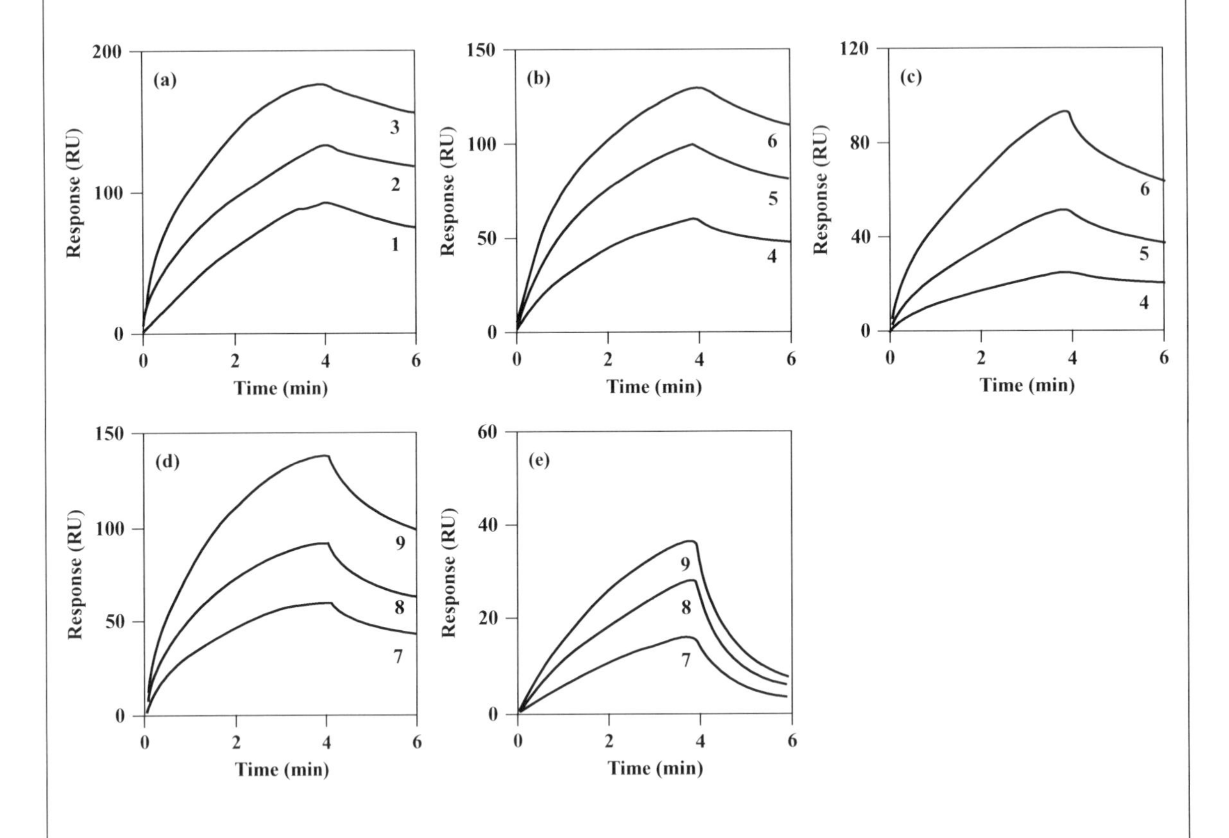

SPR association/dissociation curves of plasmin binding to factor VIII components. (a) Factor VIII (1 = 12.5 nM, 2 = 25 nM, 3 = 50 nM), (b) Intact heavy chain (4 = 20 nM, 5 = 40 nM, 6 = 80 nM), (c) Intact light chain (4 = 20 nM, 5 = 40 nM, 6 = 80 nM), (d) A2 subunit (7 = 40 nM, 8 = 80 nM, 9 = 120 nM), (e) A1 subunit (7 = 40 nM, 8 = 80 nM, 9 = 120 nM).

Nonlinear regression analysis of these data yielded the following binding parameters. The K_{ds} were also determined by an independent ELISA assay:

(*Continues*)

(Continued)

Binding parameters for the interacton of factor VIII(a) subunit and Ah-plasmin determined in SPR-based assay and ELISA-based assay

Factor VIII fragment	SPR-based assay k_{ass}	k_{diss}	$K_d^{app\,a}$	ELISA K_d^{app}
	$\times 10^4\ M^{-1}\ s^{-1}$	$\times 10^{-3}\ s^{-1}$	nM	nM
Factor VIII	26.3 ± 0.8	0.8 ± 0.1	3.1	6.7 ± 2.0
Heavy chain	40.2 ± 6.9	2.3 ± 0.1	5.6	13.0 ± 3.1
Light chain	2.9 ± 0.4	2.0 ± 0.7	68.2	115 ± 11
A2	14.3 ± 6.2	3.2 ± 1.3	22.6	40.7 ± 8.7
A1	5.2 ± 0.4	10.9 ± 0.9	208	N.d.[b]

From this it was concluded that A2 is preferentially bound to plasmin compared to A1. Since a monoclonal antibody recognizing residues 484–509 abolished plasmin binding, mutant forms of A2 were generated in which Lys or Arg residues were changed in this region and a panel of 17 mutants and wild type were assayed by SPR as above. This revealed that Arg 484 Ala displayed ~2000-fold lower k_{ass} ($250\,M^{-1}\,s^{-1}$) and ~250-fold increased K_d (5800 nM) compared to wild type confirming that residue 484 plays a key role in plasmin binding.

(Reproduced with permission from Nogami *et al.* (2008) Biochimica et Biophysica Acta **1784**, 753–763).

x_d the thickness of the boundary layer. We can define a mass transfer coefficient, k_m, as D/x_d. Thus:

$$J_d = k_m(A_b - A_s) \qquad (3.54)$$

In general for a flow cell shown in Figure 3.62:

$$k_m = \frac{D}{x_d} = 1.86\frac{(D^2 v)^{1/3}}{h \cdot L} \qquad (3.55)$$

where v is the flow rate, L is the distance between the flow cell inlet and sampling area and h is the height of the flow cell. Initially the kinetics are dominated by mass transfer and are diffusion-controlled but, as adsorption proceeds, the process becomes limited by the intrinsic kinetics of binding. From Equations (3.54) and (3.55) we could predict that the mass transfer rate can be increased by reducing L or increasing flow rate, v, (which both would increase k_m) or by increasing the concentration gradient of the analyte ($A_b - A_s$).

When measuring binding of a ligand (B) to a ligate (A) such that A + B → AB, the overall rate of AB complex formation is:

$$\frac{d[AB]}{dt} = \text{formation of AB} - \text{dissociation of AB} \qquad (3.56)$$

$$= K_a[A] \cdot [B] - K_d[AB] \qquad (3.57)$$

where K_a is the association constant, K_d the dissociation constant, [A] the total ligate concentration, [B] is the concentration of free ligand and [AB] the concentration of the ligand-ligate complex. The maximum change in the SPR angle (R_{max}; Figure 3.63) is dependant directly on the total ligand concentration ([B] + [AB]). The term ($R_{max} - R$) is dependant on [B] and [A] and can be regarded as a constant, C, since it is constantly replaced under flow conditions (Figure 3.62). Based on these relationships we can rewrite Equation (3.57) as:

$$\frac{dR}{dt} = K_a \cdot C \cdot (R_{max} - R) - K_d R \qquad (3.58)$$

$$= K_a \cdot C \cdot R_{max} - (K_a \cdot C + K_d) \cdot R \qquad (3.59)$$

A plot of dR/dt against R for a range of ligate concentrations can be used to determine rate constants (Figure 3.65). The slope of each such plot (k_s) is then plotted against C to form a straight line where:

$$-k_s = K_a \cdot C + K_d \qquad (3.60)$$

K_a is the slope of this plot and K_d the y-axis intercept. Alternatively, traces can be analysed by nonlinear regression using the software provided with the instrument.

(a)

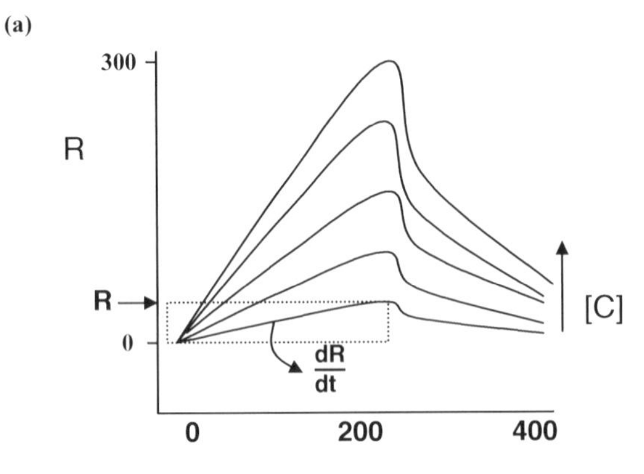

(a)

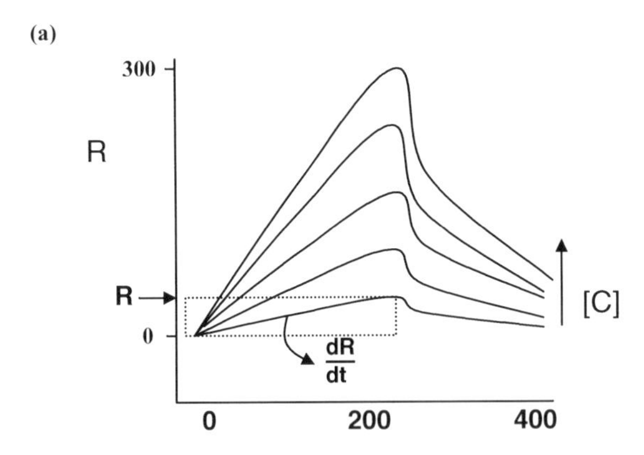

(c)

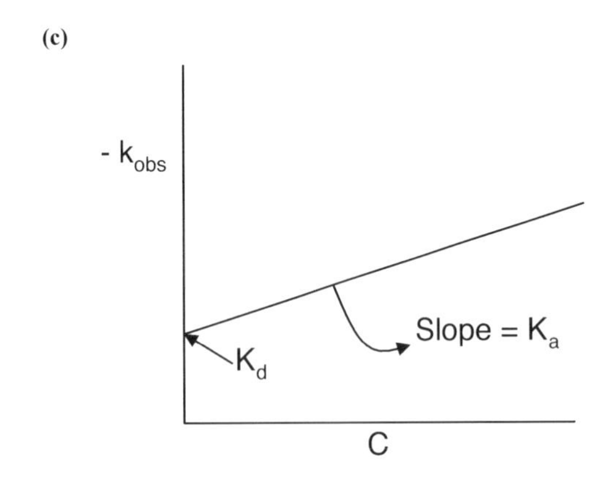

Figure 3.65. Results of SPR experiment. (a) Reflectivity is plotted versus time. (b) dR/dt determined from (a) is plotted versus R for a range of ligate concentrations. (c) The slope of each such plot (kobs) is then plotted against C to form a straight line Ka is the slope of this plot and Kd the y-axis intercept.

REFERENCES

General reading

Schmid, F.X. (1997) Optical spectroscopy to characterize protein conformational changes, in *Protein Structure: A Practical Approach* (ed. T.E. Creighton), Oxford University Press, Oxford, UK. An overview of the use of fluorescence, absorbance and CD as measures of conformational change in proteins.

Weiss, R.J. (1996) *A Brief History of Light and Those That Lit The Way*, World Scientific Publishing, Singapore. A very readable and lighthearted overview of the history of light.

Wilson, R.G. (1995) *Fourier Series and Optical Transform Techniques in Contemporary Optics: An Introduction*, John Wiley & Sons, Inc., NY, USA. A mathematical description of the application of the Fourier Transform to optical problems.

Absorbance spectroscopy

Mach, H., Volkin, D.B., Burke, C.J. and Middaugh, C.R. (1995) Ultraviolet absorption spectroscopy, in *Protein Stability and Folding* (ed. B.A. Shirley) Humana Press, Totowa, NJ, USA, pp. 91–114. A description of the use of absorption spectroscopy in protein folding studies.

Pace, C.N. and Schmid, F.X. (1997) How to determine the molar absorbance coefficient of a protein, in *Protein Structure: A Practical Approach* (ed. T.E. Creighton), Oxford University Press, Oxford, UK, pp. 253–60. An overview of common intrinsic protein chromophores and their uses.

Fluorescence and chemiluminescence spectroscopy

Brand, L. and Johnson, M.L. (eds) (1997) *Fluorescence Spectroscopy* (Methods in Enzymology), Vol. **278** (eds J.N. Abelsonand and M.I. Simon), Academic Press, San Diego CA, USA. A comprehensive collection of examples of the applications of fluorescence in biochemistry.

Cheung, H.C. (1995) Resonance energy transfer, in *Topics in Fluorescence Spectroscopy; Principles*, Vol. **2** (ed. J.R. Lakowicz), Plenum Press, NY, USA, pp. 127–76. An overview of principles of resonance energy transfer.

Cubitt, A.B., Heim, R., Adams, S.R. *et al.* (1995) Understanding, improving and using green fluorescent proteins. *Trend in Biochemical Sciences*, **20**, 448–55. A review of the biochemistry and biotechnological applications of green fluorescent proteins.

Darby, I.A. (ed.) (1999) *In Situ Hybridisation Protocols*, Humana Press, Totowa, NJ, USA. A collection of laboratory protocols for *in situ* hybridization.

Eftink, M.R. (1991) Fluorescence quenching: Theory and applications, in *Topics in Fluorescence Spectroscopy; Principles*, Vol. **2** (ed. J.R. Lakowicz) Plenum Press, NY, USA, pp. 53–126. A comprehensive overview of principles of fluorescence quenching.

LaRossa, R.A. (ed.) (1998) *Bioluminescence Methods and Protocols*, Humana Press, Totowa, NJ, USA. A collection of applications of chemiluminescence.

Mátyus, L., Sjöllφsi, J. and Jenei, A. (2006) Steady state fluorescence quenching applications for studying protein structure and dynamics. *Journal of Photochemistry and Photobiology B, Biology*, **83**, 223–36. An excellent review of fluorescence quenching.

Royer, C.A. (1995) Ultraviolet absorption spectroscopy, in *Protein Stability and Folding* (ed. B.A. Shirley), Humana Press, Totowa, NJ, USA, pp. 65–89. A description of the use of fluorescence spectroscopy in protein folding studies.

Schuler, B. and Eaton, W.A. (2008) Protein folding studied by single-molecule FRET. *Current Opinion in Structural Biology*, **18**, 16–26. A review of use of FRET in protein folding studies.

Sullivan, K.F. (ed.) (2008) Fluorescent proteins, in *Methods in Cell Biology*, Vol. **85**, Elsevier. A special issue on fluorescent proteins and their uses especially in cell biology.

Sun, C.X., Yang, J.H., Li, L. *et al.* (2004) Advances in the study of luminescence probes for proteins. *Journal of Chromatography B*, **803**, 173–90. A review of spectral probes.

Torrent, J., Alvarez-Martinez, M.T., Liautard, J.P. *et al.* (2005) The role of the 132-160 region in prion protein conformational transitions. *Protein Science*, **14**, 956–67.

Wolfbeis, O.S. (ed.) (1993) *Fluorescence Spectroscopy: New Methods and Applications*, Springer-Verlag, Berlin, Germany. An interesting collection of analytical applications of fluorescence spectroscopy.

Circular and Linear Dichroism

Capaldi, K.P., Ferguson, S.J. and Radford, S.E. (1999) The greek key protein Apo-pseudoazurin folds through an obligate on-pathway intermediate. *J. Mol. Biol.*, **286**, 1621–32. An interesting example of the use of CD in protein folding investigations.

Eriksson, S., Kim, S.K., Kubista, M. and Norden, B. (1993) Binding of 4′,6-diamidino-2-phenylindole (DAPI) to AT regions of DNA: evidence for an allosteric conformational change. *Biochemistry*, **32**, 2987–98. An example of the application of LD to a study of ligand binding to DNA.

Hanson, K.R. (1966) Applications of the sequence rule. I. Naming the paired ligands g, g at a tetrahedral atom Xggij. II. Naming the two faces of a trigonal atom Yghi. *Journal of the American Chemical Society*, **88**, 2731–48. A description of the RS convention for enantiomers.

Johnson, C.R., Morin, P.E., Arrowsmith, C.H. and Freire, E. (1995) Thermodynamic analysis of the structural stability of the tetrameric oligomerization domain of p53 tumor suppressor. *Biochemistry*, **34**, 5309–16. A good example of the application of CD to investigations of protein structure.

Kelly, S.M., Jess, T.J. and Price, N.C. (2005) How to study proteins by circular dichroism. *Biochimica et Biophysica Acta*, **1751**, 119–39. A review of the application of CD to the study of proteins.

Martin, S.R. and Schilstra, M.J. (2008) Circular dichroism and its application to the study of biomolecules, in *Methods in Cell Biology*, Vol. **84** (eds J.J. Correia and H.W. Detrich), pp. 263–93. A review of CD.

Rodgers, A. and Norden, B. (1997) *Circular Dichroism and Linear DIchroism*, Oxford University Press. An excellent introduction to the theory and practice of CD and LD.

Infrared spectroscopy

Barth, A. and Zscherp, C. (2002) What vibrations tell us about proteins. *Quarterly Review of Biophysics*, **35**, 369–430. A comprehensive review of infrared and raman spectroscopy as applied to proteins.

Cooper, E.A. and Knutson, K. (1995) Fourier transform infrared spectroscopy investigations of protein structure, in *Physical Methods to Characterize Pharmaceutical Proteins* (eds J.N. Herron, *et al.*), Plenum Press, NY, pp. 101–43. A review of the use of FTIR spectroscopy in protein structure determination.

Franck, P. Nabet, P. and Dousset, B. (1998) Applications of infrared spectroscopy to medical biology. *Cell and Molecular Biology*, **44**, 273–5. A brief overview of the potential of FTIR spectroscopy in biomedical research.

Jackson, M. and Mantsch, H.H. (1995) The use and misuse of FTIR spectroscopy in the determination of protein structure. *CRC Crit. Reviews in Biochemistry and Molecular Biology*, **30**, 95–120. A critical review of the potential of FTIR spectroscopy for protein structure determination.

Larrabee, J.A. and Choi, S. (1993) Fourier transform infrared spectroscopy. *Advances in Enzymol.*, **226**, 289–305. A review of the theory, instrumentation and some applications of FTIR spectroscopy.

Stuart, B.H. (2004) *Infrared Spectroscopy: Fundamentals and Applications*, John Wiley & Sons, Ltd, Chichester, UK. An introduction to the fundamentals of IR spectroscopy.

van Thor, J.J., Pierik, A.J., Nugteren-Roodzant, I. *et al.* (1998) Characterization of the photoconversion of green fluorescent protein with FTIR spectroscopy. *Biochemistry*, **37**, 16915–21. A good example of the use of difference FTIR spectroscopy.

Vu, D.M., Myers, J.K., Oas, T.G. and Dyer, R.B. (2004) Probing the folding and unfolding dynamics of secondary and tertiary structures in a three-helix bundle protein. *Biochemistry*, **43**, 3582–9. An application of infrared spectroscopy to measuring rates of protein folding/unfolding.

Raman spectroscopy

Wang, Y. and Van Wart, H.E. (1993) Raman and resonance raman spectroscopy. *Advances in Enzymol.*, **226**, 319–73. A review of the theory, instrumentation and applications of Raman spectroscopy.

Zhu, F.J., Isaacs, N.W., Hecht, L. and Barron, L.D. (2005) Raman optical activity: A tool for protein structure analysis. *Structure*, **13**, 1409–19. A review of applications of Raman spectroscopy to proteins.

NMR

Bushong, S.C. (1996) Nuclear magnetic resonance spectroscopy, in *Magnetic Resonance Imaging*, Mosby, St. Louis, pp. 104–17. A non-mathematical description of the basics of NMR.

Evans, J.N.S. (1995) *Biomolecular NMR Spectroscopy*, Oxford University Press, NY. A comprehensive description of the physical basis of biological NMR.

Garcia-Martin, M.L., Garcia-Espinosa, M.A., Ballasteros, P. *et al.* (2002) Hydrogen turnover and subcellular compartmentation of hepatic [2-C-13] glutamate and [3-C-13] aspartate as detected by C-13 NMR. *Journal of Biological Chemistry*, **277**, 7799–807. A good example of *in vivo* NMR.

Kernec, F., Le Tallec, N., Nadal, L. *et al.* (1996) Phosphocreatine synthesis by isolated rat skeletal muscle mitochondria is not dependent upon external ADP: A ^{31}P NMR study. *Biochem. Biophys. Res. Commun.*, **225**, 819–25. A good example of application of ^{31}P NMR to the biochemical investigation of intact mitochondria.

Reid, D.G.. (ed.) (1997) *Protein NMR Techniques*, Humana Press, Totowa, NJ, USA. Some good examples of applications of single and multi-dimensional NMR to biochemistry.

Ruan, R.R. and Chen, P.L. (1998) Nuclear magnetic resonance techniques, in *Water in Foods and Biological Materials: A Nuclear Magnetic Resonance Approach*, Technomic, Lancaster, PA, pp. 1–50. An introduction to the basics of NMR.

Teng, Q. (2005) *Structural Biology: Practical NMR Applications*, Springer, NY, USA. A description of applications of NMR to structural biology with an excellent introduction to basics of NMR.

Watts, A. (1998) Solid-state NMR approaches for studying the interaction of peptides and proteins with membranes. *Biochimica et Biophysica Acta*, **1376**, 297–318. A review of the potential of solid-state NMR in the study of membranes with particular reference to interactions with proteins and peptides.

ESR

Cruz, A., Marsh, D. and Perez-Gil, J. (1998) Rotational dynamics of spin-labelled surfactant-associated proteins SP-B and SP-C in dipalmitoylphosphatidylcholine and dipalmitoylphosphatidylcholine bilayers. *Biochimica et Biophysica Acta*, **1415**, 125–34. A good example of ESR studies in bilayer model systems.

Edwards, T.E. and Sigurdsson, S.T. (2007) Site-specific incorporation of nitroxide spin labels into 2′ position of nucleic acids. *Nature Protocols*, **2**, 1954–62. A method for labeling nucleic acids with nitroxide spin labels.

Hubbell, W.L., Cafiso, D.S. and Altenbach, C. (2000) Identifying conformation changes with site-directed spin labeling. *Nature Structural Biology*, **7**, 735–9. A review of use of site-directed spin labeling in protein studies.

Klug, C.S. and Feix, J.B. (2008) Methods and applications of site-specific spin labeling EPR spectroscopy, in *Methods in Cell Biology*, Vol. **84**, Elsevier, pp. 617–58. A review of site-directed spin labeling.

Marsh, D. and Pali, T. (2004) The protein-lipid interface: Perspectives from magnetic resonance and crystal structures. *Biochimica et Biophysica Acta*, **1666**, 118–41. An excellent review of use of ESR in studies of membrane proteins.

Millhauser, G.L. (1992) Selective placement of electron resonance spin labels: new structural methods for peptides and proteins. *Trends in Biochemical Sciences*, **17**, 448–52. A description of potential of ESR spin labelling.

Russell, C.J., Thorgeirsson, T.E. and Shin, Y.-K. (1999) The membrane affinities of the aliphatic amino acid side chains in an α-helical context are independent of membrane immersion depth. *Biochemistry*, **38**, 337–46. An elegant example of application of ESR in studies of the interactions between proteins and membranes.

Lasers

Ceurrier, C.M., Lebel, R. and Grandbois, M. (2007) Single cell transfection using plasmid decorated AFM probes. *Biochemical and Biophysical research Communications*, **355**, 632–6. A fascinating example of a novel use of AFM.

Ishijima, A. and Yanagida, T. (2001) Single molecule nanobioscience. *Trends in Biochemical Sciences*, **26**, 438–44. A review of the potential of single molecule studies.

Mehta, A.D., Rief, M., Spudich, J.A. *et al.* (1999) Single-molecule biomechanics with optical methods. *Science*, **283**, 1689–95. A review of some applications of the single-molecule approach to the study of biomechanics.

Meschede, D. (2007) *Optics, Light and Lasers: The Practical Approach to Modern Aspects of Photonics and Laser Physics*, 2nd edn, Wiley-VCH Verlag GmbH, Weinheim, Germany. An overview of the physics and applications of lasers.

Moerner, W.E. and Orritt, M. (1999) Illuminating single molecules in condensed matter. *Science*, **283**, 1670–6. A review of the problems encountered in single-molecule studies.

SPR

Boozer, C., Kim, G., Cong, S. *et al.* (2006) Looking towards label-free biomolecular interaction analysis in a high-throughput format: A review of new surface plasmon technologies. *Current Opinion in Biotechnology*, **17**, 400–5. A review of developments in SPR.

Nogami, K., Katsumi, N., Evgueni, L.S. *et al.* (2008) Identification of a plasmin interactive site within the A2 domain of the factor VIII heavy chain. *Biochimica et Biophysica Acta*, **1784**, 753–63. An excellent illustration of the power of SPR especially when combined with other techniques.

Chapter 4
Mass Spectrometry

Objectives

After completing this chapter you should be familiar with:

- *The* **physical basis** *of mass spectrometry (MS).*
- *The* **range** *of MS methods available.*
- *How MS techniques can be* **interfaced** *with other high-resolution techniques.*
- *How MS techniques can be applied to the study of biomolecules such as proteins, peptides and DNA.*

A wide range of molecular mass is found in living cells varying from very tiny ionic species such as Na^+ to molecules made up of many thousands of atoms such as proteins and DNA. These molecules possess a variety of *mass/charge* (*m/z*) *ratios*, depending on their mass, m, and their charge (positive or negative), z. If we pass such a mixture through a powerful magnetic field, its components will be deflected differently based on this ratio, giving a means of making very accurate mass measurements. A variety of ionization methods are available to impart charge and momentum to sample molecules in the gas phase, which may then pass through a magnetic field and be differentially deflected. These methods vary in the energy they give to the sample molecules and in the degree to which they may cause fragmentation of their chemical structure. Depending on the ionization method chosen, it is therefore possible to determine a highly accurate mass for an intact chemical species or, alternatively, to fragment it into smaller pieces. The pattern of fragments observed in the latter experiment may be used to deduce the complete structure of the original molecule.

The technique of *mass spectrometry* (MS) has long been used in the study of naturally volatile molecules such as lipids. In cases of non-volatile samples, it was possible to *derivatize* them (e.g. acylation or esterification) to make them amenable to analysis. Because of limitations imposed by ionization, MS has in the past therefore had quite limited applications to the study of large, complex biomolecules such as proteins. However, the development of novel MS ionization techniques has now made possible high-resolution mass determinations of large proteins (>200 kDa). This provides a particularly useful method for analysis of structural variants of proteins (e.g. mutants, post-translationally-modified proteins) at resolutions of a few atomic mass units (AMU). This chapter will describe the physical basis and practice of MS in the study of complex biopolymers in biochemistry and show how we can use this group of techniques to obtain high-resolution structural information about such biopolymers. For more details on use of MS in *proteomics*, the reader is referred to Chapter 10.

4.1 PRINCIPLES OF MASS SPECTROMETRY

4.1.1 Physical Basis

Any moving charged species of mass, m, and velocity, v, will be deflected by an applied magnetic field (Figure 4.1). The magnitude of this deflection will depend on the momentum, μ, of the species which is given by Equation (4.1).

$$\mu = m \cdot v \qquad (4.1)$$

Species with large momentum are deflected less than those with small momentum. Thus, if a stream of atoms and small molecules of identical velocity and charge but different mass in the gas phase is passed through a magnetic field, the deflection experienced by each atom or molecule depends on its mass. This deflection can therefore provide an accurate measure of mass. Larger molecules are uncharged in the gas phase so, in order to deflect these, it is necessary to confer a charge upon them. This may be achieved by irradiating the molecules with a beam of electrons and is called *electron impact* (*EI*) *ionization* (Figure 4.2). Since the beam of electrons is of quite high energy, this ionization method can cause extensive breakdown of molecular structure, splitting a single molecule into a number of *fragment ions* of different mass (Figure 4.3) each of which may be deflected differently in the magnetic field. This generates a *mass spectrum* (MS) which is characteristic of the chemical structure of the molecule. In fact, the complete structure of

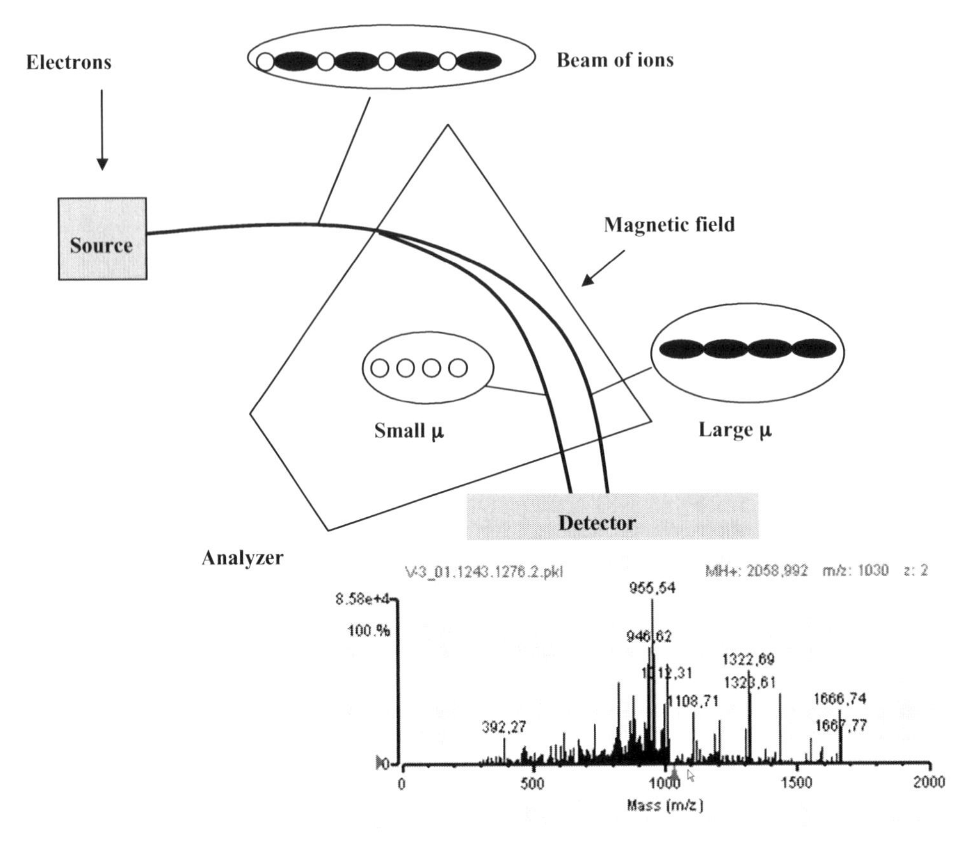

Figure 4.1. Differential deflection of charged species in magnetic field. Two species, of identical charge and velocity, will be deflected differently by a magnetic field. The species with the smallest momentum, μ, is deflected greatest.

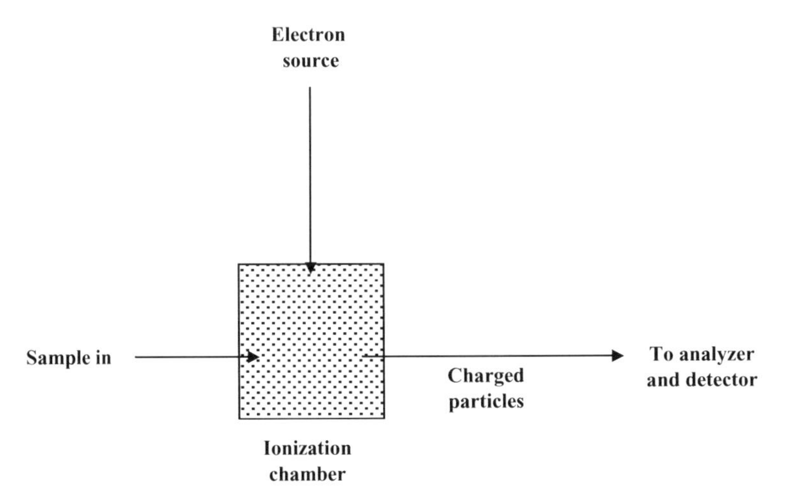

Figure 4.2. Electron Impact ionization. Sample is bombarded with a high-energy beam of electrons which causes ionization. These ions are subsequently accelerated and deflected in the analyzer.

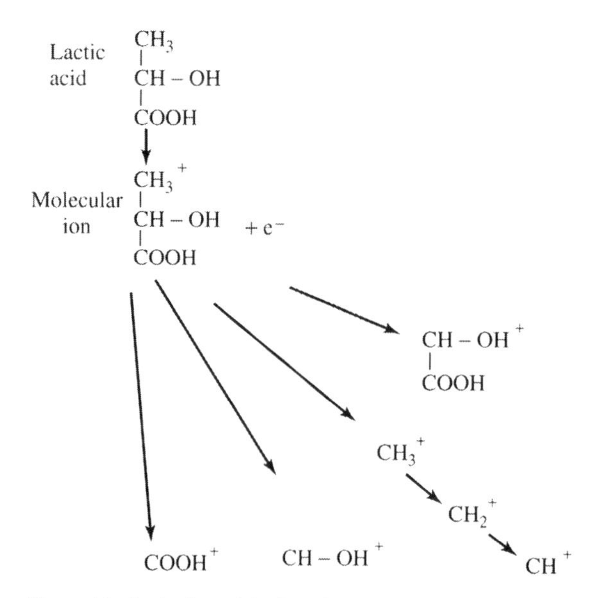

Figure 4.3. Ionization of lactic acid by electron impact. Lactic acid is ionized by electrons forming a molecular ion. This breaks down into a large number of fragment ions, each possessing a characteristic mass allowing their identification. The pattern of ions generated is characteristic of the chemical structure of the parent molecule and can therefore be used to calculate this structure

small biomolecules may be determined by electron impact MS (EI/MS) analysis alone.

For many years, EI/MS was used in determination of structures of relatively low mass species but was not directly applicable to larger biomolecules such as proteins and peptides. The principal reason for this was the inadequacy of EI for the generation of ions of sufficiently large mass which could be introduced to the gas phase. However, development of novel ionization methods combined with other technical developments has now made MS analysis of large proteins possible.

4.1.2 Overview of MS Experiment

A mass spectrometer consists of three distinct components (Figure 4.4; Section 4.1.4); a *source*, *analyzer* and *detector* all maintained under a powerful vacuum (~1 mPa). The source is an *ionization chamber* where the stream of ions is generated. In this overview of MS we will concentrate on EI/MS but other ionization methods will be described in the following section. It is the possibilities offered by these alternative ionization methods which have facilitated MS analysis of macromolecules such as proteins and peptides. The analyzer maintains either a magnetic or electric field which accelerates the stream of ions to a single velocity and then differentially deflects them so that they can be detected.

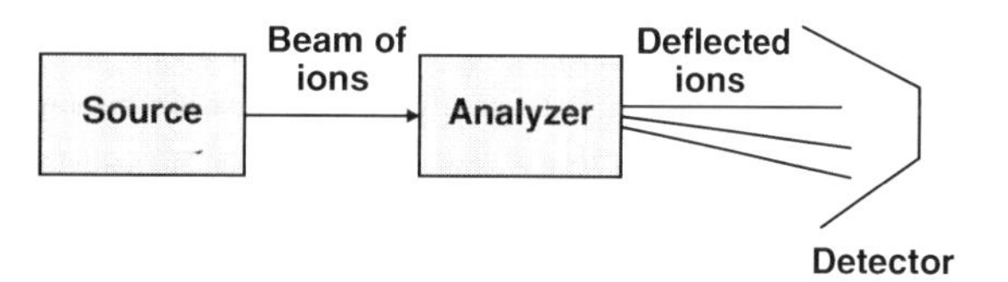

Figure 4.4. Overview of MS experiment. A beam of charged ions is generated in the source which is accelerated and deflected in the analyzer. Ions are detected by the detector. The system is maintained under a powerful vacuum.

This separates ions based on differences in their *m/z* ratio. Because the experiment takes place in a powerful vacuum, ions are encouraged to exist in the gas phase and are unlikely to encounter air molecules.

It is possible to analyse both positive and negative ions in MS systems, depending on the format of the analyzer (N.B. cations and anions must be separately analysed). These formats are called *positive ion mode* and *negative ion mode*, respectively. In EI/MS, a beam of electrons bombards the sample which has diffused into the ionization chamber. This efficiently strips the sample molecules of single electrons, converting them into positive ions which are called *molecular ions*. Since in chemical bonds electrons exist in pairs rather than singly, loss of a single electron leaves an unpaired electron behind. These ions are therefore also *radicals* and are chemically unstable. The excess energy imparted by bombardment with the electron beam leads to fragmentation of the molecular ions into *fragment ions*. This fragmentation produces an uncharged free radical and a cation (*see* Example 4.1). Every possible fragment ion can exist both in the form of an uncharged free radical and of a cation (the ratio between how much ion and radical are formed depends on fragment thermodynamic stability and will accordingly vary from fragment to fragment). The molecular ion breaks up into nearly every possible size of fragment down to individual atoms. Since only ions can be accelerated in the analyzer, free radicals play no further part in the MS experiment.

An alternative, though less common type of EI ionization is *electron capture ionization*, where the sample captures an electron from the beam to form an anion as shown in Figure 4.5.

Provided all ions carry the same charge, they will deflect in the analyzer according to their *m/z* ratio *that is* according to their different mass. A spectrum of differing mass can therefore be generated from a single chemical species as shown in Example 4.1. It is possible to generate ions with $z > 1$, thus allowing analysis of large mass species.

The *ion beam* may be detected by a *photomultiplier* detector and converted into an electric signal. This signal is amplified by a factor of up to 10^6 resulting in extremely sensitive detection. The spectrum of *m/z* ratios measured is

Example 4.1 Electron impact mass spectrometry of 2-hydroxy-5-nitrobenzyl alcohol (HNB)

This compound was purified by reversed phase HPLC from commercial dimethyl (2-hydroxyl-5-nitrobenzyl)sulphonium bromide (HNBB). On EI/MS the following spectrum was obtained. Suggested allocations of structure to the principal ions are as follows:

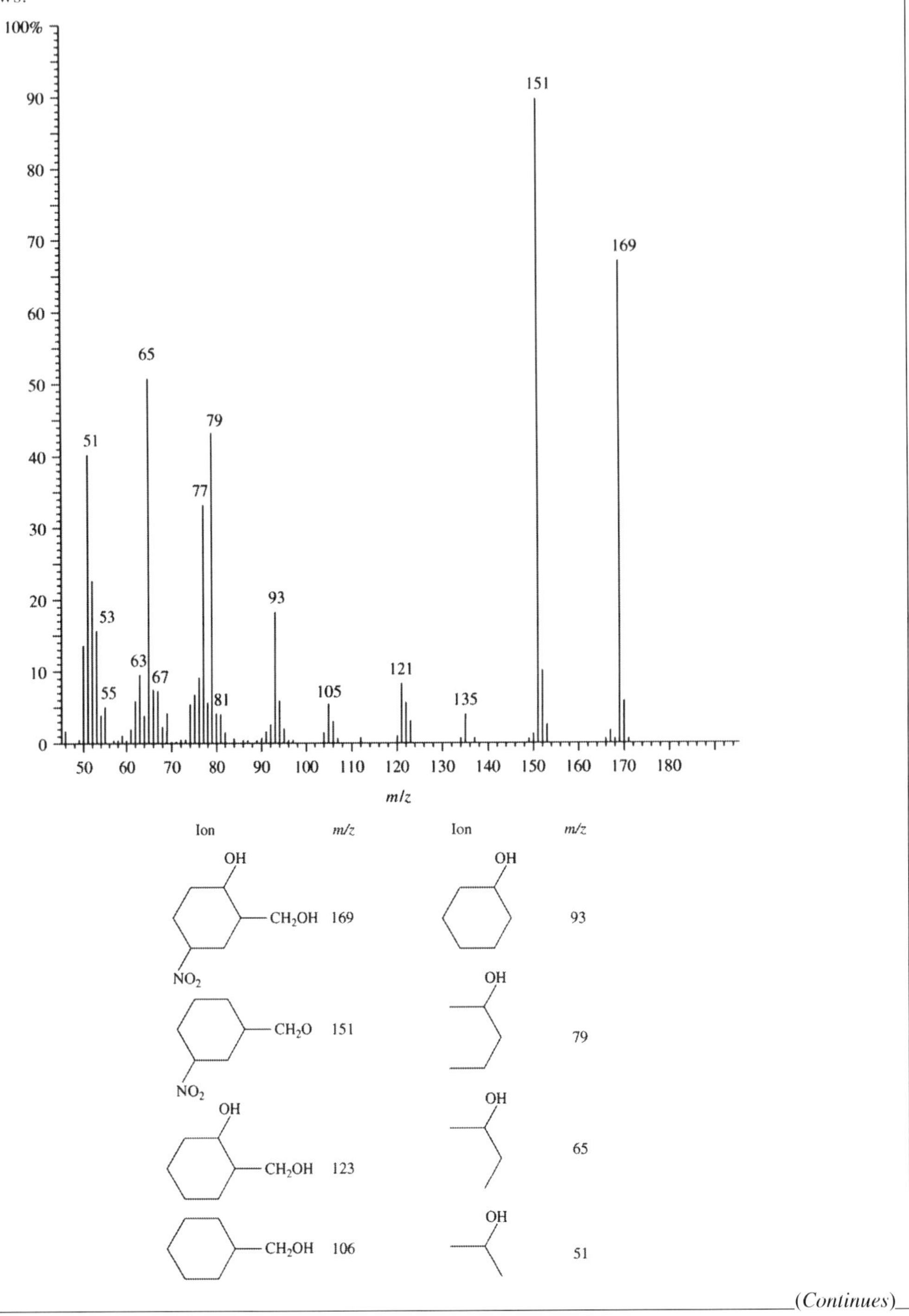

(*Continues*)

(Continued)

From this, it was concluded that the species with m/z of 169 represented the parent ion in this analysis and that the parent compound, HNB, had the following chemical structure:

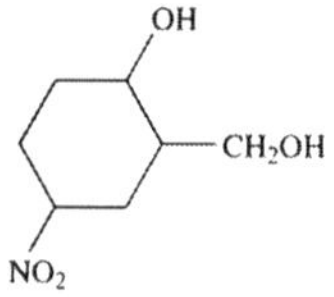

OH
CH$_2$OH
NO$_2$

This was further supported by ^{1}H NMR (Section 3.6.4), by elemental analysis and by the fact that this would be an expected product of hydrolysis of HNBB.

plotted against intensity normalized to that of the *base peak* (i.e. the peak in the spectrum with the highest intensity). All peaks in the spectrum are therefore in the relative intensity range 0–100 with the base peak at a value of 100 (Example 4.1). Peaks found in the spectrum in addition to the base peak include the *molecular ion* (i.e. the intact molecule of interest), *fragment ions* (due to breakdown of the parent ion), *isotope ions* (*see* below) and *background ions*. The latter are usually due to air-contamination, oil leaks or contaminants of the sample.

Several MS scans are carried out on a sample to yield a statistically-averaged spectrum. Especially when interfaced with other high-resolution techniques (Section 4.3), many spectra may be generated from a single sample. A large amount of data can therefore be generated in MS analysis which can be difficult to interpret manually. Modern EI-MS mass spectrometers contain a *library* of mass spectra of thousands of molecules which may be automatically compared to experimental spectra. MS spectra derived from analysis of protein peptide maps may be automatically compared to protein sequence databases facilitating identification of particular proteins in the sample. Automated comparisons such as these allow rapid identification of sample components and their fragments in MS spectra enabling highly-efficient throughput of samples.

Isotope peaks arise from the fact that some atoms in the species may exist in a variety of *isotopic forms* giving a range of possible mass. We can calculate the percent probability of occurrence of these satellite peaks from the percent occurrence of each isotope in nature (Table 4.1).

For small molecules, the presence of isotopes leads to quite small mass differences between MS fragments differing only in isotope content and these can safely be ignored for most purposes. However, for large macromolecules containing up to thousands of C, H and N atoms, the effect on calculated mass can be large. Moreover, the use of a simple atomic weight scale where C=12, H=1 and O=16 rather than the accurate monoisotopic atomic weight (Table 4.1) increasingly leads to inaccuracy in nominal molecular weight as molecular size increases. For example, a small protein with the atomic formula $C_{150}H_{310}O_{200}N_{160}S_6$ might have a nominal molecular weight of 7.742 kDa with a simple atomic weight scale but the monoisotopic mass would be 7.744 kDa when estimated with the accurate atomic mass scale given in Table 4.1. If we allow for the presence of naturally-occurring isotopes we could calculate an isotopically-averaged value of 7.747 kDa.

$$M + e^- \longrightarrow M^{-\cdot}$$

Figure 4.5. Electron capture ionization.

Table 4.1. Mass and abundance of isotopes found in biomolecules

Element	Isotope	Isotopic mass (AMU)	Abundance (%)
H	^{1}H	1.00784	99.985
	^{2}H	2.01410	0.015
C	^{12}C	12.00000	98.90
	^{13}C	13.00336	1.1
N	^{14}N	14.00307	99.634
	^{15}N	15.00011	0.366
O	^{16}O	15.99491	99.762
	^{17}O	16.99913	0.038
	^{18}O	17.99916	0.2
P	^{31}P	30.97376	100
S	^{32}S	31.97207	95.02
	^{33}S	32.97146	0.75
	^{34}S	33.96787	4.21
	^{35}S	35.96708	0.08
Cl	^{35}Cl	34.96885	75.77
	^{37}Cl	36.96590	24.23
K	^{39}K	38.96371	93.256
	^{40}K	39.96400	0.012
	^{41}K	40.96183	6.73

Table 4.2. Summary of main MS ionization modes used in the study of biomolecules

Ionization mode	Basis of ionization	Nature of sample	Principal applications
Electron impact (EI)	Molecule loses or gains electron from beam of electrons *in vacuo*	Vapour	Determination of structure of small molecules by interpreting fragmentation pattern obtained. Unsuitable for proteins
Chemical ionization (CI)	Beam of electrons passes through large excess of ammonia or methane. Molecule gains H^+ to form quasimolecular ion, M + H	Vapour	Accurate mass of intact M + H. Unsuitable for proteins
Fast atom bombardment (FAB)	High-energy beam of argon or xenon causes high-temperature spike which volatilizes and ionizes sample	Liquid matrix (e.g. glycerol)	Accurate mass of intact M + H or M − H ions. Suitable for peptides or small proteins
Plasma desorption ionization (PDI)	Beam of plasma ionizes and desorbs sample	Deposition on matrix (e.g. nitrocellulose)	Gives mass of large, intact samples such as proteins (<20 kDa). Uses TOF analyzer
Matrix-assisted plasma desorption ionization (MALDI)	High-energy laser (UV) causes sample and matrix to "sputter" into the gas phase	Sample mixed with solid matrix (e.g. sinapinic acid) which absorbs strongly at lmax of laser	Gives mass of large, intact macromolecular ions. Uses TOF analyzer samples such as proteins (<20 kDa). Uses TOF analyzer
Electrospray ionization (ESI)	A fine spray of charged droplets is generated. Matrix is evaporated leading to expulsion of ions	Liquid matrix (aqueous or aqueous/organic)	Gives mass of intact macromolecular ions. Uses Formation of multiply-charged ions facilitates very accurate mass determination. Usually uses quadrupole analyzer
Ion spray ionization (ISI)	ESI in which a flow of nebulising gas is also used		High flow-rates facilitate interfacing with HPLC. Suitable for studies with proteins

4.1.3 Ionization Modes

The description of EI/MS in the preceding section serves as an introduction to the basic theory and practice of MS. Historically, this was the first and most widespread ionization procedure used. However, in practice this technique is limited to small, stable and volatile sample components and is unsuited to macromolecules such as peptides and proteins. We will now review other ionization methods some of which allow informative MS analysis of larger mass molecules. A summary of the principal ionization methods used in MS of biomolecules is provided in Table 4.2.

Chemical ionization (*CI*) involves ionization of sample in the presence of a large excess of a reagent gas such as ammonia or methane. A beam of electrons similar to that used in EI is passed through this gas leading to the formation of radicals such as CH_4^+ and NH_3^+. Unlike EI, which generates a chemically unstable radical, CI generates a stable ion from each sample component due to proton transfer from the reagent gas to the molecule (Figure 4.6).

This ion is called a *quasimolecular* ion and it is one AMU greater than the mass of the molecule itself (often referred to as M + H). The stability of this ion is advantageous as the mass of the species generated can be accurately determined. It is disadvantageous, however, in that no fragment ions are

Methane;

$$CH_4 \longrightarrow CH_4^{+\bullet}, CH_3^{+\bullet}, CH_2^{+\bullet}$$

$$CH_4^{+\bullet} + CH_4 \longrightarrow CH_5^+ + CH_3\bullet$$

$$M + CH_5^+ \longrightarrow MH^+ + CH_4$$

Ammonia;

$$NH_3 \longrightarrow NH_3^{+\bullet}, NH_2^{+\bullet}, NH^{+\bullet}$$

$$NH_3^{+\bullet} + NH_3 \longrightarrow NH_4^+ + NH_2\bullet$$

$$M + NH_4^+ \longrightarrow MH^+ + NH_3$$

Figure 4.6. Chemical ionization of particle of mass, M, with methane and ammonia.

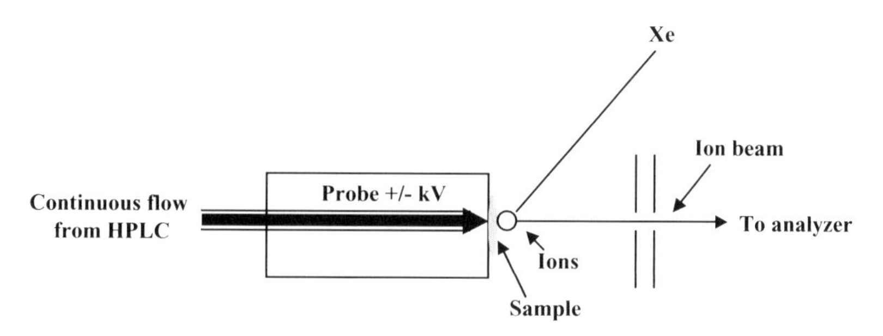

Figure 4.7. Continuous Flow Fast Atom Bombardment (FAB) Ionization. Sample, held in a liquid matrix, is bombarded with a beam of atoms such as Xenon. Ions are formed in a dense gas which may be accelerated towards the analyzer and detector. In continuous flow (dynamic) FAB, sample may flow continuously in a capillary passing through the probe as shown.

formed in the process and, therefore, little structural information is generated from this ionization method. In practice, both EI and CI spectra can be generated from a given sample generating complementary information on chemical structure and mass, respectively. As with EI, it is still necessary to vaporize the sample before ionization. This means that CI shares some of the limitations of EI, being directly useful mainly in MS analysis of small, stable and volatile molecules and unsuitable for biological macromolecules. A partial exception to this is *atmospheric pressure chemical ionization* (*APCI*) which is a specific form of CI described briefly at the end of this section. CI is also an important facet of many of the ionization procedures which allow MS analysis of large molecules such as proteins (e.g. the formation of quasimolecular ions).

These procedures differ from each other mainly in their *ionization mode*. The experimental problem is to generate intact, singly- or multiply-charged ions from macromolecules such as proteins in the gas phase. Most ionization modes used with macromolecules depend on either *bombardment* of the sample by a beam of energetic particles or else on *evaporation* of sample into the gas phase. A common feature of these methods is the use of some form of *matrix* in which sample is bombarded or from which it may be evaporated. The matrix, which is usually liquid but is a solid in some ionization modes, serves to isolate sample components and prevent their aggregation. It provides a reservoir from which ions may be gently recruited into the vapour phase without resulting in extensive fragmentation. The matrix may be regarded as a sort of 'platform' which is removed after volatilization leaving single sample ions in the gas phase.

Fast Atom Bombardment (*FAB*) *ionization* uses a high-energy (8–30 keV) beam of neutral atoms such as argon or xenon, accelerated to high velocities to cause both volatilization and ionization of the sample (Figure 4.7). The material to be analysed, dissolved in a liquid matrix such as glycerol, is placed on the surface of a *probe* and, as a result of bombardment by the beam of atoms, a dense gas forms immediately above the sample. Ionization of sample molecules in this gas is thought to be due to the generation of an extremely transient increase in temperature called a *high temperature spike* which is of sufficient duration to generate ions but too short to cause heat-induced bond breakage. The gas contains a mixture of anions, cations and neutral molecules (which can be subsequently ionized by CI interactions with charged ions in the gas, *see* below). Alternatively, ions may be preformed in the matrix and desorbed by the gas bombardment. By varying the voltage, anions or cations can be accelerated towards the analyzer.

If sample were merely deposited on the probe as a dry residue, there would be a rapid decrease in yield of sample ions. However, introducing the sample in a *liquid matrix* allows continuous replenishment of sample ions from the matrix surface. Moreover, in the gas phase the matrix behaves in a similar manner to the reagent gas in CI shown in Figure 4.6, generating quasimolecular ions (either protonated (M + H) cations or deprotonated (M – H) anions) from uncharged molecules of the sample in the gas phase. In FAB, the matrix itself generates characteristic ions which are called *matrix cluster ions* which are useful in calibration of spectra.

A limitation of FAB arises from the fact that there seems to be a significant difference in ionization efficiency between molecules that exist on the surface of the liquid matrix and those which are buried inside it (i.e. between sample components of differing hydrophobicities and surface activities). Molecules of the former type are much more readily ionized than those of the latter. This is called *selective desorption* or *suppression effect* and it complicates analysis of FAB/MS spectra since the precise composition of the sample may not be known prior to the analysis. Suppression effects are also a problem with other ionization methods and can be avoided by separating the sample before MS by some other high-resolution technique such as HPLC (Chapter 2; Section 4.3.1) or electrophoresis (Chapter 5; Section 4.3.3).

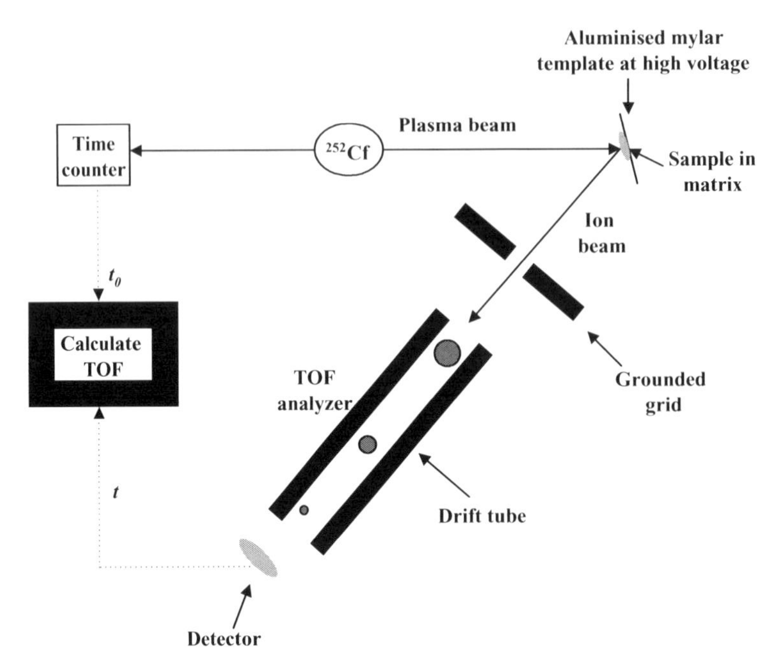

Figure 4.8. Plasma Desorption Ionization (PDI). ^{252}Cf breaks down to form two particles which travel in opposite directions at the same velocity. One of these activates a time counter (at t_0) while the other, causes vapourization and ionization of the sample. Smaller mass ions travel down the drift tube of the TOF analyzer more quickly than larger ones. When ions arrive at detector, time of flight (t) is calculated.

Continuous analysis of samples eluting from such separations, which is called *continuous flow FAB* or *dynamic FAB*, ensures that no component will be overlooked due to its different ionization efficiency (Figure 4.7).

Plasma desorption ionization (*PDI*) is similar to FAB except that it involves bombardment of sample by a beam of *plasma* (i.e. atomic nuclei stripped of their electrons (Figure 4.8). This is achieved with the radioactive isotope Californium, ^{252}Cf, which breaks down to emit a stream of emission nuclei. Typical products, which are simultaneously emitted in opposite directions at identical velocities, are $^{20}Ba^{+}$ and $^{18}Tc^{+}$. Sample is coated onto a suitable substrate such as aluminized mylar in a matrix composed of either nitrocellulose or glutathione. It is then placed in front of the source and exposed to the plasma beam which imparts sufficient energy to each component of the sample to cause ionization and to *desorb* these ions into the gas phase for analysis.

PDI is used with a *time-of-flight* (*TOF*) analyzer (*see* following section) which measures the time required for each ion to travel to the end of a tube called the *drift tube* where they are detected. A feature of the plasma source is that, simultaneous with emission of the plasma product which causes sample ionization, a second product is emitted in the opposite direction with the same velocity. This is used to begin a *time counter* which allows accurate estimation of *zero time* and, thus, accurate determination of the time-of-flight. Since ions generated in PDI have the same momentum, ions of smaller mass travel down the drift tube more quickly than those of larger mass (i.e. they have greater velocity). From the time measured for each ion, it is therefore possible accurately to measure ion mass. Like FAB this ionization technique is regarded as quite gentle and rarely results in fragmentation of sample molecules. Because of development of other ionization methods, PDI is now mainly of historical interest and has been largely superseded by *matrix-assisted laser desorption ionization* (*MALDI*).

MALDI depends on bombardment of sample (held cocrystallized in a suitable matrix) with pulses of high-energy laser radiation in the UV part of the electromagnetic spectrum (Figure 4.9). The nature of the matrix is critical to this ionization method since it absorbs much of the energy from the laser, avoiding significant breakdown of sample structure. The matrix (e.g. 2,5-dihydroxybenzoic acid) is a solid present in large excess and is composed of molecules which absorb the radiation energy at the wavelength of the coherent laser light, unlike the sample components. At ultraviolet wavelengths, conjugated structures are especially useful as matrix components. When the sample is exposed to the high-energy radiation of the laser, matrix and intact sample molecules are vapourized in a jet. They are said to '*sputter*' into the gas phase where sample molecules are ionized

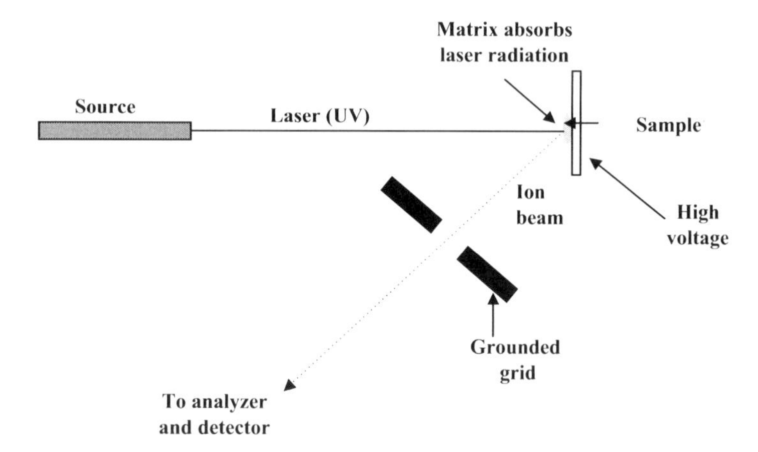

Figure 4.9. Matrix-Assisted Laser Desorption Ionization (MALDI). A pulse of laser light (UV) is directed at the sample which is cocrystallized with a suitable matrix component. This matrix component absorbs most of the energy of the laser. Both matrix and sample are vapourized by the laser. In the gas phase, proton exchange occurs between matrix and sample components producing M + H ions.

by the excited matrix molecules, presumably by a process of proton transfer similar to CI. A molecular ion (M + H) is generated by this transfer. Although singly-charged ions predominate in MALDI spectra, ions which are multi-charged are also formed. Because of the fact that laser radiation is used discontinuously as a *pulse*, this ionization mode lends itself to use of TOF analyzers. It is possible therefore to use this technique to analyse large macromolecules such as proteins, glycoproteins, oligonucleotides and oligosaccharides. The technique is especially versatile allowing MS analysis without significant sample fragmentation. A disadvantage is the difficulty of observing low molecular mass ions (e.g. <1000 Da) due to the presence of ions due to the matrix. This can be avoided by varying the choice of matrix material.

The preceding ionization modes all depend ultimately on bombardment of sample for the generation of ions. *Electrospray ionization* (*ESI*) is an evaporation-based ionization mode. It involves the generation of a fine *spray* of charged liquid droplets at atmospheric pressure from which the matrix is later evaporated (Figure 4.10). Because ionization takes place at atmospheric pressure rather than in a vacuum, this ionization mode is also sometimes called *atmospheric pressure ionization* (*API*). Sample is again introduced in a liquid matrix which is either aqueous or aqueous/organic. Typically, sample is sprayed from a very fine metal syringe

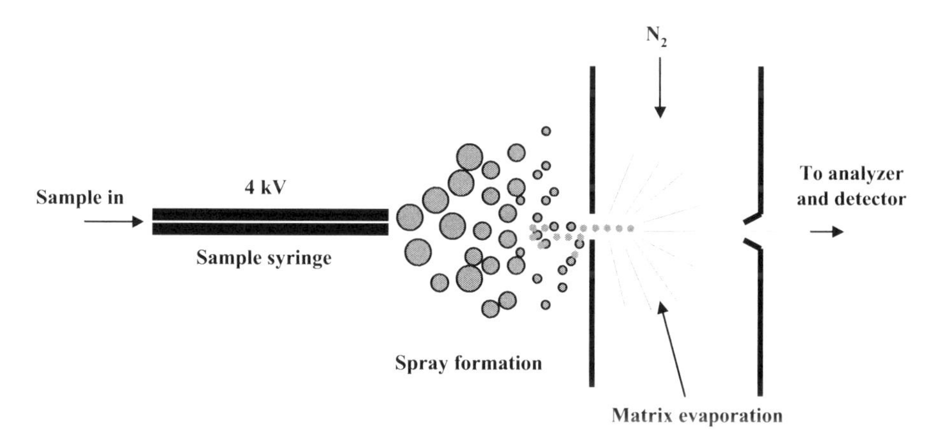

Figure 4.10. Electrospray ionization (ESI). Sample is converted to a fine spray of droplets. Matrix is evaporated from this, leading to Coulombic explosion and the liberation of ions which then are directed to the analyzer and detector. In ion spray ionization (ISI), the sample is surrounded by a sheath of gas such as N_2 which imparts greater energy and hence facilitates faster input of sample to the system.

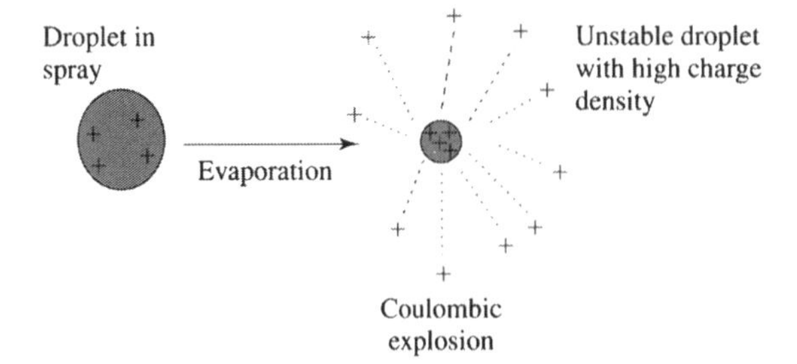

Figure 4.11. Generation of ions during ESI. Charged droplets are generated in a fine spray. When exposed to warm, drying N_2 gas, they undergo evaporation of matrix. Charge density increases as a result of decreasing droplet size, leading to an unstable droplet which explodes in a Coulombic explosion. Thus ions are liberated into the gas phase where they are accessible to MS analysis.

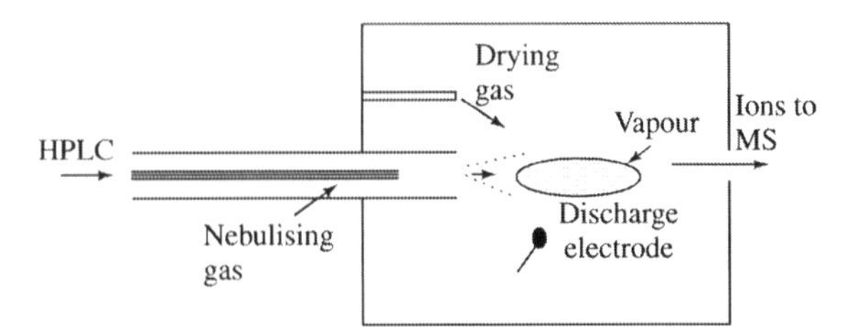

Figure 4.12. Atmospheric Pressure Chemical Ionization. Sample is vapourized by heat and nebulizing gas. The vapour contains both sample and solvent. Solvent is ionized by a corona discharge from the nearby discharge electrode. Sample molecules are in turn ionized by the excited solvent molecules in a CI process. The vapour is sampled and subsequently analysed by MS.

maintained at 4 kV. Dry gas, heat or a mixture of each is applied to the droplets which, because of the powerful electric field, are now highly-charged. This causes the solvent to evaporate, in turn, leading to a decrease in droplet size and a large increase in charge density on the droplet surface (Figure 4.11). This leads to increasing instability in the droplet resulting in a *Coulombic explosion* in which the droplet generates a number of smaller droplets. Still smaller droplets may be formed from these by an identical process until, ultimately, sufficiently small droplets are formed in which the field due to excess charge is large enough to expel ions. These ions enter the gas phase where they may be directed into a mass analyzer. This ionization method requires relatively small inputs of thermal energy and results in gentle ionization of sample components with little attendant fragmentation. Accordingly, it is especially useful in the ionization of intact large macromolecules such as proteins.

As well as generating singly-charged ions, a particular feature of ESI is the generation of *multiply-charged ions*. On average, a charge is added for every kDa of mass with protein samples. This means that, since detectors with relatively limited mass ranges such as quadrupole detectors (*see* Section 4.1.4) measure m/z ratio and not just mass alone, it is possible to determine the mass of molecules outside the normal mass range of the analyzer. For example a molecule of 20 kDa possessing 20 charges has a m/z ratio equal to that of a singly-charged molecule of mass 1 kDa. Therefore, ESI brings large macromolecules into the normal mass range of MS analyzers. Moreover, the distribution of multiply-charged peaks in the MS spectrum forms a characteristic 'fingerprint' for each ion and allows determination of an *average relative molar mass*. This is a more accurate estimate of mass than single m/z measurements of singly-charged ions.

A version of ESI in which a flow of *nebulizing* gas (usually Nitrogen) surrounds the spraying needle within a circular sheath is called *ion spray ionization* (*ISI*). This allows the input of greater energy into droplet formation facilitating the use of higher sample flow-rates (up to 1 ml/min compared to 10 μl/min for conventional ESI) and is capable of tolerating higher water levels in solvent.

In discussing CI above, it was pointed out that this ionization mode was not really very suitable for direct analysis of macromolecules since it is difficult to vapourize these without at the same time destroying their structure. *Atmospheric pressure chemical ionization* (APCI) is a process in which sample is nebulized at atmospheric pressure by a combination of a stream of gas and direct heating (Figure 4.12). This produces a vapour which may be then ionized by a corona discharge from an electrode placed immediately after the nebulizing inlet. Ions are formed by CI and are then accelerated and analysed in the usual way. Because this ionization mode can accommodate relatively large flow rates, it is conveniently interfaced with HPLC to allow MS analysis of peaks which have been chromatographically separated (Section 4.3). However, because heat is used to volatilize sample, this ionization mode is not appropriate for studies of proteins and peptides.

4.1.4 Equipment Used in MS Analysis

The formation of gas phase ions is a prerequisite for MS experiments. These ions are formed in the *source*. In EI/MS, this is an evacuated chamber into which sample may be introduced as a vapour (Figure 4.13). A thin piece of metal such as rhenium or tungsten which is called a *filament* is heated to 2000 K. At high temperature, metals can lose electrons by a process of *diffusion*. This happens more quickly if an electric potential is also applied to the heated filament. A potential of 70 V generates a beam of electrons with energy of 70 eV. These 70 eV electrons stream across the chamber

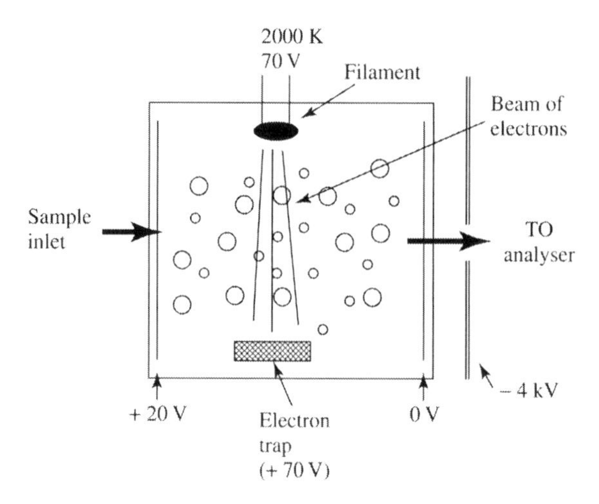

Figure 4.13. EI/MS source. Sample (shown as circles) is introduced as vapour into evacuated chamber. Electrons diffuse from the heated filament at energy of 70 eV. The electron beam ionizes sample molecules into anions or cations which are accelerated by an electric potential towards the analyzer.

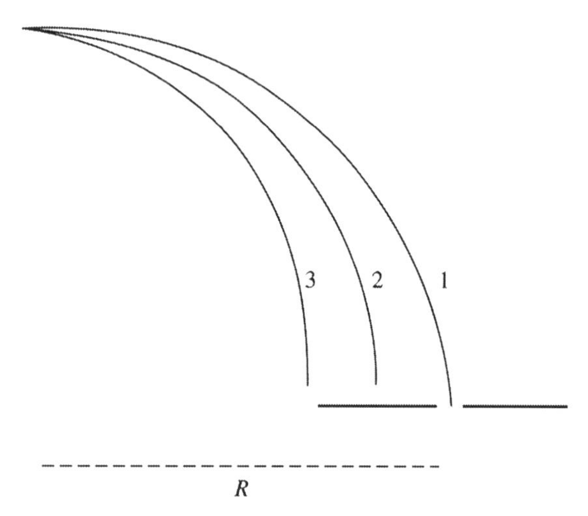

Figure 4.14. Magnetic focusing of ions in MS A detector slit is placed at a distance, R, from the centre of trajectory arc. By varying field strength or accelerating voltage, the beam may be focused through a number of possible trajectories (1 to 3). Similarly, beams of different ions may be detected in this way.

and interact with the diffusing sample molecules (which are neutral) and converts them either into anions or cations as described in Section 4.1.2. In order to increase the ease with which electrons diffuse from tungsten filaments, they are often coated with thorium oxide. A similar source is used for CI/MS but with the proviso that it requires to be more gas-tight than that used for EI/MS, to accommodate the high gas pressure (1 torr) required.

In bombardment ionization methods, a beam of high-energy atoms or ions needs to be generated. In FAB, a *fast atom gun* is used to generate a beam of high-energy (3–10 keV) argon, Krypton or xenon atoms while a higher-energy beam of caesium ions (20–30 keV) may instead be generated by a *caesium gun*. For MALDI a laser source generates a beam of UV laser light.

The *analyzer* has the function of separating ions based on their different m/z ratios. Before looking at the main types of analyzers available for MS of macromolecules, it is important to define *resolution*, R, in the context of MS This is the ability of a mass spectrometer to distinguish between ions of different m/z ratios. It may be calculated from Equation (4.2);

$$R = \frac{M}{\Delta M} \tag{4.2}$$

where M is the mass of an MS peak of interest and ΔM is it's width at 50% of it's height. A system with a resolution of 500 can distinguish between ions of m/z of 500 and 501, while one with a resolution of 4000 can distinguish between this mass and 4001.

When an ion, accelerated with a voltage V, enters a constant magnetic field of strength B, it will tend to follow a *circular* path of radius R according to Equation (4.3);

$$\frac{m}{z} = \frac{B^2 \cdot R^2}{2V} \tag{4.3}$$

If a detector slit is placed at a constant distance, R, from the centre of the trajectory arc, it is possible to focus ions on this slit by varying either the field strength B or the accelerating voltage V (Figure 4.14). The maximum value for B determines the upper limit of m/z attainable at each value of V. This technique, *magnetic focusing*, is inadequate for high-resolution work since ions of a single m/z ratio entering the magnetic field or *sector* will often possess a wide range of velocities and therefore have a range of momentum values. They would therefore describe a range of possible arcs as shown in Figure 4.14. However, if the magnetic sector is preceded by an *electrostatic analyzer* to ensure that all ions entering the magnetic field have the same velocity, high-resolution of m/z is possible. This type of analyzer is called a *double-focusing magnetic sector analyzer* and is shown schematically in Figure 4.15. The upper limit of m/z attainable with such an instrument is approx. 10 000.

An alternative analyzer suitable for work with lower m/z values (up to 4000) is the *quadropole mass analyzer* (Figure 4.16). In this analyzer, two pairs of parallel rods are arranged on either side of the beam of ions. Opposite rods are

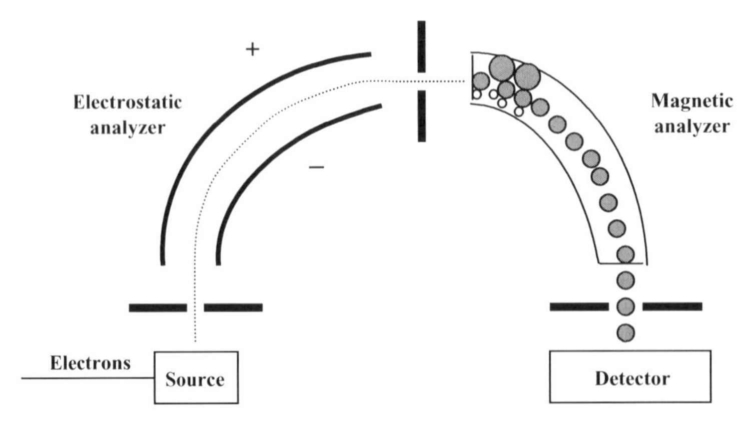

Figure 4.15. Double-focusing magnetic sector analyzer. Ions are deflected in the electrostatic sector in a manner proportional to their energy and independent of their mass. Thus a beam of ions of identical energy is selected. Deflection in the magnetic sector depends on m/z ratio.

connected electrically while adjacent rods are connected by a voltage which has radiofrequency (RF) and direct current (DC) components. By varying the radiofrequency, it is possible to select for single m/z values, since all other ions collide with the rods and are lost. Because only a small number of m/z ratios may traverse a path along the axis of the rods, this analyzer is also called a *quadropole mass filter*. Particular advantages of these analyzers are the fact that they are lightweight, cheap, can be rapidly switched between positive and negative ion modes and, since they can accommodate higher gas phase pressures than sector analyzers, they are especially suitable for interfacing with GC and HPLC systems.

Time of flight (TOF) *analyzers* (Figure 4.17) can measure m/z values up to 200–500 kDa. In this system, ions are accelerated down a long tube called the *drift tube* which is free of magnetic field. The time required for each ion to arrive at the detector (i.e. the *time of flight*) is measured and, from this, the m/z is calculated. These detectors lend themselves especially to discontinuous ionization methods such as MALDI and PDI (*see* previous section).

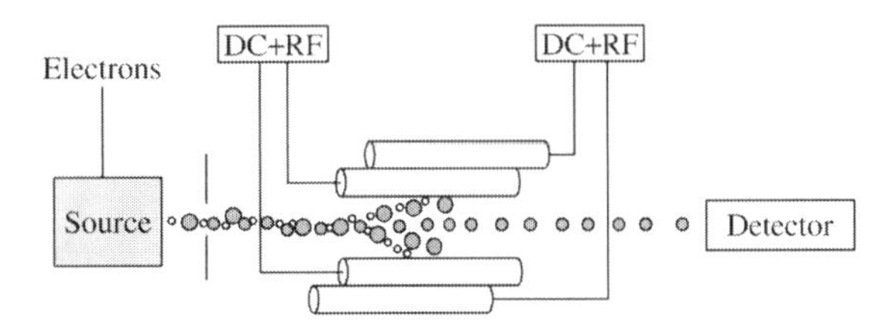

Figure 4.16. Quadropole mass analyzer. Varying the radiofrequency (RF) component of voltage allows 'tuning' for specific m/z ratios. Ions with either larger or smaller m/z collide with rods and are lost.

Ion cyclotron resonance Fourier transform MS (*FTMS*) analyzers are in some respects similar to quadropole detectors. Ions are induced to move in circular (hence *cyclotron*) orbits within an electrostatic field held within the solenoid of a powerful magnet which generates a field of strength, B. The angular frequency of motion, ω, of a particular ion is given by Equation (4.4);

$$\omega = \frac{z \cdot B}{m} \tag{4.4}$$

The ions are exposed to an alternating electric field and, when the frequency of this field is the same as ω, the ions are steadily accelerated to larger and larger radii and, simultaneously, move into phase with each other. Since ω depends on m/z, this ratio can be determined in the analyzer for individual ions to a very high degree of accuracy. The resolution attainable with this analyzer is $>1\,000\,000$. What makes this type of detector unusual is that ions are not directly detected by impacting on the detector but rather by their effect as they pass near plates within the detector which induces a current in the plates consisting of sine waves called a *free induction decay* (FID). As will also be described later (Sections 6.2.1), a continuous wave function like the FID can be converted by a mathematical tool called the *Fourier transform*

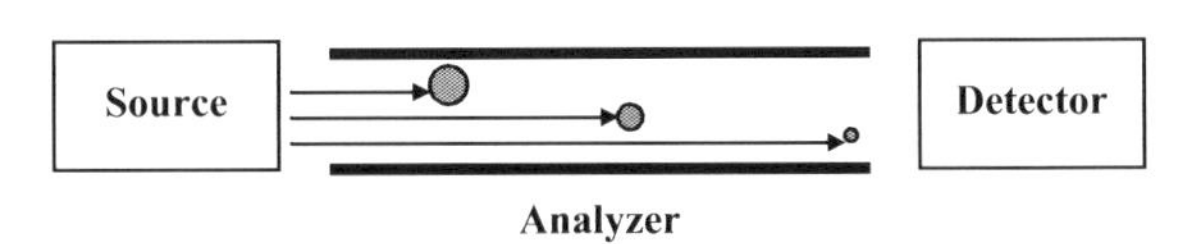

Figure 4.17. Time of flight analyzer. Ions are accelerated down a long tube (drawing not to scale) and the time required for arrival at detector is measured. The m/z ratio may be calculated from this.

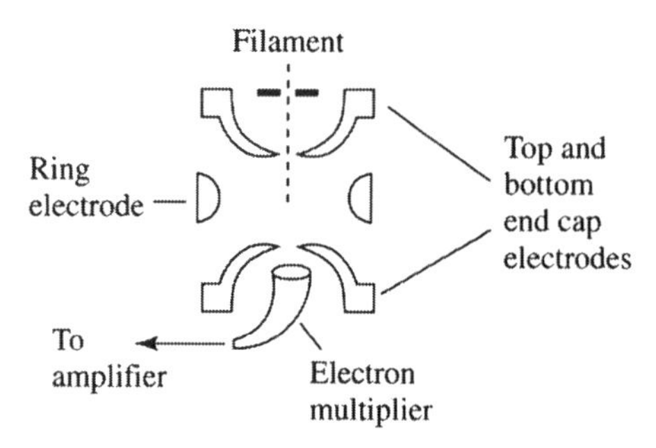

Figure 4.18. Ion Trap Analyzer. Ions are either created within or rapidly injected into the trap. The electrostatic field forces them into specific orbits, depending on the *m/z* ratio. By varying this electrostatic field, ions may be selectively ejected in order of increasing *m/z* from the trap and detected.

(Appendix 2) into a *discontinuous* function, in this case a mass spectrum.

Ion trap analyzers (Figure 4.18) are part of a family of analyzers called *quadropole ion storage systems*. This device consists of three electrodes, two of which are the *end cap electrodes* while the third is the *ring electrode* surrounding the central part of the analyzer. Ions pass into this analyzer and are held in constant orbital motion within it by the electrostatic field created by the electrodes. By variation of the precise field conditions, ions are selectively ejected in order of increasing *m/z* ratio and detected.

4.2 MASS SPECTROMETRY OF PROTEINS/PEPTIDES

4.2.1 Sample Preparation

MS has become especially popular for the structural investigation both of mixtures of proteins/peptides and of highly-purified proteins/peptides. Because of the extremely high mass accuracy of these techniques (±0.01%), it is possible to obtain detailed structural information on aspects of protein structure such as amino acid sequence, chemical and post-translational modification (e.g. N-terminal modification, glycosylation), disulfide bridge formation and to compare closely-related proteins (e.g. mutants/wild-type) by peptide mapping. It is often necessary in such studies to pretreat samples with specific chemical agents before carrying out MS analysis. Normal precautions such as prior cleavage, carboxymethylation or reduction of the sample should be carefully included in the sample preparation before MS analysis but care should be taken that reagents used in such treatments do not interfere in the analysis. Interfacing of MS with other high-resolution techniques (Section 4.3) greatly extends its analytical potential and introduction of sample to MS by HPLC or electrophoresis may often facilitate removal of contaminants such as buffer components or chemical agents used in pretreating sample. In other situations, it may be desired to introduce the sample directly to MS in a one-dimensional analysis.

Before analysing proteins or peptides by MS, it is necessary to remove salts present in the sample to avoid the formation of adduct ions with alkali metals. These would reduce the abundance of protonated quasimolecular ions. This is easily achieved either by dialysis (Section 7.6.1), by gel filtration (Section 2.4.2) or by reversed phase HPLC of the sample (Section 2.4.3). Early studies of peptides using EI/MS and CI/MS involved rendering sample molecules volatile by their extensive derivatization by acylation or esterification. However, using matrix-based ionization modes, this is no longer necessary. For ionization modes using liquid matrices, it is simply required to mix the protein (dissolved in water) with a solution of the matrix. For modes using solid matrices (e.g. MALDI), the mixture of matrix/sample is allowed to dry on the probe before acquiring the spectrum.

4.2.2 MS Modes Used in the Study of Proteins/Peptides

Information on intact protein molecular mass is readily achieved by FAB, PDI, MALDI, ESI and so on. In general, use of a TOF analyzer facilitates determination of very large masses. Fragmentation information may be obtained either by CID or by pretreating sample proteins chemically or enzymatically, for example by digesting with trypsin followed by peptide mass estimation. If the sequence of the protein is known, it may be possible to assign masses to each peptide generated in the digest. A further level of resolution is added to MS analysis by interfacing with other high-resolution techniques such as HPLC and electrophoresis.

4.2.3 Fragmentation of Proteins/Peptides in MS Systems

Proteins and peptides are *oligomers* of repeating (–NH–CHR–CO–) units, differing only in the nature of the side-chain, R. The amino acid residues are held together by peptide bonds, (–NH–CO–), which have lower bond energies than standard (–N–C–) bonds. When proteins or peptides *fragment* (e.g. due to heat or acid hydrolysis), peptide bonds are commonly broken releasing mostly intact amino acid residues. In MS experiments (e.g. FAB/MS), however, there may also be fragmentation at other locations in addition to peptide bonds resulting in a complex pattern of ions. These are related to each other by *incremental m/z* differences because only 20 R-groups of known structure are

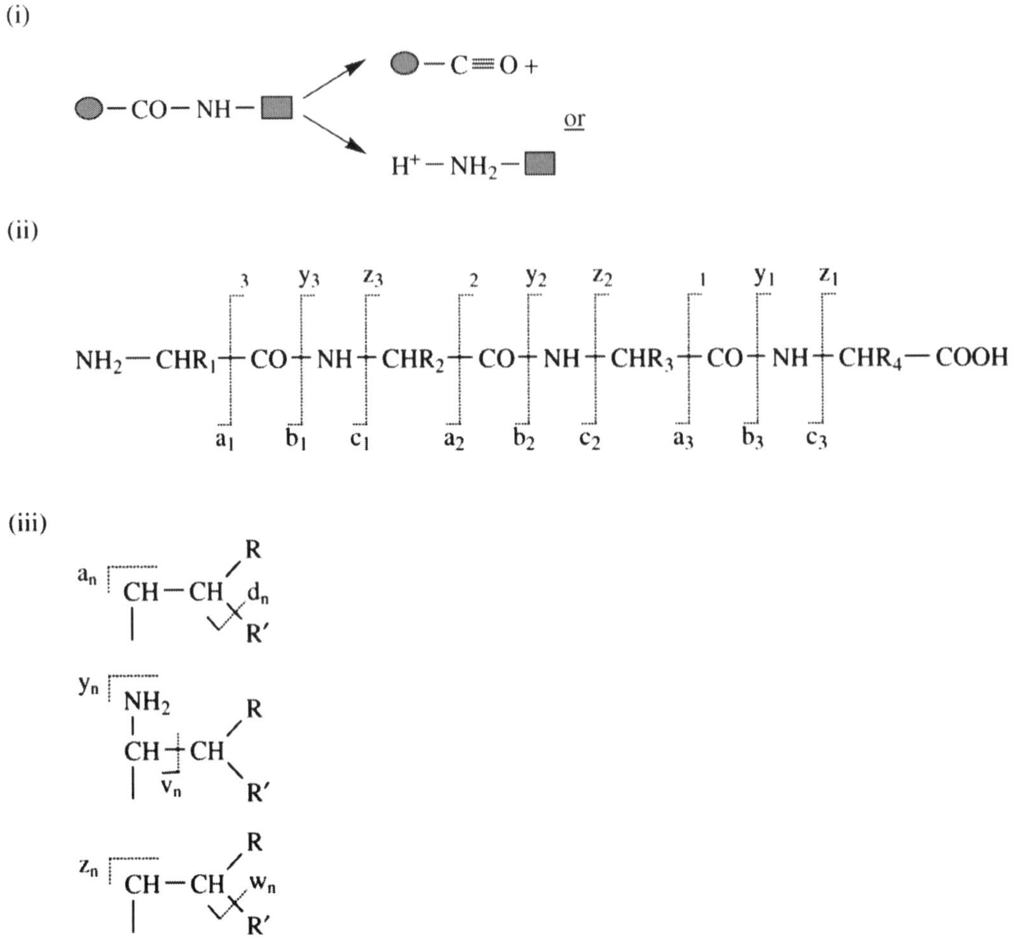

Figure 4.19. Fragmentation of protonated peptide in FAB/MS (i) Positive ions may be generated at the C- or N-terminus of peptide bonds. (ii) Main-chain fragmentation. (iii) Side-chain fragmentation. Note that d_n and w_n -type ions facilitate distinction of isomers such as Leu and Ile.

commonly found in proteins and because breakage points have fixed structural locations relative to each other. It is therefore possible to interpret complex MS spectra to learn about the structure and especially amino acid sequence of peptides and proteins (Section 4.4.6). A nomenclature proposed by Roepstorff is widely used to describe polypeptide fragmentation in MS (Figure 4.19).

Ions generated from a polypeptide during MS may retain a positive or negative charge either on their C- or N-terminus. A horizontal line pointing towards the C- or N-terminus at the breakage point is used to denote which fragment carries the charge in that particular ion (Figure 4.19). Two series of possible fragments are denoted by letters with a subscript number, n, to denote the sequence position of the residue. For the N-terminal ions these are a_n, b_n, c_n and d_n (numbered from the N-terminus) while the C-terminal ions are designated v_n, w_n, x_n, y_n and z_n (numbered from the C-terminus). The ion represented by a_1 represents the first residue in the sequence (minus the CO group) while a_2 represents the first *two* residues (minus the CO group), and so on. Identical main-chain breakage points are denoted by the pairs a/x, b/y and c/z (each member of a pair referring to a positive charge retained either on the N- or C-terminus, respectively). Breakage points denoted by d_n, v_n and w_n are due to side-chain fragmentations. These are used to distinguish residue pairs such as Leu/Ile and Gln/Lys which have identical masses (i.e. they are *isobaric*).

An alternative fragmentation method is offered by interacting samples with electrons in *electron capture dissociation* (ECD) and *electron transfer dissociation* (ETD).

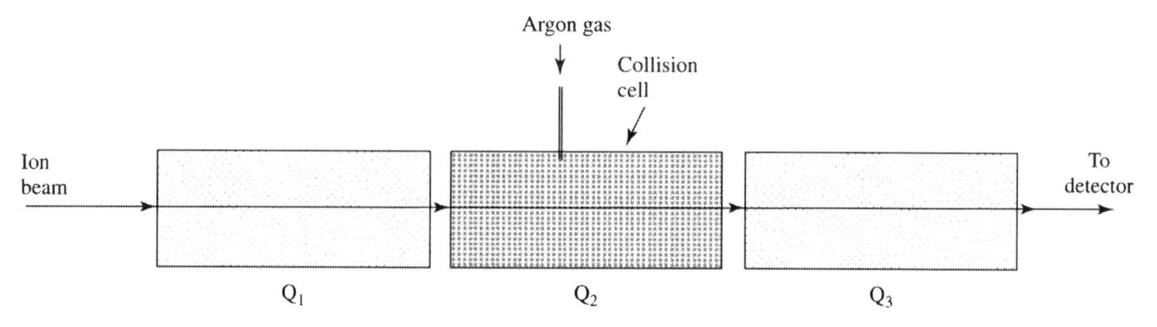

Figure 4.20. Collision-Induced Dissociation (CID) in ESI. Three quadropole analyzers (Section 4.1.4) are used to select for ions with different *m/z* ratios. Q1 selects intact ions of specific *m/z*. A collision gas (argon) is introduced into the second analyzer (Q2) called the collision cell. Collision with this gas causes fragmentation of the ion into smaller ions. These are analysed in the third quadropole analyzer, Q3. Other formats for this kind of experiment are also possible.

Both methods are based on reaction of electrons with peptide (or intact protein) cations in the magnetic field of an FTMS instrument (Section 4.1.4). In ECD, the protonated sample anion interacts directly with a low-energy electron while in ETD radical anions are the source of electrons. Peptide/protein cations with an additional electron are generated which undergo rearrangement resulting in backbone fragmentation. Importantly, there is no fragmentation of bonds in side-chains. Thus, unlike collision-based fragmentation, ECD/ETD are relatively insensitive to peptide length (collision-based fragmentation is poor with either large or small tryptic peptides) and to post-translational modifications (these are often labile in collision-based dissociations).

4.3 INTERFACING MS WITH OTHER METHODS

Interfacing MS with other high-resolution methodologies greatly extends the range of analytical problems which may be addressed in biochemistry. This topic is further explored later in this volume in the context of proteomics (Chapter 10). Interfacing may be with MS alone (e.g. FAB or ESI/MS) or with tandem MS. Major technical challenges encountered in developing such methodologies include matching analytical system flow-rates with those of MS and achieving ionization of sample presented in a continuous flow to the MS system. In the following sections, we will look at some two-dimensional analysis systems possible with MS

4.3.1 MS/MS

Fragmentation can be achieved by a process called *collision-induced dissociation* (*CID*). This involves introduction of a single-mass ion of interest from the analyzer into a second analyzer called the *collision cell* where it may collide with an inert gas such as argon (Figure 4.20). The ion fragments into a characteristic set of daughter ions which can be analysed by MS in a third analyzer. Since this experiment involves the successive use of three MS analyzers, one to select the parent ion, one to fragment it and one to analyse the daughter ions, it is called *tandem MS* (other terms for this technique include MS/MS, MS^2 and four-sector MS). It allows selective fragmentation of single ions from the first analyzer.

4.3.2 LC/MS

We have seen that MS analysis of components of complex samples may be complicated by selective desorption of sample components at rates dependent, for example, on their surface activity (Section 4.1.3). By first fractionating samples by HPLC (Chapter 2) and following this with MS analysis, we can combine high-resolution chromatographic separation with high-resolution analysis (i.e. mass determination). This is called HPLC/MS or LC/MS. In such a system, MS can be regarded as essentially a *detector* for the LC system, but one which gives useful structural information about each component of the sample. An outline of the experimental arrangement used in LC/MS is shown in Figure 4.21. Although a variety of chromatography modes are possible for LC, in practice reversed phase chromatography

Figure 4.21. Schematic of LC/MS system. The role of the interface is to transfer sample components effectively and quantitatively to the MS, prepare sample components for ionization, cope with LC solvent and bridge pressure differences between LC and MS.

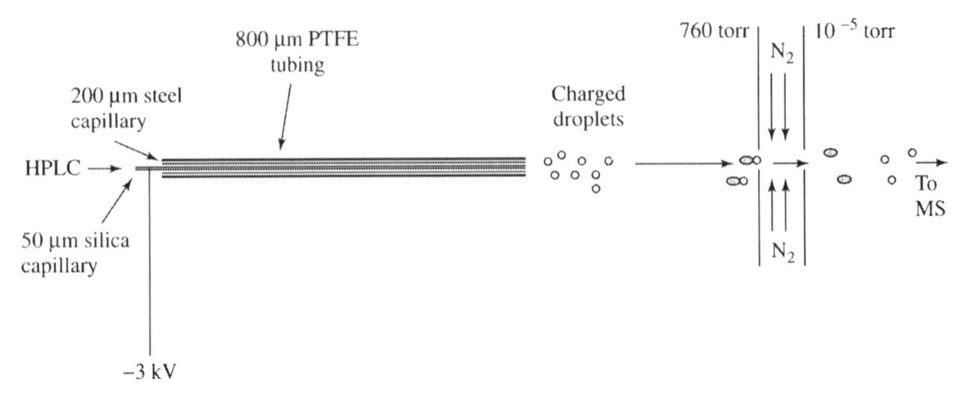

Figure 4.22. The ionspray interface. Sample flows through HPLC and is converted into a fine mist of electrically-charged droplets by electrospray (Section 4.1.3). These droplets are rapidly dried by a flow of nitrogen which removes solvent.

is used most often for protein samples. It is feasible to carry out LC/MS analysis *off-line* (i.e. by first collecting peaks from LC and evaporating solvent followed by MS analysis), but *on-line* analysis is preferable in most applications. A feature of all LC/MS experiments is the need for an *interface* between the two systems. A wide variety of interface designs are possible but not all of these are suitable for samples such as biopolymers. The *ionspray interface* (Figure 4.22) is particularly useful for protein and peptide samples because it facilitates the high flow-rates of LC systems as mentioned earlier (Section 4.1.3).

Flow-rates of most LC systems (e.g. 1–2 ml/min) are generally far higher than those of MS If the two systems were simply connected together then the vacuum component of the MS could not be maintained due to limitations in the pumping capacity of the system. There are a number of ways in which this problem can be overcome. One possibility is enlargement of MS pumping capacity. Alternatively, solvent (usually volatile organic solvents) could be removed prior to introduction of sample components into the high-vacuum region of the MS by an *analyte-enrichment interface*. This preferentially removes solvent rather than sample components leading to enrichment for the latter. Yet another approach is postcolumn *splitting* and diversion of solvent flow (Figure 4.23). In addition, the volume of the column may be reduced by the reduction of column diameter, that is the use of columns with small internal diameters (e.g. 1–2 mm) or *capillary* columns (internal diameters; approx. 50–320 μm). These possibilities are not exclusive of each other and several of them are found in LC/MS formats used for the study of proteins.

The most common ionization modes used in LC/MS are ESI and FAB although it has recently become feasible continuously to introduce sample from LC either as an aerosol or mixed with a liquid matrix to MALDI.

4.3.3 GC/MS

GC analysis (Section 2.1.4) is especially useful in study of low molecular weight, volatile samples and has no direct applications to the study of proteins or other biomacromolecules. However, the technique facilitates very high resolution separation of molecules such as drug metabolites, toxins and steroid hormones which are of biochemical interest. Just as with LC/MS, the analytical potential of GC can be greatly increased by interfacing it with MS in GC/MS. Mass estimates allow identification of the most likely chemical structure for each sample component and the identification, for example, of metabolites of a parent compound. The problem which we have seen presented by LC flow-rates is potentially even greater in the case of GC where flow-rates can exceed 20 ml/min. This is mainly overcome by the use

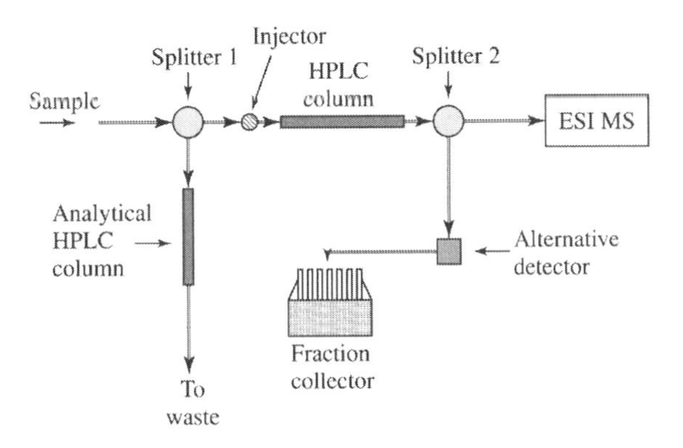

Figure 4.23. LC/MS system suitable for analysis of proteins by electrospray MS. Experimental setup featuring splitting of solvent flow and use of small-diameter HPLC. Analytical HPLC of sample is facilitated by splitter 1 while splitter 2 allows collection of sample components.

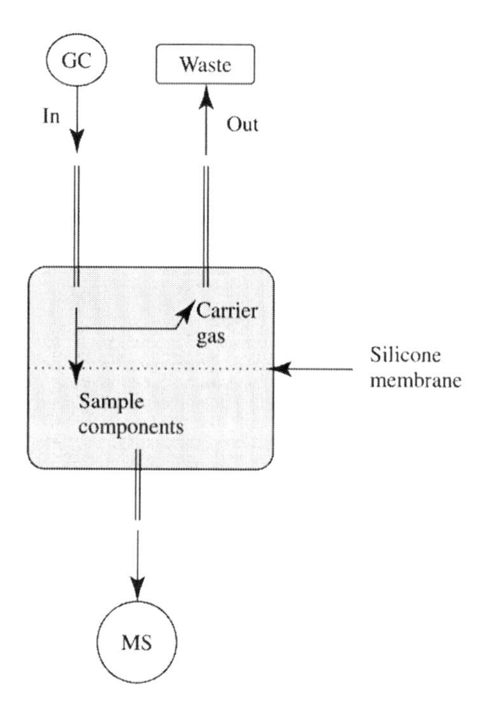

Figure 4.24. Analyte-enrichment interface for GC/MS Sample enters interface from GC. Sample components diffuse more efficiently across a dimethylsilicone membrane than the carrier gas. 10–30-fold enrichment of sample is achieved with this interface.

of GC capillary columns which have low flow-rates (e.g. 1 ml/min). An alternative approach is use of highly-efficient analyte-enrichment interfaces of the type described above. These depend on the fact that, in GC, a carrier-gas is used instead of a liquid solvent. This is usually lighter than the sample components and may have different diffusion properties. An example of such an interface is shown in Figure 4.24. When capillary columns are used, flow is usually directly into the MS instrument without the use of splitters or enrichment interfaces.

4.3.4 Electrophoresis/MS

High-resolution separation of complex samples may be routinely achieved by carrying out *electrophoresis* under a wide range of experimental conditions and in a wide variety of formats (Chapter 5). The highest-resolution analytical electrophoretic systems, *capillary electrophoresis* (CE; Section 5.10) and *capillary isotachophoresis* (CITP; the physical basis of *isotachophoresis* is described in Section 5.3.1) take advantage of *electrophoretic flow* in extremely narrow capillaries under very high voltages (10–60 kV). These capillaries efficiently dissipate heat, allowing high-resolution separation without deterioration of sample. This can be applied to a

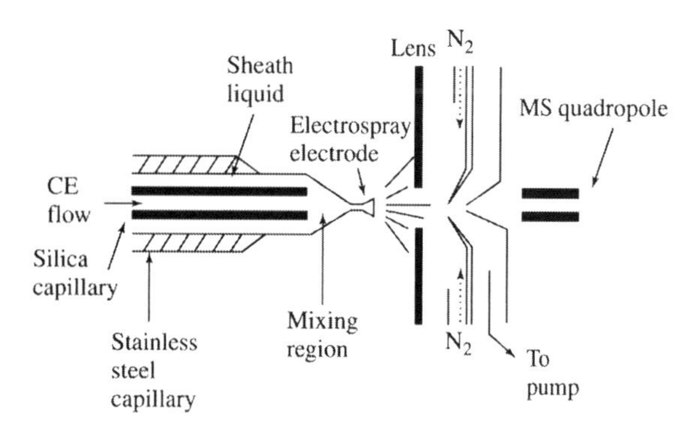

Figure 4.25. CE-MS electrospray interface. Flow of sheath liquid (approx. 5 μl/min methanol, propanol or acetonitrile) adds to CE flow, resulting in increased flow-rate to the electrospray. This system functions with aqueous buffers and low inner diameter capillaries. The electrospray electrode is stainless steel.

wide spectrum of bioseparations including enantiomeric resolution of low molecular weight isomers and separation of peptides, proteins, DNA, viruses and even whole cells. Since they are capillary-based methods with consequent small total volume, both CE and CITP use extremely low sample loadings and flow-rates compared to other electrophoretic and chromatographic methods.

In interfacing these procedures to MS, we are therefore presented with the problem of matching the very low flow-rates of CE/CITP (0–100 nl/min) with those of MS but without losing the high resolution achieved in CE/CITP by peak broadening effects. One approach to this is the addition of *make-up flow* postcapillary that is the addition of extra solvent to increase the flow volume to that of the MS system. Alternatively, a low dead-volume T-piece may be inserted between the capillary and CE/MS interface. A further problem may be presented by the low *migration times* possible in CE systems (e.g. 100–1000 s). This may be too short for sufficient scans to allow accurate characterization of MS peaks.

As with LC/MS and GC/MS, interfaces are also used in electrophoresis/MS Among the most popular of these is the *electrospray interface* (Figure 4.25). This illustrates the use of make-up flow to facilitate low CE flow-rates and also facilitates the use of CITP/MS CITP is a form of CE which results in concentration of sample components into compact bands and which allows somewhat larger loading volumes than standard CE.

4.4 USES OF MASS SPECTROMETRY IN BIOCHEMISTRY

Variations in protein structure are often of great importance biochemically. Many techniques used in the study of proteins may not adequately distinguish such variants from each other. However, the high accuracy mass determination possible by MS analysis makes it feasible to distinguish proteins which may differ from each other by only a few AMU. Moreover, since mass values are directly related to distinct chemical structures, MS analysis makes interpretations possible which might be problematic using other analytical procedures. One of the first demonstrations of the usefulness of MS analysis of proteins was the demonstration that cDNA-derived sequences of some proteins were incorrect. The following sections describe some further examples which illustrate the versatility of MS in structural studies of biomacromolecules.

4.4.1 MS and Microheterogeneity in Proteins

When expressed in heterologous expression systems (e.g. a human gene cloned into and expressed in *Escherichia coli*), many proteins are inappropriately, incompletely or otherwise incorrectly processed. This arises from the fact that the expression system may lack particular enzymes responsible for catalyzing processing or else these enzymes may simply not recognize processing sites due to species-specific sequence differences. This may happen either within or at the termini of the polypeptide chain (Figure 4.26; Table 4.3).

Eukaryotic proteins are biosynthesized with the initiation methionine intact which is post-translationally removed by enzymatic digestion. Prokaryotic proteins, by contrast, are biosynthesized with an initiation N-formyl methionine. Prokaryotic and eukaryotic systems may lack appropriate enzymes for removal of initiation residues from eukaryotic and prokaryotic proteins, respectively. A range of other chemical structures are also possible at the N-terminus (Table 4.3), which can result in N-terminal 'blocking' (i.e. an inability to sequence the protein chemically using the Edman degradation). This can be important, for example, if N-terminal sequence is required by regulatory authorities as confirmation of protein identity/purity during commercial production of recombinant proteins. Heterogeneity is also possible at the C-terminus of the polypeptide chain. Protein biosynthesis may be prematurely terminated in heterologous expression systems which can result in a variety of C-terminal sequences, called '*ragged ends*'. Moreover, bacterial cells may contain a variety of *exopeptidases* which may cause proteolysis at the N- or C-terminus of proteins/peptides during the purification procedure.

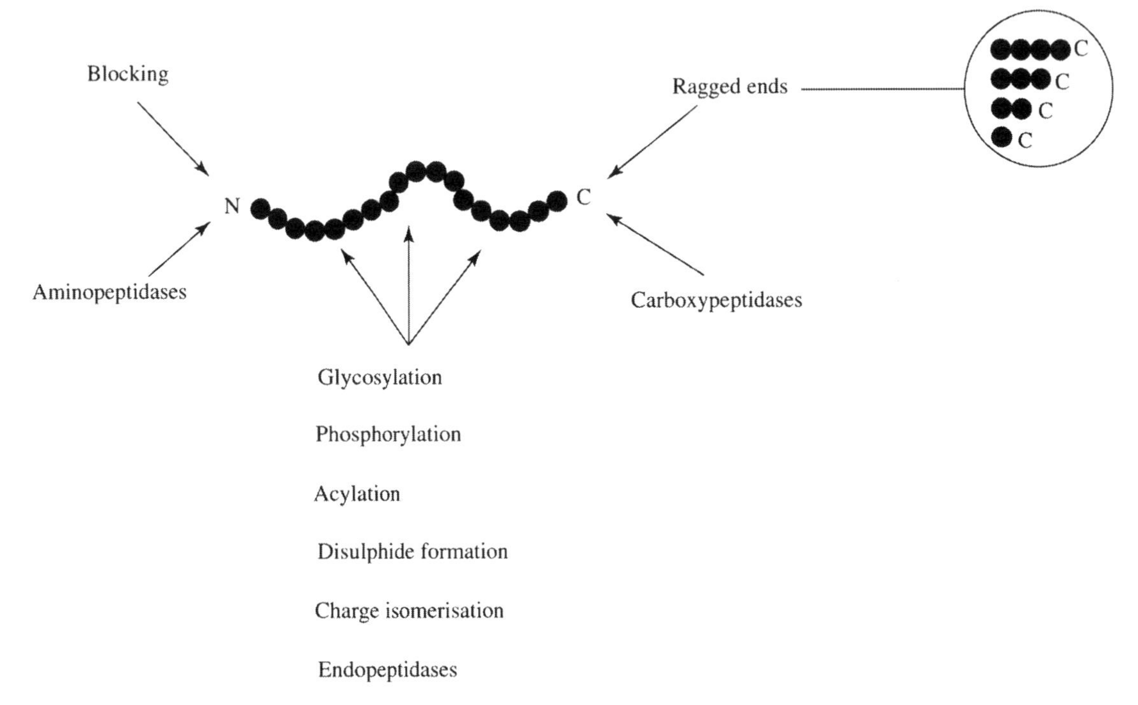

Figure 4.26. Some sources of covalent modification of polypeptides. Polypeptides may be covalently modified internally or at their termini. These modifications result in altered molecular mass detectable by MS.

Table 4.3. Some possible sources of heterogeneity within polypeptides

	Modification	Possible structures
N-terminus		
	Acylation	$-NH_2$ $-NH-CO-CH_3$
	N-formylation	$-NH_2$ $-NH-CO-H$
	N-formyl methionine	$-NH_2$ 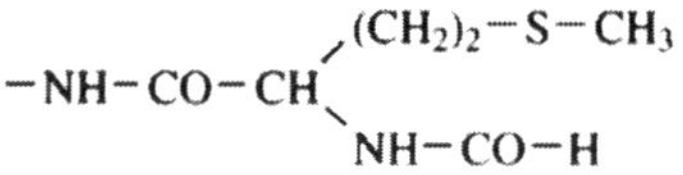 $-NH-CO-CH(-(CH_2)_2-S-CH_3)(-NH-CO-H)$
	Pyrollidone carboxylic acid (due to cyclisation of glutamine)	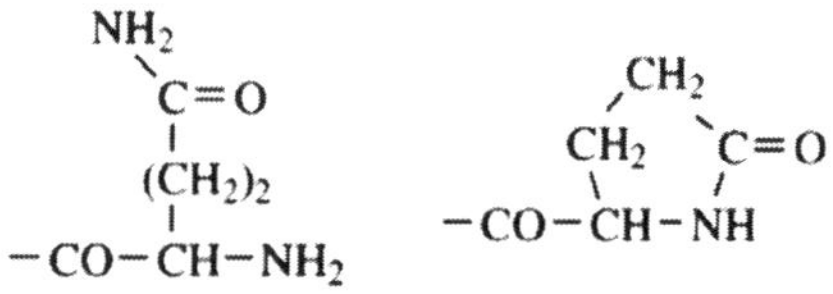 $-CO-CH(-(CH_2)_2-C(=O)-NH_2)-NH_2$ $-CO-CH-NH$ (ring: CH_2, CH_2, $C=O$)
	Aminopeptidases and aminoproteinases[d]	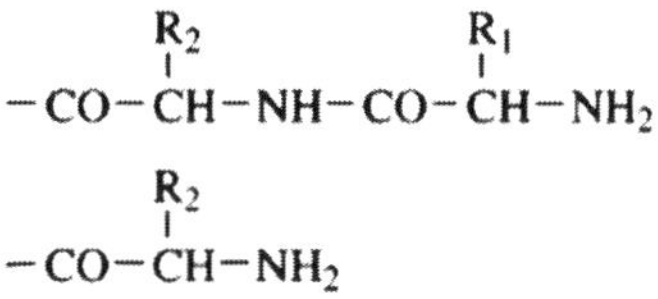 $-CO-CH(R_2)-NH-CO-CH(R_1)-NH_2$ $-CO-CH(R_2)-NH_2$
Internal:		
	Glycosylation[a]	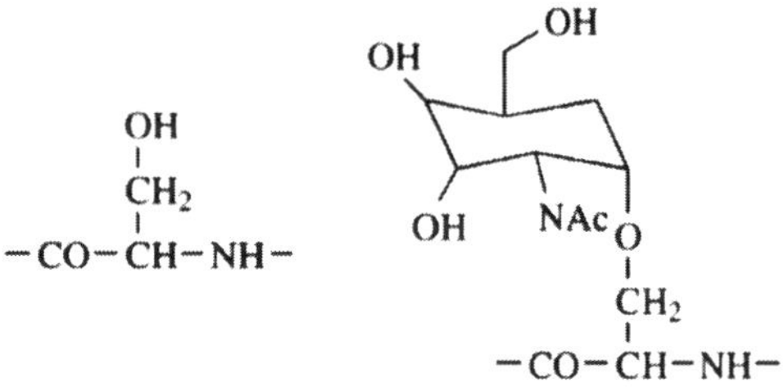 $-CO-CH(-CH_2-OH)-NH-$ $-CO-CH(-CH_2-O-\text{sugar (OH, OH, OH, NAc)})-NH-$
	Phosphorylation[a]	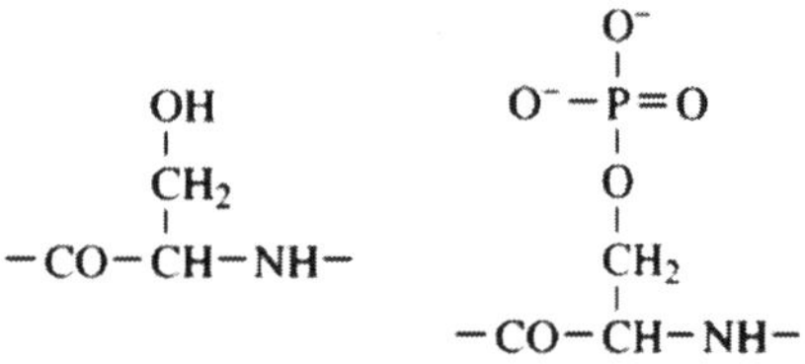 $-CO-CH(-CH_2-OH)-NH-$ $-CO-CH(-CH_2-O-P(=O)(O^-)-O^-)-NH-$
	Charge isomerisation[b] (e.g. deamidation)	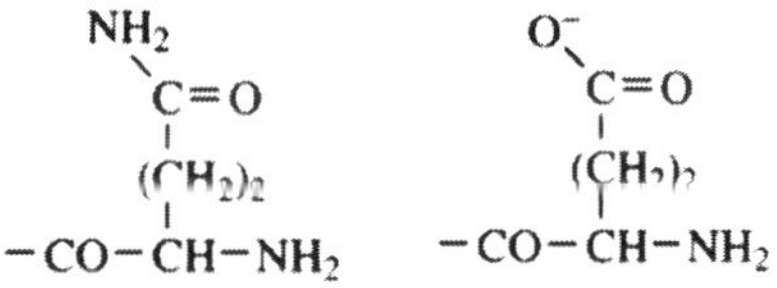 $-CO-CH(-(CH_2)_2-C(=O)-NH_2)-NH_2$ $-CO-CH(-(CH_2)_2-C(=O)-O^-)-NH_2$

(*Continued*)

Table 4.3. (*Continued*)

	Modification	Possible structures
	Disulphide formation[c]	NH / $CH-CH_2-SH$ / CO and NH / $SH-CH_2-CH$ / CO → NH / $CH-CH_2-S-S-CH_2-CH$ / CO (NH) (CO)
	Oxidation	CH_3–S–CH_2 on $-CO-CH-NH-$ **Methionine**; CH_3–S=O–CH_2 on $-CO-CH-NH-$ **Sulphoxide**; CH_3–O=S=O–CH_2 on $-CO-CH-NH-$ **Sulphone**
	Endopeptidases and endoproteinases[d]	$-CO-CH(R_2)-NH-CO-CH(R_1)-NH-$ → $-CO-CH(R_2)-NH_2$ $COOH-CH(R_1)-NH-$
C-terminus:	Carboxypeptidases[d]	$-NH-CH(R_2)-CO-NH-CH(R_1)-COOH$ → $-NH-CH(R_2)-COOH$ $NH_2-CH(R_1)-COOH$
	'ragged ends'	$-NH-CH(R_2)-CO-NH-CH(R_1)-COOH$ $-NH-CH(R_2)-COOH$

[a] Also possible at threonine as well as serine. Glycosylation is also possible at asparagine and lysine residues.
[b] Also possible at asparagine. Deamidation of asparagine and glutamine results in charge isomer formation as these amides are uncharged.
[c] MS of samples treated/untreated with reducing agents would be expected to give distinct patterns depending on number and location of disulphides.
[d] Exposure to these enzymes may occur artefactually during protein purification.

Sources of possible heterogeneity *within* polypeptide chains include glycosylation, phosphorylation, acylation, disulfide bond formation, charge isomerization and *endoprotease* activity. Some examples of these are described in more detail in the following sections.

Covalent modifications (Figure 4.26; Table 4.3) result in mass alterations measurable by MS Comparison of *m/z* values of variants of proteins allows us, therefore, readily to identify such modifications. This determination is independent of the chemical basis of heterogeneity and this is a major reason for the versatility of MS.

4.4.2 Confirmation and Analysis of Peptide Synthesis

Solid-phase peptide synthesis (Figure 4.27) has proven an invaluable tool for the generation of model peptides which can be used in a wide variety of studies (e.g. synthesis of epitopes for antibody preparation, design of novel antibiotics and identification of novel protein ligands in drug screening programs). The technique involves use of amino acids which are initially chemically protected or *blocked* at $-NH_2$ and other functional groups (e.g. ε-NH_2, –SH) while chemically *activated* at their COOH group. These amino acids will readily react with the exposed $-NH_2$ group of an amino acid immobilized on a solid phase, thus forming a peptide bond. De-blocking of the blocked $-NH_2$ group allows a second cycle of reaction leading to formation of a second peptide bond and, in this way, a polypeptide chain of 20–30 residues may be synthesized from the C-terminus. When synthesis is complete, the peptide is cleaved from the resin (exposing a C-terminal –COOH) and functional groups are *deblocked*. This procedure is readily automated and peptides may be routinely synthesized by adding commercially-available activated/blocked amino acid derivatives in the desired sequence.

Common problems which occur in peptide synthesis and which lead to heterogeneity in the final peptide product are the formation of *deletion peptides* and *incomplete deblocking*. Modifications specific to particular residues such as air-oxidation of methionine or trifluoroacetylation of serine are also possible. Deletion peptides arise due to incomplete peptide bond formation in a synthetic cycle, resulting in a proportion of peptide lacking the residue corresponding to that cycle. Since deletion peptides will lack a particular amino acid, they can be readily and sensitively identified by MS analysis. Incomplete deblocking can result in functional amino acid side-chains retaining their blocking groups, giving them an inappropriate covalent structure. Since the masses of blocking groups are known and vary between different amino acid side-chains, MS analysis allows identification of specific incomplete deblocking steps, facilitating their improvement in further rounds of synthesis, perhaps by increasing reaction times or by altering reaction conditions. MS analysis of a synthetic peptide is described in Example 4.2.

4.4.3 Peptide Mapping

Digestion of proteins with specific endoproteinases results in a unique population of peptides. Reversed phase HPLC using C-18 stationary phase (Section 2.4.3) may be used to achieve high-resolution separation of these. This is particularly useful for comparing related proteins such as haemoglobins or isoenzymes; where there is a degree of similarity in primary structure, we expect common peptides. However, in order to associate such peptides with specific regions of primary structure in the parent protein, we need to determine the individual sequence of each peptide (e.g. by the Edman degradation). MS provides a complementary method for analysis of peptide maps which, unlike HPLC analysis *per se*, gives molecular mass information about the peptides. Thus, it may be possible to associate a peptide with a particular part of the protein primary structure simply by mass comparison (Example 4.3) although care should be taken in such experiments to take the *isobaric* nature of some residues into consideration (e.g. Leu/Ile; Gln/Lys).

By interfacing HPLC with MS (Section 4.3.1) or by performing tandem MS, we can achieve a detailed analysis of peptide populations which can be used to identify cleavage points in the original sample protein.

4.4.4 Post-Translational Modification Analysis of Proteins

The post-translational modifications summarized in Figure 4.26 and Table 4.3 are important to protein function because they may be implicated in processes such as protein activation/inactivation (e.g. protein phosphorylation in signal transduction cascades involving G-proteins), recognition (e.g. glycosylation of receptors) or proteolysis (e.g. protein turnover). Chemical procedures for analysis of individual modifications such as glycosylation, acylation or phosphorylation exist. However, these techniques are laborious and only give information about a particular type of modification. The observation that all of the post-translational modifications of proteins result in altered mass makes MS particularly suitable for their analysis. MS analysis of *O*-glyosylation is described in Example 4.4.

4.4.5 Determination of Protein Disulfide Patterns

Several methods (including MS; *see* next section) are available for determination of amino acid sequences. However, the complete covalent structure of a protein is only

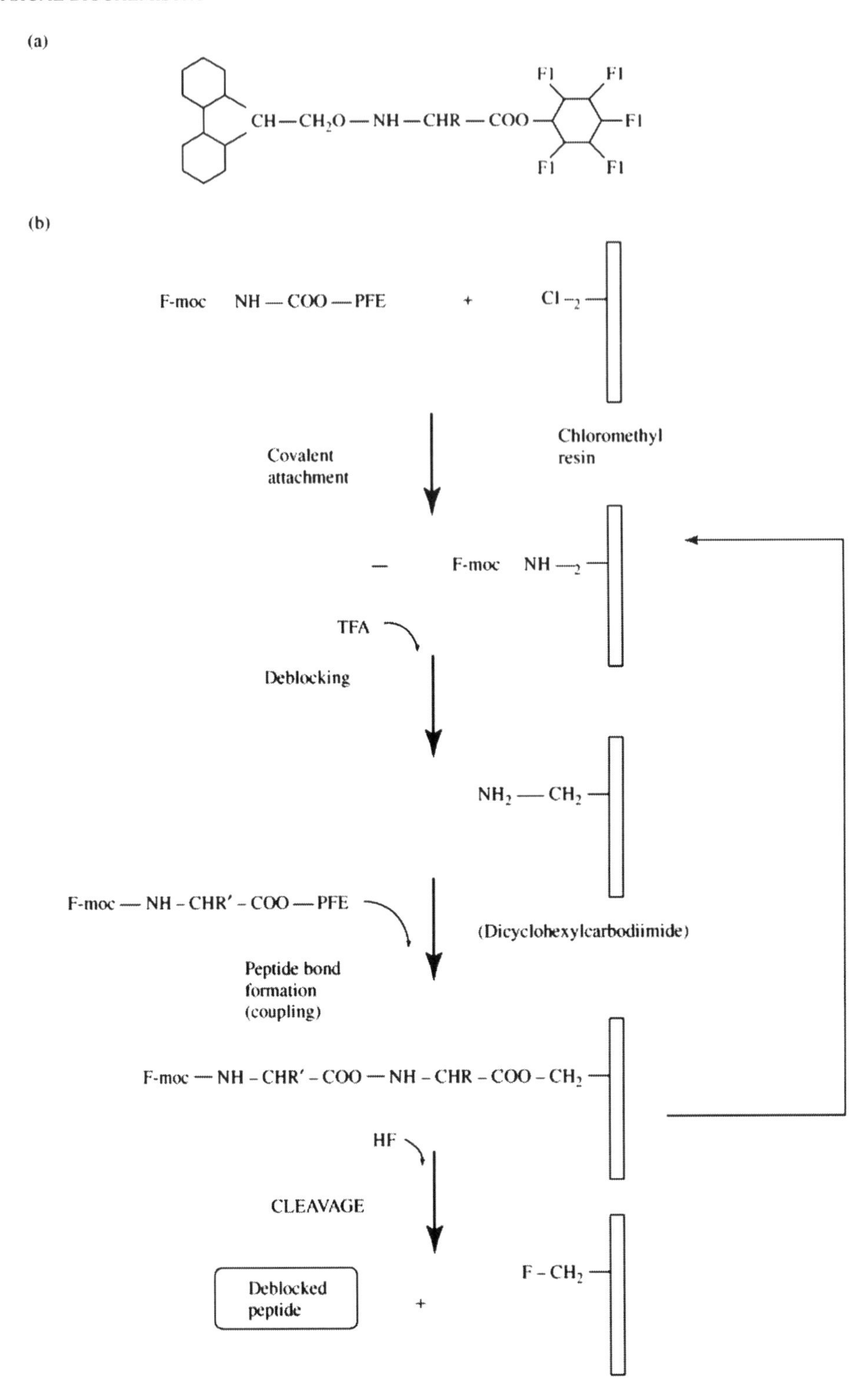

Figure 4.27. Solid phase peptide synthesis. (a) An amino acid blocked at the N-terminus (with a 9-fluoenylmethoxycarbonyl; F-moc group) and activated at the C-terminus (*via* a pentafluorophenyl ester). (b) Solid phase peptide synthesis procedure involving successive cycles of deblocking and coupling to extend the polypeptide chain from the C- to the N-terminus.

Example 4.2 Determination of microheterogeneity in synthetic peptides

A peptide was synthesized by the Merrifield method (Fmoc NH_2-blocking groups) with the following sequence:

NH_2-Ser-Asn-Tyr-Thr-Phe-Ser-Gln-Met-COOH
(predicted m/z; 977 Da).

EIS/MS was performed in a triple-quadropole instrument (Figure 4.12) and the result is shown below:

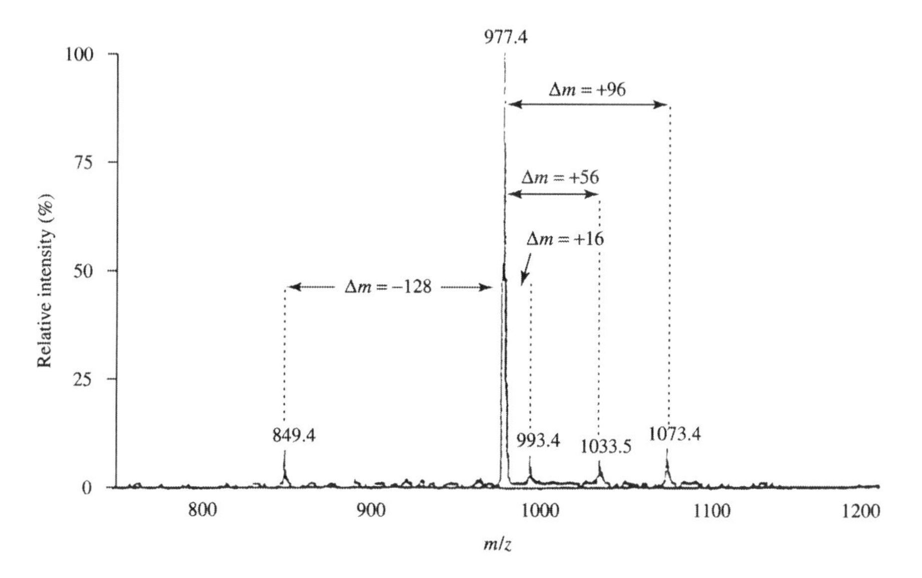

The M + H ion with a m/z of 977.4 is the major product but a number of minor peaks are also evident in the spectrum. Ions with m/z greater than this are attributed, respectively, to oxidation of methionine-8 (+16 AMU), a peptide retaining a tert-butyl blocking group (+56 AMU) and the presence of an O-trifluoroacetyl group on a serine or threonine (+96 AMU). The species with m/z of 849 is a deletion peptide lacking glutamine-7 (−128 AMU). More detailed analysis of peptides is possible by tandem MS including, for example, unique identification of the O-trifluoroacetylated residue. Of the three possible sites of trifluoroacetylation in this example (2 serines, 1 threonine), this residue was identified as serine-1 by tandem MS.

Taken from Metzger et al., Analytical Biochemistry (1994) **219**, 261–277, reproduced with permission from Academic Press, Inc.

known when the number and location of disulfide bridges involving cysteine residues have been determined. This is especially important for proteins normally secreted from their cells of origin (e.g. hormones, immunoglobulins). Disulfide bridges are used to stabilize otherwise unstable structures and also represent important intermediates in the folding pathways of such proteins. In addition, the disulfide bridged polypeptide is an oxidized form of the protein which is appropriate to the oxidizing nature of the extracellular environment. Indeed, in protein engineering, disulfide bridges are frequently introduced into proteins to enhance their stability to heat (e.g. in the design of enzymes suitable for enantioselective synthesis).

The sulphydryl groups of cysteine are among the most chemically-reactive side-chains found in proteins. Advantage may be taken of this to alkylate or oxidize free sulphydryls or to reduce disulfides to sulphydryls (Figure 4.28). Chemical treatments of this type are often used (in association with proteolytic digestion) to determine the pattern of disulfide bridge formation in proteins by MS (Example 4.5).

Digestion of protein samples of known sequence with agents of high specificity generates a characteristic mixture of peptides with easily predicted m/z values on FAB/MS The presence of disulfide bridges results in one or more unexpectedly large ions (Figure 4.29). Reduction prior to digestion leads to the disappearance of these large ions and

Example 4.3 FAB/MS analysis of peptide maps

A novel haemoglobin mutant was digested with trypsin generating a population of peptides. One of these peptides was purified and subjected to FAB/MS (A below). Instead of the expected m/z of 1833.9 found with wild-type α-globin, a new peptide with m/z of 1634.8 was detected. This is consistent with a glutamine $\rightarrow$ arginine mutation at position 54. This conclusion was confirmed by digestion with a second endoproteinase, lysylendopeptidase (Lys-C). This enzyme does not cleave at arginyl peptide bonds, unlike trypsin. The FAB/MS spectrum for this digest (B below) shows an ion with m/z of 1862, an increase of 28 AMU on the value of 1833.9 obtained for the wild type protein. This corresponds to the mass difference between glutamine and arginine.

(Taken from Structural Analysis of Protein Variants, Yoshido Wada in Protein and Peptide Analysis by Mass Spectrometry (J. Chapman ed.), 1996, Human Press, reproduced with permission.

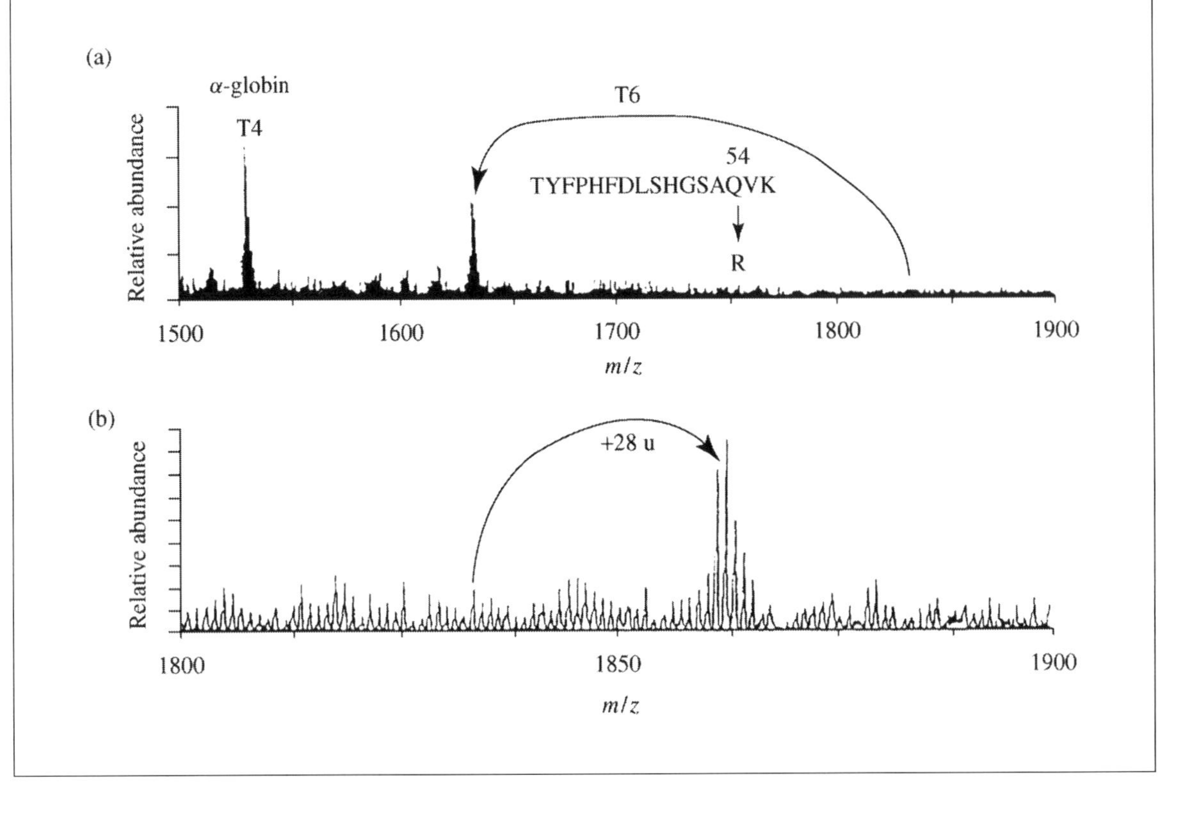

the appearance of pairs of smaller ions in the spectrum. In this way, disulfides can be identified in proteins and allocated to regions of primary structure based on *m/z* values. The total number of free sulphydryl groups and disulfide bonds in the protein may be determined by comparison of *m/z* of alkylated samples of both reduced and nonreduced protein, while oxidation with performic acid results in a characteristic increase in the mass of cysteine (and methionine)-containing peptides (Example 4.5).

Similar experiments to these can be performed using reversed phase C-18 HPLC (Section 2.4.3) but MS has the advantage that *m/z* data are measured in the experiment which may be correlated directly with the protein primary structure. In HPLC, it is necessary first to collect individual peptides before chemically determining their sequences.

4.4.6 Protein Sequencing by MS

The main strategies for obtaining protein sequence information in biochemistry are cDNA sequencing and direct chemical sequencing by the Edman degradation. The former method has the limitation that it tells us nothing about post-translational modifications of the protein such as those described in Section 4.4.4 while the latter method is only applicable to protein samples with unblocked $-NH_2$ groups and is generally limited to short sequences (e.g. <40 residues).

Example 4.4 Determination of erythropoietin glycosylation pattern by FAB/MS

Erythropoietin is a 165-residue protein hormone produced in the kidney and liver which plays a role as a physiological modulator of erythrocyte production. FAB/MS previously revealed that removal of arginine-166 was essential for activation of recombinant erythropoietin. The active hormone is extensively glycosylated and this is the basis of isoforms of the protein. When digested with lysylendopeptidase (Lys-C), a glycosylated peptide is released. FAB/MS analysis revealed the following mass spectrum:

The species with m/z 2500.6 corresponds to the following unglycosylated erythropoietin sequence;

NH_2-Glu-Ala-Ile-Ser-Pro-Pro-Asp-Ala-Ala-Ser-Ala-Ala-Pro-Leu-Arg-Thr-Ile-Thr-Ala-Asp-Thr-Phe-Arg-Lys-COOH

The pattern of glycosylation explaining the other ions may be described diagrammatically as follows:

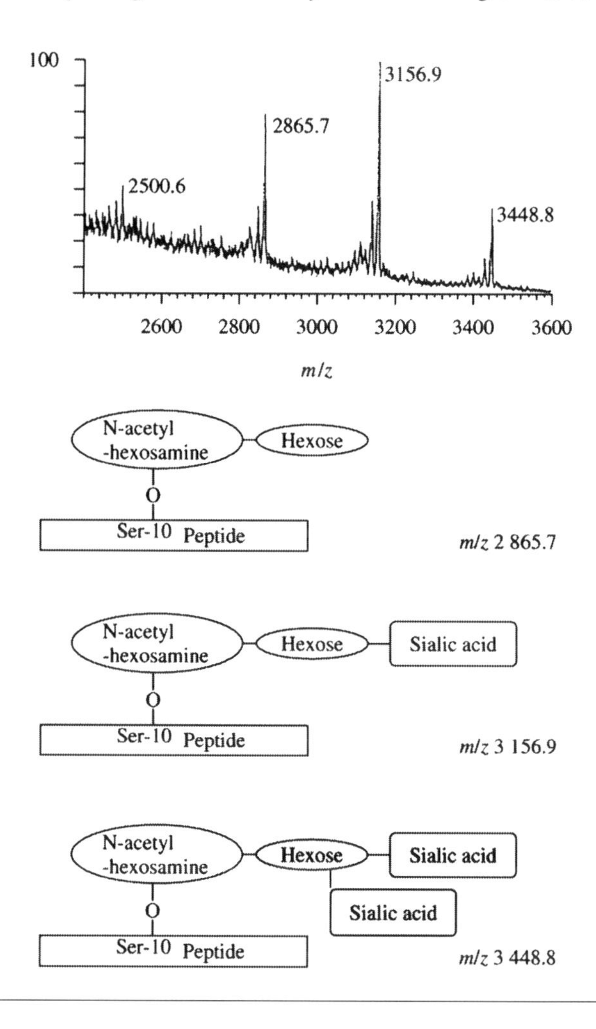

We have already seen that polypeptides fragment into sets of related ions during some MS procedures for which accurate *m/z* values can be recorded (Section 4.2.3). Comparison of such ions allows identification of short sequences of proteins and peptides. *Identities* of amino acids may be determined by measuring differences in *m/z* between consecutive peaks in the mass spectrum while the *sequence* is determined by the order of the peaks. Information on post-translational modifications can also be detected as *m/z* values of modified residues including the mass of modifying

(i) $\text{SH} + \text{I-CH}_2\text{—COOH} \longrightarrow \text{S—CH}_2\text{—COOH}$

(ii) $\text{S—S} + 2\,(\text{HS—(CH}_2)_2\text{—OH}) \longrightarrow \text{S—S—(CH}_2)_2\text{—OH} + \text{S—S—(CH}_2)_2\text{—OH}$

(iii) S—S / SH $+ \text{CHO—OOH} \longrightarrow SO_3^-$

Cysteic acid

Figure 4.28. Chemical reactions of cysteine. (i) Alkylation with iodoacetic acid, (ii) Reduction of disulfide bridge with 2-mercaptoethanol, (iii) Oxidation with performic acid (note; this treatment also oxidizes methionine to the corresponding sulphone).

groups (e.g. +80 Da for phosphate attached to a serine or threonine; *see* Example 4.6).

Due to the complexity of ion patterns generated in MS of peptides/proteins, direct *de novo* sequencing is a very challenging undertaking. However, when combined with other experimental approaches, MS can be very informative. Usually, it is carried out on samples which have been digested with a number of specific cleavage agents such as proteases before MS analysis.

Combining some elements of the Edman degradation with MALDI/TOF MS is called *protein ladder sequencing* (Figure 4.30). In this method, fragments of the protein to be sequenced are generated by partial Edman degradation which are then subjected to MALDI/MS (Example 4.6). The fragments may be generated in an automated protein sequencer by incomplete reaction with phenylisothiocyanate (PITC) or by the inclusion of a small amount of a *terminating agent*. For example if 5% phenylisocyanate is included with PITC, a small amount of phenylcarbamyl-peptide is generated in each cycle of the Edman degradation which is stable to subsequent processing. These fragments may be analysed after the desired number of sequencing cycles in a *single* MALDI/MS experiment. The mass spectrum resembles a ladder with the spaces between each 'rung' corresponding to the mass of a particular residue. It should be noted that this technique cannot distinguish between leucine/isoleucine which have identical mass (Section 4.2.3) but glutamine and lysine (which also have the same mass) may be readily distinguished since the ε-NH_2 group of lysine reacts with PITC to give a mass larger by 135 Da, while glutamine does not react with this reagent.

Example 4.5 Identification of disulfide bridges by FAB/MS

Vasopressin is a peptide hormone of the sequence Cys-Tyr-Phe-Gln-Asn-Cys-Pro-Arg-Gly, which contains a single disulfide bridge between Cysteines-1 and -6. FAB/MS of A) untreated and B) oxidized (i.e. treated with performic acid) vasopressin are shown below. The presence of a cysteic acid at positions 1 and 6 in the sequence results in a characteristic m/z increase of 98 AMU. This treatment can be used to estimate the total number of cysteines in a peptide or protein sample. Sun & Smith (1988) Analytical Biochemistry **172**, 130–138. Academic Press, Inc, reproduced with permission.

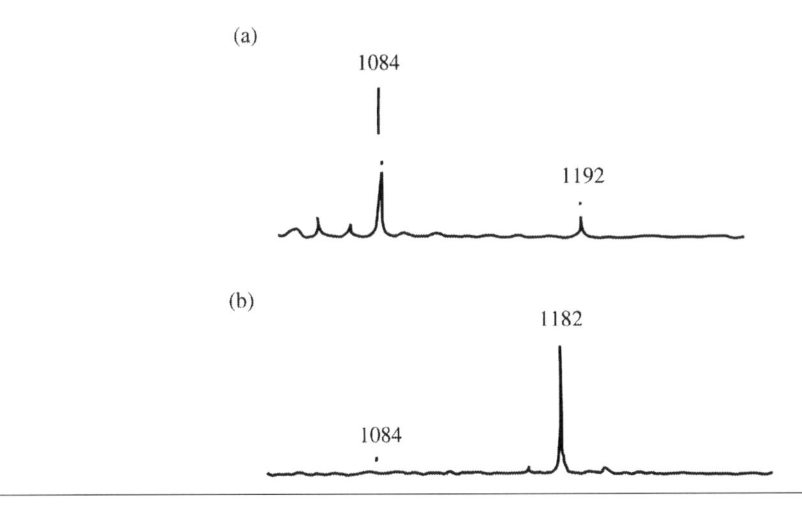

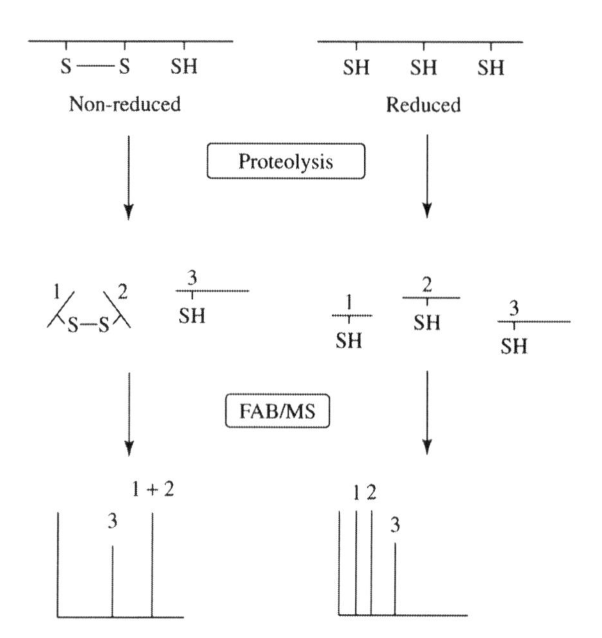

Figure 4.29. Identification of intrachain disulfide bridges in proteins. Based on the known sequence of the protein, three peptides are expected from proteolytic cleavage. One of the peptides (3) has the predicted *m/z*, while a single, unexpectedly large ion is also obtained. On reduction, this large ion disappears and is replaced with two smaller ions of predicted *m/z*. The precise location of the intrachain disulfide between cysteines of peptides 1 and 2 may be determined from this experiment.

Sequences of up to 30 residues may be measured by protein ladder sequencing which is similar to the upper limit of the Edman degradation. Since this method generates *structural* information (e.g. identification of blocking or modifying groups) in addition to *sequence* information, it neatly complements cDNA and direct chemical sequencing.

4.4.7 Studies on Enzymes

Enzymes achieve very high rate enhancements (the ratio of the reaction rate in presence of enzyme to that in its absence) which enable the very diverse spectrum of chemical reactions found in living systems to occur under relatively mild chemical conditions of pH, temperature and pressure. Enzymes bind substrates to form an enzyme-substrate complex within which a chemical reaction happens converting the substrate to product. A key component of catalysis is stabilization of a reaction intermediate called the *transition state*. This is the least stable species in the reaction pathway which has a vanishingly short half-life ($\sim 10^{-13}$ s). Since transition states are generally highly specific to particular reaction types, there is much interest in understanding more about them and molecules structurally resembling transition states are promising candidates as new drugs. Very high resolution MS methods such as fourier transform ion cyclotron resonance MS can detect tiny mass differences in enzymes and their complexes giving powerful insights into enzymatic catalysis (Example 4.7).

4.4.8 Analysis of DNA Components

Because MS is simply concerned with the measurement of *m/z* values, it is a particularly versatile technique for the analysis of nonprotein biomolecules. MS has been applied to the analysis of carbohydrates, nucleic acids and heterogeneous molecules such as proteoglycans. DNA is of particular interest in this regard since modest changes in the covalent structure of DNA components can result

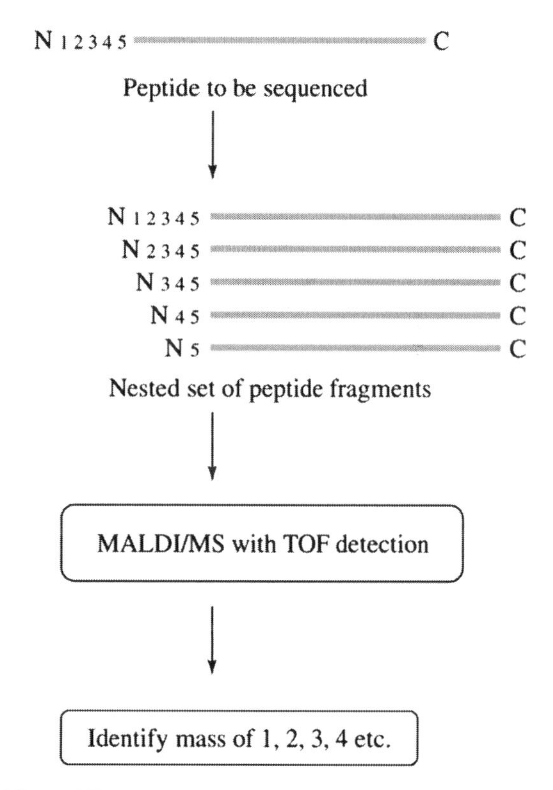

Figure 4.30. Protein ladder sequencing by partial Edman degradation. The Edman degradation results in step-wise removal of residues at the N-terminus. This generates a nested series of fragments, with consecutive fragments differing from each other by a single residue (i.e. 1–5). This series is analysed in a single MALDI/MS experiment and the 'ladder' of fragments is interpreted as an amino acid sequence.

Example 4.6 Protein ladder sequencing by MALDI/MS

A peptide which is a substrate for protein kinase C was subjected to partial Edman degradation as outlined in Figure 4.30. The MALDI/MS spectra obtained from A) nonphosphorylated and B) phosphorylated peptides are shown below.

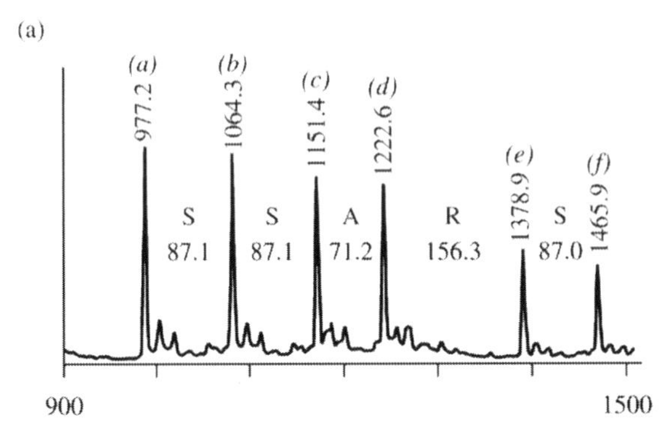

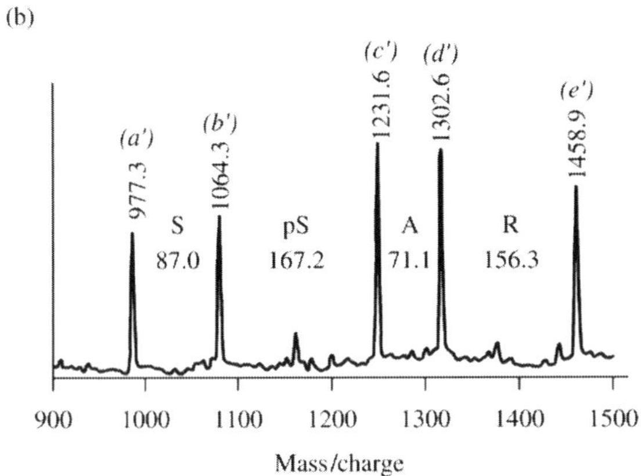

These spectra could be interpreted by measuring incremental mass differences between successive peaks a–f and a′–e′, respectively (e.g. 1064.3 − 977.2 = 87.1, the mass of a Ser residue). The following sequence was read out for the nonphosphorylated peptide; Ser-Ser-Ala-Arg-Ser. Ser-2 in this sequence is the site of phosphorylation by protein kinase C. Note the mass shift in the spectrum at the fragment corresponding to the phosphorylation site (i.e. m/z of ions c′, d′ and e′ are greater by 80.1 Da than corresponding ions c-e in A, due to the mass of the phosphate group).

in serious biological consequences. For example, many mutagens cause their cytotoxic effects by forming covalent adducts with DNA bases. ESI/MS techniques have been extensively applied to the characterization of such DNA-chemical adducts (Example 4.8). Other applications of MS to the study of DNA include the study of noncovalent binding to regulatory DNA sequences and duplex formation in oligonucleotides. A particular problem which may be observed with DNA samples is the presence of large amounts of sodium ions complexed to phosphate groups. These can be efficiently removed by use of on-line microdialysis prior to MS.

Example 4.7 Detection of a single water molecule in the transition state complex of cytidine deaminase

Cytidine deaminase (CDA) removes an amino group from cytidine to form uridine. This is a key target in stopping rapidly dividing cells (e.g. bacteria, cancer cells) meeting their requirements for new DNA. The enzyme is a dimer containing a single active site per subunit and requires a Zn atom at each site for addition of water to cytidine. Analogues of cytidine lacking $-NH_2$ bind tightly to CDA and are thought to be transition state analogues. One such analogue, 5-fluoro-3, 4-dihydrouridine (FDHU) is formed by addition of water to an inhibitor of the enzyme, 5-fluoropyrimidine-2-one ribonucleoside (FZeb), catalyzed by CDA:

A → (H₂O) → B

$C_9H_{11}FN_2O_5$ (FZeb) exact mass 246.07

$C_9H_{13}FN_2O_6$ (FDHU) exact mass 264.08

It can be predicted that the missing $-NH_2$ group in transition state analogues of CDA might leave a 'hole' in the complex formed. It is not known if this is filled by something else or is, perhaps, removed by a structural rearrangement within the protein. In this experiment CDA exposed to FZeb was subjected to fourier transform ion cyclotron resonance MS using electrospray ionization and the mass of the complex accurately confirmed:

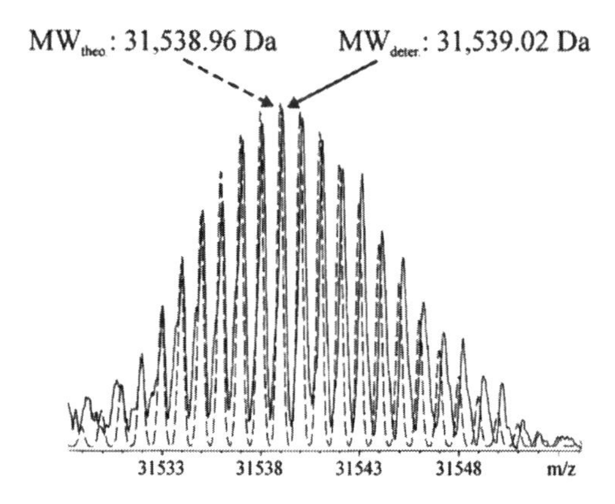

This showed that the mass of the dimer-FZeb complex observed (MW_{deter}) was within 2 ppb of that theoretically predicted from atomic composition (MW_{theor}) for the enzyme dimer plus two Zn atoms plus two FZeb molecules. The very high-resolution mass determination made possible by the fourier transform ion cyclotron detector allowed the conclusion that CDA was essentially in its native structure in the gas phase.

(Continues)

(Continued)

Careful analysis of deconvoluted CDA-FZeb mass spectra revealed three peaks separated by ~264 Da, the mass of FDHU, which were not found in spectra of the protein alone:

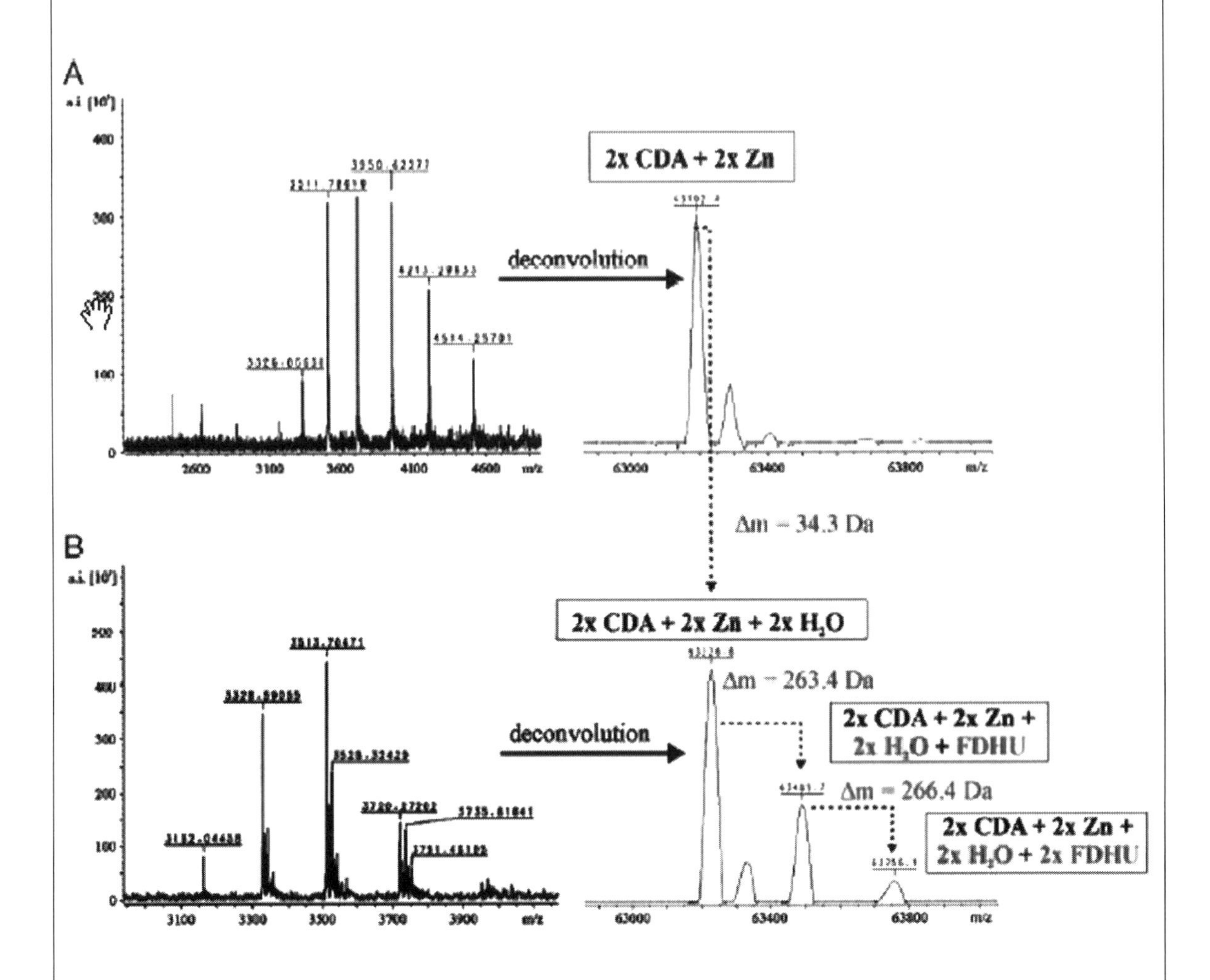

This suggested that CDA had converted FZeb into FDHU (confirming its native status). The data are interpretable in terms of a CDA-FDHU complex containing two Zn atoms and two water molecules per CDA dimer. It was concluded that a single water molecule occupies the niche vacated by the absent $-NH_2$ and thus contributes to stabilization of the complex between cytidine deaminase and FDHU. In the absence of FZeb, water is lost to the vacuum and thus does not feature in the MS of CDA alone.

(Reproduced with permission from Borchers *et al.* (2004), Fourier transform ion cyclotron resonance MS reveals the presence of a water molecule in an enzyme-transition state analogue complex. Proceedings of the National Academy of Sciences USA **101**, 15341–15345)

Example 4.8 Characterization of 7-(2-hydroxyethyl) guanine from DNA by ESI/MS

A covalent adduct is formed between guanine of DNA and ethylene oxide, a commonly-used sterilant in synthetic chemistry.

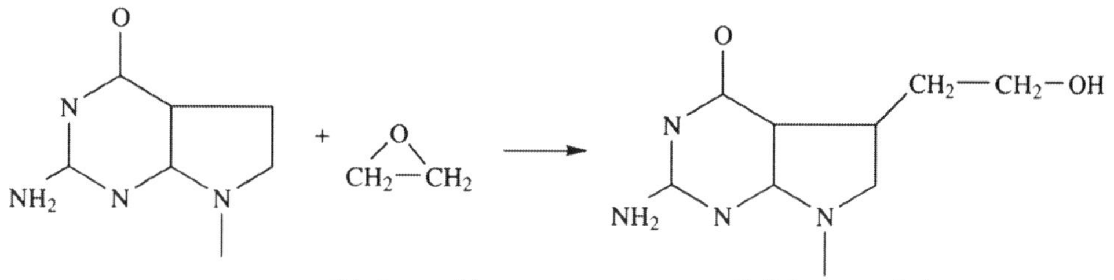

Guanine + Ethylene oxide → 7-(2-hydroxyethyl) guanine

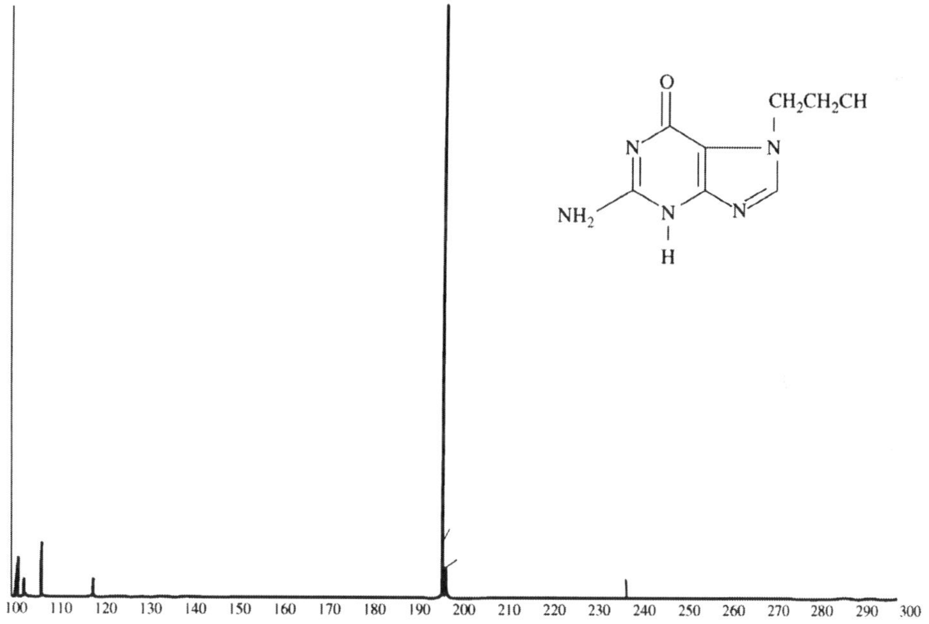

This adduct is known to be both genotoxic and carcinogenic in man and rodents. DNA was exposed to ethylene oxide in vitro and hydrolyzed to release base adducts. The sample was fractionated by C-18 reversed phase HPLC followed by ESI/MS. The spectrum obtained is shown below.

Using this HPLC/MS system, it was possible to detect hydroxyethylation of 1 in 10^6 DNA nucleotides. This type of approach could facilitate screening of exposed individuals and also screening for other types of DNA adducts.

REFERENCES

Basic theory of Mass Spectrometry

Dongre, A.R., Eng, J.K. and Yates, J.R. (1997). Emerging tandem-mass-spectrometry techniques for the rapid identification of proteins. *Trends in Biotechnology*, **15**, 418–25. An excellent overview of tandem MS.

Graham, R.L., Graham, C. and McMullan G. (2007) Microbial proteomics: A mass spectrometry primer for biologists. *Microbial Cell Factories*, **6**, article no 26. (available online at; http://www.microbialcellfactories.com/content/6/1/26). A good general introduction to MS.

Jonscher, K.R. and Yates J.R. (1997). The quadropole ion trap mass spectrometer – a small solution to a big challenge. *Analytical Biochemistry*, **245**, 1–15. A review of ion trap MS.

Mass Spectrometry of proteins/peptides

Asara, J.M., Schweitzer, M.H., Freimark, L.M. *et al.* (2007) Protein sequences from Mastodon and *Tyrannosaurus rex* revealed by mass spectrometry. *Science*, **316**, 280–5. A fascinating application of MS to study of fossil samples.

Borchers, C.H., Marquez, V.E., Schroeder, G.K. *et al.* (2004), Fourier transform ion cyclotron resonance MS reveals the presence of a water molecule in an enzyme-transition state analogue complex. *Proceedings of the National Academy of Sciences USA*, **101**, 15341–5. An excellent example of interpretations made possible by the high resolution of FTICP MS detectors.

Chait, B.T. (2007) Mass spectrometry: Bottom-up or top-down? *Science*, **314**, 65–6. A though-provoking commentary on new approaches to protein MS.

Cooper, H.J., Håkansson, K. and Marshall, A.G. (2005). The role of electron capture dissociation in biomolecular analysis. *Mass Spectrometry Reviews*, **24**, 201–22. A review of ECD.

Domon, B. and Aebersold, R. (2006) Mass spectrometry and protein analysis. *Science*, **312**, 212-7. A general review on protein MS.

Fricker, L.D., Lim, J.Y., Pan, H. and Che, F.Y. (2006) Peptidomics: Identification and quantification of endogenous peptides in neuroendocrine tissues. *Mass Spectrometry Reviews*, **25**, 327–44. A review of use of MS in peptide profiling (peptidomics) of the neuroendocrine system.

Han, X.M., Jin, M., Breuker, K. and McLaffert, F.W. (2006) Extending top-down mass spectrometry to proteins with masses greater than 200 kilodaltons. *Science*, **314**, 109–12. Describes new MS conditions for fragmenting proteins in the MS instrument which increases the protein Mr limit of the technique.

Mann, M. and Wilm, M. (1995) Electrospray mass spectrometry for protein characterization. *Trends in Biochemical Sciences*, **20**, 219–24. A review of the application of electrospray MS to the study of proteins.

McLafferty, F.W., Fridricksson, E.K., Horn, D.M. *et al.* (1999) Biomolecule mass spectrometry. *Science*, **284**, 1289–90. A brief overview of potential of MS in protein characterization.

Mikesh, L.M., Ueberheide, B., Chi, A. *et al.* (2006) The utility of ETD mass spectrometry in proteomic analysis. *Bichimica et Biophysica Acta – Proteins and Proteomics*, **1764**, 1811–22. A review of electron transfer dissociation in MS of proteins.

Nakazawa, T., Yamaguchi, M., Okamura, T. *et al.* (2008) Terminal proteomics: N- and C-terminal analyses for high-fidelity identification of proteins using MS. *Proteomics*, **8**, 673–85. A review of the potential of N- and C-terminal sequencing by MS in protein identification.

Scoble, H.A. (1989) Characterization of peptides by fast atom bombardment mass spectrometry and high performance tandem mass spectrometry, in *A Practical Guide to Protein and Peptide Purification for microsequencing* (ed. P. Matsudaira), Academic Press, London, pp. 89–109.

Spickett, C.M., Pitt, A.R., Morrice, N. and Kolch, W. (2006) Proteomic analysis of phosphorylation, oxidation and nitrosylation in signal transduction. *Biochimica et Biophysica Acta (BBA) – Proteins and Proteomics*, **1764**, 1823–41. Review on use of MS in detection of post-translational modifications.

Sun, Y. and Smith, D.L. (1988). Identification of disulfide-containing peptides by performic acid oxidation and mass spectrometry. *Analytical Biochemistry*, **172**, 130–8.

Tamvakopoulos, C. (2007) Mass spectrometry for the quantification of bioactive peptides in biological fluids. *Mass Spectrometry Reviews*, **26**, 389–402. A good review of the issues involved in MS analysis of biofluids.

Interfaced Mass Spectrometry

Cohen, S.L. and Chait, B.T. (1997) Mass spectrometry of whole proteins eluted from sodium dodecyl sulfate - polyacrylamide gel electrophoresis gels. *Analytical Biochemistry*, **247**, 257–67.

Krone, J.R., Nelson, R.W., Dogruel, D. *et al.* (1997) BIA/MS: Interfacing biomolecular interaction analysis with mass spectrometry. *Analytical Biochemistry*, **244**, 124–32.

Kelly, J.F., Ramaley, L. and Thibault, P. (1997) Capillary zone electrophoresis – electrospray mass spectrometry at submicroliter flow rates : Practical considerations and analytical performance. *Analytical Chemistry*, **69**, 51–60.

Norwood, D.L., Mullis, J.O. and Feinberg, T.N. (2007) 7 Hyphenated techniques. *Separation Science and Technology*, **8**, 189–235. A critical overview of LC-MS.

Qian, W.J., Jakobs, J.M., Liu, T. *et al.* (2006) Advances and challenges in liquid chromatography-mass spectrometry-based proteomics profilimg for clinical applications. *Molecular and Cellular Proteomics*, **5**, 1727–44. A review of LC-MS.

Sonke, J.E. and Salters, V.J.M. (2007) Capillary electrophoresis-high resolution sector field inductively coupled plasma mass spectrometry. *Journal of Chromatography Part A*, **1159**, 63–74. A review of CE-MS.

Non-protein Mass Spectrometry

Ball, R.W. and Packman, L.C. (1997) Matrix-assisted laser desorption ionization time-of-flight mass spectrometry as a rapid quality control method in oligonucleotide synthesis. *Analytical Biochemistry*, **246**, 185–94.

Bedair, M. and Sumner, L.W. (2008) Current and emerging mass-spectrometry technologies for metabolomics. *TrAC Trends in*

Analytical Chemistry, **27**, 238–250. Review on potential of MS in metabolomics.

Thomas, B. and Akouilitchev, A.V. (2006) Mass spectrometry of RNA. *Trends in Biochemical Sciences*, **31**, 173–81. MS as applied to RNA.

Useful web sites

John Wiley & Sons maintain an MS web-site: http://basepeak.wiley.com.

Virtual Mass spectrometry laboratory hosted by Carnegie-Mellon University and University of Pittsburgh: http://Massspec.chem.cmu.edu/VMSL/.

Chapter 5
Electrophoresis

Objectives

After completing this chapter you should understand:

- *The* **physical basis** *of electrophoresis.*
- *The* **range** *of electrophoresis formats available.*
- *How electrophoresis can be* **interfaced** *with other high-resolution techniques.*
- *Applications of electrophoresis to the study of biomolecules such as proteins, peptides and DNA.*

Positive or negative electrical charges are frequently associated with biomolecules. These can be important for *structural* reasons (e.g. ionic interactions leading to the formation of salt bridges in proteins) and we have already seen how they may be taken advantage of in *bioseparation* procedures (e.g. ion exchange chromatography; Section 2.4.1). When placed in an electric field, charged biomolecules move towards the electrode of opposite charge due to the phenomenon of *electrostatic attraction. Electrophoresis* is the separation of charged molecules in an applied electric field. The relative *mobility* of individual molecules depends on several factors the most important of which are net charge, charge/mass ratio, molecular shape and the temperature, porosity and viscosity of the matrix through which the molecule migrates. Complex mixtures can be separated to very high resolution by this process. A wide variety of electrophoretic formats are available for analytical, preparative and functional experiments in biochemistry. This chapter describes the physical basis and principal applications of electrophoresis. Electrophoresis applications to proteomics are described in Chapter 10.

5.1 PRINCIPLES OF ELECTROPHORESIS

5.1.1 Physical Basis

If a mixture of electrically-charged biomolecules is placed in an electric field of *field strength* E, they will freely move towards the electrode of opposite charge. However, different molecules will move at quite different and individual rates depending on the physical characteristics of the molecule and on the experimental system used (Figure 5.1). The velocity of movement, v, of a charged molecule in an electric field depends on variables described by Equation (5.1);

$$v = \frac{E \cdot q}{f} \tag{5.1}$$

where f is the *frictional coefficient* and q is the *net charge* on the molecule. The frictional coefficient describes frictional resistance to mobility and depends on a number of factors such as the mass of the molecule, its degree of compactness, buffer viscosity and the porosity of the matrix in which the experiment is performed. The net charge is determined by the number of positive and negative charges in the molecule. Charges are conferred on proteins by charged amino acid side-chains as well as by groups arising from post-translational modifications such as deamidation, acylation or phosphorylation. DNA has a particularly uniform charge-distribution since a phosphate group confers a single negative charge per nucleotide. Equation (5.1) means that, in general, molecules will move *faster* as their net charge increases, the electric field strengthens and as f decreases (which is a function of molecular mass/shape). Molecules of similar net charge separate due to differences in frictional coefficient while molecules of similar mass/shape may differ widely from each other in net charge. Consequently, it is often possible to achieve very high resolution separation by electrophoresis.

In practice, an electric field is established by applying a voltage, V, to a pair of electrodes separated by a distance, d (Figure 5.1). This results in the electrical field of strength E;

$$E = \frac{V}{d} \tag{5.2}$$

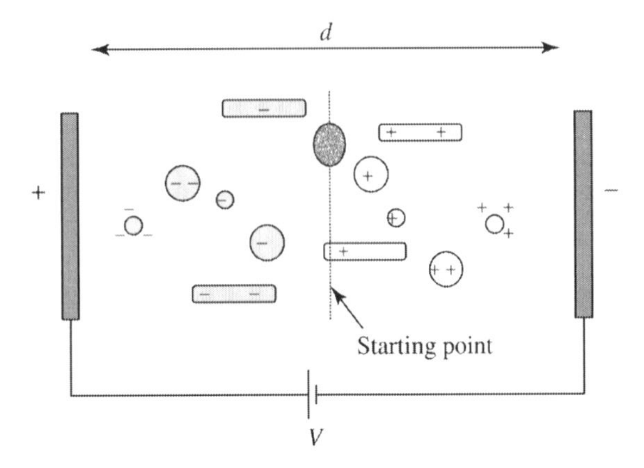

Figure 5.1. Physical basis of electrophoresis. Molecules move in an electric field of field strength E depending on their net charge, molecular mass and shape. Note that smaller and more charged molecules move more rapidly towards the electrode of opposite charge and that molecules of different shape but identical net charge have different electrophoretic mobilities.

The bulk of the current is carried between the electrodes by charged components of the buffer which also perform the function of maintaining a constant pH. This is especially important in electrophoresis since pH changes during the experiment would result in altered net charge and hence altered mobility. The most commonly-used buffer systems in electrophoresis of biomolecules are Tris-Cl or Tris-glycine. Buffers are held in reservoirs connected to each electrode and provide a constant supply of ions to the electrophoresis system throughout the separation.

Ohm's law relates V to current, I, by electrical resistance, R, as follows;

$$V = R \cdot I \tag{5.3}$$

We might predict that increasing V would result in much faster migration of molecules due to greater current. However, experimentally we find that large voltages result in significant *power* generation which is mainly dissipated in the form of heat. The power (in Watts) generated during electrophoresis is given by Equation (5.4).

$$W = I^2 \cdot R \tag{5.4}$$

Heat generation is undesirable because it leads to loss of resolution (due to convection of buffer causing mixing of separated samples), a decrease of buffer viscosity (which also decreases R) and, in extreme cases, structural breakdown of thermally-labile samples such as proteins and nucleic acids. A decrease in R means that, under conditions of *constant voltage*, I will increase during electrophoresis which, in turn, leads to further heat generation (Equation (5.3)). In practice, *constant voltage* conditions are used in most electrophoresis experiments but, for certain applications, a *constant power* supply may be used. A constant power supply allows V to change during the experiment to maintain a constant value of W (Equation (5.4)). In general, electrical conditions are selected which are adequate to separate samples in a reasonable time-frame but which will not cause extensive heating.

Because the electrical field strength, E, may vary widely between different experimental formats, the *electrophoretic mobility*, μ, of a sample is defined by Equation (5.5);

$$\mu = \frac{v}{E} \tag{5.5}$$

Combining this with Equation (5.1) shows that;

$$\mu = \frac{E \cdot q}{E \cdot f} = \frac{q}{f} \tag{5.6}$$

That is, biomolecules migrate based on the ratio of net charge to frictional coefficient. Since f is strongly mass-dependent for classes of biopolymers of similar shape (e.g. globular proteins; linear DNA), differences in μ approximate closely to differences in charge/mass ratio.

This description is adequate to understand how net charge, mass and shape underlie the separation of molecules in electrophoresis. It should be noted, however, that it is an *incomplete* description since it does not include relevant factors such as possible interaction of the sample molecules with the support medium (e.g. gels; *see* next section), charge suppression on the surface of large biomolecules or differences due to the buffer composition used in the experiment. For this reason it is important to understand that electrophoresis, as used in most situations in biochemistry, is largely an *empirical* technique. High resolution mobility data can be obtained for biomolecules by comparison with standard molecules of similar charge-density and shape and such measurements have proved useful in all areas of biochemistry. However, it is not usually possible to make *direct* measurements (as compared to *comparative* measurements) of molecular mass or shape from electrophoretic mobility alone due to lack of detailed information on variables involved in the process. While most protein and nucleic acid molecules behave predictably in electrophoresis, there are many examples of molecules of different charge/mass comigrating or of similar charge/mass separating due to differences in their electrophoretic mobility.

5.1.2 Historical Development of Electrophoresis

Early electrophoresis experiments were performed in free solutions of aqueous buffers. In 1937, the Swedish

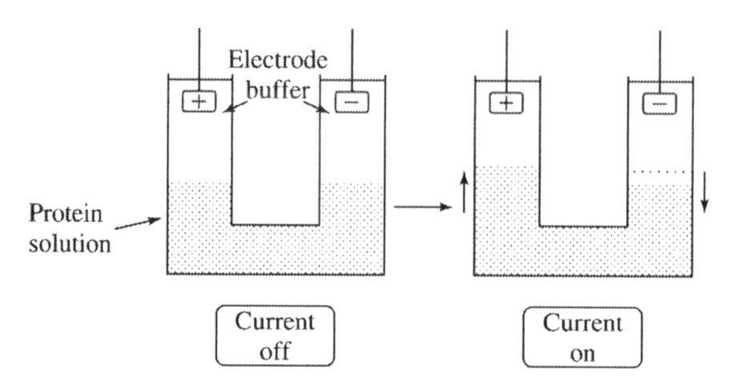

Figure 5.2. Moving boundary apparatus of Tiselius. When current is turned on, a solution of negatively charged protein moves towards the anode. Interfaces between the electrode buffer and protein solution form ascending and descending boundaries as shown by arrows. The rate of migration of these 'moving boundaries' may be measured.

scientist Arne Tiselius developed an apparatus which could be used to measure μ values of proteins (*see* Figure 5.2). In this system, proteins moved towards the electrode of opposite charge as an *ascending* and *descending* boundary in a U-shaped tank. While this could separate proteins into *moving boundaries* allowing mobility measurements to be made, the resolution of the system was extremely low, large volumes of protein were required and the apparatus used was quite cumbersome. *Zone electrophoresis* was then introduced which involved separation of biomolecules in some form of solid support. Examples of these supports include paper, cellulose and gel networks. These supports have considerable mechanical strength and allow staining of biomolecules after separation. Most modern electrophoretic separations take place in *zonal gel electrophoresis systems*. More recently, it has been found that electrophoresis in very thin capillaries allows the use of higher than normal voltages with efficient heat dissipation giving very high resolution. *Capillary electrophoresis* is a versatile analytical technique capable of separating a wide range of biological samples such as biomolecules, viruses and even whole cells (Section 5.10). A wide range of zonal and free solution systems are possible in capillary electrophoresis making an interesting historical link with the pioneering work of Tiselius.

5.1.3 Gel Electrophoresis

Hydrated gel networks have many desirable properties for electrophoresis. They allow a wide variety of mechanically-stable experimental formats such as horizontal/vertical electrophoresis in slab gels or electrophoresis in tubes or capillaries. Their mechanical stability also facilitates post-electrophoretic manipulation making further experimentation possible such as *blotting*, *electroelution* or *MS identification/fingerprinting* of intact proteins or of proteins digested in gel slices (Chapters 4 and 10). Since gels used in biochemistry are chemically rather unreactive, they interact minimally with biomolecules during electrophoresis allowing separation based on *physical* rather than *chemical* differences between sample components. Highly-controlled procedures are available for the formation of gels of a narrow range of *porosity*. This means that an individual gel can be prepared containing pores which will allow only molecules of a defined maximum mass to pass through, while excluding larger masses. Increasing or decreasing the size of these pores alters the mass which can be selected by the gel. Where the electrophoretic mobility of a biopolymer is directly related to gel porosity such gels are called *restrictive* gels. Most hydrated polymers used in electrophoresis applications in biochemistry also have good *anticonvection* properties which enhance resolution.

Gel networks may be formed from polymers which are *crosslinked* or *noncrosslinked* (the latter have proved particularly useful in *capillary electrophoresis*; Section 5.10). Carbohydrate polymeric materials such as starch were historically used to form gel networks for the separation of proteins (Figure 5.3). There is extensive hydrogen bonding in these polymers and, when boiled and then cooled, suspensions will 'set' into a gel similar to those used widely in food systems. The gel network arises from a rearrangement of the hydrogen bonding pattern to form extensive *interchain* hydrogen bonds. It is difficult to control this process accurately, however, to ensure uniformity of porosity. Moreover, the presence of carboxyl, sulfate and other chemical groups in starch can cause *electroosmosis* (Section 5.10) and ionic interactions. In starch gel electrophoresis of proteins, these effects result in low-resolution electrophoresis with broad bands. The most widely-used polysaccharide gel matrix nowadays is that formed with *agarose*. This is a polymer composed of a repeating disaccharide unit called *agarobiose* which consists of galactose and 3, 6-anhydrogalactose (Figure 5.3). Agarose gives a more uniform degree of porosity than starch and this may be varied by altering the starting concentration of the suspension (low concentrations give large pores while high concentrations give smaller pores). This gel has found widespread use especially in the separation of DNA molecules (although it may also be used in some electrophoretic procedures involving protein samples such as *immunoelectrophoresis*; Section 5.7). Because of the uniform charge-distribution in nucleic acids, it is possible accurately to determine DNA molecular masses based on mobility in agarose gels. This application is described in more detail in Section 5.8. The limited mechanical stability of agarose, while sufficient to form a stable horizontal gel compromises the possibilities for postelectrophoretic manipulation, however.

A far stronger gel suitable for electrophoretic separation of both proteins and nucleic acids may be formed by the

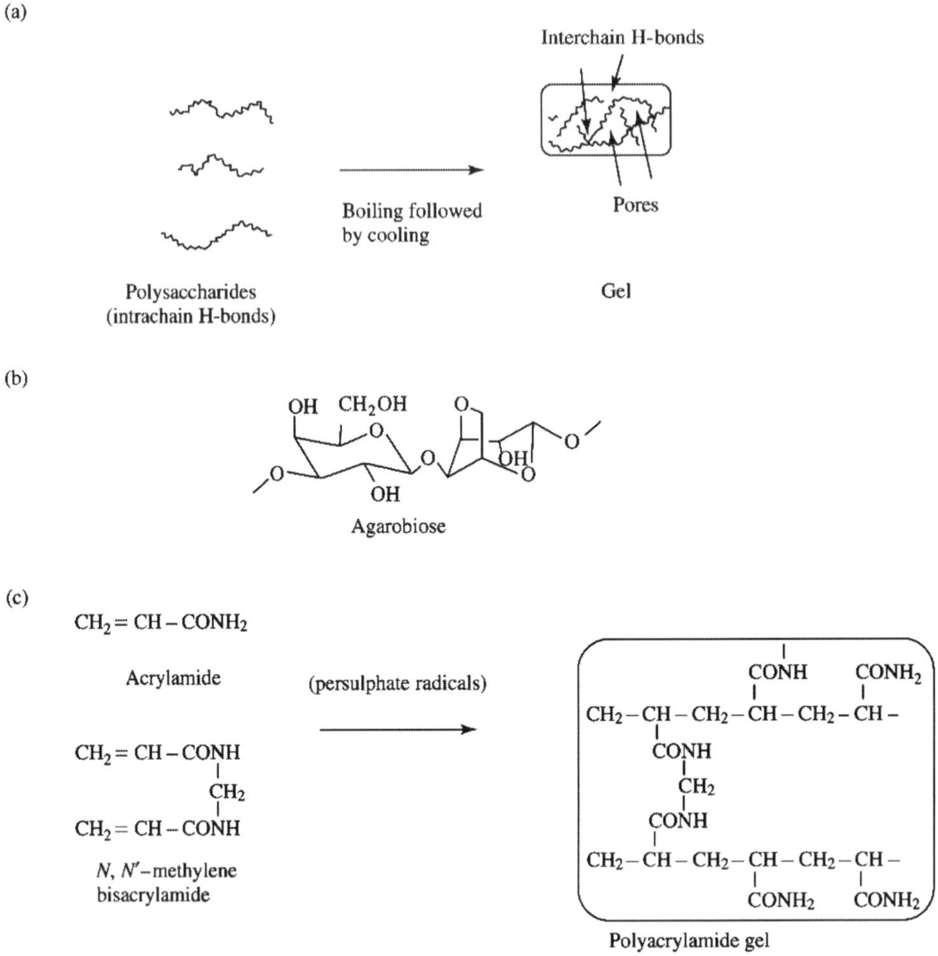

Figure 5.3. Gels commonly used in electrophoresis of proteins and nucleic acids, (a) Polysaccharide gels are formed by boiling followed by cooling. Rearrangement of hydrogen bonds gives interchain crosslinking. (b) Agarose is composed of agarobiose. (c) Polymerisation of acrylamide to form polyacrylamide gel. The polymerisation reaction is initiated by persulphate radicals and catalysed by TEMED.

polymerization of *acrylamide* (Figure 5.3). The inclusion of a small amount of acrylamide crosslinked by a methylene bridge (*N*,*N*′ methylene bisacrylamide) allows formation of a crosslinked gel with a highly-controlled porosity which is also mechanically strong and chemically inert. For separation of proteins, the ratio of acrylamide: *N*,*N*′ methylene bisacrylamide is usually 40 : 1 while for DNA separation it is 19 : 1. Such gels are suitable for high resolution separation of DNA and proteins across a large mass range (Table 5.1). A wider range of molecular mass in an individual gel is achievable by the use of *gradient* gels. These consist of a gradient of polyacrylamide (e.g. 5–20%) which is prepared as described in Figure 5.4.

Once sample molecules have separated in the gel matrix it is necessary to *visualize* their position. This is achieved by *staining* with an agent appropriate for the sample. Some of the more common staining methods used in biochemistry are listed in Table 5.2. DNA molecules are easily visualized under an ultraviolet lamp when electrophoresed in the presence of the extrinsic fluor *ethidium bromide* (Table 3.4). Alternatively, nucleic acids can be stained *after* electrophoretic separation by soaking the gel in a solution of ethidium bromide. When intercalated into double-stranded DNA, fluorescence of this molecule increases greatly (Chapter 3; this molecule also stains single-stranded nucleic acids albeit not so strongly). This is also the basis of the undesirable mutagenesis and carcinogenesis properties of ethidium which makes caution necessary when using it in electrophoretic staining. It is also possible to detect DNA with the extrinsic fluor *1-anilino 8-naphthalene sulphonate* (ANS;

Table 5.1. Range of separation of DNA and proteins in polyacrylamide gels of different polyacrylamide concentration

Acrylamide concentration % (w/v)	Range of separation	
	DNA (bp)	Protein (kDa)
3.5	1000–2000	—
5	80–500	>1000
8	60–400	300–1000
12	40–200	50–300
15	20–150	10–80
20	5–100	5–30

Note: Proteins separated in the presence of SDS (see Section 5.3.1).

Table 3.4). Protein is usually stained with the dye *coomassie blue*. Although most proteins stain with this dye, it should be noted that not all proteins take up the dye with equal affinity and care should be taken when trying to obtain *quantitative* information from such stains. Less sensitive protein dyes include *ponceau red* and *amido black*. Ponceau red has the advantage that it stains reversibly and may be removed from the protein to allow subsequent analysis (e.g. immunostaining). The most sensitive staining method for protein is *silver staining*. This involves soaking the gel in $AgNO_3$ which results in precipitation of *metallic silver* (Ag^0) at the location of protein or DNA forming a black deposit in a process similar to that used in black-and-white photography. Because of the multiple washing steps involved, however, this staining method is much more tedious than dye staining and is normally used to detect protein bands not visible by other methods. Radioactive proteins/nucleic acids can be visualized after electrophoretic separation by *autoradiography*.

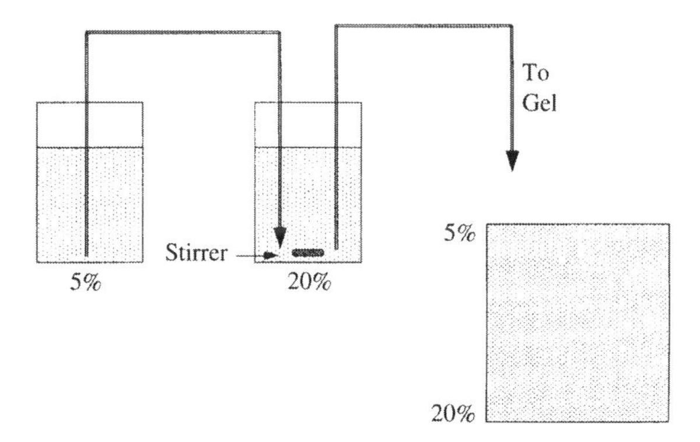

Figure 5.4. Gradient gel electrophoresis. A polyacrylamide gradient gel is prepared by mixing solutions of the two percentages required as shown. The 20% solution contains glycerol which stabilises the gradient until polymerisation is complete. These gels are used to achieve separation of proteins across a wide molecular mass range.

Table 5.2. Commonly used stains for biopolymers after electrophoretic separation in agarose or polyacrylamide gels

Stain	Use	Detection limit[a] (ng)
Amido black	Proteins	400
Coomassie blue	Proteins	200
Ponceau red	Proteins (reversible)	200
Bis-1-anilino-8-Naphthalene sulphonate (ANS)	Proteins	150
Nile Red	Proteins (reversible)	20
SYPRO orange	Proteins	10
Fluorescamine (protein treated *prior to* electrophoresis)	Proteins	1
Silver chloride	Proteins/DN	1
SYPRO red	Proteins	0.5
Ethidium bromide	DNA/RNA	10

Note: Some of these stains may also be applied to molecules after transfer to nitrocellulose or other membranes (i.e. blotting; see Section 5.11).
[a]These limits of detection should be regarded as approximate since individual proteins may stain more or less intensely than average.

The dried gel is clamped on a piece of X-ray film and placed at −70 °C (to stop band diffusion). On developing the film, bands become visible.

Despite the possibility that different sample components may stain to different extents, it is often possible to obtain *quantitative* information from electrophoresis gels. If the bands are labelled with radioisotopes, they may be cut out and counted in a scintillation counter. Alternatively, a variety of *image analysis* procedures are now possible with the advent of computers equipped with a scanner or video camera and appropriate image analysis software. These allow analysis of the gel and recording of results as a digital signal. A *densitometer* (a type of spectrophotometer; Chapter 3) may be used to measure absorbance at a particular wavelength along each track of the gel. This generates a trace similar to a HPLC chromatogram in which peaks of absorbance represent bands of protein/DNA in the gel. This process is called *densitometry* and allows comparison of one track with another and of one gel with another.

While electrophoresis at *continuous* pH is possible, improved resolution in polyacrylamide gel electrophoresis is obtained in a *discontinuous* system consisting of two gels called the *stacking* gel and *resolving* gel which are held at pH 6.9 and 8–9, respectively (Figure 5.5). The purpose of the stacking gel (sometimes also called *spacer* gel) is concentration of sample into thin bands or 'stacks' which can accumulate at the interface between the two gels prior to

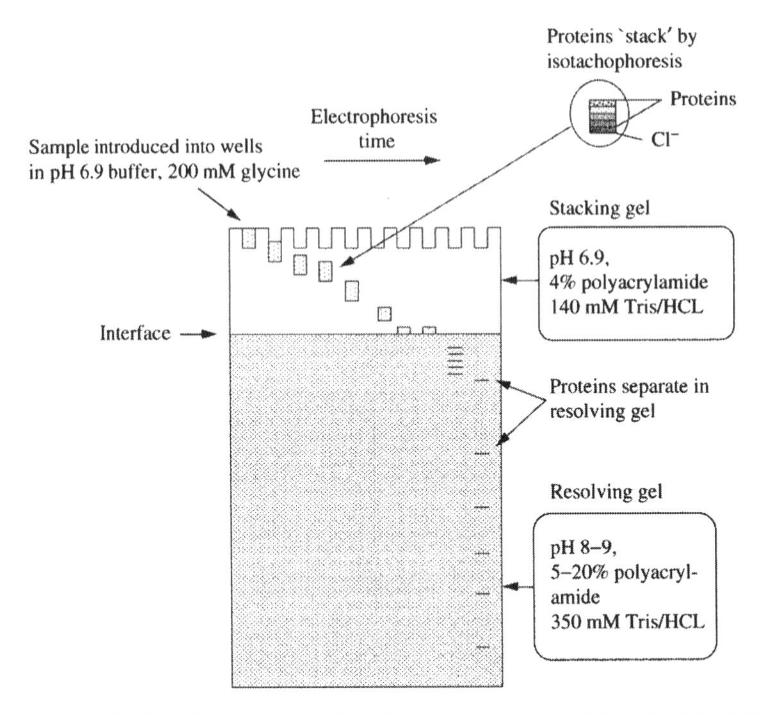

Figure 5.5. Discontinuous polyacrylamide gel electrophoresis. Sample is introduced into small (e.g. 50 μl) 'wells' in the stacking gel which has a different pH, ion strength and polyacrylamide concentration of the resolving gel. When current is turned on, protein/nucleic acids 'stack' by a process of 'isotachophoresis' as they pass through the stacking gel (see Section 5.3.1). At the interface between the stacking and resolving gels, a process of destacking begins (i.e. separation of individual protein/DNA molecules). Sample components continue to separate as they pass through the resolving gel due to differences in mass/charge, shape, etc. Adaptation of Figure 2 from Cruz *et al.* (1998) *Biochemica et Biophysica Acta*, **1415**, 125–134, with permission from Elsevier Science.

separation. This gel has a low concentration of polyacrylamide (3–5%), low ionic strength and a pH near neutrality. The resolving gel, by contrast, is composed of a higher percentage of acrylamide (8–20%), higher ionic strength and an alkaline pH. This gel achieves separation of sample molecules stacked at the interface. The lower ionic strength of the stacking gel causes higher electrical resistance and hence stronger electrical field strength in this gel compared to the resolving gel. This means that, at a given voltage, samples have higher mobility in the stacking compared to the resolving gel (Equation (5.1)).

Sample is applied to the stacking gel in the *electrophoresis* or *running buffer* composed of Tris-glycine at pH 8–9. Glycine exists in the form of a mixture of anion and *zwitterion* at this pH;

$$\underset{\text{Zwitterion form}}{NH_3^+ - CH_2 - COO^-} \rightarrow \underset{\text{Anion form}}{NH_2 - CH_2 - COO^-} + H^+ \quad (5.7)$$

As the aliquot of sample enters the stacking gel at pH 6.9, the balance of equilibrium shifts strongly towards the *uncharged* zwitterion form *which has no electrophoretic mobility*. As mentioned above, buffer ions are required to carry current in electrophoresis systems (Section 5.1.1). To maintain a constant current, a flow of anions is necessary. At pH 6.9, most proteins and nucleic acids are anionic and these, together with chloride, replace glycinate as mobile ions. In practice, protein/nucleic acid ions become 'sandwiched' between chloride (high mobility) and a small amount of glycinate ions (low mobility) as they move quickly through the stacking gel, unimpeded by the large pores due to low polyacrylamide concentration. This phenomenon of 'sandwiching' between ions of different electrophoretic mobility is called *isotachophoresis* (for a more detailed description of this *see* Section 5.10.1). Chlorine is a very strongly electronegative atom which moves towards the anode with much greater velocity than any other species present. This band of anions leaves behind it a *zone* of low conductivity. As this zone passes through the rest of the sample, molecules in the sample become sorted on the basis of their charge from most to least negatively charged. Simultaneously, they are *concentrated* based on this charge discrimination. This stacks the sample into a number of thin layers. When this front of ions reaches the interface with the resolving gel (pH 8–9), however, the glycinate ion concentration increases dramatically (Equation (5.7)) and now carries the bulk of the current. At the same time, protein/nucleic acid ions encounter a higher concentration of polyacrylamide with a narrow pore size and a more alkaline pH. The thin stacks of protein/nucleic acid therefore separate in this gel depending on their mass/charge and shape characteristics as described in Section 5.1.1.

In most of the applications described in this chapter, electrophoresis is used as a high-resolution *analytical* procedure. However, it is also possible to recover protein, peptide or DNA samples from gels after electrophoretic separation. This *preparative* procedure usually involves cutting out a slice of gel containing a single molecular species and recovering the protein/DNA by exposing the gel-slice to an electrical field. This process is called *electroelution* and it is suitable for the purification of small quantities of material for further analysis. This approach is especially appropriate for samples difficult to purify by conventional chromatographic methods (Chapter 2).

5.2 NONDENATURING ELECTROPHORESIS

5.2.1 Polyacrylamide Nondenaturing Electrophoresis

The gel network formed by polyacrylamide is a suitable environment for the electrophoretic separation of proteins in their *native* state, that is under *nondenaturing conditions*. In such conditions, the protein is regarded as being in its intact fully-active form with the correct secondary, tertiary and quaternary structure. It separates on the basis of *intrinsic* charges on amino acid side-chains and other groups located on the protein surface. Since the precise number and strength of positive and negative charges will vary from protein to protein, each will have a characteristic mobility in a nondenaturing system determined by a combination of these charges together with physical characteristics of the protein such as mass and shape.

5.2.2 Protein Mass Determination by Nondenaturing Electrophoresis

Non-denaturing electrophoresis allows the determination of protein *native molecular mass* (M_r). Since mobility is strongly affected by mass and shape, it is possible to compare the mobility of a protein of unknown mass with a series of standard proteins of similar shape but known mass. This mobility is measured in a number of nondenaturing gels each of *differing polyacrylamide concentration* called *Ferguson* gels (Example 5.1). The mobility of each protein (standards and unknown) is measured and expressed as R_f. A plot of log R_f for each standard protein versus % polyacrylamide is then made (it is necessary to use a minimum of five different polyacrylamide concentrations for accurate results). The slope of these plots is related to the *retardation coefficient*, K_r, for that protein as;

$$K_r = -(\text{slope}) \tag{5.8}$$

There is a direct relationship between K_r and molecular mass of molecules of similar general shape. In the case of globular proteins, there is a linear relationship between K_r and the radius of the globule shape, r (Stokes radius). By plotting K_r versus M_r, a standard curve may be constructed from which it is possible to estimate native molecular mass. This data complements mass estimates from techniques such as gel filtration (Chapter 2) and is especially useful when only small amounts of protein are available. By combining such measurements with M_r estimates under denaturing conditions (Section 5.3), the quaternary structure of a protein can be routinely determined.

5.2.3 Activity Staining

Since proteins are separated by nondenaturing electrophoresis in their native form, they generally retain biological activity. In the particular case of enzymes, catalytic activity can be used as a specific stain for the enzyme in a process called *activity staining*. This involves use of an artificial substrate which is converted to an insoluble, coloured product by the catalytic cycle. The location of the protein in the gel can be visualized by observation of the insoluble (and therefore precipitated) product. Since a single enzyme may go through many thousands of catalytic cycles in a few minutes, very tiny amounts of enzyme can be visualized in this way.

Although in principle an individual activity stain is required for each enzyme, in practice particular activity stains are available for groups of enzymes sharing a common catalytic process. For example, many dehydrogenases catalyze formation of NADH from NAD. This can be used to reduce *tetrazolium* to produce insoluble *formazan* (Figure 5.6). If, after nondenaturing electrophoresis, a gel is soaked with tetrazolium together with appropriate substrates and cofactors for a particular dehydrogenase, formazan will only form where the enzyme is located in the gel. By selecting substrates specific for the particular dehydrogenase under investigation (e.g. succinate for succinate dehydrogenase, malate for malate dehydrogenase) highly individual staining for dehydrogenases can be achieved using essentially the same staining procedure.

This type of experiment is useful in identifying which band among several visible on a nondenaturing gel is the band representing the enzyme of interest. It is also

Example 5.1 Determination of protein native mass by nondenaturing electrophoresis

When working with a newly-purified protein, we may need to know its quaternary structure (i.e. is it a monomer, dimer, etc.?). Non-denaturing electrophoresis is particularly useful for this determination in situations where availability of protein is limited (<500 μg). In restrictive polyacrylamide gels, the electrophoretic mobility of individual proteins (Rf) will vary in a straight-line manner with respect to the porosity of the gel (i.e. % acrylamide).

For a protein of unknown molecular mass (subunit molecular mass; 50 kDa) and some standard proteins of known mass, Rf values were determined in a nondenaturing system using gels of 5, 7.5, 10, 12.5 and 15% polyacrylamide . The standard proteins were α-lactalbumin (14.2 kDa), carbonic anhydrase (29 kDa), ovalbumin (45 kDa) and urease (trimeric form; 272 kDa). A Ferguson plot of log Rf versus % acrylamide was prepared as follows;

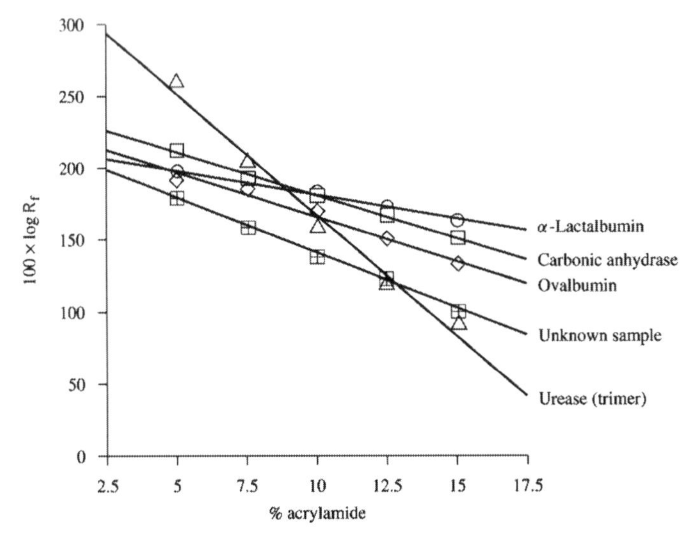

The Ferguson plot demonstrates the linear relationship between Rf and gel porosity mentioned above.

A slope value was calculated for each protein from which the retardation coefficient (K_r) is determined as – (slope). There is a direct relationship between K_r and the molecular radius (Stoke's radius) of the protein which allows us to calculate molecular mass (M_r). Since we know the K_r of a number of standard proteins (of known M_r) as well as of the unknown sample, a standard curve of K_r versus molecular mass can be plotted. From this plot, the native mass of the protein was measured as 108 kDa. Since the subunit M_r is 50 kDa, it could be deduced from this experiment that the unknown protein was a dimer.

Standard curve of retardation coefficients from Ferguson Plot.

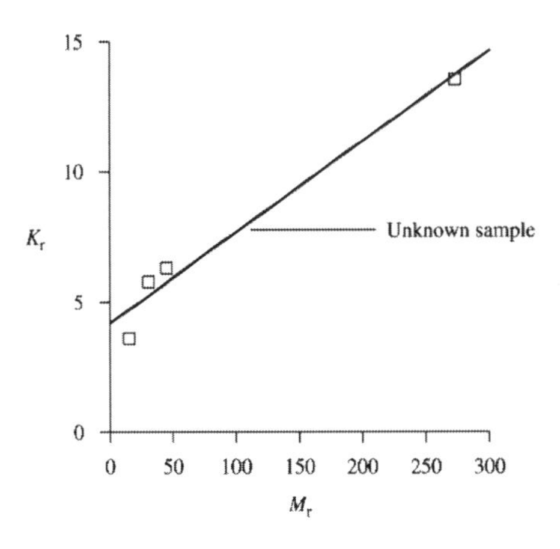

(DATA COURTESY OF DR. VIVIENNE FOLEY).

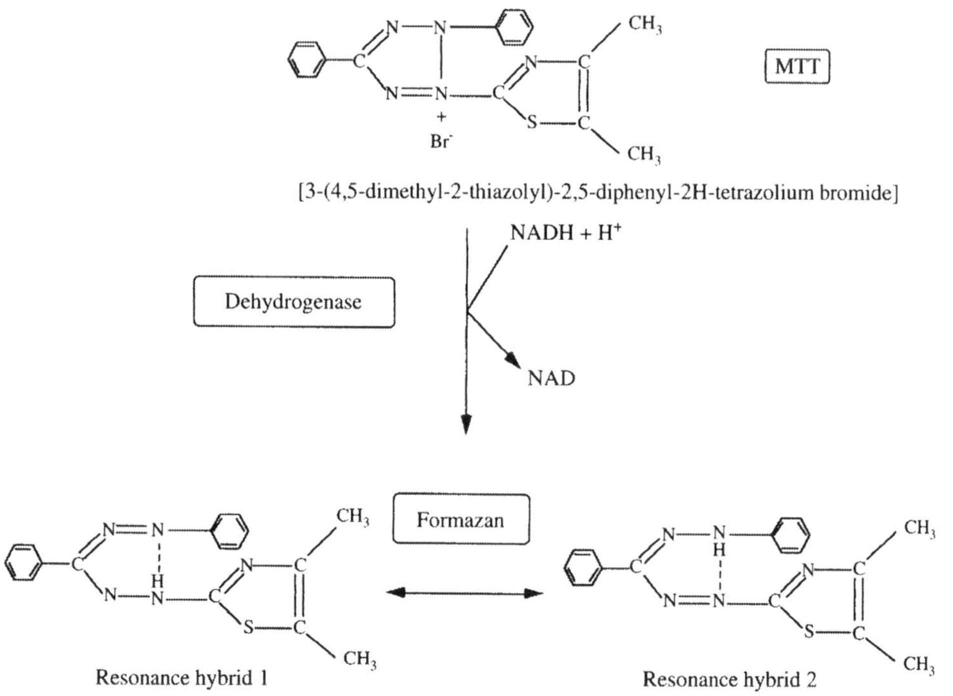

[3-(4,5-dimethyl-2-thiazolyl)-2,5-diphenyl-2H-tetrazolium bromide]

Figure 5.6. Activity staining for dehydrogenases. MTT is a tetrazolium salt which may be reduced in a reaction catalysed by dehydrogenases to an insoluble formazan product which has two resonance hybrid forms stabilised by an internal hydrogen bond (dashed line).

useful in demonstrating the presence of multiple enzymes catalyzing a particular chemical reaction called *isoenzymes* (Example 5.2).

5.2.4 Zymograms

Many enzyme activities are expressed as isoenzymes and there is often a characteristic pattern of expression in different tissues, individuals, populations or in samples taken at different stages of development. Examples of the use of these patterns include diagnosing the onset of cancer in cultured cells and the mixing of genetically-distinct biological populations such as farmed and wild salmon. Activity-stained nondenaturing gels are called *zymograms* and these may be used to characterize cell-types, tissues, individuals and populations on the basis of isoenzyme expression (Example 5.2).

5.3 DENATURING ELECTROPHORESIS

Biopolymers such as proteins and nucleic acids are folded into compact structures held together by a variety of non-covalent, ionic interactions such as hydrogen bonding and salt bridges. We have seen that, under nondenaturing conditions, such samples migrate in a manner determined largely by their *intrinsic* electrical charge. However, it is possible to disrupt these ionic interactions to *denature* the biopolymers and then to separate them electrophoretically by *denaturing electrophoresis*. Such experiments are especially useful in the study of proteins since they have a more varied range of tertiary structure than nucleic acids. The electrophoretic mobility of the denatured molecule will be changed, compared to that in nondenaturing conditions, due to altered charge and/or shape since it now migrates as an unstructured monomer through the electrical field. Any biological activity or quaternary structure associated with the sample components is lost in denaturing electrophoresis. However, important structural information can be obtained especially about proteins as described in the following sections.

5.3.1 SDS Polyacrylamide Gel Electrophoresis

The detergent *sodium dodecyl sulphate* (SDS) consists of a hydrophobic 12-carbon chain and a polar sulfated head. The hydrophobic chain can intercalate into hydrophobic parts of the protein by *detergent action*, disrupting its compact folded structure (Figure 5.7). The sulfated part of the detergent remains in contact with water thus maintaining the detergent-protein complex in a highly-soluble state. This interaction disrupts the folded structure of single polypeptides

Example 5.2 Use of nondenaturing electrophoresis for zymogram analysis

Proteins retain their intact native structure when electrophoresis is performed under nondenaturing conditions. This means that enzymes are usually fully active even after electrophoretic separation. Frequently, enzymes are present in multiple forms or isoenzymes in cells. These proteins may be variants arising from a range of post-translational modifications (e.g. glycosylation) or may be distinct products of specific genes. After electrophoretic separation, instead of staining the gel with nonspecific dyes such as coomassie blue, we can often activity stain with enzyme substrates which are specific for the group of enzymes under investigation. This is called zymogram analysis and it facilitates analysis of individual isoenzymes in cell extracts without the need for their prior purification. Isoenzyme expression patterns from such studies can give us useful comparative information on the genetic composition of natural populations of animals, differences between species, tissue-specific enzyme expression within a single individual and localized expression within a single tissue.

A good example of an enzyme present in multiple forms in cells is *lactate dehydrogenase* (LDH) which catalyses the reduction of pyruvate under anaerobic conditions. This tetrameric enzyme is composed of two types of subunit polypeptides designated H and M. LDH from heart and muscle are composed of four H and M subunits, respectively, but five isoenzymes are possible in vertebrate liver (H_4, HM_3, H_2M_2, HM_3, M_4). These isoenzymes are designated LDH-1 to LDH-5, respectively, and are separable by electrophoresis in polyacrylamide gels. Specific activity staining of these enzymes is possible by the following method;

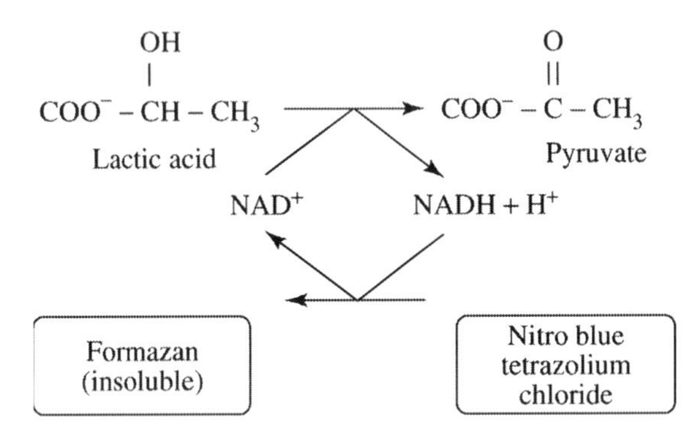

Wherever LDH isoenzymes are located in the gel, they will produce the coloured formazan product (this type of reaction is widely-used in activity staining of dehydrogenases generally). Since this compound is insoluble, it will not diffuse away from the location to which the isoenzyme has electrophoresed and this is an important feature of most activity stains. The pattern of isoenzyme expression determined for liver samples from pig (1), cattle (2), rabbit (3), rat (4) and human (5) is shown below;

Using this method with ng-sized microdissected samples, it was possible to identify differential expression of isoenzymes in the basic functional unit of liver, the acinus. The level of the M-subunit was found to be consistently higher in the periportal part of the acinus than in the perivenous part in all these species while the H-subunit showed species-specific variation. These results are important in understanding the contribution of LDH to anaerobic metabolism in this vital tissue.

Reproduced from Maly, I.P. and Toranelli, M. (1993), Ultrathin-layer zone electrophoresis of lactate dehydrogenase isoenzymes in microdissected liver samples. *Analytical Biochemistry*, **214**, 379–388.

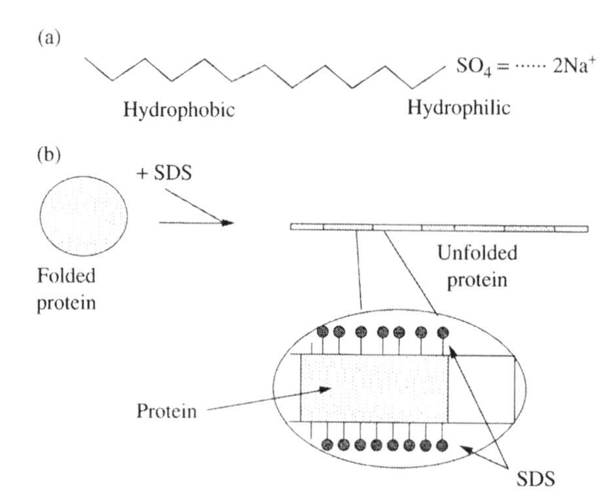

Figure 5.7. Interaction of protein with sodium dodecyl sulphate, (a) Sodium dodecyl sulphate has a hydrophilic head and a hydrophobic tail, (b) The detergent interacts with folded protein by coating hydrophobic regions of the polypeptide conferring negative charges on them. This disrupts both subunit–subunit and protein–membrane interaction.

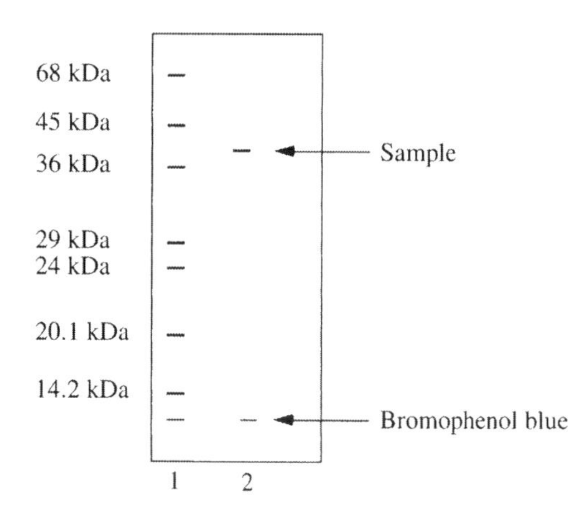

Figure 5.8. SDS PAGE of a protein of unknown molecular mass. Standard proteins of known molecular mass (M_r) were applied to a 15% SDS PAGE gel (track 1) as follows; bovine serum albumin (68 kDa), ovalbumin (45 kDa), glyceraldehyde 3-phosphate dehydrogenase (36 kDa), carbonic anhydrase (29 kDa), trypsinogen (24 kDa), soya bean trypsin inhibitor (20.1 kDa) and α-lactalbumin (14.2 kDa). A sample protein of unknown mass was simultaneously applied to track 2 and, after electrophoresis (3 h at 175 V), the resolving gel was stained with coomassie blue. This visualized both standard and sample proteins. Bromophenol blue was included as a 'tracker dye' which shows the front of electrophoresis. The distance each protein migrates into the gel is measured and M_r of sample protein may be determined from a plot of this distance versus log M_r of standards.

as well as subunit-subunit (i.e. quaternary structure) and protein-membrane interactions for most proteins. SDS coats proteins with a uniform 'layer' of negative charges which causes them to migrate towards the *anode* when placed in an electrical field, regardless of the net intrinsic charge of the uncomplexed proteins. The negative charge gives a charge-density largely independent of the primary structure or mass of the polypeptide. For this reason, there is a close relationship between the mobility of SDS-protein complexes in polyacrylamide gels and the molecular mass of the polypeptide (Figure 5.8). This type of experiment is called *SDS polyacrylamide gel electrophoresis* or *SDS PAGE* and it is used widely to obtain mass and purity estimates for polypeptides. A further application of this technique is to visualize *peptide maps* arising from proteolytic or chemical digestion of a pure sample. This latter application complements HPLC (Chapter 2) and mass spectrometry (Chapter 4) analyses.

Despite its widespread use and versatility, SDS PAGE has a number of important limitations. It is assumed that proteins electrophorese as perfect spheres with a uniform charge-distribution. If proteins deviate from a globular shape or if they bind above- or below-average amounts of SDS, they may behave in a nonideal way and inaccurate mass estimates will result. For this reason, it is important to compare unknown proteins to appropriate standard proteins in SDS PAGE (i.e. globular unknowns with globular standards and fibrous unknowns with fibrous standards). Mass estimates obtained by this technique are often referred to as the *apparent* molecular mass since they depend on comparison with other proteins rather than on direct measurement such as determination of complete amino acid sequence. Possible effects of post-translational modifications on protein mobility in SDS PAGE gels should also be considered. For example, glycoproteins migrate slower in such systems because SDS does not bind to sugar moieties.

5.3.2 SDS Polyacrylamide Gel Electrophoresis in Reducing Conditions

A number of variants of SDS PAGE are possible by appropriate chemical pretreatment of the protein sample. Many proteins, especially those secreted outside the cell such as immunoglobulins, contain disulphide bridges as a distinct structural feature. They are formed post-translationally between cysteine side-chains, adding greatly to the stability of the protein. In order to understand the complete covalent structure of a protein, it is necessary to know the number and location of these disulphide bridges. By pretreating protein samples with chemical *reducing agents* such as 2-mercaptoethanol or dithiothreitol, it is possible to reduce disulphides, breaking the links between individual polypeptides (Example 5.3). In the presence of SDS, such

Example 5.3 SDS PAGE in reducing conditions

Disulphide bridges are formed between cysteine side-chain sulphydryl groups either *within* a single polypeptide (intrachain) or *between* different polypeptides (*interchain*) in oligomeric proteins. They are important structural features of proteins because they frequently maintain the protein in a particular shape and usually add to its stability. Disulphide bridges can be reduced to sulphydryls by treatment with reducing agents such as 2-mercaptoethanol or dithiothreitol. In order to understand fully the covalent structure of a protein, we need to identify its pattern of disulphide bridging.

A 100 kDa protein is a trimer composed of two molecules of 25 kDa and a single polypeptide of 50 kDa, giving an apparent native M_r estimate of 100 kDa. Treatment with β-mercaptoethanol disrupts disulphide bridges as shown below. SDS PAGE analysis of this protein in the presence (+) and absence (−) of 2-mercaptoethanol reveals a single pattern of disulphide bridging in this protein.

SDS PAGE analysis of protein under reducing conditions

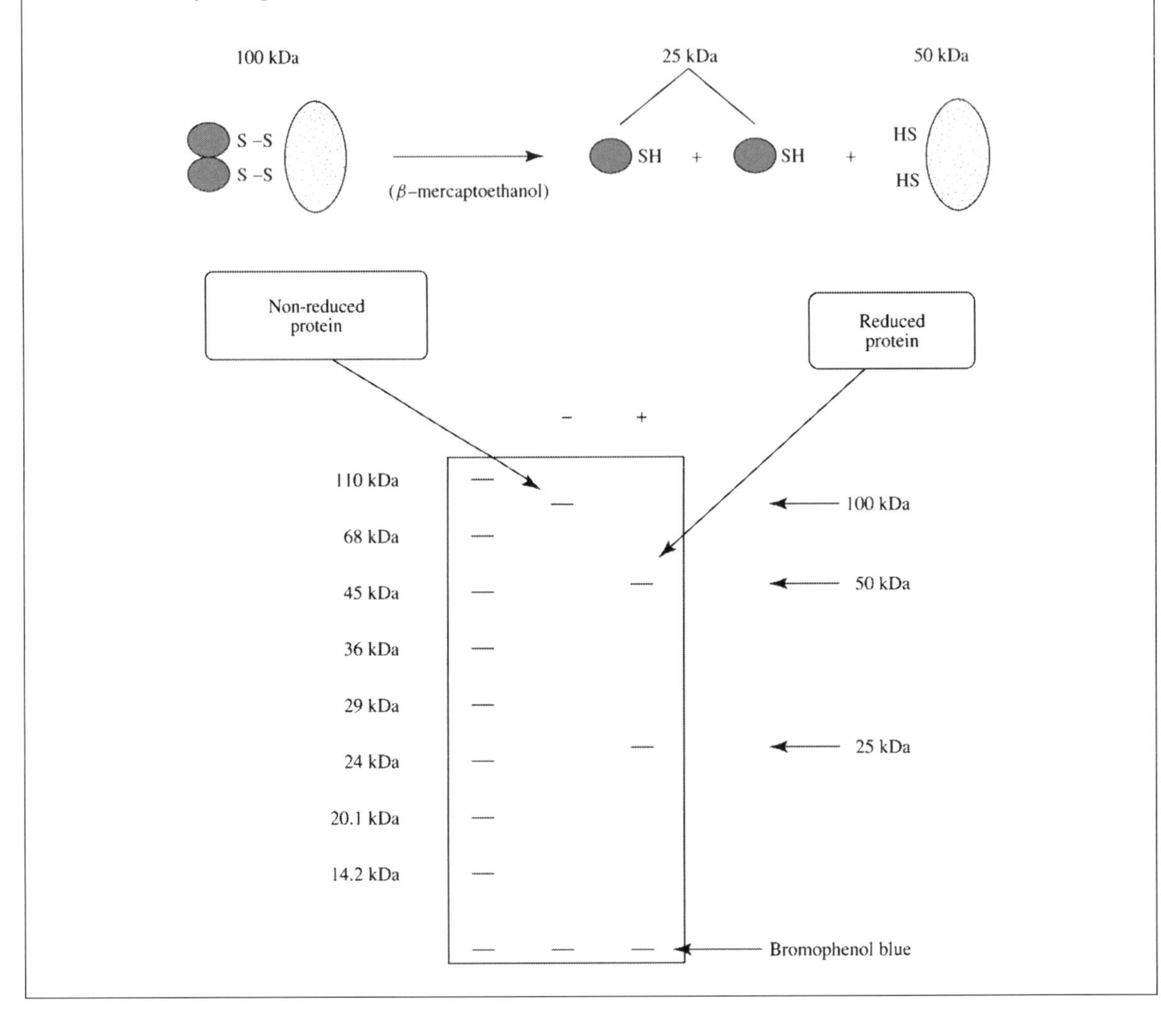

polypeptides then migrate *individually* in an electrophoresis system. Analysis of mass estimates in the presence and absence of reducing agents allows deductions about disulphide bridging in sample proteins to be made.

A related and complementary nondenaturing electrophoretic technique allows *counting* of integral numbers of cysteines in proteins by pretreatment with mixtures of iodoacetic acid and iodoacetamide (Figure 5.9). This

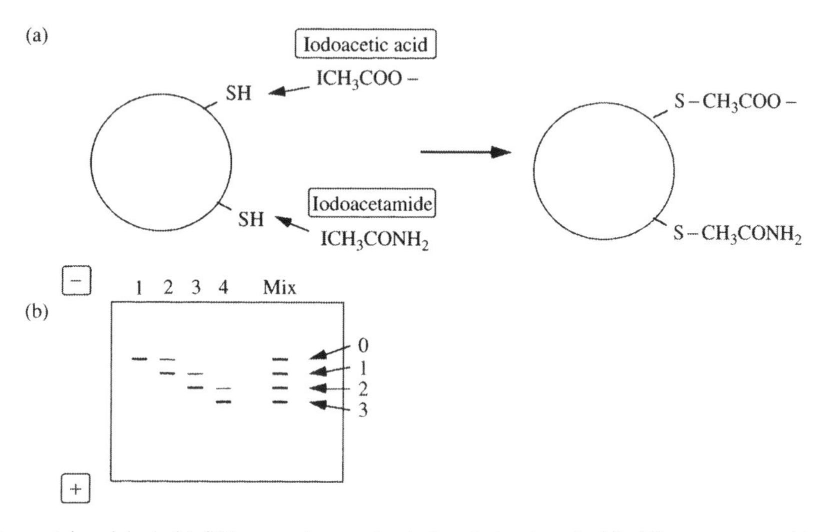

Figure 5.9. Counting protein sulphydryl (–SH) groups by non-denaturing electrophoresis. (a) –SH groups react with both iodoacetic acid (giving a negative charge) or iodoacetamide (uncharged). Negatively charged proteins migrate towards the anode, (b) Treatment of sample protein (containing three –SH groups) with an increasing ratio of iodoacetic acid/iodoacetamide (1–4), gives a series of differently charged proteins with zero to three negative charges. 'Mix' contains a mixture of aliquots from tracks 1–4. Counting of bands in this track gives the number of –SH groups in the protein.

information can be useful as a first step in revealing the disulphide distribution pattern within a protein.

5.3.3 Chemical Crosslinking of Proteins – Quaternary Structure

Quaternary structure describes the arrangement of individual polypeptide chains in *oligomeric* proteins (i.e. proteins consisting of more than one polypeptide). We have already seen that secreted proteins such as immunoglobulins often consist of several polypeptide chains or *subunits* linked together by disulphide bridges. Other oligomeric proteins are held together by noncovalent interactions, especially *hydrophobic forces*. Such structures are often quite unstable and may be difficult to study by conventional methods such as X-ray crystallography (Chapter 6). One experimental approach to this problem takes advantage of the fact that the amino acid lysine, which has a chemically-reactive ϵ-NH_2 group, is quite abundant in proteins (an average of ~7% of total amino acid residues). This group will react readily with *bifunctional* reagents such as dimethyl suberimidate (Figure 5.10). Polypeptides adjacent in the quaternary structure of a

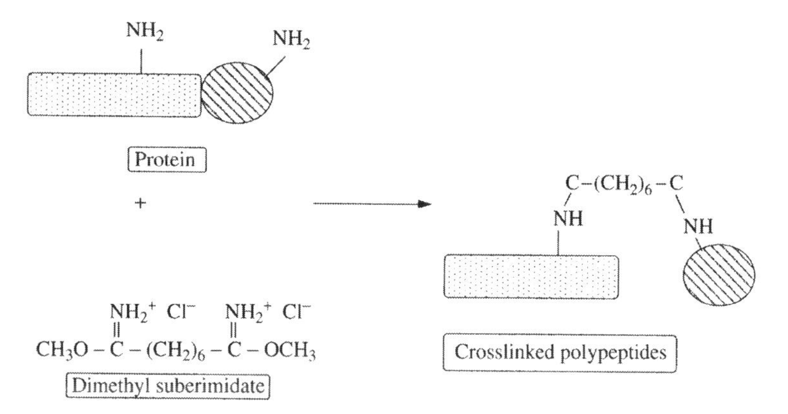

Figure 5.10. Cross-linking of polypeptides with bifunctional reagents. Lysine ε-NH2 groups of adjacent polypeptides may be crosslinked via a bifunctional reagent such as dimethyl suberimidate. Crosslinked proteins will electrophorese in SDS PAGE to an apparent mass equal to that of the sum of the crosslinked polypeptides.

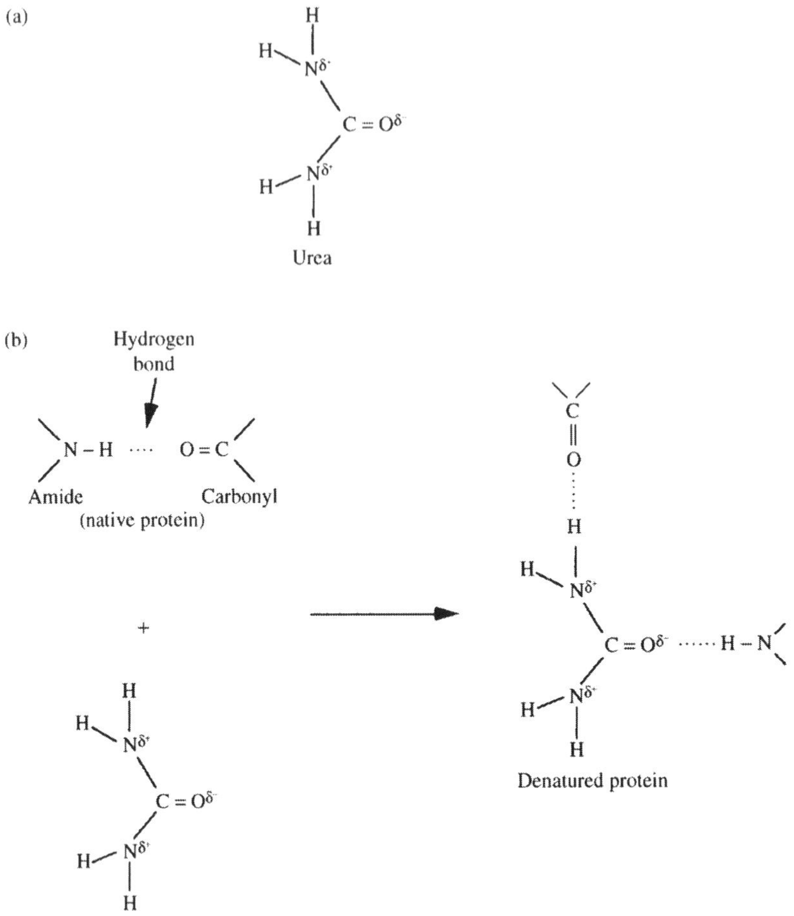

Figure 5.11. Urea as a protein denaturant. (a) Urea is a net uncharged molecule which is nonetheless polar in nature with positive and negative dipoles as shown (δ). (b) At high concentrations, urea can interrupt hydrogen bonding between amide and carbonyl groups of proteins. This leads to unfolding of the protein.

large protein or complex will be *covalently crosslinked* by this reaction and will electrophorese as a single molecule in SDS PAGE.

5.3.4 Urea Electrophoresis

One of the great disadvantages of SDS as a denaturing agent is the fact that it binds tightly to protein and can be very difficult to remove after electrophoresis. An alternative denaturation strategy is offered by the *chaotropic* agent, *urea*. We have previously seen (Chapter 2) how this molecule may be used in ion exchange chromatography of peptides. Urea (Figure 5.11) has the useful property that, although it possesses zero net charge, it is nonetheless a *polar* molecule with unequal internal charge distribution. Protein secondary structure depends on hydrogen bonding between the amide and carbonyl groups of peptide bonds. Similar hydrogen bonds are also involved in stabilization of tertiary and quaternary structure. High concentrations of urea interrupt these hydrogen bonds, leading to a complete disruption of secondary, tertiary and quaternary structure. Urea renders polypeptides highly water-soluble but, because it is uncharged, it does not migrate in electrical fields. Proteins therefore electrophorese in the presence of urea as determined by their net intrinsic charge, despite the fact that they are denatured in these conditions. Urea also features in DNA electrophoresis in situations where single strands are required. As with proteins, urea disrupts the hydrogen bonding responsible for DNA secondary structure.

5.4 ELECTROPHORESIS IN DNA SEQUENCING

DNA is a long rod-like molecule which moves through polyacrylamide gels with limited mobility. Moreover, DNA has a uniform charge-distribution which means that this mobility is directly proportional to the size of the molecule. For these reasons, electrophoresis of DNA in polyacrylamide gels allows separation of molecules differing by as little as a single nucleotide. Advantage has been taken of this to develop sequencing strategies for the study of DNA.

5.4.1 Sanger Dideoxynucleotide Sequencing of DNA

In *Sanger dideoxynucleotide sequencing*, the sequence of a single DNA strand is determined. Where double stranded DNA is used in the reaction, the strand complementary to the strand to be sequenced must be displaced in some way before sequencing. Double stranded DNA is normally prepared by the *polymerase chain reaction* (PCR) which generates many copies of a double stranded *template*. In DNA sequencing, single stranded DNA functions as a template for replication of a series of *daughter strands* using DNA polymerase.

In order to obtain large amounts of single stranded DNA the *M13 bacteriophage* system is used (Figure 5.12). This is a virus capable of infecting bacterial cells such as *E. coli*. Outside of such cells, the genome of the virus consists of approximately 7 kb single stranded DNA. When M13 infects bacteria, this DNA is replicated into a double stranded form called the *replicative form* (RF). Once approximately 100 copies of RF have accumulated, synthesis switches to production of single stranded DNA which is eventually packaged with protein and extruded from the cell. Approximately 1000 such viruses may be released into the medium in a single cell generation. This life cycle means that M13 DNA may be isolated in *either* single or double stranded form and can be handled, respectively, as a virus or as a plasmid. Various modifications have been made to wild-type M13 such as the inclusion of a *polylinker* site and of the *Lac Z* gene. These factors make M13 a particularly versatile system for replicating single stranded DNA fragments.

DNA to be sequenced is cloned into the polylinker site of the RF form of M13 and used to transform lac$^-$ *E. coli* cells in the same manner as with bacterial plasmids. Recombinants are identified as white colonies and these are grown up in broth. Several copies of the recombinant RF form of M13 are made in the *E. coli* cell at first, but then synthesis switches to the single stranded form which eventually predominates. Single stranded DNA collects at the periplasm of the cell and is combined with coat proteins to form M13 bacteriophage. This is extruded from the cells without lysis and is collected by centrifugation. Single stranded DNA is prepared from the virus by precipitation of coat proteins with polyethylene glycol followed by DNA extraction with phenol/chloroform and precipitation with ethanol.

A short oligonucleotide called a *primer*, complementary to an M13 sequence near the origin of replication, is labelled radioactively at the 5′ end which allows detection of each daughter strand and ensures that the sequence is read in one direction (5′ to 3′). A small amount of *competitive inhibitors* of *DNA polymerase*, dideoxyadenine, dideoxyguanine, dideoxythymidine or dideoxycytosine (ddNTPs) is included in each of four separate reactions, together with substrates for this enzyme; dATP, dGTP, dCTP and dTTP (dNTPs; Figure 5.13). The precise amount of dideoxynucleotide used may need to be optimized but it is usually 1 : 100 relative to the dNTPs. DNA polymerase is then added and catalyses synthesis of a new daughter-strand until a molecule of inhibitor is incorporated terminating the synthetic reaction. Since the ddNTP lacks a 3′-hydroxyl group, it cannot form a 3′,5′-phosphodiester bond and the DNA chain cannot be extended further. Thus, this technique is often referred to as the *chain termination* procedure. For example, for every G residue occurring in the sequence, at least some of the time, a ddG becomes incorporated in the strand. The experiment therefore results in four sets of DNA daughter-strands which are *truncated fragments* complementary to the DNA to be sequenced.

5.4.2 Sequencing of DNA

The four mixtures are then electrophoresed in an unusually long (up to 48 cm) and thin (0.4 mm) continuous polyacrylamide (4–6%) gels in the presence of urea (8 M) and at approx. 30–40 °C. These conditions ensure that DNA strands migrate without formation of significant secondary structure. A *shark's tooth comb* is used to create small reservoirs or*wells* at the top of the gel. These form between the teeth of the comb and the gel surface. Individual samples may be applied to specific wells thus avoiding mixing with other samples (Figure 5.14). In the four mixtures DNA synthesis will have terminated at different positions as shown in Figure 5.13.

These samples are electrophoresed side-by-side with one track for ddG, ddA, ddT and ddC, respectively. After electrophoresis, the gel is dried onto a sheet of paper which is then laid against X-ray film which detects the radiolabel which was incorporated into the primer at the beginning of the experiment. The sequence may be read from the bottom of the autoradiogram, beginning with the smallest fragment. It is important to understand that this sequencing method depends absolutely on the ability to separate DNA to a resolution of one base-pair and is only possible because of the absence of secondary structure combined with the uniform charge-distribution provided by the phosphate group of each

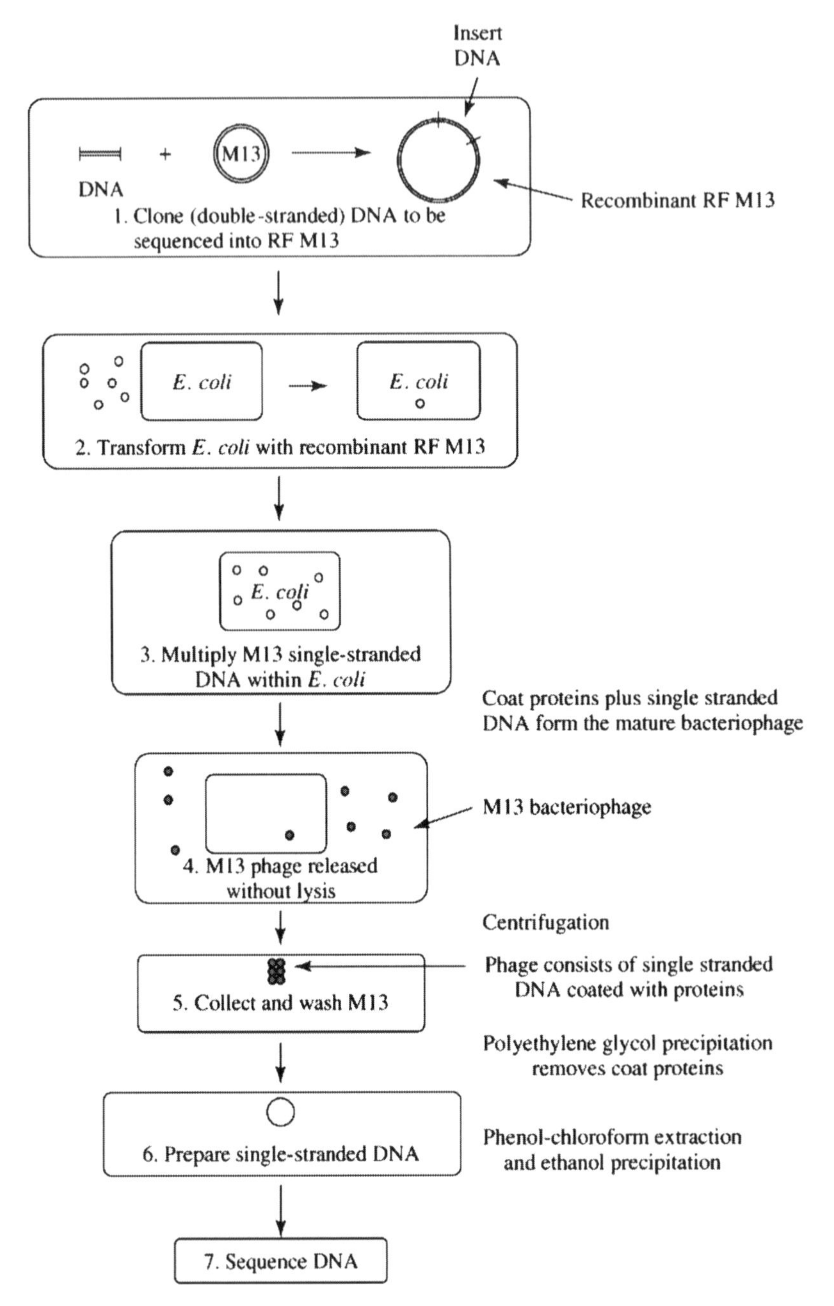

Figure 5.12. Preparation of single-stranded DNA for sequencing. DNA to be sequenced is cloned into the RF form of M13. This is transformed into competent lac- *E. coli* and replicated first into several copies of RF M13 and then into many copies of single stranded M13. Coat proteins surround this DNA to form the mature bacteriophage which is released into the medium. Single stranded DNA is collected by phenol-chloroform extraction and ethanol precipitation. This is used as template for DNA sequencing.

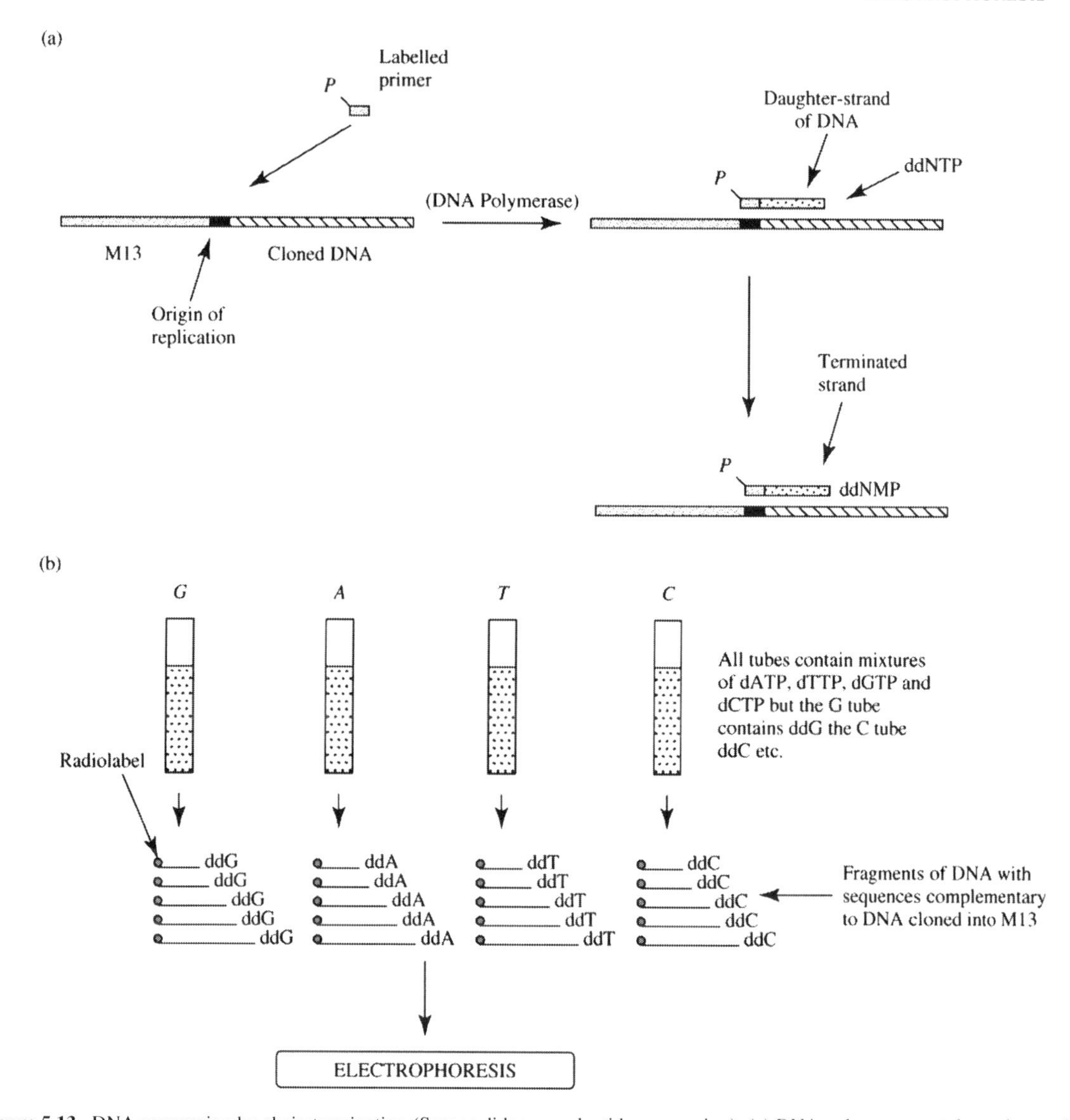

Figure 5.13. DNA sequencing by chain termination (Sanger dideoxynucleotide sequencing), (a) DNA polymerase catalyses the synthesis of a DNA strand complementary to the DNA to be sequenced. When a ddNTP is incorporated, synthesis of the daughter strand ceases. P represents a radiolabel on the oligonucleotide primer (which may be replaced by a fluorescent label, see text), (b) A separate experiment is set up for each of four ddNTPs. This generates a set of truncated fragments in each test-tube. These are then electrophoresed through a polyacrylamide–urea gel.

nucleotide. Near the top of the autoradiogram it will be difficult to resolve slight differences in electrophoretic mobility between different DNA fragments and this represents the limit of reliable sequencing in this experiment. From the rules of complementary base pairing, it is elementary to deduce the sequence of the parent strand from that of the complementary strand and, in the case of coding DNA, to translate this into a deduced amino acid sequence. Since even a single error in sequence may result in large errors in deduced amino acid sequence it is usual to carry out sequencing experiments for *both* complementary strands of parent DNA.

Using this approach, it is possible manually to sequence lengths of DNA up to 400 bp in a single experiment. By using parts of the sequence near the 3′ end to design a new primer, a second sequence overlapping at its 5′ end with the 3′ end of the first sequence may be generated (Figure 5.15). This strategy, which is called *gene walking* or *primer*

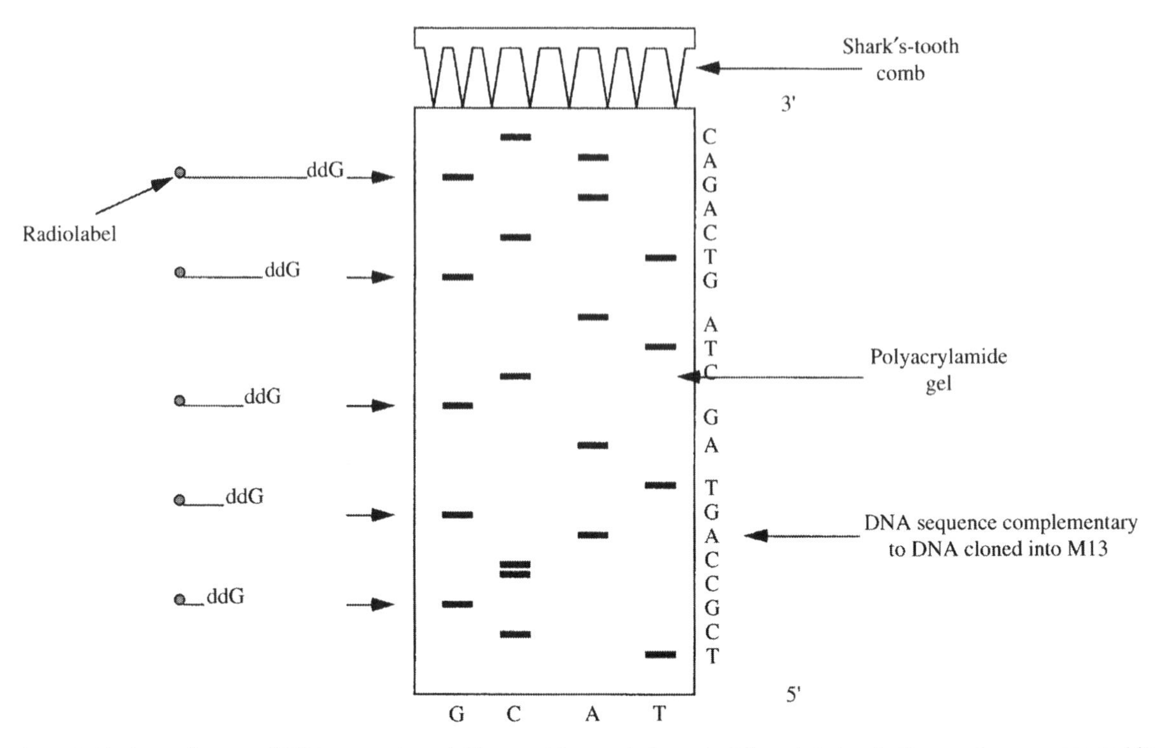

Figure 5.14. Autoradiogram of DNA sequencing gel. Truncated fragments for each of four chain termination reactions are separated in polyacrylamide gels containing 8 M urea. Smaller fragments electrophorese furthest into the gel (for clarity, only those for ddG are shown in the illustration but bands in the C, A and T lanes arise from corresponding fragment sets for ddC, ddA and ddT). Bands are visualised by autoradiography on X-ray film. The sequence obtained is complementary to the sequence of the template DNA strand.

walking, allows progressive sequencing of large amounts of DNA.

Another means of extending the amount of DNA sequence information is *automated DNA sequencing*. *Dye primerDNA sequencing* (Figure 5.16) uses four distinct fluorescent compounds to label the primer at the 5′ end and, since these labels can be distinguished from each other spectroscopically, the four sets of truncated fragments can be mixed together immediately prior to electrophoresis and separated in a single electrophoresis track. This dispenses with the need for four different tracks per sample required for manual sequencing and also avoids the use of radioactivity. Larger

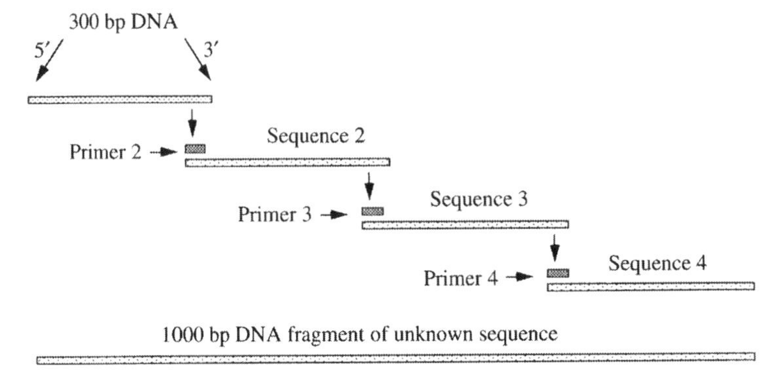

Figure 5.15. Primer walking approach for sequencing 1 kb DNA fragment. New oligonucleotide primers are designed from 3′ ends of previously sequenced region. In this way sequencing of a template can be continued until completed.

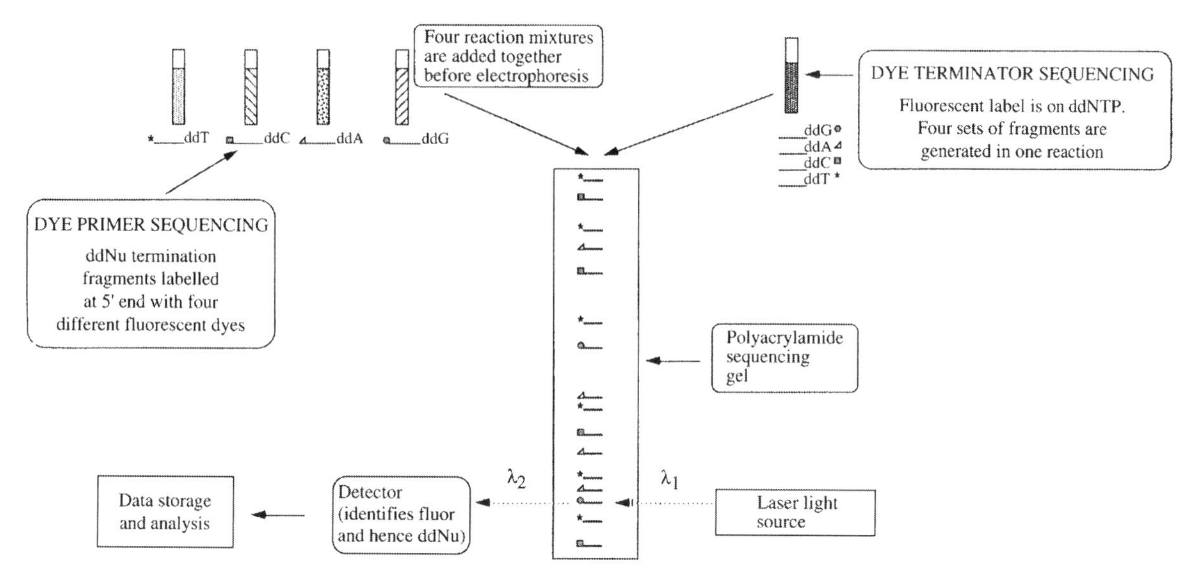

Figure 5.16. Automated DNA sequencing. Instead of a radiolabel, primer strands are labeled with a different fluor or each ddNTP. These have distinct excitation (λ_1) and emission (λ_2) wavelengths and can therefore be easily distinguished by the detector. A single track can be used for separation of all four ddNTP nested fragments.

amounts of sequence data are therefore obtainable from each electrophoresis run. Moreover, since the fluors are detected as they pass near the *bottom* of the gel, longer sequences (up to 600 bp) can be read. A modification of this method is *dye terminator DNA sequencing* which involves incorporation of fluorescent dyes in the ddNTP molecule. In this method, the four reactions may be performed in a single test-tube and electrophoresed in a single lane of the polyacrylamide gel thus giving even better sample-throughput than dye primer DNA sequencing. For some applications, it may be preferred to use a single fluorescent dye (in place of the radiolabel shown in Figure 5.13) and four electrophoresis tracks. Automatic sequencers have inbuilt facilities for storing and analyzing sequence data and may be interfaced with robots in large-scale sequencing projects such as the *human genome project*.

5.4.3 Footprinting of DNA

In eucaryotes, most DNA sequences are *noncoding*, that is they do not code for protein. Examples include *satellite DNA*, *intervening sequences* within genes (*introns*) and *regulatory* sequences. The latter control gene expression in eucaryotes and, for this reason, is widely studied. Good examples of these are *promotor*, *enhancer* and *operator* sequences which are targets for highly sequence-specific DNA-binding proteins (frequently called *factors*). Binding of protein at these sites is responsible for a regulatory effect involving turning expression of a specific gene on or off.

A widely-used experiment allowing the identification of regulatory DNA sequences is *DNA footprinting*. This experiment (Figure 5.17) involves mixing genomic DNA (or smaller fragments derived from genomic DNA) with an extract containing the protein factor under investigation followed by digestion with a small amount of DNA-ase I. Regions of DNA to which protein is bound are protected against DNA-ase digestion. The DNA is then separated from protein, denatured with urea and analysed in polyacrylamide gels. Comparison of samples digested in the presence and absence of DNA-binding protein allows identification of DNA fragments for which the protein is specific. These can then be electroeluted, cloned and sequenced. Studies with a range of DNA-binding proteins has allowed the identification of *consensus* recognition sites for such proteins and these allow screening of new DNA sequences to find putative regulatory sequences. An adaptation of DNA footprinting allows simultaneous identification and sequencing of the regulatory site (Figure 5.18).

5.4.4 Single Strand Conformation Polymorphism Analysis of DNA

If double stranded DNA is first denatured into single stranded molecules and then placed in nondenaturing conditions, DNA molecules can adopt specific *conformations* as a result of sequence-specific *intramolecular* hydrogen bonding. Single stranded molecules of similar mass may show slight differences in their electrophoretic mobility as a

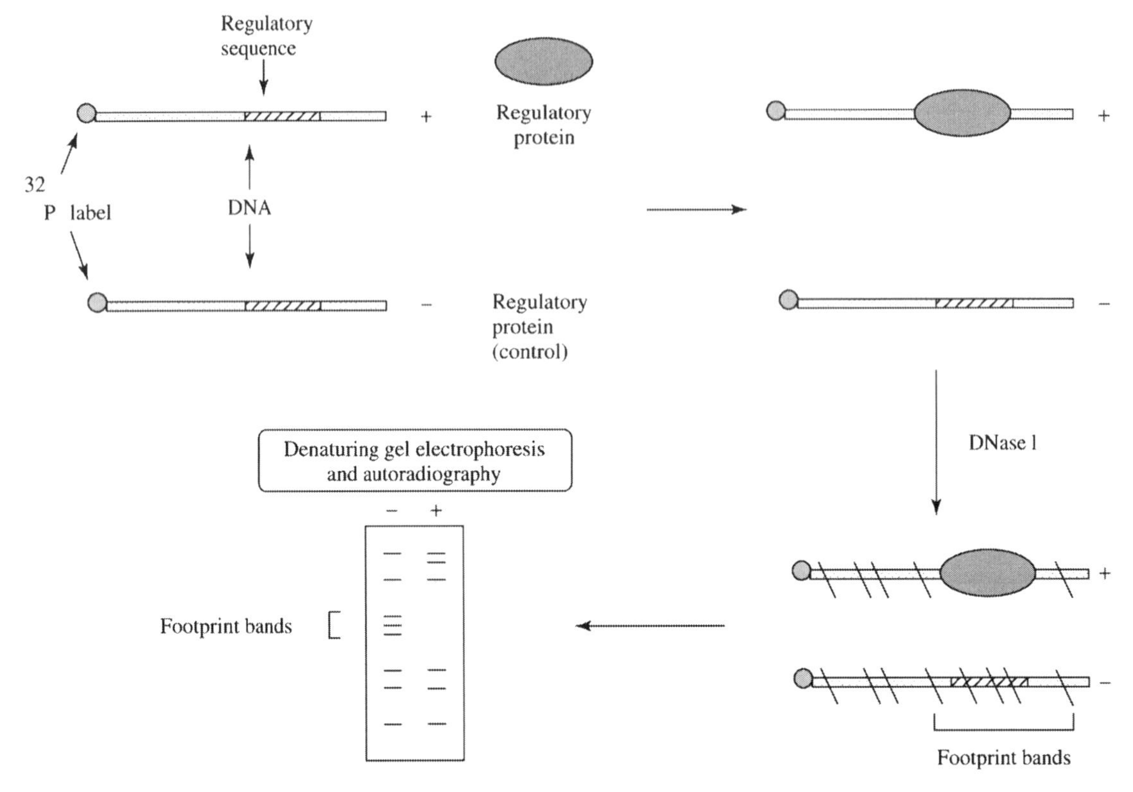

Figure 5.17. DNA footprinting. Genomic DNA is incubated with DNA-binding, regulatory protein followed by treatment with DNase I. Sufficient DNase I is added to give approximately one cleavage (\) per DNA strand. Protein is then removed and the DNA is denatured and electrophoresed in polyacrylamide gels. Sample containing the DNA-binding protein (+) is missing bands present in the control (−) sample not exposed to DNA-binding protein. These are called *footprint bands* and represent DNA sequences which are protected against DNase I digestion by the DNA-binding protein.

result of *mutations* or *polymorphisms* which cause small differences in conformation. Indeed, two molecules of DNA differing in only a single base can be distinguished electrophoretically by a *band-shift* in polyacrylamide gels. This forms the basis of a very useful and widely-used screening method for the identification of mutations in DNA called *single strand conformation polymorphism (SSCP) analysis* (Figure 5.19; Example 5.4).

The procedure involves heating DNA fragments (200–400 bp) of a single gene from a range of samples (e.g. different human individuals) in either NaOH or formamide to denature double-stranded DNA. This is then chilled and loaded onto a nondenaturing polyacrylamide gel. During electrophoresis, variants of a single DNA molecule will adopt slightly different conformations resulting in slightly different electrophoretic mobilities. This procedure may be performed with a number of samples simultaneously and is more convenient than direct sequencing for screening of samples for mutations (although the result is later confirmed by sequencing). This technique has come to be widely-used in the screening of human DNA for mutations responsible for genetically-based diseases.

5.5 ISOELECTRIC FOCUSING (IEF)

We have seen that proteins have a range of electrically-charged groups on their surface which contribute to protein solubility and which underlie the different behaviour of individual proteins in experimental systems such as non-denaturing electrophoresis (Section 5.2) and ion exchange chromatography (Section 2.4). The net charge on a protein is heavily influenced by pH since the groups on the protein surface are *titratable*. At acidic pH most proteins are positively charged while at alkaline pH they are negatively charged (Figure 2.9). However, in between these extremes, the net charge on individual proteins can vary considerably. This variation is a reflection of differences in amino

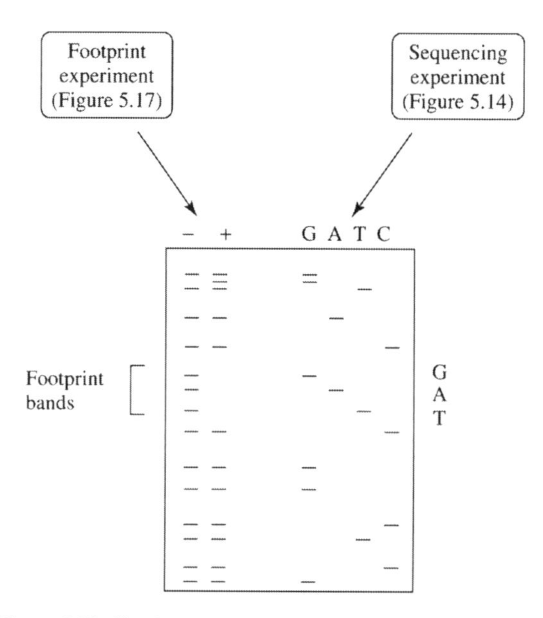

Figure 5.18. Simultaneous identification and sequencing of regulatory sites in DNA. If identical samples are used for both DNA footprinting and sequencing (see Section 5.4.2), these can be electrophoresed on the same polyacrylamide gel. The regulatory sequence determined for the example shown is ATC (complementary to TAG), although in reality, regulatory sequences are usually longer than this.

acid sequence and/or post-translational modification. The pH value at which the protein has no net charge is the *isoelectric point* (pI). Under standard experimental conditions of temperature, buffer composition and in the absence of extensive chemical or post-translational modification, pI may be regarded as a constant property of a protein (Table 5.3). Since this is an important biophysical parameter underlying the pH-dependent behaviour of proteins, we frequently wish to determine pI experimentally. In this section, we shall look at methods used in the determination and application of pI values in proteins.

5.5.1 Ampholyte Structure

The usual approach to determination of pI involves the formation of a *stable pH gradient*. It is technically difficult to achieve such a gradient with buffer components described in Chapter 1 since they would simply diffuse together in free solution. *Ampholytes* are synthetic, low Mr heteropolymers of oligoamino and oligocarboxylic acids (Figure 5.20). The various combinations of amino and carboxylic acid groups possible allow synthesis of a wide range of polymers each possessing a slightly different pI. When a mixture of ampholytes is placed in an electric field, each migrates to its individual pI value thus forming a *gradient* of pI. Since ampholytes also have good buffering

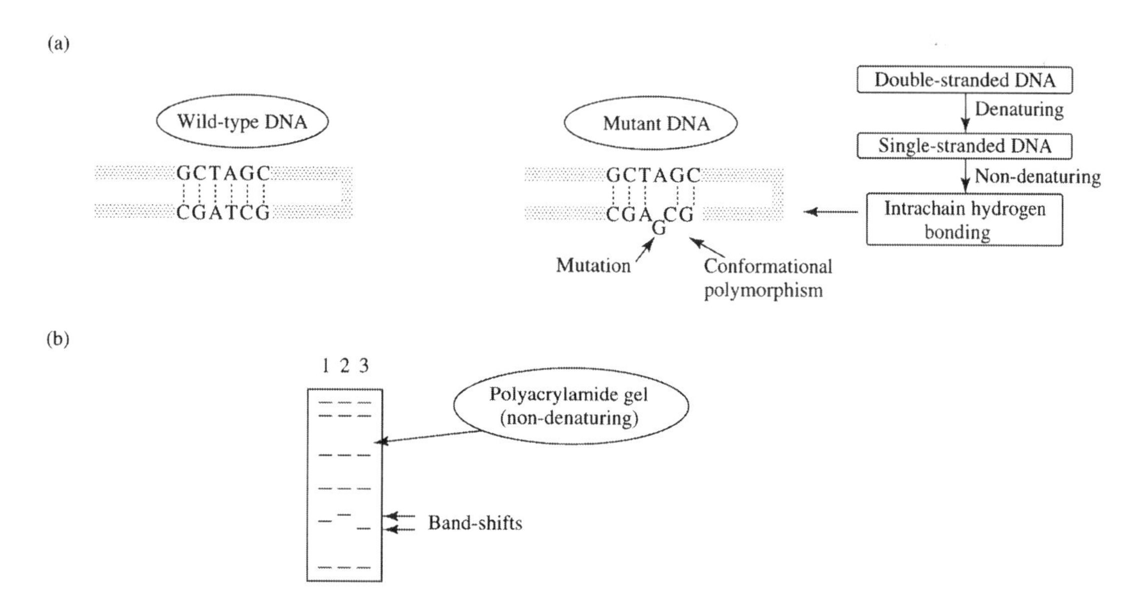

Figure 5.19. Detection of mutations in DNA by single-strand conformation polymorphism (SSCP) analysis. (a) DNA is denatured to make it single-stranded. When placed in non-denaturing conditions, intrachain hydrogen bonding can occur. Mutations in single-stranded DNA may alter the intrachain hydrogen bonding pattern giving a *conformational polymorphism*. (b) Wild-type and mutant DNA have altered mobilities in fragments containing mutations. These are visualised as *band-shifts* in polyacrylamide non-denaturing gels. Bands can either be shifted up (sample 2) or down (sample 3).

Example 5.4 Detection of point mutations in β-globin DNA using single-strand conformation polymorphism.

When denatured DNA is electrophoresed under nondenaturing conditions, intrastrand hydrogen bonds can form which cause the molecule to adopt a number of distinct shapes or conformations. These are visualized in polyacrylamide gels as a series of bands. If the DNA varies from the wild type, some intrastrand hydrogen bonds are altered which, in turn, alters the pattern of conformations formed. This conformational change changes the pattern of bands visible in polyacrylamide gels. This experiment is called *single-strand conformation polymorphism* (SSCP) analysis.

Mutations were randomly introduced into a 193 bp fragment of the promotor region of the β-globin gene in a plasmid. Large variation was observed in the pattern of fragments visualized in a 5% polyacrylamide gel as shown below;

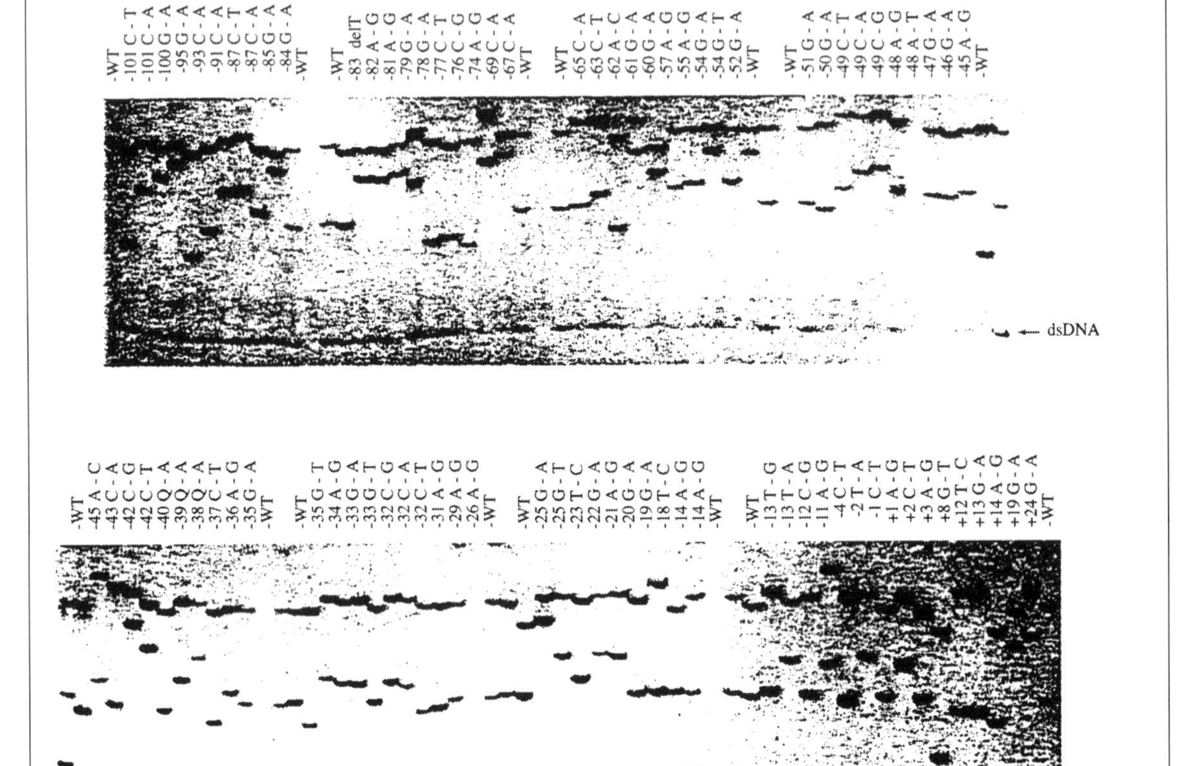

This experiment shows that all mutations were detectable by SSCP analysis. Interestingly, neither the precise position of a mutation nor its nature (i.e. whether it was a purine → pyrimidine, pyrimidine → purine or deletion) determined how different the conformation was from the pattern of the wild type. Because of its exquisite ability to distinguish otherwise very similar DNA sequences, SSCP is now the major method for detection of mutations. However, once an SSCP is detected, it is necessary to sequence the DNA displaying the SSCP in order to identify the actual mutation.

Reproduced from Glavac, D. and Dean, M. (1993) Optimization of the single-strand conformation polymorphism (SSCP) technique for detection of point mutations. *Human Mutation*, **2**, 404–414, with permission of John Wiley & Sons Ltd.

capacity, a gradient of pH is thus stabilized across the electrical field. Commercially available ampholytes may be selected to cover a wide (e.g. 3–9) or narrow (e.g. 6–7) pH range.

An alternative to the use of free ampholytes is provided by *immobilized pH gradients*. This involves the use of derivatives of acrylamide which contain weak acid or base groups called *immobilines* (Figure 5.21). The acid groups are carboxylic acids with pK values of 3.6 or 4.6 while the basic groups are tertiary amino groups with pKs of 6.2, 7.0, 8.5 or 9.3. At least one each of the acid and alkaline immobilines are mixed together in the presence of acrylamide monomers to form a polyacrylamide gel. By varying the identity and number of immobilines from the two categories available, a variety of pH gradients may be generated. A particular advantage of these gradients over those formed with free ampholytes is that they cover a very narrow pH-range allowing finer resolution between similar pI values.

Immobilized pH gradients are formed by mixing two solutions, one containing acidic immobiline(s) plus glycerol, the other containing alkaline immobiline(s). The acidic immobiline(s) will be concentrated near the end of the gradient while the alkaline immobiline(s) will be concentrated near the top. In between, there will be a continuous gradient of decreasing carboxylic acid groups and increasing amino groups. These groups act as local buffers resulting in a pH

Table 5.3. pI values of selected proteins at 10 °C

Protein	pI
Pepsin	1.00
Pepsinogen	2.80
Amyloglucosidase	3.50
Glucose oxidase	4.15
Soyabean trypsin inhibitor	4.55
Ovalbumin	4.60
Bovine serum albumin	4.90
Urease	5.00
β-Lactoglobulin A	5.20
Carbonic anhydrase B	5.85
Myoglobin (acidic band)	6.85
Myoglobin (basic band)	7.35
Trypsinogen	9.30
Cytochrome C	10.70
Lysozyme	11.00

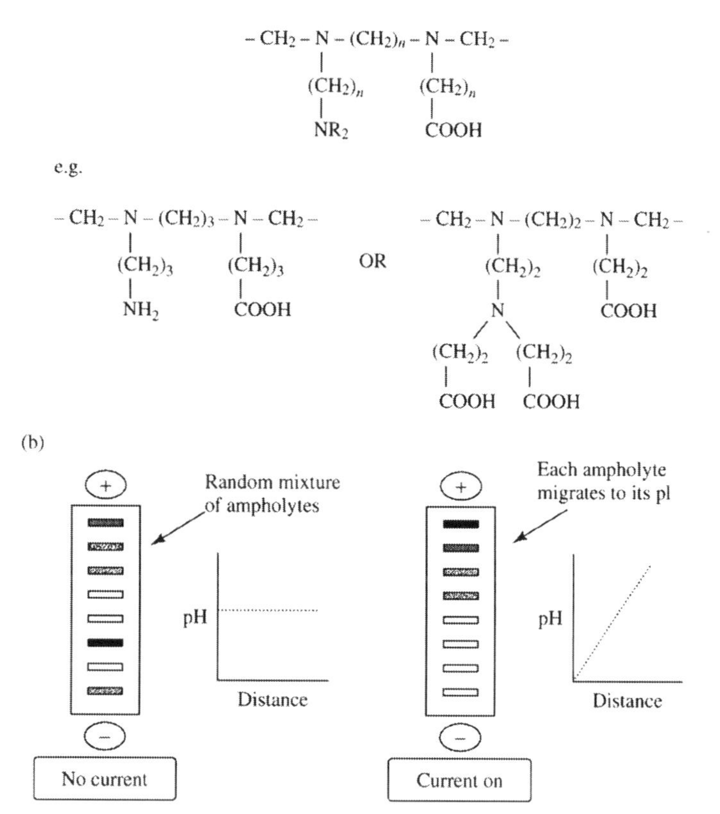

Figure 5.20. Formation of stable pH gradient with ampholytes. (a) Ampholytes are short polymers containing both NH_2 and –COOH groups. Two example structures are shown although many different individual structures will be present in a mixture ($n = 2$–3; $R = H$ *or* (CH_2)–COOH). (b) When mixed together in the absence of an electric field (left), a single net pH results. However, in the presence of an electric field (right), individual ampholytes migrate to their pI values. Here they act as buffers, resulting in a pH gradient.

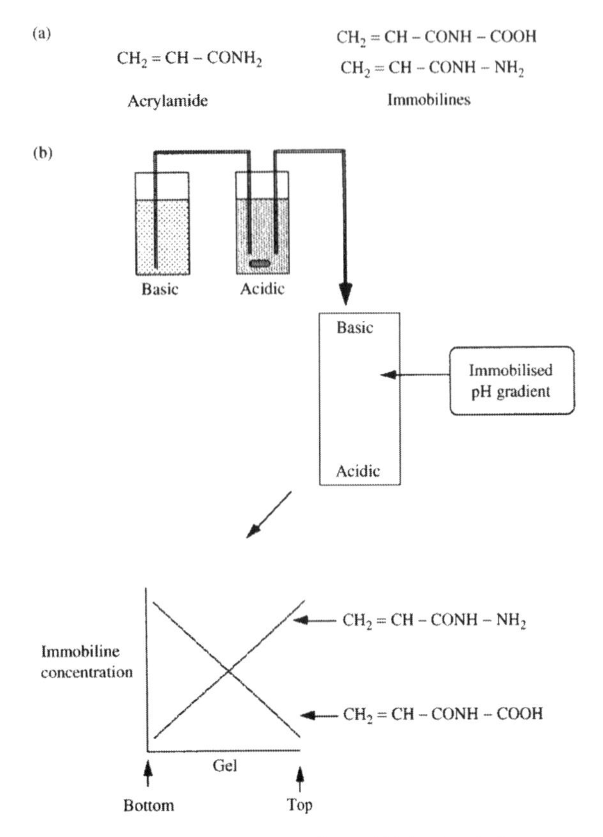

Figure 5.21. Immobilized pH gradients, (a) The chemical structure of immobilines. These are derivatives of acrylamide and may be incorporated into polyacrylamide gels essentially as described in Figure 5.3. (b) Formation of immobilized pH gradient. Glycerol is mixed with acidic ampholines which predominate at the bottom of the polyacrylamide gel. A gradient of decreasing acidic and increasing basic ampholines is formed. Once the gel is polymerized this gradient is immobilized.

gradient which, because they are covalently linked to the polyacrylamide gel, is highly-stable.

5.5.2 Isoelectric Focusing

When a protein is placed in a pH gradient, it will migrate towards its pI (Figure 5.22). At pH values above pI, the protein will possess a net negative charge which will make it migrate towards the anode. At pH values below pI, the protein will have a net positive charge and will migrate towards the cathode. Only at the pI will the protein have zero net charge and, hence, no electric mobility (Section 5.1). This experiment is called *isoelectric focusing* (IEF) because it involves *focusing* of molecules at their individual pI values.

IEF may be performed in *glass tubes* or in a *horizontal flat-bed* format (Figure 5.23). Pre-cast IEF gels are commercially available widely. Alternatively, ampholytes may be mixed with components necessary for the formation of a polyacrylamide gel. Once the gel has polymerized, an electric current is applied forming a pH gradient. Protein sample is then loaded (as an aqueous solution) and electrophoresis is resumed. When focusing is complete, the gel is fixed with trichloroacetic acid. In the case of flat-bed horizontal gels, pI is determined by comparison with standard proteins of known pI run on the same gel. In tube gels, the gel is cut into small pieces along its length and the pH of each piece is then measured. A graph of distance along the gel *versus* pH may then be plotted. The migration distance of the protein under investigation allows estimation of pI.

5.5.3 Titration Curve Analysis

Knowledge of pH-dependent effects on net charge of proteins is helpful in many experimental situations such as, for example, designing ion-exchange protocols (Chapter 2). Since the complement and pH-dependent behaviour of amino acid side-chains is highly individual, we find that a characteristic *titration curve* can be obtained for each protein. Such curves can be determined with the use of ampholytes in IEF gels (Figure 5.24). A pH gradient is first established in a low-percentage (e.g. 4–5%) polyacrylamide gel. Then the gel is rotated 90° and protein sample is applied in a trough at the centre of the gel. Proteins migrate across the pH gradient in a manner dependent on their charge while ampholytes, since they are at their pI and uncharged, remain in position and maintain a stable pH gradient. Near pH 3, most proteins have a net positive charge while, near pH 9, most have a net negative charge. Since the gel is of low percentage acrylamide, the main factor determining mobility at and between these extremes is net charge rather than protein mass or shape. This experiment allows the identification of specific pH values where two proteins may have different charge, thus facilitating rational design of an ion exchange protocol.

5.5.4 Chromatofocusing

If an ion exchange resin (Section 2.4.1) were equilibrated at pH 9.0 and then exposed to a second buffer at pH 7.0, a pH gradient would be formed between these two buffers. However, this gradient would be *narrow* because of the limited buffering capacity of the two buffers across the whole range 9.0–7.0 and it would be quite *unstable*. A chromatographic technique called *chromatofocusing* (Figure 5.25) based on the same principle as IEF uses the high buffering capacity of an ion-exchange resin coupled with *amphoteric* buffers to generate a stable pH gradient. Amphoteric buffers are

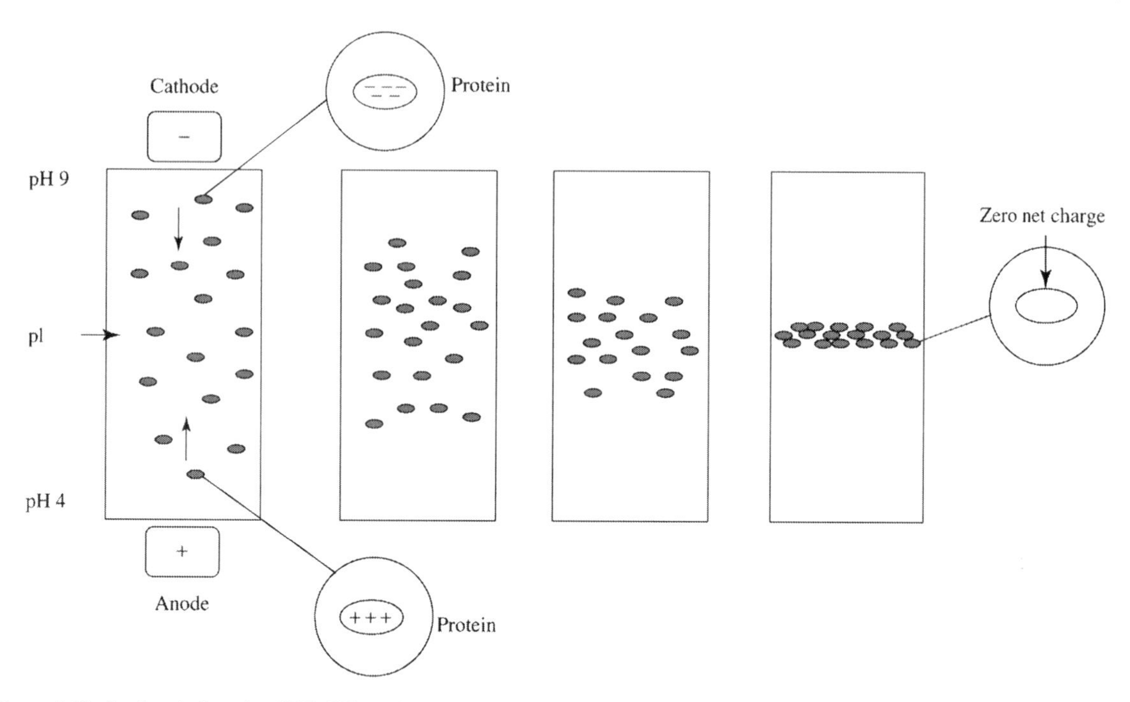

Figure 5.22. Isoelectric focusing (IEF). When placed in a pH gradient, proteins with positive net charge will migrate towards the cathode while those with a negative net charge migrate towards the anode (arrows). Proteins at their pI have no net charge and hence no electrophoretic mobility.

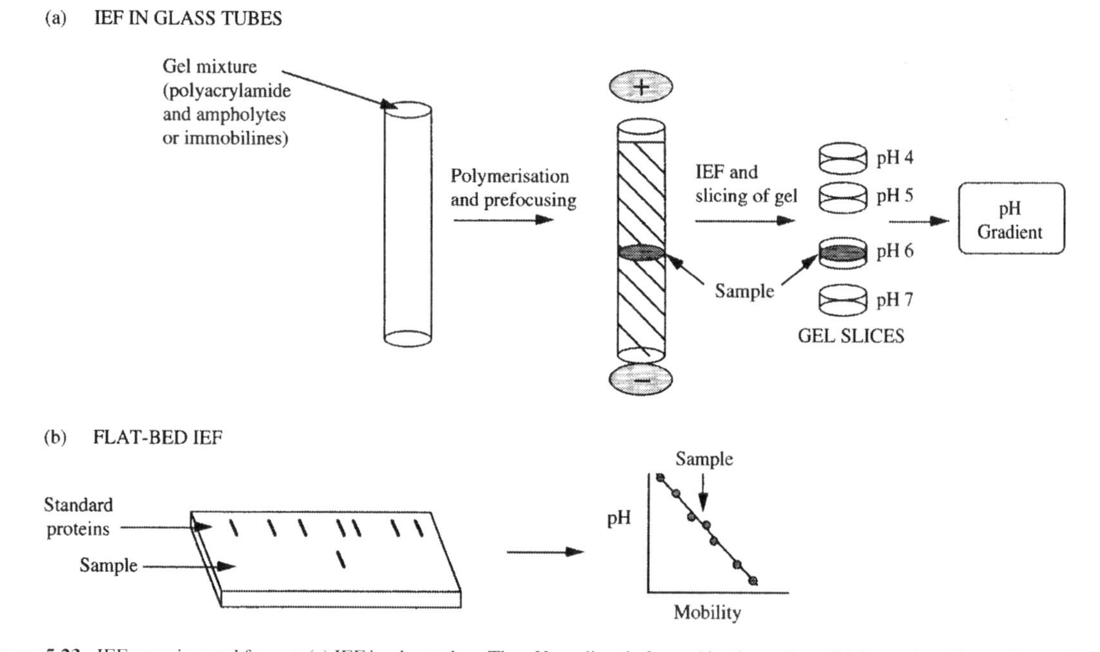

Figure 5.23. IEF experimental format, (a) IEF in glass tubes. The pH gradient is formed in glass tubes which are placed in an electrophoretic device. One sample is loaded per tube. After IEF, the gel is cut into thin slices and the pH of each slice is determined. From this a graph of mobility *versus* pH may be generated. From the mobility of protein band, pI of protein may be determined, (b) The flat-bed format allows IEF of standard proteins of known pI alongside sample protein. Comparison with standards allows determination of pI.

(a) Pre-electrophoresis establishes a stable pH gradient

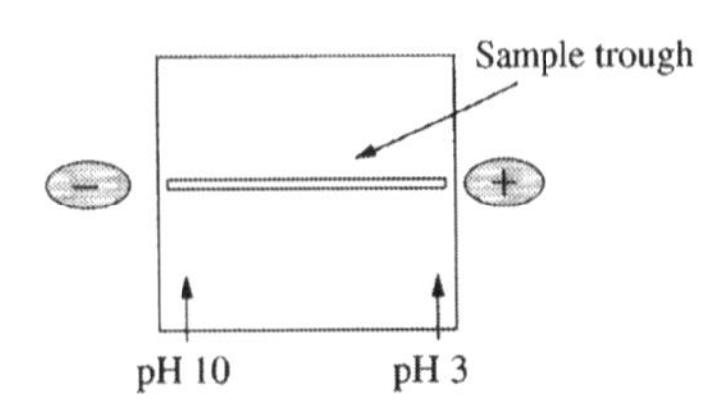

(b) Gel rotated 90°, sample is loaded in trough and electrophoresed

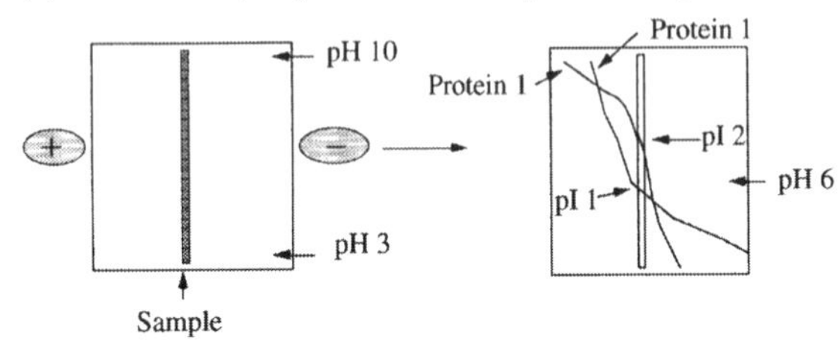

Figure 5.24. Titration curve analysis of proteins, (a) A stable pH gradient is established in a 5% polyacrylamide gel containing ampholytes by pre-electrophoresis. (b) Gel is rotated 90° and sample is loaded in central trough. Electrophoresis causes proteins 1 and 2 to migrate differently due to their differing net charge across the pH range. Ampholytes do not move since they have no net charge at their pI. Proteins 1 and 2 give different titration curves and their pIs may be measured as the point where the curve and sample trough intersect. From this experiment we could conclude that, at pH 6, protein 1 has a net positive charge while protein 2 has a net negative charge.

similar to ampholines in that they have good and even buffering capacity across a broad range of pH. In this experiment it is possible to achieve selective elution of proteins very close to their pI values.

An anion-exchange resin is first equilibrated at a particular pH (e.g. pH 9.4), followed by amphoteric buffer. This buffer has good buffering capacity across a wide pH range (e.g. 9.0–6.0) and automatically forms a pH gradient through the column in the range 8.3–6.4. Sample proteins are now applied. At the top of the column (pH 6.4), the protein will be positively-charged and will pass freely through the resin without binding but encountering a gradually increasing pH. Eventually, the protein will reach its pI and, soon after this, acquire a negative charge which will allow it to bind to the anion exchange resin. As elution buffer continues to be applied to the column, the pH at the binding site of the protein would be gradually lowered giving the protein again a negative charge and causing it to dissociate from the resin. The protein would elute further down the column until a pH is reached which would again give it a positive charge causing it to bind to the resin once more. In this way, individual proteins move down the column being continuously desorbed and adsorbed as the pH gradient progresses through the resin bed. If a second batch of sample protein were added to the column sample proteins would behave similarly to those of the first batch and quickly 'catch up' at pH values just below protein pI. This *focusing* effect gives chromatofocusing a high capacity. Eventually, each protein in the mixture applied elutes from the resin at a pH value slightly above its pI.

5.6 IMMUNOELECTROPHORESIS

Antibodies are proteins which are highly evolved for specific recognition of immunological determinants called *antigens*. These antigens may be low molecular mass molecules such as sugars or else more complex structural components of biopolymers such as proteins. The specificity of this recognition makes antibodies, especially *immunoglobulins* of the G class (IgGs), particularly useful molecular probes for the structure of proteins, macromolecular complexes and viruses. The principle sources of antibodies are either experimental animals such as rabbits (which raise *polyclonal* antibodies in their serum in response to injected antigen) or mouse hybridoma cells (which produce *monoclonal* antibodies in the culture medium in which cells are grown). We have seen how, once electrophoretically separated, we often wish to achieve *specific recognition* of individual proteins in complex mixtures (Section 5.2.3). Antibodies have found a number of uses in such specific detection.

When immunodetection is combined with electrophoretic separation of proteins, this technique is called *immunoelectrophoresis* and it will be described in Sections 5.6.2–5.6.5. It should be noted that immunoelectrophoresis techniques are generally performed in *nonrestrictive* gels with very large pores relative to the size of the antigen and antibody molecules. For proteins, these gels are composed usually of agarose but for low molecular mass molecules such as peptides, they may be polyacrylamide. The purpose of nonrestrictive gels is to encourage separation based only on differences in charge between proteins rather than differences in mass or shape. They also allow very fast separations which may take less than one hour compared to the days required for immunodiffusion procedures. The techniques described in this section are based on recognition of antigens possessing *native structure*. Some other popular immunodetection methods (e.g. western blotting; Section 5.10.2) involve recognition of *denatured antigens*.

5.6.1 Dot Blotting and Immunodiffusion Tests with Antibodies

Before carrying out immunoelectrophoresis, it is first necessary to confirm that antibodies are present which are specific for an antigen of interest and which have a sufficiently high

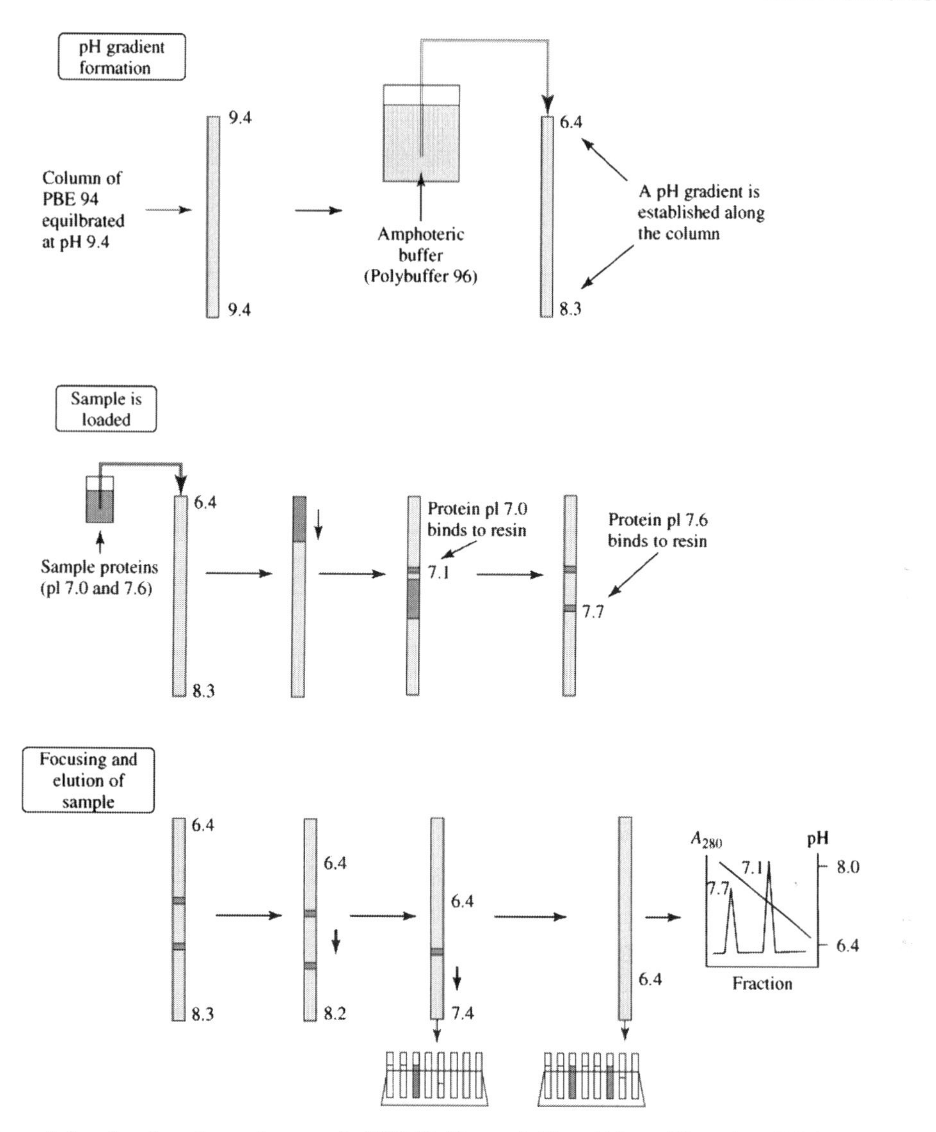

Figure 5.25. Chromatofocusing. An anion exchange resin (PBE 96; Pharmacia Biotech) is equilibrated at pH 9.4. An amphoteric buffer (Polybuffer 96) is applied, resulting in the formation of a stable pH gradient (8.3–6.4). A protein sample consisting of two components (pi 7.0 and 7.6) is applied. These bind to resin when negatively charged (i.e. at pH 7.1 and 7.7, respectively). As polybuffer 96 continues to be loaded, the pH gradient moves down the column, continuously desorbing and re-adsorbing the two proteins. They elute at pH values near their pIs.

affinity or *avidity* for this antigen to give sensitive detection. In this section two *nonelectrophoretic* experiments will be described in which antigens and antibody are brought together to allow detection and quantification of antibody concentration. Measurement of presence and level of antibody is necessary since there is considerable variability in the *immunogenicity* of different antigens, some antigens giving a strong immunological response in individual experimental animals but not in others. Conversely, some antigens are highly immunogenic while other antigens give little detectable immunological response.

When a protein and antibody encounter each other, a specific binding event occurs between the *variable region* of the antibody and a region of the antigen called the *epitope*. In the case of polyclonal antibodies (i.e. antibodies raised to a range of epitopes) a number of binding sites may exist on the surface of a single antigen. Monoclonal antibodies represent a population of identical antibodies which recognize

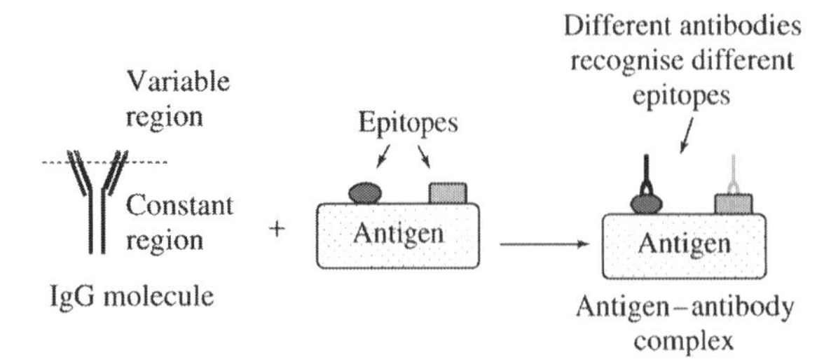

Figure 5.26. Binding of antibody to antigen. The variable region of the antibody binds specifically to a region of the antigen called the epitope. Different epitopes in protein structures may be recognised by specific antibodies. Polyclonal antibodies recognise many epitopes in a single antigen while monoclonal antibodies recognise only one.

a single epitope. If an excess of either antibody or antigen is used, staining of the complex formed between them can be used to measure levels of antigen or antibody, respectively. In *dot blotting*, antigen is applied to a nitrocellulose membrane to which it binds (Figure 5.26). Additional binding sites are saturated with nonantigen protein such as *bovine serum albumin* (BSA). A solution of antibody may then be added allowing specific binding of the antibody to the antigen. The complex formed between these molecules may be *quantified* by staining with a second antibody which has two important properties. Firstly, it is specific for the *species* in which the first antibody was raised, usually recognizing the *constant region* of the IgG molecule (e.g. antigoat IgG or antirabbit IgG). Secondly, an enzyme (e.g. horseradish peroxidase) is *covalently attached* to it which facilitates quantification. For every molecule of antigen, one antibody molecule will bind. The second antibody, in turn, will specifically bind to this IgG. Since an enzyme is attached to this second antibody, many thousands of molecules of substrate will be converted by this enzyme into a product which is coloured. Considerable *amplification* of staining is therefore achieved compared to that which might be expected, for example, with a protein dye which would only stain the original antibody-antigen complex. This factor makes dot blotting a highly-sensitive detection procedure for antibodies. It is usual to carry out the experiment with a range of dilutions of the antibody solution, thus giving an indication of the antibody concentration or *titre* in serum taken from an experimental animal or in hybridoma cell culture medium.

A second frequently-used, nonelectrophoretic procedure to demonstrate the presence and specificity of antibodies in various sources is *Ouchterloney double-immunodiffusion* assay in agarose gels. At low ratios of antigen to antibody, initially soluble antibody-antigen complexes are formed. As this ratio is increased, however, extremely large macromolecular assemblies of repeating antibody-antigen-antibody-antigen units are formed. This occurs at an optimum antigen/antibody ratio which is called the *equivalence point*. This complex is insoluble and will *immunoprecipitate* to form *precipitin*. In double immunodiffusion, a small amount (3–5 μl) of antiserum is placed in a central well and a range of dilutions of antigen is added to similar wells surrounding it (Figure 5.27). Both antigen and antibody solutions freely diffuse through the agarose (hence *double immunodiffusion*). If there is specific binding of antibody to the antigen, then precipitin forms. This is visible to the naked eye but may also be stained with coomassie blue after washing away nonprecipitated protein. If two different antigens (e.g. two different proteins) recognize the same antibody, a *common* precipitin line is formed, that is the two lines fuse (Figure 5.28). These proteins are said to have *immunological identity*. If the proteins are *nonidentical*, the precipitin lines cross. In the case of *partial identity* between the antigens, a *spur* is formed. This type of experiment gives useful evidence for structural similarity between different antigens.

5.6.2 Zone Electrophoresis/Immunodiffusion Immunoelectrophoresis

In *zone electrophoresis/immunodiffusion* immunoelectrophoresis, it is first necessary to perform electrophoretic separation of antigens in an agarose gel containing a trough as shown in Figure 5.29. After electrophoresis, a solution of antibody is poured into this trough. As with Ouchterloney double immunodiffusion, antigens and antibodies diffuse towards each other through the agarose and form precipitin *arcs*. This technique is useful for *qualitative* detection of antigens in serum, cell extracts or other preparations.

5.6.3 Rocket Immunoelectrophoresis

At pH 8.0, most proteins possess a net negative charge and will electrophorese towards the anode. IgG molecules possess little net charge at this pH, however, and consequently experience little *electrophoretic mobility*. However, because of movement of water towards the cathode called *electroosmotic flow* which will be described in detail later, (Section 5.10.1) IgGs migrate towards the cathode. Advantage of this is taken in *rocket immunoelectrophoresis* (Figure 5.30). Antigen is placed in wells cut in an agarose gel containing antibodies. During electrophoresis, most antigens migrate towards the anode while IgG molecules migrate towards the cathode (electroosmotic flow). At first, soluble antibody-antigen complexes are formed at low antibody concentration and these continue to move towards the anode. However, as electrophoresis proceeds and all the antigens enter the gel, insoluble precipitin forms. When the gel is

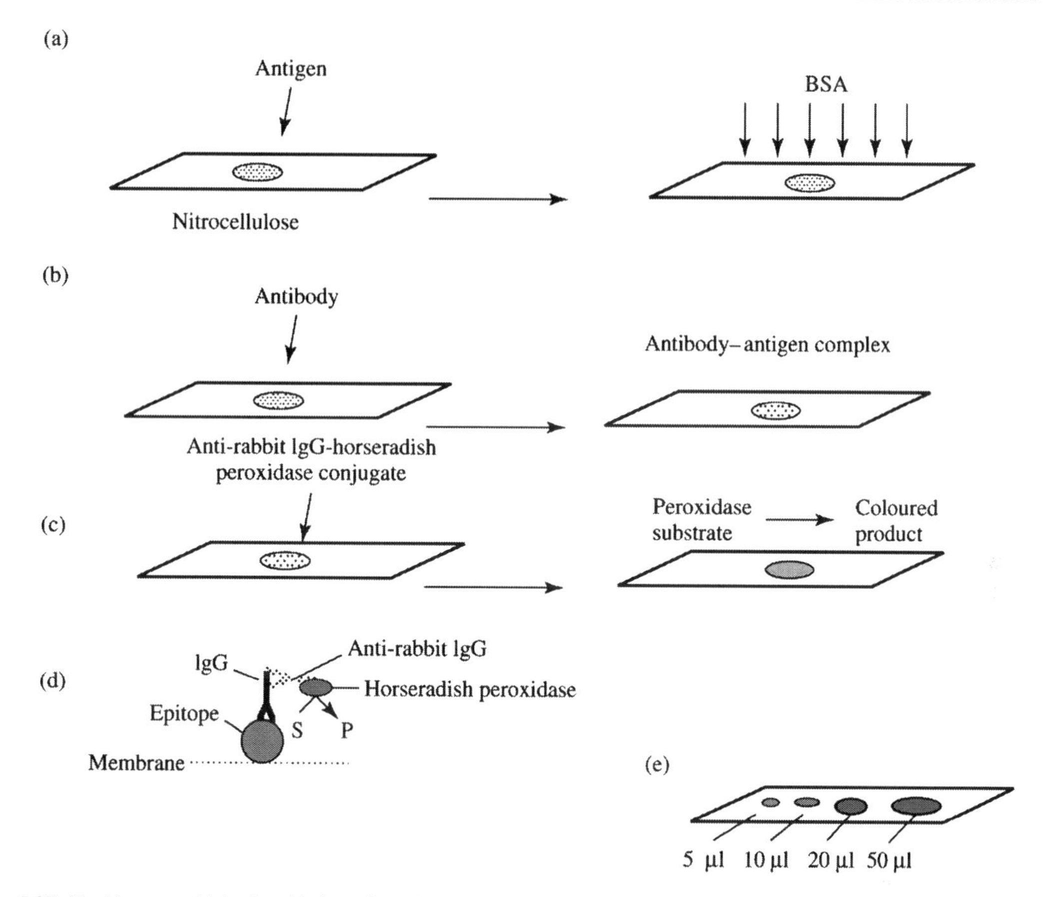

Figure 5.27. Dot blot assay, (a) Antigen binds to nitrocellulose membrane and bovine serum albumin (BSA) is added to saturate non-specific binding sites, (b) Antibody is added to membrane forming an antibody–antigen complex, (c) A second antibody is added which binds to the constant region of the IgG molecule. This antibody is conjugated to the enzyme, horseradish peroxidase. (d) Schematic of arrangement on nitrocellulose. (e) A dot blot of increasing volumes of antiserum.

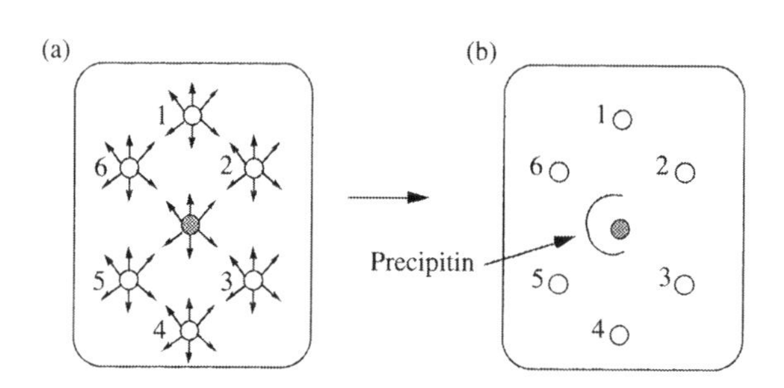

Figure 5.28. Ouchterloney double immunodiffusion. (a) Antibody is placed in a central well (shaded) of an agarose gel. A range of dilutions of antigen is placed in wells 1–6 (e.g. 1 : 10 000 to 1 : 100 in phosphate buffered saline). Both antibody and antigen solutions diffuse outwards in all directions (arrows), (b) When equivalence is reached, precipitin is formed which is visible as a white arc against a clear background.

stained, this precipitin is found to form characteristic *rocket* shapes around each antigen well. The area under this shape is proportional to the antigen concentration in that well since greater amounts of antigen will migrate further towards the anode. Therefore, using standard antigen solutions of known

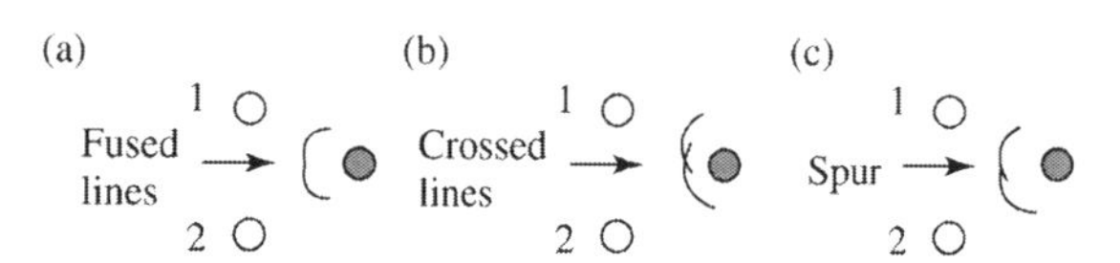

Figure 5.29. Patterns of reaction in Ouchterloney double immunodiffusion assays, (a) Fused precipitin lines mean that the antigens in wells 1 and 2 have immunological identity, (b) Crossed precipitin lines mean that the antigens in wells 1 and 2 are non-identical immunologically. (c) A spur between precipitin lines mean that the antigens in wells 1 and 2 are partially identical immunologically.

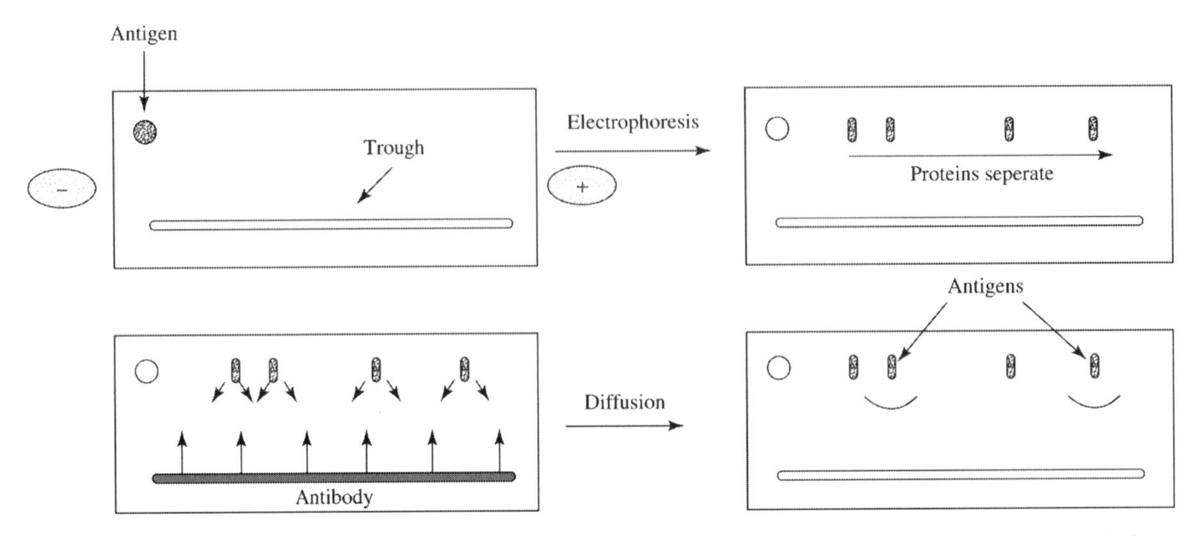

Figure 5.30. Zone electrophoresis/immunodiffusion immunoelectrophoresis. Protein antigens are first separated electrophoretically in an agarose gel. Antibody is added to the trough in the gel. Antibodies and antigen diffuse through the gel and, at equivalence, form precipitin arcs. In this experiment, two antigens are identified in the protein mixture.

concentration, a calibration curve may be generated and the antigen concentration of a range of samples conveniently determined.

5.6.4 Counter Immunoelectrophoresis

In *counter immunoelectrophoresis,* antigen is placed in a well at the cathode end of an agarose gel while antibody is placed in a well at the anode end (Figure 5.31). The antigen migrates towards the anode (electrophoretic flow) while the antibody migrates towards the cathode (electroosmotic flow). When they meet, a precipitin band is formed.

5.6.5 Crossed Immunoelectrophoresis (CIE)

A fourth form of immunoelectrophoresis is *two-dimensional* or *crossed immunoelectrophoresis* (CIE). This technique involves initial (i.e. first dimension) electrophoretic separation of antigens in an agarose gel. This is followed by cutting out a section of gel containing the separated proteins (Figure 5.32). A second agarose gel containing antibodies (i.e. second dimension) is formed against this section and electrophoresis is performed at 90° relative to the axis of the first gel. Precipitin arcs form similar to those of rocket immunoelectrophoresis and these allow quantification of several antigens in a single sample.

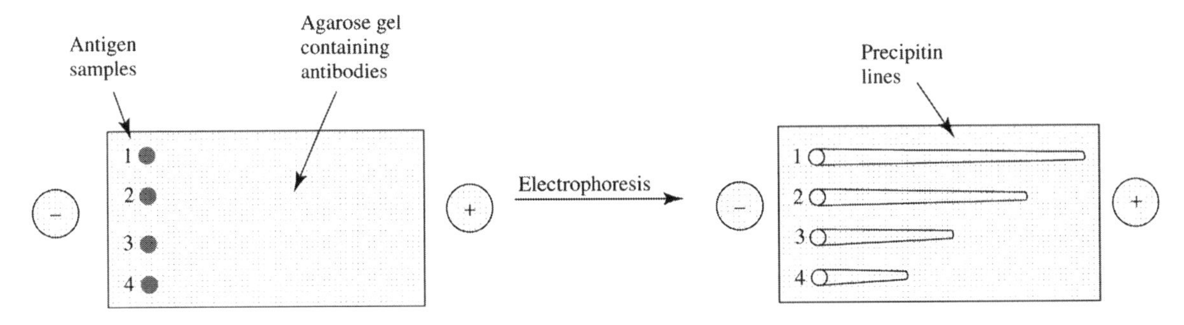

Figure 5.31. Rocket immunoelectrophoresis. An agarose gel is prepared containing antibodies. Decreasing concentrations of antigens are placed in wells 1–4. As antigens electrophorese towards the anode, precipitin is formed at the equivalence point. The area of the rocket shapes formed is proportional to the concentration of antigen in the well.

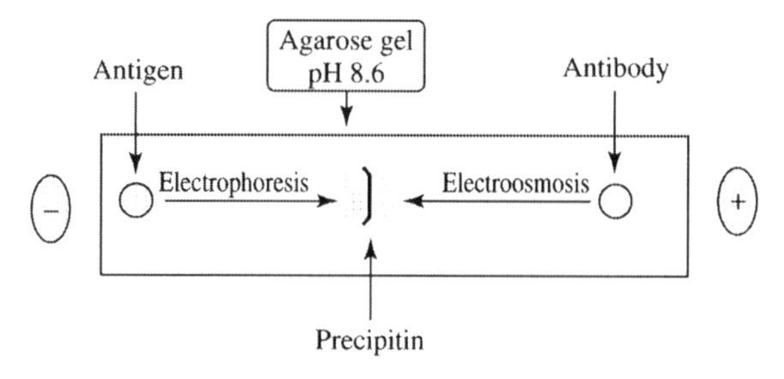

Figure 5.32. Counter-immunoelectrophoresis. Antigen moves towards the anode by electrophoretic flow. At pH 8.6, IgG is uncharged but is carried towards the cathode by electroosmotic flow. At the equivalence point, precipitin forms.

5.7 AGAROSE GEL ELECTROPHORESIS OF NUCLEIC ACIDS

5.7.1 Formation of an Agarose Gel

Agarose gels are formed by first heating a suspension of agarose and then allowing it to cool with consequent formation of a gel. Heating disrupts the largely *intramolecular* hydrogen bonding pattern of agarose while cooling allows the reformation of hydrogen bonds (Figure 5.33). At least some of the time, these reformed bonds will be *intermolecular* in nature and it is these bonds which are responsible for gel formation and stabilization. Unlike other polysaccharide gels, agarose is largely uncharged and does not give secondary ionic interactions with nucleic acids or proteins. It is nonetheless a hydrated polymer due to extensive hydrogen bonding to water. Pores formed in agarose gels are much larger than those in polyacrylamide. This has important implications for the mass and shape of molecules for which agarose gel electrophoresis is suitable (Section 5.8.3).

5.7.2 Equipment for Agarose Gel Electrophoresis

Although both vertical and horizontal gels are possible in principle, in practice agarose gels are usually formed in horizontal plastic moulds (Figure 5.34). This is sometimes called the *submarine* gel electrophoresis technique. Sample wells are formed in the gel by insertion of a plastic comb into the warm gel suspension before setting. This comb is removed once the gel has set producing a set of wells into which individual samples may be loaded. The gel is placed in a flat-bed electrophoresis tank for electrophoretic separation. This experiment allows side-by-side separation of a number of samples which facilitates direct comparison between them. For preparative electrophoresis, a single large well is formed by use of a *continuous* comb.

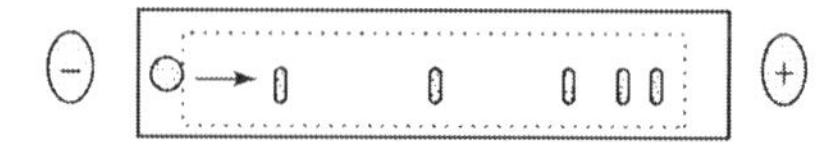

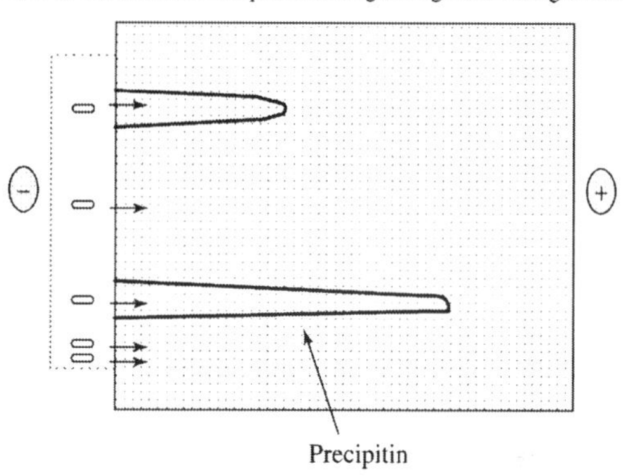

Figure 5.33. Crossed immunoelectrophoresis (CIE). (a) Antigens are separated by electrophoresis in agarose gels, (b) A slice is taken from this gel and a second gel is poured against it which contains antibodies. Electrophoresis is then performed at 90° to the direction of gel in a). At the equivalence point, precipitin forms. Two of the proteins in the original mixture are antigens for the antibody shown and their relative amounts may be calculated from the area under the peaks.

5.7.3 Agarose Gel Electrophoresis of DNA and RNA

Electrophoresis of nucleic acids is dependent on molecular mass in the range 1–25 kb. For DNA masses *less* than 1 kb it would be necessary to use gels of high percentage agarose but gels of concentration greater than 1% exhibit

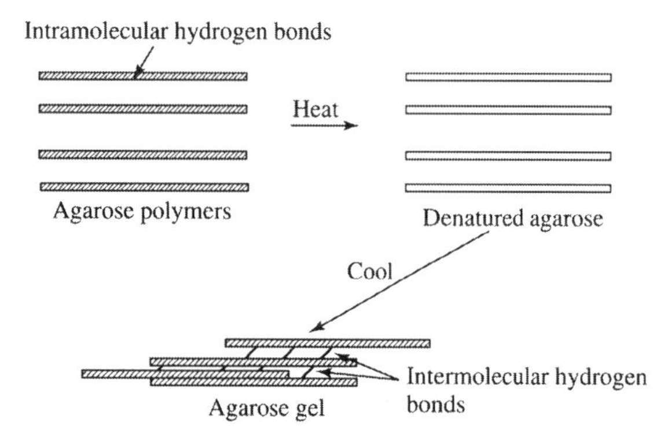

Figure 5.34. Formation of agarose gel. A suspension of agarose is heated to 70 °C which disrupts intramolecular hydrogen bonds between –OH and O groups in agarobiose. On cooling, some intermolecular hydrogen bonds form (dashed lines) which result in a hydrated gel network. The greater the percentage agarose, the smaller the average pore size of the gel.

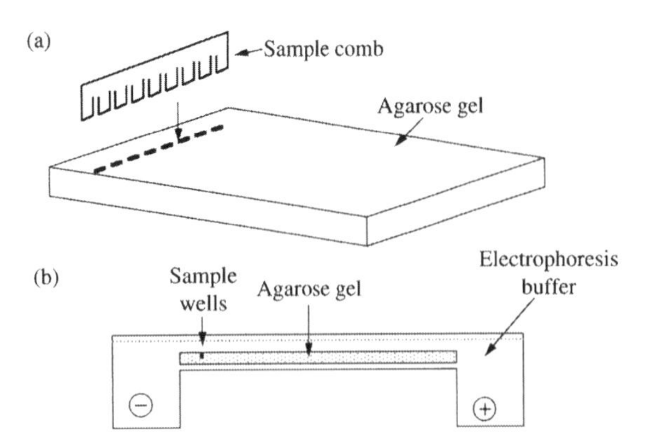

Figure 5.35. Submarine electrophoresis in agarose gels, (a) Gel is formed in horizontal mould. Sample wells are formed in the gel during cooling by insertion of a *comb*, (b) Electrophoresis is performed under electrophoresis buffer (i.e. *submarine*) in a horizontal (i.e. flat-bed) format.

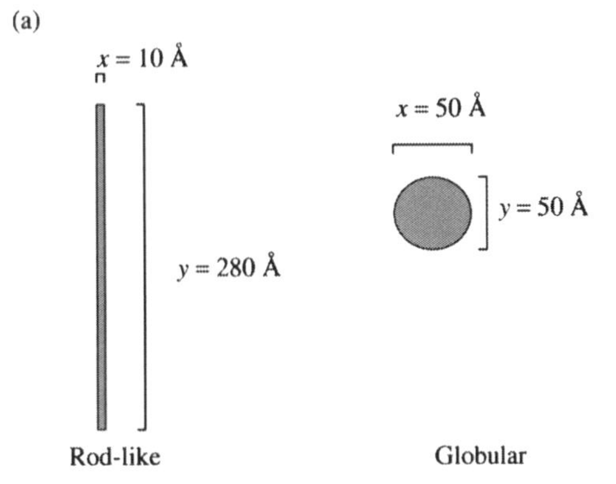

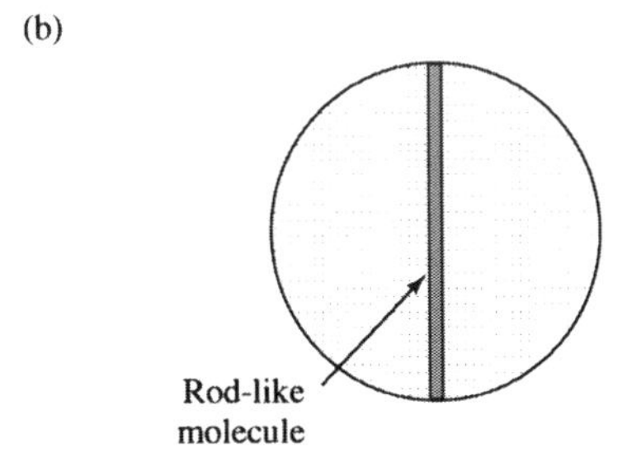

Figure 5.36. Effect of axial ratios on electrophoresis of biopolymers. (a) Rod-like molecules such as nucleic acids and fibrous proteins have a large axial ratio (y/x). The example shown has a ratio of approximately 28 while the globular molecule of similar mass has a ratio of 1 (for simplicity, these representations take no account of the third dimension of molecular size, z). (b) Rod-like molecules migrate as if they have a mass many times greater than their actual mass. The rodlike molecule shown migrates with an apparent mass equivalent to that of the shaded molecule.

high electroosmotic flow (Section 5.10.1) which places an effective *lower* limit on the mass of nucleic acid it is possible to separate. While newer commercially-available agaroses are almost capable of matching their impressive resolving power, in practice polyacrylamide gels are normally used to study nucleic acids of smaller mass (e.g. <200 bp).

We have seen in Section 5.7 that agarose gels are largely nonrestrictive for proteins. However, compared to most proteins, nucleic acids are extremely long molecules with a large *axial ratio* (ratio of longitudinal to transverse axes of the molecule). Since, during migration through the gel the nucleic acid is free to rotate through all possible orientations, it migrates as if it has an extremely large molecular mass (Figure 5.35). The same is true for proteins but, because their axial ratios are much *smaller* than nucleic acids, the *apparent* mass is much closer to the *actual* mass. This means that agarose gels, while not restrictive for compact proteins, are restrictive for nucleic acids and should give separation based on differences in charge and mass. As nucleic acids have a fairly uniform charge-distribution, this means that they separate in practice according to differences in mass. Agarose gels are normally used in the concentration range 0.8–1.5% allowing separation of molecules up to 50 kb (although 20 kb is the upper limit for high-resolution separations).

Above 20 kb, separation of DNA becomes increasingly *independent* of molecular mass. This has been explained by two principal models (Figure 5.36). In the *biased-reptation* model, a DNA molecule larger than the average pore-size of an agarose gel is proposed to display the motion of a crawling snake (i.e. *reptation*) adopting elongated shapes orientated in the direction of the electric field. Large DNA molecules will therefore 'slither' through the agarose pores in a mass-independent manner. An alternative model is supported by later work with both fluorescently-labelled DNA and computer simulations which show that large DNA molecules move through agarose gels undergoing cycles of elongation and contraction along the axis of the electric field. The leading part of the DNA molecule becomes concentrated into a compact head which moves through the gel at a constant rate independent of molecular mass. During transition between contracted and extended structures, the DNA often adopts an overall U-shape around agarose fibres with the tail of the molecule gradually catching up on the head before taking over as the leading part of the molecule. In this model, the DNA behaves somewhat like a caterpillar with the leading and trailing parts of the U-shaped molecule alternately being sieved through the agarose pores in a manner independent of molecular mass. This latter model depends on DNA behaving as an *elastic* molecule during electrophoresis.

Table 5.4. Some examples of separations performed using capillary electrophoresis

Sample component	CE mode	Detection
Metals	Free solution	UV
Catecholamines	Free solution	Electrochemical
Propranolol (chiral separation)	Free solution	UV
Amino acids	Free solution	UV
Peptides	Free solution	UV
Insulins	SDS PAGE	UV
Serum proteins	Discontinuous free solution	UV
DNA	Free solution	Fluorescence
DNA (mutation detection)	Pulsed field in ultradilute sieving solutions	UV
DNA (restriction fragments)	Agarose	Fluorescence
Oligonucleotides	Free solution	UV
Virus particles	Free solution	UV
Bacterial	Free solution	UV

5.7.4 Detection of DNA and RNA in Gels

Once separated, nucleic acids may be detected with the aid of fluorescent dyes such as ethidium bromide (Table 3.4). This intercalates into double-stranded DNA to give strong fluorescence. Although weaker fluorescence is found in single-stranded molecules such as RNA, it is still possible to visualize bands under a UV lamp. Usually, this dye is incorporated into the electrophoresis buffer which means that staining and electrophoresis occur *simultaneously*. Electrophoresis may be halted while the gel is inspected under UV light and, if bands are not adequately separated, the gel may be replaced in the electrophoresis system and separation continued. Because dyes such as ethidium bromide are also mutagens, great care must be taken in handling them. Newer, nonmutagenic dyes which equal or exceed the sensitivity of ethidium bromide are now widely available.

5.8 PULSED FIELD GEL ELECTROPHORESIS

DNA *in vivo* is organized in extremely large single molecules with high axial ratios (*see* previous section) which are arranged in chromosomes. It is therefore necessary to process DNA by *restriction digestion* to obtain sufficiently small molecules for separation and study in agarose or polyacrylamide gels. Sometimes though, we wish to study *intact* chromosomes or large DNA fragments of chromosomes which exceed the resolution-range of these systems. An electrophoretic technique called *pulsed field gel electrophoresis* has been specifically developed for separations in the mass range 200–3000 kb (Example 5.5).

5.8.1 Physical Basis of Pulsed Field Gel Electrophoresis

All the electrophoretic techniques we have so far encountered in this chapter use a *constant* and *unidirectional* electrical field. However, in pulsed field gel electrophoresis, the *direction* of the field is varied continuously during electrophoresis (Figure 5.37). This is achieved by alternately turning on and off the current or *pulsing* the electrical field in short time intervals called the *pulse time*, t_p. Intact chromosomal DNA or large fragments of chromosomal DNA are obtained by first lysing cells in an agarose matrix by treatment with a detergent or with enzymes such as lysozyme. This treatment often generates fragments of chromosomal DNA due to *mechanical shearing*. Large fragments may also be generated, if required, by digestion with unusual restriction enzymes recognizing rarely-occurring 8-nucleotide restriction sites (e.g. *Asc* I, *Not* I). Slices of agarose containing large DNA molecules are then loaded on 0.6–1.5% agarose gels and electrophoresis is initiated. An array of electrodes is arranged around the gel such that the direction of electrophoresis may be continuously varied (*see* next section). *Pulses* of electricity are passed through these electrodes for times in the range 0.1 s to 90 min before changing the direction of the field and passing pulses of electricity through the field in the new direction (Figure 5.37). The direction of electrophoresis of DNA is constantly alternated between the two electric field directions as the molecules *reorientate* themselves through a *reorientation angle*.

The precise reason why large DNA fragments separate so well on pulsed field gels is not yet fully understood. It is thought that, during reorientation, the helical structure is stretched and then compressed. The time required for this to occur is the *visceoelastic relaxation time*, t_r, and it is dependent on molecular weight. Small molecules reorient themselves to the changed electric field much more quickly than larger molecules which means that the latter have less time available to electrophorese during a given electrical pulse. Accordingly the larger the molecule, the slower it will migrate through the gel. The ratio of t_r to t_p is crucial for good separation in pulsed field electrophoresis. If t_p is much *shorter* than t_r then there is insufficient time for the molecule to reorientate itself in response to the pulse. If t_p is much *larger* than t_r, then the molecules will behave as in conventional agarose gel electrophoresis. Under both of these conditions large DNA molecules will not separate according to their mass. When the t_r/t_p ratio is between 0.1 and 1, optimal separation on the basis of mass is achieved. Since t_r is much longer for larger molecules, this means that

Example 5.5 Determination of chromosome size in bacteria by pulsed field gel electrophoresis.

Actinobacillus pleuropneumoniae causes a contagious pulmonary disease in pigs which is of major worldwide agricultural importance. A number of varieties or *serotypes* of this bacterium have been identified and these differ widely from each other in their pathogenicity and geographical distribution. *Agar blocks* were made with cells from a range of serotypes and these were digested with lysozyme and proteinase K. *Pre-electrophoresis* was then performed to remove any degraded or extrachromosomal (e.g. plasmid) DNA. The blocks were then digested with restriction enzymes such as Asc I (5′-GGCGCGCC-3′). These were subjected to *contour-clamped homogeneous electric field gel electrophoresis* (Figure 5.41) with pulse times of 5–20 s for 24 h at 200 V. The pattern of medium sized (20–400 kb) restriction fragments obtained for 13 serotypes (S1 to S12) is shown below (M is a standard made from polymers of bacteriophage λ genome);

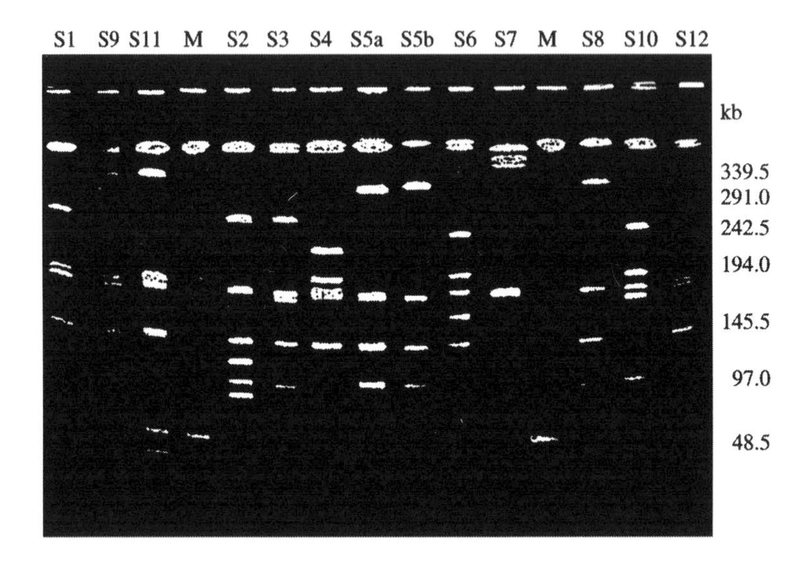

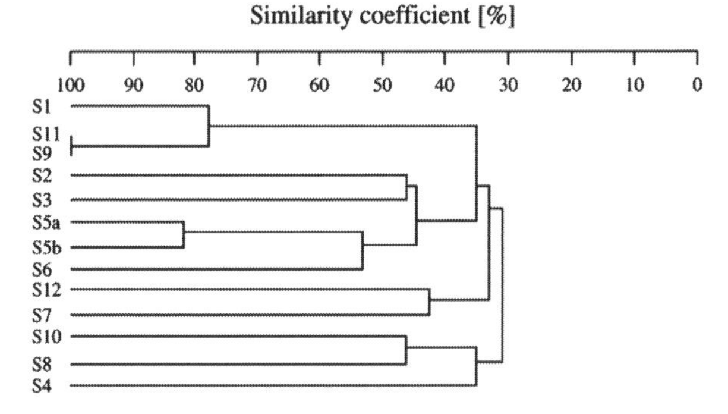

Analyses of this type showed a fragmentation of the *A. pleuropneumoniae* genome into 6–12 fragments of molecular mass 11–1217 kb. Summing the sizes of all fragments for each serotype gave a total genome size of 22 283 to 2409 kb for this bacterium. Moreover, since significant polymorphism was evident in the digestion patterns of 12 of the 13 serotypes (see above). It was found possible to compare the degree of genetic relatedness between the serotypes using pulsed field gel electrophoresis by analysing digestion patterns obtained with Apa I. This is summarized in the above dendogram comparing percentage similarity of digestion patterns with Apa I for each serotype. This analysis shows, for example, that serotypes 1, 9 and 11 are closely-related to each other while distantly related to serotype 4;

This type of study gives useful information on genome size and the degree of relatedness between different strains of pathogenic bacteria. It is particularly appropriate in studying the geographical spread of such organisms and for monitoring the appearance and likely origins of new serotypes.

Images reproduced from Chevallier *et al.*, (1998) Chromosome sizes and phylogenetic relationships between serotypes of *Actinobacillus pleuropneumoniae*. *FEMS Microbiol. Letts.* **160**, 209 16, by permission of the Federation of European Microbiological Societies.

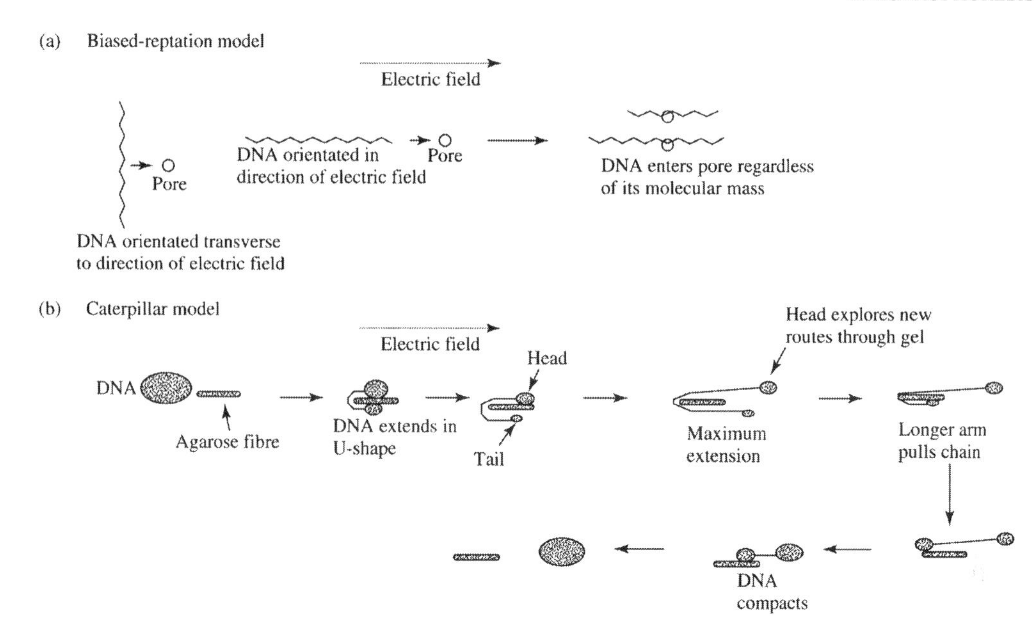

Figure 5.37. Some models for migration of large DNA molecules through agarose gels, (a) Biased-reptation model. DNA transverse to direction of electrophoresis cannot enter pores in gel. DNA orientated along axis of electric field can enter pores, (b) 'Caterpillar' model. DNA forms a compact mass which becomes retarded on an agarose fibre. It extends into a U-shape around the fibre. The head explores new routes through the gel. The tail is pulled along behind the head until compact mass is again formed. The tail can act as head in the next cycle of extension and contraction. Migration of head region through gel is independent of mass.

pulse times of as little as 0.1 s are adequate for separation of molecules of 10 kb while pulse times of 1000 s are required in the mass range approaching 10 Mb.

Any factors affecting t_r will alter the resolution of pulsed field gel electrophoresis. Experimentally these have been found to include; DNA molecular mass (*see* above); electrical field strength, E (reorientation is faster in stronger electrical fields); gel concentration (affects pore size); temperature (affects buffer viscosity); reorientation angle (larger angles give longer t_r).

5.8.2 Equipment Used for Pulsed Field Gel Electrophoresis

In a *homogeneous* electrical field a DNA molecule experiences a uniform field strength throughout electrophoresis. An *inhomogeneous* field arises due to variations in the electrical field strength. Both types of field arise during pulsed field electrophoresis. There is an upper limit to the electrical field strength which may be used in the experiment since too high values of E result in *trapping* of DNA in the gel. This limitation results in longer t_p values which can push the time theoretically required for individual experiments from one day (1 Mb, t_p 61 s, 1000 pulses) to 14 months (20 Mb, t_p 38 000 s, 1000 pulses). In practice, electrophoresis is usually performed for 12–24 h.

Specialized apparatus featuring an array of electrodes is required for pulsed field gel electrophoresis. The shape or *geometry* of this array differs between particular types of pulsed field electrophoresis experiments (Figures 5.39–5.42). A unique electric field orientated in a particular direction is achieved by varying which electrodes carry current, for how long and in what direction. Flow of current to the electrodes and pulse-times are electronically controlled allowing selection of optimized separation conditions. Agarose gels are poured and electrophoresed in a flat-bed, horizontal format as shown in Figure 5.37 and the gel may be arranged in a variety of orientations relative to the array of electrodes.

In *orthogonal field alternating gel electrophoresis* (Figure 5.38), pairs of electrodes are arranged at 90° relative to each other, resulting in creation of an inhomogeneous electrical field. This inhomogeneity has the disadvantage that lanes are distorted outwards at the bottom of the gel. In *contour-clamped homogeneous electric field gel electrophoresis* (Figure 5.39), point electrodes are arranged hexagonally around the gel. This results in a homogeneous electrical field in all lanes of the gel. *Crossed-field* or *rotating-gel electrophoresis* (Figure 5.40) takes advantage of a constant electrical field with the gel is held on a *rotating platform*. This is limited to separation of DNA >50 kb since smaller molecules require t_p less than one second which is

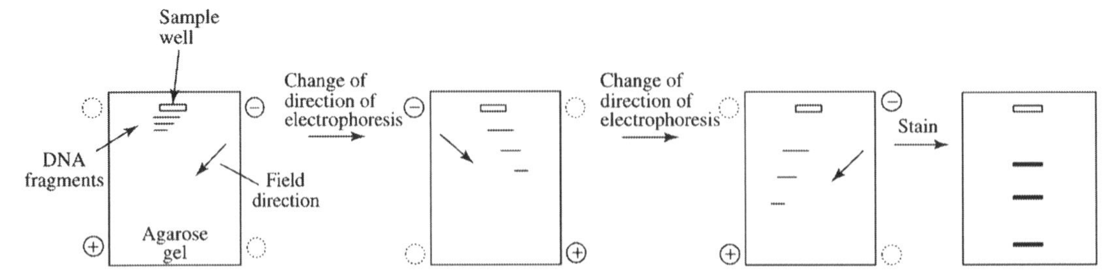

Figure 5.38. Pulsed field gel electrophoresis. DNA fragments are exposed to an electric field which alternates in different directions as a result of electrical pulses. The DNA moves through the gel in response to this field. Small molecules move more quickly than large ones due to their shorter visceoelastic relaxation times. For simplicity, an array of only four electrodes carrying equal charge is shown. In reality, the array may consist of many electrodes arranged in various geometries (Figure 5.40–5.43) and carrying different proportions of the current thus generating a complex electric field. Separation is optimised by varying the pulse-time, homogeneity and geometry of the field.

the minimum time required to rotate the gel. These three methods give a simple linear dependence of mobility on molecular mass. *Field-inversion gel electrophoresis* (Figure 5.41) uses a periodic *inversion* of the electric field in one direction (i.e. a reorientation angle of 180°). The forward pulse time is longer than that in the reverse direction thus resulting in net forward motion of DNA. Unlike the other pulsed field methods described, the dependence on DNA molecular mass in this technique is near-parabolic. Particularly large or small molecules progress rapidly through the gel with little resolution while intermediate-sized molecules move more slowly and are well-resolved.

For a given sample it is necessary to vary the pulse-time, field-strength, homogeneity and geometry of the electric field created across the gel to optimize separation.

5.8.3 Applications of Pulsed Field Gel Electrophoresis

Because this technique facilitates analysis of DNA fragments on the scale of individual chromosomes, it has found extensive uses in large-scale mapping of chromosomes. This has particular importance in bacterial taxonomy allowing the identification of relationships between existing and novel strains of bacteria. In eukaryotes, pulsed field gel electrophoresis of yeast chromosomes has revealed widespread length polymorphisms in both sexual and asexual species. The technique has also been applied to studies on strand-breaks in human chromosomes as a result of exposure to toxic chemicals.

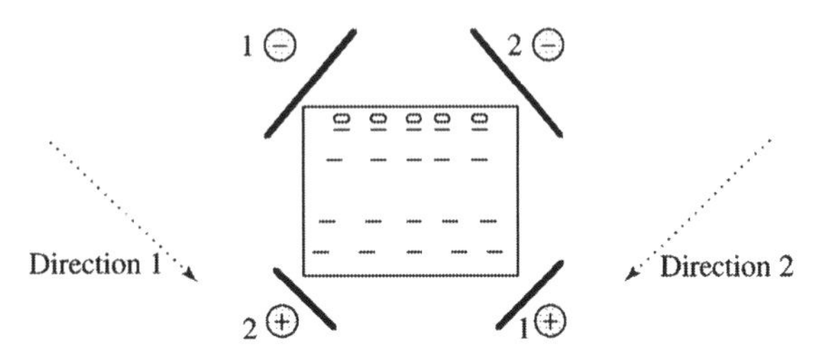

Figure 5.39. Orthoganol field-alternating gel electrophoresis. The electric field is alternated between two directions (arrows) by pulsing between two pairs of electrodes, the reorientation angle is that formed between directions 1 and 2. Note the outward distortion of bands near the end of gel.

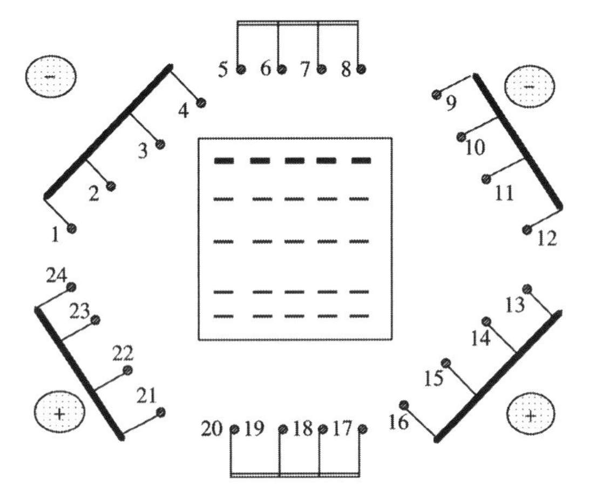

Figure 5.40. Contour-clamped homogeneous electric field electrophoresis. Point-electrodes (labelled 1–24) are arranged in a hexagonal array around the gel as shown. Electrodes 1–4 are held at a negative potential while electrodes 13–16 are held at the same, though positive, potential. Intermediate potentials are applied to all other electrodes in the array. To change the direction of the electric field by 105°, the negative potential is then applied to electrodes 9–12 while positive potential is applied to electrodes 21–24 and other electrodes again adopt intermediate potentials. This system achieves a homogeneous electric field throughout the gel.

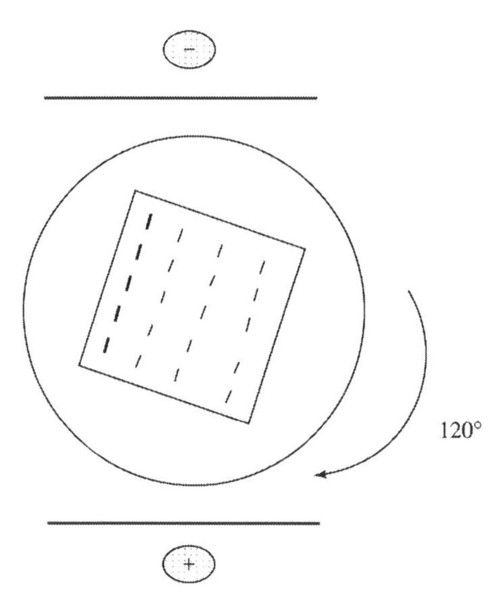

Figure 5.41. Rotating-gel electrophoresis. Gel is rotated backwards and forwards through a reorientation angle (e.g. 120°). Since the minimum rotation time is approximately 1 s, this technique cannot be used for separations requiring pulse times less than 1 s (i.e. DNA of mass <50 kb).

5.9 CAPILLARY ELECTROPHORESIS

When describing the basic theory of electrophoresis (Section 5.1.1) it was pointed out that use of high voltages leads to considerable *heat generation*. As well as affecting the structural integrity of biopolymers, heat is likely to affect the resolution attainable with electrophoresis since convection due to heat will *decrease* resolution of electrophoretically-separated bands. This places an effective *upper limit* on the voltage which may be used in most electrophoresis experiments at 300–500 V. If electrophoresis is performed in very thin *capillaries*, with *small internal volumes* and *large surface areas*, however, it is to be expected that it will be possible to use higher voltages than normal since heat generated would be efficiently *dissipated*. This is the basis of *capillary electrophoresis* (CE) which is now a major analytical technique for the study of biomolecules, macromolecular assemblies and even intact cells (Example 5.6).

5.9.1 Physical Basis of Capillary Electrophoresis

Use of CE in mapping of peptides is described in Figure 5.42. The glass capillary used in CE has an inner-diameter of 10–100 μm, an external diameter of 300 μm and a typical length of 10–100 cm. This gives an *included volume* for separation which is approximately a thousand times smaller than that of a slab gel or a HPLC column. The capillary is usually filled with a buffer (*free solution CE*) although it may also be filled with a gel such as agarose, polyacrylamide or a noncrosslinked linear polymer (*gel electrophoresis CE* or *capillary gel electrophoresis*). All of the electrophoresis applications so far described in this chapter such as isoelectric focusing, nondenaturing electrophoresis, SDS PAGE and pulsed field gel electrophoresis are possible in capillary systems. The main difference between electrophoresis in capillaries and that in rod or slab gels is the very high voltages (5–50 kV) possible in capillary systems. If such voltages were used in slab or rod gels, the heat generated would make successful electrophoresis impossible. Use of capillaries facilitates highly-efficient *heat dissipation*. A further difference between capillary and slab/rod gel electrophoresis is the small sample volumes (nL) and loadings (Attomoles) necessary for capillary electrophoresis (*see* Example 5.7). This is a consequence of the small volume of the capillary used and means that CE is principally an *analytical* technique suitable for highly-sensitive detection of as little as 10^{-11} M sample components.

The resolution attainable is also extremely high and CE has found particular uses in the high-resolution separation of *chiral mixtures*. The number of theoretical plates in a capillary used in CE is given by Equation (5.9);

$$N = \frac{\mu \cdot V}{2D} \qquad (5.9)$$

where μ is electrophoretic mobility, V is the applied voltage and D is the diffusion coefficient of the molecule in the electrophoresis medium. High voltages in CE give a high number of theoretical plates and hence, higher efficiency of separation.

The *time*, t, required for a molecule to pass through a capillary of length L is given by Equation (5.10);

$$t = \frac{L^2}{\mu \cdot V} \qquad (5.10)$$

Equations (5.9) and (5.10) mean that column length makes *no contribution* to the efficiency of separation but does influence separation time while higher voltage gives better separation efficiency. Since μ and D are properties of the sample component, they are not amenable to extensive experimental variation. Therefore the most efficient CE system would be expected to include the *shortest* capillary possible (to shorten separation time) and the *highest* voltage possible (to maximize separation efficiency). In practice, the ability of the capillary efficiently to dissipate heat limits both of these variables leading to use of maximum voltages of 50 kV and minimum capillary lengths of 10 cm.

Since a major component of the glass used in capillaries is negatively-charged silica, a layer of *counter-ions* such as

Example 5.6 Detection of prion protein by *capillary electrophoresis* (CE)

Transmissible spongiform encephalopathies cause progressive degeneration of the central nervous tissue in mammals. While scrapie has been known in sheep for over two centuries, bovine spongiform encephalopathy (BSE or 'mad cow disease') and associated human conditions such as Creuzfeld–Jakob disease and Kuru have only relatively recently been described and understood to form a family of related diseases. The mode of transmission of these illnesses is thought to be unique in that they may not be primarily caused by infectious agents such as bacteria and viruses. Protein tangles have been observed to accumulate in the brain tissue of infected animals which are protease-resistant aggregates of a normal protein (the prion protein) from which the N-terminal 30–50 residues have been proteolytically removed. The normal protein is approx. 40% α-helix while the abnormal form is 45% β-pleated sheet with considerably less α-helix and it is thought that these two forms can interconvert. This conformational interconversion is stimulated by the presence of small amounts of abnormal protein and even by short peptides derived from this abnormal protein. It is hypothesized that small amounts of abnormal protein in the diet can avoid proteolysis in the gut, cross the blood-brain barrier and infect brain cells promoting conformational change in the normal protein and resulting in the formation of tangles.

Prions may be identified by western blot analysis but this method of detection is probably not sensitive enough to detect the early stages of infection in sheep or cattle. Moreover, large amounts of brain tissue are required for such analyses and the results are not completely quantitative. SDS PAGE CE in a 57 cm capillary at a voltage of 15 kV was therefore assessed as a method for detection of the abnormal prion protein in samples from infected sheep brain. Separations obtained for A) Normal and B) scrapie-infected individuals are shown below;

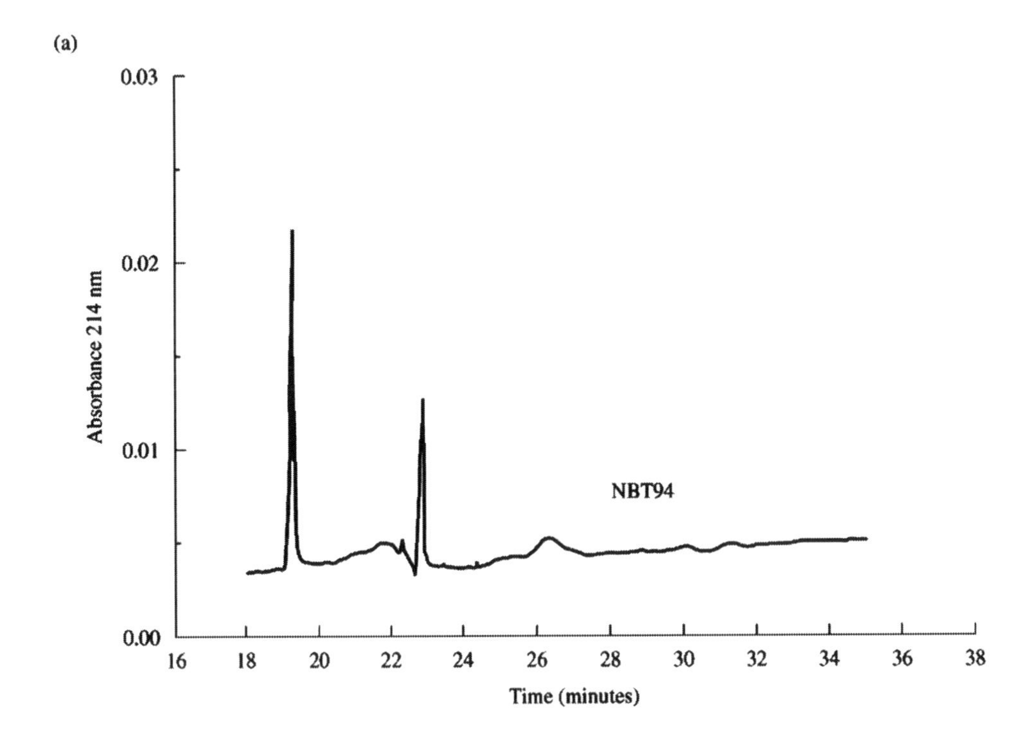

A peak eluting from the capillary at 26 min absent in all controls, was identified in all infected sheep. This separation could be performed in <40 min. and is sensitive and robust enough to detect the prion protein with a minimum of sample. While some 20 mg of brain tissue is required for a single western blot as little as 50 μg is necessary for each CE experiment.

(*Continues*)

(*Continued*)

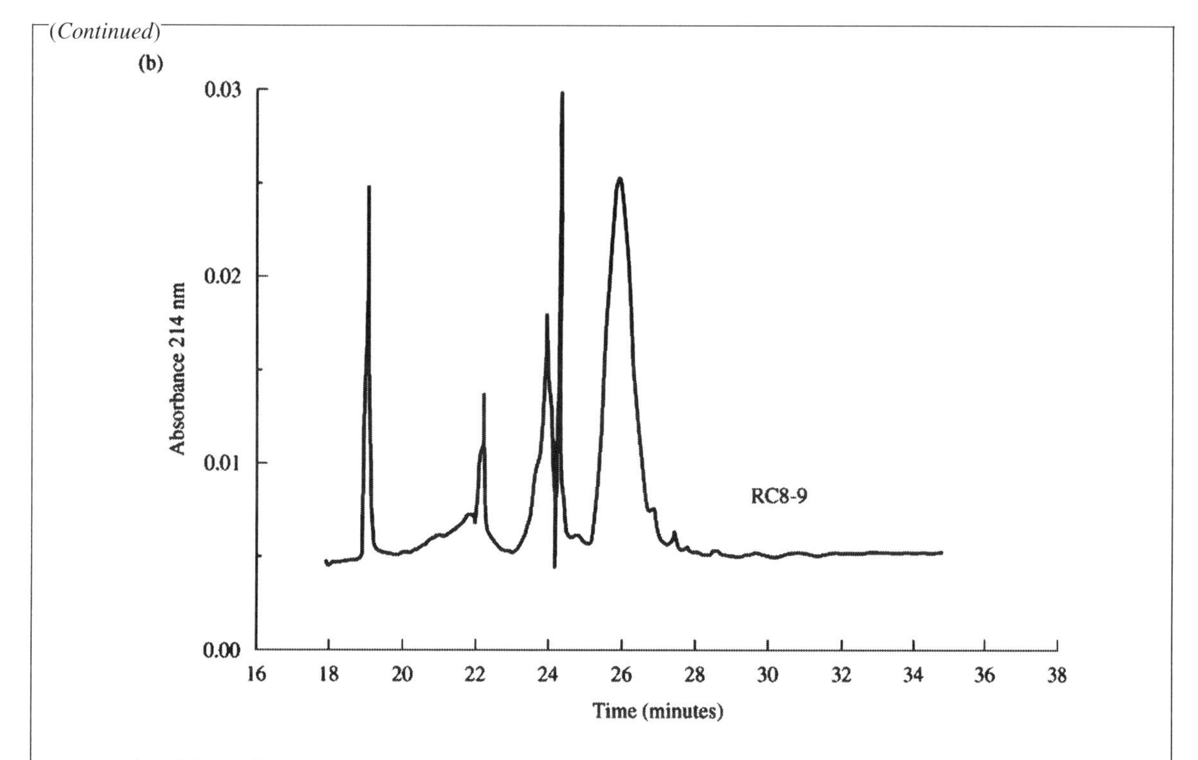

Reproduced from Schmerr, M.J., Jenny, A. and Cutlip, R.C. (1997) Use of capillary sodium dodecyl sulfate gel electrophoresis to detect the prion protein extracted from scrapie-infected sheep. *Journal of Chromatography B*, **697**, 223–9, with permission of Elsevier.

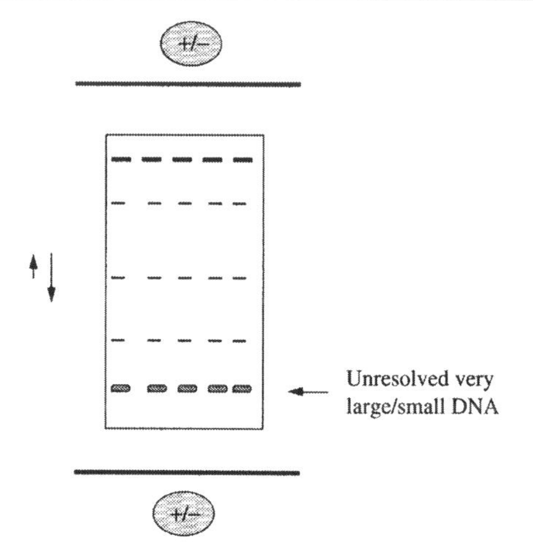

Figure 5.42. Field-inversion gel electrophoresis. Electrodes alternate between being cathode or anode. The pulse time or field strength in the forward direction is represented by the downward-facing arrow while that for the backward direction is represented by the upward facing arrow. The ratio between the forward and reverse pulses depends strongly on the molecular mass of sample DNA and must be chosen to yield sizedependent resolution. Very small and very large DNA migrate quickest through the gel.

Na^+ and H^+ assemble along the interior surface of the capillary (Figure 5.43). This layer of ions, in turn, attracts a *hydration layer* of water. During electrophoresis, the positively-charged Na^+/H^+ ions are electrophoretically *attracted* towards the cathode and carry the hydration layer of water with them. This phenomenon is called *electroosmotic flow*. This is a common occurrence during electrophoresis and we have previously encountered this term in a number of applications of electrophoresis described in this chapter. Electroosmotic flow has a particular significance in CE, however, since the internal volume of the capillary is so *small*. The thin hydration layer (together with other water molecules to which this layer is hydrogen bonded) represents a major fraction of the water contained within the capillary. Electroosmotic flow is therefore *quantitatively* particularly important in CE. Electroosmotic flow is strongly pH-dependent being up to ten times faster at low than at high pH.

In addition to electroosmotic flow, sample components also experience *electrophoretic flow* (i.e. positively-charged ions are attracted to the cathode, negatively-charged ions to the anode). The precise mobility of an individual ion will therefore be the result of a combination of these two flows (Figure 5.44). Positively-charged ions will tend to flow

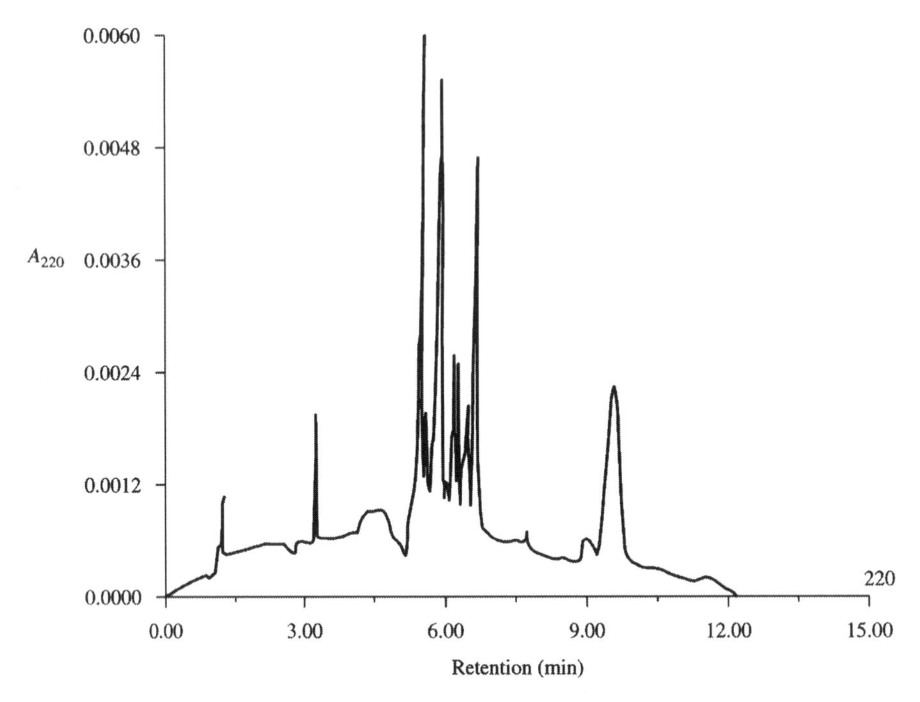

Figure 5.43. Free solution capillary electrophoresis of peptides in SDS. A typical CE analysis of peptides generated by digestion with a specific protease. In this example, the peptides arise from an insoluble antigen (HC-31) of the hepatitis C virus which was digested with *Staphylococcus aureus* V8 protease (Winkler *et al.*, 1996).

toward the cathode as a result of *both* electroosmotic flow and electrophoretic flow. Uncharged sample components will move towards the cathode in response to electroosmotic flow *only* (since they will experience no electrophoretic flow). Negatively-charged ions will move either towards the cathode or the anode *depending on the relative strength of the two flows*.

This gives rise to some apparent paradoxes since *uncharged* molecules move in a CE system (in most electrophoresis systems they experience no movement) and negatively-charged ions can move towards the *cathode* in such systems (which seems to contradict electrostatic attraction). By varying pH and hence the strength of electroosmotic flow, the separation of individual components can be optimized in CE. If it is desired to carry out separations on the basis of differences in electrophoretic flow *alone*, it is possible to *coat* the inner surface of the capillary to remove ionic interactions. Such interactions may be undesirable, for example, in separation of proteins.

A major limitation of CE is the limited volume of sample which may be applied. In many cases, the components of the sample may be quite dilute and difficult to detect despite the high-sensitivity detectors used. *Isotachophoresis* is an electrophoretic technique involving separation in a *discontinuous* buffer system which has found wide application in CE. In *capillary isotachophoresis* a sample ion is electrophoresed under constant current behind a *leading ion* of higher mobility and in front of a *terminating ion* of lower mobility in a capillary (Figure 5.46). The relationship between the mobilities of these ions is given by;

$$\mu_L > \mu_A > \mu_B > \mu_T \tag{5.11}$$

where A and B represent sample ions. When an electrical field is applied at constant current, the field strength is *greater* in the vicinity of lower-mobility ions and *lower* in the region of higher-mobility ions and the product of field strength and mobility is constant for each ion so that all ions migrate at the same *velocity*.

$$\mu_L \cdot E_L = \mu_A \cdot E_A = \mu_B \cdot E_B = \mu_T \cdot E_T \tag{5.12}$$

If this were not the case then a 'gap' could develop between the ions which would stop the current being carried altogether. The relationship described in Equation 5.12 results in a *zone sharpening effect*. Ions which diffuse into a zone with higher mobility will be slowed because of the weaker electrical field. Conversely, ions diffusing into a zone of

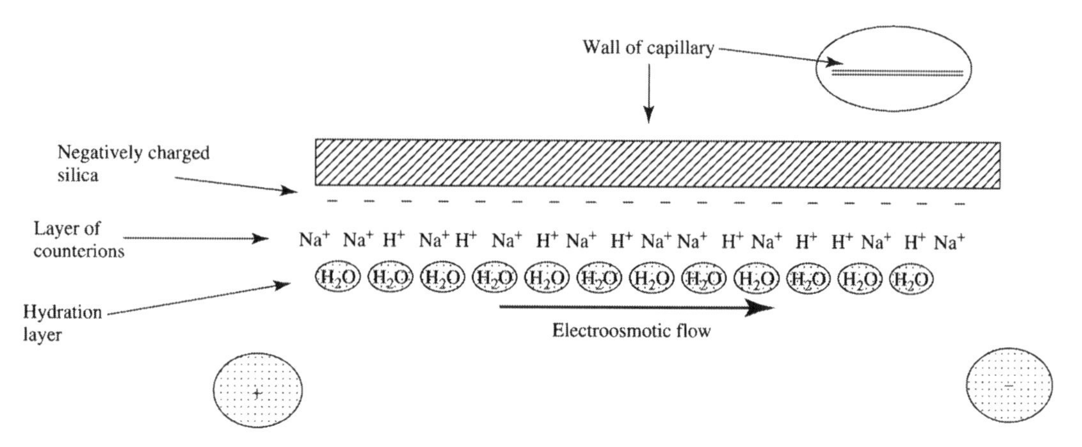

Figure 5.44. Electroosmotic flow in capillary electrophoresis. The silica of which glass capillary is composed is negatively charged. This attracts a layer of positively charged counterions (Na^+/H^+) which, in turn, is coated with a hydration layer of water. During electrophoresis, the counterion layer migrates towards the cathode, bringing the hydration layer with it. Since this layer is, in turn, hydrogen bonded to bulk water and since the capillary is very narrow, this means that there is a net flow of water towards the cathode.

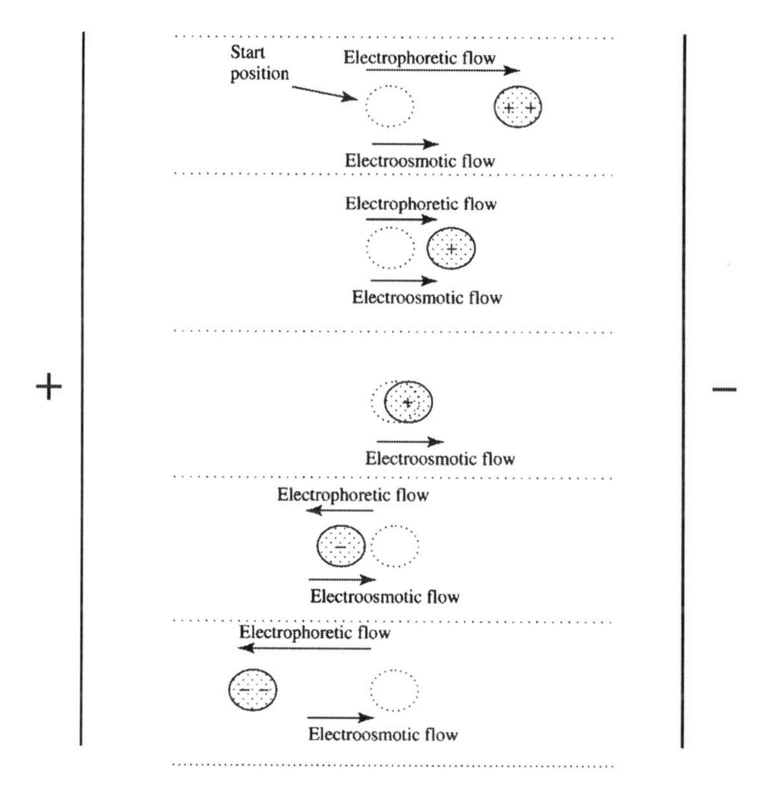

Figure 5.45. Interaction between electrophoretic and electroosmotic flow in capillary electrophoresis. Cations move towards the cathode in response to both electroosmotic and electrophoretic flow. Uncharged molecules move towards the cathode in response to electroosmotic flow only. Anions move dependent on the balance between electroosmotic and electrophoretic flows. In the example shown, electrophoretic flow is sufficiently strong to overcome electroosmotic flow giving net mobility towards the anode. However, if electroosmotic flow were strong enough, electrophoretic flow could be overcome and net mobility to the cathode could result. Note that electrophoretic mobility is stronger in multiply charged molecules than in those which are singly charged.

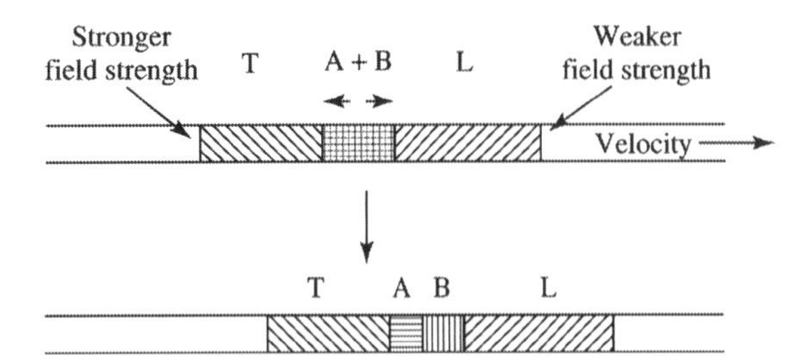

Figure 5.46. Capillary isotachophoresis. Sample components A and B are electrophoresed in front of a terminating ion (T) and behind a leading ion (L). Ions A and B concentrate into distinct 'stacks' due to a combination of local differences in the field strength and their differing electrophoretic mobilities. Ion A has a lower electrophoretic mobility than B which, in turn, has a mobility less than L. The stacks move with a constant velocity through the capillary.

lower mobility will be accelerated due to the stronger electric field. In isotachophoresis, sample ions therefore form *stacks* during electrophoresis. The concentration of each sample component in these stacks is a fixed function of that of the leading ion. For sample component A, the concentration of A, c_A, is given by;

$$c_A = f(c_L) \tag{5.13}$$

By varying the concentration of the leading ion, we can vary the concentration of each sample component, *regardless of their actual concentrations in the original sample* (Figure 5.47).

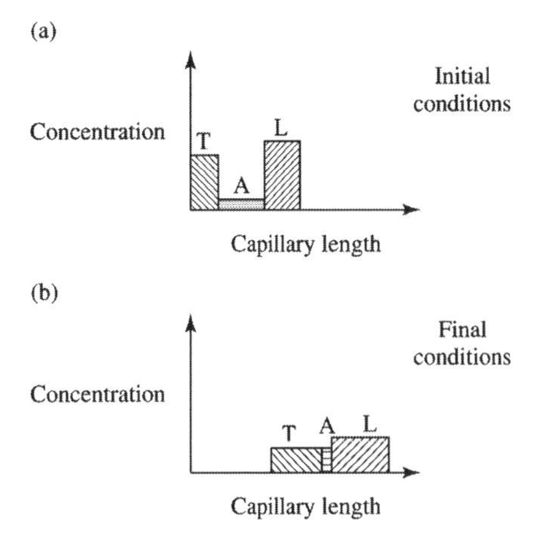

Figure 5.47. Concentration effect during capillary isotachophoresis. The concentration of sample component A is related directly to that of the leading ion L. During isotachophoresis, the length of the zone occupied by A alters to achieve this concentration. In this way isotachophoresis can be used to concentrate dilute samples prior to CE.

Isotachophoresis is performed in teflon or quartz capillaries at voltages up to 30 kV and may be used to concentrate samples for application to CE. The limited loading volumes possible with CE are therefore overcome.

5.9.2 Equipment Used in Capillary Electrophoresis

The experimental apparatus used in CE is described in Figure 5.48. A 5–50 kV DC source provides the electrical potential difference between the electrode buffer tanks or *electrolyte chambers*. These are connected by a narrow capillary of variable length (~50–100 cm). *Detection* is possible by a number of different methods such as conductance and electrochemical detection. MS (Chapter 4) can also be interfaced with CE and used as a detector allowing a two-dimensional characterization of sample components. For proteins, peptides and nucleic acids *laser-induced fluorescence* in the UV-visible part of the spectrum is especially popular for high-resolution work. Because the voltages used in this equipment are potentially lethal, care is necessary when carrying out CE experiments. Sample throughput can be greatly improved by the use of an autosampler, data acquisition/storage systems and the recent development of *capillary array electrophoresis* which allows simultaneous analysis in an *array* of up to 96 capillaries.

Since heat dissipation is critical for the success of CE it is essential that *heat gradients* between the interior wall of the capillary and the outside be as *low* as possible. This depends on the *rate of heat transfer* between the outermost layer of the capillary and the surrounding surface. Where the surrounding surface is simply air, the temperature gradient between the innermost and outermost layer of the capillary can be as large as 125 °C/mm. Forced air cooling can reduce this to 51 °C/mm while arranging contact with a temperature-controlled aluminium plate with high thermal conductivity reduces it further to 7 °C/mm. Effective *thermostatting* is therefore a common feature of CE instruments.

5.9.3 Variety of Formats in Capillary Electrophoresis

CE offers a number of distinct advantages for the analysis of biological samples. The technique facilitates *rapid* separations of complex mixtures in minutes with *attomole* sensitivity. The small volume of the capillary allows analysis of sample loadings in the nL and ng range and very small amounts of reagent are consumed (in free solution CE flow rates of μl/min are used). Moreover, since a wide variety of electrolyte buffers and gels can be used in CE, a range of separation conditions can be assessed to optimize separation. Table 5.4 summarizes some CE separations ranging from atoms to whole cells.

A range of CE separation conditions may be assessed by varying the contents of the capillary. The pH can be changed

Example 5.7 Detection of mutant proteins by western blotting.

Protein kinases and phosphatases (PPs) combine to play a major role in the regulation of cell function by, respectively, phosphorylating and dephosphorylating proteins. Three groups of phosphatases (which hydrolyze phosphate groups from phospho-amino acids) have been described in eukaryotes differing on substrate specificity (i.e.; protein serine/threonine, tyrosine and serine/threonine/tyrosine phosphatases). One of the protein serine/threonine phosphatase enzymes is found in two distinct gene products; PP2Cα and β. Five cDNA clones for the latter have been identified which differ only in their C-terminal sequence and which also show tissue-specific expression. Crystallographic data and point mutations, however, identified a possible N-terminal catalytic domain. It is thought that PP2Cβ isoenzymes may have distinct physiological roles in each tissue but the relative importance of the N- and C-terminal domains to this are unclear.

This question was probed by selectively deleting parts of the PP2Cβ sequence from both the N- and the C-terminus (A). These mutants were purified from a GST fusion expression system (Chapter 2; The residues Gly - Ser- Pro originate from the GST portion of the fusion) and analysed by western blotting (B) and catalytic assay of dephosphorylation (C).

The western blot (B) carried out with antiserum to PP2Cβ showed a single band except in the case of ΔC(313–390) and ΔC(291–390) where two bands were observed. The blot is an important part of the experiment as it allows confirmation of deletion of large parts of the N- and C-terminal domains as evidenced by the greater electrophoretic mobility of mutants. Moreover, the presence of two bands for ΔC(313–390) and ΔC(291–390) allowed identification of inappropriate thrombin digestion of the GST-PP2Cβ fusion (Figure 2.19). It is clear from the results in C that the first 12 N-terminal residues are essential for catalytic activity while mutations at the C-terminal end had much more modest effects. The inactivation observed for ΔC(291–390) may well be due to gross structural changes in the protein.

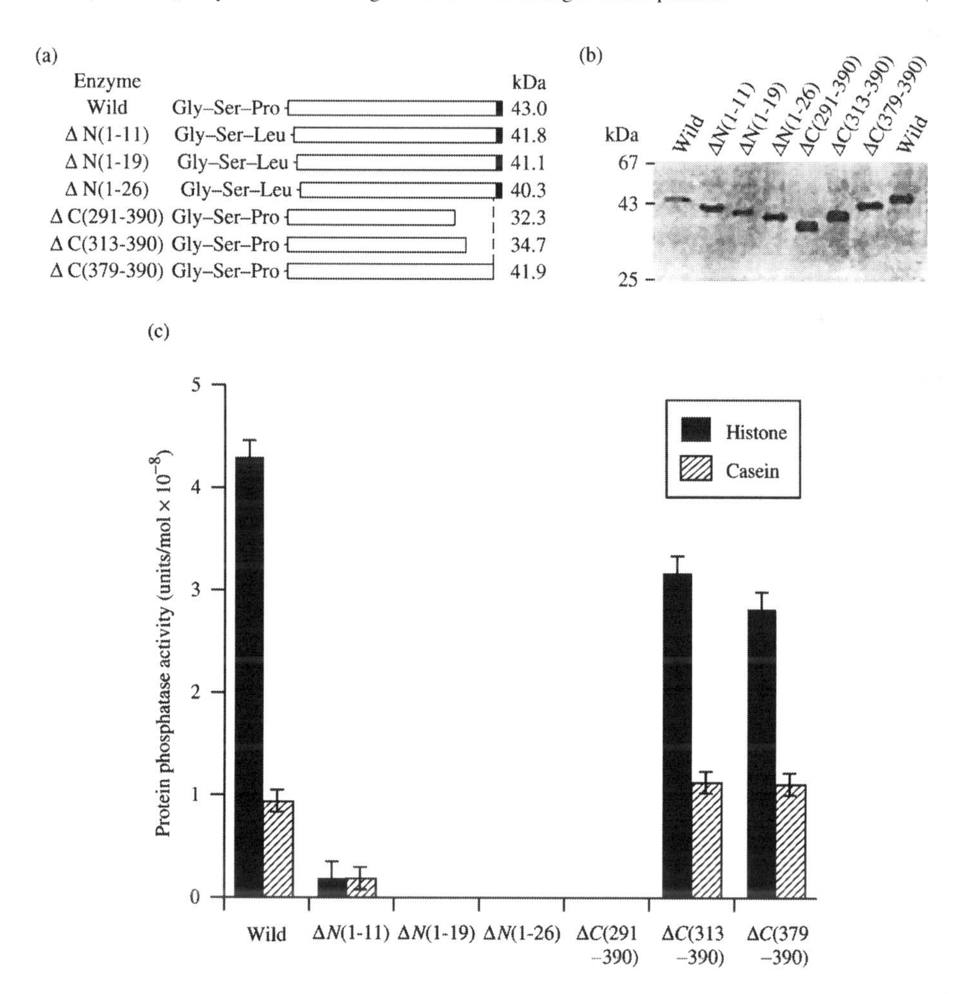

(Continues)

(Continued)

Western blots facilitate the identification of structurally-similar proteins in SDS PAGE separations and are important structural probes in studies of proteins.

Images reproduced from Kusuda *et al.* (1998) Mutational analysis of the domain structure of mouse protein phosphatase 2Cβ. *Biochemical Journal*, **332**, 243–50, by permission of Portland Press.

to alter electroosmotic flow while changes in the voltage can alter both electrophoretic and electroosmotic flow. Agents such as ampholytes, SDS and urea can be included in the electrophoresis buffer to create denaturing conditions. Incorporation of *ligands* within the capillary can be used to estimate *ligand affinity* of proteins or DNA since the electrophoretic mobility of a ligand-protein complex will differ from both uncomplexed ligand and protein, allowing quantification of all three species. Incorporation of proteins such as BSA in electrophoresis buffers allows *chiral separations* in CE because each of a pair of enantiomers will differ in binding affinity for the protein thus giving them different electrophoretic mobility.

CE electrophoretic separation may be combined with chromatography (Chapter 2) in a number of ways. CE may be interfaced with HPLC chromatography systems to facilitate two-dimensional analysis of complex samples. In such systems, samples eluting from HPLC are analysed immediately by CE. An alternative approach is to pack the CE capillary with a suitable stationary phase and to combine electrophoretic and chromatographic separation within the capillary. Anionic detergents such as SDS may be included in the capillary in *micellar electrokinetic* separations. The detergent forms *micelles* which act as a stationary phase with which sample components may interact in different ways. In *capillary electro-chromatography* C-18 reversed phase stationary phase is packed into the capillary and electroosmotic flow pumps sample through this phase. Again, sample components separate based partly on individual affinity for C-18 and partly on individual electrophoretic mobility.

CE provides high-resolution analysis which complements techniques such as HPLC and MS and is versatile across the whole range of structural complexity found in biochemistry.

5.10 ELECTROBLOTTING PROCEDURES

Electrophoresis allows separation of complex samples using equipment which, in most cases, is cheap and universally available and procedures which have been shown to be robust and versatile. The resolution attainable in electrophoresis is often better than that obtained in HPLC and MS analyses and is far superior to techniques such as low performance column chromatography. It is sometimes possible to take advantage of the high-resolution nature of electrophoresis to isolate samples difficult to purify by other means for further study. One approach to this is to *transfer* separated sample components onto a mechanically-stable *membrane* which may be manipulated in further analyses. This transfer process is called *blotting* and, where electrophoresis is the basis of transfer, it is termed *electroblotting*. A method for blotting DNA restriction fragments was developed by Prof. Ed. Southern and called *Southern blotting*. Subsequently, other techniques were developed for blotting proteins (Example 5.7) and RNA and named after other points of the compass to denote the fact that they were all based on a similar experimental approach (Figure 5.48). Even though some of these blotting methods are not electrically-based, they will be described together in this section to clarify the relationship between them.

5.10.1 Equipment Used in Electroblotting

An outline of an electroblotting experiment is shown in Figure 5.49. Electrophoresis may be performed in a variety of vertical or horizontal gel formats. The separated biomolecules are then electrophoresed onto a membrane in an *electroblotting apparatus*. This is achieved by laying the gel flat on a piece of membrane and ‘sandwiching’ this between two flat electrodes which apply an electrical field *at right angles* to the plane of the gel. The sample components electrophorese from the gel and attach irreversibly to the membrane by hydrophobic interaction. The electroblotting apparatus may consist of a *tank* in which the gel-membrane sandwich is completely immersed in buffer or it may be a *semidry* system where buffer is provided by buffer-soaked filter papers which bracket the sandwich. The latter system allows more economic use of buffer. The membrane used most widely in electroblotting is *nitrocellulose*. However, where further chemical analysis is desired (e.g. N-terminal microsequencing of proteins by the Edman degradation), more robust membranes such as *polyvinylidenedifluoride (PVDF)* may be used.

5.10.2 Western Blotting

If proteins are separated by SDS PAGE, they can be transferred to nitrocellulose by electroblotting and stained with

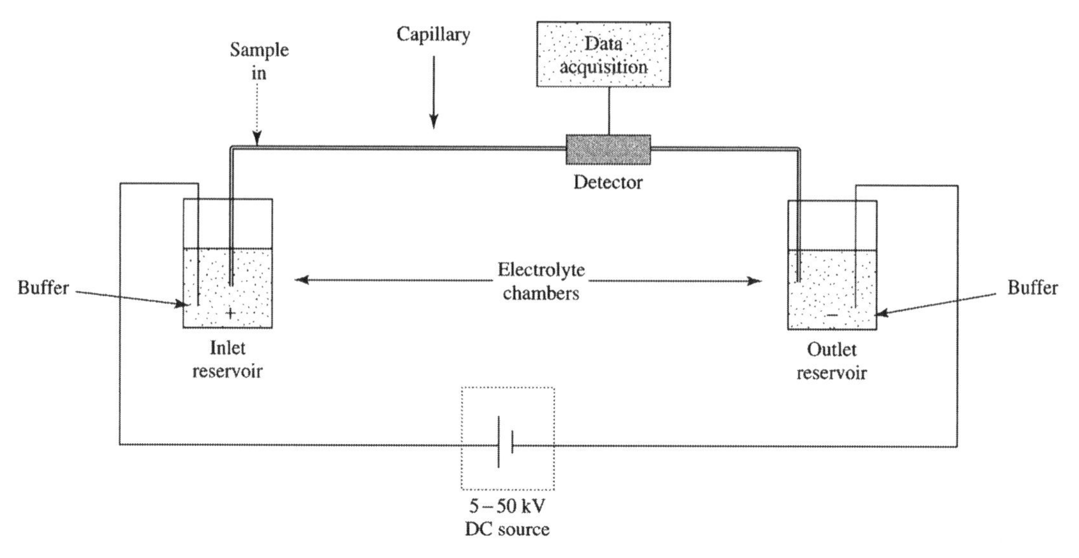

Figure 5.48. Apparatus used for capillary electrophoresis. Sample is applied to sample inlet near the inlet reservoir and electrophoresis takes place towards the outlet reservoir. The capillary contains buffer (free solution CE) or a gel (gel electrophoresis CE). Because of the narrowness of the capillary, heat is dissipated very efficiently allowing the use of very high voltages.

antibodies essentially as described for the dot blot assay in Figure 5.27. This experiment is called *western blotting* (Figure 5.50). Proteins which have been separated by SDS PAGE are transferred to a membrane to which they bind. The membrane is then washed with a protein solution called the blocking solution, which blocks any nonspecific antibody binding sites. This is followed by washing with an antibody solution (e.g. rabbit IgG). Antibodies bind to epitopes on proteins. Bound antibodies are visualized by washing with a second antibody specific for the class of antibodies to

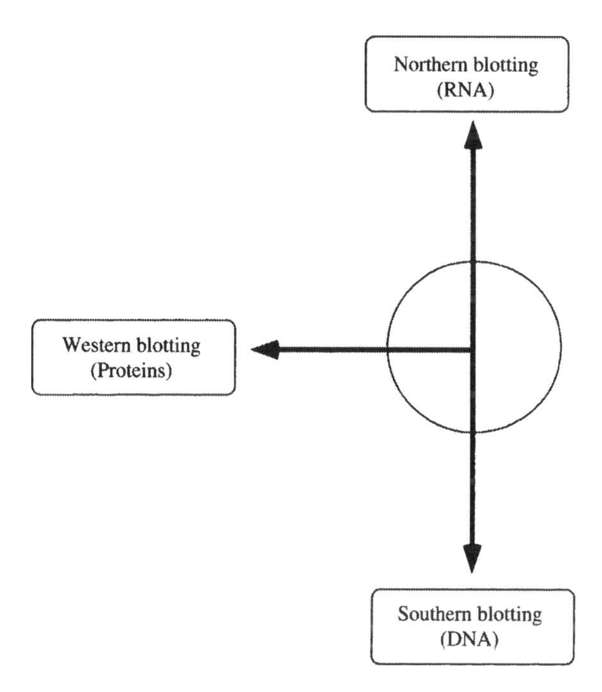

Figure 5.49. Blotting techniques widely used in biochemistry. Southern blotting was originally named after Professor Ed. Southern. The other techniques were named after other points of the compass to reflect their overall similarity of approach.

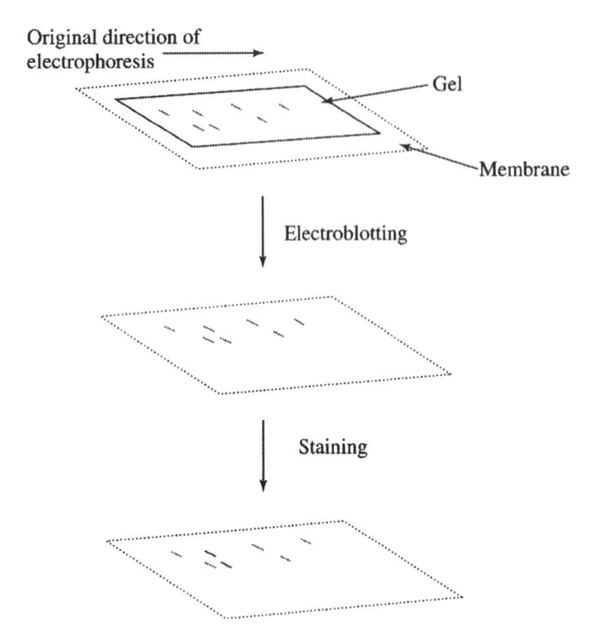

Figure 5.50. Outline of electroblotting experiment. Sample is electrophoresed in a polyacrylamide or agarose gel. It is then transferred to a membrane such as nitrocellulose by electrophoresis at right angles to the original direction. The membrane is then stained to identify particular sample components.

which the first antibody belongs (e.g. antirabbit IgG). These second antibodies, which are available commercially, are raised to the constant part of the IgG structure and are therefore useful for all antibodies of that class. They are usually attached to an enzyme such as horseradish peroxidase and their site of binding may be visualized by washing the membrane with a suitable chromogenic substrate for this enzyme. It will be recalled that this visualization approach is similar to that described for *in situ* immunostaining (Section 3.4.6) and to dot blotting (Section 5.6.1).

Because *selective staining* is achieved with antibodies, this technique often helps in clarification of structural relationships between proteins. It is useful, for example, in comparing structurally-related proteins such as haemoglobins and serum albumins and also in identifying new strains of bacteria and viruses. Since we can identify an immunoblotting protein in a complex mixture separated by SDS PAGE, we know the molecular mass of the polypeptide as well as its immunological identity.

Protein preparations used for western blotting need not be pure and the experiment can therefore be performed on crude preparations such as serum, cell lysates and cell-wall extracts. A single protein preparation can be screened with a panel of antibodies or, alternatively, a single antibody may be used to screen a variety of protein preparations. In the latter experiment, the various preparations can be electrophoresed side-by-side to allow direct comparison between them. An important variable in western blotting is *protein transfer efficiency*. Small and soluble proteins transfer best while larger and more fibrous proteins may transfer with very low efficiency. For this reason, it is usual to stain the gel after blotting to establish how much of each component of the sample has successfully transferred to the membrane.

Since transfer between gel and membrane covers a very short distance, it should be complete in a short time (<1 h). In cases where a particular protein does not transfer efficiently, it may be wise to vary the transfer conditions rather than extending the transfer time to several hours. Useful points of variation include the presence/absence of methanol, SDS concentration, ionic strength of transfer buffer and current density (mA/cm^2).

5.10.3 Southern Blotting of DNA

Because of its large size, DNA is often digested with *restriction enzymes* to generate smaller *restriction fragments* suitable for separation on agarose gels. Since restriction enzymes are *sequence-specific*, the pattern of fragments obtained is usually characteristic for a particular piece of DNA and is called a *restriction digest map*. It is possible to sequence restriction fragments as described in Section 5.4.2. However, where many fragments have been generated, this would be very time-consuming. It would be better if sequence relationships could be identified between fragments either within a digest or between digests from different samples *before* carrying out direct sequencing.

Southern blotting (Figure 5.51) allows such comparisons by blotting the DNA fragments onto a membrane followed by washing with DNA probes *complementary* to the sequence of interest. We have already seen examples of the use of such probes in *in situ* hybridization (Chapter 3). Nucleotide bases in the probe hydrogen bond to bases in the blotted DNA fragments according to the rules of Watson-Crick base pairing (i.e. A-T and G-C). The specific recognition offered by base pairing may be regarded as analogous to the recognition of individual epitopes by antibodies in western blotting.

After electrophoretic separation of restriction fragments, the agarose gel is exposed to alkali to *denature* the DNA. Fragments are transferred to nitrocellulose by capillary action (although electroblotting and *vacuum-transfer* are also possible) as described in Figure 5.52. The DNA is bonded to the membrane and then hybridized to a labelled probe (which may be an oligonucleotide or a DNA fragment). Only DNA fragments with sequences complementary to the probe will hybridize. This process is heavily dependent on the temperature and salt concentration in which the hybridization is performed. Where a very high degree of sequence

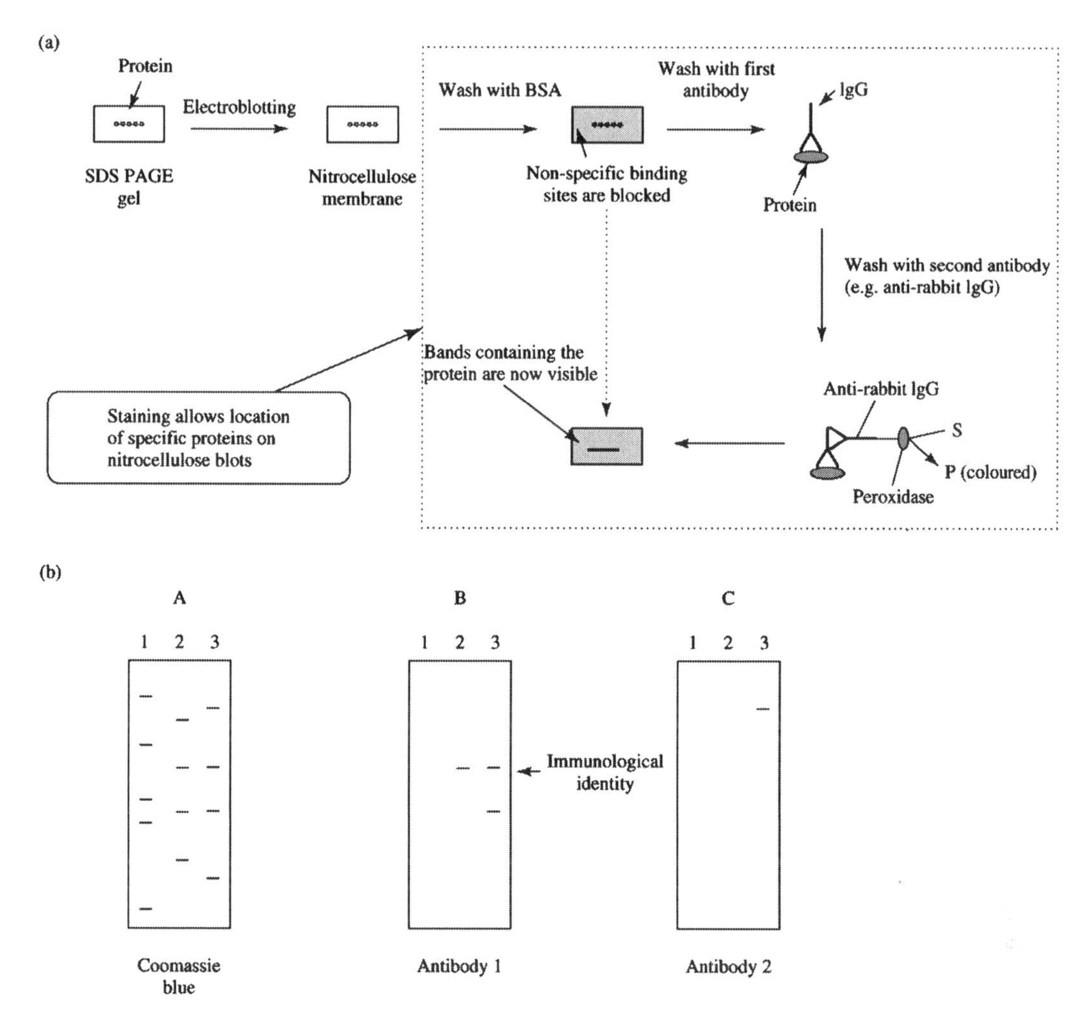

Figure 5.51. Western blotting, (a) Proteins are electroblotted onto nitrocellulose and stained. Non-specific binding sites are first saturated with BSA or some other protein. The nitrocellulose is then washed with an antibody specific for one or more of the electrophoretically-separated proteins. This antibody binds specifically to these proteins on the surface of the nitrocellulose. The membrane is then washed with a solution of a second antibody (specific for the first antibody) which is conjugated to an enzyme such as peroxidase. Peroxidase converts a substrate into a coloured product allowing visualisation of the location of target proteins on the blot. Note that more than one molecule of second antibody can bind to each IgG. (b) SDS PAGE was carried out in three separate, identical gels on a mixture of proteins of known identity and mass (track 1) and two individual samples (tracks 2 and 3). Gel A was stained with coomassie blue making all protein bands visible. Western blotting was carried out with two separate antibodies on the other two gels. The blot obtained for these are B and C, respectively. An immunologically identical band was found in both samples (arrow in B) and a second band also immunoblotted which was present only in sample 3. This could be a breakdown product of the common immunoblotting band but is clearly distinct from the co-migrating band in sample 2. A different protein is detected by antibody 2.

complementarity is expected, *high stringency conditions* (i.e. high temperature and/or low salt concentration) are used. Where the degree of complementarity is low, *low stringency conditions* (i.e. low temperature and high salt concentration) are used. Under the latter conditions imperfect hybrids containing some mismatches can form. The length and composition of the probe also affects the stringency with longer sequences requiring lower stringency. The degree of stringency required in a given experiment is a matter for experiment and frequently varies between identifying too many positives and no positives at all. More detailed information will be found in the references cited in the bibliography but hybridization is usually performed 12 °C below the theoretical melting temperature of the probe/target

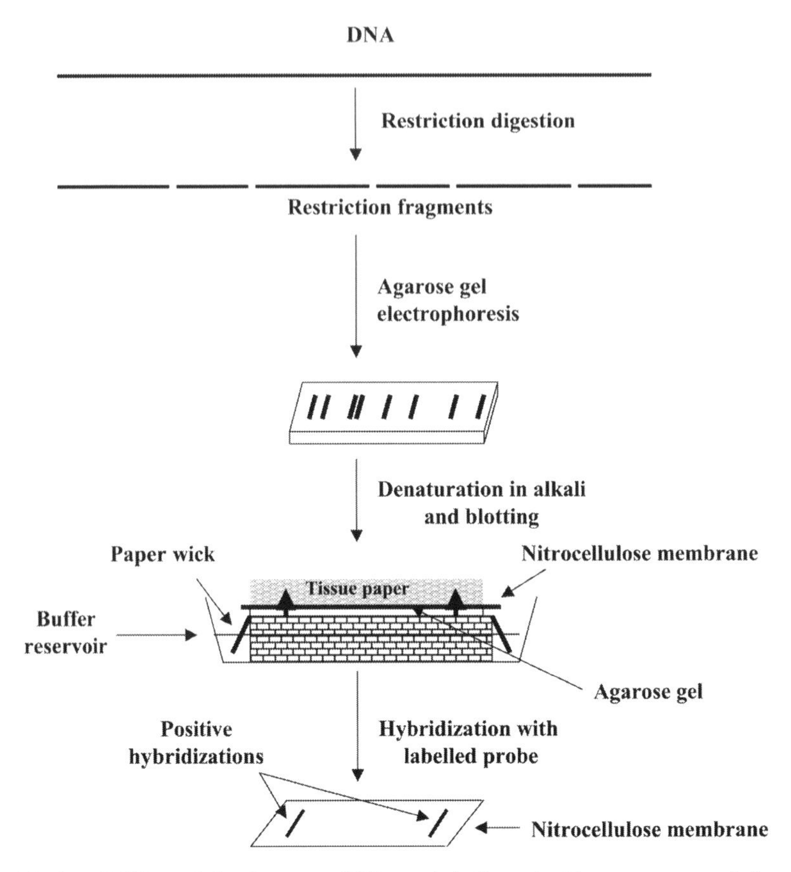

Figure 5.52. Southern blotting of DNA restriction fragments. DNA sample is digested with one or more restriction enzymes to generate a series of restriction fragments. These are separated on an agarose gel. The fragments are then transferred by capillary action to a nitrocellulose membrane (upward arrows). This is hybridized with a labeled probe (which may be an oligonuclcotide or DNA fragment). Positive bands are identified and related to bands visible in the original gel. This allows recognition of restriction fragments complementary to the sequence of the probe.

hybrid, T_m, which is calculated from the length (l), and percentage G and C content ($\% \, G + C$) of the probe as follows;

$$T_m = 69.3^\circ + 0.41[\%(G + C)] - 650/l \tag{5.14}$$

A typical hybridization temperature is 68 °C but this can be reduced to 43 °C by the inclusion of formamide.

The principal use of Southern blotting is the comparison of DNA preparations from different sources. Under conditions of high stringency, only fragments which are almost identical will hybridize to the same oligonucleotide probe. Thus very similar pieces of DNA can be identified in samples from related bacteria, different tissues (in mammals) or different individuals. If a cDNA molecule is used as a probe instead of an oligonucleotide, restriction digests of genomic DNA can be studied to locate fragments corresponding to the cDNA. *Polymorphisms* in the pattern of fragments generated by restriction enzymes are due to interindividual differences in DNA sequence and are often encountered in genetic diseases. These are called *restriction fragment length polymorphisms* (RFLPs).

5.10.4 Northern Blotting of RNA

RNA may also be separated on agarose gels and blotted in the same way as DNA fragments and such blots are called *northern blots*. The main difference between these experiments is that RNA is quite unstable and would hydrolyze in the presence of alkali. Accordingly, RNA is denatured *before* electrophoresis by treatment with formaldehyde followed by blotting. Nylon membranes or membranes to which RNA may be covalently attached are also used in this procedure

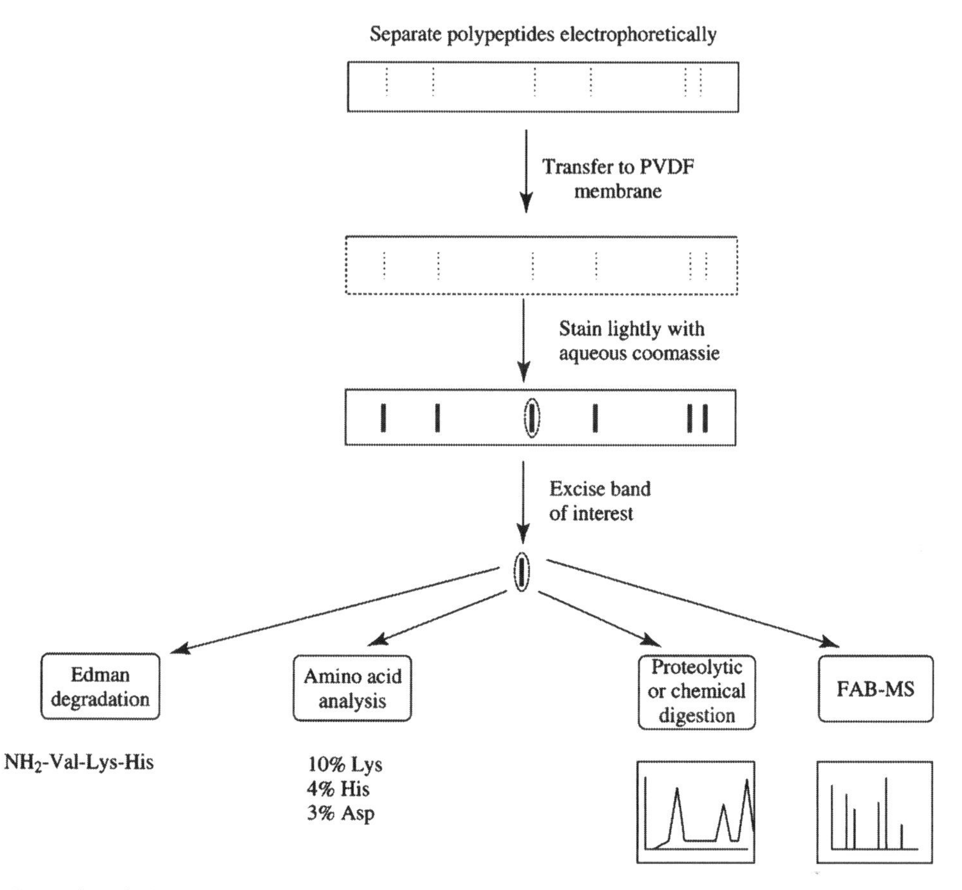

Figure 5.53. Preparation of polypeptides for further analysis by electroblotting. Polypeptides separable by electrophoresis may be transferred to a PVDF membrane. This is stained for protein in aqueous dye. Once bands become visible, the polypeptide of interest is excised and subjected to further analysis.

rather than nitrocellulose. This technique is routinely used to demonstrate and quantify expression of particular mRNA molecules in cells and may, for example, give size estimates for specific mRNAs or reveal how many different types of the mRNA may exist.

5.10.5 Blotting as a Preparative Procedure for Polypeptides

Most of the applications of electrophoresis techniques are *analytical* allowing separation, visualization and sometimes quantification of complex mixtures. Because of the high-resolution separations possible in electrophoresis, we occasionally identify polypeptides separable by electrophoresis which prove difficult or impossible to purify by conventional techniques (Chapter 2). In such cases, we can use electrophoresis as a *preparative* procedure. It is possible to *electroelute* DNA and proteins from gels and, in this way, to collect sufficient sample for further experimentation. In the case of DNA we can clone such molecules into suitable vectors while, in the case of proteins, we can perform direct chemical and structural analysis. Blotting of polypeptides onto membranes (Figure 5.53), however, offers a particularly attractive way of preparing these molecules for further analysis since the membrane may act as a *solid phase* allowing high-efficiency removal of contaminants such as buffer components. This also minimizes loss of sample during handling and facilitates analyses such as N-terminal microsequencing and MS analysis (Chapter 4) on as little as a few μg of material. Because nitrocellulose has limited chemical stability to corrosive chemicals such as trifluoroacetic acid, it may be replaced by vinyl-based membranes such as PVDF. Another advantage offered by PVDF is the fact that this membrane does not bind protein stains such as coomassie blue while nitrocellulose does. It is consequently possible to visualize stained protein bands against a white background using PVDF (Figure 5.53) while nitrocellulose membranes stain *nonspecifically* making such visualization impossible.

Using electroblotting, therefore, we can obtain detailed chemical and structural information on polypeptides without the prior necessity of obtaining them quantitatively as highly-purified preparations.

5.11 ELECTROPORATION

Electric fields can have important effects on charged biomolecules. Most applications of such fields in biochemistry involve *in vitro* separations. There are relatively few examples of the use of electric fields on living cells. *Electroporation* is a technique which takes advantage of the effects of high voltages on cell membranes to facilitate the uptake of foreign genetic material. This process is called *transformation.*

5.11.1 Transformation of Cells

When a gene is cloned into a suitable vector, it is often desired to insert this vector into a suitable bacterial, fungal, plant or animal host-cell. This would generate a genetically-altered cell which is said to be *transformed*. If the vector is simply mixed with the host-cell, very few, if any, of the cells take up vector because the cell membrane acts as an efficient barrier against such foreign material. A wide variety of biological, chemical and physical methods have been developed to increase the *uptake efficiency* of transformed cells (Figure 5.54). *Biological* approaches include the use of bacterial, animal and plant viruses or parts of viruses in the vector. Viruses recognize *receptors* on the surface of cells and are internalized from these receptors. The most widely-used *chemical* method involves making cells *competent* to take up vectors by treatment with $CaCl_2$. *Physical* methods for introduction of vectors include *microinjection*, firing microscopic tungsten pellets coated with genetic material as *microprojectiles* into blocks of cells and *electroporation*.

5.11.2 Physical Basis of Electroporation

Electroporation involves exposing cells to a pulse of high voltage electricity. It is thought that this causes reorientation of molecules in the cell membrane making it transiently porous to vectors. The technique is especially useful in transforming eucaryotic cells. When dealing with plant cells, the cell wall is first removed by enzymatic digestion to form a *protoplast*.

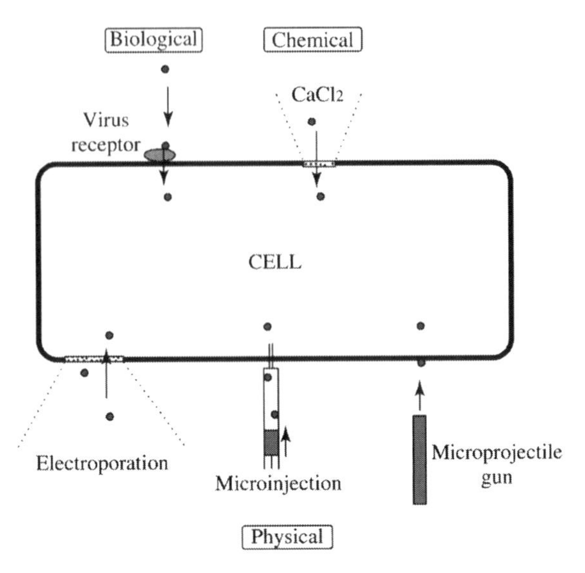

Figure 5.54. Transformation of cells. Cells may be transformed by a variety of biological, chemical and physical methods to increase the uptake efficiency for foreign DNA. Electroporation and $CaCl_2$ treatment make the cell membrane transiently porous to vectors (circles).

REFERENCES

General reading

Adamson, N.J. and Reynolds, E.C. (1997) Rules relating electrophoretic mobility, charge and molecular size. *Journal of Chromatography B*, **699**, 133–48. General review of theory of electrophoresis.

Cruz, A., Marsh, D. and Pérez-Gil, J. (1998) Rotational dynamics of spin-labelled surfactant-associated proteins SP-B and SP-C in dipalmitoylphosphatidylcholine and dipalmitoylphosphatidylglycerol bilayers. *Biochemica et Biophysica Acta*, **1415**, 125–34.

Issaq, H.J., Conrads, T.P., Janini, G.M. and Veenstra, T.M. (2002) Methods for fractionation, separation and profiling of proteins and peptides. *Electrophoresis*, **23**, 3048–61. A good general review of electrophoresis methods.

Rathore, A.S. (2002) Theory of electroosmotic flow, retention and separation efficiency in capillary electrochromatography. *Electrophoresis*, **23**, 3827–46. A review of the theory of behaviour of buffers, charged and uncharged particles in open channels and porous media.

Ji, L.T. and Choi, J.K. (2004) Usefulness of visible dyes for the staining of protein or DNA in electrophoresis. *Electrophoresis*, **25**, 2429–39. Review of visible stains.

Landers, J.P. (ed.) (2008) *Handbook of Capillary and Microchip Electrophoresis and Related Techniques*, 3rd edn, CRC Press, Boca Raton, FL, USA. Describes a broad range of applications of CE and related techniques.

Westermeier, R. (2006) *Electrophoresis in Practice: A Guide to Methods and Applications of DNA and Protein Separations*,

4th edn, Wiley-VCH Verlag GmbH, Weinheim. An introduction to the theory and practice of electrophoresis.

Winkler, M.A., Kundu, S., Robey, T.E. and Robey, W.G. (1996) Comparative peptide mapping of a hepatitis C viral recombinant protein by capillary electrophoresis and matrix-assisted laser desorption time-of-flight mass spectrometry. *Journal of Chromatography A*, **744**, 177–185.

Nondenaturing electrophoresis

Goldenberg, D.P. (1997) Analysis of protein conformation by gel electrophoresis, in *Protein Structure: A Practical Approach* (ed. T.E. Creighton), IRL Press, NY, USA, pp. 187–218.

Maly, I.P. and Toranelli, M. (1993) Ultrathin-layer zone electrophoresis of lactate dehydrogenase isoenzymes in microdissected liver samples. *Analytical Biochem.*, **214**, 379–88. Description of zymogram analysis of LDH.

Moriyama, T. and Ikeda, H (1996) Hydrolases acting on glycosidic bonds: chromatographic and electrophoretic separations. *Journal of Chromatography B*, **684**, 201–16. A review of amylase detection including activity staining.

Rivoal, J., Smith, C.R., Moraes, T.F. *et al.* (2002) A method for activity staining after native polyacrylamide gel electrophoresis using a coupled enzyme assay and fluorescence detection: Application to the analysis of several glycolytic enzymes. *Analytical Biochemistry*, **300**, 94–9. An elegant activity staining method suitable for dehydrogenases.

See, Y.P. and Jackowski, G. (1989) Estimating protein molecular weights of polypeptides by SDS gel electrophoresis, in *Protein Structure; a Practical Approach* (ed. T.E. Creighton), IRL Press, NY, USA, pp. 1–21. Determination of native protein masses by non-denaturing electrophoresis.

Denaturing electrophoresis

Laemmli, U.K. (1970) Cleavage of structural proteins during the assembly of the head of bacteriophage T4. *Nature (London)*, **227**, 680–5. The most widely-used SDS PAGE method.

Makowski, G.S. and Ramsby, M.L. (1996) Calibrating gelatin zymograms with human gelatinase standards. *Analytical Biochem.*, **236**, 353–6. Describes activity staining in SDS PAGE gels.

Makowski, G.S. and Ramsby, M.L. (1997) Protein molecular weight determination by sodium dodecyl sulfate polyacrylamide gel electrophoresis, in *Protein Structure: A Practical Approach* (ed. T.E. Creighton), IRL Press, NY, USA, pp. 1–28. A good practical description of SDS PAGE, including detailed protocols.

Patel, A.C. and Matthewson, S.R. (1998) *Molecular Biomethods Handbook* (eds R. Rapley and J. Walker), Humana Press, Totowa, NJ, USA, pp. 401–11. Describes the preparation of radiolabelled proteins and peptides.

Electrophoresis in the study of DNA

Anderson, P.S., Jespersgaard, C., Vuust, J. *et al.* (2003) Capillary electrophoresis-based single strand DNA conformation analysis in high-throughput mutation screening. *Human Mutation*, **21**, 455–65. Review of use of CE in SSCP analysis.

Aquino de Muro, M. (1997) Gene probes, in *Biomolecular Biomethods Handbook* (eds R. Rapley and J. Walker), Humana Press, Totowa, NJ, USA, pp. 59–72. Describes the preparation of radiolabelled oligonucleotides.

Kleparnik, K. and Bocek, P. (2007) DNA diagnostics by capillary electrophoresis. *Chemical Reviews*, **107**, 5279–5317. An exhaustive review of use of CE in DNA diagnostics.

McCreery, T.P. and Barrette, T.R. (1997) in *Biomolecular Biomethods Handbook* (eds R. Rapley and J. Walker), Humana Press, Totowa, NJ, USA, pp. 73–6. Describes the preparation of non-radioactively labelled oligonucleotides.

Orita, M., Suzuki, I., Sekiya, T. and Hayashi, K. (1989) Rapid and sensitive detection of point mutations and DNA polymorphisms using the polymerase chain reaction. *Genomics*, **5**, 874–9. Description of SSCP analysis for identification of mutations.

Righetti, P.G., Gelfi, C. and D'Acunto, M.R. (2002) Recent progress in DNA analysis by capillary electrophoresis. *Electrophoresis*, **23**, 1361–74. A review of DNA CE.

Sanger, F. (1981) Determination of nucleotide sequences in DNA. *Science*, **214**, 1205–10. Original report of dideoxynucleotide DNA sequencing method.

Slater, G.W., Guillouzic, S., Gauthier, M.G. *et al.* (2002) Theory of DNA electrophoresis (similar to 1999-2002 (1)/(2)). *Electrophoresis*, **23**, 3791–816. Review of developments in DNA electrophoresis.

Zhou, S.F. (2003) Separation and detection methods for covalent drug-protein adducts. *Journal of Chromatography B*, **797**, 63–90. Review of methods for protein-drug conjugate isolation and detection.

Immunoelectrophoresis

Parodi, A., Drosera, M., Barbieri, L. and Rebora, A. (1998) Counterimmunoelectrophoresis, ELISA and immunoblotting detection of anti-Ro/SSA antibodies in subacute cutaneous lupus erythematosus. A comparative study. *British Journal of Dermatology*, **138**, 114–7. A comparison of the three methods for clinical diagnosis.

Nichols, E.J. and Kenny, G.E. (1984) Immunochemical characterisation of a heat-stable surface antigen of *Mycoplasma pulmonis* expressing both species-specific and strain-specific determinants. *Infection and Immunity*, **44**, 355–63. An application of CIE.

Pulsed field gel electrophoresis

Bustamante, C., Gurrieri, S. and Smith S.B. (1993) Towards a molecular description of pulsed-field gel electrophoresis. *Trends in Biotechnology*, **11**, 23–30. A review of DNA mobility in agarose gel systems including pulsed field gel electrophoresis.

Chevallier, B., Dugourd, D., Tarasiuk, K. *et al.* (1998) Chromosome sizes and phylogenetic relationships between serotypes of *Actinobacillus pleuropneumoniae*. *FEMS Microbiology Letters*, **160**, 209–16. A good example of the application of pulsed field gel electrophoresis

Cole, S.T. and Saint Girons, I. (1994) Bacterial genomics. *FEMS Microbiol. Rev.*, **14**, 139–60. Use of pulsed field gel electrophoresis in bacterial taxonomy.

Lahti, C.J. (1996) Pulsed field gel electrophoresis in the clinical microbiology laboratory. *Journal Clin. Lab. Analysis*, **10**, 326–30.

A review of use of pulsed field gel electrophoresis in the identification of new pathogenic strains of bacteria.

Nevas, M., Lindstrom, M., Hielm, S. *et al.* (2005) Diversity of proteolytic *Clostridium botulinum* strains determined by a pulsed-field gel electrophoresis approach. *Applied and Environmental Microbiology*, **71**, 1311–7. An application of pulsed field electrophoresis to strain typing.

Capillary electrophoresis

Bao, J.J., Fujima, J.M. and Danielson, M.D. (1997) Determination of minute enzymatic activities by means of capillary electrophoresis. *Journal of Chromatography B*, **699**, 481–97. Use of CE in immunoassay.

Camilleri, P. (1998) *Capillary Electrophoresis; Theory and Practice*, 2nd edn, Springer-Verlag, London. An overview of the theory and applications of CE.

El Rassi, Z. (1997) Selectivity and optimization in capillary electrophoresis. *Journal of Chromatography A*, **792**, 1–519. A collection of reviews by various authors on different aspects of CE.

Hage, D.S. (1997) Chiral separations in capillary electrophoresis using proteins as stereoselective binding agents. *Electrophoresis*, **18**, 2311–21. Review of chiral separation using CE.

Krivankova, L. and Bocek, P. (1997) Synergism of capillary isotachophoresis and capillary zone electrophoresis. *Journal of Chromatography B*, **689**, 13–34. A review of capillary isotachophoresis.

Rickard, E.C. and Towns, J.K. (1996) Applications of capillary zone electrophoresis to peptide mapping. *Methods in Enzymology*, **271**, 237–64. A review of application of CE in polyacrylamide gels to peptide mapping.

Schmerr, M.J., Jenny, A. and Cutlip, R.C. (1997) Use of capillary sodium dodecyl sulfate gel electrophoresis to detect the prion protein extracted from scrapie-infected sheep. *Journal of Chromatography B*, **697**, 223–29.

Shimura, K. and Kasai, K.-I. (1997) Affinity capillary electrophoresis: a sensitive tool for the study of molecular interactions and its use in microscale analyses. *Analytical Biochemistry*, **251**, 1–16. Review of determination of binding constants using CE.

Wu, S.L. (1997) The use of sequential high-performance liquid chromatography and capillary zone electrophoresis to separate the glycosylated peptides from recombinant tissue plasminogen activator to a detailed level of microheterogeneity. *Analytical Biochemistry*, **253**, 85–97. Two-dimensional analysis using HPLC and CE.

Electroblotting

Bolt, M.W. and Mahoney, P.A. (1997) High-efficiency blotting of proteins of diverse sizes following sodium dodecyl sulfate-polyacrylamide gel electrophoresis. *Analytical Biochemistry*, **247**, 185–92. Blotting of proteins.

Dunn, M. J. (1996) Electroblotting of proteins from polyacrylamide gels, in *Protein Purification Protocols*, (ed. S. Doonan), Humana Press, Totowa, NJ, USA, pp. 363–70. A detailed method for electroblotting of proteins.

Kusuda, K., Kobayashi, T., Ikeda, S. *et al.* (1998) Mutational analysis of the domain structure of mouse protein phosphatase 2Cb. *Biochemical Journal*, **332**, 243–50. An example of the use of western blotting.

Schagger, H. (2006) Tricine-SDS-PAGE. *Nature Protocols*, **1**, 16–22. A description of Tricine-SDS-PAGE which is especially useful for electroblotting of smaller and more hydrophobic proteins.

Southern, E. (1975) Detection of specific sequences among DNA fragments separated by gel electrophoresis. *Journal of Molecular Biology*, **98**, 503–17. Original description of southern blotting.

Electroelution and preparative electrophoresis

Atmeh, R.F., Massad, T.T., Kana'an, B.M. and Abu-Alrob, A.A. (2008) Separation of biomacromolecules by electrofiltration through gel layers. *Analytical Biochemistry*, **373**, 307–12. A good example of electroelution.

Dunn, M. J. (1996) Electroelution of proteins from polyacrylamide gels, in *Protein Purification Protocols* (ed. S. Doonan), Humana Press, Totowa, NJ, USA, pp. 357–62. A detailed method for electroelution of proteins.

LeGendre, N. and Matsudaira, P.T. (1989) Purification of proteins and peptides by SDS-PAGE, in *A Practical Guide to Protein and Peptide Purification For Microsequencing* (ed. P.T. Matsudaira), Academic Press, San Diego, pp. 52–72. Description of electroblotting and electroelution of proteins from SDS PAGE gels.

Sequence analysis of electrophoretically-separated proteins

Aebersold, R. (1989) Internal amino acid sequence analysis of proteins after *in situ* protease digestion on nitrocellulose, in *A Practical Guide to Protein and Peptide Purification for Microsequencing* (ed. P.T. Matsudaira), Academic Press, San Diego, pp. 73–90. A guide to peptide mapping of electroblotted proteins.

Aebersold, R. (1991) High sensitivity sequence analysis of proteins separated by polyacrylamide gel electrophoresis, in *Advances in Electrophoresis*, Vol. **4** (eds A. Chrambach, M.J. Dunn and B.J. Radola), Wiley-VCH Verlag GmbH, Weinheim, Germany, pp. 81-168. Sequencing of proteins separated by electrophoresis.

Patterson, S.D. (1994) From electrophoretically separated protein to identification: strategies for sequence and mass analysis. *Analytical Biochemistry*, **221**, 1–15. Sequence determination of proteins separated by electrophoresis.

Chapter 6

Three-Dimensional Structure Determination of Macromolecules

Objectives

After completing this chapter you should be familiar with:

- *The* **protein folding problem**.
- *Determination of three dimensional structures of biomolecules by* **multi-dimensional NMR**.
- *How* **crystals** *of biomolecules may be obtained.*
- *Determination of three dimensional structures of biomolecules from* **X-ray diffraction experiments**.

The chemical and biological properties of biomolecules (especially biomacromolecules such as proteins and DNA) are determined mainly by their *molecular structure*. This underlies important processes such as enzyme catalysis, binding of ligands at protein receptors and correct expression of genetic information at the level of protein. Even modest changes in structure (e.g. arising from point mutations or from interaction with environmental chemicals) can result in molecules with drastically altered biological activity. Several techniques have been presented elsewhere in this text which yield information on the approximate structure of biomolecules such as their quaternary structure, overall shape/mass and secondary structure (e.g. Chapters 3, 4, 5, 7 and 10). These techniques often give us crucial information on the importance of particular structural domains for function and allow comparison of normal with aberrant molecules. However, in order to understand the detail of how biomolecules function, we need *high-resolution* structural information (i.e. at ~2 Å resolution). In practice, this information is particularly difficult to obtain for the two most important classes of biomacromolecules (proteins and nucleic acids) and macromolecular complexes composed of mixtures of these such as ribosomes and viruses.

Since proteins adopt a much wider range of *tertiary* (i.e. three-dimensional) structures than other biomacromolecules and frequently mediate the functioning of other classes of biopolymers, this chapter will concentrate in particular on protein structure. We will first examine how proteins come to have a particular shape and then describe the two main experimental techniques which give us high-resolution insights into protein structure; multi-dimensional NMR and X-ray crystallography. These techniques are independent of each other and have the major difference that the applicability of NMR is in practice limited to relatively small proteins. However, as will be described later (Section 6.7) these methods also *complement* each other because NMR gives information relevant to molecular structure in solution while X-ray crystallography gives structural information on molecules in the solid (i.e. crystalline) phase. It is important to understand that processes similar to those which determine protein structure also underlie structure formation in nucleic acids and macromolecular complexes. Moreover, both X-ray crystallography and multi-dimensional NMR are just as applicable to structural investigations in other classes of biomolecules as in proteins.

6.1 THE PROTEIN-FOLDING PROBLEM

Proteins fold up into compact, bioactive shapes or conformations largely as determined by their *primary structure* (amino acid sequence). In computer parlance, we could regard primary structure as a sort of 'program' encoding a particular three-dimensional structure. For this reason proteins are *informational* biomolecules in which alterations in primary structure (i.e. mutations) often result in an altered three-dimensional structure lacking normal function. Understanding the consequences of such mutations has led to knowledge of the biochemical basis of many genetic diseases as well as the prospect of eventual gene therapy for treating such conditions. Moreover, because we can now readily alter primary structure, the possibility exists of creating completely new proteins such as enzymes or antibodies designed for specific biotechnological applications. A major limitation to advance in this new field of *rational protein design* is our lack of understanding of precisely how primary structure determines tertiary structure. This is referred to as the *protein-folding problem*. This section describes the process of protein folding leading to a particular tertiary structure.

6.1.1 Proteins are only Marginally Stable

Proteins show great structural and functional complexity. They vary widely in size ranging from a few thousand to several million Da and are found in both hydrophobic and hydrophilic environments (e.g. biological membranes and cytosol, respectively). Proteins are also chemically complex and are specialized for a wide variety of cellular functions such as transport, defense and catalysis. This structural and chemical diversity arose by variation of the sequence and composition of only twenty amino acids which are the building blocks of proteins. In the early years of the twentieth century it was believed that proteins were essentially *random colloids* which acted as hosts for smaller bioactive components thought to be responsible for processes such as catalysis. The application of X-ray crystallography to biomolecules such as myoglobin and pepsin, however, revealed that proteins have a definite three-dimensional or *tertiary* structure which underlies their function.

In the 1950s Christian Anfinsen (Nobel prize in Chemistry, 1972) and his colleagues demonstrated that tertiary structure is a consequence of protein *primary* structure and that small proteins fold up into compact, bioactive structures in a manner determined largely by amino acid sequence alone. Anfinsen showed that, when ribonuclease is denatured by treatment with moderate heat or urea, it spontaneously renatures on subsequent cooling or removal of urea. Later, Bruce Merrifield (Nobel prize in Chemistry, 1984) chemically synthesized a polypeptide with identical sequence to ribonuclease and found that it folded up spontaneously such that its physical and catalytic properties were indistinguishable from the natural enzyme. Thus, primary structure does not alone determine the ultimate tertiary structure but also in some way apparently *directs the folding* of the protein into that structure. We now know (see below) that some large proteins often require assistance from other proteins to fold into their correct structure *in vivo* but Anfinsen's insight reveals that a particular three-dimensional shape is a natural consequence of sequence.

Proteins are *compact* structures with the interior having a density similar to that of a protein crystal. We can regard the interior as being composed mostly of a tightly folded polypeptide chain from which mainly hydrophobic amino acid side-chains radiate to occupy any available space. Proteins are also heavily solvated (40–60%) with water and, in cases where hydrophilic residues are enclosed in the interior of the protein, they are usually hydrogen-bonded to molecules of *structural water*. These are water molecules which form an integral part of the structure of the protein. For example, in the case of bovine pancreatic trypsin inhibitor, four water molecules are invariably found in specific locations and steric arrangements in the interior of the protein. We can think of structural water as providing a low-energy option for internalizing hydrophilic side-chains thus extending folding possibilities available to the polypeptide chain. The folding of even a small protein is a highly efficient packaging process which takes place remarkably quickly with, in the majority of cases, minimum intervention by other molecules.

This folding process is accompanied by extensive *desolvation* of the polypeptide backbone in which hydrogen bonding contacts between water and amide C=O and N–H groups are lost. A consequence of this is that such groups then interact with other amide groups by intrachain or interchain hydrogen bonds such as (C=O. . .H–N) (Chapter 1; Figure 6.1). This is probably the reason why, in proteins for which three-dimensional structures have been determined by X-ray crystallography, some 50% of amino acids (except glycine and alanine) adopt dihedral angle values close to those of α-helix and some 40% adopt values close to those of β-sheet. These backbone geometries facilitate desolvation by providing nonwater hydrogen bond partners for amide groups.

The surfaces of proteins are more mobile than the interior and consist mainly of hydrophilic side-chains protruding from the protein. These interact with solvent water by hydrogen bonding forming a *solvation shell* around the protein (Chapter 1). The *accessible surface area* of globular proteins (A_s) is approximately proportional to its molecular weight (M) as follows:

$$A_s = 11/12\, M^{2/3} \tag{6.1}$$

This value is nearly *twice* that which would be calculated from the geometry of a smooth sphere showing that the surface of a protein is in fact quite *rough*. Studies with extrinsic fluors such as ANS (Table 3.4; Section 3.4.4) reveal that many proteins have pockets on their surface to which such fluors can bind which are more hydrophobic than the average for the surface. Such pockets may provide ligand binding sites for the protein and they may also interact with hydrophobic stationary phases in hydrophobic interaction chromatography (Chapter 2).

Although proteins from microorganisms specialized for extreme environments such as thermophilic bacteria are particularly heat-stable, in general proteins are of only marginal stability. Comparatively modest increases of temperature or pressure can result in complete unfolding or *denaturation* of

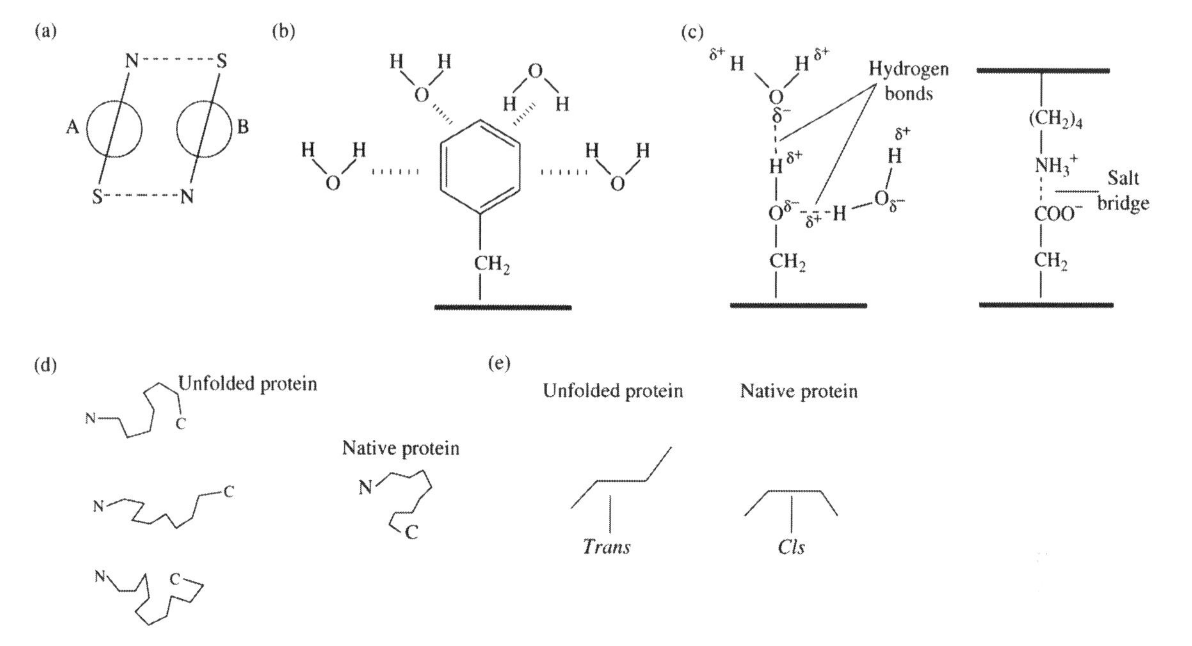

Figure 6.1. The main forces involved in protein folding. Stabilizing forces include (a) van der Waals interactions. These are short-range attractive forces between any pair of atoms due to transient magnetic dipoles which they induce in each other. North (N) and south (S) poles of the dipoles are shown. If two atoms come too close together, repulsion between electron clouds can overcome this attraction. Each atom has a van der Waals radius and the sum of these radii for a pair of atoms is the van der Waals contact distance. (b) The hydrophobic effect. Hydrophobic amino acids such as phenylalanine 'prefer' to be buried in the interior of a protein rather than be exposed to solvent water. Repulsion of water by phenylalanine is shown (''''). (c) Electrostatic interactions. Hydrogen bonds stabilize interactions between polar side-chains and solvent water. Salt bridges arise from electrostatic attraction between side-chains of opposite charge. The main destabilizing forces are (d) Chain entropy. Many more conformations are available for the protein in the unfolded state than in the folded state. (e) Bond strain due to unfavourable interactions in folded state (e.g. constraint into a *cis* rather than the more favoured *trans* configuration).

the polypeptide chain. For reasons which will be explained in the next section, for many single-domain small proteins we can represent protein denaturation (and, by analogy, protein folding) as a two-state system:

$$N \rightleftarrows U \tag{6.2}$$

where N and U represent the native and unfolded (i.e. denatured) forms of the protein.

Adoption of a particular folded structure seems to be the result of a delicate balancing act between large stabilizing forces and almost as large destabilizing forces (Figure 6.1; Table 6.1). A relatively weak stabilizing force is provided by *van der Waals interactions* (also sometimes called *London dispersion forces*) which occur between pairs of nonbonded atoms brought close together in space as a result of attraction between transient magnetic dipoles which they induce in each other. These forces are *additive* and, while not quantitatively important in small molecules, they represent a significant contribution to stabilization of large molecules such as proteins. If two atoms come closer together than the sum of their individual *van der Waals radius*, (referred to as their *van der Waals contact distance*) repulsive forces exceed attraction. In fact, most atoms in the interior of proteins

Table 6.1. Estimates of net magnitudes of stabilizing and destabilizing forces in protein folding.

Interaction (protein of 100 residues)	ΔG at 25 °C (kcal/mole)
Destabilizing	
Chain entropy	330–1000
Unfavourable folding interactions	200
Stabilizing	
Increased van der Waals bonds due to close packing	−227
Hydrogen bonds	−49 to −719
Hydrophobic effect	−264
Disulfide bridges	−4

Note that stabilizing forces contribute a negative ΔG while destabilizing force ΔG values are positive.

are in van der Waals contact with other atoms and cannot be packed more densely due to the limitation imposed by their van der Waals contact distance.

The *hydrophobic effect* is another important stabilizing force in proteins. This arises from folding of the protein to minimize exposure of hydrophobic residues to solvent water. Strongly hydrophobic side-chains (e.g. phenylalanine) can only be dissolved in water at considerable energy cost because they would strongly disrupt the hydrogen-bonded structure of the solvent. Moreover, the hydrophobic effect is largely *entropic* (rather than enthalpic) since removal of a hydrophobic residue from water releases solvent water from the constraints it had imposed leading to a net increase in solvent entropy. Packing of the polypeptide chain to remove such residues from contact with water is therefore strongly thermodynamically favoured.

The last important group of stabilizing forces are electrostatic interactions which arise from *electrostatic attraction* between either partial charges arising from the differing electronegativity of atoms (e.g. δ^-, δ^+; Chapter 1) or full charges arising from ionization of residue side-chains at physiological pH (e.g. acidic and basic residues). Electrostatic attraction between partial charges (i.e. $\delta^- --- \delta^+$) permit formation of low-energy, noncovalent bonds through hydrogen atoms called hydrogen bonds. Such bonds facilitate binding of the solvation shell of water at the protein surface and are also responsible for binding of structural water. Attraction between oppositely charged side-chains (e.g. Asp which is negatively charged and Lys which is positively charged) allows formation of *salt bridges* which are frequently found to stabilize folding of sequentially distant parts of the polypeptide chain. (This is a way of 'burying' charged side-chains at lower thermodynamic cost to the protein).

Further stability can be introduced by post-translational formation of *disulfide bridges*. These are covalent disulfide (i.e. –S–S–) bonds formed between sulphydryl (–SH) side-chains of cysteine residues which may be on the same or different polypeptides. They are frequently found in proteins destined for secretion from the cell and stabilize otherwise unstable folded structures by decreasing conformational entropy.

The principal destabilizing forces in proteins are due to *chain entropy* and *unfavourable structures* adopted by the polypeptide as a result of folding. Chain entropy represents the energy 'cost' in constraining the folded polypeptide into a single or small number of conformations rather than the greater number of conformations possible for the unfolded protein. Unfavourable interactions are structural features adopted within the folded protein which are of higher energy than the arrangement which would normally occur in the unfolded polypeptide. These may represent electrostatic repulsion between side-chains or constraining the polypeptide chain into an unfavourable arrangement (e.g. *cis* peptide bonds may occasionally occur rather than *trans*).

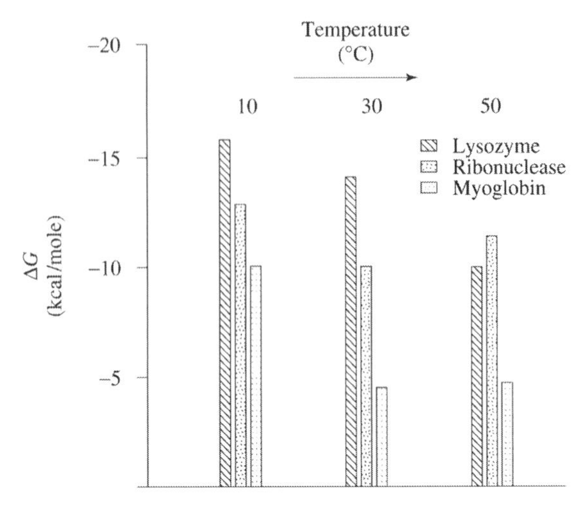

Figure 6.2. Folded proteins are highly-sensitive to temperature. ΔG values (i.e. $G_{\text{folded}} - G_{\text{unfolded}}$) for some small proteins in the temperature-range 10–50 °C. Temperature affects individual stabilizing/destabilizing interactions in a complex way which may result in the protein being more or less stable at higher temperature (e.g. compare ribonuclease and myoglobin at 30° and 50 °C). However, most proteins are unstable at high temperatures. Note the small magnitude of the energy difference which maintains the protein in a folded state.

The relative size of stabilization and destabilization forces results in a difference in Gibb's free energy (ΔG) between N and U which is slightly negative, that is the native structure is *thermodynamically favoured* (Figure 6.2). Relative to the magnitude of stabilizing and destabilizing forces, however (Table 6.1), ΔG values for small proteins are quite modest (−5 to −20 kcal/mole at room temperature). Variation of temperature can alter the delicate balance between stabilization and destabilization, resulting in an unfolded structure having a lower value of G at high temperature than the folded structure. This is the reason why a protein which is structurally stable at 50 °C may become denatured when exposed to a temperature of 80 °C for even a few minutes.

Protein folding is therefore a *thermodynamically spontaneous* process (i.e. folding causes a decrease in G) which allows the bulk of the molecules of a protein to adopt a single or small number of stable structures under physiological conditions. Preference for the folded state over the unfolded is marginal and is dominated by low-energy interactions (hydrophobic effect, salt bridges, etc.). The most likely biological rationale for this is that it allows proteins to be assembled and dismantled without a requirement for input or release of large amounts of energy. A consequence of this is that the bioactive conformations of proteins are fragile structures compared, for example, to synthetic polymers or low molecular weight organic/inorganic compounds. Methods for determination of three-dimensional structure may be severely constrained by this fragility.

6.1.2 Protein Folding as a Two-State Process

The model of protein folding shown in Equation (6.2) suggests that protein folding may be represented as a simple two-state process where essentially all the protein exists either as a completely unstructured polypeptide (U) capable of adopting many thousands of individual conformations *or* as a highly structured native molecule with a single or small number of conformations (N). This model simplifies the process of protein folding significantly because it takes no cognisance of the possibility of stable intermediates being formed during folding. *Molten globule intermediates* in which stable, native-like secondary structure is present are known to form during folding of cytochrome C and ribonucleases and such proteins have been designated *class I*. Proteins which fold by a two-state process as shown in Equation (6.2) and which do not form measurable, stable, intermediates such as chymotrypsin inhibitor 2 and cold shock protein B have been designated *class II*. It seems that class I proteins fold by a *hierarchical* process in which structural information is delocalized over the protein as a whole and in which tertiary structure interactions with parts of the protein distant in sequence stabilize secondary structure in measurable intermediates. This may explain why mutations often introduce instability to proteins but only rarely alter their overall fold. Class II proteins, by contrast, fold very rapidly based largely on local interactions such as the formation of secondary structure (Example 6.2).

The model proposed in Equation (6.2) gives a good quantitative approximation of the bulk of the population of a small single-domain class II protein at any stage of the folding process. What evidence do we have for such a simple model? Protein folding may be studied by first denaturing the protein (e.g. with heat or a chaotropic agent such as urea) and then allowing it to renature after removal of the denaturing treatment (e.g. cooling or removal of urea). Alternatively, we can study protein unfolding (i.e. $N \rightarrow U$) rather than directly studying protein folding ($U \rightarrow N$) since it is likely that a protein will follow similar energetic and kinetic pathways to unfold as to fold. Protein unfolding may be induced by a wide variety of treatments such as exposure to extremes of heat, pH or pressure, or treatment with denaturants such as urea or detergents and can be monitored with a range of spectroscopic and electrophoretic methods (Chapters 3 and 5). A typical denaturation profile for a protein is shown in Figure 6.3 and one of the interesting observations we can make from such experiments is that, regardless of the denaturation treatment employed, most small proteins appear to denature completely over a remarkably narrow range of temperature, pH or denaturant concentration. This suggests that protein unfolding is a rapid and highly cooperative process.

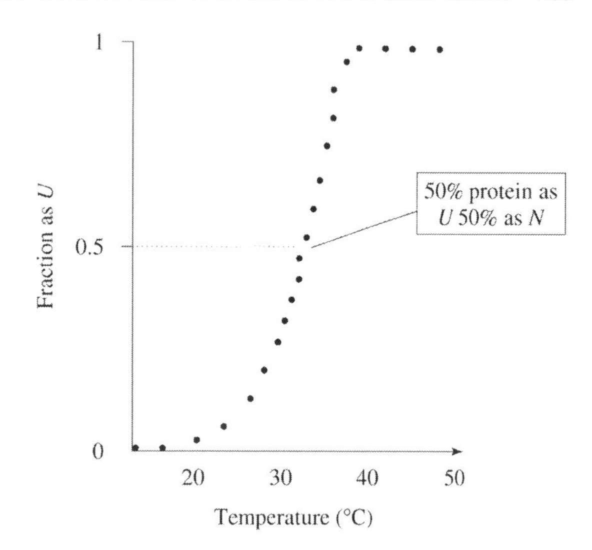

Figure 6.3. Protein denaturation profile of ribonuclease A. Temperature-induced denaturation was determined spectroscopically and is expressed as the fraction of unfolded (U) protein present. The situation when this fraction is 0.5 is described in the box.

Spectroscopic techniques such as absorbance, fluorescence and circular dichroism can be used to monitor folding/unfolding and to calculate the *fraction* of protein present in the U state at any particular temperature/pH/denaturant concentration. This fraction will be 0 when the protein is fully folded, 0.5 when 50% denatured and 1 when fully denatured. An equilibrium constant for 6.2 can be calculated from such measurements since

$$K = \frac{[N]}{[U]} \tag{6.3}$$

We would expect this equilibrium constant to be related to ΔH, the enthalpy change on unfolding by the *van't Hoff relationship* as follows:

$$\Delta H = RT^2 \frac{\mathrm{d} \ln K}{\mathrm{d}T} \tag{6.4}$$

where R is the universal gas constant and T is the absolute temperature of the sample. When enthalpy changes associated with unfolding of small proteins are directly measured by *calorimetric* measurements (Sections 8.3 and 8.4), it is found that they agree well with those arising from the van't Hoff relationship. In fact, the ratio between the two ΔH values is 1.05 suggesting that only some 5% of protein molecules are in a state other than the N or U states at any temperature. This body of evidence strongly suggests that protein folding is described well by a simple two-state model such as that shown in Equation (6.2) although a small amount of the protein exists in states other than N or U. What is this small amount of protein doing during protein folding?

6.1.3 Protein-Folding Pathways

A possible thermodynamic interpretation of protein folding is to think of the folded conformation as occupying a lower free energy value than any of the unfolded conformations. It could be imagined that the protein might *sample* all possible conformations until the lowest energy conformation was randomly arrived at. This is known as *Levinthal's paradox* and it may be explained by examining folding of a 100-residue protein. If it was assumed that each residue could adopt ten possible conformations (which is much less than the number of likely conformations available to most residues) then;

$$\text{Total number of conformations} = 10^{100} \quad (6.5)$$

If the time allowed for transition between each individual conformation was assumed to be 10^{-13} s then;

$$\text{Time required to sample all conformations} = 10^{77} \text{ years} \quad (6.6)$$

Since we know that proteins in fact fold on a time-scale of 10^{-1} to 10^{3} s, it is not likely that proteins randomly sample all possible conformations.

Thermodynamics does not offer a full description of reaction processes. We also have to consider the *pathway* through which the reaction moves, that is the kinetics of the process. Levinthal's paradox is explained by the fact that, although there may be an astronomically large number of possible conformations each with an individual value of G, not all of these are *kinetically accessible* and therefore the search for a folded structure would be expected to be biased towards conformations which are accessible. The protein could therefore be expected to fold in a *biased* way along a definite pathway rather than by *unbiased* sampling. Protein folding may be viewed as the movement from a relatively extended and disordered structure of a given G value to a far more compact and ordered structure of lower G through a *folding pathway* (Figure 6.4). This pathway might contain local energy minima which would represent *folding intermediates* as well as higher-energy structures which we

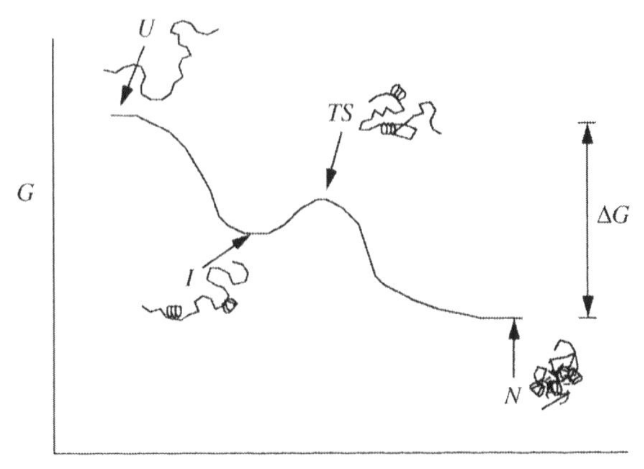

Figure 6.4. A protein folding pathway. As protein folds, the compactness of the structure increases and free energy (G) decreases. The pathway is defined by intermediate structures (I) which have intermediate compactness and free energy, occupying a local energy minimum. Although only one is shown, there may be several of these in a given pathway. Transition states (TS) are high-energy structures which provide energy barriers to the folding pathway. The final folded structure (N) occupies a *global* energy minimum and is the most compact structure. The ΔG between the U and N states provides the thermodynamic driving force for folding.

can regard as *transition states*. The local energy minima of folding intermediates will be of intermediate G values between the U and N conformations of the protein. The small amount of protein not in the N or U states at any given time during protein folding is present mainly as folding intermediates along the folding pathway.

Based on a range of experimental studies, several models have been proposed to explain protein folding. Usually these studies have focused on small proteins such as lysozyme, ribonuclease and bovine pancreatic trypsin inhibitor. Three classical mechanisms for folding have been proposed (Figure 6.5). The *framework model* postulates that secondary structure forms *independently* of tertiary structure and that diffusion and coalescence of regions of secondary structure then go on to form tertiary structure (this is also sometimes called the *diffusion-collision model*). The *hydrophobic collapse model* suggests rapid collapse of the protein around hydrophobic side-chains and that secondary and tertiary structures form in the consequently restricted conformational space. Both of these models are consistent with the formation of measurable intermediates in the folding pathway. The *nucleation model* postulates that localized regions of native secondary structure act as *nuclei* from which correct tertiary structure is propagated. Like the framework model, this proposes formation of tertiary structure as a

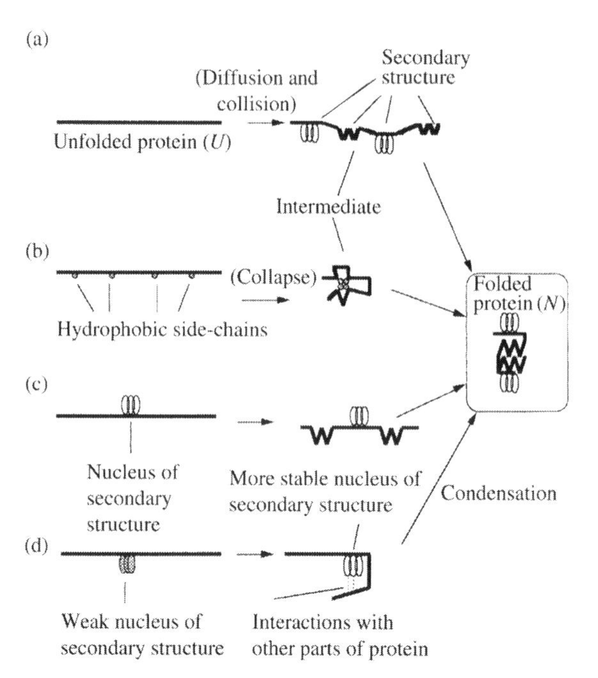

Figure 6.5. Models of protein folding. (a) Framework model; local elements of secondary structure form independently of tertiary structure. (b) Hydrophobic collapse model; protein chain rapidly collapses around hydrophobic side-chains. Secondary and tertiary structure form in restricted conformational space around this. (c) Nucleation model; secondary structure elements act as nuclei around which tertiary structure forms. (d) Nucleation-condensation model; A weak central nucleus is first formed followed by stabilization with sequentially distant residues leading to a larger and more stable nucleus. This condenses with the rest of the structure until folding is complete.

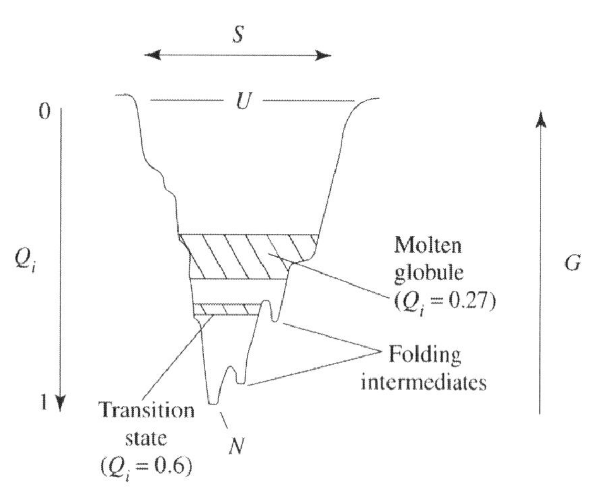

Figure 6.6. A protein folding funnel. The funnel represents incremental changes in conformation as the protein folds from U to N. Entropy (S) decreases as the number of conformations available reduces. The parameter Q_j reflects gradual forming of the correct structural contacts in N. Experimentally identifiable species such as molten globules, folding intermediates and transition states are shown.

consequence of secondary structure but it does not necessitate formation of intermediates.

More recent work involving a combination of mutational analysis, computer simulations using molecular dynamics and two-dimensional NMR (Section 6.2) has led to the proposal that a *nucleation–condensation model* may explain the folding of small proteins (Figure 6.5). This postulates that a weak central nucleus is formed consisting primarily of adjacent residues. The nucleus is stabilized by long-range interactions with residues distant in the sequence to form a progressively larger and more stable nucleus. This nucleation process is closely coupled to condensation of the rest of the structure around the stable nucleus until the complete structure is formed.

It is unlikely that any one model of protein folding fully explains all cases. Every protein is an individual molecule and the details of folding are likely to be unique to each. It is probable, though, that formation of secondary structure, nucleation and condensation are frequent occurrences in most protein-folding processes.

An alternative to the classical image of the folding pathway is the *folding funnel* (Figure 6.6). This moves away from the concept of a folding pathway and instead represents the folding of a protein along a funnel formed by two variables; entropy, S, which is expected to *decrease* on folding and a parameter Q_i (which is not linear with free energy, G) which is given by

$$Q_i = \frac{\text{Number of pairwise contacts in state } i}{\text{Number of contacts in state } N} \qquad (6.7)$$

State i is any individual conformation along the folding funnel. Q_i varies between the unfolded state, U (where it has a value of 0), to a maximum value of 1 in the native state, N. Thus, this parameter is expected to increase on folding at the same time as the value of S decreases. In this representation, localized energy minima which in a pathway are designated as folding intermediates instead are represented as *folding traps*. In other words *some* proteins in the population may occupy this localized minimum but others may not. Folding traps are therefore not obligatory in the sense that intermediates in a folding pathway are.

The power of this representation of protein folding is that it allows us to explain Levinthal's paradox. Distinct structures for which experimental evidence have been obtained such as molten globules, folding traps and transition states occupy distinct parts of the funnel but progression along the funnel arises from small conformational changes across a large continuum of values for Q_i and S. This representation has the advantage that we can imagine the search for a lower free energy as a driving force through the funnel but that aspects of the kinetics of the process such as formation of intermediate and transition states are also readily visualized.

The protein folding problem is to understand how primary structure directs folding of a protein into a particular tertiary structure. The main practical difficulty in solving this problem is that we lack sufficient experimental tools to 'freeze' all forms of the protein along the folding pathway since many of them have only a transient existence. For this reason we are limited to constructing models of folding which are based largely on quasi-stable structures such as folding intermediates and molten globules for which we have direct experimental evidence. We need to infer the existence of unstable species such as transition states and steps along the pathway by interpolating between such quasi-stable structures. Even though most studies on protein folding have concentrated on small proteins, it is thought that the same physical principles govern folding of larger proteins. Because of the size and complexity of the folding problem in larger polypeptides it has been discovered that some 10% of proteins require additional proteins to fold correctly *in vivo*. These are referred to as *chaperones* because they 'chaperone' the protein into the correct folded structure.

6.1.4 Chaperones

Chaperones are required *in vivo* because exposure of hydrophobic residues on nascent or partially unfolded polypeptide chains to the aqueous environment of the cytosol could lead to aggregation and formation of disordered protein precipitates likely to be toxic to the cell. This is especially likely to be important in eukaryotes where proteins are significantly larger and often consist of several domains. The 'crowding' effect of high cellular protein concentration must also contribute to a tendency towards mis-folding. Formation of aggregates or incorrectly folded proteins could also have serious consequences for the rate of protein synthesis (since they would slow it down) and its efficiency (since undesirable variability would be introduced). Mis-folding can occur with nascent polypeptide chains while still attached to the ribosome. Once synthesized, some large proteins require a finite amount of time to fold correctly and aggregation between unfolded proteins is a possibility during this period. In any case, it is known that unfolded polypeptides do not survive very long in the cell as they are rapidly degraded by proteases. Therefore, the cell requires a system which will allow sufficient time for correct folding whilst simultaneously protecting unfolded proteins from aggregation or proteolysis. Two general categories of proteins prevent protein mis-folding. Chaperones are generally small proteins that protect nascent polypeptide chains from aggregating before the chain is completely synthesized. *Chaperonins* are larger oligomeric proteins which often form large cylinders within which completely synthesized though unfolded proteins can complete folding.

Chaperones and chaperonins were originally discovered as *heat shock proteins* in eukaryotic cells, that is proteins expressed in response to short periods of heating. It is most likely that they are induced because of the risk of thermal denaturation of proteins in the cell which would tend to lead to exposure of hydrophobic residues and consequent aggregation. They were designated *Hsp* followed by their apparent kDa molecular mass (e.g. Hsp 10) and their role in protein folding was only discovered subsequently. Chaperonins were first identified in bacteria as essential host factors necessary for assembly of bacteriophages. We now know that chaperonins are a widespread family of highly conserved proteins which are present in bacteria and eukaryote cytosol, mitochondria, endoplasmic reticulum and chloroplasts. The following description focuses on the main chaperones and chaperonins (exemplified by Hsp60 and Hsp 70, respectively) of *E. coli* but there are several eukaryotic homologues showing close sequence, structural and functional similarities to these proteins.

Hsp 60 and Hsp 70 are mechanistically distinct and have been extensively studied. These proteins are not substrate-specific, that is they bind to a wide range of different unfolded proteins with approximately the same affinity. They both work in conjunction with other proteins called *cochaperones* which are specific for each chaperone but which can distinguish between ATP-bound and ADP-bound structural variants of them. Both chaperones are ATP-ases and ATP-hydrolysis provides energy for allosteric transitions within Hsp 60 and Hsp 70 which the unfolded proteins experience as alternating hydrophobic and hydrophilic surfaces. This close coupling of ATP-hydrolysis to allosteric alterations in chaperones is crucial to their correct function. The effect of these changes in structure is gradually to ease the unfolded protein into its correct folded structure by successive cycles of binding to the chaperone. Not all the details of the precise mechanism of chaperones have yet been completely established but the following is an overview of the main features of Hsp 70 and Hsp 60.

Hsp 70 is also called *DnaK* and is specialized for binding to short sequences enriched in hydrophobic residues such as are found in nascent polypeptides being synthesized on ribosomes (Figure 6.7). This chaperone can therefore function *cotranslationally*, that is at the same time as translation. DnaK can exist in two structurally different forms; an ATP-bound form which has high affinity for unfolded substrate protein and an ADP-bound form which has low affinity.

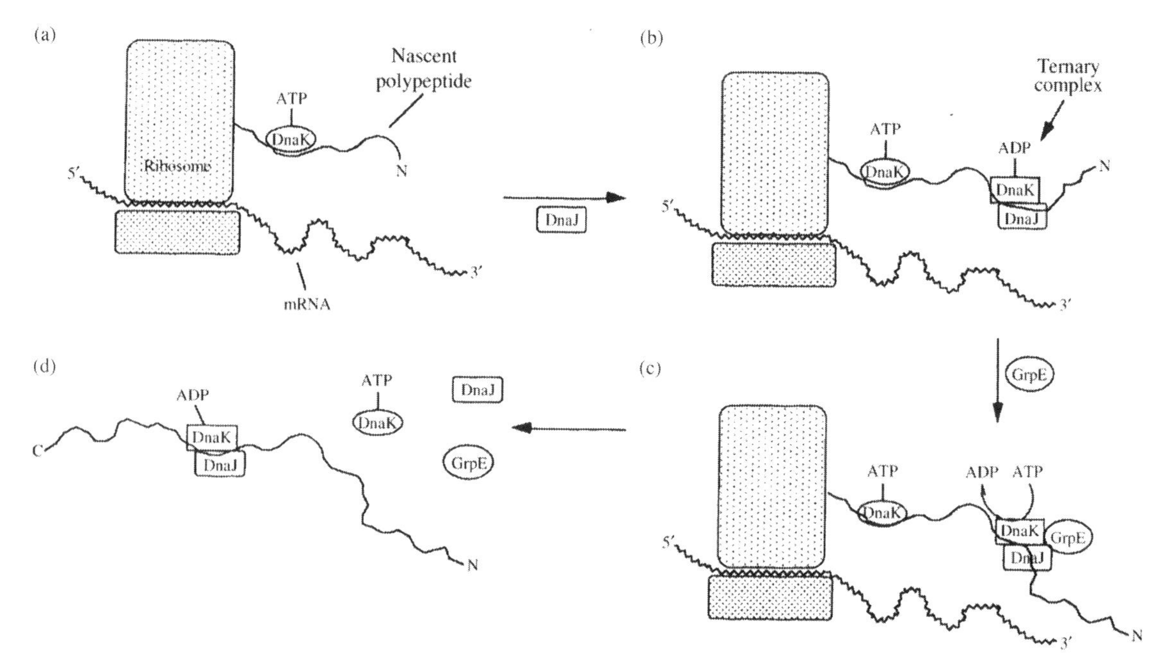

Figure 6.7. Functioning of Hsp 70 (DnaK) in folding of nascent polypeptides. (a) DnaK binds to hydrophobic stretches of nascent polypeptide chains during protein synthesis. (b) The co-chaperone, DnaJ, binds to DnaK, forming a ternary complex. This stimulates the ATP-ase activity of DnaK resulting in ATP being hydrolyzed to ADP. Note that the structure of DnaK-ADP differs from that of DnaK-ATP (c) A further protein, GrpE binds to the ternary complex, causing release of ADP and binding of ATP at DnaK. (d) The complex dissociates from the polypeptide. Note that several cycles of binding and dissociation can be gone through before peptide synthesis is complete. Moreover, several hydrophobic sites on a single polypeptide can act as substrates for the Hsp70 system.

It forms a ternary complex with hydrophobic stretches of the as yet incomplete polypeptide chain and another protein called *DnaJ* (Hsp 40) which stimulates the ATP-ase activity of DnaK. DnaJ is in fact a chaperone in its own right and has folding activity even in the absence of DnaK. A further protein (*GrpE*) can bind in turn to this complex causing the release of ADP from DnaK followed by binding of ATP. The complex now dissociates from the polypeptide chain and several cycles of binding and dissociation can be gone through before the polypeptide is complete and released from the ribosome. Most proteins fold independently of chaperones while some only require the aid of the Hsp 70 system to fold completely. A third group of proteins, however, do not fold into their native structure even with the help of the Hsp 70 system.

This third group of polypeptides now passes to the second main class of chaperones, the chaperonins, the principal component of which is Hsp 60 (also known as *GroEL*). This is composed of 58 kDa subunits of which 14 are arranged in the form of a hollow cylinder made up of two rings of seven (Figure 6.8). These rings are arranged head-to-head and enclose a cavity calculated to have an average diameter of 45 Å and a length of approximately 146 Å which would be capable of accommodating a folded polypeptide of maximum mass 147 kDa (assuming perfect packing). In practice, the maximum unfolded structure likely to be accommodated is approximately 55 kDa. Like Hsp 70, GroEL has ATP-ase activity and, in the absence of ATP has high affinity for unfolded polypeptides. Although not visible in

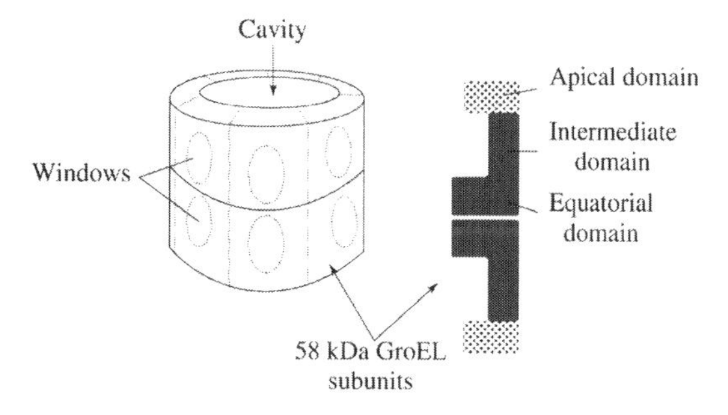

Figure 6.8. Organization of Hsp 60 chaperonin. Fourteen 58 kDa GroEL subunits are arranged in two circular rows of seven as shown. This forms a cavity within which folding can occur. Both the individual subunits and the complex they form are referred to as GroEl. Each subunit consists of three distinct structural domains; apical, intermediate and equatorial (which contain the ATP-ase site). Each equatorial domain also contains a 'window' sufficiently large for small molecules to diffuse freely through.

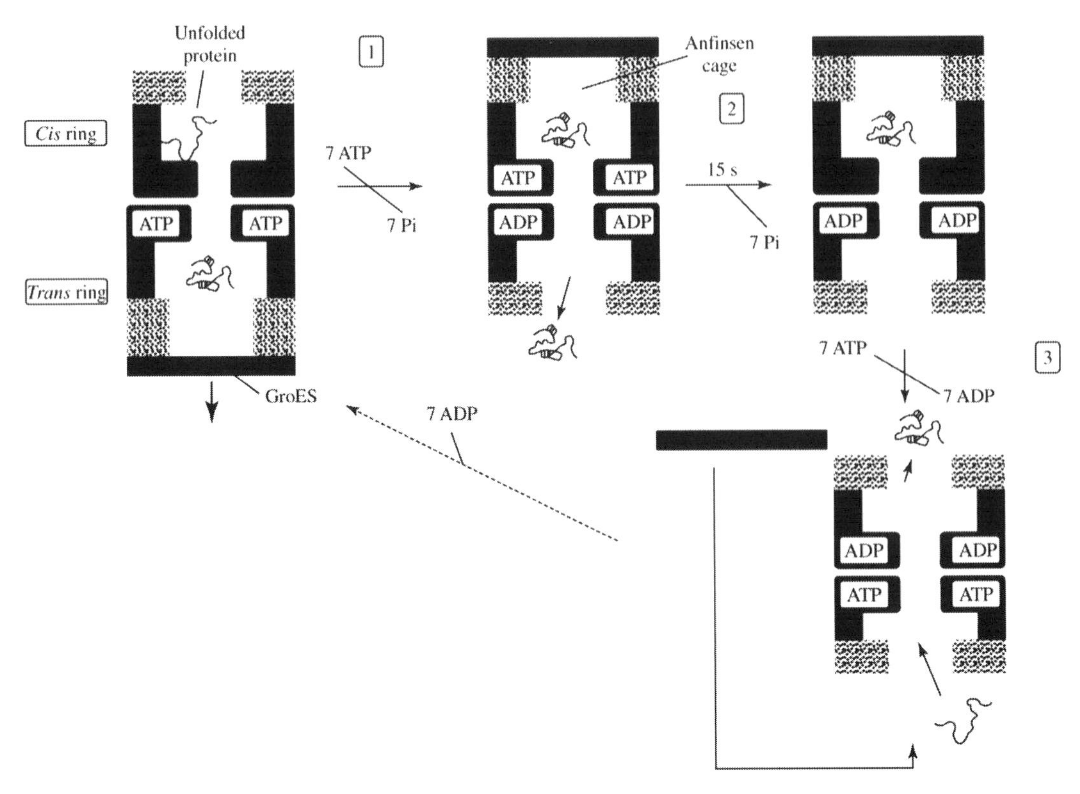

Figure 6.9. A vertical view of structural events in the GroEl/GroES system during protein folding. A transverse section of the upper (*cis*) and lower (*trans*) rings of seven 58 kDa subunits are shown (for clarity, two subunits from each ring are represented). (1) Unfolded protein binds strongly to GroEL (see also Figure 6.10). ATP hydrolysis in the *trans* ring causes binding of GroES to the apical domain and ATP binding to the seven equatorial domains of the *cis* ring of 58 kDa subunits (note enlarged cavity in *cis* ring). Simultaneously, GroES and folded protein are released from the *trans* ring. Conformational changes result in unfolded protein being released into the hydrophilic 'Anfinsen cage' allowing an attempt at folding. (2) After 15 s, ATP is hydrolyzed in the *cis* ring destabilizing the GRoEL/GroES complex. (3) ATP binds to the *trans* ring of 58 kDa subunits causing release of GroES and either release of the folded protein or another cycle of binding/release if the protein is not yet fully folded.

the crystal structure, the interior cavity is likely to be divided at the equatorial region of the complex between the upper and lower ring of subunits by flexible parts of the N- and C-termini of the subunits. Flexible regions of polypeptides are not visible in electron density maps from which such structures are determined (Section 6.5). The cavity of the ADP–GroEL complex is capable of accommodating several substrate protein molecules at a time and presents a nonpolar face made up of hydrophobic regions or 'patches' on the surface of each subunit.

The 58 kDa subunits consist of three distinct structural domains the *equatorial* domain is the largest of these and forms the bulk of the interface between the two rings of subunits and is also the site of ATP-ase activity. The *apical* domain provides most of the interactions with *GroES* (see below) and is connected to the equatorial domain by an *intermediate* domain which transmits allosteric structural changes between the other two domains.

Exposed hydrophobic regions of unfolded protein bind at the hydrophobic face lining the interior cavity of GroEL, leading to ATP-ase catalyzed ATP hydrolysis in the distal (*trans*) ring of subunits (Figures 6.9 and 6.10). ATP can now bind to the *cis* ring of subunits which results in major structural changes in GroEL. The ATP-bound form of GroEL has an enclosed cavity volume which is twice that of the ADP-bound form (the diameter increases to 70 Å). Moreover, ATP-GroEL has low affinity for unfolded protein due to the exposure of polar side-chains on the previously hydrophobic interior and also due to adjacent 58 kDa subunits moving further apart, thus disrupting binding of

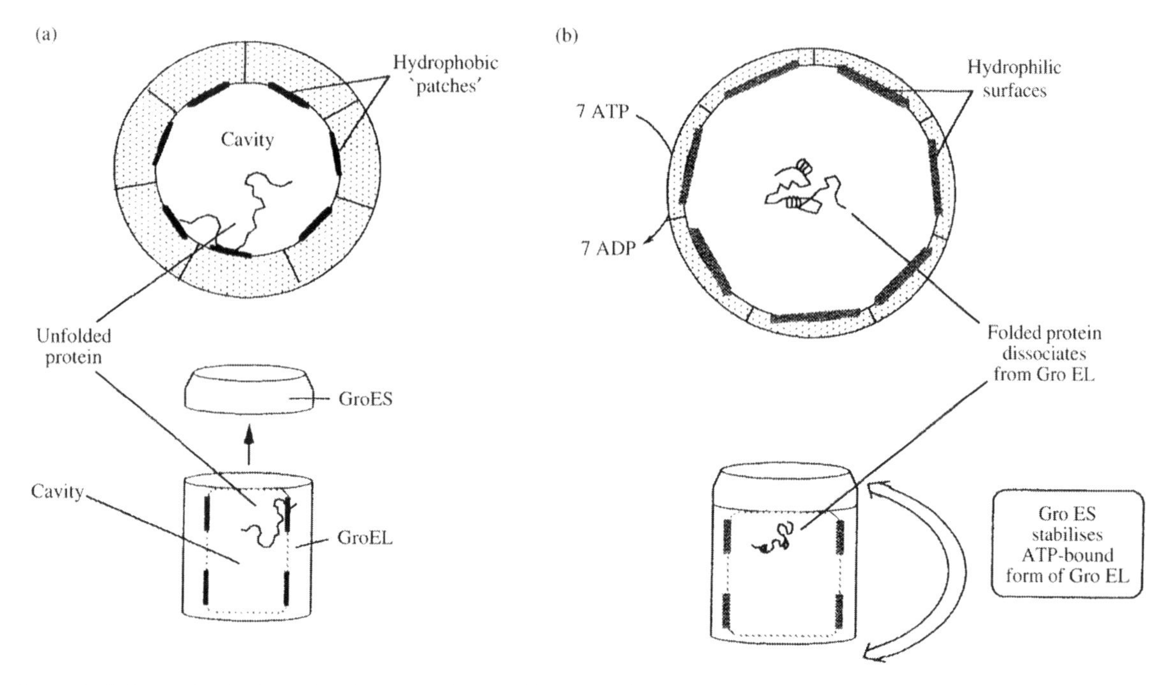

Figure 6.10. Protein folding in the Hsp 60 chaperonin. Views from above (top) and horizontal with (bottom) the Gro EL cylinder are shown after unfolded protein enters the cavity. (a) Hydrophobic stretches of unfolded protein bind to hydrophobic 'patches' of the interior of Gro EL. The co-chaperone, Gro ES (Hsp 10) has low affinity for this form of Gro EL. (b) ATP hydrolysis in the *trans* ring (see Figure 6.9) of Gro EL together with Gro ES binding causes an allosteric alteration in Gro EL. 58 kDa subunits move apart and hydrophilic residues (grey bars) replace hydrophobic side-chains on the interior of the cavity which significantly enlarges. The unfolded protein is released into the cavity where it can fold. Gro ES stabilizes this form of Gro EL by binding at one end of the cylinder to the apical domains of the 58 kDa subunits. It can alternate between opposite ends of the cavity as shown by arrow in successive cycles of binding and dissociation of unfolded protein.

substrate proteins which interact with several GroEL subunits at a time. The unfolded protein consequently dissociates and is released into the cavity where it attempts to fold.

In vivo, GroEL requires the co-chaperone GroES which is also a heat shock protein (Hsp 10) and is assembled into a heptamer. This acts as a 'lid' which covers one end of the GroEL cylinder (Figures 6.9 and 6.10). The initial binding of unfolded protein to the *cis* ring leads to dissociation of GroES from the *trans* ring. GroES can bind to the *cis* ring of the GroEL cylinder at the same time as ATP causing the cavity formed by the adjacent (*cis*) ring of 58 kDa subunits to enlarge relative to the distal (*trans*) ring by structural effects on the apical domain of the *cis* ring subunits (Figure 6.10). This forms a hydrophilic structure which is sometimes referred to as the *Anfinsen cage* (Figure 6.10). ATP binding to the *trans* ring of 58 kDa subunits causes release of GroES which opens the cage allowing folded protein either to escape or, if incompletely folded, to undergo another cycle of binding and folding. On each cycle of binding and release of unfolded protein, GroES binds at *alternate* ends of the GroEL cylinder. When the protein has folded correctly, it is released from the cavity into the cytosol. The structure of the GroEL-GroES complex has now been solved by X-ray crystallography and its structure and function can be viewed on the Internet at the address given in the Bibliography.

The GroEL/GroES system provides a dynamic surface environment of alternating hydrophobicity/hydrophilicity which is sequestered from the cytosol (Example 6.1). It increases the rate of individual steps in the protein folding pathway and protects the unfolded protein from proteolysis. The system also has an *unfolding activity* and, in the event of a protein folding incorrectly, can promote unfolding of the incorrect structure and give the protein another opportunity to fold. While many proteins do not need this system to fold correctly, most do fold faster in its presence. Such fast-folding proteins will often fold even before a cycle of ATP-hydrolysis is complete, thus avoiding waste of energy by the cell. It has been suggested that the ATP-ase activity of GroEL may act as a 'gatekeeper' providing a selection mechanism for slow-folding proteins which may require one or several cycles of ATP-hydrolysis.

Example 6.1 Forced Unfolding of RUBISCO in the GroEL-GroES Complex

Ribulose-1,5-bisphosphate carboxylase-oxygenase (RUBISCO) is a protein which cannot fold spontaneously into its native structure in the absence of chaperonins. It becomes blocked in an unfolded state capable of exchanging most of its hydrogen atoms with hydrogens from solvent water. This process can be followed by radioactively labeling the protein with tritium and following tritium-hydrogen exchange. It was observed that some twelve protons persist in unfolded RUBISCO which are not available for rapid exchange. The half-life for exchange of these protons is some 30 min. as compared to the more usual half-life of approximately 10 ms. The most likely reason for this is that they are amide hydrogens involved in secondary structure such as α-helix or β-sheet and thus protected from rapid exchange.

These hydrogen atoms provide a specific probe to follow events in the GroEL-GroES complex. It was observed that, when RUBISCO is mixed with GroEL alone, GroES alone or GroEL plus various nucleotides, the time-course of exchange is similar to that of the unfolded protein. However, when GroES and GroEL are added in the presence of either Mg-ATP or β,γ-imidoadenosine 5′-triphosphate (AMP-PNP) rapid exchange occurred with ten of the twelve hydrogens as shown in (1).

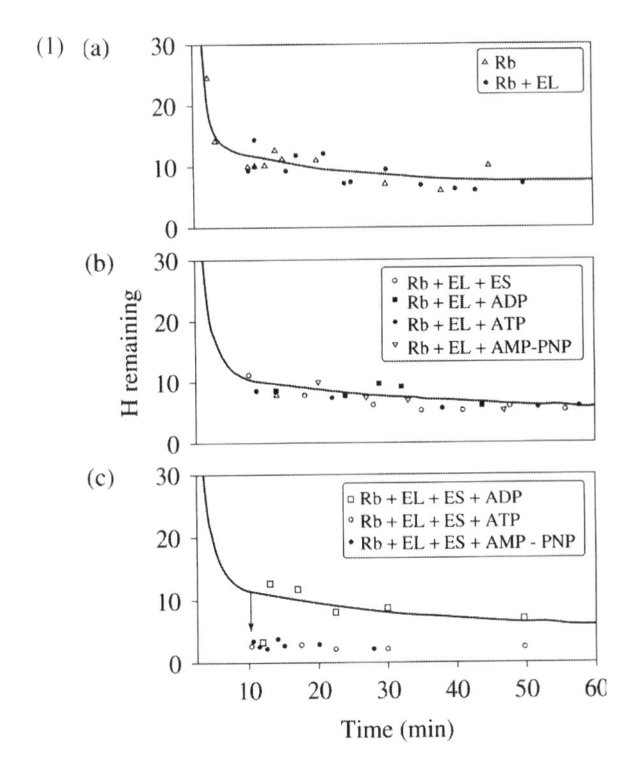

Number of tritium atoms remaining as a function of time. Note the rapid exchange (arrow) in panel C.

The most likely reason for this is that the energy of binding of the adenine nucleotide (rather than its hydrolysis) stimulates formation of the GroEL–GroES complex which disrupts whatever secondary structure is protecting most of the twelve protons from exchange. This is consistent with a model in which the substrate protein is 'stretched' to disrupt secondary structure, thus allowing further attempts at correct folding. The process is complete within 5 s which is well within the 13 s required for each unfolding cycle.

This is an example of forced unfolding which is thought to be a general mechanism for overcoming the slowness of unfolding of proteins which have become blocked in the folding process. The structure of GroEL is shown in (2) together with a sketch of the stretching model.

(*Continues*)

(Continued)

(2)

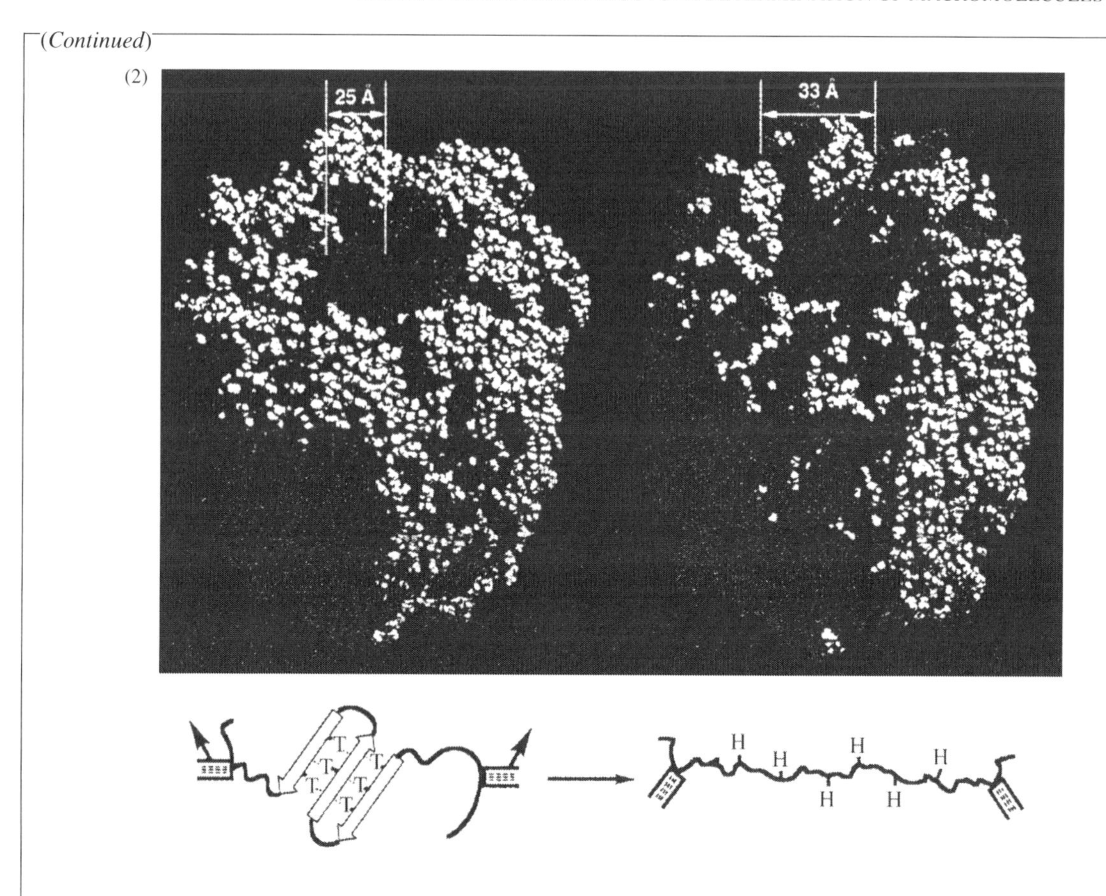

Forced unfolding in GroEL. The crystal structure of the GroEL part of the GroEL-GrES complex determined by X-ray crystallography is shown above while a sketch of the model is shown below (T = tritium; H = hydrogen). The unfolded structure binds to GroEL in the apical domains shown on the left with the distance between binding sites of 25 Å. When GroES binds, these domains are forced further apart which stretches the substrate protein and widens the gap between binding sites to 33 Å. Reprinted with permission from Shtilerman *et al.*, Chaperonin Function: Folding by Forced Unfolding *Science* **284**, 822–825 (1999) American Association for the Advancement of Science.

A functional distinction between the Hsp 70 and Hsp 60 systems has been drawn in that the former system seems to function largely to *maintain the folding competence* of proteins in the cell frequently operating co-translationally. In other words, Hsp 70 may not necessarily achieve complete folding of the protein into its native structure but rather may primarily *prevent it aggregating* which would halt the folding process. The Hsp 60 system, on the other hand, seems to be designed mainly to fold proteins completely into their native state and functions *post-translationally*. While some proteins pass sequentially from the Hsp 70 to the Hsp 60 system as described above, some *in vitro* evidence suggests that this may not necessarily be a *unidirectional* pathway, that is some proteins (e.g. firefly luciferase) may pass to the Hsp 70 system after prior exposure to GroEL. Luciferase also provides an example of a group of Hsp 60 substrates which are ejected from the chaperonin in unfolded form.

Membrane proteins are not substrates for chaperonins so this system is mainly concerned with folding of cytosolic proteins. GroEL substrates are typically in the mass range 10–55 kDa with a clear cut-off at 55 kDa, corresponding to the volume of the interior cavity. It is estimated that 15–30% of *E. coli* proteins may be GroEL substrates. The majority of small substrates (<25 kDa) are released within a single cycle which corresponds closely to the time required for synthesis of an average *E. coli* protein while substrates in the mass range 25–55 kDa require multiple chaperonin cycles and are typically multi-domain proteins. Of the 4300 or so proteins in *E. coli* some 2600 are cytosolic with an average size of

35 kDa. Only 13% of these proteins exceed the GroEL mass cut-off (55 kDa). In eukaryotes, the average protein size is larger (53 kDa) and 38% of yeast's 5800 proteins exceed the GroEl cut-off. Coupled with the fact that eukaryote protein synthesis is significantly slower than bacteria, this suggests that post-translational chaperonin activity may not be as important in these organisms. A possible reason for this may be a preference for cotranslational domain-wise folding in eukaryotes (i.e. folding of individual domains while translation of other domains is still in progress) compared to a preference for post-translational folding in prokaryotes. Indeed it has been suggested that, during evolution, domains which had a predisposition to fold cotranslationally may have been selected and this may explain why eukaryote proteins heterologously expressed in bacteria so often aggregate to form *inclusion bodies*.

Chaperones make use of ATP-hydrolysis coupled to allosteric structural changes to promote protein folding in cells. But many of the physicochemical forces driving folding are expected to be largely independent of chaperones and these proteins may merely provide a supportive microenvironment where these forces can operate. Two main models for the contribution of chaperones to folding of slow-folding proteins have been proposed. The *Anfinsen cage* model suggests that *intermolecular* aggregation could provide a limit to folding and that the main role of chaperones is to avoid this. An alternative view is the *iterative annealing* model which proposes that the rate-limiting step in slow folding may be the *intramolecular* formation of incorrectly folded structures and that the role of chaperones is to unfold such polypeptides and give them an opportunity to fold correctly. These two models are not mutually exclusive and it is possible that chaperones contribute to the folding of individual proteins in different ways depending on the particular rate-limiting step in their folding pathways.

6.2 STRUCTURE DETERMINATION BY NMR

In our previous discussion of NMR spectroscopy (Chapter 3; Section 3.7) we saw that the chemical shift associated with a nucleus possessing magnetic spin may be altered by its precise chemical environment. Interactions between magnetic dipoles in different nuclei can arise as a result of either the bond structure of the group in which the nucleus is involved or the presence of nearby nuclei. These interactions provide a means by which atoms in a protein can be connected to one another and *multi-dimensional NMR* can be used to detect these interactions. NMR is therefore a major method for structure determination of biomolecules in solution and for his work in this area Kurt Wütrich was awarded the Nobel Prize in Chemistry in 2002. This section will outline the physical basis, advantages and limitations of this important group of techniques.

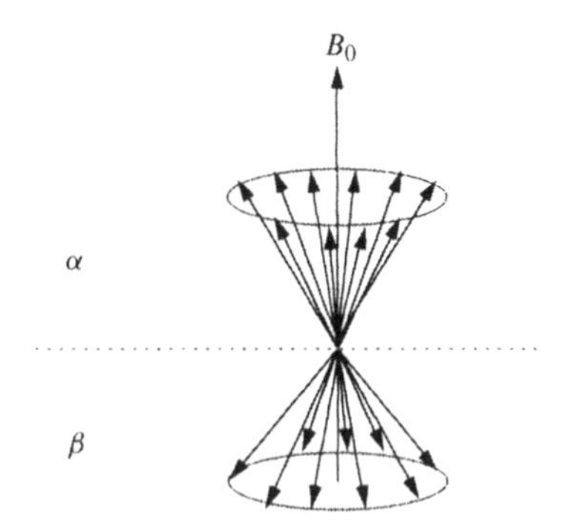

Figure 6.11. Molecular magnetization. The two spin states possible for a nucleus of spin $I = 1/2$ are shown. The lower-energy α state is slightly more heavily populated than the higher-energy β state.

6.2.1 Relaxation in One-Dimensional NMR

One-dimensional NMR depends on the fact that the energy of nuclei which have magnetic spin (i.e. $I \neq 0$) becomes *orientation-dependent* in an applied external magnetic field. In biological NMR the most useful nuclei are ^{1}H, ^{13}C, ^{15}N, ^{31}P and ^{19}F (Table 3.6). Because the nucleus is subatomic, the number of energy levels possible is constrained by quantum mechanics to $2I + 1$. For nuclei of $I = 1/2$, there are two possible energy levels which may be regarded as precessing around the applied external magnetic field, B_0, as shown in Figure 6.11. As previously described (Figure 3.49), there is only a slight preference for the low-energy spin state over the high-energy one in the absence of the external magnetic field. Figure 6.11 is a representation of *molecular magnetization*.

Since NMR spectra are actually determined at the macroscopic level, we could alternatively represent this process as *macroscopic magnetization* (Figure 6.12). This represents magnetization, M as a vector quantity which is subjected to a *torque* as a result of the applied magnetic field:

$$\text{Torque} = M\,B_0 \qquad (6.8)$$

This may be regarded as exposing an *equilibrium magnetization* vector, M_z, parallel to B_0, to a *longitudinal* component so that the magnetization now processes about the axis of the external magnetic field. When the sample is exposed to radiofrequency radiation (which has a small magnetic field associated with it), the resultant magnetic field B_{total} also exposes M_z to a *transverse* component which is

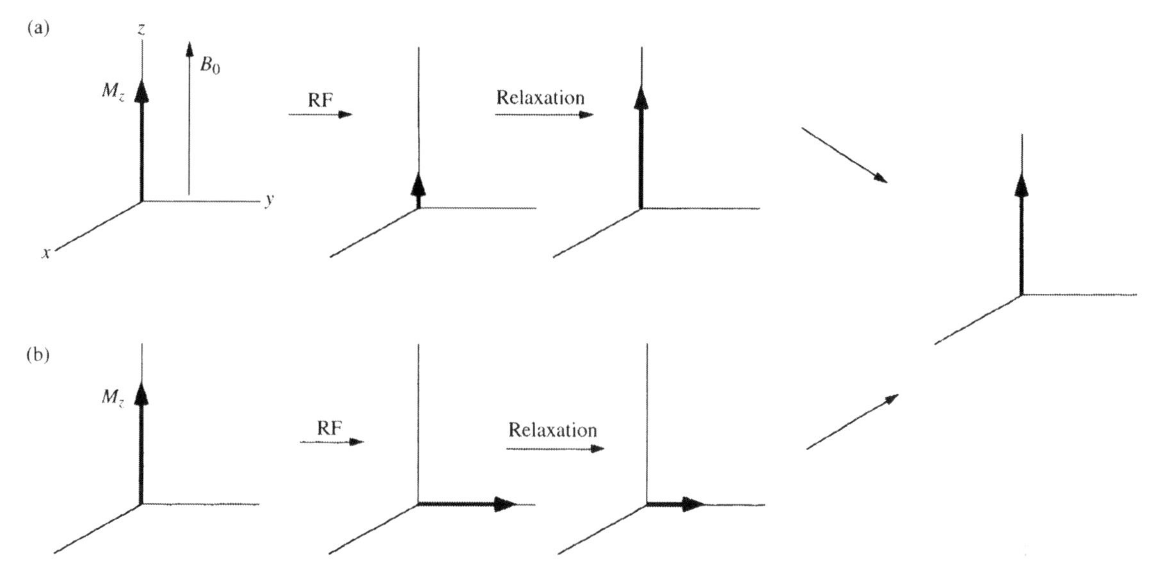

Figure 6.12. Macroscopic magnetization. Magnetization is represented as an equilibrium vector quantity (M_z) parallel to the direction of applied magnetic field (B_0). Exposure to radiowave radiation (RF) under resonance conditions causes perturbation of this vector to a value M. This perturbation has two components; (a) longitudinal and (b) transverse. Relaxation from this perturbed state back to the equilibrium state is shown.

perpendicular to the z-axis and which precesses around it at the Larmor frequency (this term was previously introduced in Chapter 3, Section 3.7.1).

After exposure to a pulse of radiofrequency radiation, the longitudinal magnetization must return to the value of the equilibrium magnetization vector, M_z, while the transverse magnetization must decay to zero. The processes by which this happens are called, respectively, *longitudinal and transverse relaxation* and are quite slow relative to the time-scale of other spectroscopic methods (Figure 6.12). M is a *bulk property* of a magnetized sample which is evenly distributed throughout the solution but which nonetheless has direction. Many individual resonances contribute to M and, after exposure to radiowave radiation, each will precess at its own characteristic Larmor frequency.

This induces a tiny current in the receiver coil of the NMR spectrometer. This current has characteristics described by the *free-induction decay* (*FID*), namely a wavelength and a decay time which may be represented as a plot of intensity versus time. This plot is converted through the *Fourier transform* (Appendix 2) into a plot of intensity versus frequency; the NMR spectrum (Figure 6.13). A molecular species containing several nuclei ($I \neq 0$) will produce an FID which is the sum of individual FIDs, each with a characteristic

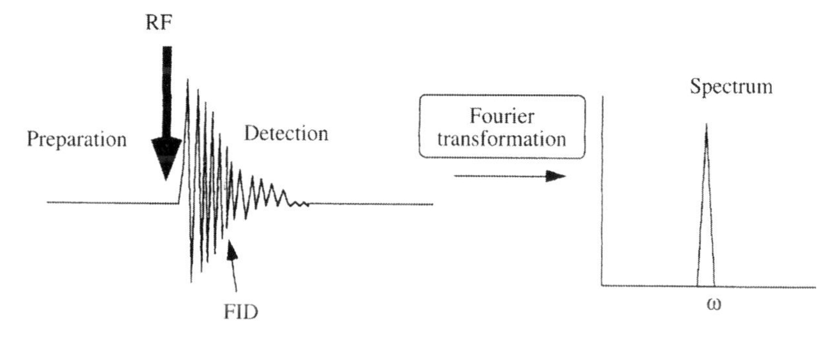

Figure 6.13. Free induction decay (FID). When sample is magnetized, a component of M remains in the xy plane where it generates radio signals. These produce the free induction decay (FID) which decays exponentially with time. This function is directly related to the frequency domain (ω) by the Fourier transform.

wavelength and decay time arising from each nucleus. The one-dimensional NMR spectrum of small molecules is structurally informative since resonances resulting from the experiment are well-resolved and arise from individual protons. However, large biomolecules such as proteins and nucleic acids yield complex spectra with extensive overlap between individual peaks which makes structural interpretation of these samples from one-dimensional NMR spectra alone impossible.

Two-dimensional NMR (2-D NMR) overcomes this difficulty by extending measurements of these effects into a second dimension and, as will be described later, three- and four-dimensional NMR are also possible. These measurements can be used to detect interactions between magnetized protons which are near together in space and are of two general types; *through-space or through-bond*. Through-space interactions are called *nuclear overhauser effects* (NOEs) while through-bond interactions result in *correlation spectroscopy*. The two most important types of 2-D NMR spectroscopy are therefore referred to as *NOESY* (nuclear overhauser effect spectroscopy) and *COSY* (correlation spectroscopy). Both of these effects are due to interactions between magnetic dipoles in a pair of nuclei which are spatially near to each other and both types of experiment are used in the determination of structure in large biomolecules by NMR.

6.2.2 The Nuclear Overhauser Effect (NOE)

Using the molecular magnetization representation (Figure 6.11), we can draw an energy level diagram for two protons, i and j, near together in space (Figure 6.14). Four spin states are possible which may be represented as; $\alpha\alpha$, $\alpha\beta$, $\beta\alpha$ and $\beta\beta$. Transitions between these states may occur which can be followed by relaxation back to the ground state. Magnetization can be exchanged between these protons by a form of relaxation called *cross-relaxation* but, at equilibrium, the rate of this process in both directions is the same. The cross-relaxation rate, σ, between i and j is directly dependent on two parameters:

$$\sigma = \frac{\text{constant}\ \tau_{\text{eff}}}{r^6} \tag{6.9}$$

where r is the distance between the protons and τ_{eff} is the *effective rotational correlation time* of the interaction between the protons. When the magnetization of i is perturbed, however, the magnetization of j will change during a time-period immediately after the perturbation. This change in magnetization of j on perturbation of i is the NOE and the initial rate of build-up of this *is equal to the cross-relaxation rate*, σ, and is hence *proportional to* $1/r^6$.

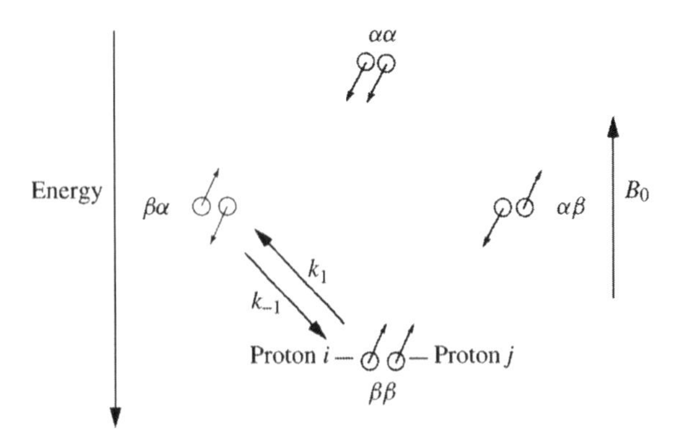

Figure 6.14. The nuclear overhauser effect (NOE). Two protons (i and j) may each exist in a low-energy state (α) or a high-energy state (β) in a magnetic field (B_0) as previously described (Figure 6.11). Transitions between these states at equilibrium (arrows) occur by cross-relaxation with rate constants which are equal in opposite directions (for clarity, only k_1 and k_{-1} between $\beta\beta$ and $\beta\alpha$ are shown). However, when this situation is perturbed by exposure to radiofrequency radiation, the equilibrium is disturbed ($k_1 \neq k_{-1}$). A change in magnetization of i changes the magnetization of j for a time period after the perturbation. This is the nuclear overhauser effect (NOE) and the rate of build-up of this is equal to the cross-relaxation rate.

This dependence of NOE on the distance between protons, r, is the reason why they are structurally informative since measurement of an NOE gives direct indication of the magnitude of r. However, the presence of internal motion in a protein will affect σ in Equation (6.9). Therefore τ_{eff} can vary in a manner dependent on the location of a proton in the protein and this may place a limitation on the direct measurement of r values from NOEs. Moreover, NOEs are determined with a finite experimental accuracy and may be subject to quenching effects due, for example, to the presence of paramagnetic metal ions as contaminants in the sample. For these reasons, it is more difficult to obtain reliable *lower limit* distance estimates from NOEs than *upper limit* distance estimates. In practice, the van der Waals radius is taken as the lower limit for estimates of r while upper limits can be calculated safely from NOEs. Because of the strong dependence of NOEs on r, they only arise between protons located within 5 Å of each other and are classified into strong (1.8–2.7 Å), medium (1.8–3.3 Å) and weak (1.8–5 Å) effects.

Clearly, the variation of r between two protons in the range between the van der Waals radius (approximately 2 Å) and the upper limit of an NOE (5 Å) would allow a relatively large amount of conformational space for these protons to arrange themselves relative to each other. These values represent a *distance constraint*. However, both i and j are also involved in NOEs with other protons (e.g. h, k, l, etc.) and

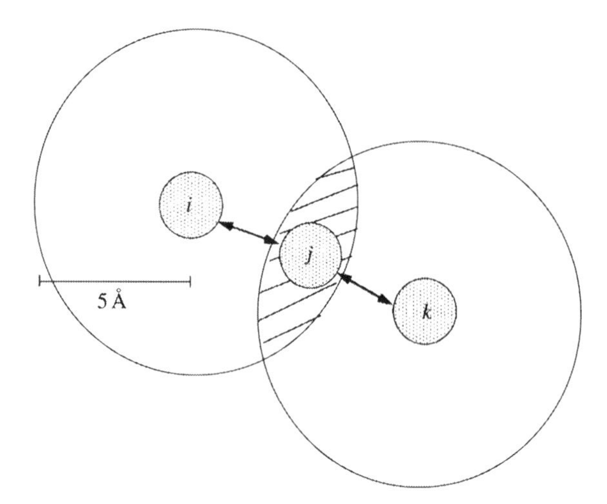

Figure 6.15. Distance constraints determined by NOEs. A NOE (double arrow) between protons i and j would be equally consistent with j occupying any location within the large circle (dashed) around proton i. However, identification of a second NOE between protons j and k suggests that j is likely to occupy the hatched area. Note that there is no NOE between protons i and k since they are more than 5 Å apart. In reality, many NOEs between many protons are used to produce a list of distance constraints which operate in three dimensions rather than the simplified two-dimensional representation shown here.

these *narrow down* the amount of space which would be consistent with NOEs between individual pairs of protons (Figure 6.15). NOESY spectra produce a list of interproton distance constraints for many individual protons in the protein. Since there are only a small number of structural arrangements which would satisfy this list of distance constraints, an accurate indication of three-dimensional structure is possible from NOESY spectra (see below).

6.2.3 Correlation Spectroscopy (COSY)

In Section 3.7.4 we saw that peak-splitting in one-dimensional NMR can result from a process variously called *spin–spin coupling*, *J-coupling* or *scalar coupling*. This is due to spin-pairing between pairs of nuclei (e.g. i and j) and electrons located in the bonds between them and are called *correlation effects*. Unlike the NOE, these operate through *covalent bonds*. The energy of this interaction, E, is given by

$$E = hJ\,I_i I_j \tag{6.10}$$

where h is Planck's constant and J is the *coupling constant* (see Section 3.7.4). Karplus calculated from theory that the three-bond coupling constant ^3J between two protons i and j in (^{i}H–C–C–H^j) has a form dependent on the dihedral angle θ as follows:

$$^3J = A + B\cos\theta + C\cos^2\theta \tag{6.11}$$

where A, B and C are coefficients dependent on substituent electronegativity. This *Karplus equation* can be used to estimate the ϕ angle in polypeptides (see below) using measurements of coupling constants derived from 2-D-NMR.

All 2-D-NMR experiments follow the general scheme outlined in Figure 6.16. The result of this sequence is a series of FIDs which are detected and form the data input subsequently analysed to produce a structure. The effects and significance of the various time-periods into which the experiment is divided on the magnetization vector are described in more detail in Figure 6.17.

After a preparation period, sample is exposed to a radiofrequency pulse of radiation which magnetizes each proton in the sample. This has the effect of forcing M from the z-axis into the x–y plane as previously described in Figure 6.12. This is called the *evolution period* and various times (t_1) are allowed for this varying from 0 to a maximum value ($t_{1\,\max}$) which may be several seconds. Sample is then exposed to a *second* radiofrequency pulse which rotates the plane of the magnetization vectors into the x–z plane. This is called the *mixing period* (τ_{m}). The magnitude of the vector rotated in this step will vary depending on the duration of t_1. During τ_m, z-components of the magnetization vectors are selected which vary depending on the length of the evolution period (i.e. value of t_1). These z-components are now said to be *frequency-labeled* since their precession frequency will vary in a manner which depends directly on t_1. A third radiofrequency pulse now allows us to 'read' these frequencies by inducing transverse relaxation which results in an FID which is detected and analysed.

This experiment results in a data matrix, $s(t_1, t_2)$. 2-Dimensional Fourier transformation of this matrix yields a 2-D frequency spectrum $S(\omega_1, \omega_2)$ where ω corresponds to chemical shifts (i.e. PPM) in one-dimensional NMR. A key point to understand is that the diagonal of this spectrum corresponds to the one-dimensional spectrum for the sample where $\omega_1 = \omega_2$.

COSY spectroscopy allows identification of through-bond coupling effects in an experiment distinct from NOESY as *off-diagonal peaks* (sometimes referred to as *cross-peaks*; $\omega_1 \neq \omega_2$). Like NOEs, spin–spin coupling becomes weaker with distance and correlation effects are only observed when two nuclei are linked by up to a maximum of three bonds. Correlation effects are in principle possible between any pair of nuclei possessing spin but, for the

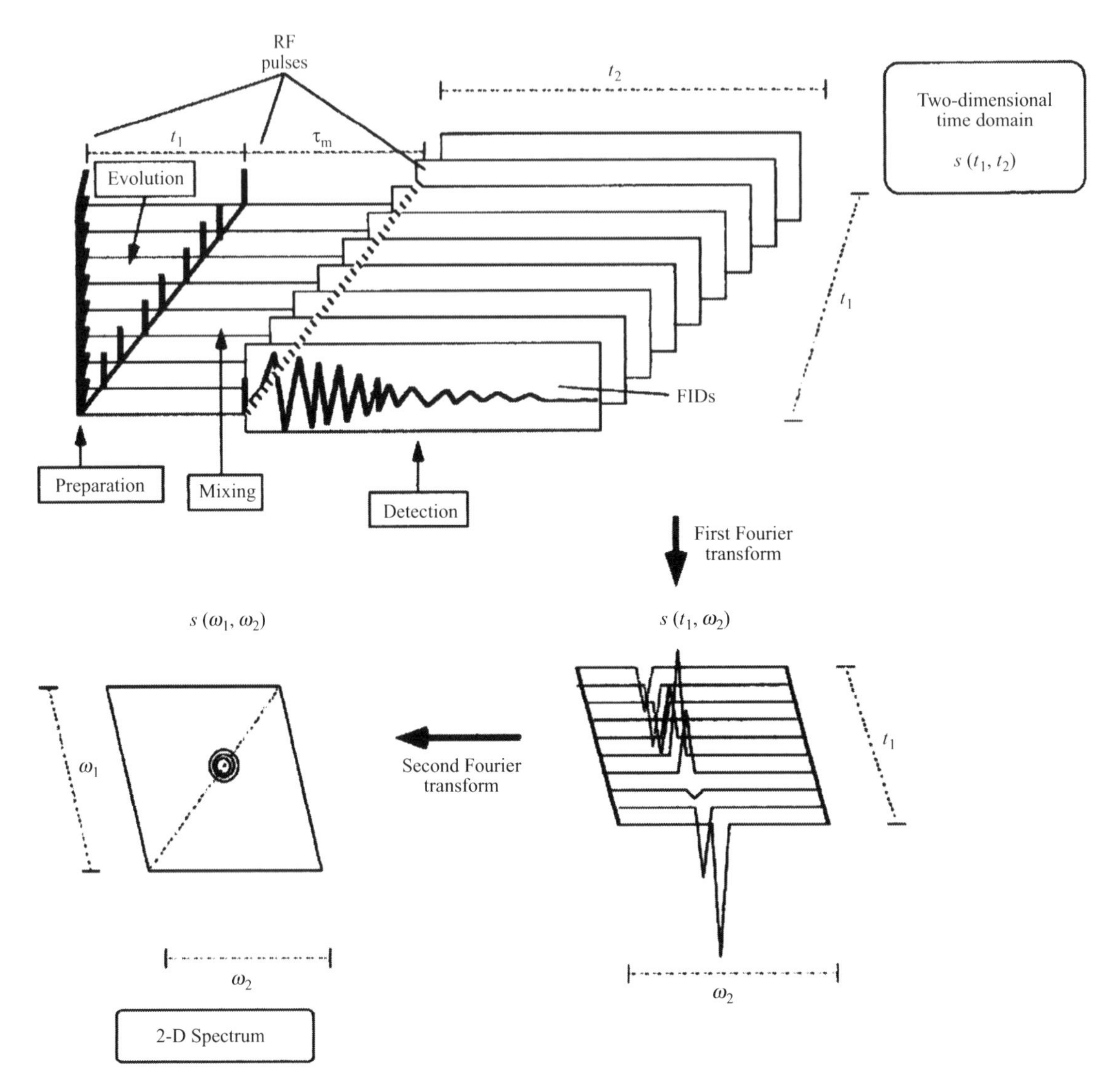

Figure 6.16. Diagrammatic overview of COSY 2-D NMR experiment. After a preparation period, sample is exposed to three pulses of radiofrequency (RF) radiation with various time delays. The first pulse is followed by an evolution period (t_1) which varies from 0 to $t_{1\,max}$. The second pulse is followed by a mixing period (τ_m), while a detection period (t_2) follows the third pulse. FIDs are collected for each value of t_1 and produce a two-dimensional time domain data matrix, $s(t_1, t_2)$. This is Fourier transformed first with respect to t_2. This produces a one-dimensional spectrum in which the peak amplitude varies regularly as a cosine function of t_1. A second Fourier transform is then performed with respect to t_1 to produce a two-dimensional spectrum (ω_1, ω_2). The diagonal of this plot is identical to a one-dimensional NMR spectrum. For simplicity, the spectrum from a single NMR resonance is presented. If multiple resonances were present, correlation effects would be detected as spots off the diagonal ($\omega_1 \neq \omega_2$).

purposes of protein structure determination in small proteins ($M_r < 15$ kDa), the most widely used COSY experiment is that involving *homonuclear* interactions between protons (i.e. $^1H–^1H$). For larger proteins ($M_r = 15–30$ kDa) *heteronuclear* interactions are determined while other methods are required for proteins of $M_r > 30$ kDa.

Amino acid residues, by definition have unique covalent bonding patterns, they represent distinct spin systems which can be identified by COSY spectroscopy (Figure 6.18). Moreover, the dependence of the coupling constant, J, on the dihedral angles between groups linked by a covalent bond allows determination of ϕ using the Karplus equation

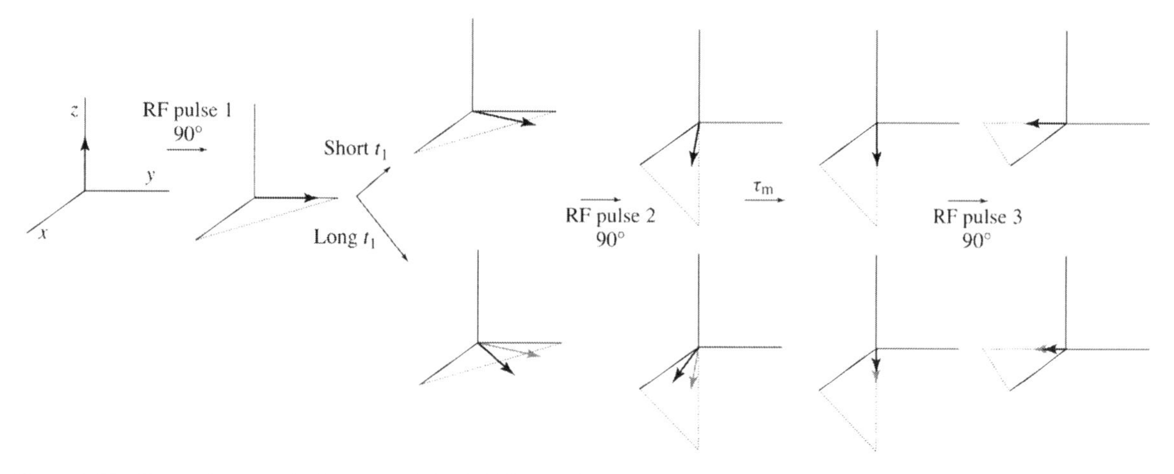

Figure 6.17. Changes in magnetization vector during NOESY. The first radiofrequency (RF) pulse forces the magnetization vector, M, into the x–y plane (dashed line). The vector precesses around the z-axis and is sampled at various times (t_1) during the evolution period. After t_1, sample is exposed to a second RF pulse and M is again forced through 90° into the x–z plane. A time-period called the mixing period (τ_m) is now allowed at the end of which longitudinal components (i.e. those along the z-axis) of M are selected by a third RF pulse and again rotated through 90°. Cross-relaxation happens during τ_m as a result of NOEs. These are evident in the two-dimensional spectrum as $\omega_1 \neq \omega_2$ (Figure 6.20). Note that shorter t_1 results in an outcome from the series of pulses which differs from that from longer t_1 values as illustrated by the faint arrow (in lower panel).

(Equation (6.11)). Dihedral angles are limited to only certain ranges in proteins and are crucial for defining the conformation of the overall polypeptide chain (Figure 6.19). COSY spectroscopy can yield a list of *torsion angle restraints* for the protein. Together with the list of distance constraints generated by NOESY, these form the major data inputs in determination of protein three-dimensional structure by multidimensional NMR.

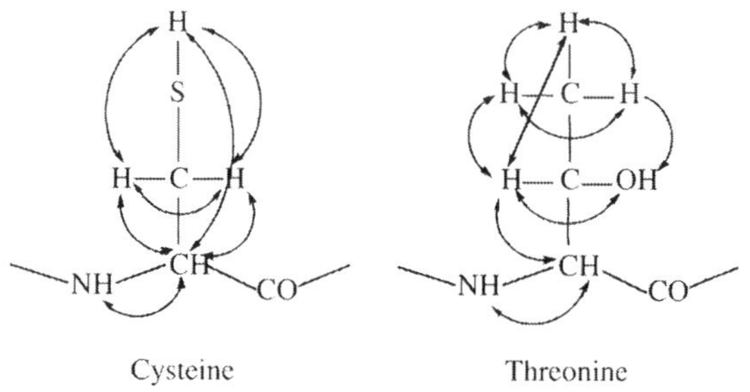

Figure 6.18. Amino acid residues are distinct spin systems. J-coupling between protons is shown for cysteine and threonine residues. These couplings operate through a maximum of three covalent bonds and form the basis of off-diagonal peaks in COSY spectroscopy. In addition to ^1H–^1H homonuclear coupling, heteronuclear coupling is also possible. Each amino acid is a distinct spin system which gives a characteristic 'fingerprint' for each residue type.

6.2.4 Nuclear Overhauser Effect Spectroscopy (NOESY)

NOESY spectroscopy uses generally the same sequence of RF pulses and time-periods as shown for COSY in Figure 6.16. The diagonal of a NOESY spectrum consists of resonances from protons which have **not** undergone cross-relaxation during τ_m (i.e. $\omega_1 = \omega_2$) while off-diagonal peaks ($\omega_1 \neq \omega_2$) represent through-space NOE exchanges of magnetization between pairs of protons arising from cross-relaxation during the mixing period (τ_m).

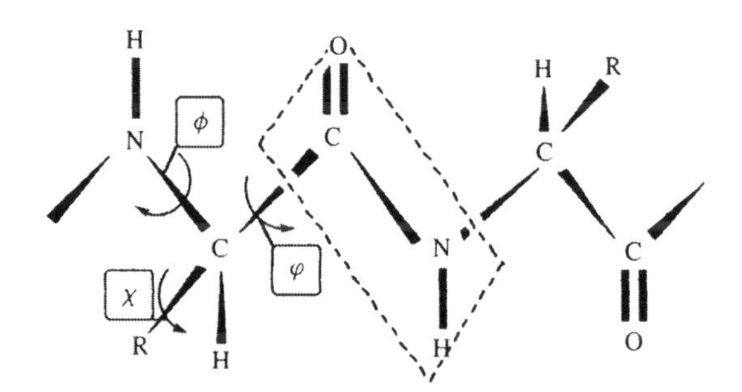

Figure 6.19. Torsion angles in polypeptides. Rotation is possible around three types of bonds in polypeptides since the peptide bond is essentially planar (dashed plane). ϕ is the C–N bond, φ is the C–(C=O) bond while χ is rotation around side-chain bonds. Not all values are sterically allowed for these torsion angles. Their values are related to coupling constants which may be determined from COSY spectra by simple geometric relationships. Note that some residues may have more than one χ bond (e.g. valine).

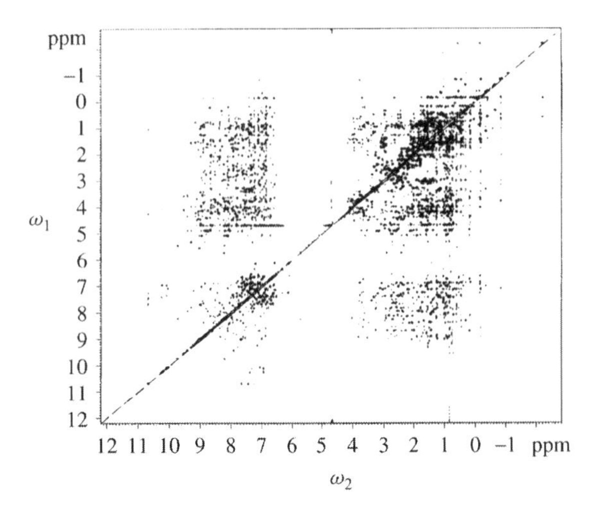

Figure 6.20. NOESY spectrum of lysozyme. A typical spectrum of 2 mM hen lysozyme in 90% H_2O/10% D_2O at 298 K, pH 6.0. Note that the diagonal ($\omega_1 = \omega_2$) represents many resonances while NOEs between protons appear off the diagonal as ($\omega_1 \neq \omega_2$). (Courtesy of Dr L.-Y. Lian, Biological NMR Centre, Leicester, UK).

In a real NOESY experiment thousands of NOEs may be visible in such a spectrum but, because of the two-dimensional presentation of the data, it is possible to resolve individual resonances in a 2-D NMR experiment which would be undetectable in a one-dimensional spectrum due to spectral overlap (Figure 6.20).

6.2.5 Sequential Assignment and Structure Elucidation

The overall strategy for determining structure from 2-D NMR for proteins up to approximately 100 residues was developed by Kurt Wütrich in the early 1980s. More sophisticated NMR experiments suitable for larger proteins based on this overall approach were developed later and these are briefly described below. We can summarize the approach in four main steps (Figure 6.21):

1. Resonances in 2-D NMR spectra are assigned to individual amino acids in the sequence in a process called *sequential resonance assignment*. This involves a combination of COSY and NOESY experiments allowing the identification on the diagonal of such spectrum (i.e. $\omega_1 = \omega_2$) of resonances associated with specific individual amino acids in the sequence. NOE interactions between adjacent residues in the sequence (e.g. –NH ($i + 1$) with αH (i)) are identified in this process and, for example, allow unambiguous identification of a specific residue as being adjacent to two others, one on either side of it, in the primary structure.
2. Bond-angle restraints are identified from COSY measurements and intraresidue and sequential inter-residue NOE data allowing determination of conformation limits on the polypeptide chain.
3. Through-space inter-residue NOEs between individual residues which are far apart in the sequence are identified.
4. Based on data inputs from steps 1–3 a three-dimensional structure consistent with the 2-D NMR data is calculated using a variety of computer algorithms. This structure is then refined to produce a high-resolution structure.

To make this clearer, let us consider a protein of known primary structure (Figure 6.22). There are three glycine residues in this sequence (at positions 10, 16 and 35). Glycine 16 is in a particular sequence in which it is flanked by leucine and tyrosine residues. COSY experiments allow us to identify positions corresponding to the glycine 'fingerprint' on the 2-D NMR diagonal ($\omega_1 = \omega_2$) but at this stage we do not know which position corresponds to residue 10, 16 or 35. However, because of their proximity within the primary structure, we expect the NOESY spectrum to reveal off-diagonal resonances corresponding to an interaction between glycine-16 and leucine-15 and between glycine-16 and tyrosine-17. This allows us to identify the diagonal position of resonances due to glycine-16 in an unambiguous way which distinguishes it from the other two glycines in the sequence. We also now know the diagonal positions of leucine-15 and tyrosine-17. In principle, we can proceed to identify NOEs between tyrosine-17 and the residue at position 18 and between leucine-15 and the residue at position 14 by simply repeating this process in an iterative way along the sequence. At the end of this process we should know the diagonal positions of each residue in the sequence (see Example 6.2).

Bond angles along the polypeptide chain are now determined by a combination of COSY and NOESY. These define value-ranges for ϕ, φ and χ angles shown in Figure 6.20. This is followed by inspecting the NOESY spectrum for through-space interactions between resonances which are *far apart in the sequence but within 5 Å of each other in conformational space*. This generates a list of distance constraints which may be analysed by computer algorithms. The most usual approach is the use of *distance geometry* which defines a family of three-dimensional conformations consistent with the observed NOEs based on the type of analysis previously shown in Figure 6.15. Since there are only a limited number of ways in which a polypeptide of known sequence can fold up to produce a particular pattern of NOEs while at the same time staying within the torsion angle ranges defined by COSY and without violating the well-established geometrical principles governing protein structure, 2-D NMR usually results in a unique family of conformations.

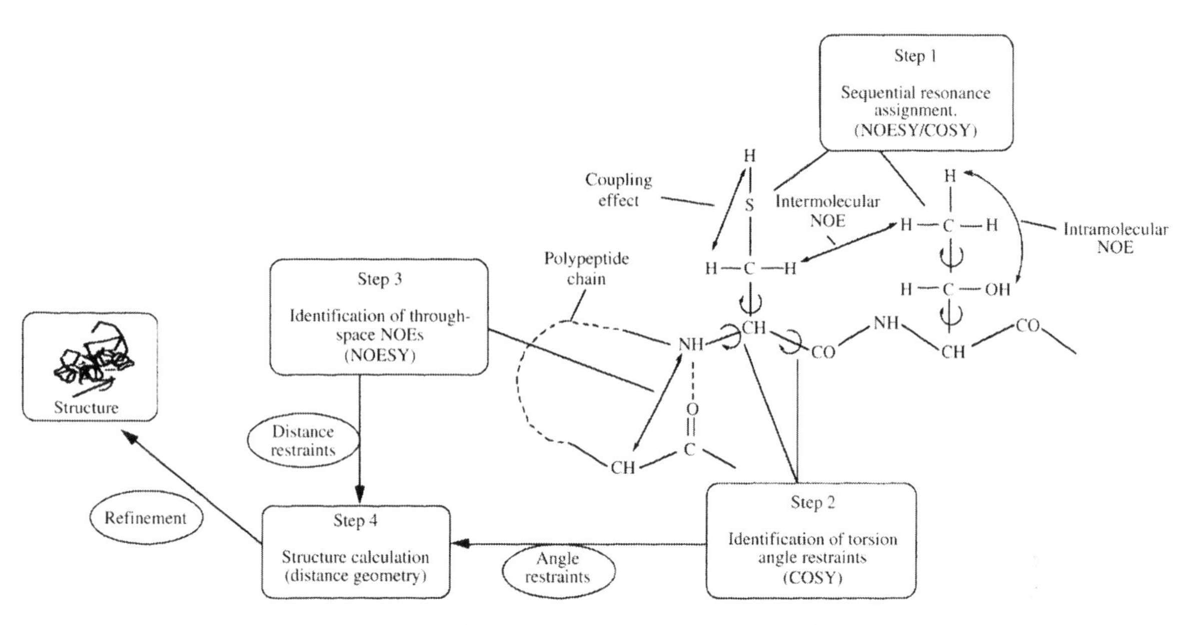

Figure 6.21. Overall approach to structure determination by 2-D NMR. *Step 1.* Based on a combination of NOESY and COSY experiments, resonances are assigned to individual residues in the sequence (e.g. cysteine–threonine). Each will have a distinctive 'fingerprint' in COSY (correlation effects) as well as characteristic intramolecular NOEs as shown. Moreover, intermolecular NOEs between adjacent residues are also identifiable by NOESY. *Step 2* COSY spectroscopy identifies value ranges for torsion angles since these are directly related to the coupling constant, J. This produces a list of torsion angle restraints for each residue in the sequence. *Step 3.* Through-space intermolecular NOEs identify interactions between residues which may be far apart in the primary structure (dashed line). Two sequentially distant parts of the polypeptide brought together by a hydrogen bond are shown. This produces a list of distance restraints. *Step 4.* Using distance geometry or other methods, the list of distance and angle restraints are used to calculate a structure for the polypeptide. This is refined to produce a high-resolution structure.

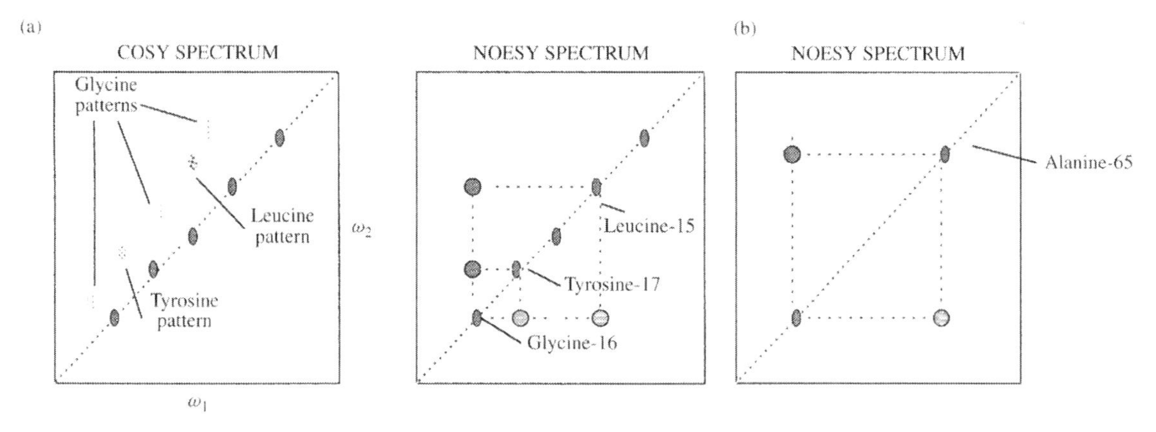

Figure 6.22. Diagrammatic representation of sequential resonance assignment in 2-D NMR. A protein containing the sequence Leu–Gly–Tyr is under investigation. This protein contains three glycines (at positions 10, 16 and 35). (a) Resonances on the diagonal correspond to individual amino acid residues. Each type of amino acid gives a distinct COSY spectrum. NOEs between adjacent residues may be identified in the NOESY spectrum and this identifies the diagonal position of Glycine-16. This process is repeated for every residue in the sequence. (b) Intermolecular NOEs between sequentially distant parts of the protein are then identified from NOESY spectra since diagonal positions of each residue are now known.

Example 6.2 Structure of a Protein Determined by Multi-dimensional NMR

Δ^5-3-Ketosteroid isomerase (KSI) is a dimeric enzyme which catalyzes highly efficient isomerization of various β,γ-unsaturated oxo-steroids. Despite the fact that crystals of this protein were obtained in 1976 and a low-resolution structure determined to 6 Å resolution was published in 1984, determination of a high-resolution structure had proved elusive. For this reason heteronuclear, multi-dimensional NMR was carried out to determine the structure of this protein.

^{15}N- and ^{15}N^{13}C-labelled KSI samples were obtained from bacteria grown in media rich in ^{15}N and ^{13}C. Ten percent dioxane was included in solvent to inhibit aggregation at concentrations greater than 80 μM. Four-dimensional ^{15}N, ^{13}C- and ^{13}C, ^{13}C NOESY spectra were generated and used to assign backbone and side-chain signals. A list of 1865 distance constraints ((1) below) was used as an input to generate 20 individual distance geometry structures which are shown in (2).

Disorder at the dimer interface is responsible for the greater convergence of the family of monomeric structures compared to that for the dimer. The model agreed well with previously published structure–function work on this enzyme and allowed identification of important active-site residues.

(1)

Distance restraints generated by multi-dimensional, heteronuclear NMR	
Intraresidue	49
Sequential	306
Medium range ($\lvert i - j \rvert = 2$–5 residues)	351
Long range ($\lvert i - j \rvert > 5$ residues)	545
Intermolecular	60
Hydrogen bond restraints	554
Total NMR-derived restraints	1865
Mean restraints/residue	15.2

(2)

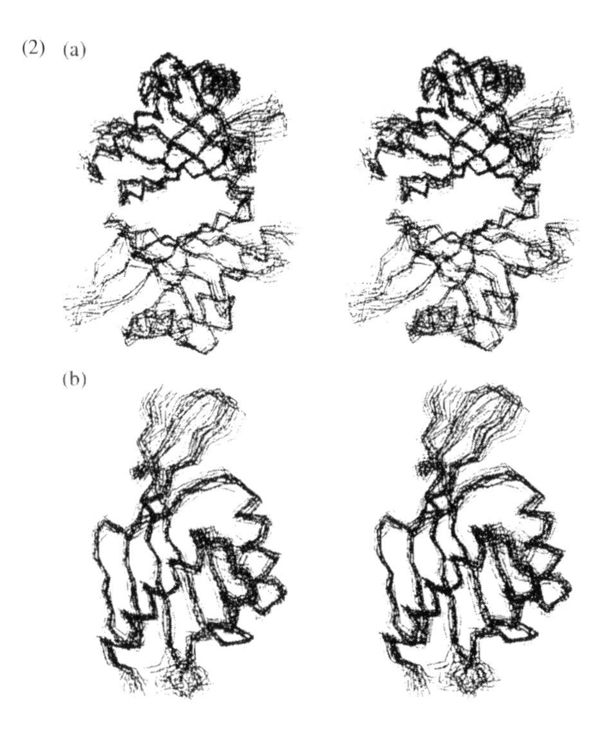

Superposition of the family of 20 structures generated for the dimer (A) and the monomer (B) of KSI. Reprinted with permission from Wu *et al.*, Solution Structure of 3-oxo-Δ^5-Steroid Isomerase, *Science* **276**, 415–418 (1997) American Association for the Advancement of Science.

At this stage, the overall fold of the polypeptide chain and secondary structure features such as alpha-helices and beta-sheet are clearly defined. The final step in the procedure is refinement of the structure to a resolution of a few Å. This will often involve the use of statistical information obtained from high-resolution databases of structures derived from crystallographic measurements as well as more sophisticated analysis of NMR data. More information on protein structure refinement is given below in Section 6.5.8.

6.2.6 Multi-Dimensional NMR

The approach to structure determination which has just been described has been successfully used to determine three-dimensional structures to a resolution comparable to 2.5 Å crystal structures for proteins up to M_r of 15 kDa. However, because of problems in assigning resonances unambiguously, spectral overlap and the large number of resonances obtained in 2-D NMR, the technique is not generally appropriate for larger proteins (>15 kDa). Moreover, some proteins smaller than 15 kDa may produce spectra of such complexity that their structures cannot be determined using 2-D homonuclear proton NMR alone. For such difficult samples we may require NMR spectroscopy in a third or fourth dimension, that is *3-D and 4-D NMR* (Example 6.2).

These experiments may be simply regarded as linear combinations of 2-D NMR sequences of the type previously described (Figure 6.16) with the proviso that there is a single preparation period at the beginning and a single detection period at the end of the sequence. Thus, 3-D NMR consists of two successive 2-D NMR sequences while 4-D NMR consists of three (Figure 6.23). The evolution (t_2) and mixing (τ_m) periods of each 2-D NMR sequence are varied independently of (t_2, τ_m) values of other sequences.

3-D and 4-D NMR experiments make use of *heteronuclear* correlation effects along the polypeptide backbone (Example 6.3). In these experiments, the sample protein is *isotopically labeled uniformly* with ^{15}N and ^{13}C. These isotopes are introduced into the protein by overexpression in bacteria grown on $^{15}NH_4Cl$ and ^{13}C-glucose as their exclusive sources of nitrogen and carbon. For larger proteins, *specific isotopic labeling* may be used in which isotopes are *selectively* introduced at known sites. These isotopes allow 2-D proton homonuclear NMR to be followed by heteronuclear NMR in such a way that 3-D NMR generates a (ω_1, ω_2, ω_3) data-set while 4-D NMR generates (ω_1, ω_2, ω_3, ω_4) (Figure 6.24). These experiments facilitate resonance assignments to much higher resolution than 2-D NMR alone thus allowing structure determination of larger proteins or of smaller proteins for which 2-D NMR is insufficient. This approach has been successfully used to determine protein structure up to 30 kDa.

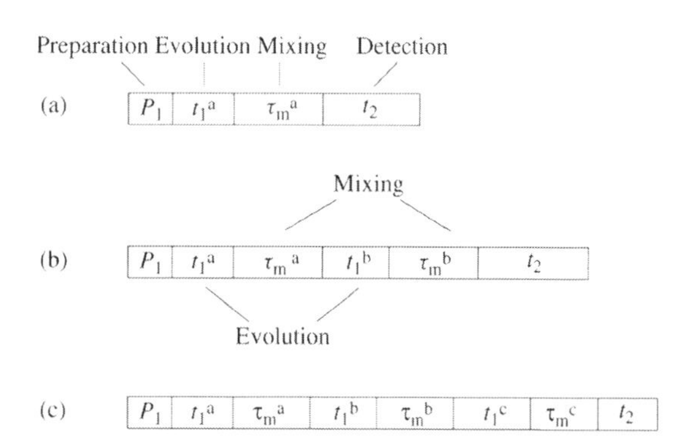

Figure 6.23. Multi-dimensional NMR. Pulse-sequences responsible for (a) 2-D NMR, (b) 3-D NMR and (c) 4-D NMR are built up by successive 2-D NMR pulses. Successive evolution (t_1) and mixing (τ_m) time-period ranges are denoted by superscripts a, b and c. Note that time-period ranges in successive 2-D NMR pulses are varied independently of each other..

6.2.7 Other Applications of Multi-Dimensional NMR

Our discussion of multi-dimensional NMR has thus far focused almost exclusively on the determination of three-dimensional structures of proteins (Example 6.2). However, this group of techniques is also useful *as a general probe for protein structure* which can also shed light on aspects of protein dynamics such as enzyme catalysis, protein–protein complex formation, ligand binding and protein folding (Example 6.3). In principle, any biomolecule containing atoms possessing spin can give an NMR spectrum and resonances due to individual atoms can be followed by multi-dimensional methods. Although, high-resolution structural information may not always be obtainable from such spectra, they can nonetheless yield important clues to shape or function. The technique is therefore generally applicable to other classes of biomolecules such as DNA and carbohydrates and a similar approach can be taken to analysis of structure and function in these molecules as for proteins (Table 6.2).

In the case of DNA, multidimensional NMR studies have tended to concentrate on short oligonucleotides (10–20 bp) rather than large structures. This can yield structural information on functionally important parts of DNA such as promoters and operators. Since there are only four bases commonly found in DNA (compared to twenty amino acid residues in proteins), proton NMR spin systems are more easily assigned to bases than to amino acids. The major interproton distances measured from NOESY in DNA samples are between sequentially adjacent bases rather than the more long-distance interactions which we have come across in protein structure. NOEs are also measurable between protons on separate, individual strands of double-stranded DNA which happen to be brought close together by hydrogen

Example 6.3 Using Two-dimensional NMR to Follow *in vitro* Evolution of a Protein Fold

Comparison of related protein sequences suggest that mutation of one amino acid residue for another is a major factor in the evolution of novel proteins. When such changes are deliberately made to proteins by site-directed mutagenesis, they often result in altered biological activity and stability. However, mutation rarely results in change in the overall fold of the protein. The most likely reason for this is delocalization of information specifying a particular fold over the whole protein rather than merely being specified by the sequence alone (Section 6.1.1). A mutagenesis study of the Arc repressor provides an interesting exception to this rule.

Residue-by-residue differences in chemical shift show most changes concentrated between residues 9 and 14

Portion of NOESY spectrum showing cross-peaks between Leu-21 and other residues (numbered) in mutant Arc. NOEs shown as solid lines are *not* consistent with wild-type structure while those shown with dashed lines are.

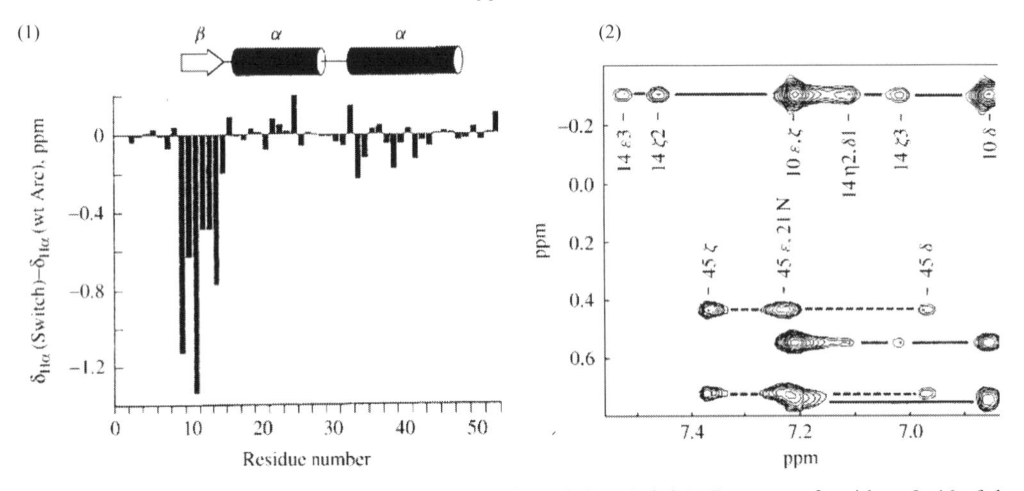

Representation of the structure of Arc from two perspectives (left and right). Structure of residues 8–46 of the wild type is shown in panels A and B while that of residues 15 to 46 of the mutant is shown in C and D;

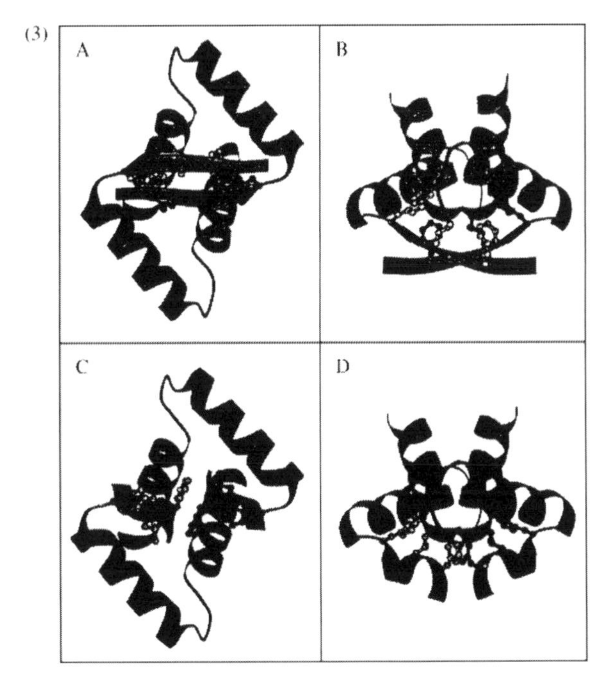

(Continues)

(Continued)

Arc is a homodimer each subunit of which contains a crucial sequence (–Gln^9–Phe–Asn–Leu–Arg–Trp^{14}–). This sequence contributes to an antiparallel β-sheet in the dimer. A feature of this sequence is that even-numbered side-chains are hydrophobic and buried in the core of the protein where they contribute to folding and stability while odd-numbered side-chains are polar and located on the surface of the protein where they can bind to operator DNA.

Mutations were introduced into this protein which resulted in an interchange of residue 11 for residue 12 (i.e. resulting in the sequence; –Gln^9–Phe–Leu–Asn–Arg–Trp^{14}–). Spectroscopic investigations such as fluorescence and circular dichroism (Chapter 3) showed that this mutation had disrupted the structure of the protein while only slightly affecting its thermal stability. Heteronuclear two-dimensional NMR was used to study NOEs between ^{1}H and ^{15}N and are shown in (1) and (2).

Analysis of the chemical shifts for each residue revealed that the bulk of the differences between the NMR spectra of the mutant and wild-type proteins is concentrated between residues 9 and 14. This suggests a changed structure in this region while the rest of the protein is largely unchanged.

A family of three-dimensional structures of the mutant Arc determined from 99 NOEs was determined and the averaged structure is shown in (3). This reveals a localized conformational change in the region of residues 9–14 from β-sheet (wild type) to helix (mutant). This was rationalized in terms of the amphiphilicity of the mutant sequence PHHPPH (where *P* is polar and *H* is hydrophobic) which is consistent with a helical periodicity (3–4 residues per turn) rather than the PHPHPH periodicity of the wild type (consistent with a β-strand in which *H* is on one side and *P* is on the other).

Images reprinted with permission from Cordes *et al.*, (1999) Evolution of a protein fold *in vitro*. *Science* **284**, 325–327, American Association for the Advancement of Science.

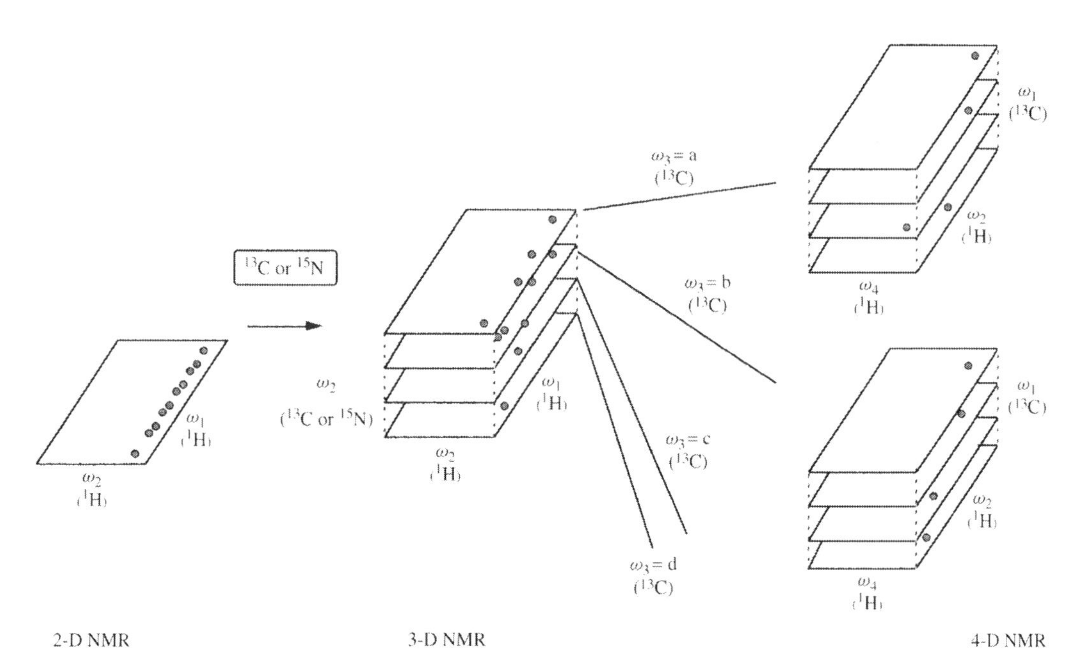

Figure 6.24. Multidimensional NMR. In a homonuclear 2-D NMR experiment a number of resonances are observed which have the same value of ω_2. 3-D heteronuclear NMR allows further resolution of these resonances depending on the identity of the heavier atom (nitrogen or carbon) to which one of the protons is attached. This data set consists of (ω_1, ω_2, ω_3) resonances due, respectively to (^{1}H, ^{15}N or ^{13}C, ^{1}H) 4-D heteronuclear NMR allows unique resolution of all resonances producing a data-set of values (ω_1, ω_2, ω_3, ω_4) corresponding respectively to (^{13}C, ^{1}H, ^{13}C, ^{1}H). For clarity, only two sets of values (a and b) are shown for ω_3. Resonances due to each proton of the homonuclear proton pair are uniquely defined in terms of the chemical shifts of the heavier atoms to which they are attached.

Table 6.2. Selected applications of NMR to the study of biomolecules.

Type of NMR	Sample
1-D proton NMR	Small molecules (e.g. amino acids)
Homonuclear proton	Proteins (<15 kDa)
2-D NMR	
Heteronuclear	Proteins up to 30 kDa
3-D and 4-D NMR	
Heteronuclear	DNA up to 20 bp
3-D and 4-D NMR	
Solid-state multidimensional NMR	Membrane proteins (up to 30 kDa)
Solid-state ^{31}p	
2-D magic angle NMR	DNA fibres (to distinguish conformation)

bonding. Important contributions of NMR to structural studies of DNA include the identification of novel secondary structures such as the triple-helix, novel higher-order structure such as hairpins and monitoring of drug–DNA and protein–DNA interactions.

NMR has also been used in structural investigations of RNA which can adopt a much wider range of three-dimensional structures than DNA and for which very few crystal structures have been determined. This is important because knowledge of structural alterations in solution is essential to understanding interactions between RNA and proteins or DNA. The advent of solid phase oligonucleotide synthesis has allowed the production of large quantities of oligonucleotides capable of adopting a wide range of secondary structures suitable for NMR investigation.

Biological membrane components are not amenable to investigation by crystallography but are of major biochemical interest. In particular, heteronuclear NMR has been used to investigate the structure of phospholipid bilayers and membrane-bound proteins as well as interactions between them. Recent improvements in *solid-state NMR* are of particular importance to these investigations. This technique is applicable to phospholipid bilayers, DNA fibres, crystals, amyloid fibrils and membrane-bound samples such as receptors. It depends on the fact that these samples are *anisotropic*, that is they have a definite orientation and are not tumbling freely in solution in a random manner (Figure 6.25).

Solution-state NMR depends on a resonance interaction in which angle-dependent mathematical terms are normally averaged to zero due to random tumbling. Samples investigated by solution-state NMR are said to be *isotropic* which means that the angle between the magnetic field, B_0 and the sample rotation axis does not affect the spectrum obtained. When anisotropic samples are analysed by solution-state NMR, however, they produce very noisy spectra with broad peaks which are difficult to interpret. If the sample is instead analysed in a solid-state, an angle can be selected which removes the noise from the spectrum yielding a high-resolution result. This angle is sometimes called the *magic angle* and it has a value of 54.7°. This means that solid-state heteronuclear multidimensional NMR of anisotropic samples can be used to determine high-resolution structures of samples which are impossible to analyse structurally either by crystallography or by solution-state NMR. This approach has determined structures of membrane-bound proteins up to M_r 30 kDa and allowed structural and functional analysis of heterogenous noncrystalline samples including membrane-bound peptides and proteins.

6.2.8 Limitations and Advantages of Multi-Dimensional NMR

Table 6.2 gives a list of selected applications of NMR to the study of biomolecules. From this it will be clear that this group of techniques facilitates analysis of a wide range of size (up to 30 kDa), type (building-block molecules to proteins and macromolecular complexes) and state (solution-state/solid-state; homogeneous/heterogeneous; soluble/membrane-bound). It is likely that new methods of data analysis and resonance assignment will increase the protein mass-range amenable to study by NMR into the region 30–50 kDa in the foreseeable future. This will bring the subunits of many oligomeric proteins of interest within reach of NMR.

In addition to the versatility of the technique, it has the distinct advantage that it is gentle and nondestructive facilitating analysis under conditions which are close to physiological. It is also possible to include ligands and other biopolymers such as DNA or protein with the sample to follow effects on the sample molecule and to incorporate radioisotopes into the molecule to study individual atoms. Analysis can also be carried out along a time-course to follow dynamic processes such as protein folding. The main advantages relative to crystallography are that molecules for which it is impossible to grow crystals (e.g. membrane proteins) can be analysed by NMR and that the measurement takes place under conditions which are significantly closer to physiological. Moreover, flexible regions of proteins are not apparent in electron density maps determined by crystallography but are detectable by NMR. Direct comparisons between NMR and crystallography suggest that the resolution of NMR is close to that of crystallography in the region of 2.5 Å.

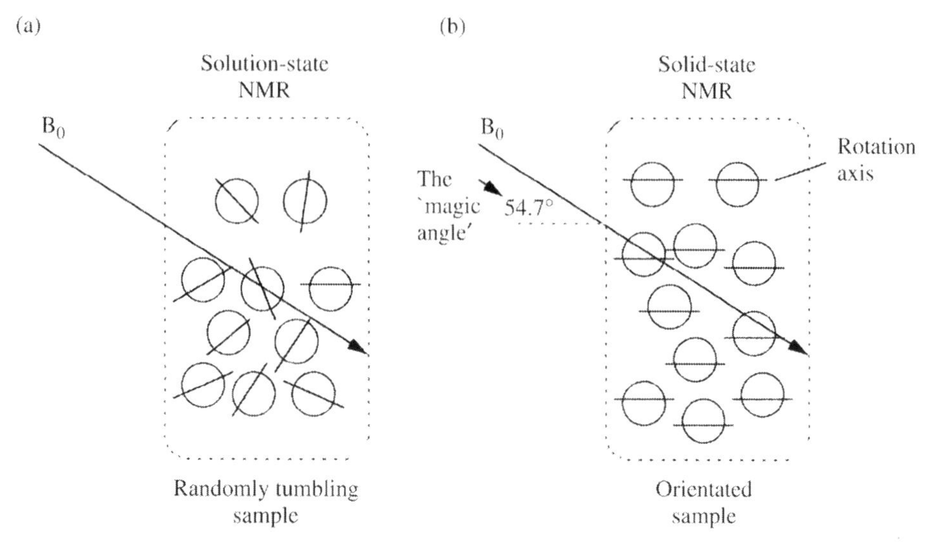

Figure 6.25. Solid-state NMR. Freely tumbling samples (a) are isotropic while orientated samples (b) are anisotropic. Solid-state NMR involves analysis of solid-state, orientated samples at the magic angle of 54.7°. At this angle, high-resolution spectra are obtainable with solid-state samples.

NMR measurements have a number of important limitations, however, which place limits on their general applicability. Many of these relate to the size and type of sample which can be analysed as discussed above. Protein NMR requires a high sample concentration (0.5–2 mM) at temperatures above 15 °C to allow detection of resonances with a high signal: noise ratio. Some proteins may aggregate under these conditions which precludes their analysis by this group of techniques. Since it is usually proton resonances which are measured in the analysis, it is essential that the sample is fully protonated for the duration of the experiment. At high pH values, the rate of *proton exchange* with solvent is high relative to the time-scale of the experiment. In practice, this means that multidimensional NMR measurements are performed in the pH range 4–8. This precludes the study of pH effects in the alkaline range and also places a limitation on the pH-stability of the sample since some proteins may not be fully stable or fully bioactive below pH 6.0. Moreover, the refined structure resulting from the experiment outlined in Figure 6.21 represents a *family* of equally possible structures and it can be difficult to distinguish between these statistically. In comparison to X-ray crystallography, it is particularly difficult accurately to quantify error through the structure and this error may vary from one part of the sequence to another and from one structure to another.

Information obtained from NMR studies is particularly important because it is independent of, though complementary to, data from crystallographic experiments, thus facilitating detection of structure from molecules which may not readily crystallize. We will compare the two techniques at the end of this chapter but we must now turn to a description of the main method for structure determination in the solid state; X-ray diffraction.

6.3 CRYSTALLIZATION OF BIOMACROMOLECULES

The second widespread technique for determination of three-dimensional structure of biomacromolecules is based on the study of X-ray diffraction patterns by crystals of protein/DNA called *X-ray crystallography*. The basic outline of this experiment is shown in Figure 6.26. The technique is equally applicable to structure determination of small molecules such as those encountered in organic/inorganic chemistry and is based on the fact that atoms diffract X-rays in a pattern which is dependent on their location in three-dimensional space. The reason X-rays are used is that their wavelength range is of the same order of magnitude as chemical bonds thus allowing us to 'see' at a resolution equivalent to interatomic distances (i.e. 0.8–2.5 Å). To detect diffracted X-rays with high enough sensitivity, it is essential that many atoms contribute to the diffraction pattern obtained. In practice, this means that the molecule under investigation must be present in the ordered three-dimensional array of a *crystal* so that many equivalent atoms in different molecules contribute to the diffraction pattern.

In the case of biomacromolecules, it is often difficult to obtain crystals of adequate quality and this is a major

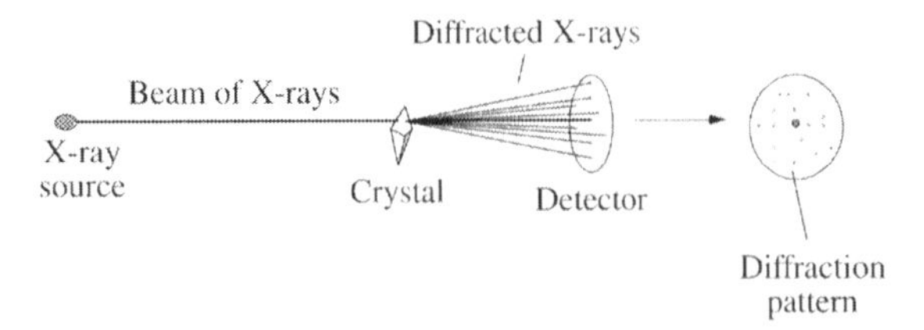

Figure 6.26. X-ray diffraction experiment. A beam of X-rays is passed through a crystal. Most of the beam passes straight through but some of the beam is diffracted by the systematic arrangement of atoms in the crystal. When a monochromatic X-ray beam is used, it is necessary to rotate or oscillate the crystal slightly during the experiment (Section 6.3.2). Diffracted beams are detected as a two-dimensional array of spots of particular position and intensity forming the diffraction pattern. The crystal is then rotated through a small angle (e.g. 1°) and the measurement is repeated. At each angle some spots disappear, others appear and individual spot intensities change.

limitation of this approach to structure determination. In particular, it means that biomolecules which *do not form crystals* (e.g. intrinsic membrane-bound proteins) are not immediately amenable to study by this technique. Notwithstanding this limitation, X-ray crystallography has resulted in determination of three-dimensional structures for thousands of proteins and is still far more widely used than multi-dimensional NMR for high-resolution structure determination. It has made a number of major contributions to biochemistry such as the demonstration that enzymes are protein in nature, the Watson–Crick model of DNA (although diffraction in this case was by DNA *fibres*) and understanding the molecular basis of phenomena such as allosterism and protein/DNA interactions. Moreover, structures derived from X-ray crystallography provide the major experimental link between structure and function underlying most modern biochemistry and molecular biology.

In this section we will describe the symmetry properties of crystals and how crystals of biomacromolecules suitable for X-ray diffraction work may be obtained. In following sections we will look at the diffraction experiment and how structure may be calculated from diffraction patterns.

6.3.1 What are Crystals?

A crystal (from the Greek *krustallos* meaning *clear ice*) can be defined as a regularly repeating three-dimensional array of atoms or molecules. Some crystals are formed naturally such as table salt (NaCl), graphite and diamonds (crystalline forms of carbon) and quartz (SiO_2). These crystals are held together by strong ionic interactions (e.g. Na^+–Cl^-) which make them quite mechanically stable. Crystals of protein and DNA are far more fragile than these materials because they are held together by weak interactions such as hydrogen bonds and van der Waals forces. Moreover, protein/DNA crystals are far more difficult to grow than such naturally occurring organic and inorganic crystals. Notwithstanding this, much of the terminology used in biomolecular crystallography derives from early work on organic/inorganic crystals and, in principle, the same overall experimental approach is followed in protein/DNA X-ray diffraction experiments as those with small molecule crystals.

Diffraction is the process whereby electromagnetic radiation is scattered by matter. The phenomenon is not unique to X-rays. It also occurs with visible light and it is the *focusing* of diffracted visible light by the lens in optical microscopes which allows its recombination to form an image of the object. If it were possible to do the same thing with diffracted X-rays, then we could simply build a microscope to view the atomic structure of crystals directly. However, X-rays cannot be focused and, for this reason, atomic arrangements within the crystal must instead be determined by *calculation* based on the X-ray diffraction pattern obtained (Section 6.4). This calculation is possible or helped through the use of geometry to describe the arrangement of atoms or molecules within the crystal.

6.3.2 Symmetry in Crystals

The basic building block of a crystal is the *unit* cell which is a microscopic repeating *unit* within the crystal (Figure 6.27). This may be thought of as a 'box' of six *faces* or sides, defined by three lengths a, b, c (one for each *edge* of the box) and three angles α, β, γ (between the axes b–c, a–c and a–b, respectively) which are collectively referred to as

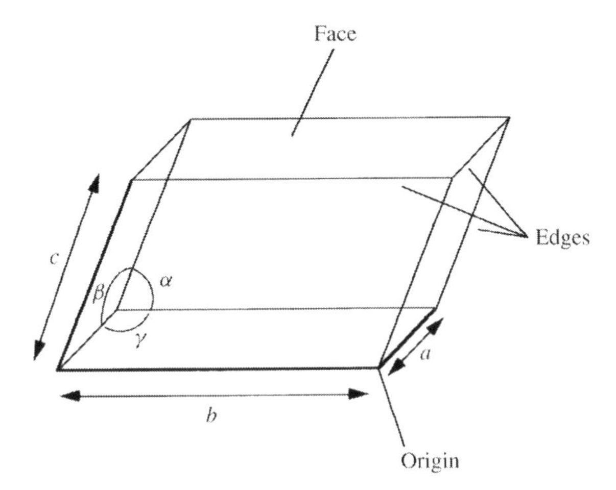

Figure 6.27. The unit cell. A unit cell is determined by lattice constants consisting of three edges (a, b and c) and three angles between them; $\alpha(c - b)$, $\beta(c - a)$ and $\gamma(a - b)$. This also forms six faces. A crystal consists of unit cells which are repeated in three dimensions. Many different unit cells could be chosen for a crystal but it is usual to choose the unit cell with shortest edge lengths, angles nearest to 90° and with highest symmetry (Table 6.4).

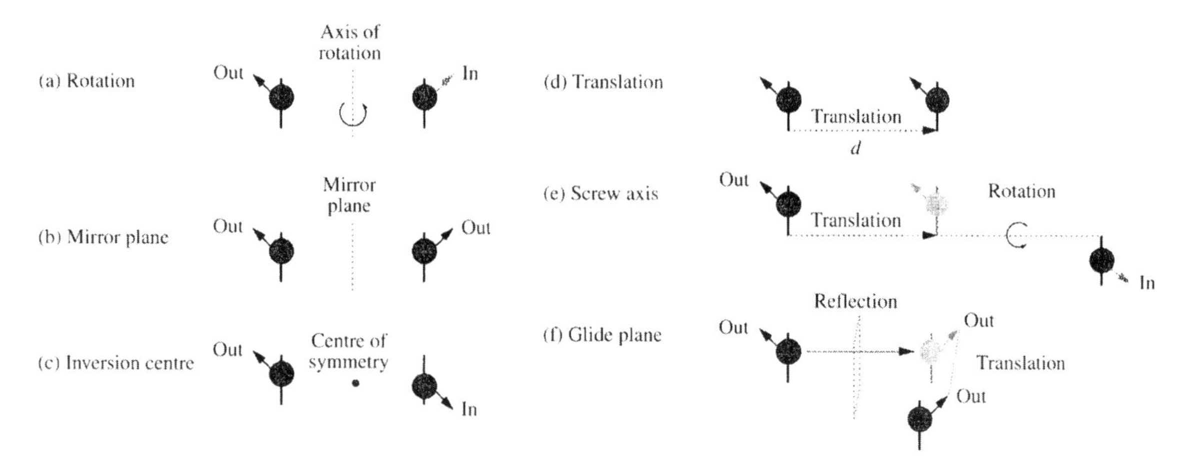

Figure 6.28. Symmetry operations and symmetry elements in crystallography. A symmetry operation relates the location of one object (left) to that of another (right) in the crystal array. Six distinct symmetry operations are shown (a–f). The points, planes axes, and so on about which these operations occur are the symmetry elements. (a) Rotation. A counterclockwise rotation of $360°/n$ where n is 1, 2, 3, 4 or 6. A twofold axis where $n = 2$ resulting in a 180° rotation is illustrated. 'Out' and 'in' denote that arrows are protruding out of and into the page, respectively. (b) Mirror plane. A reflection through a plane is shown by which one object is converted to its mirror image. (c) Inversion centre; objects are related through a point called the centre of symmetry. (d) Translation; objects are related by translation through a distance, d, in a single direction. (e) Screw axis; objects are related by a translation p/n (p is an integer $< n$) along the axis of rotation followed by a rotation as in (a) (f) Glide plane; objects are related by reflection as in (b) followed by a translation in a single direction as in (d) through a distance which is half the unit cell dimension parallel to the mirror plane. Because of the chirality of biomolecules, (b), (c) and (f) are not applicable and do not occur in crystals of protein and DNA.

lattice constants. The unit cell may contain part of a molecule, a single molecule or several molecules of the species under investigation. To arrive at a final structure, however, it is essential that *the volume of the unit cell is equal to or greater than the volume of one molecule*. The unique part of the unit cell is called the *asymmetric unit*. This is the smallest part of the unit cell which lacks internal symmetry. Symmetry simply means that the location of an atom or molecule when subjected to a *symmetry operation* around a point, plane or axis may be converted into that of another atom/molecule in the same unit cell. The point, plane or axis is referred to as a *symmetry element*. The main symmetry operations and elements encountered in crystallography are summarized in Figure 6.28. In order to describe a crystal, several symmetry elements may be combined and a particular combination of symmetry elements is called a *space* group.

The unit cell will frequently contain more than one protein/DNA molecule. These may be related to each other by *noncrystallographic symmetry*. This is not part of the symmetry defined by the space group which is referred to specifically as *crystallographic symmetry*.

It is important to understand that unit cells, symmetry operations, space groups, and so on are imaginary mathematical devices allowing us to describe a crystal uniquely and to simplify calculations. The crystal is in reality made up of hundreds of thousands of individual atoms and our goal in X-ray diffraction is to identify and quantify the three-dimensional arrangement of these atoms. A single crystal may potentially be described by several unit cells/space groups. This choice is to some extent arbitrary. In the example shown in Figure 6.29, it is clear that the simplest unit cell contains one object while a more complex possible unit cell for the same array contains two objects and may have *higher symmetry*. This is because the latter unit cell is a more 'unique' representation of the pattern in the crystal. In general, a unit cell is chosen which has the shortest possible lengths for a–c, the closest angles possible to 90° for the angles α–γ and the highest possible symmetry.

The crystal visible at the macroscopic level is composed of atoms/molecules repeated regularly in three dimensions according to a mathematical pattern called the *crystal lattice*. A lattice is defined as *an array of points repeated periodically through three-dimensional space such that the arrangement of points around any one point is identical to that around any other point in the same array and in the same orientation in space*. In fact, a crystal may be thought of as a succession of symmetry operations around symmetry elements on the asymmetric unit of the unit cell to produce a crystal lattice. Theoretical investigations of crystals in the nineteenth century arrived at the conclusion that there are a limited number of possible geometric arrangements in crystalline arrays.

Depending on the values and relationships of their lattice constants, seven main types of unit cell are possible which are called the *crystal systems*. In order of increasing

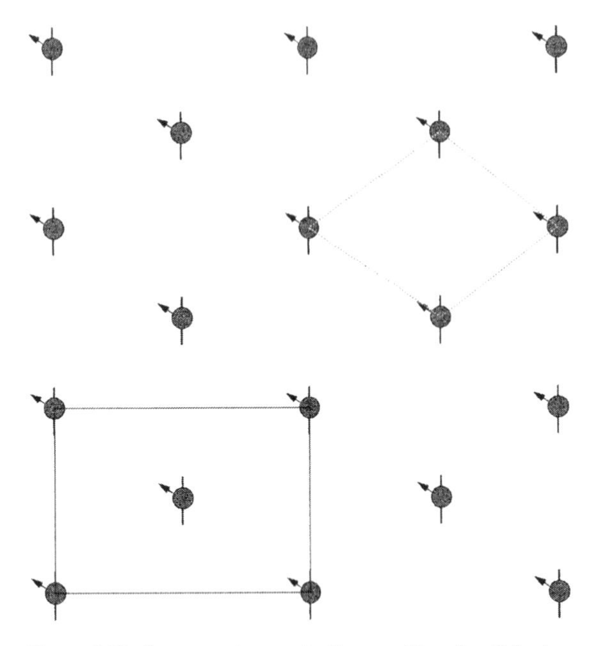

Figure 6.29. Symmetry in crystals. One possible unit cell (broken lines) contains one object per unit cell and is described as 'primitive'. Another possible unit cell shown with solid lines contains two objects and therefore has higher symmetry. Both unit cells could be used to represent this two-dimensional array of objects but, in three-dimensional crystallography we select the unit cell with highest symmetry.

symmetry these are triclinic, monoclinic, orthorhombic, tetragonal, trigonal/rhombohedral, hexagonal and cubic. Six of these can form primitive lattices which are denoted by (P) (Table 6.3, Figures 6.29 and 6.30). The word 'primitive' (from the Latin *primus* for first) refers to the fact that the cell contains only one lattice point as illustrated in Figure 6.29. In addition to the primitive lattices, seven further lattices are possible. Two of these feature an extra point at the *centre of one of the six faces*. By convention, this face is denoted as the *a–b* face and such lattices are referred to as *C-centred (C)*. Extra points on each of the six faces result in a *face-centred* lattice (*F*) of which two are possible. An extra point in the *centre* of the primitive unit cell is possible in three lattices which are known as a *body-centred* and denoted as (*I*; from the German *innenzentrierte* for *interior*). In all, there are a total of fourteen *Bravais lattices* the characteristics of which are summarized in Table 6.3 and which are illustrated in Figure 6.30. It may be further calculated that there are 230 mathematically possible space groups. However, since proteins and nucleic acids are usually *enantiomorphic* (Section 3.5.2), only 65 of these are relevant to crystallography of biomolecules.

Diffraction of X-rays by a crystal results in a pattern which is mathematically related to the pattern of the crystal lattice. One of the first steps in analyzing diffraction patterns is to assign the crystal to its specific space group in a Bravais lattice with *maximum* symmetry as shown in Table 6.3. This analysis also determines the shape and dimensions of the unit cell which are important parameters in calculation of structure from crystallographic data.

An important point of difference between crystal lattices of biomacromolecules such as proteins and small molecules like NaCl is that real crystals of large molecules frequently consist of *mosaics* of many *microcrystals* aligned more or less in the same direction. This *mosaicity* has the result that diffracted X-ray beams from crystals of large molecules actually emerge as narrow cones rather than fine beams which means that they need to be collected over a very small angle in practice rather than at a single, fixed angle.

6.3.3 Physical Basis of Crystallization

The crystalline state is achieved when molecules are recruited in a *highly ordered* manner from the liquid phase

Table 6.3. The seven crystal systems.

Crystal system	Possible Bravais lattices	Minimum symmetry	Restrictions on lattice constants
Triclinic	P	Single onefold symmetry axis	$a \neq b \neq c; \alpha \neq \beta \neq \gamma$
Monoclinic	P, C	One twofold axis (// to b)	$a \neq b \neq c; \alpha = \gamma = 90°; \beta \neq 90°$
Orthorhombic	P, C, I, F	Three mutually Perpendicular twofold axes	$a \neq b \neq c; \alpha = \beta = \gamma = 90°$
Tetragonal	P, I	One fourfold axis (// to c)	$a = b \neq c; \alpha = \beta = \gamma = 90°$
Trigonal	P	One threefold axis (// to c)	$a = b \neq c; \alpha = \beta = 90°; \gamma < 120°$
Rhombohedral	R	One threefold axis (// to c)	$a = b = c; \alpha = \beta = \gamma \neq 90°$
Hexagonal	P	One sixfold axis (// to c)	$a = b \neq c; \alpha = \beta = 90°; \gamma = 120°$
Cubic	P, I, F	Four threefold axes (along cube diagonals)	$a = b = c; \alpha = \beta = \gamma = 90°$

Rhombohedral and Trigonal are closely related and usually regarded as a single. Trigonal, crystal system. Trigonal R, shown in Fig. 6.30, is Rhombohedral. A Trigonal P form (with a 'primitive' lattice) would resemble Hexagonal P except the γ angle would be $<120°$ rather than $=120°$.

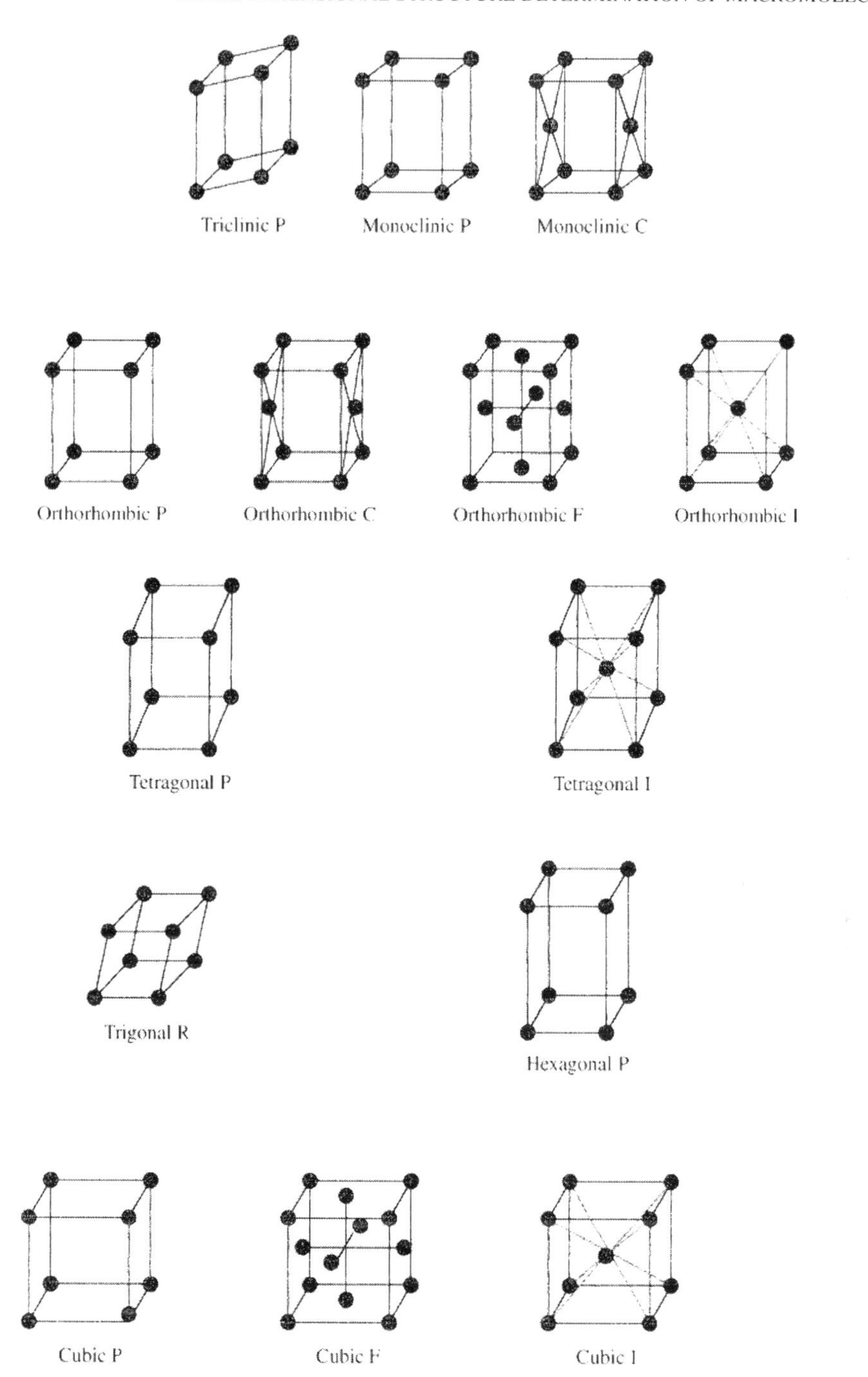

Figure 6.30. The Bravais lattices. There are fourteen types of lattices possible, depending on the lattice constants (Table 6.4). These are designated as primitive (P), C-centred (C), face-centred (F) body-centred (I) or rhombohedral (R).

(i.e. solution) to the solid phase (i.e. crystal). This contrasts with phenomena such as *precipitation* in which molecules leave solution in a *random* manner forming an amorphous and disordered aggregate. Inorganic materials such as NaCl form crystals very readily but more complex molecules such as proteins and DNA will only crystallize under a comparatively small range of chemical circumstances. In fact, we have a very limited understanding of biomacromolecular crystallization and it is difficult to predict before a crystallization trial whether or not crystals will be successfully obtained. For this reason the normal practice is to screen *empirically* many different sets of crystallization conditions to identify the very small number which will produce suitable crystals in a given case.

Crystallization of biomacromolecules is highly dependent on the availability of *pure sample*. While techniques described elsewhere in this volume may be used to assess purity of protein and other samples (e.g. SDS PAGE of proteins, Section 5.3.1; reversed-phase HPLC of proteins or DNA, Section 2.6) these may not detect *microheterogeneity* within the sample. This could be due, in the case of proteins, for example, to variable patterns of glycosylation or disulfide bridge formation, N- or C-terminal heterogeneity or the possibility of several structural conformations for the sample molecule. Microheterogeneity due to differences in covalent structure can usually be detected by mass spectrometry or electrophoresis (Chapters 4 and 10).

A process likely to result in microheterogeneity within the sample will tend to hinder crystal formation; heterogeneity interferes with orderly recruitment into and efficient packing within the crystal. Frequently, it is easier to crystallize a protein in the presence of a low molecular weight ligand as this reduces conformational variability within the population of sample molecules. This process is referred to as *cocrystallization*. Good examples are cocrystallization of enzymes with inhibitors or of hormone receptors with hormones. The necessity for ultrapure sample also leads to a requirement for particularly low batch-to-batch variation during protein purification or oligonucleotide synthesis for crystallographic work.

The general approach taken to crystallization is to expose a concentrated aliquot of a sample of protein/DNA to a concentration of *precipitant* which is *slightly below* that necessary to precipitate the sample. The most widely used precipitants are low molecular weight poly-alcohols (e.g. hexane diol), salts (e.g. ammonium sulfate), high molecular weight polymers (e.g. polyethylene glycol) and organic solvents (e.g. methanol). At the beginning of the experiment, the sample solution, though concentrated, is still unsaturated (see the phase diagram in Figure 6.31). During the experiment the protein and/or precipitant concentration is increased slowly such that eventually the *saturation point* of the phase diagram is reached. This is the point where the solid and liquid states of the protein exist *in equilibrium*

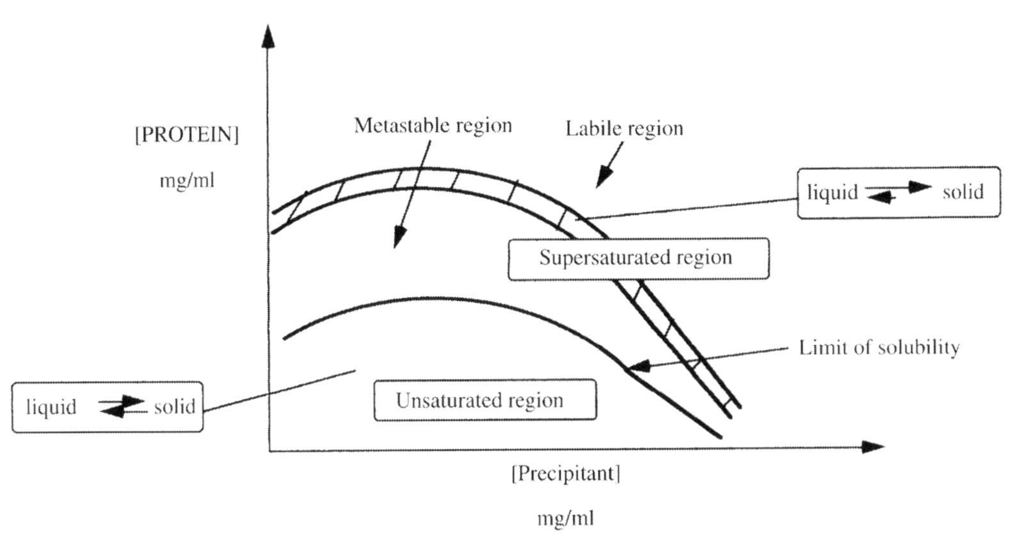

Figure 6.31. Phase diagram for protein. The region demarcated by the lower curve represents the limit of solubility of a protein. This is characterized by equilibrium between the solid and liquid states of the protein. The region above this represents supersaturation which may be divided into a metastable and labile region. In the metastable region, crystals can grow while in the labile region stable nuclei can both form and grow. Supersaturation is characterized by nonequilibrium between the liquid and solid states of the protein which drives crystal formation. The hatched area shows the optimum part of the diagram for growth of crystals suitable for diffraction.

with each other (Figure 6.31). Any increase in solid protein here tends to be counterbalanced by dissolution of solid protein with the consequence that *crystals cannot grow at the saturation point*.

Formation of a crystal depends on the formation of a *stable nucleus*. This is initiated by two or more sample molecules assembling together to form a suitable 'platform' onto which other sample molecules can assemble. Ordered successive addition of sample under the nonequilibrium conditions of supersaturation leads to growth of the crystal in three dimensions until equilibrium is re-established. The driving force behind this process is formation of the maximum number of intermolecular bonds such as hydrogen bonds and van der Waals interactions. Of course, there are many more ways for a protein/DNA molecule *not to form* a crystal since entropy in chemical systems tends to the maximum. Therefore, it is much more common for a nucleus to grow to a certain point and then stop growing. Such a nucleus is called an *unstable nucleus*. Alternatively, sample molecules may add together in a disordered manner leading to precipitation. Conditions leading to formation of a crystal of suitable size (minimum 0.5 mm) and quality for X-ray diffraction are rare and are usually only found after exhaustive screens of hundreds or thousands of individual conditions.

If the protein concentration is increased beyond saturation, that is to *supersaturation*, a *nonequilibrium state* is achieved. At supersaturation, the tendency of the system to return to equilibrium forces protein into the solid phase, occasionally resulting in crystal formation. In practical terms, the supersaturated state may be achieved by evaporation of solvent or by manipulation of temperature (Section 6.3.4). The supersaturated region of the phase diagram may, in turn, be divided into metastable and labile regions (Figure 6.31). The metastable region is defined as a region where stable nuclei cannot form. If, however, stable nuclei are introduced into the metastable region, they can continue to grow. This is distinguished from the labile region where stable nuclei can *both* form and continue to grow.

The importance of this is that, if the supersaturated system is very far from equilibrium (e.g. far into the labile region), then very many stable nuclei can rapidly form in the early stages of crystallization but crystal growth then slows and halts as equilibrium is re-established. This results in a large number of very small and often imperfect crystals which are not suitable for diffraction (these are called 'showers' of *microcrystals*). Conversely, the nearer the system is to the metastable region, the fewer stable nuclei are formed and the slower crystallization proceeds in the early stage of the experiment. This is the situation desired in biomacromolecule crystallization as it results in a small number of large crystals of the high quality required for diffraction (**Figure 6.32, see colour plate section**). The behaviour of each individual protein or DNA molecule in the phase diagram is unique which is why hitting on optimum conditions resulting in the formation of a small number of large crystals is so difficult.

Any variable affecting protein solubility will affect crystallization. For this reason, it is usual to establish a discrete set of chemical conditions which include a specific pH, temperature, buffer and precipitant. Very small differences in pH (e.g. 0.1 pH units) can sometimes make the difference between obtaining crystals or not. Similarly, crystallization is particularly sensitive to temperature and, since several months may be necessary for crystals to form, it is essential that samples are not exposed to fluctuations of temperature. This is achieved by storing samples under conditions where temperature is tightly controlled (e.g. in temperature-controlled rooms, refrigerators or incubators). It has even been discovered that the earth's gravitational field may hinder the growth of crystals and protein crystallization trials in the near-zero gravity of space are now a common feature of space shuttle missions. Unlike inorganic molecules, protein and DNA structures are large, complex and sometimes labile and their charge and shape characteristics may alter during the crystallization process in highly-individual ways. This is why crystallization of biomacromolecules is such a slow and somewhat inexact process.

In addition to nongrowth of crystals, a number of other problems can be encountered during crystallization trials. *Twinning* can occur when two crystals grow together. Because they consist of two distinct lattices, they cannot be used to determine structure. This phenomenon can be avoided by inclusion of organic solvents such as 1,4 dioxan in the crystallization buffer. Growth of larger single crystals can be encouraged by increasing the volume of the sample or by manipulating the conditions to slow crystallization (e.g. by decreasing the protein concentration).

6.3.4 Crystallization Methods

Because of the vast number of possible crystallization conditions (pH, precipitant concentration/identity, ion strength, temperature) and the limited amount of protein normally available, it is not realistic to assess every possibility. Instead, a process called *sparse matrix sampling* has come to be widely used. This consists of a set of 25–50 individual conditions which have been chosen based largely on previously successful crystallizations. These are now commercially available as kits and are often used in initial crystallization screens. Once one or a small number of conditions producing some crystalline material have been identified, further secondary screens using incrementally different conditions are then systematically carried out with the object of obtaining larger and better-quality crystals.

Crystallization methods provide a means to expose protein/DNA samples to the supersaturated state usually by one or more of the following processes:

1. Facilitating *gradual evaporation* of solvent (e.g. organic solvents).
2. Achieving *gradual desolvation* of sample by polyalcohols or salts. This is also achieved by the presence of polyethylene glycol or organic solvents which lower the dielectric constant of solvent and thus also alter the bulk properties of water.
3. Variation of pH and temperature to *reduce protein solubility*. These processes must occur under conditions which do not significantly alter the structure or stability of the sample molecule.

We will now describe three of the principal crystallization methods currently in use.

Microdialysis is a low-volume form of *dialysis* (Chapter 7) which is a process of *diffusion* across semipermeable membranes (Figure 3.33). A solution containing the desired crystallization condition is placed on one side of the membrane while the protein solution (5–50 µl) is placed on the other. Equilibration across the membrane gradually alters the properties of the protein solution to that of the crystallization condition. The properties of the crystallization condition can be gradually changed by altering the solution on that side of the membrane. Two formats of this technique are illustrated in Figure 6.33. One of these features a specially designed commercially available plexiglass 'button' into which the protein sample is introduced. The other involves use of glass capillaries with dialysis membrane held in place across one end of the capillary. Both of these can be placed in the external solution to allow equilibration to occur. Microdialysis is especially suited to crystallization because diffusion across semipermeable membranes is relatively slow.

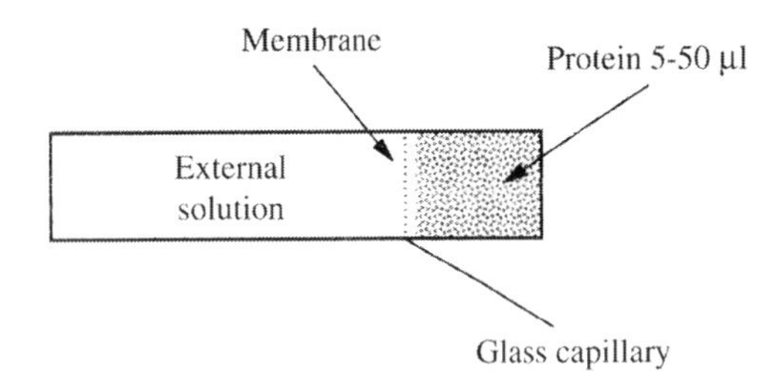

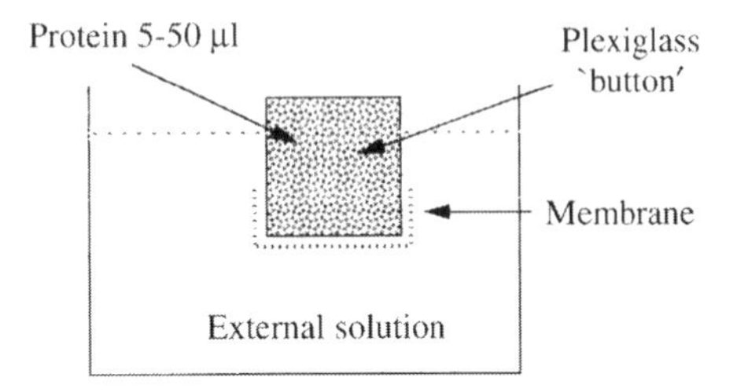

Figure 6.33. Microdialysis. Solution containing desired conditions (external solution) diffuses across a semipermeable membrane into the protein solution (liquid–liquid diffusion). Properties of the external solution can be gradually varied leading to crystallization.

(a)

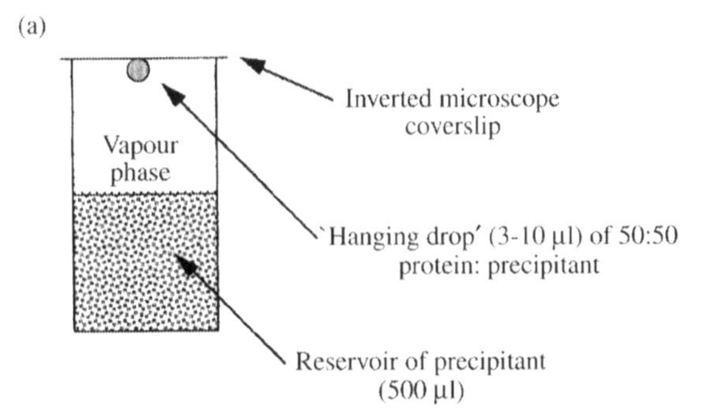

(b)

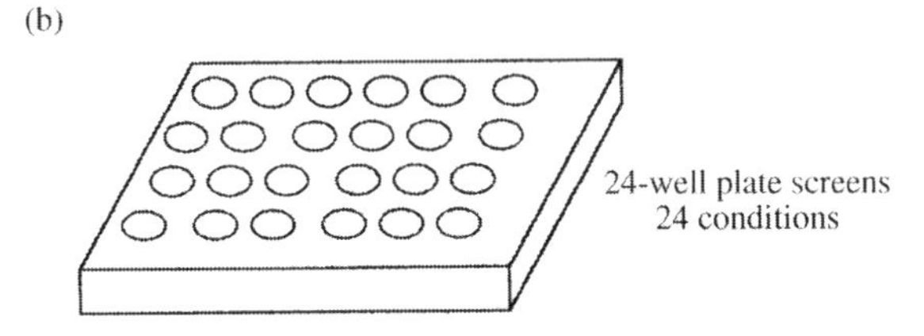

Figure 6.34. The 'hanging drop' vapour diffusion method. (a) Sample is suspended as a 'hanging drop' (3–10 µl) over a much larger volume (e.g. 500 µl) of the desired crystallization condition. This equilibrates very slowly (days, weeks, years) with the drop through the vapour phase (hence 'vapour diffusion'). (b) A 24-well plate may be used to screen simultaneously 24 individual conditions.

Vapour phase diffusion features slow equilibration between a bulk crystallization solution and a small volume (<10 µl) of concentrated (10–50 mg/ml) sample through the vapour phase. The most popular form of this technique is the 'hanging drop' method (Figure 6.34). A small volume of sample solution mixed 1 : 1 with crystallization solution is pipetted onto a glass microscope slide. This is then inverted over a bulk (500 µl) crystallization solution. Equilibration occurs between the bulk and drop solutions until eventually the chemical conditions in the drop are identical to those of the crystallizing solution. The availability of 24-well linbro plates allows simultaneous screening of 24 conditions using as little as 24 µl of sample solution per plate.

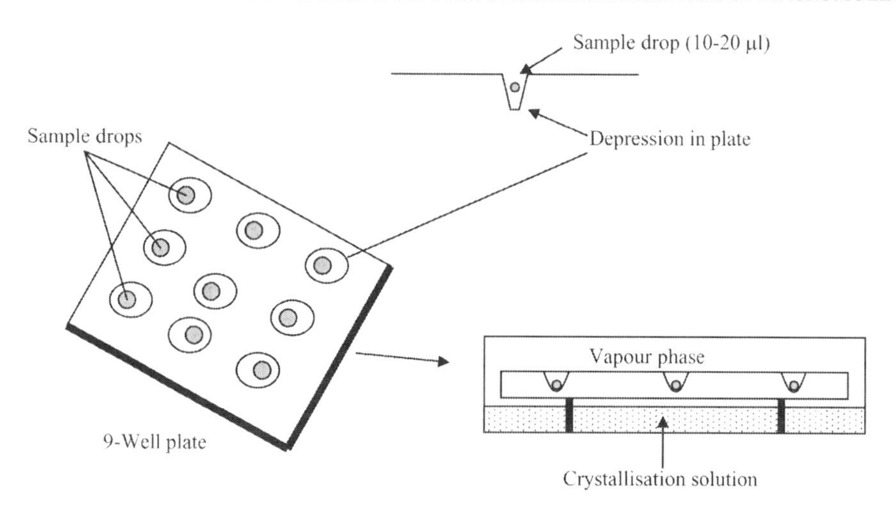

Figure 6.35. The 'sitting drop' vapour diffusion method. Sample (10–20 μl) is placed in a depression formed in a 9-well plate. This is supported over the crystallization solution in a sealed chamber (e.g., on an inverted petri dish). Sample equilibrates with the crystallization solution through the vapour phase.

Closely related to this is the *'sitting drop' technique* in which a drop of sample is placed in a depression on a plate which is then placed in the vapour phase of a crystallizing solution (Figure 6.35). Again, equilibration is through the vapour phase and the conditions of the drop slowly become identical to those of the bulk crystallizing solution. This technique is useful for assessing a single crystallization condition with several protein/DNA samples but is not as useful as the hanging drop procedure for screening a large number of crystallization conditions.

The third major crystallization technique is *free interface diffusion* which is similar to microdialysis except no semipermeable membrane is used (Figure 6.36). A crystallization solution is layered gently on the sample solution and diffusion slowly generates concentration gradients at the interface. This creates localized supersaturation with the possibility of crystal formation. This technique is especially useful in the microgravity of space.

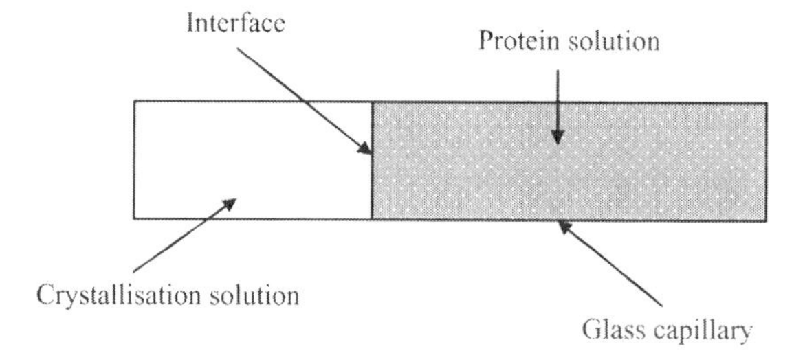

Figure 6.36. Free interface diffusion. This is generally similar to microdialysis except no membrane is used. Concentration gradients are generated at the interface which gives local supersaturation. The technique is particularly useful in microgravity.

6.3.5 Mounting Crystals for Diffraction

The above discussion makes clear that crystals of biomacromolecules are difficult to obtain in a size and quality suitable for diffraction. A further step is necessary before collection of diffraction data however. This is *mounting* the crystal in a format suitable for exposure to the X-ray beam. Crystals of protein or DNA are far less stable than inorganic crystals. Unintended collision of the crystal with a solid surface (e.g. a needle or the edges of plasticware) can be sufficient to cause it to crack or fragment. Exposure to mild heat (e.g. from fingertips transmitted through glassware), air or any chemical condition different to those in which the crystal was grown can introduce distortions into the crystal making it useless for diffraction work and wasting the time expended in growing it.

Protein/DNA crystals are usually mounted in thin glass capillaries suitable for assembly and rotation in the X-ray beam (Figure 6.37). The small size of the crystal necessitates availability of a dissecting microscope to follow the manipulations required. A single crystal is introduced into the capillary suspended in a small volume of the solution in which the crystal was grown (the *mother liquor*) by aspiration from the reservoir in which crystallization occurred. The crystal is then dried by removing this liquid by capillary action through wicks of paper. A small amount of mother liquor is then placed on either side of the crystal and

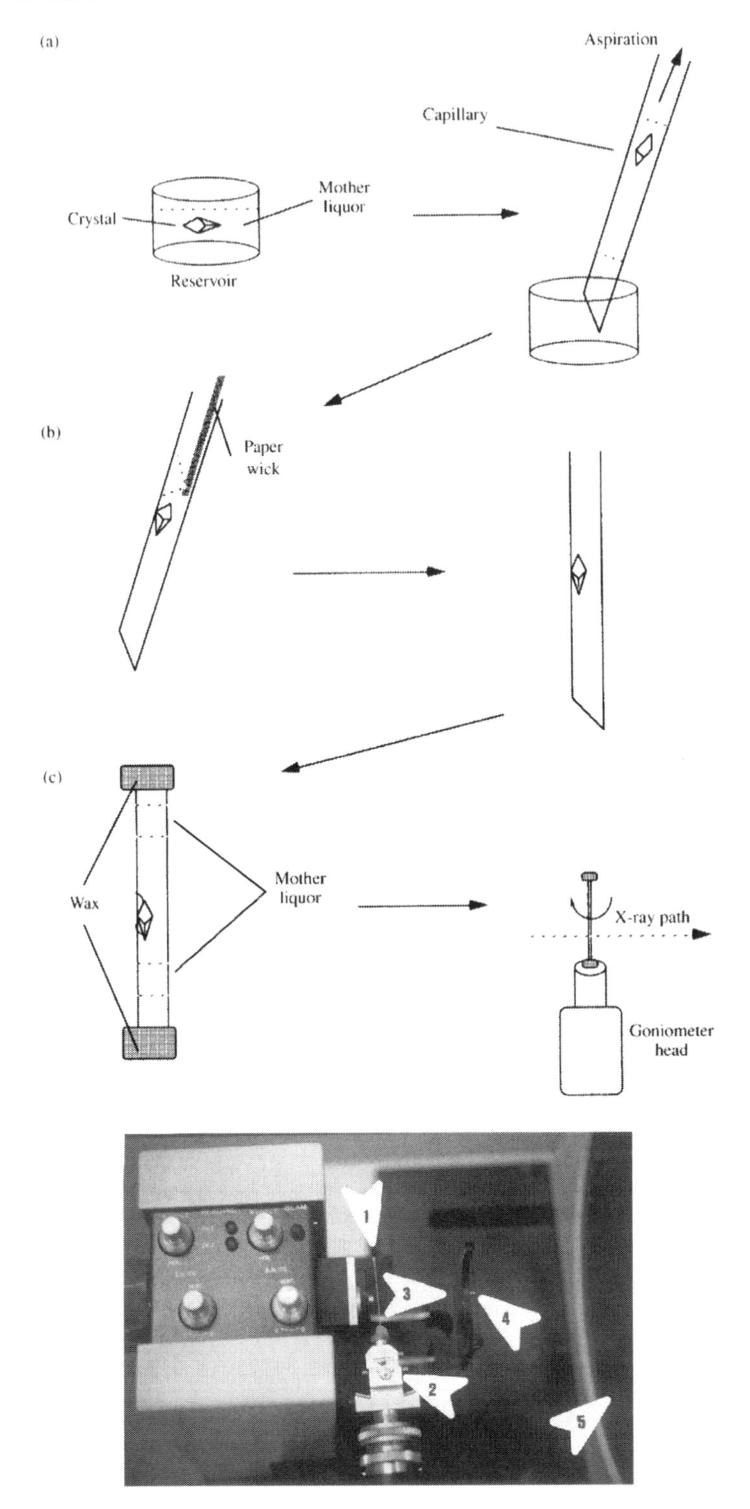

Figure 6.37. Mounting crystals. (a) A single crystal (minimum 0.5 mm) is transferred to a glass capillary by aspiration. (b) The crystal is dried by removing mother liquor with paper wicks. (c) A small plug of mother liquor is applied on either side of the crystal before sealing capillary ends with wax. Capillary is then mounted on the goniometer head. This holds the crystal in the X-ray beam at all angles of rotation (arrow). (d) View of a crystal mounted in a capillary (1) attached to a goniometer head (2) in an X-ray beam path (3). The lead beam stop (4) and area detector (5) are also visible.

the ends are sealed in wax. This creates an environment in which the crystal is constantly exposed to a vapour phase from the mother liquor and is protected from exposure to air and attendant contaminants.

The mounted crystal can now be assembled on a *goniometer head* in the apparatus used for diffraction. This is a device which allows arc and lateral adjustments to the three-dimensional position of the crystal such that, at any rotation angle, the crystal remains in the X-ray beam. The crystal is also aligned in such a way that one of the axes a–c is orientated along the axis of the X-ray beam. This point is discussed further in Section 6.4.3 below.

Some crystals deteriorate rapidly at room temperature and it may be necessary to freeze them before exposing them to the X-ray beam. This is achieved by *flash freezing* the crystal in the presence of a *cryoprotectant* to a temperature of −196 °C in cooled liquid propane. This is carried out very quickly to freeze water molecules in the crystal lattice into amorphous ice rather than into ice crystals (which would disrupt the lattice). Frozen crystals are mounted on loops of fine nylon rather than in capillaries and are exposed to the X-ray beam in a cold stream at −196 °C.

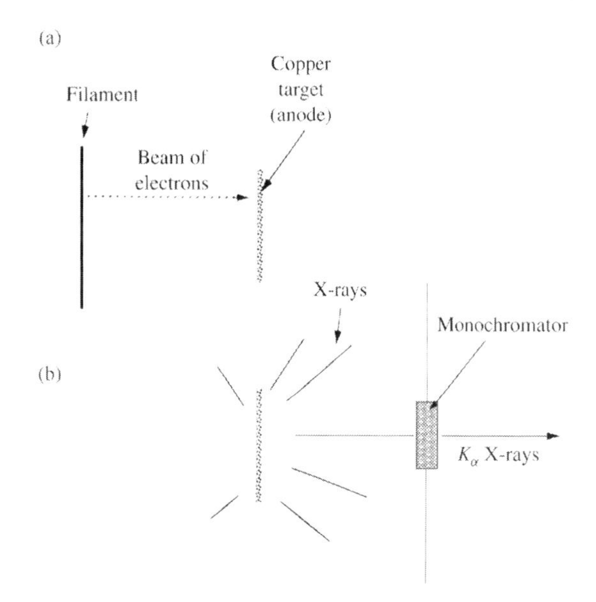

Figure 6.38. Source of X-rays. (a) Bombardment of a copper anode target by a beam of electrons excites copper electrons to outer atomic shells. (b) Return to the ground state results in emission of X-rays of a number of wavelengths. K_α rays ($\lambda = 1.542$ Å) are specifically selected by passage through a graphite monochromator or a nickel filter.

6.4 X-RAY DIFFRACTION BY CRYSTALS

Having described crystals, some of their symmetry properties and how we can prepare them for X-ray diffraction work we will now turn to the other important components of the experiment by examining the nature of X-rays and their diffraction by crystals. This leads to a data-set consisting of thousands of individual diffraction spots each with a particular intensity, location and phase. In Section 6.5 we will see how this data is used to generate a three-dimensional structure for the sample molecule.

6.4.1 X-Rays

X-rays are a form of high-energy electromagnetic radiation with wavelengths in the range $0.1–100 \times 10^{-10}$ m (Chapter 3; Figure 3.7) which were originally called *Roentgen rays* in honour of their discoverer, Wilhelm Roentgen. They may be generated by bombarding a copper or molybdenum *target* (anode) with a beam of accelerated electrons at a potential of some 10 000 eV (Figure 6.38). The energy of this bombardment is absorbed by the dislocation of electrons from the inner K and M atomic shells to outer atomic shells. When these electrons return to the ground state, X-ray radiation is emitted. For crystallographic studies, radiation of $\lambda = 1.542 \times 10^{-10}$ m (=1.5 Å; also-called K_α rays) are selected by passage through a graphite monochromator or a nickel filter. X-rays having a single wavelength are desirable in crystallography because they give a single pattern of strong reflections.

Two main laboratory-scale X-ray sources are widely used. *Sealed tube generators* feature an evacuated glass tube which contains a filament (source of the electron beam) and a hollow target. They have the important advantages of ease of maintenance and replacement but produce beams of limited energy output (up to 3 kW) because too-powerful an electron beam could melt the target. *Rotating anode generators* consist of a target anode which can be rotated relative to the filament. This means that, since the electron beam impacts on a constantly-varied cool part of the filament, higher energy electron beams can be used which results in higher-power output (up to 12 kW). In addition to these, *synchrotron* sources are large-scale facilities which use a particle accelerator to produce a continuous spectrum of high-energy X-rays and these are described in more detail in Section 6.5.9.

6.4.2 Diffraction of X-Rays by Crystals

In order to determine molecular structure it is necessary to use X-rays because the wavelength of this radiation is of a similar order of magnitude as atoms and covalent bond-lengths. The diffraction pattern obtained contains the information we use to calculate this molecular structure. However, the diffraction pattern is a two-dimensional array of 'spots' of particular position and intensity (Figure 6.26). In order to convert this into a three-dimensional structure it is necessary to collect a *data set* of many such diffraction patterns from a single crystal (the precise number required depends on the symmetry and other characteristics of the crystal). Repeated exposure to the high-energy X-ray beam can sometimes lead to deterioration of the crystal during the experiment which can preclude the collection of a complete data set. In such cases it is often necessary to rescreen in an attempt to obtain more robust crystals, perhaps in a different crystal form.

The electrons of atoms are responsible for diffraction of X-rays. This is due to an interaction between the X-ray electromagnetic wave and the electron. The electric vector of the electromagnetic wave (Chapter 3; Figure 3.1) induces an *oscillation* in the electron of a frequency identical to that of the wave (Figure 6.39). This results in emission of secondary radiation of a wavelength and frequency identical to that of the incident X-ray but 180° out of phase with it which is known as *coherent scattering*. A related process called *anomalous scattering* can occur if the frequency of the incident beam happens to be near that of a naturally-occurring oscillation in the electron. This latter phenomenon can be useful in determining the phase of scattered beams as discussed below (Section 6.6).

Since electrons are responsible for diffraction, there is a close relationship between the intensity with which incident X-rays are diffracted and the atomic number (Z) of the atom which diffracts them. This may be quantified by the *scattering factor*, f, which is related to the angle of incidence

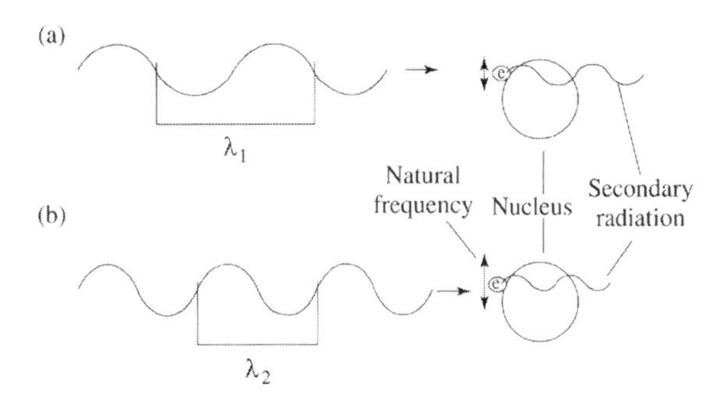

Figure 6.39. Scattering of X-rays by electrons. (a) Coherent scattering. The beam of X-rays induces an oscillation in the electron (e) which has the same frequency as the incident radiation. This results in emission of secondary radiation which is 180° out of phase with incident radiation. (b) Anomalous scattering. When the frequency of the incident beam (v) corresponds or is close to a naturally-occurring frequency in the electron, this results in emission of radiation known as anomalous scattering.

(θ) between the incident beam and the plane containing the diffracting atom and the X-ray wavelength. f is in fact a function of sin θ/λ and has a maximum value of Z when $\theta = 0$.

The scattering factor is a ratio between the intensity of scattering of X-rays by an atom and that of a single electron in the same experimental circumstances. Relative values (in brackets) for some atoms of biological interest are; ^{1}H (1), ^{12}C (6), ^{16}O (8) and ^{54}Fe (26). This means that heavy metals diffract with much greater intensity than other atoms and that hydrogen diffracts so weakly that it is usually not detectable. For this reason the actual structure we calculate from X-ray diffraction patterns is an *electron density map* representing the location of C, N, S, O and other nonhydrogen atoms. Because of the similar scattering factors for N, O and C it can be difficult to distinguish side-chains of aspartate, glutamine and threonine from each other in electron density maps of proteins.

The detection system for scattered X-rays should record both the position of the diffraction spot and its intensity as well as the overall diffraction pattern. X-ray diffraction patterns may be detected on photographic film since X-rays, like visible light, interact with film causing deposition of metallic silver. The intensity of each spot can be estimated relative to standard spots or by using a *densitometer*. The film is mounted in a camera which is orientated in a known angle relative to the incident beam and crystal. Alternatively, diffracted beams may be detected using *electronic detectors* such as the *area detector*. These convert spot intensities and position to an electrical charge which is recorded (Figure 6.40). Electronic detectors have a number of advantages over film; they are *more sensitive* which shortens the time necessary to collect a diffraction pattern (and hence shortens the exposure time of the crystal to the X-ray beam). Moreover, the data can be *stored electronically* in a computer-readable form which eliminates any requirement for a densitometer and again saves time. Once the crystal has been exposed to the X-ray beam, the diffraction pattern is collected and stored electronically. The detector is then erased, the crystal rotated slightly and a second diffraction pattern collected. In this way, a data-set of diffraction patterns is built up for the crystal. Because most of the incident beam passes straight through the crystal without being scattered, this would saturate the detector limiting its sensitivity. Accordingly, the detector is shielded from the direct incident beam by a circular piece of lead called a *beam stop*. More recently, detectors with an X-ray-sensitive phosphor screen coupled to a charge coupled device have become available.

6.4.3 Bragg's Law

As with all electromagnetic radiation waves, diffraction can result in interference between diffracted waves which results

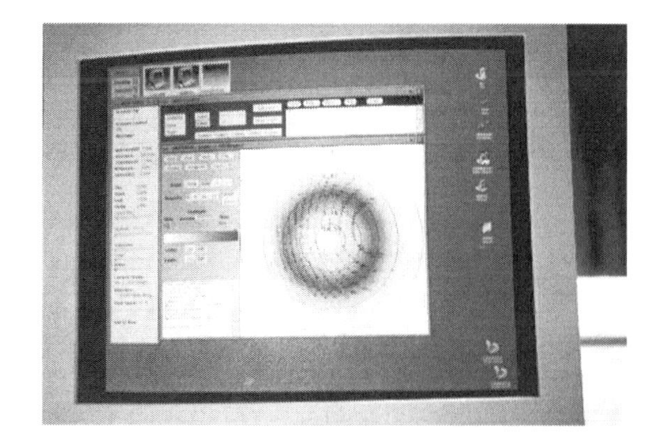

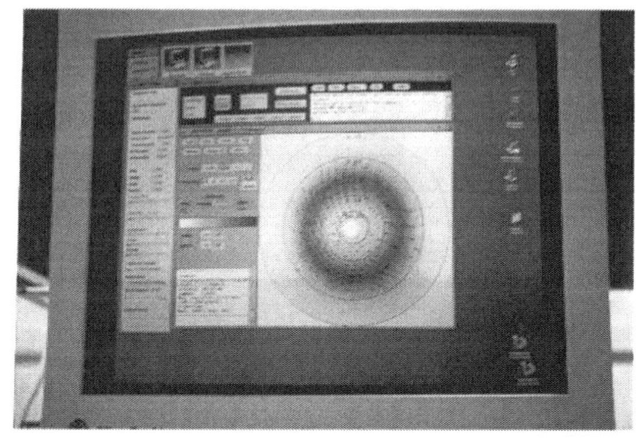

Figure 6.40. X-ray diffraction pattern collected on an area detector. A diffraction pattern collected for lysozyme crystals on an electronic area detector is displayed on a silicon graphics computer. Note the different intensities and location of diffraction spots forming a unique pattern. These intensities are converted into structure factors. After rotation through a small angle (e.g. 1°), a further diffraction pattern is collected. Some spots increase in intensity while others decrease. A series of such diffraction patterns forms the data-set used in calculation of the electron density map.

(a)

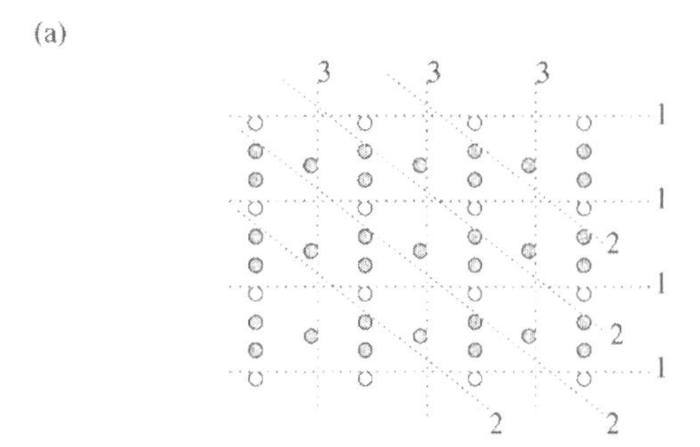

(b)

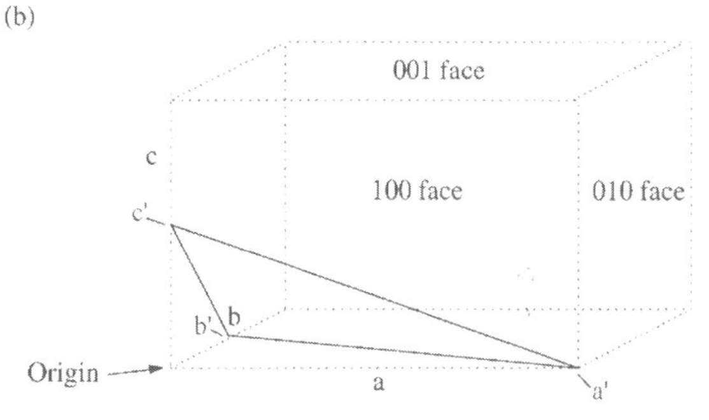

Figure 6.41. Lattice planes. (a) Three lattice planes (dashed lines, 1–3) are shown for a two-dimensional array. These planes link corresponding atoms through the crystal lattice. (b) Miller indices allow identification of individual planes. They are designated in terms of (hkl, where $h = a/a'$, $k = b/b'$, $l = c/c'$ representing points where the plane transects the unit cell (dashed lines). The plane shown is designated 122 since the axes are cut at $a/1$, $b/2$ and $c/2$. The faces of the unit cell may similarly be designated as 100 (a), 010 (b) and 001 (c).

in the waves either reinforcing or weakening each other depending on their relative phases (Chapter 3; Figure 3.44). These phenomena are characteristic of wave functions and are known as *constructive* and *destructive interference*, respectively. In an X-ray diffraction experiment, a complicated pattern of scattering of X-ray beams is observed in which *both of these processes occur simultaneously*. The pattern of interference depends on the distribution of atoms in the sample through which X-rays pass. Since crystals are highly ordered arrays of atoms, systematic interference patterns occur which may be thought of as 'encoding' information on relative three-dimensional atomic locations.

In 1913 Bragg proposed that a crystal may be regarded as a series of planes which behave as 'mirrors' reflecting X-rays. In a real crystal, these *lattice planes* cut through the crystal lattice in three dimensions (a two-dimensional representation is shown in Figure 6.41). *Miller indices* provide a convenient way of identifying specific three-dimensional locations for such planes in the lattice. They are defined as the three intercepts (h, k, l) that a plane makes with the cell axes, in units of the cell edge. For example, if a plane intersects the cell axes a–c at points a', b' and c', the indices are given by $h = a/a'$, $k = b/b'$ and $l = c/c'$ (Figure 6.41). Values for h, k and l are usually less than six for most crystals.

Consider two such parallel lattice planes, each containing equivalent diffracting atoms and separated by a distance, d (Figure 6.42). Constructive interference between beams diffracted from successive planes occurs when there is an *integral* difference in λ between them (i.e. $n\lambda$, λ, 2λ, 3λ, etc.). In Figure 6.42 this condition is met when

$$PQ + QR = n\lambda \tag{6.12}$$

where PQ and QR represent the distances shown. The relationship between these distances and the angle of incidence

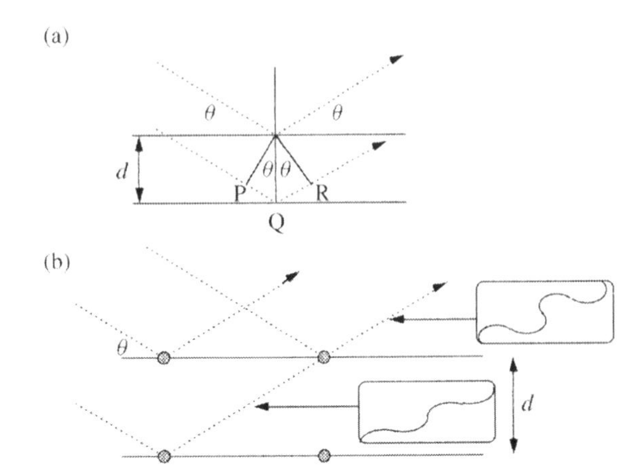

Figure 6.42. Bragg's law of diffraction. (a) X-rays incident on successive lattice planes at an angle θ are reflected when $n\lambda = 2d\sin\theta$, where d is the distance between planes. (b) This is because successive lattice planes which obey Bragg's law give rise to constructive interference. Note that the diffracted wave from the upper lattice plane has greater amplitude than that from the lower plane as a result of constructive interference.

of the X-ray beam (θ), is geometrical and is given by

$$PQ = QR = d\sin\theta \tag{6.13}$$

Substituting Equation (6.12) into (6.12) gives us *Bragg's law of diffraction*:

$$2d\sin\theta = n\lambda \tag{6.14}$$

In practice this means that the diffracted beams detected in an X-ray diffraction pattern (sometimes called *Bragg reflections*) are those originating from successive lattice planes which obey Bragg's law and reinforce each other by constructive interference with a single reflection arising from each equivalent lattice plane (*hkl*) in the crystal. All other diffracted beams undergo some destructive interference and, since this would happen through many equivalent lattice planes in a real crystal, are so weakened in intensity that they are not detectable. The only experimental variable in the Bragg equation is the angle of incidence, θ (since d is a fixed property of the crystal for a given hkl and λ is in most circumstances a fixed property of the X-ray beam). Variation of θ is achieved by slight rotation of the crystal (approximately. 1°) between exposures to the X-ray beam allowing detection of a distinct pattern of Bragg reflections for each angle of rotation. In a complete data-set, most atoms in the crystal lattice will contribute some Bragg reflections to some diffraction patterns.

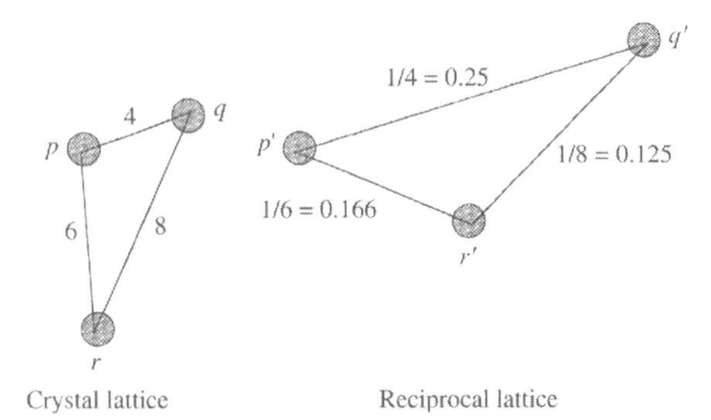

Figure 6.43. Reciprocal space. Objects in the crystal lattice (p, q, r) are related reciprocally to objects in the reciprocal lattice ($p'q'r'$). The triangle shown on the left therefore gives rise to a different-shaped triangle as shown above.

6.4.4 Reciprocal Space

Bragg's law makes clear that there is a fixed relationship between the pattern of Bragg reflections obtained during X-ray diffraction and the spacing of atoms within the crystal lattice. This relationship can be clarified further using the concept of the *reciprocal lattice* which is widely used in the study of diffracting systems. This is a real-space lattice (often referred to as *reciprocal space*) which is related *reciprocally* to points in the microscopic crystal lattice. The reciprocal relationship means that macroscopic measurements made at a distance of 10–100 cm from the crystal may be directly related to *microscopic* spacings in the crystal lattice (Figure 6.43). We can imagine a diffracting crystal as generating a macroscopic lattice in three dimensions around it. In our collection of data as diffraction patterns we *sample* this reciprocal lattice in two dimensions (since detectors such as photographic film or electronic detectors are two-dimensional). As the crystal is rotated, the reciprocal lattice also rotates and, since the detector is fixed during collection of a single data-set, this means that we can sample a considerable amount of the crystal lattice by rotating the crystal through 45–180°.

A consequence of Bragg's law is that there is an inverse relationship between the angle θ and d (i.e. sin $\propto 1/d$). This means that large spacing between lattice planes (which we find in the large unit cells of protein/DNA crystals) result in small values for θ while small spacing (of the type we find in small unit cells of small molecule crystals) give large values for θ. Because the unit cells for protein/DNA crystals are much larger than those for small molecule crystals, the distance between Bragg reflection spots is significantly smaller in diffraction patterns obtained from crystals of biomacromolecules. This problem can be overcome by setting a larger crystal-detector distance with protein/DNA crystals than is

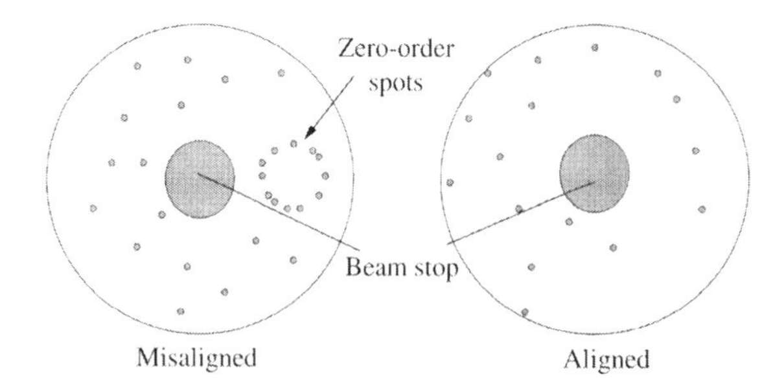

Figure 6.44. Alignment of crystal. The inner ring of diffraction spots (zero-order spots) are shown for a mounted crystal. A single diffraction pattern is collected to check alignment (left). This shows misalignment since the zero-order spots are visible. The position of the crystal is altered to bring zero-order spots behind the beam stop (inner circle). This procedure is not necessary if the crystal is oscillated during collection as described in Figure 6.26.

used with crystals of smaller molecules even though this results in less intense spots in the resulting diffraction pattern.

In order to maximize the information obtainable from X-ray diffraction while minimizing the number of diffraction patterns required, it is necessary to position the crystal in the beam such that its unit cells are orientated in a known way along the X-ray beam. This process is called *indexing* the crystal. Since the shape of the crystal is usually related to the dimensions and orientation of the unit cell, it is sometimes possible to align one of the crystal's (and hence the unit cell's) main axes $a-c$ along the beam. However, macroscopic inspection is not always sufficient to identify the orientation of the axes (Figure 6.27). In practice, the crystal is first mounted in the path of the X-ray beam as described in Figure 6.37 and a single diffraction pattern is collected (Figures 6.40 and 6.44). The position of the crystal is adjusted such that the inner ring of diffraction spots is not visible in the pattern because it is incident on the beam stop. This means that one of the unit cell axes, $a-c$, is now orientated along the X-ray beam although, at this stage, we do not know which one. Inspection of the diffraction pattern allows calculation of the dimensions of the reciprocal lattice and, hence tells us the orientation and dimensions of the crystal lattice axes.

The *Ewald construction* can be used to understand this concept better (Figure 6.45). This describes the geometrical relationship between the orientation of the crystal, the direction of X-ray beams diffracted from it as Bragg reflections and the reciprocal lattice. An X-ray beam incident at an angle θ on a lattice plane (hkl) obeying Bragg's law at point C will be diffracted as shown. A circle centered at point C may be drawn with a radius of $1/\lambda$. This is called the *Ewald circle* but, for a real three-dimensional crystal lattice, a three-dimensional *Ewald sphere* would be generated. If the crystal is rotated such that a reciprocal lattice point touches the surface of this sphere at a point corresponding to that where the diffracted beam intersects the circle (e.g. P in Figure 6.45), the origin of the reciprocal lattice is defined as point O on the circumference of the circle. This construction therefore allows us to relate points in the crystal lattice to corresponding points in the reciprocal lattice and each lattice plane (hkl) in a diffracting crystal to a specific Bragg reflection.

Determining how many reflections we need to collect in a given experiment depends on the properties of the crystal. As described earlier (Section 6.3.2), crystals contain considerable internal symmetry. A consequence of this is that there is identical though reciprocal symmetry in reciprocal space. This means that identical reflections may result at different crystal rotation angles. This has three useful consequences:

1. Symmetry in the pattern of reflections allows identification and measurement of the lattice constants.
2. All reflections do not need to be collected for every crystal.
3. Averaging of equivalent reflections gives more accurate estimates of intensities than measurement of single reflections alone.

It would be possible to rotate the Ewald sphere through a larger sphere in reciprocal space (radius $2/\lambda$) called the *limiting sphere*. This contains all the Bragg reflections from the crystal which might amount to 2×10^6 or more. In practice, even a very large protein rarely gives rise to more than 10^5 unique reflections because the diffraction pattern is limited to θ values around 20–25° for Kα rays. The diffraction pattern fades strongly at angles outside this range.

6.5 CALCULATION OF ELECTRON DENSITY MAPS

We have now seen how we can grow crystals and, by passing a beam of X-rays through them, collect a data-set consisting of a number of individual diffraction patterns each, in turn, consisting of spots of intensity I_{hkl}. Because the atoms in the crystal are regularly arranged relative to each other within the lattice, diffracted beams are scattered in a systematic way which is directly related to their distribution within the crystal lattice. As this scattering is in fact caused by electrons, there is therefore a clear relationship between the pattern of Bragg reflections obtained in the data-set and the density distribution of electrons within the crystal lattice and, hence, within each individual molecule of the sample under investigation. Calculation of this *electron density* distribution is the goal of X-ray crystallography since it allows us to visualize and quantify molecular

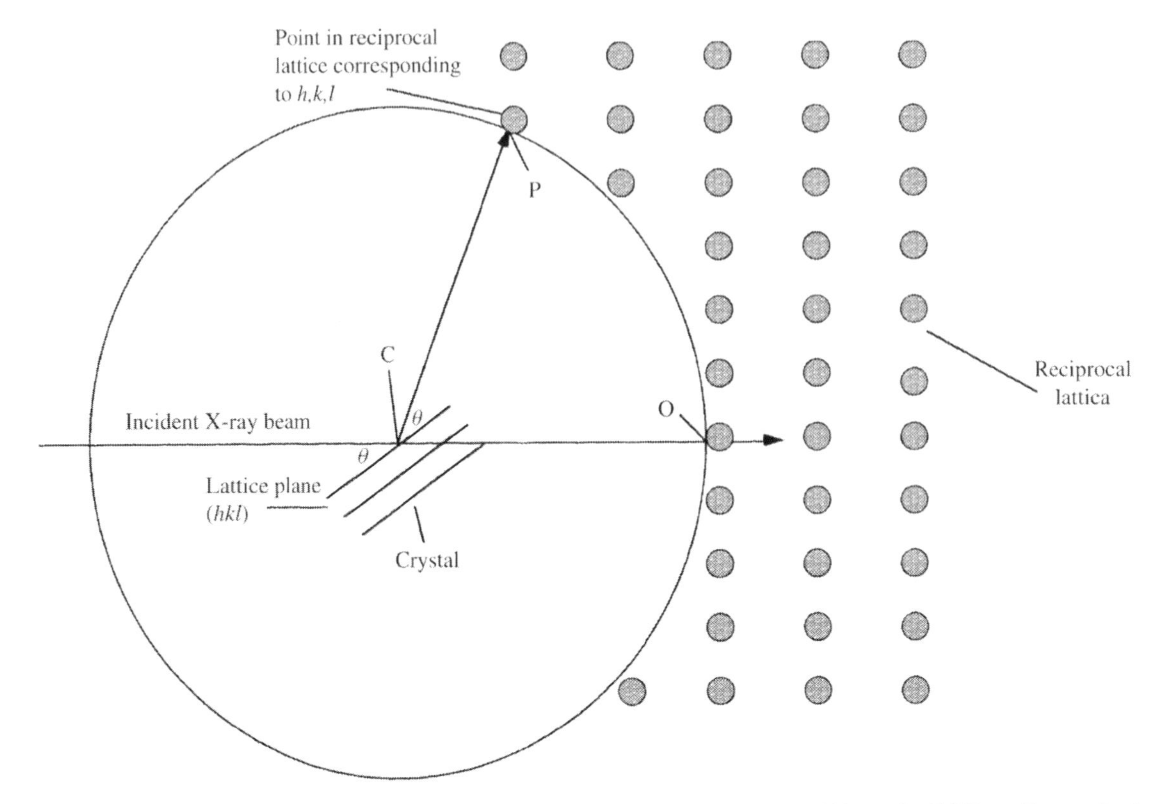

Figure 6.45. The Ewald construction. The incident X-ray beam is diffracted from lattice plane (*hkl*) at point C. When a Bragg reflection corresponding to this intersects with a circle (radius of $1/\lambda$) where a reciprocal lattice point touches the circumference of the circle, the origin of the reciprocal lattice is defined as O. The three-dimensional equivalent to this is the Ewald sphere. This allows us to predict detection of Bragg reflections and to orientate the crystal lattice with respect to the reciprocal lattice.

structure. The three-dimensional arrangement of electrons within the molecule is represented as an *electron density map*. In this section we will describe how Bragg reflections may be used to calculate an electron density map for the crystallized molecule.

A large molecule such as a protein can generate hundreds of thousands of individual Bragg reflections, much more than crystals of low molecular mass organic/inorganic molecules. This is because there are thousands of individual atoms in a protein. Moreover, the intensity with which large molecules diffract X-rays is significantly *lower* than that for small molecules which means substantially more weak reflections. The calculation of an electron density map from a large molecule such as a protein therefore represents a significantly more difficult task than a similar calculation for a small molecule, even though the individual steps which need to be taken are substantially similar. In current practice, calculation of electron density maps is carried out with the aid of powerful computer programs the availability of which is largely responsible for the speed with which macromolecular structures can now be determined compared to a decade or two ago.

Three distinct pieces of information act as inputs to these programs. (1) The individual intensity of each spot corresponding to a Bragg reflection (I_{hkl}); (2) the overall pattern of spots obtained; (3) the *phase* of the X-ray waves corresponding to each spot. We will now look at how each of these inputs contributes to generation of the electron density map.

6.5.1 Calculation of Structure Factors

In order to calculate an electron density map the intensity of each Bragg reflection, I_{hkl}, is converted into a *structure factor*, F_{hkl} (*note*; since this is a vector quantity on the Argand diagram, it is conventionally written hereafter in bold type). This has an associated *amplitude* (Figure 6.46; Section 3.1.1) denoted hereafter by $|F_{hkl}|$. These intensities, I_{hkl}, are directly measured by the detector (*or* recorded on photographic film and then converted into intensity as

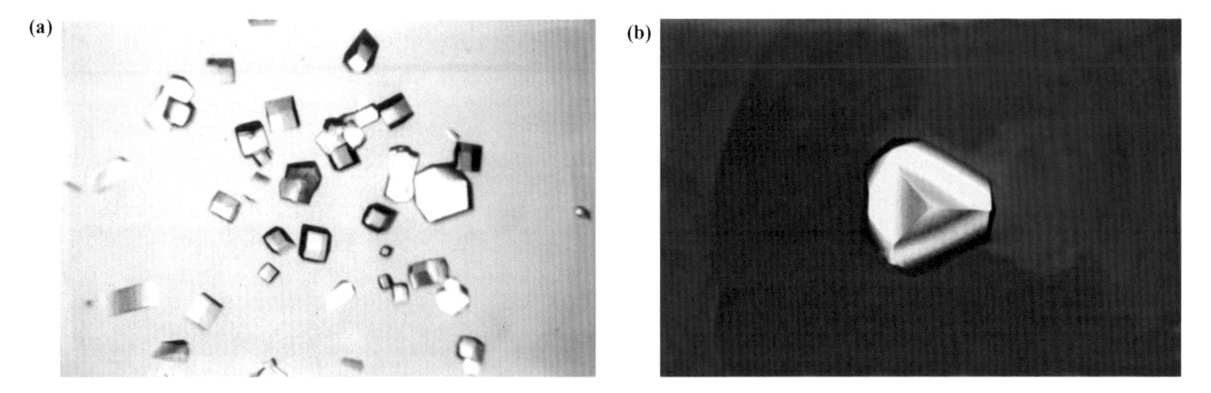

FIGURE 6.32. Crystals of lysozyme. Depending on the crystallization conditions used a large number of smaller crystals (a) or a small number of larger crystals (b) can be obtained.

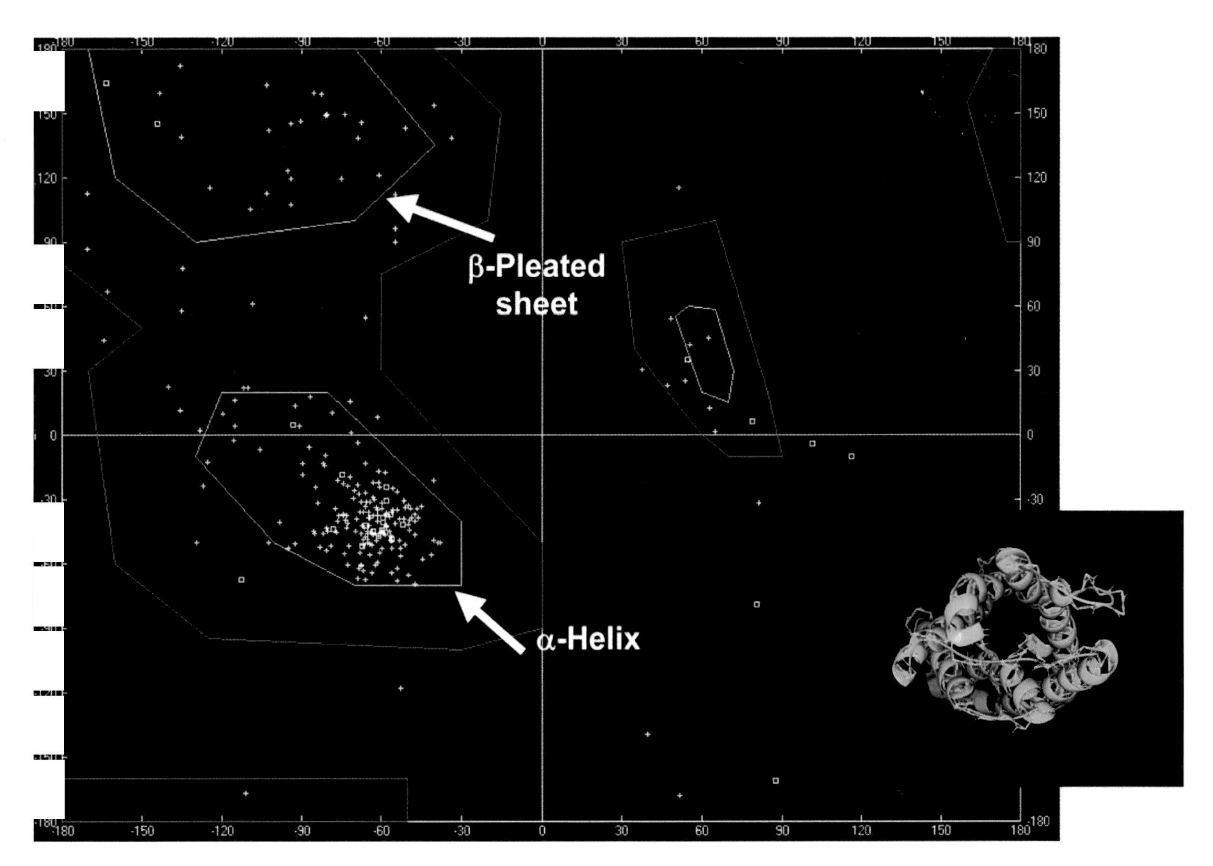

FIGURE 9.11. The Ramachandran diagram. The Ramachandran diagram represents all possible values for α-C to carbonyl C (Psi) and α-C to amide N (Phi) bond angles as −180 to +180°. The plot shown is for aquaporin 1 (structure as inset) which is exclusively α-helical. The region where β-sheet would be found if present is shown

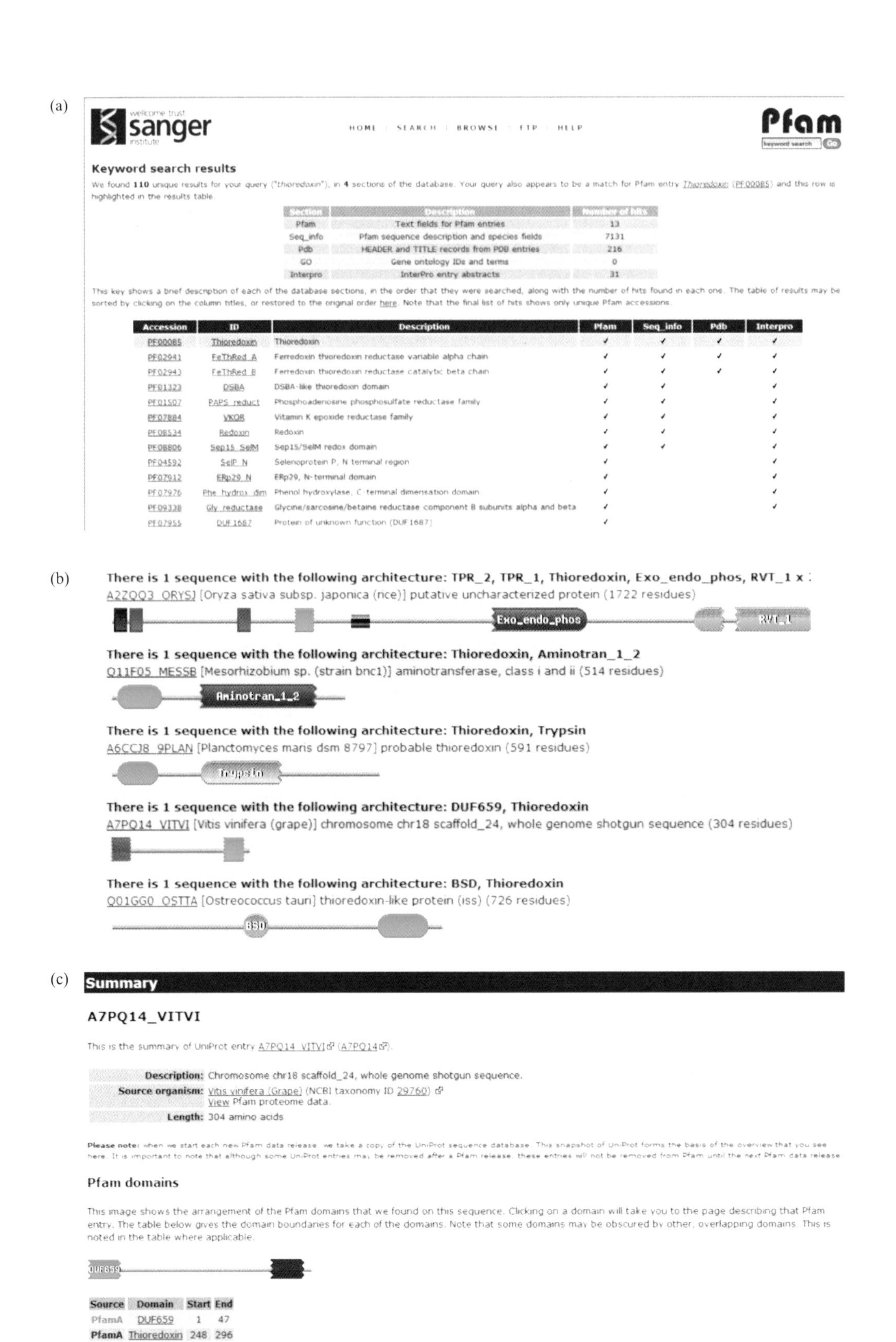

FIGURE 9.14. The Pfam database. (a) The screen generated when the search-term 'thioredoxin' is entered. Searches may also be made by sequence. (b) Detail of listing of thioredoxin family members. Note how easy it is to visualize shared domains. (c) Detail of a single entry in listing. This specifies exactly where domains begin and end.

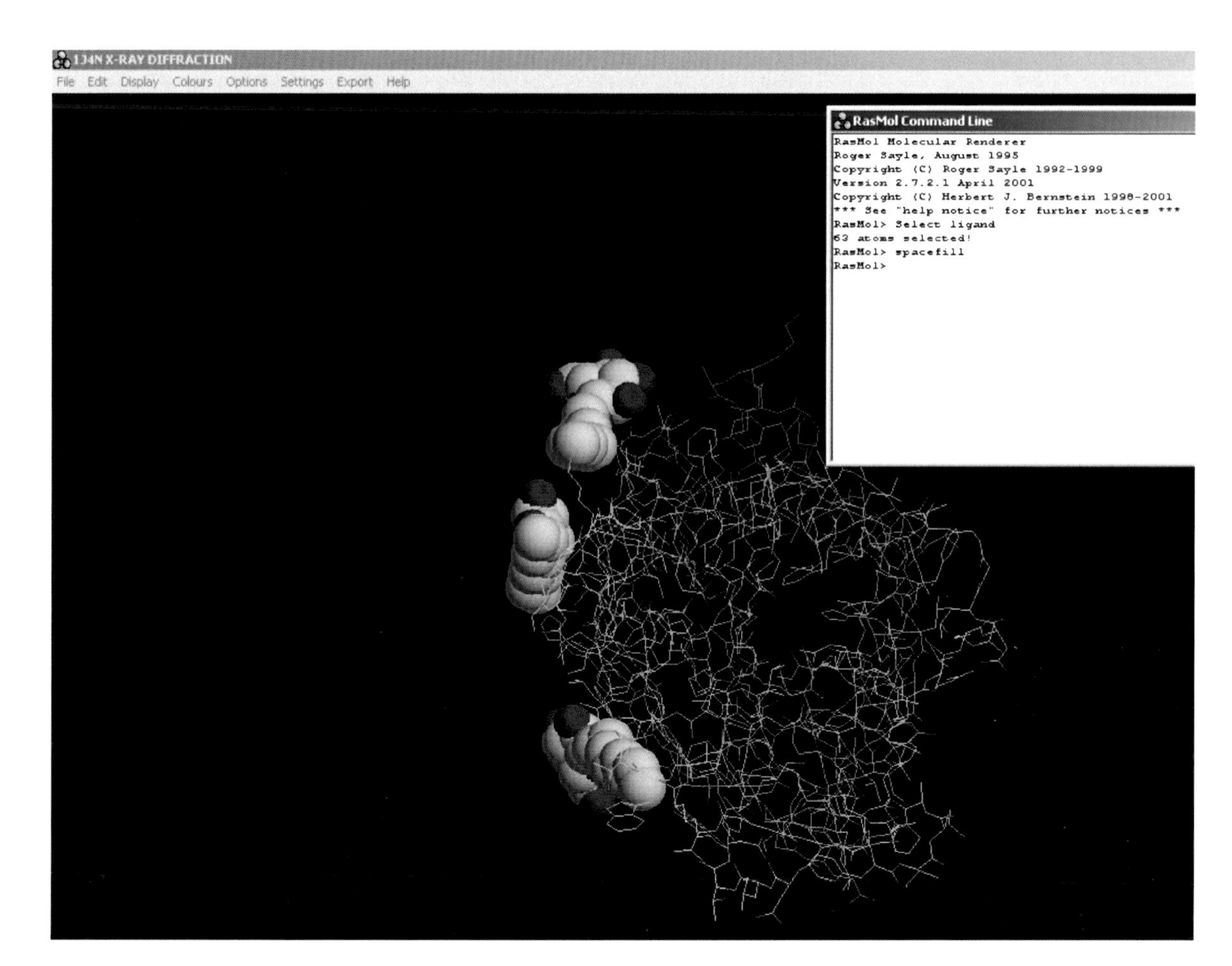

FIGURE 9.24. Viewing structures with RasMol. The PDB file 1J4N for aquaporin 1 was opened as a default wireframe model. The commands 'select ligand' and 'spacefill' were typed into the command line to reveal the location of 3 molecules of β-nonylglucoside with which the protein was cocrystallized.

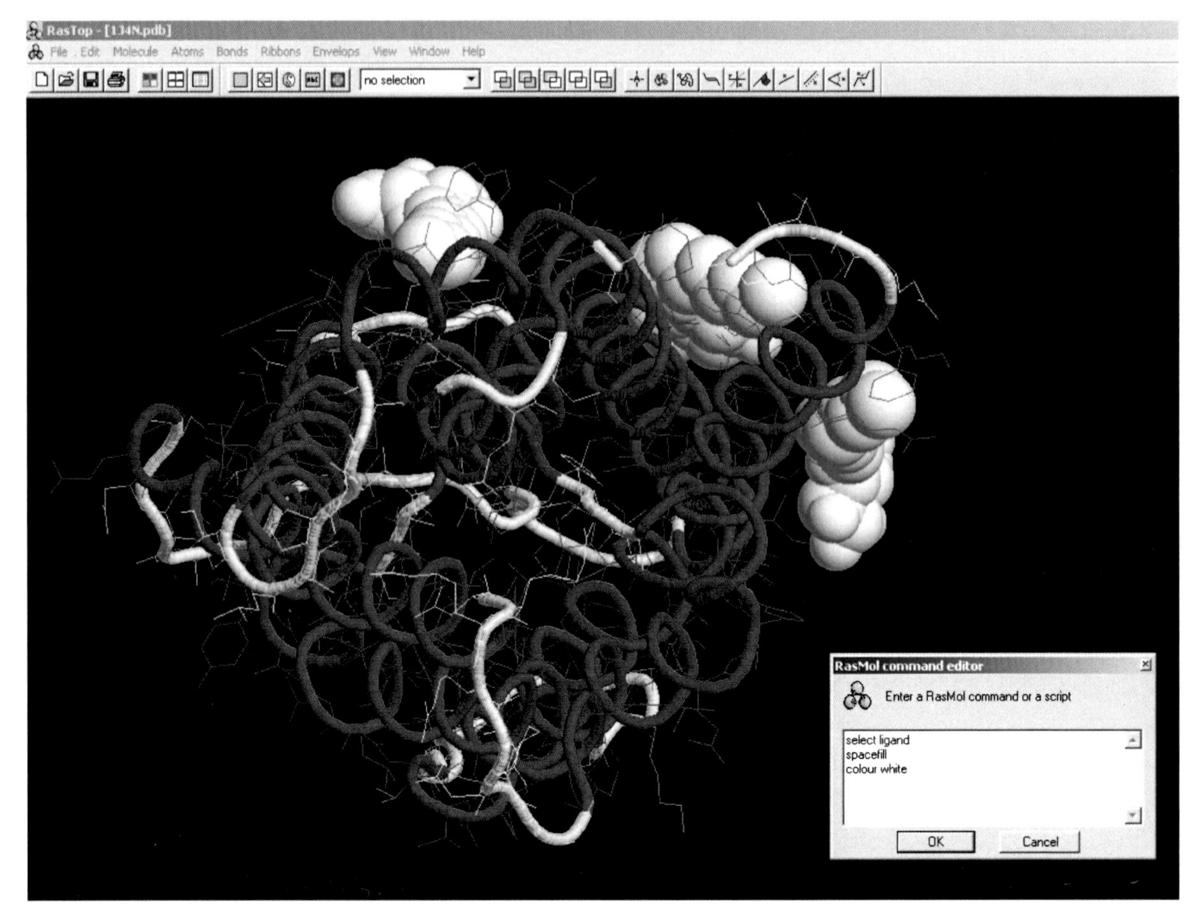

FIGURE 9.25. Viewing structures with RasTOP. The file 1J4N was opened and represented in the *colour* pulldown as 'structure' and in the *ribbon* pulldown as 'trace' .The command box was used to apply the RasMol commands: 'select ligand, spacefill, colour white' to reveal β-nonylglucosides.

(a)

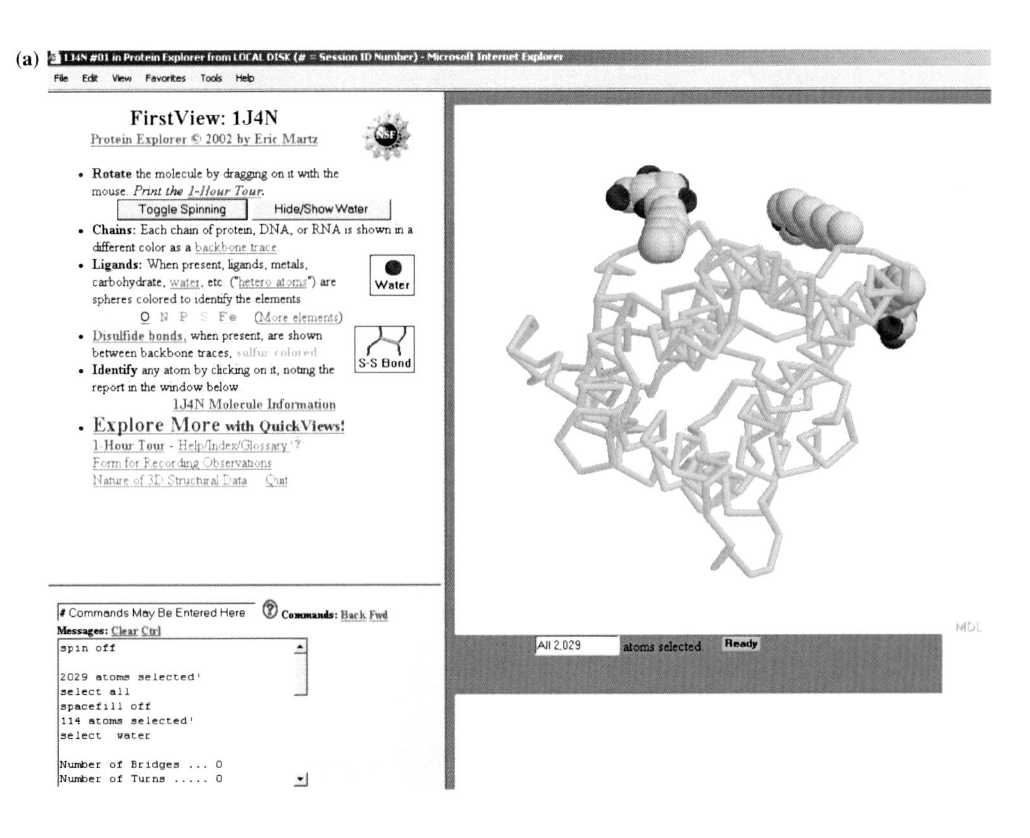

(b)

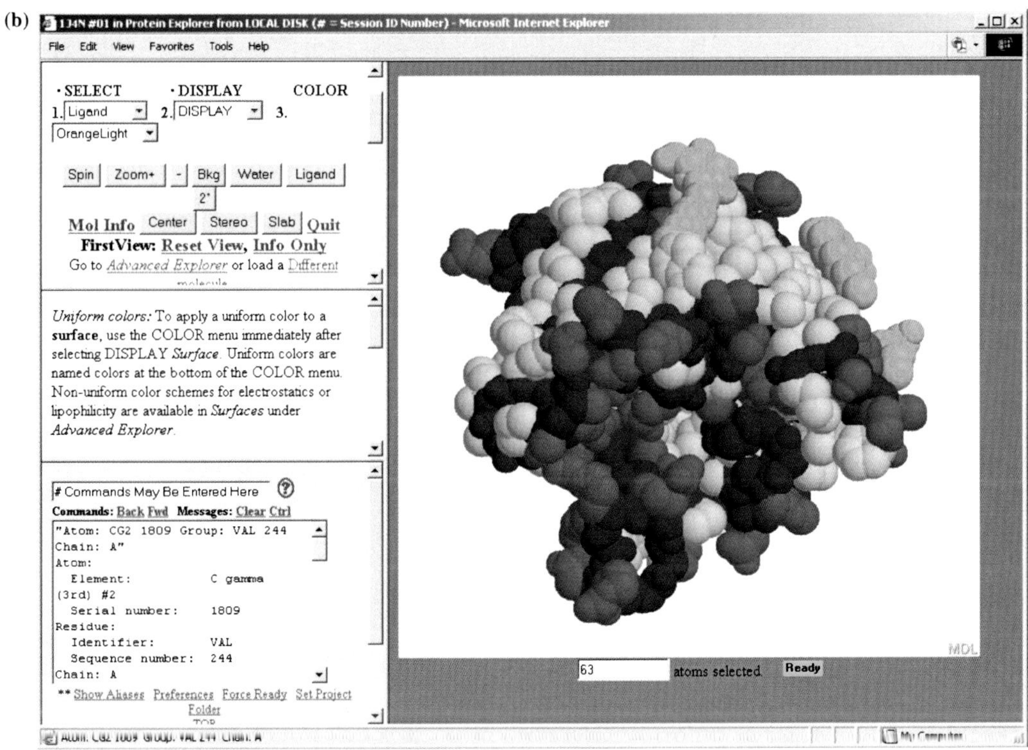

FIGURE 9.26. Viewing structures with Protein Explorer. (a) PDB file 1J4N was opened with Protein Explorer and *hide water* and *toggle spinning off* were selected. RasMol commands can be typed in the *command* box on lower left of screen. (b) Selecting *explore more* allows us to modify the image with three pulldowns. Selections are *spacefill* in *Display* pulldown and *polarity 3* in *Colour* pulldown. Note that hydrophobic residues are lighter-coloured than hydrophobic in this representation.

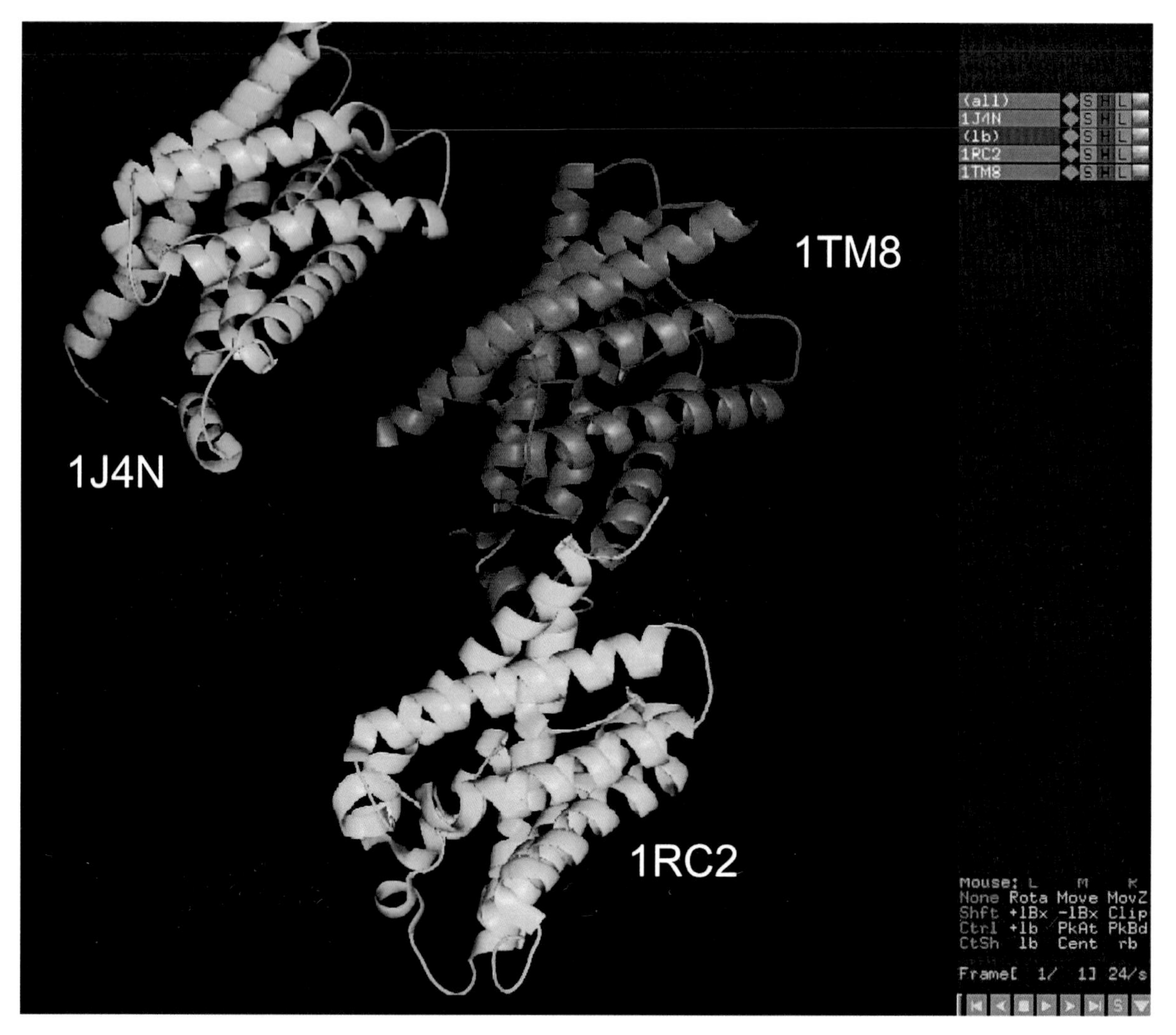

FIGURE 9.27. Viewing structures with PyMOL. The structures of human aquaporin 1 (1J4N), an *Escherichia coli* nodulin-like intrinsic protein (1RC2) and an aquaporin found as a major intrinsic protein in bovine lens fibres (1TM8) are shown in a single window. These are all members of the aquaporin family.

(a)

ESyPred3D Web Server 1.0

Christophe Lambert and Eric Depiereux

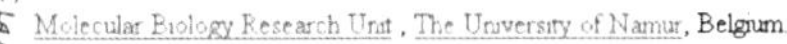

Molecular Biology Research Unit, The University of Namur, Belgium.

Project supported by the F.R.I.A.

ESyPred3D - ESyPred3D is a new automated homology modeling program. The method gets benefit of the increased alignment performances of a new alignment strategy using neural networks. Alignments are obtained by combining, weighting and screening the results of several multiple alignment programs. The final three dimensional structure is built using the modeling package MODELLEI

Please cite:

Lambert C, Leonard N, De Bolle X, Depiereux E
ESyPred3D: Prediction of proteins 3D structures.
Bioinformatics. 2002 Sep;18(9):1250-1256

Data and parameters input for the ESyPred3D server:

Your email address

Description (optional)

Browse... **You can upload a sequence file**

(Characters others than the standard 20 amino acids codes will be removed)

```
masefkkklf wravvaefla milfvfisig salgfnypvr nnqtagaaqd
nvkvslafglsiatlaqsvg hisgahlnpa vtlglllscq isilravmyi
iaqcvgaiva tailsgitsslpdnslgrne lapgvnsgqg lgieiigtlq
lvlcvlattd rrrrdlggsg plaiglsval ghllaidytg cginparsfg
ssvithnfkd hwifwvgpfi ggalavliyd filaprssdl
tdrvkvwtsg qveeyeldgd dinsrvemkp k
```

or paste your sequence

(b)

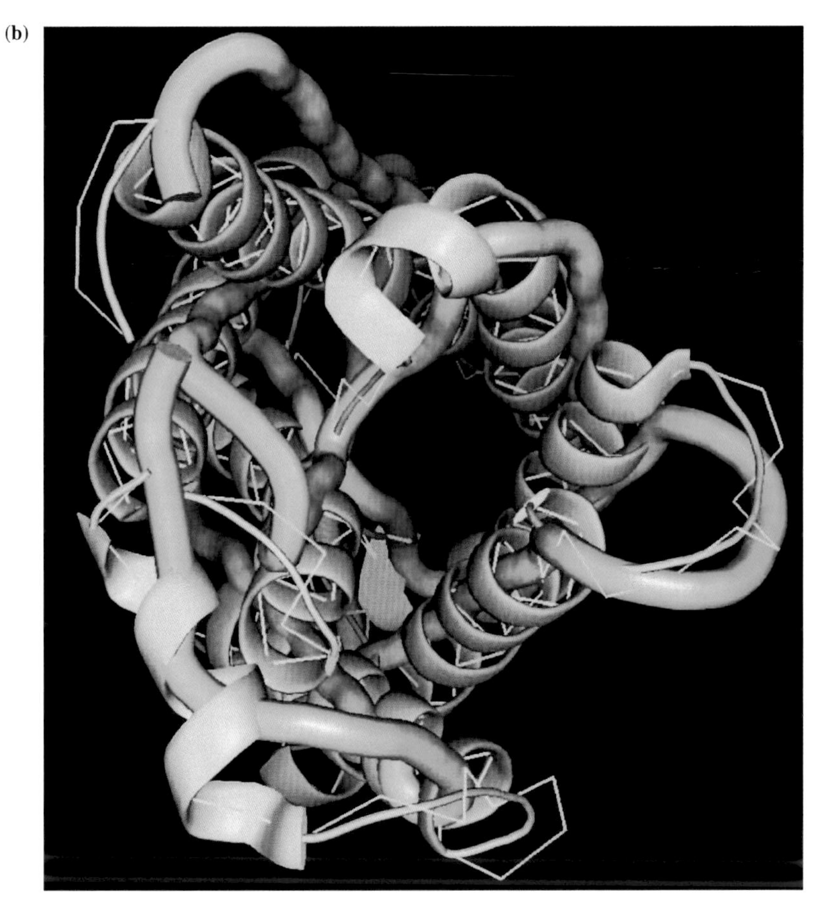

FIGURE 9.31. On-line homology modelling with ESyPred3D. (a) The sequence from NCBI of canine aquaporin is entered in ESyPred3D. (b) The model returned by e-mail is viewed with PyMol together with the structure which gave greatest sequence alignment (94%), human aquaporin 1. The model is represented as a solid cartoon while the superimposed 1J4N is in *ribbon*. Note that the structures deviate in some places while they overlay closely in others.

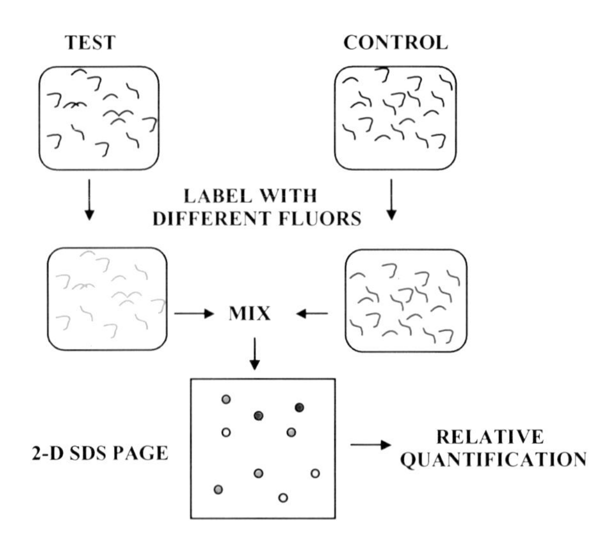

FIGURE 10.5. Difference gel electrophoresis (DIGE). Two samples (test and control) to be compared are labelled separately with distinct fluorescent dyes (e.g. green and red). They are then mixed and separated on a single 2-D SDS PAGE gel. Intensity of spots shows if they are equal in both samples (yellow), up-regulated (green) or down-regulated (red).

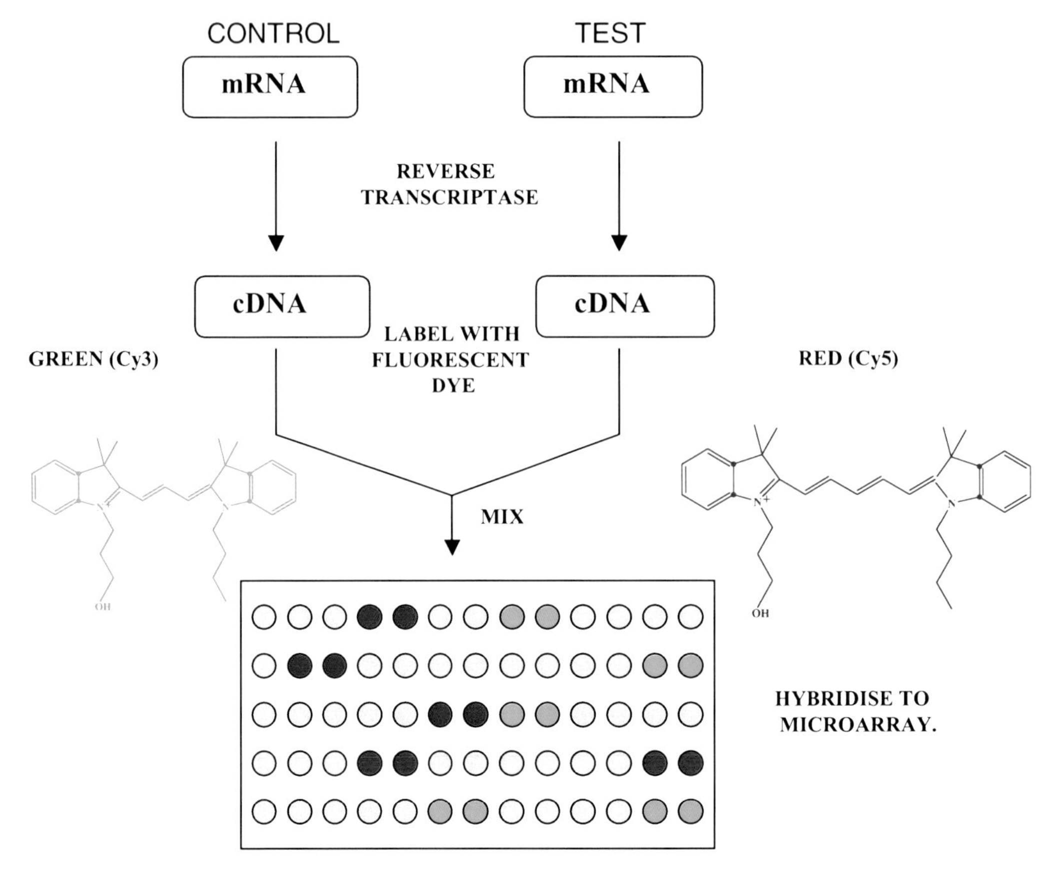

FIGURE 10.16. Detecting gene expression with DNA Microarrays. mRNA is isolated from matched test and control cells and converted to cDNA by reverse transcriptase. Each cDNA population is tagged with a fluorescent dye (red for test, green for control) and washed over the microarray. cDNA hybridises to its complementary gene. Un-hybridized cDNA is removed by washing. Scanning the microarray with a laser reveals up-regulated (green), down-regulated (red) and equal gene expression (yellow) at each spot of the microarray. Spots are in duplicate and the array contains both positive and negative controls.

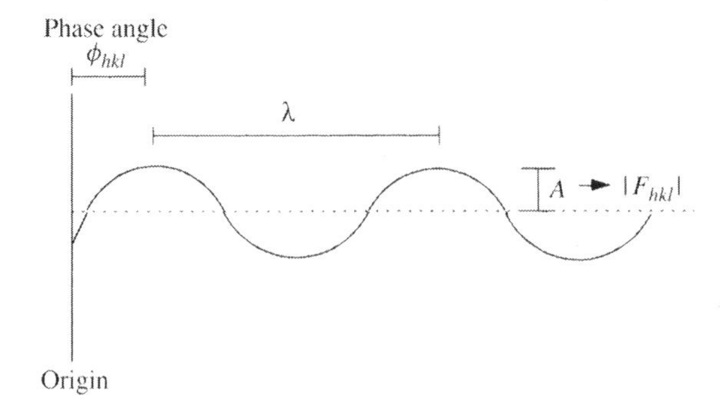

Figure 6.46. Representation of diffracted X-ray beam. The wave function is defined by wavelength (λ) and amplitude (A) but is related to a point of origin by the phase angle (ϕ_{hkl}). The amplitude is related to the square root of the intensity (I_{hkl}) for each Bragg reflection and directly to the structure factor, $\mathbf{F_{hkl}}$.

described in Section 6.4.2) and converted using the following relationship:

$$|F_{hkl}|^2 = \frac{K I_o}{LP} \quad (6.15)$$

where K is a *scaling factor* dependent on properties such as beam intensity, crystal size and other fundamental constants. L is the *Lorentz factor* which corrects for the geometry of the detection system. P is a *polarization factor*, which corrects for differences experienced by electric components of the X-ray wave parallel to the lattice plane (hkl) compared to those perpendicular to the plane (this phenomenon was previously encountered in our description of *plane polarized light*; Section 3.5).

In practice, an observed estimate of the structure factor, $F_{hkl\ \mathrm{obs}}$, for *each* Bragg reflection is first made. Once a model of the structure has been proposed, this is then employed through many iterations of a series of computer programs in a process of *refinement* (Section 6.5.8) each of which generates a calculated estimate of the structure factor, $F_{hkl\ \mathrm{calc}}$, based on the current model. The discrepancy between these two, Δ_{Fhkl}, is given by;

$$\Delta F_{hkl} = F_{hkl\ \mathrm{obs}} - F_{hkl\ \mathrm{calc}} \quad (6.16)$$

A quantity called the *R factor* represents a summation of all the structure factors for the proposed structure giving an indication of how acceptable it is:

$$R = \frac{\sum \Delta |F_{hkl}|}{\sum \Delta |F_{hkl}|_{\mathrm{obs}}} \quad (6.17)$$

The R-factor is an important criterion for the quality of the final structure and allows comparisons of goodness-of-fit between crystal structures obtained for different molecules or different crystal-forms of a single sample molecule. It is usually in the range 3–5% for small molecules but may be as high as 10–20% for protein structures.

6.5.2 Information Available from the Overall Diffraction Pattern

A number of useful and important pieces of information can be gleaned from the symmetry and relative intensities of spots observed within the overall pattern of diffraction. This is analysed by a computer program which gives an estimate of the lattice constants and the space group of the structure. We can estimate from this how many asymmetric units are in the unit cell and the volume of each. From this information in turn, we can usually obtain the number of sample molecules in the asymmetric unit. The ratio of volume of asymmetric unit: molecular weight contained in it is known as V_{m} and, for protein samples, this ratio is usually in the range 1.8–3.0 A^3/Da. Assuming protein crystals have a density consistent with 40–60% solvent content, an integral number of molecules per asymmetric unit usually results from this calculation. If the precise number of molecules per asymmetric unit is still ambiguous however, the density of the crystal can be directly measured in a *density gradient* (Chapter 7) although this is technically difficult for protein crystals because of their mechanical fragility.

Because of the greater number of calculations required with each extra biomacromolecule per unit cell it is desirable that this number is as small as possible for rapid structure determination (e.g. 1–4). Occasionally, high-quality diffracting crystals may contain 6–8 protein molecules per unit cell making calculation of a final structure almost impossible with current computational facilities. In such cases, rescreening may be necessary to obtain crystals of the same molecule with a smaller number of molecules per unit cell.

6.5.3 The Phase Problem

In Chapter 3 (Section 3.1.1) we saw how a wave function is defined by two terms; the amplitude and wavelength. Because in X-ray crystallography we are interested in the three-dimensional location from which a diffracted beam *originates* (i.e. the location of the diffracting atom) we need a third term fully to describe the diffracted X-ray wave called the *phase angle*, ϕ_{hkl} (Figure 6.46). This is defined relative to an origin (i.e. the point where $a = b = c = 0$) which, for a complete data-set, is of necessity chosen arbitrarily. Because each Bragg reflection results from a combination of wavelets scattered by all the atoms in the molecule, different reflections have different phase angles which must be determined in order for an electron density map to be calculated.

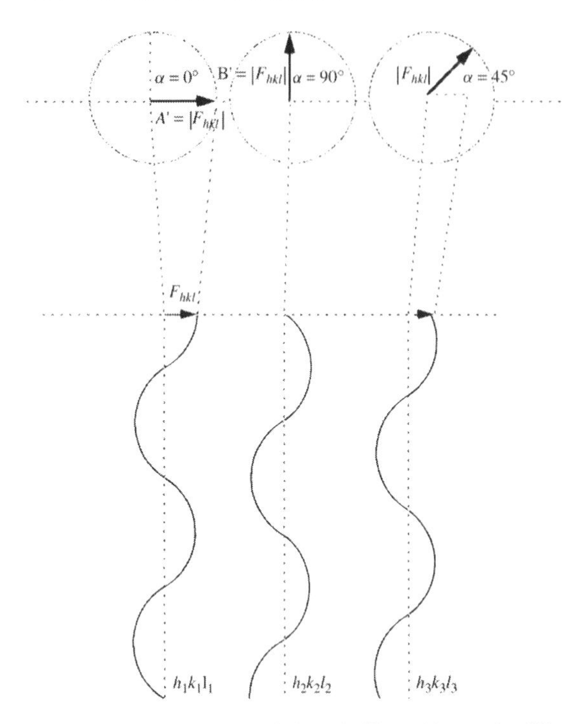

Figure 6.47. Geometric description of diffracted beams for different hkl. Three beams are shown which are out of phase and have different phase angles. The structure factor F_{hkl} is a resultant vector between the cosine and sine vectors $\boldsymbol{A}'$ and $\boldsymbol{B}'$. The angle formed between $|F_{hkl}|$ and the horizontal is α_{hkl}. The relationship between the wave description (below) and geometric description (above) is shown.

An alternative way to represent the wave function is as shown in Figure 6.47. The wave is represented as a line the length of which is proportional to the amplitude and equal to $|F_{hkl}|$. A circle with a radius equal to $|F_{hkl}|$ represents one complete wavelength. The angle formed by the line with the horizontal is called the *relative phase angle*, α_{hkl}. It is clear from this representation that $|F_{hkl}|$ can be considered as a resultant vector between two vectors A'_{hkl} and B'_{hkl} representing *cosine* and *sine* waves, respectively. The phase angle is equal to the ratio of these vectors:

$$\phi = \tan \alpha_{hkl} = B'_{hkl}/A'_{hkl} \tag{6.18}$$

The intensity is also geometrically related to their magnitude as

$$I_{hkl} = |A'_{hkl}|^2 + |B'_{hkl}|^2 = |F_{hkl}|^2 \tag{6.19}$$

While structure factors F_{hkl} may directly be obtained from the intensities of Bragg reflections detected in the diffraction pattern, we cannot obtain phase angles *directly* from such reflections. In other words, the diffraction data-set from a crystal contains the amplitude information but *has lost phase information*. Since we need to know *both* the phase and the amplitude of each Bragg reflection to calculate the electron density map this lack of phase information provides a barrier to structure calculation called the *phase problem*. A number of strategies have been developed for the recovery of phase information which we now describe in Sections 6.5.4 to 6.5.6.

6.5.4 Isomorphous Replacement

This technique was developed for proteins by Kendrew (working on myoglobin in 1960) and Perutz (working on haemoglobin in 1968). It involves the introduction of a heavy metal atom such as mercury, platinum, uranium, and so on at a *unique location* within the molecule called *single isomorphous replacement*. This may be achieved by soaking a crystal with a solution of a reactive chemical containing the heavy metal which then attaches to the biomacromolecule (Figure 6.48). It is important that this reaction does not result in an altered structure, crystal packing or unit cell size (hence the term *isomorphous* meaning having the same

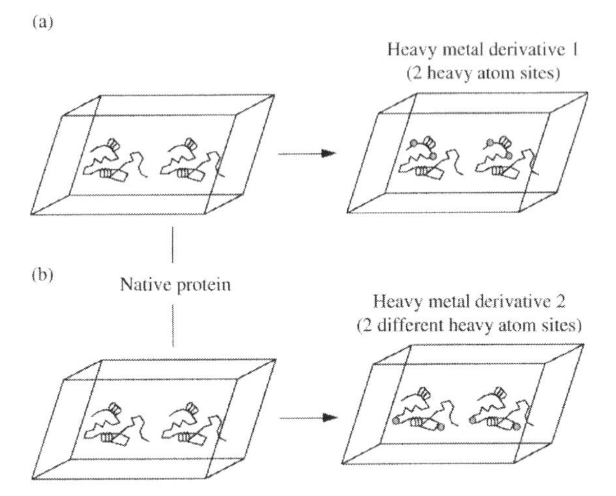

Figure 6.48. Isomorphous replacement in proteins. A protein crystal containing two molecules per unit cell is shown. Isomorphous replacement involves soaking these crystals in a solution of a reagent containing a heavy metal which attaches to the protein. (a) A heavy metal or its chemical derivative is introduced in the protein. Several sites can be occupied as shown. (b) Heavy metal atoms or their chemical derivatives are introduced in a second experiment. The sites occupied should be different from those in (a). If only one heavy metal derivative is available (e.g. (a) or (b)) the method of *single isomorphous replacement* is used. If two or more derivatives are available, *multiple isomorphous replacement* is used. The latter gives better phasing.

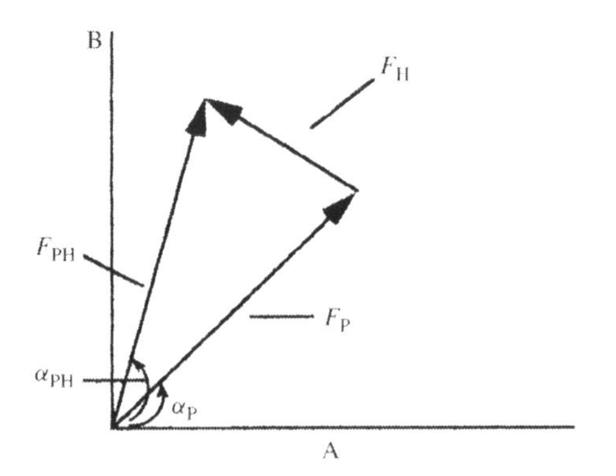

Figure 6.49. Single isomorphous replacement of a protein. The structure factor vectors for a native protein (F_P), heavy-metal modified protein (F_{PH}) and heavy metal alone (F_H) are shown together with the corresponding relative phase angles (α_P and α_{PH}).

shape). X-rays are diffracted through this crystal and, since a heavy metal contains many electrons and therefore would be expected to have high scattering factors (see above), the location and phase of Bragg reflections from metal atoms can be identified by comparison with a data-set from the crystal lacking heavy metal.

Figure 6.49 illustrates the relationship between structure factors for native protein ($F_{hkl\,P}$), heavy-metal modified protein ($F_{hkl\,PH}$) and heavy metal alone ($F_{hkl\,H}$) to each other as vector quantities:

$$F_{hkl\,PH} = F_{hkl\,P} + F_{hkl\,H} \tag{6.20}$$

Each of these structure factors has a relative phase angle (α_{hkl}) associated with it. Provided that we know the three-dimensional locations of the heavy metal (i.e. both $\alpha_{hkl\,H}$ and $|F_{hkl}|_H$), the relative phase angle for the protein and both $|F_{hkl}|_P$ and $|F_{hkl}|_{PH}$ (which are experimentally measurable from diffraction intensities), $\alpha_{hkl\,P}$, can be calculated from the following equation:

$$\alpha_{hkl\,P} = \alpha_{hkl\,H} \pm \cos^{-1}[(|F_{hkl}|_{PH^2} - |F_{hkl}|_P - |F_{hkl}|_H)/ 2|F_{hkl}|_P|F_{hkl}|_H] \tag{6.21}$$

The location of the heavy metal is arrived at using *difference Patterson maps*. These are created using a *Fourier synthesis* (Appendix 2) denoted $\Delta P(uvw)$, using the Miller indices (hkl) and the observed structure factor amplitude differences $\Delta|F_{hkl}|_H = |F_{hkl}|_{PH} - |F_{hkl}|_P$;

$$\Delta P(uvw) = \frac{1}{V}\sum\sum\sum \Delta|F_{hkl}|_{H^2} \times \cos 2\pi(hu + kv + lw) \tag{6.22}$$

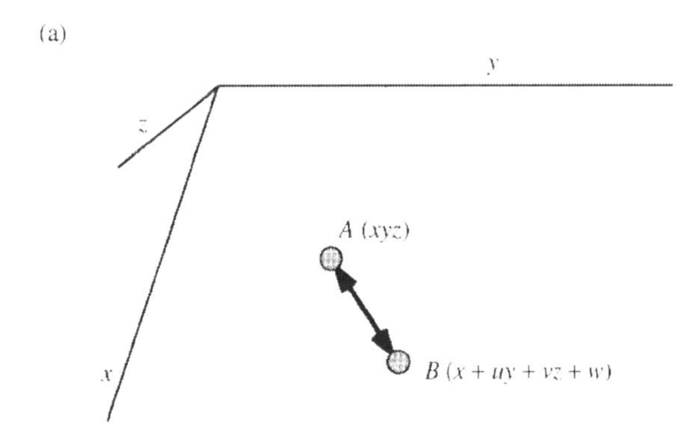

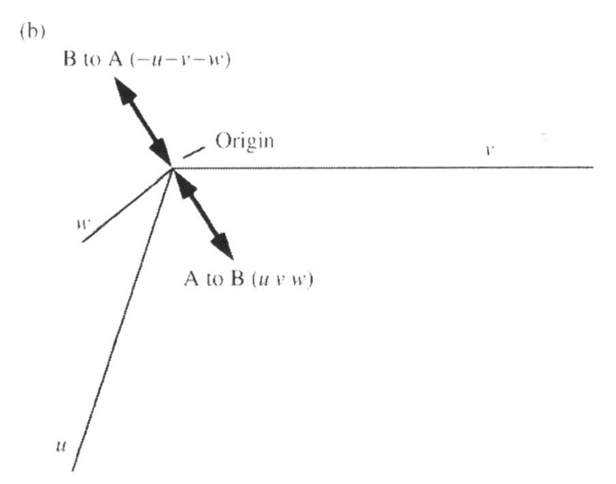

Figure 6.50. Patterson maps. (a) The relative locations of two atoms, A and B are represented in terms of their cartesian coordinates (xyz). (b) A Patterson map represents their relative location in terms of vectors with the location of one atom of the pair placed at the origin. Note that two vectors are possible (uvw) and ($-u - v - w$).

We shall see later that the form of this equation is very similar to that used for calculation of electron density maps (Equation (6.29) below) with the exception that *it does not include* a term for relative phase angle (α_{hkl}).

If two atoms in a unit cell are separated by a vector (uvw), there will be a peak in the Patterson map at (uvw). In terms of real space this represents a vector between two atoms, one at (xyz) and the other at ($x + u$, $y + v$, $z + w$) (Figure 6.50). Patterson maps represent one end of the vector at the origin and the other at (uvw). We can therefore measure the *magnitudes* of u, v and w directly from Patterson maps although we do not know x, y or z. The height of the peak is approximately proportional to the product of the atomic numbers of the two atoms. Thus, the Patterson map of a protein modified with a heavy metal is dominated by vectors between the metal and other atoms.

A crystal containing a single heavy metal in each unit cell which diffracts X-rays systematically will generate a Patterson map dominated by peaks representing vectors to and from the heavy metal atom. However, as shown in Figure 6.50, any pair of atoms yields two such vectors; (uvw) and $(-u-v-w)$. In the case of a protein, this means that Patterson maps of a protein can generate two relative phase angles, $\alpha_{hkl\,PA}$ and $\alpha_{hkl\,PB}$ as shown in Figure 6.51. One of these phase angles is true and the other imaginary. In order to distinguish which of the pair is true, it is necessary to include a second (and sometimes a third) heavy metal at another unique location in the unit cell. The effect of this may be seen using a geometrical construction called the *Harker construction* (Figure 6.51). This experiment is called *multiple* isomorphous replacement. It results in a single value for the term α_{hklP} (Equation 6.21), the relative phase angle, which is included in the equation for calculation of the electron density map (Section 6.5.7).

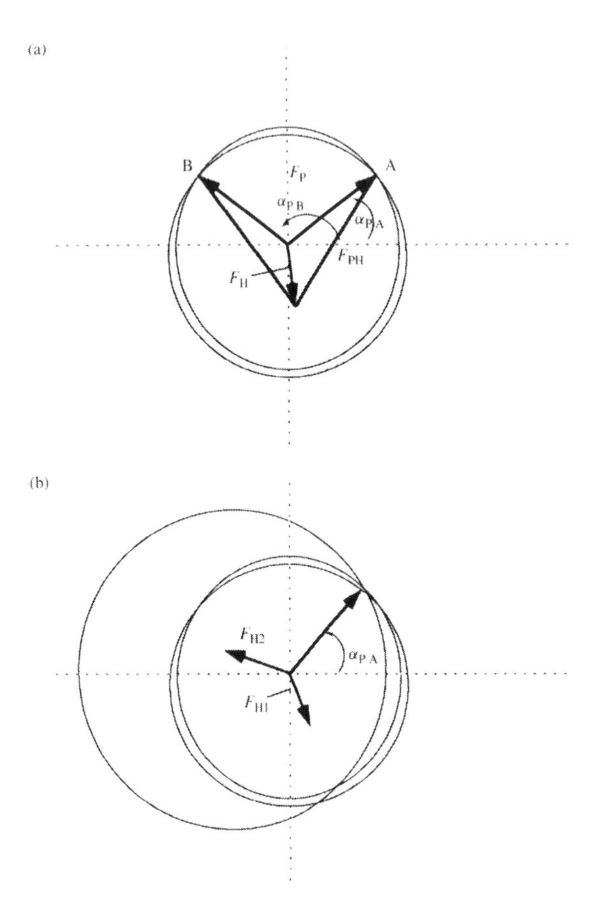

Figure 6.51. The Harker construction. (a) Single isomorphous replacement is represented as two circles, one with a radius of the structure factor for the heavy metal-derived protein (F_{PH}) and the other with a radius of that of the native protein (F_P). Note that two possible relative phase angles are possible (α_{PA} and α_{PB}). (b) This ambiguity may be resolved by using a second heavy metal derivative which will generate a third circle of radius F_{PH2}. The point where all three circles intersect represents the true F_P and hence identifies α_{PA} as the true α_P for the protein data-set.

6.5.5 Molecular Replacement

Sometimes the structure of a protein or DNA molecule similar to that under investigation will have been previously determined by X-ray diffraction. Such cases arise, for example with mutants/wild-type proteins generated by SDM experiments, families of structurally related isoenzymes, proteins cocrystallized with different ligands and functionally related protein families. Even where identity between pairs of proteins is limited or localized, *molecular replacement* can be carried out on fragments of their structures providing an alternative means to solve the phase problem.

Patterson maps contain two distinct types of vectors. *Self vectors* are vectors between pairs of atoms within each individual molecule of protein or DNA while *cross vectors* are those connecting pairs of atoms in different molecules in the unit cell. In general, these two groups are readily distinguished from each other because self vectors are *significantly shorter* than cross vectors. If vector maps are calculated for a pair of molecules for which the structure of one is known (the *model*), then the set of self vectors will be identical in the vector map of the unknown molecule (the *target*) compared to that of the model. Usually, this relationship involves rotation around a set of three angles with each individual rotation creating a new set of (x, y, z) axes. Eventually, these rotations bring many of the peaks in one Patterson map into agreement with those of the second. A faster way to do this is to rotate the Fourier transform of the model structure relative to the observed structure factors of the data-set from the target. The mathematical tools to carry out these operations are known as *rotation functions*. Their effect is to orientate the two models correctly relative to each other.

The second stage in molecular replacement is to position the fragment with respect to the crystallographic axes and the origin of the unit cell. This is achieved with a *translation function* which moves the Patterson map for the target relative to that for the model by translation but without changing their relative orientation. As with the rotation function, the translation function is defined by three translations in the three dimensions. In all, six parameters therefore define molecular replacement. The best translation function is identified by calculation of structure factors at various points in the asymmetric unit and comparing these with observed values using the R factor described in Section 6.5.1 (Equation (6.17)).

Once preliminary estimates of the relative phases are available for the target and model structures, these can be improved using standard refinement procedures. If noncrystallographic symmetry is present in the crystal, that is symmetry not described by the crystal space group, this can help or influence the refinement procedure For example, two subunits of a protein may be related by a symmetry which is not crystallographic. If a phase has been determined for two molecules related by noncrystallographic symmetry within the space group, then their electron density maps will also be symmetrically related. Any differences in electron density between them may be due to an inaccurate phase determination. If the electron density associated with the two molecules is *averaged*, however, then this must result in a more accurate set of phases. This process is known as *real space averaging*. A similar approach can be used to relate electron densities calculated for the same molecule crystallized in two different crystalline forms. Provided that they can be related by a translation function, their electron densities can similarly be averaged and the phase improved. This approach has particular applications for very large structures such as viruses where extensive noncrystallographic symmetry exists and many averaging steps can be gone through to improve phase determination.

6.5.6 Anomalous Scattering

The physical basis of anomalous scattering of X-rays has already been described in Section 6.4.2 (Figure 6.39). If the frequency of incident X-rays corresponds to a naturally occurring frequency of oscillation in the electrons of a diffracting atom, then anomalous diffraction can occur. In practice, most atoms in crystals of biomacromolecules will not undergo anomalous diffraction with fixed wavelength K_α X-rays. However, some X-ray sources (e.g. synchrotron sources; Section 6.5.9. below) are *tunable* in that they can produce X-rays across a range of wavelengths and hence frequencies (although see Example 6.5 for a case where this was achieved with a rotating anode source). Such sources can be used in conjunction with a very limited range of atoms which can undergo anomalous diffraction with wavelengths in the range 0.5–3.0 Å. Some of these are biologically occurring atoms such as Fe and Ca but most are heavy metals in the atomic number ranges 20 (Ca) to 47 (Ag) and 50 (Sn) to 92 (U) which can be introduced into the crystal by soaking. Multiple wavelength anomalous scattering can therefore be regarded as analogous to isomorphous replacement in that one or a small number of atoms are introduced into the crystallized molecule to allow determination of relative phase angles. A popular method is to replace a residue in the protein of interest with an analogue containing a heavy metal. Good examples of this are replacement of (in proteins) methionine by selenomethionine (in which Se replaces S) and (in DNA) replacement of cytosine by 1-iodocytosine or of thymine by 5-iodouracil (Examples 6.4 and 6.5). This may be achieved by replacing methionine with selenomethionine in the media on which bacteria are grown to produce the protein or, in the case of DNA, by introducing the iodonucleotide in one step of the oligonucleotide synthesis cycle. Each of the metals capable of anomalous scattering has particular advantages and limitations. Among the most important of these is the protein relative mass range in which they are useful. Most are limited to 10–14 kDa while Hg is useful up to 77 kDa.

The rather narrow range of X-ray frequencies currently possible is principally responsible for the limited number of atoms capable of anomalous diffraction. Combined with the M_r limits of the technique, this means that anomalous diffraction is used less frequently than either isomorphous or molecular replacement as a means of determining relative phase angles.

The scattering factor (f_{anom}) for an atom scattering X-ray radiation anomalously is more complex than that for coherent scattering previously described in Section 6.4.2:

$$f_{anom} = f + f' + if'' \quad (6.23)$$

where f' and f'' vary with the wavelength of the incident radiation. f is the *unperturbed scattering factor* and is, conversely, largely independent of wavelength.

A major difference between anomalous and coherent scattering is the phase change between incident and scattered X-rays. In the case of coherent scattering, this is 180° (Figure 6.39) which is the reason that diffraction patterns are *centrosymmetric*. This means that for every Bragg reflection (hkl) there is an equally intense reflection ($-h-k-l$). This is formalized in *Friedel's law* which equates the intensity (I) of this pair of Bragg reflections sometimes referred to as *Friedel mates*:

$$I_{hkl} = I_{-h-k-l} \quad (6.24)$$

By contrast, anomalous scattering results in the diffracted radiation from each Friedel mate *having a different phase* with the result that Friedel's law does not hold for anomalous diffraction (Figure 6.52). This is due to the term f'' in Equation (6.23) which affects the structure factor of each Friedel mate differently (Figure 6.53).

Because anomalous diffraction is dependent on wavelength, we can define a wavelength-dependent structure factor ${}^{\lambda}F_{hkl}$. This receives contributions from both coherent and anomalously scattering atoms and thus is composed of two vector components, ${}^{\circ}F_{hkl\,\mathrm{T}}$ (coherent scattering) and ${}^{\circ}F_{hkl\,\Lambda}$ (anomalous scattering) each of which has a

Example 6.4 X-ray Crystallographic Determination of Structure of a Protein–RNA Complex

Ribosomes are complex macromolecular structures which involve extensive interactions between a number of individual proteins and rRNA. A functionally critical part of rRNA which is heavily conserved consists of a 58-nucleotide sequence from 23S rRNA in *E. coli* which interacts structurally with a ribosomal protein L11. This is the site at which elongation factors EF-Tu and EF-G bind and it is a known target for antibiotics which confirms its functional importance.

In order to know more about this part of the ribosome, a complex consisting of the RNA-binding domain (76 residues) of L11 with a rRNA fragment containing the 58 nucleotide fragment was crystallized and its structure determined by X-ray crystallography using anomalous diffraction to solve the phase problem. For this purpose selenomethionine was incorporated into L11 in place of methionine by growing *E. coli* on media containing selenomethionine. Equimolar amounts of rRNA and L11 were used in crystallization trials.

Cubic crystals of both native and selenomethionyl complexes (dimensions; 0.1 × 0.1 × 0.2 mm) were grown successfully by vapour diffusion with the sitting drop method (Section 6.3.4) in 50 mM sodium cacodylate (pH 6.5), 15% polyethylene glycol 600, 80 mM magnesium acetate, 100 mM potassium chloride and 0.2 mM $Co(NH_3)Cl_3$ at 37 °C. The unit cell dimensions were a = b = 150.68 Å, c = 63.84 Å and the asymmetric unit contained two complexes. Diffraction data were collected from the synchrotron source at Brookhaven National Laboratory, Brookhaven, New York.

The data collected is summarized in the table shown in (1) below. Note that a number of different X-ray wavelengths were used in collecting data from the selenomethionyl crystals. The number of unique reflections is always less than the total because of symmetry in the crystal. Approximately 70 000–080 000 individual reflections were collected in each data set of which approximately 10 500–517 500 were unique.

Refinement produced an overall *R* factor of 24% for all reflections in the resolution range 8–2.8 Å and a model was built into the electron density map which is shown in (2).

(1)

Data-sets	Native	Selenomethionyl L11-rRNA complex			
1 (Å)	1.13951	0.97914	0.97871	0.96675	0.98557
Range of resolution (Å)	20–2.8	10–3.2	10–3.2	10–3.2	10–3.2
Measurements	89 078	72 499	72 879	72 186	72 579
Unique reflections	17 542	10 481	10 480	10 509	10 437
Completeness	94.5	88.3	88.3	88.7	88.2
R	24%				

Data collected from diffraction of L11-rRNA complex crystals.

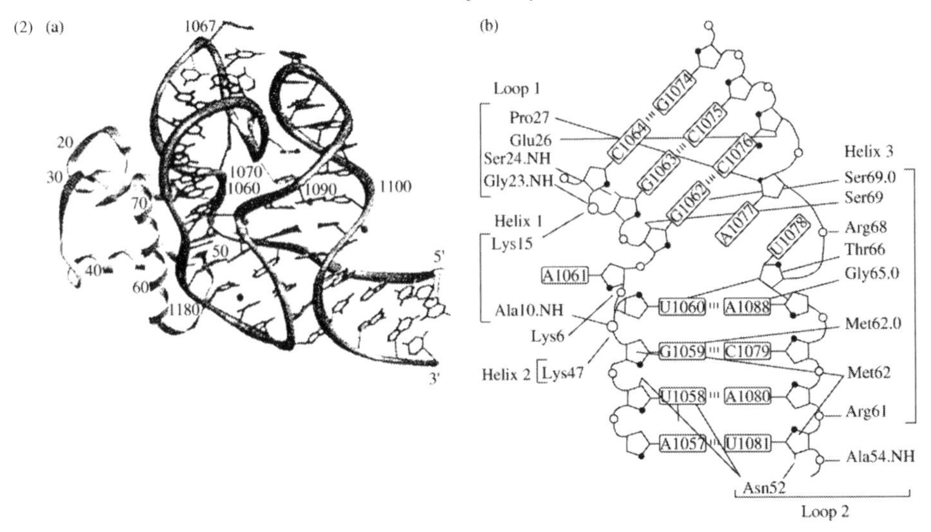

(a) Structure of L11-rRNA complex. (b) Schematic diagram showing the interactions between individual amino acid residues and individual bases

(Continues)

(Continued)

The structure determined for the complex is consistent with previous studies using site-directed mutagenesis, sequence alignments between different species and NMR. It shows a number of points where individual amino acid residues contact the rRNA fragment and explains the stabilization effect of L11 on the 58 nucleotide fragment of rRNA. A rRNA hairpin bulges into the major groove of another rRNA helix and this interaction is locked in place by a direct contact of L11 with A1088.

Reprinted with permission from Conn *et al.*, Crystal Structure of a Conserved Ribosomal Protein-RNA Complex. *Science* **284**, 1171–1174 (1999), American Association for the Advancement of Science

Example 6.5 Determination of the Structure of a Binding Domain of the Human Editing Enzyme ADAR1 Bound to DNA

Almost three decades after the structure of right-handed double helix B-DNA was elucidated by Watson and Crick, a separate left-handed form of DNA called Z-DNA was discovered which was of unknown biological function. Unlike the smooth helix traced by the phosphates of B-DNA which form relatively shallow major and minor grooves, the phosphates of Z-DNA trace a jagged helix which forms a single deep groove extending to the axis of the helix.

An enzyme called RNA adenosine deaminase (ADAR1) was found to bind to Z-DNA quite strongly and specifically by band-shift assays (Chapter 5). This binding activity was localized in a specific portion of the protein (the Z_α domain). Z-DNA is formed transiently in the wake of a moving RNA polymerase and is thought to be stabilized by negative supercoiling. ADAR1 is thought therefore to target actively transcribing genes. In order to understand this biological function in greater detail, an investigation of the detailed structural interactions between this protein and DNA was carried out using X-ray diffraction.

Cubic crystals of the Z_α domain (residues 133–209 of ADAR1) and an oligonucleotide of sequence $(TCGCGCG)_2$ were obtained with ammonium sulfate as precipitant by vapour diffusion using the hanging drop method. The cell dimensions were a = b = 85.9 Å, c = 71.3 Å. Single isomorphous crystals were obtained by including 5-iodouracil for thymine-1 (iodo-1) and 5-iodocytosine for 6-iodocytosine (iodo-6). Data was collected on a rotating anode X-ray source and the data collection details are summarized in (1) below.

(1)

	Native	Iodo-1	Iodo-6
Resolution range (Å)	40–2.4	20–2.1	20–2.8
Number of unique reflections	10 896	29 702	12 566
Completeness	99.2	99.9	99.9

The phase problem was solved by anomalous diffraction and refinement of the data-set for iodo-1 (which diffracted to higher resolution than the other two) produced an electron density map which was then used to build a model of the complex which was refined to an *R* value of 22.3%. An overview of this model is shown in (2) below. From this it is clear that two Z_α domains bind to the double-stranded Z-DNA without contacting each other.

(2)

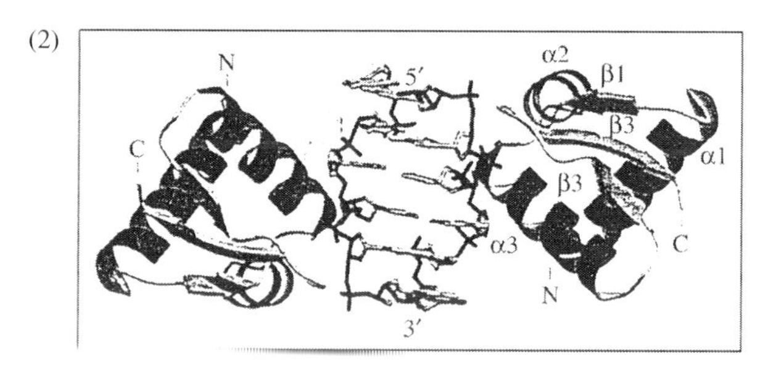

Overview of the complex formed between the Z_α domain and Z-DNA. The α-3 helix and part of the β-hairpin are responsible for direct contact with Z-DNA

(Continues)

(Continued)

The detail of the interactions between individual Z_α domains and Z-DNA is shown in 3(a) below while 3(b) is a schematic of these interactions.

(3)

(a)

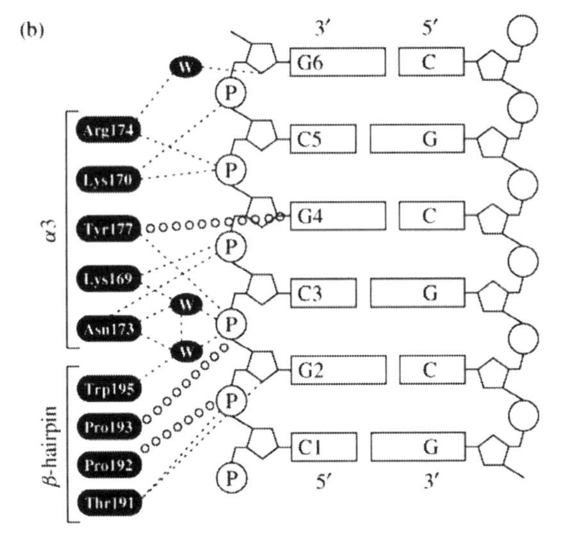

(a) Detail of protein–DNA interactions. (b) Schematic diagram of interactions

The structure makes clear that the complex is stabilized by an extensive network of hydrogen bonds which are either directly between protein and DNA or else are mediated through water molecules. The structure is consistent with previous site-directed mutagenesis and NMR studies and the overall fold is generally similar to functionally related proteins.

Reprinted with permission from Schwartz *et al.*, Crystal Structure of the Za Domain of the Human Editing Enzyme Adari Bound to Left-handed Z-DNA. *Science* **284**, 1841–1845 (1999), American Association for the Advancement of Science.

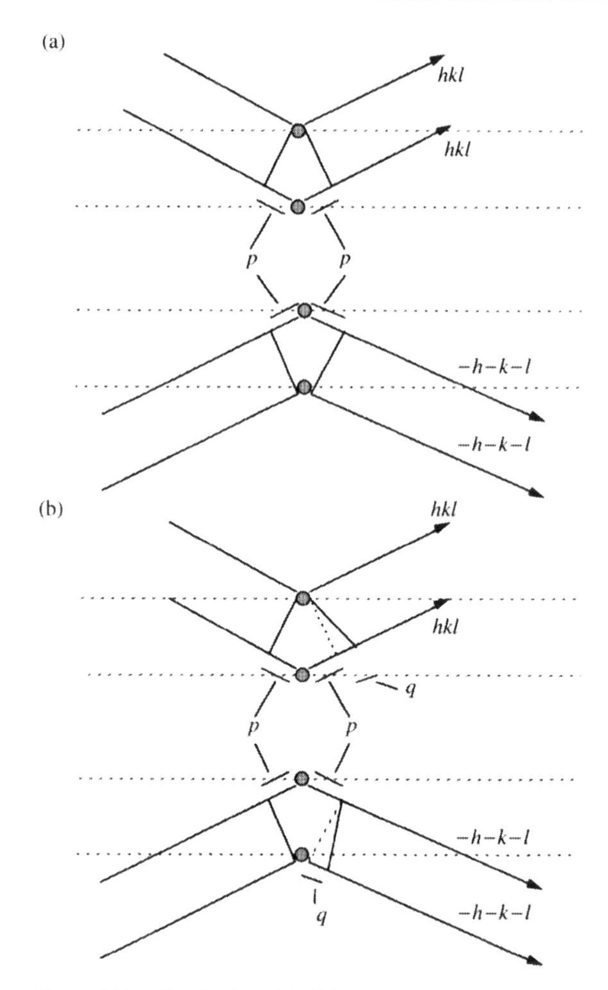

Figure 6.52. Friedel's law. (a) Coherent scattering results in a path-difference of $2p$ for both the (hkl) and $(-h-k-l)$ reflections. These give rize to equally intense Bragg reflections ($I_{hkl} = I_{-h-k-l}$) in accordance with Friedel's law. (b) Anomalous scattering results in a path-difference of $(2p+q)$ for the (hkl) relection and $(2p-q)$ for the $(-h-k-l)$ reflection. Thus $I_{hkl} \neq I_{-h-k-l}$. Friedel's law therefore does not hold for anomalous scattering.

relative phase angle, α_{hkl} and $\alpha_{hkl\,A}$, respectively, as shown in Figure 6.54. α_{hkl} is the phase of the data-set under investigation which is needed to solve the phase problem and compute an electron density map.

The objective of the anomalous scattering experiment is the determination of a value for α_{hkl}. The wavelength-dependent, coherent and anomalous structure factors are related as

$$ {}^{\lambda}F_{hkl} = {}^{\circ}F_{hkl\,T} + \sum[(f'/f) + (if''/f)]\,{}^{\circ}F_{hkl\,A} \quad (6.25) $$

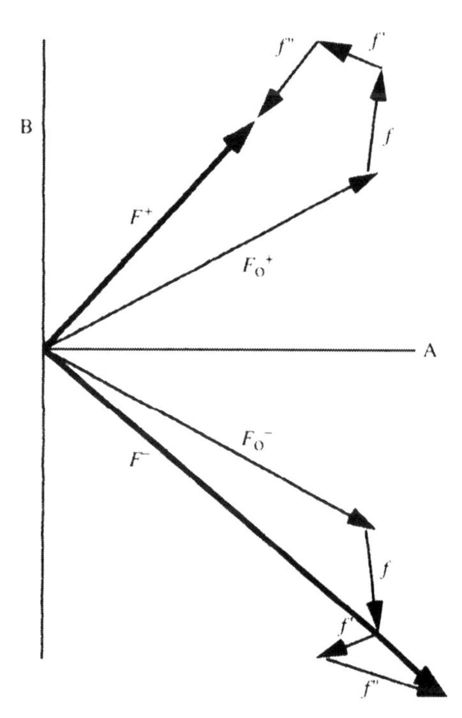

Figure 6.53. Non-equality of Friedel mates in anomalous diffraction. The structure factors for (hkl) and $(-h-k-l)$ reflections in anomalous diffraction are shown as F^+ and F^-, respectively. F_0 is the diffraction component due to nonanomalously diffracting atoms. The contribution of each of the three components of the scattering factor shown in Equation (6.23) are shown as f, f' and f'', respectively. Note that F_0, f and f' of the Friedel pair are simply symmetrical reflections of each other and thus would obey Friedel's law. The nonequality in intensity arises from differences in f''.

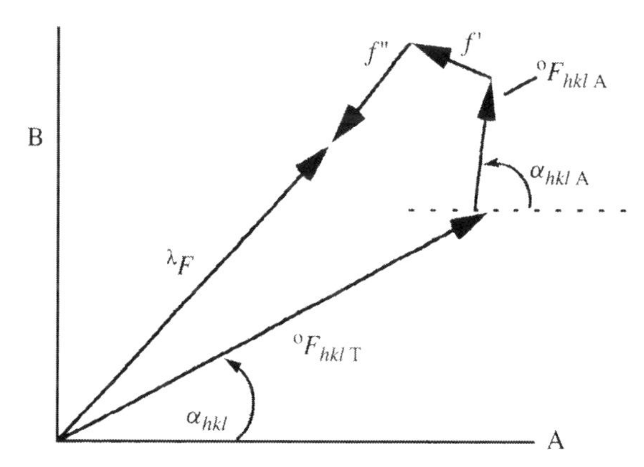

Figure 6.54. Structure factors in anomalous diffraction. The structure factor determined at any wavelength (${}^{\lambda}F$) is composed of coherent scattering (${}^{\circ}F_{hkl\,T}$) and anomalous scattering (${}^{\circ}F_{hkl\,A}$) component structure factor. The phase angles associated with these are α_{hkl} and $\alpha_{hkl\,A}$, respectively. The objective of the experiment is to determine α_{hkl}.

With a single kind of anomalous diffractor introduced into the crystal, the relationship between the vector and phase angle terms at any wavelength, λ, is

$$\begin{aligned} {}^{\lambda}F_{hkl} &= {}^{\circ}F_{hkl\,T^2} + a^{\circ}F_{hkl\,A^2} \\ &+ b({}^{\circ}F^{\circ}_{hkl\,T}F_{hkl\,A}\cos\Delta a_{hkl}) \\ &+ c({}^{\circ}F^{\circ}_{hkl\,T}F_{hkl\,A}\sin\Delta a_{hkl}) \end{aligned} \quad (6.26)$$

where a, b and c are wavelength-dependent coefficients as follows:

$$\begin{aligned} a &= \Delta f/f \\ b &= 2(f'/f) \\ c &= 2(f''/f) \end{aligned} \quad (6.27)$$

and

$$\begin{aligned} \Delta f &= (f'^2 + f''^2)^{1/2} \\ \Delta\alpha_{hkl} &= (\alpha_{hkl} - \alpha_{hkl}\mathrm{A}) \end{aligned} \quad (6.28)$$

The anomalous scattering experiment consists of collecting diffraction patterns from a crystal containing atoms capable of anomalous diffraction at a number of wavelengths. This allows estimation of the coefficients a, b and c since they consist of f' and f'' which can be experimentally measured and f which can be calculated. Experimental determination of f'' involves using the relationship shown in Figure 6.53 and measuring the different intensities of pairs of Friedel mates. f' may be estimated because of its wavelength-dependence by comparison of intensities at a chosen pair of wavelengths. This leaves three unknowns to be determined; ${}^{\circ}F_{hkl\,\mathrm{T}}$, the structure factor for coherent scattering, ${}^{\circ}F_{hkl\,\Lambda}$, the structure factor for anomalous scattering and the $\Delta\alpha_{hkl}$ term $(\alpha_{hkl} - \alpha_{hkl\Lambda})$. Equation (6.27) can be simultaneously solved for these three unknowns using data collected at different wavelengths. The value for ${}^{\circ}F_{hkl\,\Lambda}$ allows determination of the position of anomalously diffracting atoms in the crystal which, in turn, allows determination of $\alpha_{hkl\Lambda}$. Since we know $\Delta\alpha_{hkl}$, we can therefore estimate α_{hkl}.

6.5.7 Calculation of Electron Density Map

The intensity of each spot in the diffraction pattern generates a structure factor F_{hkl} for each Bragg reflection (Equation (6.15)). The relative phase angle, α_{hkl}, is estimated using the methods outlined in Sections 6.5.4–6.5.6. These data are then used as inputs for a computer program which includes a mathematical tool called the *Fourier synthesis* (Appendix 2) to analyse the waves of all the reflections. This converts the Bragg reflections from the lattice planes (hkl) to *electron density waves* which are perpendicular to (hkl) and gives a *summation* of them (Figure 6.55). The electron density at any point (x, y, z) in the unit cell may be expressed as $\rho(xyz)$ where $x - z$ are expressed as fractions of $a - c$:

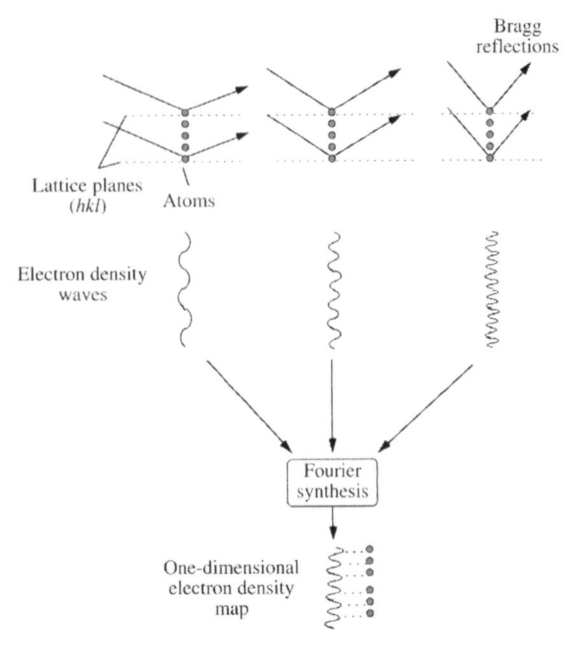

Figure 6.55. Use of Fourier synthesis in calculation of electron density map. Three Bragg reflections are shown for equivalent atoms in lattice plane (hkl). Depending on the angle of incidence (θ), an electron density wave perpendicular to the lattice plane is generated from each Bragg reflection. Fourier synthesis provides a summation of these electron density waves from many Bragg reflections which gives a one-dimensional electron density map (bottom).

$$\rho(xyz) = 1\sum_h\sum_k\sum_l |F_{hkl}|e^{2\pi i\alpha}hkl\,e^{-2\pi i(hx+ky+lz)} \quad (6.29)$$

where $i^2 = -1$ and V is the unit cell volume.

Note that this expression includes the amplitude, $|F_{hkl}|$, and phase, α_{hkl}, of each Bragg reflection as well as three-dimensional positions of points in both crystal lattice and reciprocal space $(hx + ky + lz)$. Moreover, this equation makes clear that *all* atoms in the unit cell contribute to each and every Bragg reflection (Σ). A simplified one-dimensional example of this is shown in Figure 6.55 but, for a biomacromolecule such as a protein, the electron density map generated is three-dimensional (Figure 6.56).

A consequence of Equation (6.29) is that every Bragg reflection in the data-set results from contributions from every atom in the unit cell. In fact, the structure factor F_{hkl} corresponding to each Bragg reflection is a Fourier series

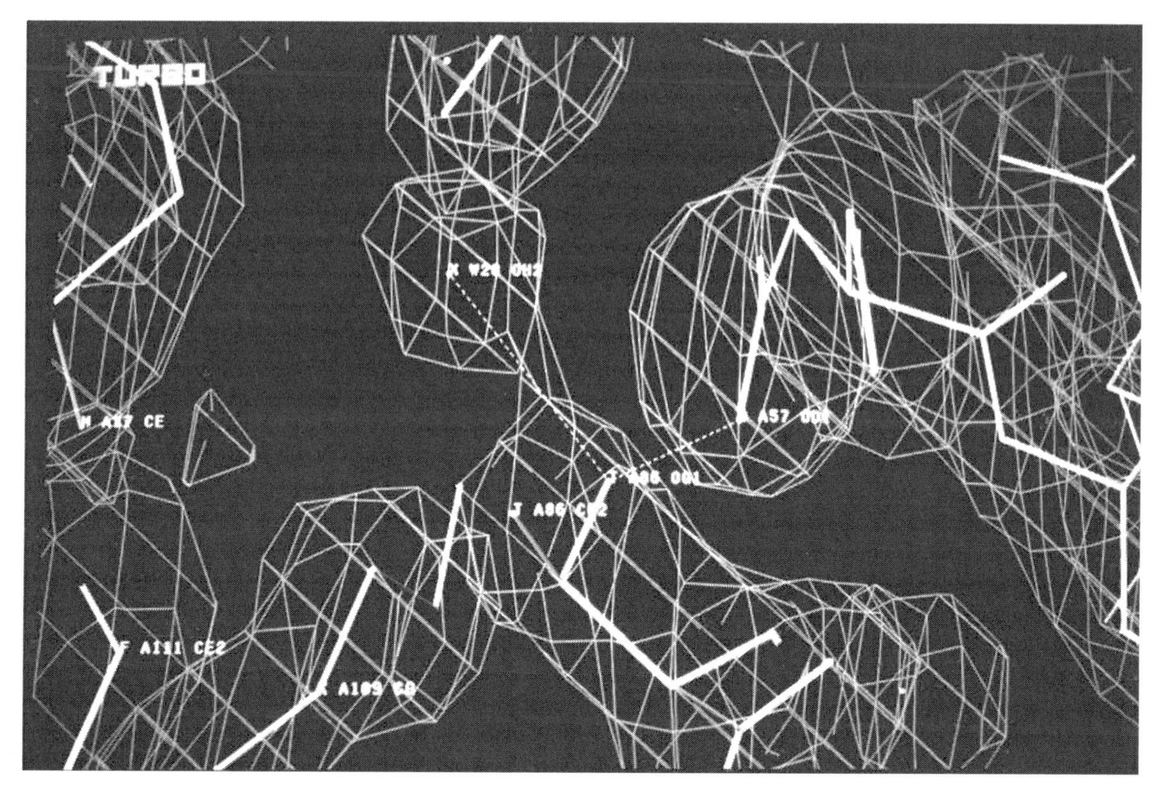

Figure 6.56. Portion of an electron density map of a protein. Amino acid side-chains (solid lines) are modelled through the electron density map of a portion of a protein. Distances due to noncovalent bonds such as hydrogen bonds can be traced as dashed lines. Note the presence of a molecule of structural water (top middle) involved in such a hydrogen bond.

consisting of individual wave functions from each atom in the unit cell (Appendix 2). This means that there is an 'all-or-nothing' aspect to structure determination by X-ray crystallography since the electron density map for the *whole* molecule is calculated from the *whole* diffraction data-set. In practice, the Fourier synthesis may be terminated once a large number of the reflections in the limiting sphere have been included.

The next step in structure determination is to *model* the structure of the biomolecule (e.g. protein primary structure) by tracing it through the electron density map to get as good a fit as possible. This involves matching atoms in the molecule to peaks in the electron density map. Difficulties can sometimes arise in distinguishing structural water molecules from metal atoms or other parts of the biomacromolecule's covalent structure such as amino acid side-chains (Section 6.1.1). In the case of proteins and nucleic acids, extensive knowledge of the normal range of bond lengths and bond angles has been built up and this is useful in the modelling process. The result is a *preliminary model* of the protein or DNA molecule in which the outline shape of the molecule and the overall arrangement of its covalent backbone should be visible.

6.5.8 Refinement of Structure

All the inputs to the calculation necessary to produce the electron density map are subject to some experimental error which consequently results in error in the map. Moreover, the level of detail visible in individual electron density maps (known as the structure *resolution*) may vary due to factors such as the number of reflections included and disorder in the crystal which results in 'smearing' of electron density. For biomacromolecules, a resolution in the range 2–3 Å is usual in a high-resolution structure although some structures of better than 1.5 Å resolution have been reported. The preliminary model is usually of significantly lower resolution, however. To arrive at a high-resolution structure we need to improve the model. This is achieved by cycling through a series of computer programs which compare the model to the diffraction data until any discrepancy between the two has reached a minimum in a process called *structure refinement*.

Two types of information act as inputs to refinement. The first is *empirical knowledge* about the chemical components of the sample molecule. In the case of proteins, this includes known ranges of bond lengths and angles, dihedral angles, preferred arrangements around chiral centres, planarity of aromatic rings and knowledge of noncovalent interactions such as van der Waals, hydrogen and electrostatic bonds. For DNA molecules empirical factors would include base-pair hydrogen bonding, base planarity and sugar puckering. This empirical knowledge can be used in two distinct ways during refinement to limit the number of calculations necessary while maximizing the accuracy of the resulting structure. One is to introduce *constraints* in the refinement which limits the number of parameters which need to be included in each set of calculations. For example, a chemical group or structure such as a benzene ring can be refined as a single entity with fixed geometry rather than as six individual carbon atoms. This significantly reduces the number of parameters required to compute a refined model location for this group. A second approach is to use *restraints* or limitations on the values that individual bond angles and lengths can have. A range of values centred on an average is used for each common bond angle and length which allows adoption of a nonideal value if this contributes to an improved structure.

The second source of information is the preliminary model derived from the X-ray diffraction experiment. The refinement process can therefore be thought of as adjusting the model of the biomacromolecule within the database of constraints provided by empirical knowledge to minimize any discrepancies between the model and experimentally observed crystallographic data. The molecular structure is represented as a collection of energy functions and the refinement process is continued until a *global energy minimum* is reached (Figure 6.57).

In refinement the model is changed in each iteration by altering the *position* (x, y, z) and *mobility* of each atom and recalculating structure factors. The mobility of each individual atom in the crystal affects its contribution to the overall diffraction pattern in that *the more mobile an atom is, the less strongly it contributes to the electron density map*. Some atoms in the structure of a protein may be quite rigid while others may be remarkably free to move. In fact, flexible regions are essentially 'invisible' in an electron density map. A consequence of this is that we need to include in our refinement a weighting factor for each atom to quantify its mobility. This is called the *temperature factor* and is denoted by B. Although temperature is indeed a source of vibration and hence mobility of atoms, this parameter is really a measure of the total mobility of the atom resulting either from thermal vibrations or disorder in the crystal and is also known by the less specific term *displacement parameter*.

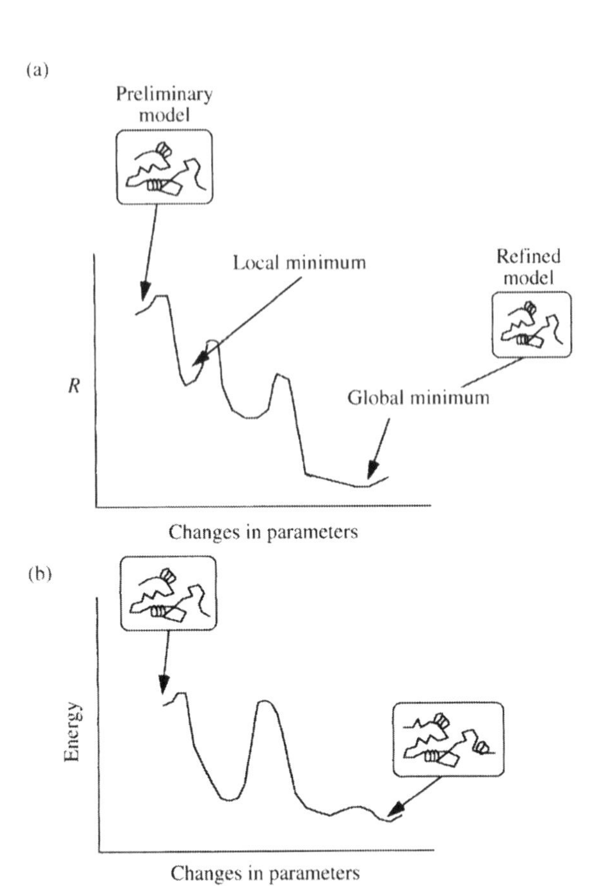

Figure 6.57. Refinement of a protein structure. (a) Least squares refinement leads to a decrease in the R factor as the preliminary model is altered with changes to the parameters describing the position and mobility of each atom. If the preliminary model is substantially correct, this leads to a refined structure. However, the model can become trapped in a local energy minimum. (b) Simulated annealing provides for a greater range of conformational parameters which is useful if the preliminary model is somewhat different from the true structure.

Because the effect of atomic vibration is to increase the apparent size of the atom, this leads to a decrease in the extent of the diffraction pattern which is a consequence of the reciprocal relationship between crystal lattice space and reciprocal space. This effect can be quantified as an exponential function of the scattering factor:

$$f = f_0 e^{B-}(\sin^2 \theta/\lambda^2) \quad (6.30)$$

where θ is the angle of incidence of X-rays of wavelength λ and f_0 is the scattering factor of a stationary atom. In crystals of salts and minerals B factors are quite low (1–3 Å^2) but they are higher in organic crystals (2–6 Å^2) and crystals of biomacromolecules (20–40 Å^2). In protein crystallography the variation of the temperature factor as a function of primary structure is frequently presented as a plot which can be useful in distinguishing flexible regions from more rigid parts of the structure.

A measure of the progress of the refinement process is provided by the R factor which was defined in Section 6.5.1 (Equation (6.17)). This represents a summation of the difference between observed structure factors and those calculated from the model. However, the R factor is not an indication of the precision of the structure in that two models with the same R factor may have bond angles and lengths determined with quite different precision.

In situations where the preliminary model is close to the true structure, the comparison may be made using a least squares procedure. This minimizes the sum of the squares of deviation of $|F_{hkl\ \mathrm{calc}}|$ from $|F_{hkl\ \mathrm{obs}}|$ for the whole structure.

However if the model, though largely correct, is incorrect in parts, the least squares procedure can become 'trapped' in a *local energy minimum* (Figure 6.57). This is because the computer algorithms which carry out the least squares refinement generally only make small alterations to the model on each iteration. To rescue the model from a local energy minimum may require extensive revision of the model. In such situations a procedure called *simulated annealing* is used. This simulates the effect of heating the molecule to give each atom much greater freedom of movement. The net effect of this is that the model can explore a much greater range of possible conformations and thus arrive at a global energy minimum.

The range of R factors estimated for biomacromolecules are significantly larger (12–20%) than are obtained in small molecule crystallography (3–5%). A major reason for this is the difficulty of modelling the relatively high water content of protein and DNA crystals (which can be up to 50%). One successful approach to this problem is to model water in layers approaching the biomacromolecule with the solvation shell having the highest order and lowest mobility.

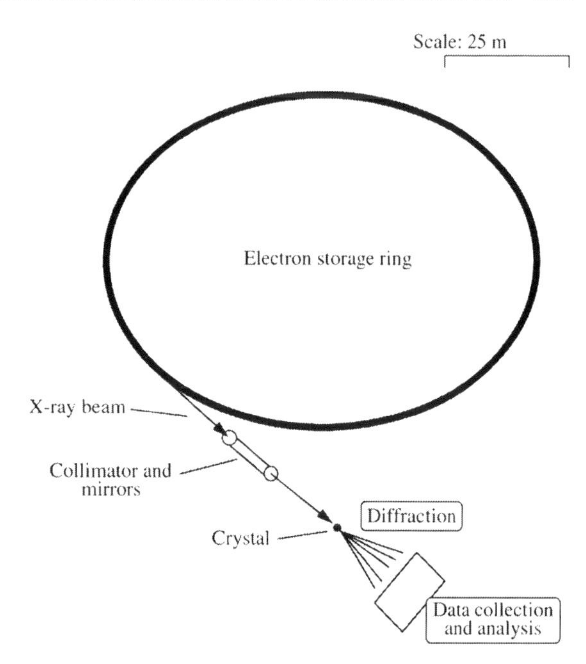

Figure 6.58. A synchrotron X-ray source. If electrons which have been accelerated in the electron storage ring are suddenly decelerated, they emit a powerful beam of X-rays across a range of wavelengths. These are directed through a collimator which, together with metal mirrors, produces a fine intense beam which is directed through the crystal for diffraction. Note the size of the electron storage ring which is similar to that of a football field.

6.5.9 Synchrotron Sources

If electrons are accelerated to near the speed of light in a ring-shaped particle accelerator called an *electron storage ring* and then quickly decelerated by forcing them into a curved trajectory, they emit their excess radiation in the form of a powerful stream of X-rays tangential to the ring which is called *synchrotron* radiation (Figure 6.58). The particle accelerators necessary to perform this function are large, extremely expensive and are located in a small number of large-scale facilities internationally such as at Brookhaven (USA), Grenoble (France), Harwell (England), Tsukuba (Japan), Trieste (Italy) and Hamburg (Germany). A web site containing links to the main synchrotron sources internationally is given in the Bibliography. Crystallographers can arrange 'beam time' to collect data-sets in these large-scale facilities and most of the high-resolution structures published in the scientific literature are determined from data collected in this way.

The beam of X-rays generated from a synchrotron source has a number of characteristics which make it extremely useful in diffraction experiments. Compared to the beam of X-rays which can be generated from the laboratory sources described in Section 6.4.1, the beam from a synchrotron has some thousand-fold greater energy and is more highly collimated which means that the diffraction spots measured are both smaller and more intense, thus enhancing the resolution and precision of the diffraction pattern. It also means that data-sets consisting of thousands of Bragg reflections can be collected in minutes rather than the hours required with laboratory sources. The synchrotron beam is extremely *narrow* and intense allowing collection of data from much smaller crystals than can be used with a laboratory-scale instrument. Moreover, the beam is *tunable* and can generate a wider range of wavelengths than laboratory-scale sources which may be used in *multiple wavelength anomalous diffraction* experiments for phase angle calculations as we have previously seen in Section 6.5.6.

6.6 OTHER DIFFRACTION METHODS

X-ray diffraction is the main means currently available for high-resolution structure determination of biomacromolecules. However, other types of diffraction are also possible which have particular applications likely to be important in the future. The principles involved in these experiments are generally similar to X-ray diffraction although each has particular attributes and limitations which we will now briefly describe.

6.6.1 Neutron Diffraction

Just as electrons diffract X-rays, so a beam of neutrons can be diffracted by the nucleus of the atom. This diffraction also results in a 180° change of phase in the diffracted beam but, because nuclei have a fixed location in space compared to the relatively diffuse electron cloud, greater structural detail can be observed than in X-ray diffraction. *Neutron diffraction* studies can be complementary to X-ray diffraction in many ways as a result. It provides a means to locate the position of light atoms such as hydrogen which are not visible in electron density maps and also allows us to distinguish individual isotopes from each other. Moreover, since electron distribution around atoms is not necessarily symmetric, neutron diffraction can give more accurate estimates for bond lengths than those apparent in the electron density map and, for example, can help locate hydrogen bonds more precisely than X-ray diffraction. Neutron diffraction can also detect *thermal vibration* in crystals which is a major source of disorder and hence scatter in the electron density map. Major constraints on the widespread use of this technique, however, are the need for much larger crystals than required for X-ray or electron diffraction (~1 mm^3) and the requirement of time at a nuclear reactor to access a neutron beam.

6.6.2 Electron Diffraction

Because the availability of three-dimensional protein crystals is a prerequisite for structure determination of proteins by X-ray diffraction, membrane-bound proteins are not directly accessible to the technique. Notwithstanding the fact that approximately 30% of eukaryotic genes encode integral membrane-bound proteins, the high-resolution structures of only a handful have been determined to date compared with tens of thousands nonmembrane protein structures. Multi-dimensional NMR is also limited as a means of structure determination because of the large weight of detergent in detergent-solubilized membrane-bound protein combined with the M_r limit of this technique (Section 6.2.8). This means that most of our knowledge of high-resolution protein structure comes from water-soluble proteins.

Electron diffraction provides an alternative means to study the structure of membrane proteins. It takes advantage of the fact that, while membrane proteins do not readily form three-dimensional crystals, they sometimes will crystallize in two dimensions (Example 6.6). While *two-dimensional crystals* cannot be mounted in an X-ray beam they can diffract electrons systematically and, when combined with suitable image analysis, can be used to determine three-dimensional structure.

In a few instances two-dimensional crystals occur naturally such as those formed by bacteriorhodopsin of *Halobacterium halobium*. However, in most cases it is necessary to screen a variety of conditions to grow crystals. Usually this involves exposing detergent-solubilized protein to a variety of lipids until suitable conditions are identified or, alternatively, growing the crystals on *lipid monolayers*. The crystals are then mounted on a grid and exposed to a beam of electrons in an electron microscope at very low temperature and at low magnification. These conditions are necessary because too strong an electron beam can damage the structure of biological molecules. The diffraction image is then collected by a *microdensitometer* and analysed by Fourier synthesis (Appendix 2). Because of the limitations imposed by too weak an electron beam and imperfections in the crystal, the resolution possible with this technique is currently in the region of 7 Å which is significantly lower than that possible with X-ray diffraction. However, since membrane-bound proteins tend to be large macromolecular complexes, this level of structural information can be highly informative about overall shape, dimensions and organization. More than 2000 structures have been determined in this way.

A related method is single particle cryo-electron microscopy (cryo-EM). This analyses electron micrographs from single particles of large structures such as ribosomes, viruses, ion channels and chaperonin complexes which do not readily crystallize. In excess of 100 such structures are presently known. The overall resolution is ~4.5–12 Å which allows the main features to be visualized. Where higher-resolution structures at the level of domain or protein components are available from X-ray diffraction or NMR, these can be combined with cryo-EM to generate a high-resolution picture of the particle. An advantage of cryo-EM is that it makes possible viewing of large biological machines in a variety of functional states, for example by collecting time-lapse images. A disadvantage is that the beam of electrons used can damage the protein as mentioned above.

Example 6.6 Electron Diffraction by Two-dimensional Crystals of Aquaporin 1

Rapid water transport across plasma membranes is facilitated by specialised water channels which transport water selectively to the exclusion of small ions and solutes. Aquaporin 1 (also called CHIP28) is a channel-forming integral glycoprotein of 28 kDa molecular mass which is expressed in erythrocytes and in cells of water-transporting tissues.

In order to study the structure and organization of aquaporin 1, two-dimensional crystals were obtained for electron diffraction studies. The crystals (dimensions 1.5–2.5 μm diameter) were obtained by reconstituting 2–4 mg/ml solubilized deglycosylated protein into synthetic lipid bilayers (dioleyl phosphatidyl choline) by dialysis in a lipid:protein ratio of 1 : 1 to 1 : 3 (Chapter 7). The crystals were attached to an electron microscope grid stained with uranyl formate and dried. Electron diffraction was then performed and structure factors calculated. Five individual diffraction images containing reflections up to 12 Å were aligned to a common phase origin and these phases and amplitudes were merged and used to reconstruct a projection density map at 12 Å resolution of aquaporin 1 shown below.

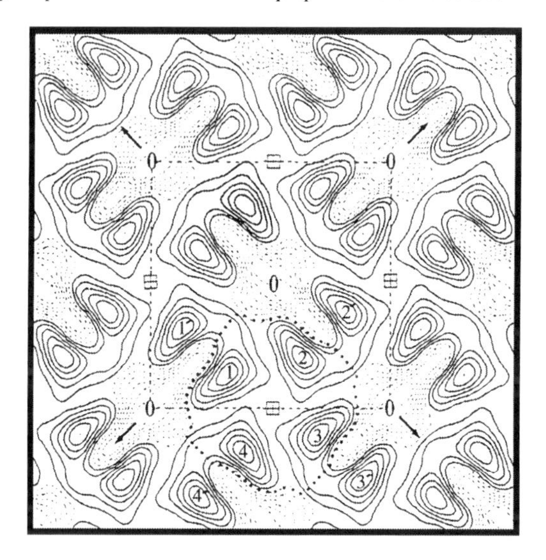

Averaged projection density map of aquaporin 1

The unit cell (light dashed line) contained eight individual aquaporin molecules arranged in four dimers and with dimensions of a = b = 99.2 Å. The tetramer (heavy dashed line) contains monomers orientated in the same direction and each monomer is in contact with a neighbouring tetramer (e.g. 1 with 1′). This structure is consistent with electron micrographs of freeze-fractured samples and allowed definition of the molecular boundaries and packing of aquaporin monomers. However, secondary structure features are not visible at this level of resolution and higher-resolution data was necessary for this. In later publications, structural data to a resolution of 6 Å was obtained by electron diffraction which allowed a more detailed model to be constructed of secondary structure in the trans-membrane domains.

Reproduced from Mitra *et al.* (1994) Projection structure of the CHIP28 water channel in lipid bilayer membranes at 12 Å resolution. *Biochemistry*, **33**, 12735–40, with permission of the American Chemical Society.

6.7 COMPARISON OF X-RAY CRYSTALLOGRAPHY WITH MULTI-DIMENSIONAL NMR

X-ray crystallography and multi-dimensional NMR are now the main experimental approaches to determination of three-dimensional structure of biomacromolecules. *These techniques are independent of each other* in both preparation of sample and the physical parameter used to calculate structure. Having described each of these techniques in some detail, we will now briefly look at the relationship between them.

6.7.1 Crystallography and NMR are Complementary Techniques

In terms of sample accessibility and of the physical basis for structure determination, NMR and X-ray crystallography

Table 6.4. Comparison of points where NMR and X-ray crystallography are complementary approaches to structure determination.

Point of comparison	Multi-dimensional NMR	X-ray crystallography
Phase of sample	Liquid	Solid
Component of atom involved	Nucleus	Electrons
Mobility		
Rigid parts of structure	Weak signal	Strong signal
Mobile parts of structure	Strong signal	Weak signal
Time-resolution	Dynamic	Fixed
Final structure	Family of related structures	Single refined structure

are largely *complementary techniques* (Table 6.4). NMR requires sample molecules capable of existing at high concentration under acidic pH conditions and therefore uses highly water-soluble samples present in the *liquid phase*. X-ray crystallography, by contrast, requires samples capable of forming large single crystals in the *solid phase*. Where proteins have had their structure determined by both methods, identical structures have usually been found although the reliability of NMR-derived structures decreases markedly with increasing sample mass. This fact, together with the knowledge that many proteins (e.g. enzymes) retain their biological activity in the crystalline state and that the same protein structure is arrived at from different crystal-forms, gives us confidence that the structural form of biomacromolecular samples present in crystals is biologically meaningful and is not merely an artefact of crystallization.

The two techniques 'report' on different parts of atomic structure since NMR is a phenomenon which depends on *nuclei* while X-ray diffraction is due to *electrons*. In NMR the location of *protons* which are largely invisible in electron density maps is identified. Moreover, lack of mobility in species responsible for an NMR signal frequently leads to a broadening of the signal similar to that observed for ESR signals (Section 3.8.3). This means that *more flexible* parts of structure produce relatively stronger NMR signals which is exactly opposite to the situation in X-ray crystallography where flexible parts of structure are much less visible than static parts. The two techniques therefore provide complementary information on structure which gives a much fuller picture than either method on its own.

Time-resolved experiments in which dynamic processes such as protein tumbling, protein folding, domain movement and ligand binding can be followed as a function of time are particularly appropriate to multi-dimensional NMR. Movement in crystals can result either from *static disorder* in the crystal (i.e. an equivalent atom occupying a range of locations in different molecules) or from *dynamic movement* (i.e. movement of the atom during the time-frame of the diffraction experiment). This may be detected by low electron density in the electron density map. In principle, rapid data collection and analysis using synchrotron sources can also provide time-resolved information from crystals. However, in most cases, X-ray crystallography provides a single static structure because intermolecular interactions necessary for crystallization may severely limit the mobility of groups on the surface of the biomacromolecule. This is an artificial situation since, in aqueous solution, surface groups are far more mobile than groups buried in the interior of the protein. NMR is therefore more generally useful for providing a 'moving image' of structure while the image provided by X-ray diffraction is closer to a single 'snapshot'.

Notwithstanding these points where the techniques are complementary it should always be borne in mind that only certain samples are suitable for both multi-dimensional NMR and X-ray diffraction. In the former case, samples must be generally small and highly soluble at acid pH without forming aggregates while, in the latter case, samples must crystallize readily. However, many proteins are not readily amenable to study by either technique (e.g. membrane-bound proteins) while some may be studied by only one of them. This sample nonaccessibility is a major bottleneck in biomacromolecule structure determination in biochemistry.

6.7.2 Different Attributes of Crystallography- and NMR-derived Structures

Even though high-resolution structures result from both techniques, there are some important points of detail in which the structures differ from each other. Multi-dimensional NMR provides a 'family' of closely related structures which are all equally possible statistically. This contrasts with X-ray crystallography which, after an extensive process of statistical analysis, results in a single refined structure. Moreover, within an individual structure arrived at by multi-dimensional NMR, there is no indication of the reliability within the structure. By contrast, X-ray crystallography has robust and well-established methods for assessment of the reliability of each individual part of the structure. This means that, while the outline of the polypeptide chain may be clear in such structures, care may need to be taken in drawing detailed conclusions on atomic locations from multi-dimensional NMR.

6.8 STRUCTURAL DATABASES

Structural information is normally disseminated to the international scientific community by means of published papers in learned journals such as the *Journal of Molecular Biology, Acta Crystallographica* and *Structure* which are accessible in most university and institutional libraries and on-line. Such papers give detailed information on the origin and preparation of the sample, the method used to determine the structure and especially interesting aspects of the structure such as its biological significance. However, structures are also included in a *structural database* which is accessible through the worldwide web. The great advantage of this is that, with the aid of a suitable computer graphics program, the structure may be viewed in three dimensions rather than the two dimensions possible on paper thus giving a more realistic impression of the structure and allowing focus on a part of the structure of specific interest. Many *sequence databases* exist which allow searching, comparison and downloading of primary structure data (Chapter 9). The main repository of three-dimensional structural information, however, is the *protein database* (PDB). Because the reader may wish to view particular structures of biomolecules we will now briefly describe the PDB and how structures may be downloaded from it and viewed.

6.8.1 The Protein Database

Entries in the protein database arise from four main experimental techniques; X-ray diffraction, multi-dimensional NMR, electron diffraction and cryo-EM. At the time of writing (Dec. 2008) there are almost 50 000 structures in the database of which some 84% are structures derived from X-ray diffraction (Table 6.5). From this table it is also clear that, in addition to protein structures, the database contains extensive entries for viruses, nucleic acids, carbohydrates and macromolecular complexes. Some 80% of the protein structures include closely related groups of structures such as mutants, protein families and individual proteins (e.g. there are over 1043 individual entries for lysozyme) which means that only some 20% of the structures are responsible for the range of some 50 000 protein structures covered.

Table 6.5. Total entries in the Protein Database (Dec. 2008).

	Technique		
Type of sample	X-ray Diffraction	NMR	Cry-EM
Proteins, peptides and viruses	43 520	6624	142
Protein/nucleic acid complexes	2011	140	50
Nucleic acids	1101	827	12
Carbohydrates and others	24	7	0
Total	46 656	7598	204

Since July 1999 the database has been managed by a consortium consisting of the National Institute of Standards and Technology (NIST, MD, USA) and the Universities of Rutgers (NJ, USA) and California (San Diego, CA, USA) called the *Research Collaboratory for Structural Bioinformatics*. Prior to this, it was operated by the Brookhaven National Laboratory and was previously referred to as the *Brookhaven structural database*.

6.8.2 Finding a Protein Structure in the Database

A tutorial program is provided to allow users to familiarize themselves with routine operations available. To visualize a structure from the database it is necessary to download the three-dimensional coordinates for the protein of interest. These may be located in three ways. Each structure is designated with a specific identification code called a *PDB ID code* (e.g. a 1.9 Å structure for glucose oxidase is denoted; 1GAL). Alternatively, a *keyword* option allows searching for the protein by name (and/or species of origin) while an *author* option allows searching by author(s) of peer-reviewed publications. A list of structures is generated from which the desired structure is selected. The coordinates are retrieved from the database and may be saved in a file on the personal computer (or attached peripheral) in use. To visualize the structure, it is necessary to use a suitable graphics program such as *Rasmol* which is available free on the internet. This allows presentation of the structure in various orientations and from various angles. The structure may be represented as *wireframe*, *space-filling* or *ribbon* models (Figure 6.59). Moreover, specific parts of the structure (e.g. from residue 10 to 30) can be highlighted in different colours or in a different representation as can nonprotein ligands and structural water.

The PDB provides a major resource to the scientific community which makes the hard-won structural data gained by the successors to Perutz, Kendrew and their colleagues increasingly readily accessible. Because of improvements in methods for generating and solving three-dimensional structures of biomacromolecules and for disseminating such data electronically, there has been a perceptible acceleration in the pace at which the structural versatility and beauty of biomacromolecules is being revealed.

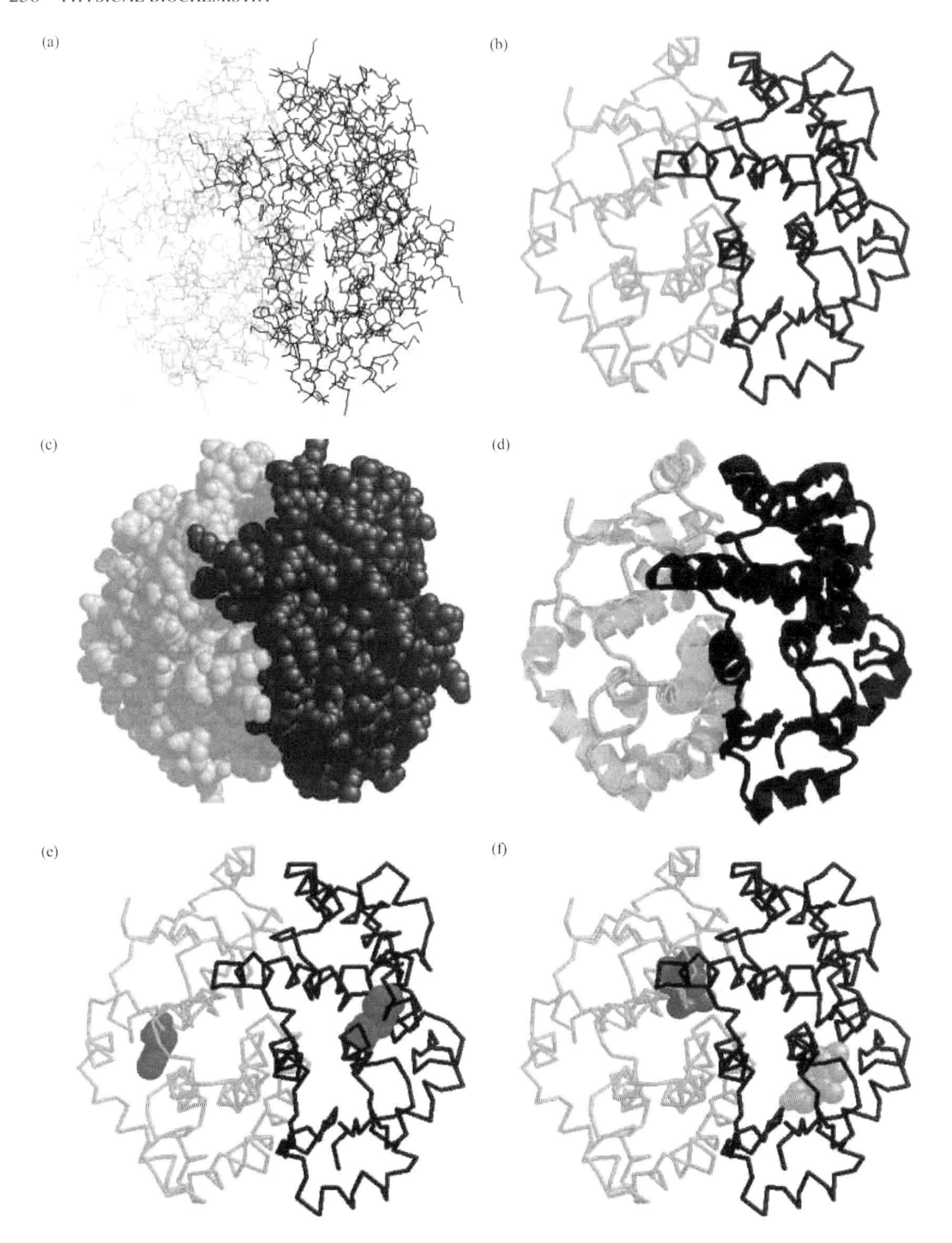

Figure 6.59. Representations of glutathione transferase 1-1 by RASMOL. The coordinates for GST 1-1 were downloaded from the protein database and used to construct the images shown. One subunit is shown in black while the other is in grey. The various representations are (a) Wireframe, (b) backbone, (c) space-filling, (d) ribbon. (e) Single tryptophans (Trp-20) are represented by space-filling. (f) The ligand ethacrynic acid with which the protein was co-crystallized is represented by space-filling.

REFERENCES

General reading

Branden, C. and Tooze, J. (1999) *Introduction to Protein Structure*, 2nd edn, Garland Press, New York. A very readable overview of protein structure and methods used in its analysis.

Campbell, I.D. (2002) The march of structural biology. *Nature Reviews Molecular Cell Biology*, **3**, 377–81. A review of the development of structural biology.

Fersht, A. (1999) *Structure and Mechanism in Protein Science: A Guide to Enzyme Catalysis and Protein Folding*, Freeman, New York. An overview of the relationship between protein structure and function with a particularly good description of protein folding.

Kleywegt, G.J., Henrick, K., Dodson, E.J. and van Aalten, D.M.F. (2003) Pound-wise but penny-foolish: How well do micromolecules fare in macromolecular refinement? *Structure*, **11**, 1051–9. A study of how structures of small molecules are often less reliable than those of macromolecules during structural refinement.

Lesk, A. (2004) *Introduction to Protein Science: Architecture, Function and Genomics*, Oxford University Press, Oxford, UK. An excellent description of protein architecture with special emphasis on information flow in Biology and on protein evolution.

Rhodes, G. (2006). *Crystallography Made Crystal Clear*, 3rd edn, Academic Press, New York. A brief but extremely well-written explanation of the basics of X-ray crystallography of biomacromolecules and how to read crystallography papers critically.

The protein folding problem

Baldwin, R.L. and Rose, G.D. (1999). Is protein folding hierarchic? I Local structure and peptide folding. *Trends in Biochemical Sciences*, **24**, 26–33. A review of class I and II protein folding featuring simulations and experimental evidence from peptides and small proteins.

Baldwin, R.L. and Rose, G.D. (1999). Is protein folding hierarchic? II Folding intermediates and transition states. *Trends in Biochemical Sciences*, **24**, 77–83. A review of hierarchical protein folding focusing on folding intermediates in class I proteins and transition states in class II.

Chiti, F. and Dobson, C.M. (2006) Protein misfolding, functional amyloid, and human disease. *Annual Review of Biochemistry*, **75**, 333–66. A detailed review on the role of protein folding in human disease.

Gordon, M.H.J., Walsh, N.P., McKnight, C.J. and Sauer, R.T. (1999) Evolution of a protein fold *in vitro*. *Science*, **284**, 325–7. An unusual example of disruption of overall protein fold by a small mutation.

Greenfield, N.J. (2006) Determination of the folding of proteins as a function of denaturants, osmolytes or ligands using circular dichroism. *Nature Protocols*, **1**, 2733–41. A review of the application of CD to the study of protein folding.

Mattos, C. and Clark, A.C. (2008) Minimizing frustration by folding in a crowded environment. *Archives of Biochemistry and Biophysics*, **469**, 118–31. A study of the role of water in protein folding.

Munoz, V. and Eaton, W.A. (1999) A simple model for calculating the kinetics of protein folding from three-dimensional structures. *Proceeding National Academy Sciences USA*, **96**, 11311–6. A simple model for calculating rates of 2-state folding from three-dimensional structure alone is described.

Royer, C.A. (2008) The nature of the transition state ensemble and the mechanisms of protein folding: A review. *Archives of Biochemistry and Biophysics*, **469**, 34–45. A review on transition states in protein folding.

Schroder, M. and Kaufman, R.J. (2005) The mammalian unfolded protein response. *Annual Review of Biochemistry*, **74**, 739–89. A review of links between protein mis-folding in the ER and signal transduction with the unfolded protein response.

Murphy, R.M. and Tsai, A.M. (eds) (2006) *Misbehaving Proteins: Protein (Mis)folding, Aggregation and Stability*, Springer, Berlin. A description of protein aggregation including computational and experimental methods and implications for engineered proteins.

Silva, J.L., Cordeiro, Y. and Foguel, D. (2006) Protein folding and aggregation: Two sides of the same coin in the condensation of proteins revealed by pressure studies. *Biochimica et Biophysica Acta – Proteins and Proteomics*, **1764**, 443–51. A study of protein mis-folding under high pressure conditions leading to stabilization of folding intermediates.

Soto, C. (2003) Unfolding the role of protein misfolding in neurodegenerative diseases. *Nature Reviews Neuroscience*, **4**, 49–60. A review of protein mis-folding in neurodegenerative diseases.

Stanley, A.M. and Fleming, K.G. (2008) The process of folding proteins into membranes: Challenges and progress. *Archives of Biochemistry and Biophysics*, **469**, 46–66. A review of folding of membrane proteins.

Thomas, P.J., Qu, B.-H. and Pedersen, P.L. (1995). Protein folding as a basis of human disease. *Trends in Biochemical Sciences*, **20**, 456–9. A review of clinical conditions possibly arising as a result of incorrect protein folding.

Chaperonins

Corrales, F.J. and Fersht, A.R. (1996) Towards a mechanism for GroEL-GroES chaperone activity: An ATPase-gated and -pulsed folding and annealing cage. *Proceedings National Academy Sciences USA*, **93**, 4509–12. A description of a proposed 'gating' mechanism allowing coupling of ATP-hydrolysis to folding of slow-folding proteins.

Lin, Z. and Rye, H.S. (2006) GroELmediated protein folding: Making the impossible possible. *Critical Reviews in Biochemistry and Molecular Biology*, **41**, 211–39. A detailed review of GroEL-mediated folding.

Macario, A.J.L. and de Macario, E.C. (2007) Molecular chaperones: Multiple functions, pathologies, and potential applications. *Frontiers in Bioscience*, **12**, 2588–600. Clinical relevance of chaperones.

Young, J.C., Barral, J.M. and Hartl, F.U. (2003) More than folding: Localized functions of cytosolic chaperones. Chaperone-assisted protein folding in the cytosol. *Trends in Biochemical Sciences*, **28**, 341–7. An excellent review of cytosolic protein folding.

Richardson, A., Landry, S.J. and Georgopoulos, C. (1998) The ins and outs of a molecular chaperone machine. *Trends in*

Biochemical Sciences, **23**, 138–43. A review of structure–function relationship in GroEL.

Shtilerman, M., Lorimer, G.H. and Englander, S.W. (1999). Chaperonin function: folding by forced unfolding. *Science*, **284**, 822–5.

Xu, Z.H., Horwich, A.L. and Sigler, P.B. (1997). The crystal structure of the asymmetric GroEL-GroES-(ADP)(7) chaperonin complex. *Nature*, **388**, 741–50. The crystal structure of the GroEL/GroES system.

Multi-dimensional NMR

Bonvin, A.M.J.J., Boelens, R. and Kapstein, R. (2005) NMR analysis of protein interactions. *Current Opinion in Chemical Biology*, **9**, 501–8. A review of how larger Mr samples, including complexes, are becoming amenable to multi-dimensional NMR investigation.

Cavanagh, J., Fairbrother, W.J., Palmer, A.J. *et al.* (2007). *Protein NMR Spectroscopy: Principles and Practice*, 2nd edn, Elsevier. A detailed description of the physical basis of biomolecular multi-dimensional NMR.

Glockshuber, R., Hornemann, S., Riek, R., Wider, G., Billeter, M. and Wuthrich, K. (1997) Three-dimensional NMR structure of a self-folding domain of the prion protein PrP(121–231). *Trends in Biochemical Sciences*, **22**, 241–2. An example of structure determination by NMR.

Ishima, R. and Torchia, D.A. (2000) Protein dynamics from NMR. *Nature Structural Biology*, **7**, 740–3. A review of use of NMR in study of protein dynamics.

Krishna, M.M.G., Hoang, L., Lin, Y. and Englander, S.W. (2004) Hydrogen exchange methods to study protein folding. *Methods*, **34**, 51–64. A description of hydrogen exchange methods for study of protein folding.

Mittermaier, A. and Kay, L.E. (2006) Review – New tools provide new insights in NMR studies of protein dynamics. *Science*, **312**, 224–8. A review of recent method improvements for study of protein dynamics by NMR.

McDermott, A. and Polenova, T. (2007) Solid-state NMR: New tools for insight into enzyme function and protein dynamics. *Current Opinion in Structural Biology*, **17**, 617–22. A review of solid state NMR in the study of enzymes.

Neuhaus, D. and Williamson, M. (1989) *The Nuclear Overhauser Effect in Structural and Conformational Analysis*, VCH Publishers, New York. A comprehensive text on the NOE.

Rattle, H. (1995). *An NMR Primer for Life Scientists*, Partnership Press. An introduction to NMR for life scientists.

Sanders, J.K.M. and Hunter, B.K. (1993). *Modern NMR Spectroscopy: A Guide for Chemists*, 2nd edn, Oxford University Press, Oxford, UK. A non-mathematical introduction to multi-dimensional NMR.

Tugarinov, V., Kanelis, K. and Kay, L.E. (2006) Isotope labeling strategies for the study of high-molecular-weight proteins by solution NMR spectroscopy. *Nature Protocols*, **1**, 749–54. A review of use of isotope labeling to allow solution NMR of high Mr samples.

Wu, Z.R., Ebrahimian, S., Zawrotny, M.E. *et al.* (1997). Solution structure of 3-oxo-Δ^5-steroid isomerase. *Science*, **276**, 415–8. A good example of structure determination by heteronuclear, multi-dimensional NMR.

X-ray crystallography

Chayen, N.E. and Saridakis, E. (2008) Protein crystallization: From purified protein to diffraction-quality crystal. *Nature Methods*, **5**, 147–53. A review of protein crystallization.

Conn, G.L., Draper, D.E., Lattman, E.E. and Gittis, A.S. (1999). Crystal structure of a conserved ribosomal protein-RNA complex. *Science*, **284**, 1171–4. A good example of structure determination of a protein–RNA macromolecular complex.

Darcy, P.A. and Wiencek, J.M. (1999). Identifying nucleation temperatures for lysozyme via differential scanning calorimetry. *Journal of Crystal Growth*, **196**, 243–9. Differential scanning calorimetry (Section 8.3) can be combined with light scattering to identify a temperature range within which crystal nucleation is maximally possible.

Dauter, Z. (2006) Current state and prospects of macromolecular crystallography. *Acta Crystallographica D – Biological Crystallography*, **62**, 1–11. A comprehensive review of the current state of macromolecular crystallography.

Dodson, E. (2008) The befores and afters of molecular replacement. *Acta Crystallographica D – Biological Crystallography*, **64**, 17–24. A review of molecular replacement.

Glusker, J.P., Lewis, M. and Rossi, M. (1994). *Crystal Structure Analysis for Chemists and Biologists*, VCH, New York. A superbly readable, comprehensive and accessible description of X-ray crystallography.

Karplus, P.A. and Faerman, C. (1994) Ordered water in macromolecular structure. *Current Opinion in Structural Biology*, **4**, 770–6. A mini-review of structural water at protein surfaces.

Ke, H.M. (1997). Overview of isomorphous replacement phasing. *Methods in Enzymology*, **276**, 448–61. A review of isomorphous replacement.

McPherson, A. (1990). Current approaches to macromolecular crystallisation. *European Journal of Biochemistry*, **189**, 1–23. A review of the process of protein crystallisation.

McPherson, A. (2004) Introduction to protein crystallization. *Methods*, **34**, 254–65. An excellent introduction to crystallization of proteins.

McPherson, A. and Cudney, B. (2006) Searching for the silver bullets: An alternative strategy for crystallizing macromolecules. *Journal of Structural Biology*, **156**, 387–406. A fascinating development in increasing the success of crystallization screens.

Messerschmidt, A. (2007) *X-ray Crystallography of Biomacromolecules*. Wiley-VCH Verlag GmbH, Weinheim. A comprehensive description of current practice in X-ray diffraction.

Qian, B., Raman, S., Das, R. *et al.* (2007) High-resolution structure prediction and the crystallographic phase problem. *Nature*, **450**, 259–64. A new modelling method which offers improved solution to the phase problem including *de novo* structure prediction.

Schwartz, T., Rould, M.A., Lowenhaupt, K. *et al.* (1999) Crystal structure of the Zα domain of the human editing enzyme ADAR1 bound to left-handed Z-DNA. *Science*, **284**, 1841–5. A good example of structure determination of a DNA–protein macromolecular complex.

Stewart, I. (1997) Crystallography of a golf ball. *Scientific American*, **276**, 96–8. An entertaining description of the mathematics of symmetry taking golf balls as a concrete example!

Vrielink, A. and Sampson, N. (2003) Sub-Angstrom resolution enzyme X-ray structures: Is seeing believing? *Current Opinion in Structural Biology*, **13**, 709–15. A review emphasizing how high-resolution structural data contribute to our understanding of enzyme mechanism.

Synchrotron X-ray sources

Ealick, S.E. and Walter, R.L. (1993). Synchrotron beamlines for macromolecular crystallography. *Current Opinions in Structural Biology*, **3**, 725–36. A review of the main synchrotron sources available internationally.

Hendrickson, W.A. and Ogata, C.M. (1997) Phase determination from multiwavelength anomalous diffraction measurements. *Methods in Enzymology*, **276**, 494–523. Review of anomalous diffraction in structure determination.

Hendrickson, W.A. (2000) Synchrotron crystallography. *Trends in Biochemical Sciences*, **25**, 637–43. A review of synchrotron radiation and its uses in structural biology.

Neutron diffraction

Fitter, J.J., Gutberlet, T. and Katsaras, J. (eds) (2006) *Neutron Scattering in Biology:Techniques and Applications*, Springer, Berlin. A collection of applications of neutron diffraction to biological problems including crystallography.

Niimura, N. and Bau, R. (2007) Neutron protein crystallography: Beyond the folding structure of biological macromolecules. *Acta Crystallographica A*, **64**, 12–22. An excellent overview of the development and current status of neutron diffraction.

Electron diffraction

Lacapere, J.-J., Pebay-Peyroula, E., Neumann, J.-M. and Etchebest, C. (2007) Determining membrane protein structures: Still a challenge! *Trends in Biochemical Sciences*, **32**, 259–70. A review of structure determination of membrane proteins.

Mitra, A.K., Yeager, M., Vanhoek, A.N. *et al.* (1994) Projection structure of the CHIP28 water channel in lipid bilayer membranes at 12 Å resolution. *Biochemistry*, **33**, 12735–40. A good example of the application of electron diffraction.

Walz, T. and Grigorieff, N. (1998) Electron crystallography of two-dimensional crystals of membrane proteins. *Journal of Structural Biology*, **121**, 142–61. A fascinating review of electron diffraction as a means to study the structures of membrane-bound proteins.

Weirich, T.E., Lábár, J.L. and Zou, X. (eds) (2006) *Electron Crystallography: Novel Approaches for Structure*, Springer, Berlin. A good overview of applications of electron diffraction methods.

Cryo-EM

Topf, M., Lasker, K., Webb, B. *et al.* (2008) Protein structure fitting and refinement guided by cryo-EM density. *Structure*, **16**, 295–307. An article on combining high-resolution and cryo-EM data.

Chiu, W., Baker, M.L., Jiang, W. *et al.* (2005) Electron cryomicroscopy of biological machines at subnanometer resolution. *Structure*, **13**, 363–72. A review of cryo-EM.

Some useful web sites

A movie of the GroEL ATPase cycle may be accessed at: http://www.cryst.bbk.ac.uk/~ubcg16z/cpn/elmovies.html.

Sequence databases: http://www.ncbi.nlm.nih.gov.

Protein database: http://www.rcsb.org.

Cambridge Crystallographic Data Centre: http://www.ccdc.cam.ac.uk.

List of synchrotron facilities worldwide: http://www.lightsources.org/cms/?pid=1000098.

Database of hydrogen and hydration in proteins (HHDB): http://hhdb01.tokai-sc.jaea.go.jp/HHDB/.

Chapter 7
Hydrodynamic Methods

Objectives

After completing this chapter you should be familiar with:

- *The principal effects of biomacromolecules and particles such as viruses/cells on flow characteristics of solutions and suspensions.*
- *The basis and main applications of centrifugation.*
- *Analytical and preparative aspects of filtration.*
- *Dialysis techniques and their uses.*
- *The basis and applications of flow cytometry.*

Because of the importance of water as the universal biological solvent, biochemical samples are usually handled and studied as aqueous solutions or suspensions. Some techniques described in this book involve interactions between biological samples present in solution/suspension and some sort of immobile stationary phase. These experiments may be critically affected by the flow characteristics of the solution/suspension (e.g. column chromatography, Chapter 2; capillary electrophoresis, Chapter 5). Moreover, common experimental manipulations in biochemistry such as centrifugation, dialysis and filtration are strongly influenced by flow interactions arising from *hydrodynamic properties* of the sample. These properties arise from physical interactions between molecules or particles and aqueous solvent. They are even more important in large-scale industrial situations encountered in biotechnology such as *downstream processing* of protein products (Chapter 2; Section 2.5.2). In order to understand such experimental situations more fully, it is helpful to consider the physical basis of hydrodynamic properties. This chapter will summarize the principal hydrodynamic properties of biomacromolecules and will describe their influence on flow-based experimental systems such as centrifugation, filtration, dialysis and flow cytometry. For simplicity, biomacromolecules (such as DNA and proteins), macromolecular complexes (such as ribosomes and viruses) and large biological entities (such as organelles and cells) are referred to as *particles* throughout this chapter.

7.1 VISCOSITY

Solutions/suspensions of particles are more viscous than the solvent alone. This alters the flow characteristics of the solution/suspension in a manner which depends on the concentration, shape and mass of the particle. This can have important consequences for the handling of such solutions/suspensions and may be used experimentally to learn about the shape and mass of the particle.

7.1.1 Definition of Viscosity

Movement of a particle across a stationary surface is inhibited by *friction*. The effect of this is to *slow* the movement of the particle. If the particle is present as a *fluid* (i.e. a liquid or gas), this friction gives rise to the phenomenon called *viscosity*. The molecular effects underlying viscosity depend partly on intermolecular interactions within the solution (e.g. van der Waals, hydrogen bonds, electrostatic interactions) and partly on the overall size and shape of solvent and solute molecules. Liquids are far more viscous than gases since intermolecular interactions are more common in the liquid than the gas phase. In the experimental situation described in Figure 7.1, flow of a liquid between two parallel plates separated by a distance, y, is shown. In 7.1(a), one of the plates is moving at a velocity, v, while the other plate is stationary and in 7.1(b) the liquid is moving with velocity, v, but *both* plates are stationary. Situations very similar to these could be expected to arise in many electrophoretic and chromatographic experimental formats using water as the main solvent. Since the velocity vectors of successive layers of water near to stationary plates would trace a parabolic shape, this phenomenon is called *parabolic flow*.

Newton observed that the layer of liquid immediately adjacent to the moving plate in 7.1(a) would have a velocity very close to v while that adjacent to the stationary plate would have a velocity close to zero. Thus, a *velocity gradient*

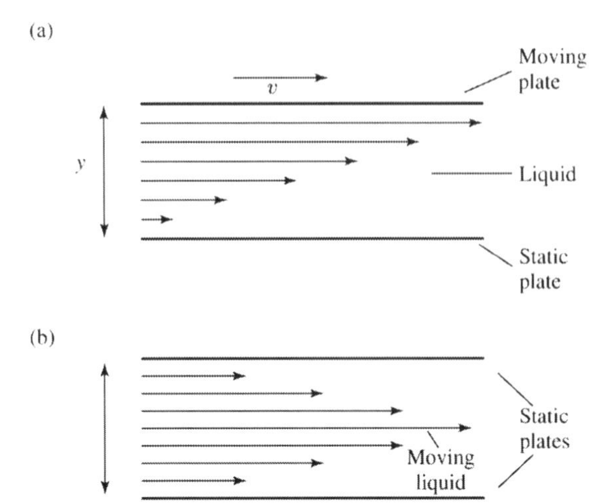

Figure 7.1. Viscosity, (a) A linear shear gradient dv/dy is established between two plates one of which is moving at speed, v, while the other is immobile, (b) A parabolic shear gradient is formed when liquid flows between two immobile plates.

(dv/dy) would exist across the solution between the two plates. He demonstrated that the *shear force*, f, between layers of liquid is proportional to *both* the area of the layers *and* the velocity gradient between them;

$$f \propto A \cdot \frac{dv}{dy} \tag{7.1}$$

The constant of proportionality in Equation (7.1) is the *viscosity*, η, of the solution;

$$f = \eta \cdot A \cdot \frac{dv}{dy} \tag{7.2}$$

In 7.1(a), friction is encountered at the stationary plate giving rise to a *linear* velocity gradient while, in 7.1(b), it is encountered at both plates giving rise to a *parabolic* velocity gradient. The term (f/A) is called the *shear stress* while dv/dy is the *shear gradient*. When η is a constant the fluid is said to be *Newtonian* while, if η is a function of either the shear stress or shear gradient, it is said to be *non-Newtonian*.

7.1.2 Measurement of Viscosity

Viscosity measurements are performed by measuring the time taken for a solution (volume, v) of a particle (e.g. a virus, organelle or cell), dissolved in a solvent of viscosity η_o to flow through a narrow capillary tube of radius, r and length, L under hydrostatic pressure. The most popular instrument is the *Ostwald viscometer* (Figure 7.2). The viscosity of the

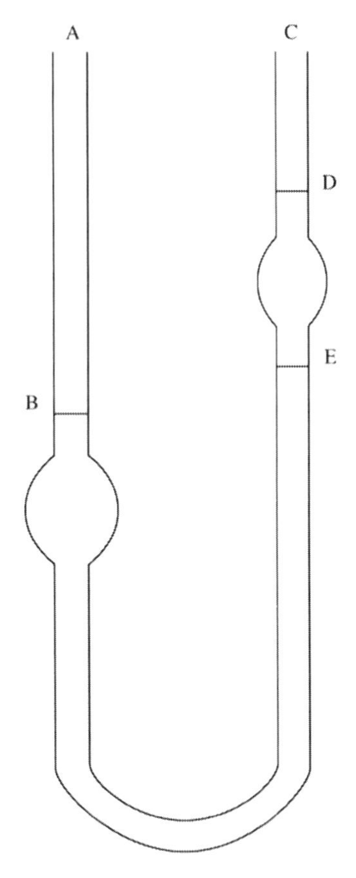

Figure 7.2. The Ostwald viscometer. Solution is added at opening A until the meniscus reaches point B. Suction is applied at point C until the liquid level is above point D. The time required for the meniscus of the liquid to flow from point D to point E is measured.

sample solution, η, is given by;

$$\eta = \frac{\pi \cdot h \cdot g \cdot \rho \cdot r^4 \cdot t}{8LV} \tag{7.3}$$

where h is the average liquid height, g the gravitational constant and ρ the density of the solution. Since viscosity measurements of the solvent alone (η_o) and of various concentrations of the particle (η) are carried out under identical experimental conditions, they may be conveniently expressed as *relative viscosity*, η_r, of the solution/suspension;

$$\eta_r = \frac{\eta}{\eta_o} = \frac{t \cdot \rho}{t_o \cdot \rho_o} \tag{7.4}$$

Thus, knowledge of the density of the different solutions under study is the only prior information required for calculation of relative viscosity using the Ostwald viscometer.

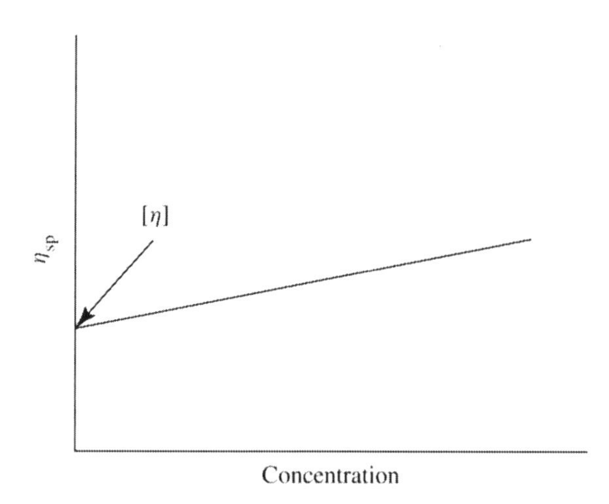

Figure 7.3. Measurement of intrinsic viscosity. Specific viscosity is measured at a range of concentrations of macromolecule/particle. Intrinsic viscosity ($[\eta]$) is the specific viscosity (η_{sp}) at a concentration of zero.

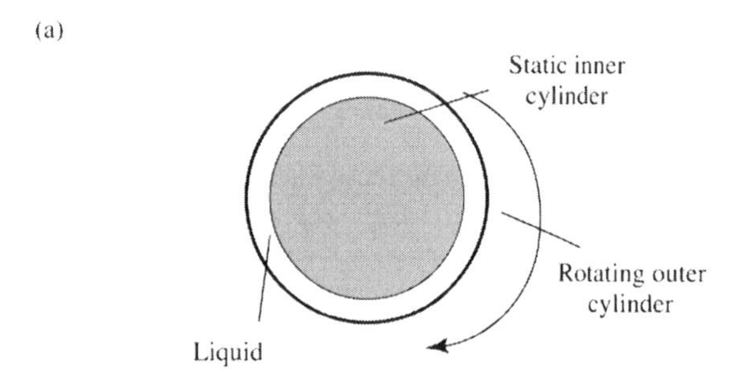

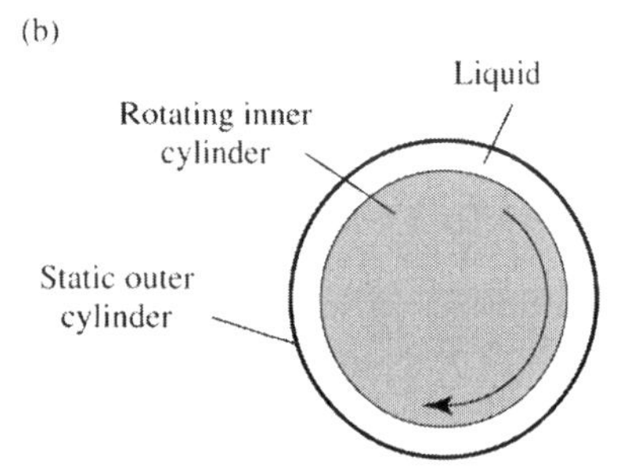

Figure 7.4. Couette-type viscometers. (a) Couette viscometer features a static inner cylinder inside a mobile outer cylinder. In this design the shear gradient is fixed while the shear stress may be experimentally varied, (b) The floating rotor viscometer in which the outer cylinder is static and the inner cylinder is free to rotate while floating in the liquid. In this design, the shear stress is fixed while the shear gradient may be experimentally varied.

It has been experimentally observed that this parameter is dependent on the shape, concentration and volume of the particle under investigation. Consequently, viscosity provides a simple means of determining aspects of the overall shape and mass of particles encountered in biochemistry.

7.1.3 Specific and Intrinsic Viscosity

An alternative, related measure to relative viscosity is *specific viscosity*, η_{sp};

$$\eta_{sp} = \frac{\eta - \eta_o}{\eta_o} = \eta_r - 1 \tag{7.5}$$

To relate specific viscosity to intrinsic molecular characteristics such as shape and volume, it is necessary to determine specific viscosity at a concentration of zero which is referred to as the *intrinsic viscosity*, $[\eta]$. This is achieved by measuring specific viscosity at a range of concentrations and extrapolating to zero (Figure 7.3).

The Ostwald viscometer has the disadvantage that the shear gradient, G, cannot be varied. This is a limitation for non-Newtonian fluids where it is necessary to measure η under conditions where $G = 0$ (since in such fluids η varies with G). More sophisticated experiments are possible with the *Couette viscometer* which consists of two concentric cylinders with a narrow space between them for liquid (Figure 7.4). This has previously been described in our discussion of experimental formats for the measurement of linear dichroism (Chapter 3; Figure 3.40). The value of G for such an instrument is;

$$G = \frac{\pi \cdot R \cdot v}{30d} \tag{7.6}$$

where R is the average radius of the cylinders, v is the rotor speed in revolutions per minute and d is the distance between the cylinders. The shear gradient is fixed while the shear stress may be calculated from;

$$F = \frac{T}{2\pi R^2 h} \tag{7.7}$$

where h is the height of the cylinder and T is the torque necessary to maintain the rotor speed, v.

A modification of this basic design is the floating rotor viscometer in which the outer cylinder is static and the inner cylinder is free to rotate while floating in the liquid. In this design, the shear stress is fixed and the shear gradient is varied.

7.1.4 Dependence of Viscosity on Characteristics of Solute

The relationship between $[\eta]$ and molecular mass and shape is complex. However, generally speaking, the larger the molecule and the more extended its shape, the greater the value of $[\eta]$. An important factor in overall shape is the *axial ratio* (i.e. the ratio if the longest to the shortest axis for the molecule). Fibrous molecules have larger axial ratios than globular molecules of the same mass and therefore give more viscous solutions (Example 7.1). For double-stranded DNA the relationship between mass and $[\eta]$ has been empirically determined as;

$$\log([\eta] + 5) = 0.665 \cdot \log M - 2.863 \qquad (7.8)$$

while for random coil proteins of n residues dissolved in 6N guanidinium hydrochloride it is;

$$[\eta] = 0.716 \cdot n^{0.66} \qquad (7.9)$$

Changes in viscosity may be used to follow processes in which the overall shape or elasticity of the molecule is altered. Examples of this include intercalation of drugs into DNA and protein denaturation (Example 7.2).

7.2 SEDIMENTATION

When particles are forced through a solution, they experience resistance to movement which depends on properties of the particle such as mass, shape and density and properties of the solvent such as its temperature, viscosity, density and composition. A commonly-used experimental format where this occurs is *sedimentation* in a centrifugal field which is also called *centrifugation*. This can be used as a *preparative* strategy to separate complex mixtures present in biological samples. Alternatively, it can also be used *analytically* to determine the mass, shape or density of particles.

7.2.1 Physical Basis of Centrifugation

Consider a particle of mass, m_o, suspended in a solvent of density, ρ. This particle would experience an upthrust equivalent to the weight of displaced liquid in accordance with *Archimedes's principle*. The *buoyant mass*, m, of the particle is therefore m_o *minus* a correction factor for this upthrust;

$$m = m_o - m_o \cdot \nu \cdot \rho \qquad (7.10)$$

where ν is the partial specific volume of the particle. This is equal to the volume displaced by the particle and is approximately equal to the reciprocal of the density of the solute although the degree of particle hydration can affect it. For example, charged particles (which *attract* solvent molecules, compacting them in its vicinity) give smaller values of ν than might be anticipated from particle density alone.

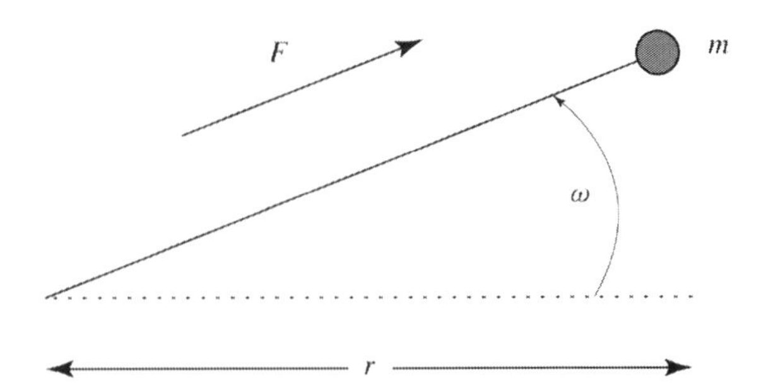

Figure 7.5. Centrifugal force. A particle of mass, m, experiences a centrifugal force, F, when centrifuged at an angular velocity, ω, around a point from which it is distant by a radius, r.

If this particle is exposed to a *centrifugal field* (Figure 7.5), it experiences a *centrifugal force*, F;

$$F = m \cdot \omega^2 \cdot r \qquad (7.11)$$

where ω is the *angular velocity of rotation* (in units of radians/sec. with one revolution $= 2\pi$ radians) and r is the distance of the particle from the centre of rotation (cm). Combining Equations (7.10) and (7.11) gives us;

$$F = m_o(1 - \nu \cdot \rho) \cdot \omega^2 \cdot r \qquad (7.12)$$

This equation means that, as the mass, angular velocity or distance from the centre of rotation increases, so does the centrifugal force experienced by a particle

As the particle passes through the solvent it experiences resistance due to a *frictional force*, F^*, operating in the opposite direction;

$$F^* = f \cdot v \qquad (7.13)$$

where f is a *frictional coefficient* dependent on the shape and mass of the particle and the viscosity of the solvent and v is the *velocity* of the particle. At the beginning of centrifugation, the particle accelerates through the solvent but eventually the frictional force decreases this acceleration such that a constant velocity called the *sedimentation velocity* is achieved. At this point;

$$\begin{aligned} &F = F^* \\ &\text{i.e.} \quad m_o(1 - \nu \cdot \rho) \cdot \omega^2 \cdot r = f \cdot v \end{aligned} \qquad (7.14)$$

Example 7.1 Use of viscometry in the study of actin polymerization

Chemical modification of actin in the lens of the eye is considered a major source of cataract formation leading to blindness. A primary source of the modification known as carbamylation is exposure to high levels of urea in the blood (uremia) which can be caused by certain clinical conditions (e.g. kidney failure, diarrhaea, dehydration). A possible mechanism of actin carbamylation involves isomerization of urea into ammonium cyanate followed by carbamylation of actin lysines:

UREA: $NH_2{-}C({=}O){-}NH_2$ ⟶ AMMONIUM CYANATE: $NH_4^+ \; N{\equiv}C{-}O^-$ (drawn as $NH_4^+{-}N{=}C{=}O$)

Ammonium cyanate + lysine residue ($—NH—CH(CO—)—(CH_2)_4—NH_2$) ⟶ carbamylated lysine ($—NH—CH(CO—)—(CH_2)_4—NH—C({=}O)—N—NH_4^+$)

This hypothesis was tested by exposing actin to the powerful carbamylating agent methylisothiocyanate and studying the effect of this on actin polymerization by viscometry using the Ostwald viscometer. Since polymerization results in a much larger protein assembly, it is to be expected that a solution containing such an assembly would be more viscous than one containing unpolymerized actin. Samples of various mixtures of carbamylated and uncarbamylated actin taken at various time-points over 24 min were analysed in the Ostwald viscometer and data for specific viscosity are presented in the following graph.

Specific viscosity of control and carbamylated actin as a function of time.

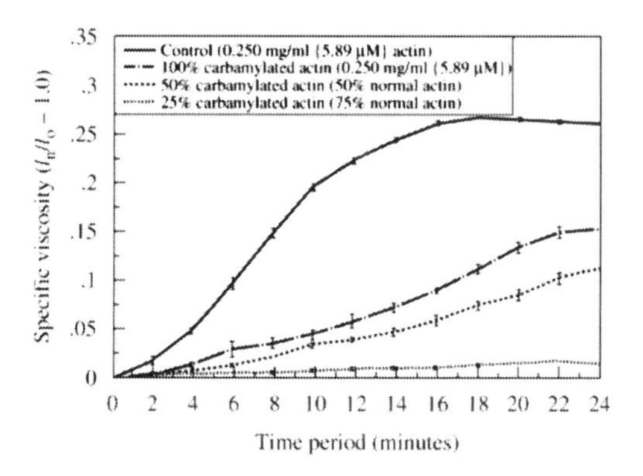

(Continues)

(Continued)

This confirms that carbamylation abolishes polymerization and thus demonstrates that lysines are essential for this process. Moreover, inclusion of carbamylated actin with unmodified protein inhibits polymerization which raises the possibility that small amounts of carbamylated protein can seriously alter rates and extent of actin polymerization in the cytoplasm.

The sedimentation velocity is given by;

$$v = \frac{m_o(1 - \nu \cdot \rho) \cdot \omega^2 \cdot r}{f} \qquad (7.15)$$

Sedimentation velocity is dependent on the shape of the particle but, for a perfect sphere, this quantity is related to physical properties of the particle and solvent by;

$$v = \frac{a^2[\rho_p - \rho_m]}{18\eta} \cdot \omega^2 r \qquad (7.16)$$

where a is the particle diameter, ρ_p and ρ_m are the densities of the particle and solvent, respectively and η is the viscosity of the solvent. The inverse relationship with viscosity means that sedimentation velocity *decreases* markedly with increase in solvent viscosity and (since viscosity is dependent on temperature) with temperature. For this reason, sedimentation velocities are usually corrected for different solvent composition and temperature to standard conditions of pure water at 20 °C.

The relationship described by Equation (7.16) underlies all centrifugation experiments in biochemistry and is the reason why different particles (i.e. with differing partial specific volumes and mass) can achieve individual sedimentation velocities in a single solvent at a single angular velocity. It provides the basis for *differential centrifugation* in which particles may be conveniently separated from each other.

It is difficult to determine ν, ρ and f accurately so a useful measure of differential behaviour in a centrifugal field is provided by the *sedimentation coefficient*, s;

$$s = \frac{v}{\omega^2 \cdot r} = \frac{m_o(1 - \nu \cdot \rho)}{f} \qquad (7.17)$$

Since both $m_o(1 - \nu \cdot \rho)$ and f are inherent properties of the particle and solvent, respectively, s is a constant for any given particle/solvent. The range of values observed in biochemistry varies widely from as low as 1×10^{-13} s for a protein to $10\,000 \times 10^{-13}$ s for intracellular organelles. A more useful notation is provided by the *Svedberg*, S, with one Svedberg equivalent to 1×10^{-13} s. There is a loose relationship between sedimentation coefficient and molecular mass in that a protein such as cytochrome C (1S) has a smaller value than bacterial ribosomes (70S), tobacco mosaic virus (400S) or lysosomes (10 000S). However, this relationship is not a simple or linear one since, for example, the 70S ribosome is composed of particles of 50S and 30S, respectively. Sedimentation coefficients are nonetheless useful indicators of relative size especially in massive particles such as viruses.

7.2.2 The Svedberg Equation

The frictional coefficient is related to a number of other physical parameters;

$$f = \frac{R \cdot T}{N \cdot D} \qquad (7.18)$$

where R is the universal gas constant, N is Avogadro's number (Appendix 1), T is the absolute temperature and D is the *diffusion coefficient*. Combination of Equations (7.17) and (7.18) gives us the relationship between s and these physical parameters;

$$s = \frac{m_o(1 - \nu \cdot \rho)}{R \cdot T/N \cdot D} \qquad (7.19)$$

This gives us;

$$R \cdot T \cdot s = N \cdot D \cdot m_o(1 - \nu \cdot \rho) \qquad (7.20)$$

But molecular mass, M_r, is the product of m_o and N so;

$$M_r = \frac{R \cdot T \cdot s}{D \cdot (1 - \nu \cdot \rho)} \qquad (7.21)$$

This is the *Svedberg equation* which relates molecular mass to intrinsic physical properties of the particle (s and ν), experimental conditions (D, T and ρ) and a physical constant (R). Since many of these terms are experimentally measurable, this equation provides us with an *a priori*

Example 7.2 Use of viscometry in the study of ligand binding to DNA

Noncovalent binding of ligands to DNA may occur by a number of distinct mechanisms. One of the most important of these is intercalation into the double-helix as exemplified by some planar compounds such as the fluor ethidium bromide (Chapter 3) and the dye aniline. Intercalation can have major biological consequences such as the introduction of mutations which underlies the antibacterial activity of molecules like aniline.

The monovalent cationic complex bis(1,10-phenanthroline) copper (I) [abbreviated to; $(Phen)_2Cu^I$] has the unusual property that it binds noncovalently to DNA and this process results in cleavage which shows some degree of sequence specificity. This makes it interesting as a probe for local DNA conformation.

Intercalation into DNA makes the molecule far more rigid and thus significantly increases the viscosity of DNA solutions. The reason for this is that intercalation results in greater distance between base-pairs leading to an apparent increase in length of the molecule. Viscometry is therefore a useful measure of intercalation.

Measurements of specific viscosity were performed in a capillary viscometer on 800 bp lengths of DNA in the presence of the following reagents; (1) Ethidium bromide; (2–4) phenanthrolene mixed with $CuSO_4$ in ratios 1 : 1, 2 : 1 and 4 : 1, respectively; (5) 2 : 1 2,9 dimethyl-1,10-phenanthrolene:$CuSO_4$; (6) phenanthrolene alone. The mixture of phenanthrolene (in the ratio 2 : 1) and 2, 9 dimethyl-1,10-phenanthrolene with $CuSO_4$ generates the $(Phen)_2Cu^I$ and bis(2,9 dimethyl-1,10-phenanthrolene) copper(I) complexes, respectively.

The results are expressed as ratios of DNA specific viscosity (hsp) in the presence (Cu) and absence (Cu = 0) of ligand.

These data demonstrate that ethidium bromide and phenanthrolene-copper complexes increase DNA specific viscosity as a result of intercalation. The1:1 ratio of phenanthrolene: copper generates mainly $(Phen)Cu^I$ while the 2 : 1 ratio generates $(Phen)_2Cu^I$. The 4 : 1 ratio contains considerable free phenanthrolene which inhibits $(Phen)_2Cu^I$ binding to DNA. Both phenanthrolene alone and bis(2,9 dimethyl-1,10-phenanthrolene) copper(I) complex resulted in negligible intercalation. Spectroscopic measurements confirmed that neither of these reagents binds to DNA. The presence of two methyl groups on each phenanthrolene substituent of the $(Phen)_2Cu^I$ complex appears to be sufficient to abolish intercalation which is thought to be due to structural rigidity in the molecule and the difficulty of oxidizing copper(I).

Reprinted with permission from Veal and Rill, Noncovalent Binding of Bis (1, 10-phenanthrolene) Copper(i) and Related Compounds, *Biochemistry* **30**, 1132–1140, © (1991) American Chemical Society.

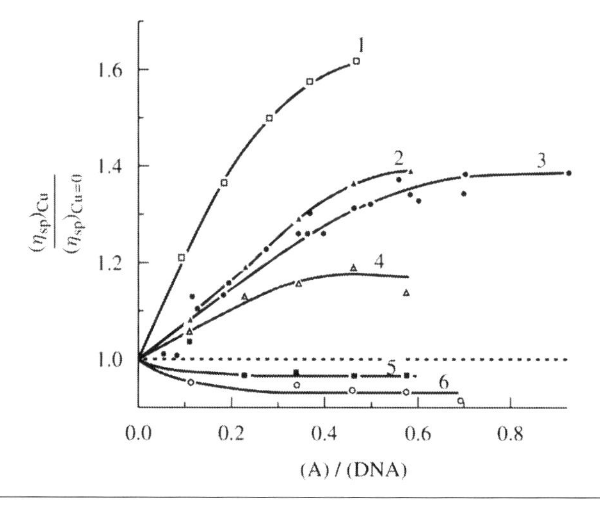

means of *directly* determining molecular mass. This contrasts with *empirical* methods of mass determination using approaches such as electrophoresis (Chapter 5), chromatography (Chapter 2) or mass/charge ratio (Chapter 4) which rely on *comparison* with molecules of known molecular mass.

7.2.3 Equipment Used in Centrifugation

Centrifugation is performed in a *centrifuge* which provides a system for rotating sample at high speed around a central point so as to develop significant centrifugal force (Figure 7.6). The sample is placed in a reinforced plastic tube which is then held in a *rotor* which rotates around a *spindle*.

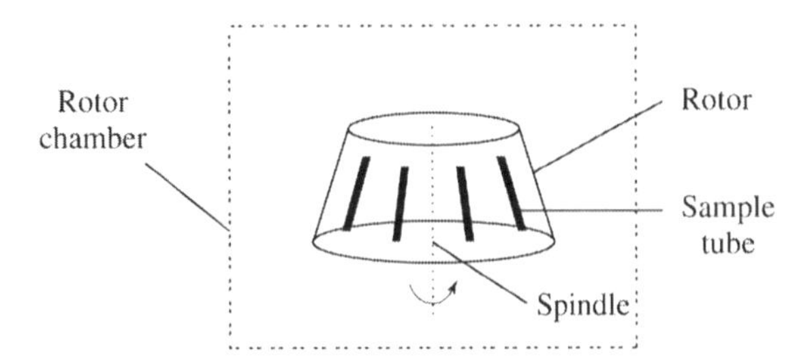

Figure 7.6. Outline design of a centrifuge. Samples contained in sample tubes (only four are shown for clarity) are placed in a rotor which is fixed on a spindle. Centrifugation takes place in the rotor chamber (dashed lines) which is often armoured, refrigerated and evacuated.

Additional possible features include refrigeration (which usually also involves evacuation of the rotor chamber), handling different volumes of samples, ability to achieve extremely high speeds and the possibility of analysis of the sample during the experiment. High-speed centrifuges are usually heavily armoured around the rotor chamber to protect against accidents which might involve the spindle breaking and release of the rotor (for this reason it is essential to ensure that sample tubes placed on opposite sides of the spindle are evenly-balanced by weighing them before the experiment).

The rotor may be of two general designs (Figure 7.7). In *fixed angle rotors* the tube is placed in a machined hole in the metal rotor which is at a fixed angle relative to the vertical axis of the spindle. This angle does not change during the centrifugation experiment and the sample is centrifuged against the side-wall of the tube. In *swinging bucket rotors*, the tube is placed in a holder which is suspended from the rotor. During the experiment, the holder swings out to become horizontal with the horizontal axis of the rotor and sample is centrifuged towards the bottom of the tube. Because rotor design varies so much between manufacturers and the distance between the spindle and the sample (r in Equation (7.15) above) can differ widely, it is often convenient to express centrifugal force as a value relative to that of the Earth's gravitational field called the *relative centrifugal force*, g;

$$g = 1.119 \cdot 10^{-5} \cdot (\text{rpm})^2 \cdot r \tag{7.22}$$

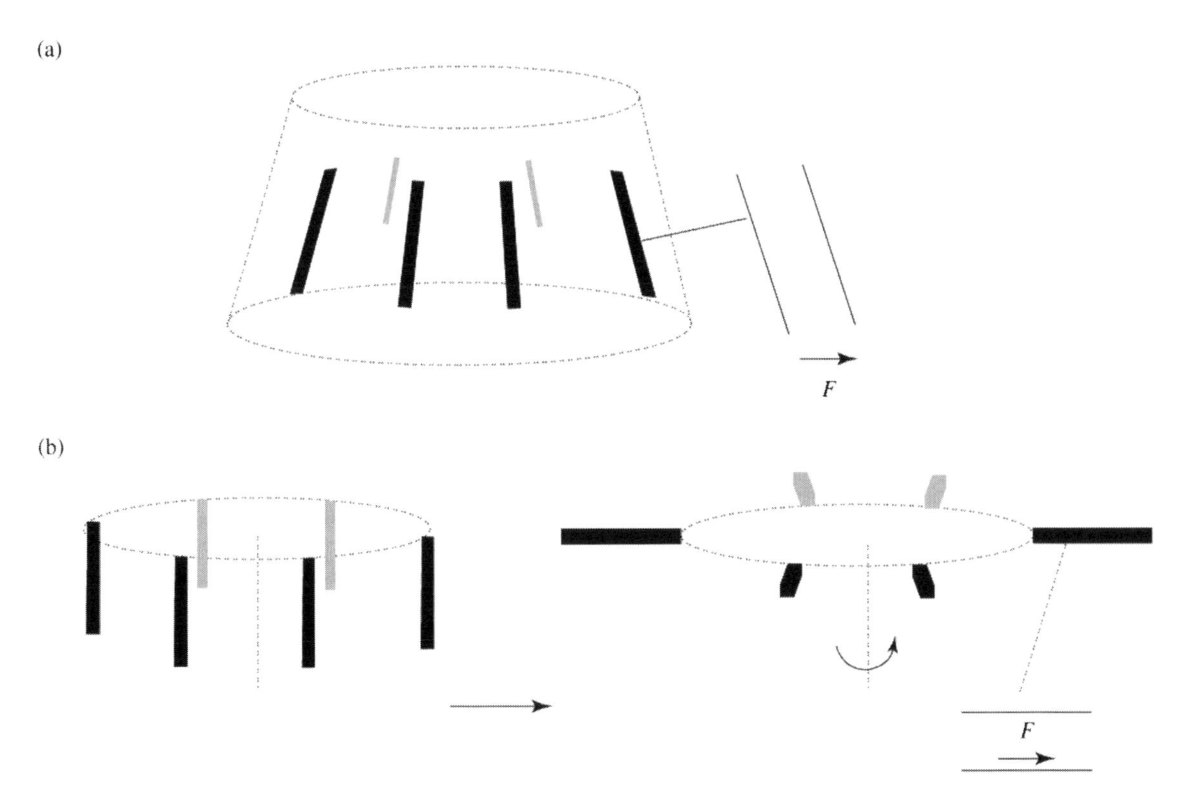

Figure 7.7. Rotors used in centrifugation. (a) Fixed-angle rotors. The rotor is solid steel with openings machined in it for sample tubes. The angle of the tubes does not change during the experiment and particles are sedimented against the back of the tube, (b) Swinging bucket rotor. The sample tubes are suspended on the rotor and swing out during the experiment. Particles sediment towards the bottom of the tube.

where rpm is the number of *revolutions per minute* of the rotor. Manufacturers usually provide a scale for each rotor by which rpm values on the instrument can be converted into g values in which centrifugation conditions should be reported. Different rotors accept sample tubes of different size allowing a range of volumes of sample to be centrifuged.

Three levels of centrifuge can be described in order of increasing sophistication (and price!). The first two would be encountered by most practicing biochemists and molecular biologists while the third is normally only used by specialists.

The most widespread type of centrifuge is the *benchtop centrifuge*. This may lack a refrigerated rotor chamber and can accept small volumes (e.g. 1–2 ml) of sample for centrifugation at relatively low speed (e.g. 6000 g). This is useful for collecting precipitated protein or DNA as a compact pellet from the bulk sample (e.g. from ammonium sulfate or ethanol, respectively). Higher speed versions need to be refrigerated because high speed heats up the rotor due to air friction. Small volume benchtop centrifuges are sometimes called *minifuges* and these can achieve relative centrifugal forces of some 12 000 g with 0.5–1.5 ml sample volumes.

High-speed large-volume *ultracentrifuges* can reach relative centrifugal force values of up to 150 000 g. These may accept swinging bucket or fixed angle rotors, have refrigerated (and evacuated) temperature-controlled rotor chambers and can accept sample tubes with volumes from 10 to 250 ml. They would be routinely used in centrifugation of tissue extracts and in recovery of precipitated protein from ammonium sulfate precipitation.

Analytical ultracentrifuges achieve relative centrifugal forces of up to 600 000 g and facilitate analysis of the behaviour of sample particles *during* the centrifugation experiment. The development of computer-controlled XL-A and XL-I centrifuges by Beckman Instruments combined with developments in data analysis has led to a resurgence of interest in analytical centrifugation. Such measurements allow the determination of M_r value with the Svedberg equation (Equation (7.21)) provided the other parameters in that equation are known. A feature of analytical centrifuges is the presence of an optical system for detection of particle movement during the experiment. These depend on either direct absorbance measurements (for the XL-A in the range 190–650 nm; Chapter 3) or on comparison of refractive index of solvent alone with that of sample which results in *interference* (the XL-I has both absorption and interference optics). These measurements follow the absorbance/refractive index of particles as a function of position during centrifugation by means of a window in the cell containing sample which passes through the detection system once during each revolution (Figure 7.8). Fluorescence detection brings

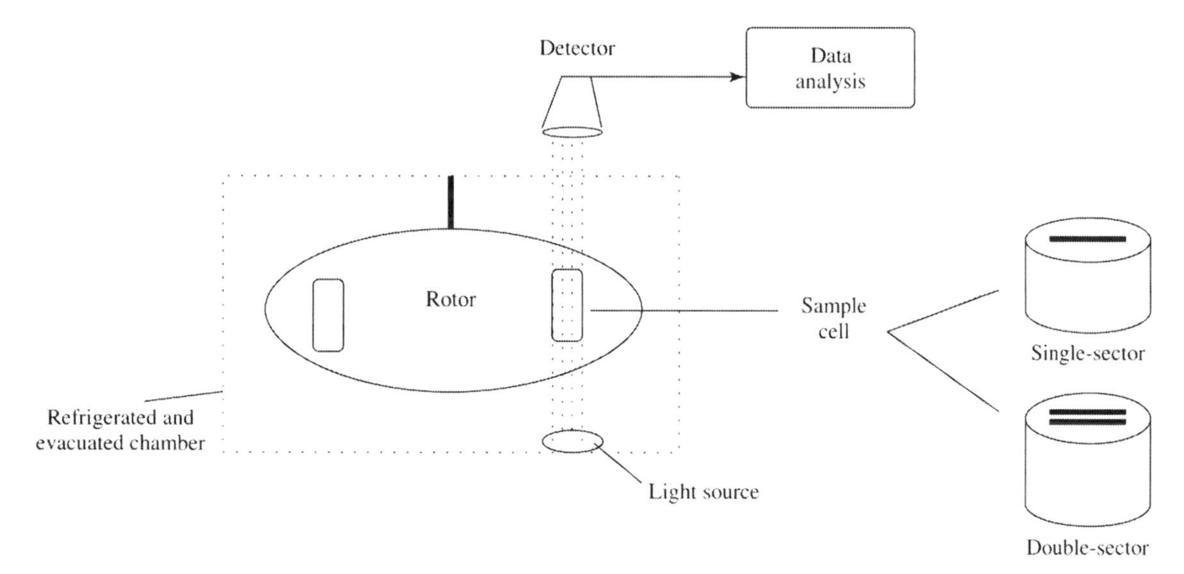

Figure 7.8. Outline design of an analytical ultracentrifuge. Sample is placed in a sample cell in a rotor contained in an evacuated and refrigerated chamber. Centrifugation is carried out at very high speed during which the behaviour of the sample is analysed. The cell is scanned with parallel light as it passes through the detection system on each revolution. Transparent windows at the top and bottom of sample cells allow light to pass through the sample. Optical systems use absorbance, fluorescence or refractive index to detect sample concentration as a function of position in the cell. Refractive index measurements of sample is continuously compared to that of solvent alone in a double-sector sample cell shown on the right. These data are continuously collected during the experiment, stored in a microcomputer and subsequently analyzed.

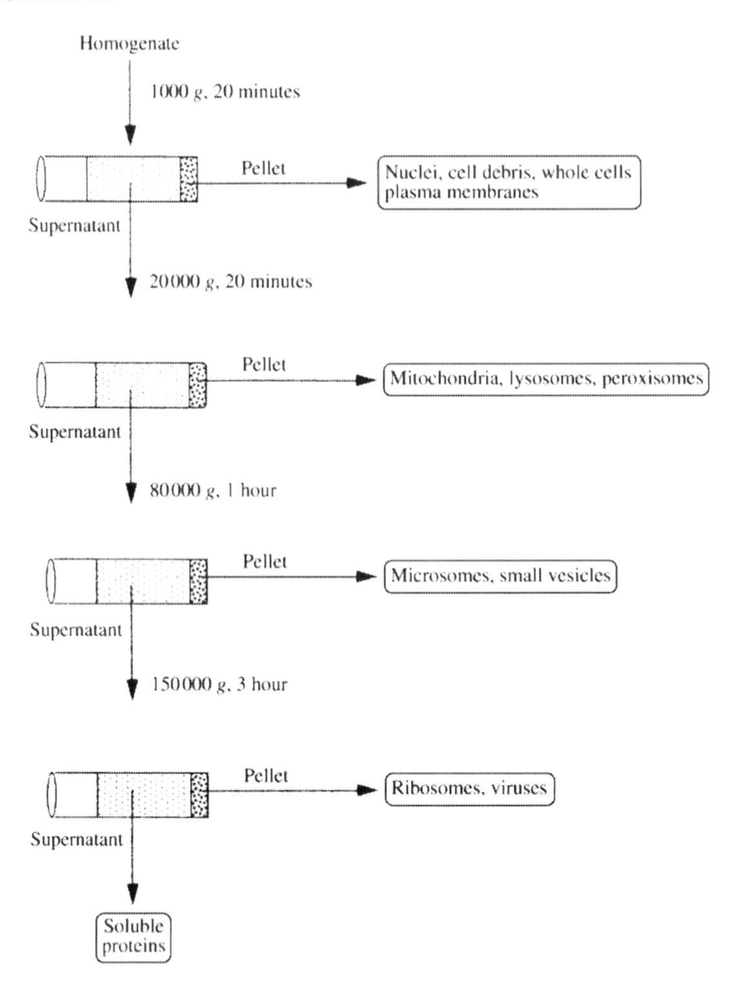

Figure 7.9. Subcellular fractionation. A tissue homogenate may be fractionated into nuclear, mitochondrial, microsomal and soluble fractions. These are enriched for biomolecules normally associated with the organelles/complexes shown. Centrifugation conditions between those described are also possible to fractionate the homogenate further. Cross-contamination between fractions can be assessed with suitable biochemical markers (Table 7.1).

the detection limit as low as 10^{-10} M for proteins and DNA labeled with extrinsic fluors.

The following (Sections 7.2.4 to 7.2.6) describe some applications of centrifugation in biochemistry.

7.2.4 Subcellular Fractionation

Individual subcellular organelles of eukaryotic cells may be regarded as individual types of particles with characteristic ranges of mass and density. Centrifugation of cell homogenates therefore allows preparation of highly-purified fractions enriched in one or other cell compartment such as the nucleus, endoplasmic reticulum or mitochondria in a process called *subcellular fractionation* (Figure 7.9). This is a useful experiment to identify the most likely subcellular location of a specific biomolecule or process. In carrying out such experiments it is important to assess the quality of separation of individual fractions and this is achieved by using marker molecules (usually enzymes) known to be expressed in a compartment (Table 7.1).

Table 7.1. Marker molecules useful in subcellular fractionation studies

Compartment	Marker
Nucleus	DNA
Cytosol	Lactate dehydrogenase
Mitochondrion	Succinate dehydrogenase
Lysosomes	Alkaline phosphatase
Microsomes	Cytochrome P-450

Ideally, 100% of the marker present in the extract should be detected in the subcellular fraction corresponding to its known subcellular location with none being detectable in other fractions. Greater and lesser values than 100% can be found as a result of the removal of some inhibitory molecule to another fraction (greater) or loss of activity due to denaturation or proteolysis (lesser). Moreover, in practice some contamination between fractions usually occurs due to handling errors but measurement of the markers allows this to be quantified. If a protein under investigation is found, for example, to mirror the distribution of succinate dehydrogenase across the subcellular fractions, then it is reasonable to conclude that its subcellular location is the mitochondrion. Isolation of high-purity subcellular organelles is a key step in *organelle proteomics* (Chapter 10).

7.2.5 Density Gradient Centrifugation

We can make use of the dependence of sedimentation velocity on solvent and particle density (Equation (7.16)) by centrifuging particles through a medium of *gradually increasing* density. This is achieved by establishing a *density gradient* between a region of high density at the bottom of the centrifugation tube and a region of low density at the top (Figure 7.10). The density gradient can be established either before particle centrifugation (*zonal centrifugation*) or during centrifugation (*isopycnic centrifugation*).

The zonal centrifugation experiment involves creation of a density gradient before centrifugation by mixing high and low density solutions of materials such as sucrose and glycerol (Figure 7.11). This gradient is created in a centrifuge tube prior to centrifugation and sample is then layered on top. Particles sediment through the gradient with gradually decreasing velocity. This is because the term $(\rho_p - \rho_m)$ is large at low density giving a relatively high velocity (Equation (7.16)). As the density of the medium increases (ρ_m) on passage through the gradient, this term gradually decreases. Eventually a point in the gradient is reached where $\rho_p = \rho_m$, resulting in sedimentation velocity values of zero.

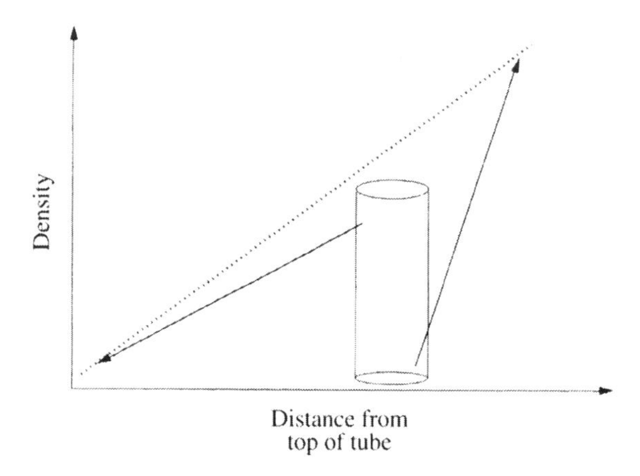

Figure 7.10. Density gradient centrifugation. A density gradient (dashed line) can be formed either during or prior to centrifugation due to a concentration of high-density material at the bottom of the centrifuge tube and correspondingly lower density near the top. An ideal, linear density gradient is shown although other gradients with other shapes are also possible.

This means that, because different particle populations have different densities, they sediment at different rates and separate out from the sample mixture to different densities. If the gradient is collected in individual fractions, highly-purified particles may be obtained in this way.

In isopycnic centrifugation, the metal salt $CsCl_2$ is used because of the high density possible with this material (up to 1.8 g/ml). A solution of $CsCl_2$ is mixed with the sample and then centrifuged which forms a $CsCl_2$ density gradient from low density at the top of the centrifuge tube to high density at the bottom (Figure 7.12). Sample particles in the region of highest density at the bottom of the tube are less dense than the surrounding medium. They therefore tend to float towards the top of the tube to a region of lower density. The driving force for this is that centrifugal force drives salt molecules towards the bottom of the tube thus exerting upward pressure on particles of lower density. Conversely, some of the same type of particle in the region of lowest density at the top of the tube will sediment towards the bottom of the tube in accordance with Equation (7.16). Eventually each particle collects and concentrates in a narrow band at its *isopycnic density*. A classic application of this is the *Meselson-Stahl experiment* which demonstrated the semiconservative nature of DNA replication by separating 'heavy' (^{15}N-containing) DNA from 'light' (^{14}N-containing) DNA in a density gradient.

With the exception of lipoproteins (which have quite low density), the densities of protein molecules are generally similar and this technique is not very useful for separating proteins from each other (the solutions that would be required would also not be compatible with native protein structure). It has been widely-used, however, for separating different forms of DNA (i.e. plasmid from chromosomal; supercoiled from uncoiled) and for preparation of macromolecular assemblies (e.g. viruses) and individual cell types. For DNA experiments, it is usual to add the extrinsic fluor ethidium bromide to the solution prior to centrifugation. This intercalates into the double-stranded DNA allowing its convenient visualization under a UV lamp (Section 3.4; Table 3.4). Materials such as ficoll and percol are particularly useful for separation of individual organelles or cell types by density gradient centrifugation since they do not subject particles to high osmotic pressure or shear forces. Density gradients (whether prepared by centrifugation or

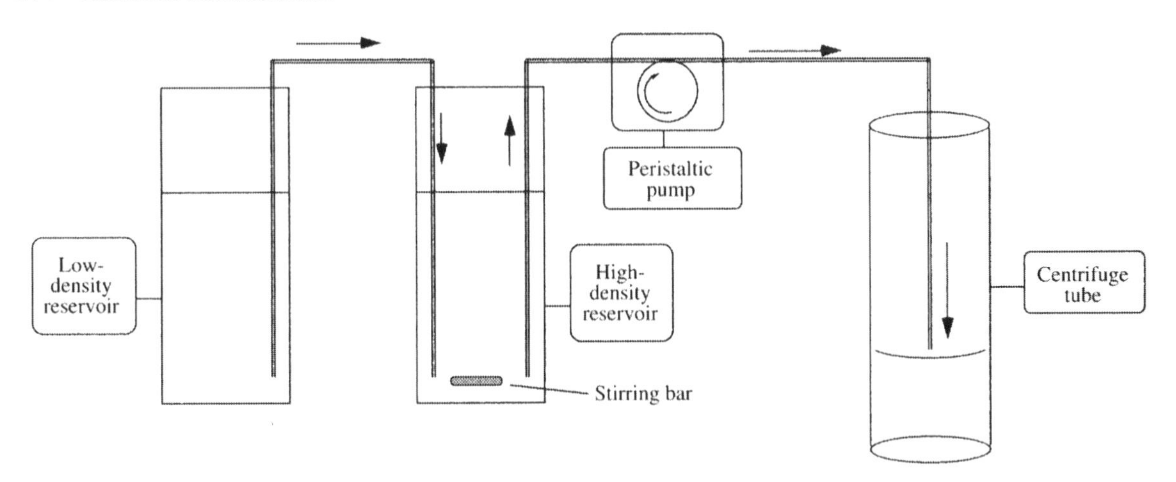

Figure 7.11. Formation of a density gradient prior to centrifugation. Two solutions of sucrose or glycerol corresponding to the lowest and highest density required are placed in separate reservoirs as shown. The low density solution is connected to the high density solution by tubing filled with low density solution. The high density solution is similarly connected to the bottom of the centrifuge tube by tubing filled with high density solution. A peristaltic pump is used slowly to fill the centrifuge tube (direction of flow shown by arrows). As the level of the high density reservoir drops, that of the low density reservoir also drops in response to hydrostatic pressure. Constant stirring of the high density solution ensures gradual lowering of solution density and formation of a smooth gradient.

otherwise) provide an extremely useful general method to determine the density of a wide range of particles such as viruses, organelles, cells and crystals of protein/DNA (Chapter 6).

7.2.6 Analytical Ultracentrifugation

So far we have described two general types of centrifugation experiment. Differential centrifugation allows separation of organelles and other biochemical compartments from each other on the basis of differential sedimentation in a centrifugal field. Density gradient centrifugation allows separation of particles in density gradients based on their differing density. Despite their obvious differences in methodology, both of these experiments involve analysis *after* centrifugation has been carried out.

It is also possible to combine analysis with the process of sedimentation and to learn about certain physical properties of particles from analysis of their dynamic behaviour in the centrifugal field. This is called *analytical ultracentrifugation* and, due mainly to improvements in equipment and data analysis, this field is currently undergoing a renaissance.

Analytical ultracentrifugation offers a number of major advantages in that it is possible to study the behaviour of biomacromolecules under a wide range of solvent conditions where they *will not interact with support materials* (which is a common problem with electrophoresis and chromatography systems). Moreover, sedimentation analysis allows behaviour of macromolecules across a *wide concentration range* to be studied in a single experiment whereas most other methods described in this volume require individual measurements to be performed at each sample concentration. Thirdly, samples in a *size-range* up to tens of millions of daltons which are difficult to study by other methods (e.g. viruses) are amenable to sedimentation analysis. In current practice, two distinct types of analytical centrifugation experiment are commonly performed and these are described in the following two sections.

7.2.7 Sedimentation Velocity Analysis

Sedimentation velocity analysis involves following the behaviour of sample particles as they move through the solvent during sedimentation at relatively high centrifugal speeds. Particles become distributed in a concentration gradient of characteristic shape which changes during the experiment (Figure 7.13). At the beginning of the experiment, sample is distributed uniformly through the sample cell. The lamella of molecules at the interface between this solution and air is referred to as the *boundary*. During centrifugation, these molecules move towards the bottom of the sample cell which causes a change in the boundary position and shape which may be followed by absorbance, fluorescence or refractive index. This is because *sedimentation* forces the sample particles to move through the solvent away from the centre of rotation while *diffusion* causes spreading of the boundary resulting in a change in boundary shape during the experiment. Computer methods for analysis of these boundaries have been greatly improved in recent years making a

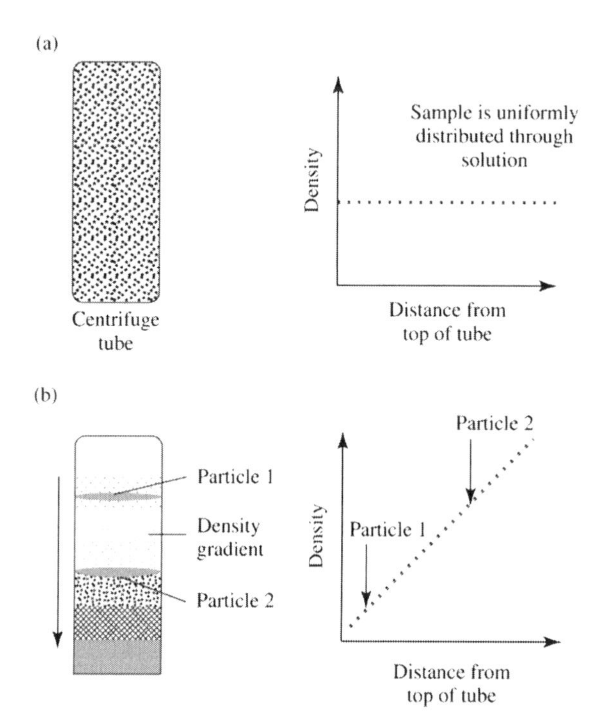

Figure 7.12. Formation of a density gradient during centrifugation. (a) A solution of uniform density of a material such as $CsCl_2$ is mixed with sample and placed in a centrifuge tube, (b) During centrifugation centrifugal force (arrow) concentrates $CsCl_2$ towards the bottom of the tube leading to greater density at the bottom and lower density at the top. Sample particles fractionate as they concentrate at their individual isopycnic densities. The shape of the density gradient obtained is shown in the graph (right), together with isopycnic densities of two particles from the sample.

number of physical and chemical parameters accessible to determination by this method. These include the *buoyant mass* $[m_o(1 - \nu \cdot \rho)]$, sedimentation and diffusion coefficients of individual molecular species.

The sedimentation coefficient, s, is the ratio of velocity to acceleration of the centrifuged particle which is related to physical and hydrodynamic properties of the sample and solvent (Equation (7.17)). If we could measure the position, r, of a boundary formed by a dilute solution of the particle at any time, t, we could experimentally determine the *apparent sedimentation coefficient*, s^*, by sedimentation velocity. This is determined by measuring the rate of movement of the boundary through the column of solvent in the centrifugation cell;

$$s^* = \frac{\mathrm{d} \ln r}{\omega^2 \,\mathrm{d}t} \tag{7.23}$$

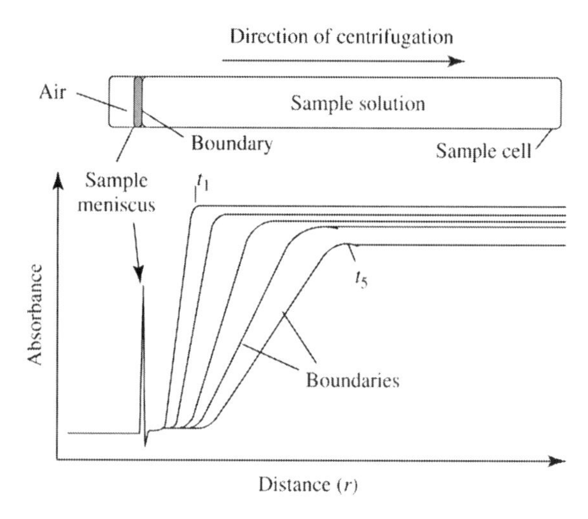

Figure 7.13. Sedimentation velocity. Sample is dissolved in solvent in sample cell. It is then centrifuged at high speed as shown and absorbance (or fluorescence or refractive index) are measured along the cell at various times during the experiment. Data for t_1 to t_5 are shown. The sample meniscus (arrow) is the interface between sample and air. This is referred to as the boundary. The shape and position of the boundary changes with time and this data is analyzed to determine s, D and M_r.

Which is related to the position of the meniscus at the top of the solution, r_m, by;

$$s^* = \frac{\ln r/r_m}{\int \omega^2 \,\mathrm{d}t} \tag{7.24}$$

This relationship may be used to convert boundary positions relative to the meniscus to values of s^*.

Diffusion is responsible for *boundary spreading* during the experiment and may be described by *Fick's first law* which relates the *diffusive flux*, J_D to the concentration of material, c, at any point, r, by the diffusion coefficient, D. This may be experimentally determined as the rate of boundary spreading $\partial c/\partial r$;

$$J_D = -D \cdot \frac{\partial c}{\partial r} \tag{7.25}$$

D can therefore be determined *simultaneously* with s^* by sedimentation velocity in relatively long centrifugation cells.

For more concentrated particle solutions, a number of corrections need to made to s^* and D to allow for factors such as hydrodynamic and thermodynamic *nonideality*. For example, solvent displaced by one particle can have an effect on sedimentation of another particle in concentrated solutions. These corrections result in a value for the parameters which would apply at *standard conditions* of near-zero concentration, at 20 °C, in pure water; $s^{o}_{20,w}$ and $D^{o}_{20,w}$.

These parameters have a number of practical uses. The buoyant molecular mass can be estimated from the values of $s^{o}_{20,w}$ and $D^{o}_{20,w}$ using the Svedberg equation (Equation (7.21)). If the molecular weight of the particle is already known, it is possible to estimate its degree of *solvation* (e.g. protein hydration) and overall *shape* during sedimentation because $s^{o}_{20,w}$ depends strongly on the frictional coefficient, f, which in turn depends on the *Stoke's radius*, R_s;

$$f = 6\pi\eta\, R_s \tag{7.26}$$

Both degree of solvation and deviation from a compact spherical shape *increase* R_s and hence f so that unexpected values for $s^{o}_{20,w}$ can be used to detect and quantify solvation and shape effects.

The analysis of more complicated experimental systems such as interacting systems (e.g. macromolecule-ligand) and multi-component systems (in which several boundaries are formed during sedimentation) is more complex but is facilitated by the availability of computer programs for data analysis which incorporate mathematical models for such systems. Fitting of experimental data to these models allows rapid estimation of quantities such as stoichiometry and binding constants for ligands (Example 7.3). More information on the analytical approaches applicable to such complex systems may be obtained from the excellent reviews cited in the bibliography at the end of this chapter.

7.2.8 Sedimentation Equilibrium Analysis

The sedimentation conditions for *sedimentation equilibrium analysis* differ from those used in sedimentation velocity in that much *lower* centrifugal speeds are used in much *shorter* columns of solvent. An equilibrium is formed between sedimentation and diffusion which results in formation of a smooth concentration gradient rather than fixed boundaries (Figure 7.14). Because formation of a stable equilibrium is quite slow, sedimentation equilibrium experiments may require days of centrifugation making them also much *slower* than sedimentation velocity.

If we consider a plane of area, A, the diffusive flux, J_D (Equation (7.25)), is equal to the sedimentation flux, J_S under these conditions;

$$J_D = J_S = \frac{c \cdot s \cdot \omega^2 r}{A} \tag{7.27}$$

where c is the concentration of the particle. Combining Equations (7.25) and (7.27) gives us;

$$\frac{s\omega^2}{D} = \frac{1}{r \cdot c} \cdot \frac{\mathrm{d}c}{\mathrm{d}r} = \frac{\mathrm{d}\ln c}{\mathrm{d}r^2/2} = \sigma \tag{7.28}$$

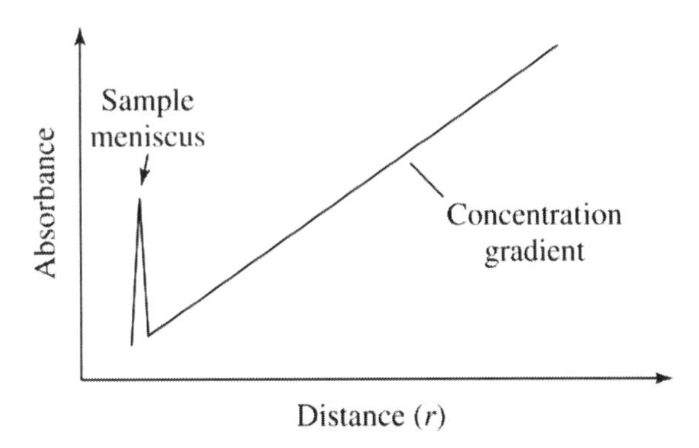

Figure 7.14. Sedimentation equilibrium. Centrifugation achieves an equilibrium between the rate of diffusion and of sedimentation. The time required for this is usually 70–100 h depending on the simplicity of the system, particle characteristics, solvent viscosity and other factors. Achievement of equilibrium can be confirmed by superimposing scans collected several hours apart. The position of the equilibrium (i.e. the shape of the concentration gradient) depends on thermodynamic properties of system components. The gradient shown is idealised for simplicity as real concentration gradients have characteristic and not necessarily linear shapes.

or;

$$\sigma = \frac{M \cdot \omega^2}{RT} \tag{7.29}$$

The concentration at any point of position r is related to the concentration, c_o, at a reference radius, r_o by;

$$c(r) = c_o\, e^{\sigma\xi} \tag{7.30}$$

where ξ is $(r^2 - r_o^2)/2$.

Rotor speeds (i.e. ω) are selected such that the value for σ lies in the range 2 to 15 cm^{-2}. The ratio of concentration at the base of the cell is 2–10 times that at the meniscus in a typical 3 mm centrifugation cell. A range of concentrations are used in different cells so that, in a typical experiment, a 100-fold range of concentration is used.

Provided that appropriate solvent conditions are chosen so that it does not contain very high concentrations of individual molecules (e.g. 8 M urea) or molecules that bind to proteins in significant quantities (e.g. detergents), the concentration gradient developed in a sedimentation equilibrium experiment is directly related to the chemical potential of the particle which, in turn, is related to its Gibb's free energy, G. In other words, while hydrodynamics may affect the time required to achieve equilibrium, the actual concentration gradient achieved is determined by thermodynamics. If the concentrations at two points in the gradient, r_1 and r_2, were expressed as c_1 and c_2, the potential energy

Example 7.3 Effects of *Vinca* alkaloids on tubumin polymerization studied by sedimentation velocity

Certain substances originally derived from the Vinca plant called the Vinca alkaloids are used extensively in cancer chemotherapy. *Vincristine* and *vinblastine* inhibit assembly and dynamics of microtubules and in this way act selectively against rapidly-dividing cells such as cancer cells. *In vitro* studies demonstrate that these molecules result in extensive microtubule depolymerization and promote formation of coiled spiral aggregates of tubulin heterodimers instead of microtubules. There is some interest therefore in knowing more about the interaction of *Vinca* alkaloids with tubulin and, in particular, on the relative effects of GTP and GDP on this interaction since the presence of GDP enhances vinblastine-induced spiral formation.

Because tubulin self-association leading to spiral formation results in a larger protein assembly, one way of following the process is by sedimentation velocity. This technique also allows determination of equilibrium constants governing binding of alkaloid to tubulin heterodimers (K_1), of alkaloid-tubulin heterodimers to spiral aggregates (K_2), of free alkaloid to tubulin polymers (K_3) and of unliganded tubulin heteropolymers to spiral aggregates (K_4) under a variety of experimental conditions (see model).

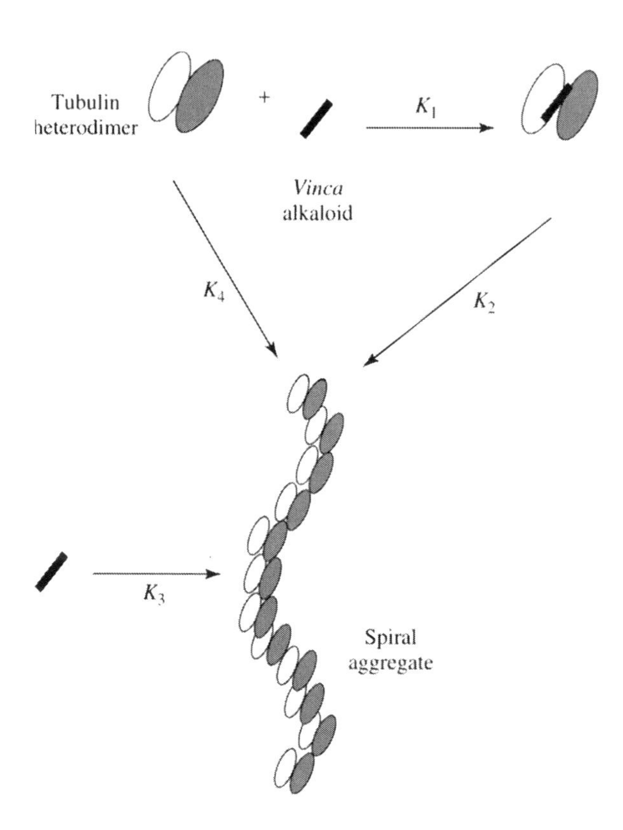

Sedimentation velocity experiments were performed with self-associating tubulin in the presence of vincristine, vinblastine and a synthetic analogue, vinorelbine. The effects of adding either GDP or GTP was assessed for these three *Vinca* alkaloids at a range of temperatures (5, 25 and 30 °C) and at either 30 000 or 40 000 RPM. Protein was quantified in the boundaries by measuring A_{280}. Fitting of sedimentation data allowed estimation of $S_{20,w}$ and equilibrium binding constants.

Sedimentation data for tubulin in the presence of GTP for A) vincristine, B) vinblastine and C) vinorelbine are shown below. These data were obtained at 5 °C (solid line), 25 °C (dashed line) and 35 °C (dotted line). The increase in $S_{20,w}$ with temperature suggests a largely entropically driven process.

(Continues)

(Continued)

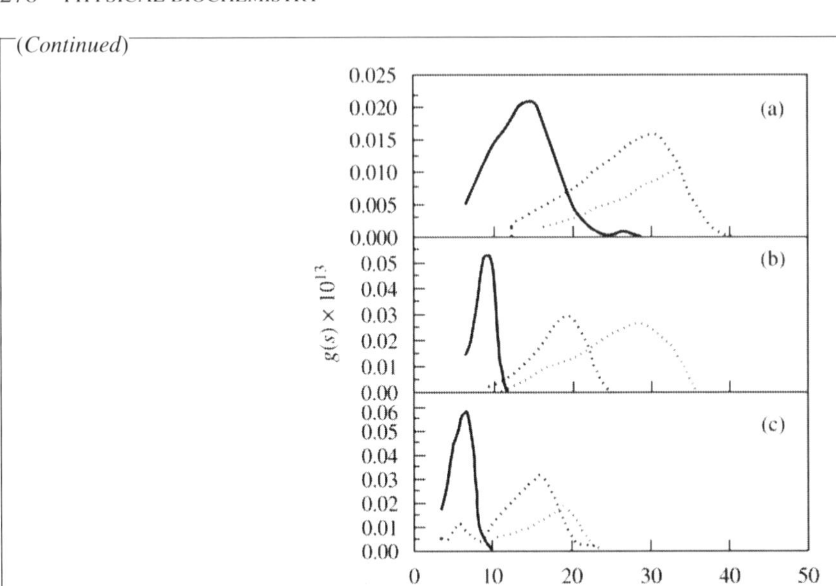

Variation in $S_{20,w}$ as a function of alkaloid concentration in the presence of GTP (open boxes) or GDP (filled boxes) are shown below. These confirm that the maximum $S_{20,w}$ in the presence of GDP is higher for all three alkaloids compared to GTP. Moreover, vincristine promotes formation of a larger species than vinblastine with vinorelbine producing a species only half as large under the same experimental conditions. These data confirm that GDP enhances tubulin assembly relative to GTP for all three *Vinca* alkaloids.

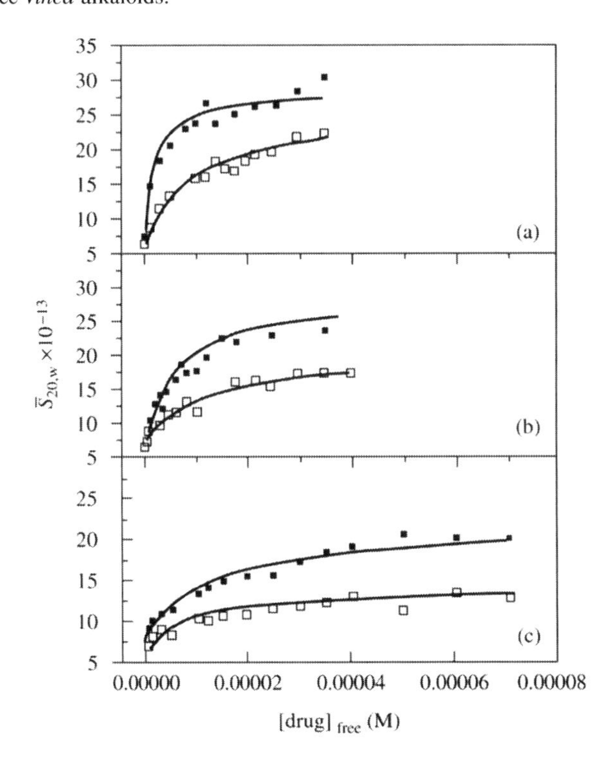

(Continues)

(Continued)

Fitting of sedimentation data allowed estimation of equilibrium constants for the model shown above. Selected data obtained for vincristine and vinorelbine at 5 °C in the presence of GTP and GTP are as follows;

Drug	GXP	K_1	K_2	K_3
		$M^{-1} \times 10^5$	$M^{-1} \times 10^5$	$M^{-1} \times 10^5$
vincristine	GTP	2.7	2.9	77
GDP	2.2	6.4	140	
vinorelbine	GTP	0.78	0.95	7.3
GDP	0.86	1.9	16	

This demonstrates how sedimentation velocity facilitates study of a wide concentration range involving simultaneous analysis of a large number of individual samples. This allows determination of detailed affinity data for complex biochemical processes involving proteins such as, in this case, assembly into large complexes.

Reprinted with permission from Lobert, Vulevic and Correla, Interaction of Vinca Alkaloids with Tubulin: A comparison of Vinblastin, Vincristine and Vinorelbine, *Biochemistry* **35**, 6806–6814, © (1996) American Chemical Society.

distribution in the gradient could be expressed as a *Boltzmann distribution*;

$$\frac{c_1}{c_2} = e^{-(E_1 - E_2)/k \cdot T} \quad (7.31)$$

where k is the *Boltzmann constant* (Appendix 1), T the absolute temperature and E_1 and E_2 the potential energy of the solute at points r_1 and r_2, respectively. The *potential energy difference* between molecules at r_1 and r_2 is given by;

$$(E_1 - E_2) = 1/2 \cdot [m_o(1 - \nu \cdot \rho] \cdot \omega^2 \left(r_2^2 - r_1^2\right) \quad (7.32)$$

Combining Equations (7.31) and (7.32) and replacing k by the Universal Gas Constant R (Appendix 1) gives us an expression for molecular mass, M_r, in units of daltons which may be determined from equilibrium centrifugation data;

$$M_r = \frac{2R \cdot T}{(1 - \nu \cdot \rho) \cdot \omega^2} \cdot \frac{\ln c_r}{c_a} \cdot \frac{1}{r_2 - a_2} \quad (7.33)$$

where c_r is the concentration of solute at a distance r from the axis of rotation c_a is the concentration at the meniscus and a is the distance of the meniscus from the axis of rotation. This relationship allows calculation of M_r from the equilibrium distribution of molecules through the concentration gradient formed in the experiment because a plot of ln c_r *versus* r^2 should yield a slope of $M_r(1 - \nu \cdot \rho)/2RT$. Note that, unlike M_r determination by sedimentation velocity (*see* previous section), knowledge of the diffusion coefficient is not required.

A number of other useful measurements can be determined from sedimentation equilibrium data. These include accurate estimates of particle density and thermodynamic information about association constants, binding stoichiometry and solution nonideality (Example 7.4). Deviation from the ideal behaviour described in Equation (7.33) can be analysed using models which simulate the presence of multiple components, hetero-/self-assembly and other forms of non-ideal behaviour. More detailed information on further uses of sedimentation equilibrium can be obtained from literature cited in the bibliography at the end of this chapter.

7.3 METHODS FOR VARYING BUFFER CONDITIONS

Biological samples are highly-sensitive to their immediate chemical environment (Chapter 1) and are best handled in an aqueous buffer at an appropriate ion strength. This helps to maintain the sample in a structure which retains maximum biological activity. Frequently, this activity may also be dependent on the presence of smaller species such as metals, coenzymes, allosteric modulators or stabilizing agents (e.g. reducing agents) and we may need to be able to change the buffer composition in a controlled way to vary the population of small molecules in contact with the sample. Alternatively, we may want to perform a *series* of procedures in which each step may require distinct conditions (e.g. column chromatography steps in protein purification; Chapter 2).

Example 7.4 Sedimentation equilibrium studies of a complement serine protease

The first component of the complement pathway, C1, consists of three interacting subcomponents, C1q, C1r and C1s. The two latter components are homologous serine protease zymogens which form a Ca^{2+}-dependent tetramer, C1s-C1r_2-C1s which associates with C1q to form C1. The zymogens are activated to functional serine proteases C1s* and C1r* in a process mediated by Ca^{2+} which triggers the first step in complement activation. C1s forms a dimer when Ca^{2+} binds at an N-terminal domain and the stabilization of the C1s-C1r2-C1s tetramer is also facilitated by Ca^{2+} coordination to homologous N-terminal domains. Interaction with Ca^{2+} is therefore a major feature of both complement structure and activation. Several binding sites of varying affinity have been identified by methods such as equilibrium dialysis (Section 7.3.2).

Detailed analysis of Ca^{2+}-binding requires a technique suitable for studying a wide range of Ca^{2+} and protein concentrations at equilibrium. Sedimentation equilibrium lends itself in particular to this type of analysis. Protein was equilibrated by dialysis with either 5 mM or 1 nM Ca^{2+}. It was then centrifuged at 6500–7000 rpm in swinging bucket rotors (90 μl volume) in a preparative centrifuge for 24 hours. Note that, in contrast to the previous example, this centrifugation experiment was not carried out in an analytical centrifuge. Sample was collected as individual fractions and the equilibrium distribution of protein under these conditions was stabilized by inclusion of dextran-70 with the protein sample.

A_{280} was plotted as a function of radial distance for the two samples as shown below. The molecular mass, Mr, was estimated using essentially the equations described in Section 7.2.8 and was found to be 79 (1 nM Ca^{2+}) and 143 kDa (5 mM Ca^{2+}), respectively. This suggests that Ca^{2+} promotes dimer formation between C1s* monomers.

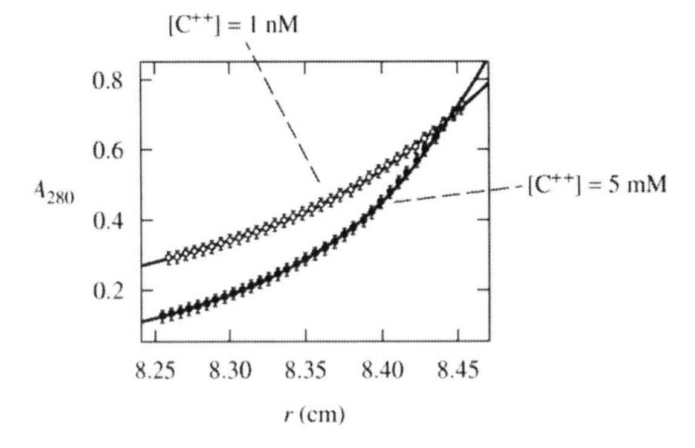

Effect of Ca^{2+} on equilibrium centrifugation of C1s*.

These data for equilibrium distribution were analysed using several models and the simplest model fitting the data was one in which the C1s* monomer contains a single Ca^{2+}-binding site ($K_s = 3 \times 10^5\,M^{-1}$) and the C1s* dimer contains three independent Ca^{2+}-binding sites two of which have low affinity ($K_s = 3 \times 10^4\,M^{-1}$) while the third ($K_s = 1 \times 10^8\,M^{-1}$) is formed by the two N-terminal domains and provides the driving force for dimer formation (see diagram below).

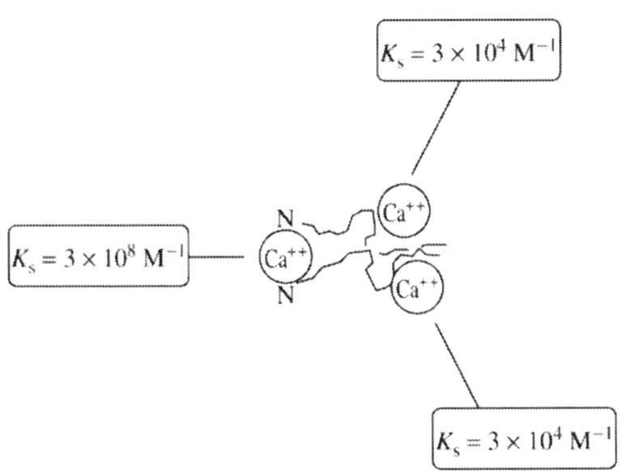

Reprinted with permission from Rivas *et al.*, Calcium-linked self-association of Human-complement C1(S-). *Biochemistry* **31**, 11707–11712, © (1992) American Chemical Society.

This section will describe the principal methods of introducing small molecules to solutions containing biomacromolecules and, conversely, of removing small molecules from such solutions while retaining biomacromolecule quantity and activity. We will also describe techniques for recovery of biomacromolecules from solution.

7.3.1 Ultrafiltration

A variety of porous membrane materials are available for filtration of solutions which selectively retain large biomacromolecules. The solution can be forced through these membranes under pressure which results in *concentration* of the protein/DNA molecules since solvent passes freely through the membrane while protein/DNA do not. This is called *ultrafiltration*. Commercially-available membranes contain a range of specified pore-sizes which give the membrane an average M_r cutoff (e.g. 10 kDa). Most molecules *smaller* than this pass freely through the membrane (usually referred to as the *permeate*) while larger molecules are retained (the *retentate*). It is important to understand that these average cutoffs are usually specified for globular proteins and may differ from the actual M_r values of fibrous proteins or DNA.

An outline design of the apparatus in which the membrane is mounted is described in Figure 7.15. A stirring bar allows constant, gentle mixing during concentration while an inert gas such as nitrogen can be used to create a pressurized atmosphere over the solution. A problem sometimes encountered with ultrafiltration is that the very fine pores through the membrane can become clogged by denatured protein or other sample debris over time. It is therefore wise to clarify the sample by centrifugation before ultrafiltration. The membrane can also be cleaned by treatment with trypsin which proteolytically degrades any protein accumulated in the pores. Care should be taken that the ultrafiltration apparatus remains cold during concentration and that filtration does not proceed to the point where all solvent is lost since this can lead to denaturation of biomacromolecules.

Ultrafiltration is useful in three main types of applications. The most obvious is *concentration of biomacromolecule solutions* prior to some further experimentation. Since rapid concentration is usually achieved, this can greatly shorten the time required for processes such as protein purification. For example, if a 200 ml solution of dilute protein is concentrated to 20 ml in an hour, then this 20 ml can be applied to a chromatography column ten times faster than would the unconcentrated solution. The inclusion of the concentration step therefore speeds up the overall process and often improves recovery of activity since many proteins lose biological activity when dilute.

A series of concentration and dilution steps can be used to *swap macromolecules from one buffer to another*. The identity/concentration/composition of the buffer can all be

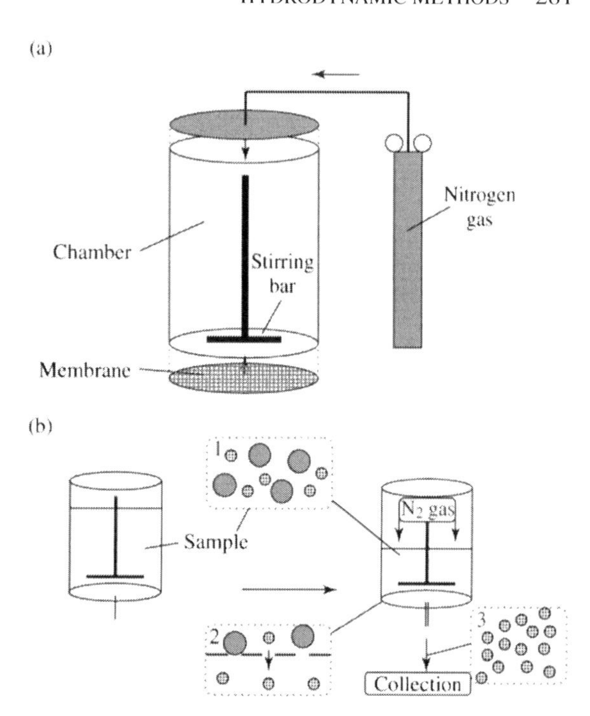

Figure 7.15. Ultrafiltration. (a) Outline of apparatus used. Sample is placed in a chamber where it can be stirred continuously. Nitrogen gas forces sample through the membrane which will have a characteristic M_r cutoff (e.g. 10 kDa). Molecules smaller than the cutoff pass freely through the membrane (permeate) while larger molecules are retained (retentate). (b) More detailed description of ultrafiltration. (1) Sample contains molecules of a range of masses (only two are shown for simplicity). (2) Molecules smaller than the cutoff pass freely through pores in the membrane while larger molecules cannot. (3) This achieves size-fractionation of the sample in that the retentate is enriched for larger mass molecules while the permeate is enriched for smaller molecules.

changed in this way. If, for example, 200 ml of buffer A is concentrated to 20 ml, and then diluted to 200 ml with buffer B, reconcentrated and the process repeated 3–4 times, then the macromolecules will eventually be surrounded by buffer B. Again, this type of procedure is very useful where successive procedures require quite different buffer conditions.

A third type of application is *size fractionation* by use of different membranes in successive ultrafiltration steps. For example, if a protein solution which passes through a 50 kDa cutoff membrane is then passed through a 10 kDa membrane, a fraction rich in molecules of mass-range 10–50 kDa can be selected. This is a comparatively easy way to achieve a significant enrichment during protein or peptide purification or to decide if a particular biological activity is associated with small, intermediate or very large molecules (Example 7.5).

Example 7.5 Use of ultrafiltration to form a size-excluded peptide bank

Many small peptides have pharmacologically-interesting properties and are major targets for drug discovery. Techniques such as biological assay or orphan receptor screening may be used to identify fractions containing promising peptides. In this example, a collection of endogenous peptides comprising a peptide bank was selected from pig brain.

The first step in this process was ultrafiltration through membranes with a 50 kDa cutoff. This was achieved by preparing 18 l of extract by homogenization, centrifugation and filtration. A pressure of 1 bar was maintained at a temperature of 5 °C and a feed-rate of 100 l/h. Collection of permeate was stopped when 43 l had been collected and it was acidified to pH 3.0 to prevent bacterial growth. This allowed the removal of proteins, nucleic acids and cell debris and produced a heterogeneous mixture of peptides suitable for further fractionation. This was achieved using a combination of cation exchange and reversed phase chromatography.

Reproduced from Seiler *et al.* (1999) Application of a peptide bank from porcine brain in isolation of regulatory peptides. *J. Chromatog. A*, **852**, 273–283, with permission from Elsevier.

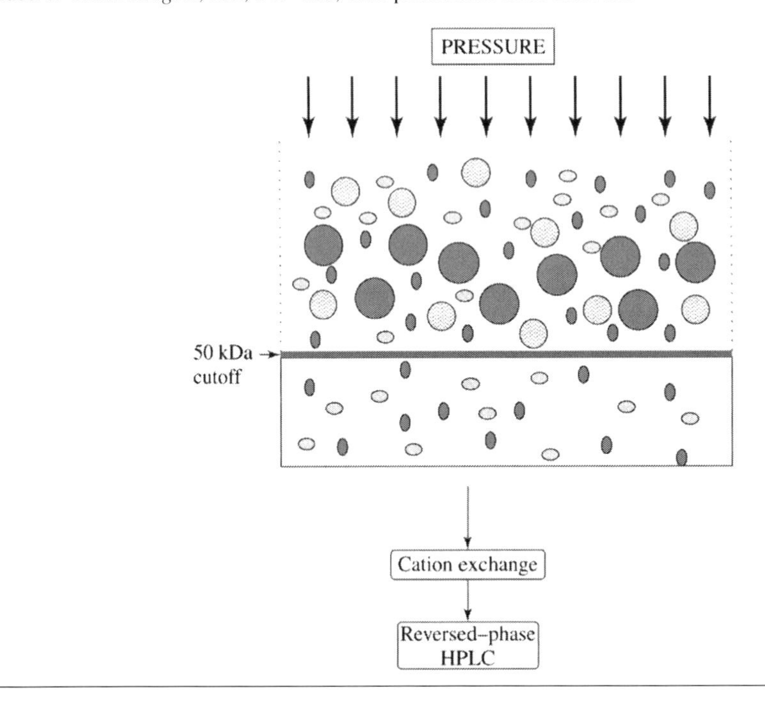

Another format for filtration (Figure 7.16) is useful where it is necessary to sterilize solutions which would not withstand heat sterilization. This is called *sterile filtration* and involves passage of the solution through a 2 μ filter (i.e. a filter with 2 μ pores) into a sterile container. This filter retains bacteria, fungi and spores thus resulting in a sterile solution. A common application for sterile filtration is where it is necessary to add heat-sensitive growth factors or antibiotics to cell culture media.

Small scale ultrafiltration may be performed in commercially-available *centrifugal concentrators*. An outline design of the apparatus used for this is shown in Figure 7.17. The sample to be concentrated (approx. 0.1–1.5 ml) is centrifuged through a filter which retains molecules larger than the M_r cutoff for that filter. As with large-scale ultrafiltration, a range of membranes with differing M_r cutoff values are available. Centrifugal force achieves the same effect as gas pressure in ultrafiltration by forcing the bulk solution through the filter. The filter cartridge is designed to retain a small minimum volume (~10 μl) which prevents macromolecules from losing all solvent and thus becoming denatured. As with large-scale ultrafiltration, centrifugal concentrators can also be used to desalt or otherwise alter the buffer composition of macromolecule solutions.

7.3.2 Dialysis

Dialysis involves equilibration of two solutions across a semipermeable membrane. If the volume of one solution massively exceeds that of the other, it is possible to

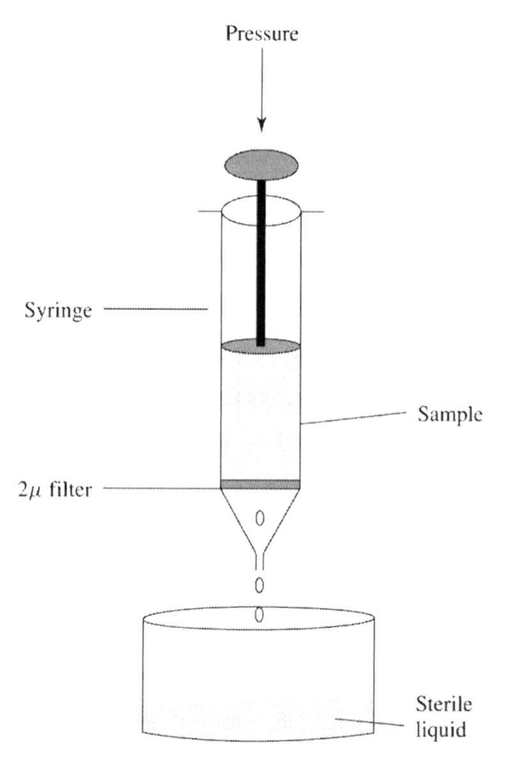

Figure 7.16. Sterile filtration. Filtration of buffers and other liquids through 2μ filter retains bacteria and facilitates sterilisation of the solution.

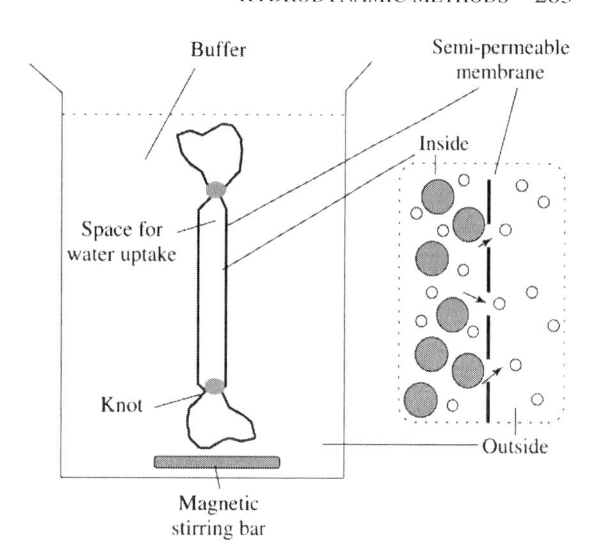

Figure 7.18. Principle of dialysis. A solution of macromolecules (e.g. protein or nucleic acids) is placed in a dialysis bag and this is tied at each end with a knot or clip, allowing some space for water uptake as a result of osmosis. This is placed in a large volume of buffer and stirred for approximately 4–6 h. Buffer is changed at least twice. As shown in the detail presented in dashed box (right), molecules larger than the M_r cutoff for the membrane (usually 10 kDa) are retained inside the bag while buffer components and other molecules smaller than the cutoff equilibrate between the inside and the outside.

convert the buffer conditions of the smaller solution to that of the larger by repeated changes of the larger solution. Thus, dialysis provides a means to desalt or otherwise change the buffer composition of solutions of biomolecules too large to pass through the membrane. In practice, commercially-available dialysis tubing with a M_r cutoff of 10 kDa is used (Figure 7.18). One end of the tubing is tied into a knot or compressed with a clip, the macromolecule solution is placed inside the tubing and the other end of the tubing is tied in a knot or clipped, taking care to squeeze air from the tubing and to allow some space for extra water to enter the tubing by osmosis. The dialysis tubing is then placed in a much larger volume of a solution of desired composition and allowed to equilibrate for 4–6 h. During this time, the buffer components of the small volume solution equilibrate with the larger volume and *vice versa*. The bulk solution is changed at least twice which means that the small volume solution eventually becomes identical with the large volume solution. Macromolecules, however, are retained by the membrane which allows recovery of protein/nucleic acids in a solution of predetermined composition.

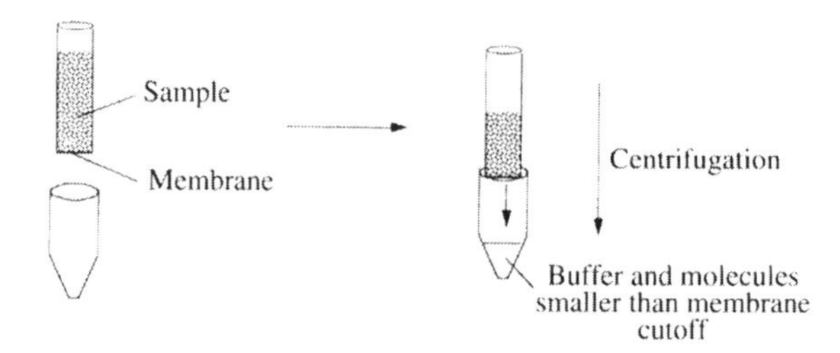

Figure 7.17. Centrifugal concentrators. Sample (approximately 0.1–1.5 ml) is placed in a tube which contains a permeable membrane at one end. This membrane has a defined cutoff value (e.g. 3 kDa; 30 kDa). Centrifugation forces buffer through the membrane, retaining molecules larger than the cutoff within the sample tube. Repeated dilution and concentration can achieve rapid buffer exchange.

This approach has the advantages that it is applicable to a range of sample volumes and allows gentle alteration of the composition of macromolecular solutions. It has the disadvantage that, since equilibration across the membranes used is slow, the process is quite time-consuming which can often lead to loss of biological activity.

Various adaptations of the basic dialysis technique are possible. *Electrodialysis* involves speeding up the exchange of charged buffer components by exposing the

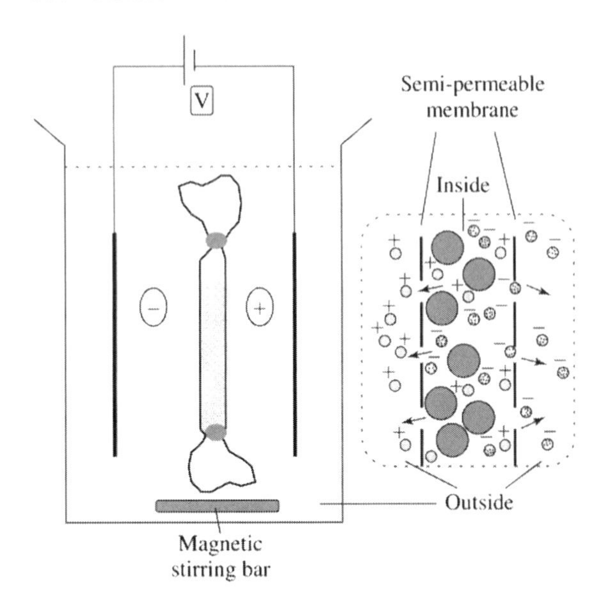

Figure 7.19. Principle of electrodialysis. If electrodes are placed in the buffer solution and a voltage applied, ions migrate towards the electrode of opposite charge. As shown in the detail presented in dashed box (right), cations move towards the cathode while anions move towards the anode (arrows). Molecules larger than the membrane cutoff may migrate within the dialysis membrane but still cannot pass through its pores. This speeds up desalting of proteins and is widely used in industrial scale protein purification.

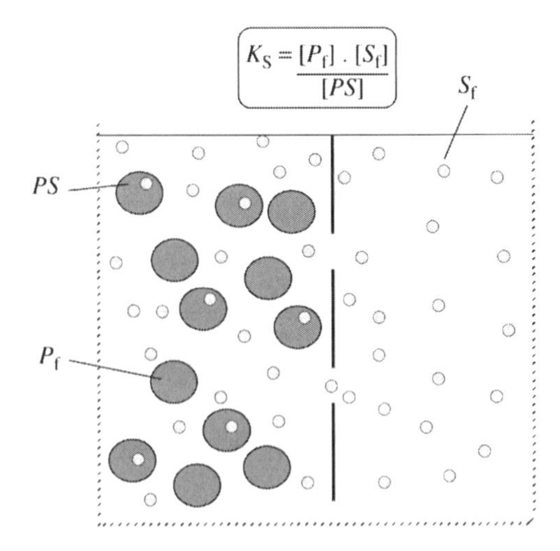

Figure 7.20. Equilibrium dialysis. The binding constant (K_s) of a protein (large circles) for a small ligand (small circles) may be determined by equilibrium dialysis. Two chambers are separated by a semi-permeable membrane which has a M_r cutoff smaller than the protein but larger than the ligand. Free protein and ligand are denoted as [P_f] and [S_f], respectively, while the protein-ligand complex is denoted by [PS]. The total ligand concentration (= [S_f] + [PS]) and the total protein concentration (= [Pf] + [PS]) are known at the beginning of the experiment. [S_f] may be determined by measuring in the chamber as shown (right). Therefore, [PS] and, hence, [P_f] can be determined. Ligands are usually labelled (e.g. with radioactivity) to facilitate quantitation.

macromolecule solution to an electric field (Figure 7.19). Charged molecules migrate out of the solution contained in the dialysis chamber while macromolecules are retained. A similar type of approach underlies some electroelution experimental formats (Chapter 5). *Microdialysis* involves small-scale dialysis in which a large number of samples can be dialyzed simultaneously.

In addition to its use in desalting, dialysis can also be used to estimate binding constants for small ligands. This experiment is called *equilibrium dialysis*. A protein (or a macromolecular species capable of binding, e.g. a membrane receptor on a vesicle) is placed inside one of two equal-sized chambers separated by a dialysis membrane (Figure 7.20). The M_r cutoff for the membrane should be smaller than the M_r of the macromolecule and larger than the M_r of the ligand. Equilibration occurs across the membrane such that the free ligand concentration ([S_f]) becomes the same in both chambers although some of the ligand is bound to the protein (as the PS complex). If the total ligand concentration is known, the fraction of ligand bound to the protein ([PS] can be calculated by subtracting [S_f] measured at equilibrium from the total ligand concentration ([S_t]). The total protein concentration ([P_t]) is also known and subtracting [PS] from this gives us the concentration of free protein (i.e. with no ligand bound, [P_f]). We can determine K_S for the protein for this ligand from the relationship $K_S = ([P_f] \cdot [S_f])/[PS]$.

7.3.3 Precipitation

The procedures outlined in the two previous sections facilitate changing the buffer conditions of a macromolecular solution by taking advantage of the fact that small molecules can pass through pores which are too small to allow passage of large biomacromolecules. It is also possible selectively to *precipitate* biomacromolecules and thus remove them from solution (Figure 7.21). Some precipitation procedures are *biocompatible* in that they retain the biological activity of the macromolecule allowing them to be redissolved in a small volume of a different buffer. Even where precipitation conditions are chemically severe and lead to loss of biological activity, they may nonetheless be useful in concentrating sample for further analysis (e.g. precipitation of protein for SDS PAGE; Chapter 5). All precipitation procedures affect interactions between the biomacromolecule and the buffer responsible for keeping it in solution.

Proteins are readily precipitated by treatment with a high concentration of a salt such as ammonium sulfate

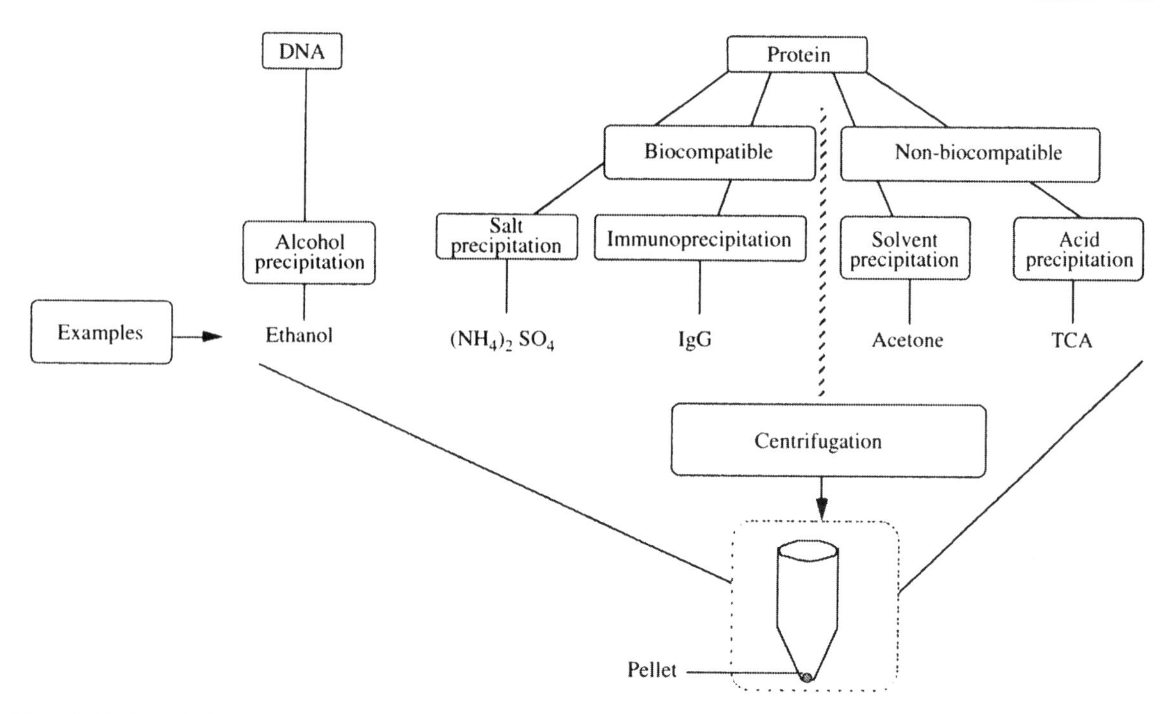

Figure 7.21. Precipitation methods useful with biomacromolecules. An overview of biocompatible (left) and non-biocompatible precipitation methods for proteins and DNA is shown. Precipitated material is collected by centrifugation. In the case of biocompatible methods, fully active DNA/protein is usually recovered. The pellet is normally washed followed by further centrifugation.

$[(NH_4)_2SO_4]$ which interferes with hydrogen bonds between the protein surface and the solvation shell of water. This process is know as *salting out* and the precipitate may be collected by centrifugation and resolubilized in an alternative aqueous buffer which usually results in recovery of fully bioactive protein. Since the volume of buffer used for protein solubilization is chosen by the experimenter, it is possible to concentrate protein by this method. It is important to understand that a minimum protein concentration is required for ammonium sulfate precipitation. If the protein solution is too dilute, no precipitate is generated and it is necessary to concentrate the solution by some other method such as ultrafiltration before repeating the precipitation.

A second means of biocompatible precipitation is immunoprecipitation. This requires that antibodies to one or more antigens on the sample are available. Binding of antibodies to antigens in the protein solution forms a complex of low solubility called *immunoprecipitates* which can be collected by centrifugation. This is a useful method to purify a single protein or group of structurally-related proteins from a crude mixture.

If subsequent analysis or experimentation does not require fully active protein, precipitation may be achieved with an organic solvent such as acetone. This alters the physical properties of the bulk solvent, particularly its dielectric constant and polarity. These effects completely disrupt protein structure, leading to denaturation and precipitation of the protein. Acetone precipitation has the advantage that it is effective at much lower protein concentrations than ammonium sulfate. The protein solution is mixed with an excess of acetone which has been cooled to $-20\,^{\circ}\mathrm{C}$ and precipitated protein is collected by centrifugation. Precipitation with a concentrated acid such as trichloroacetic acid (TCA) causes extremely low pH resulting in precipitation. The minimum protein concentration required for TCA precipitation is higher than that for acetone precipitation. To minimize the risk of acid hydrolysis of peptide bonds, the experiment is performed on ice and care should be taken to remove residual acid by washing the protein pellet with acetone.

A similar approach to precipitation of nucleic acids involves precipitation with ethanol. DNA and RNA precipitate readily from ethanol solutions allowing their recovery by centrifugation. However, in handling RNA, it should be borne in mind that ribonucleases are ubiquitous on laboratory plastic- and glassware and fingertips. Test-tubes and other materials used in the handling of RNA should therefore be baked in *diethyl pyrocarbonate* before use and gloves should be worn at all stages.

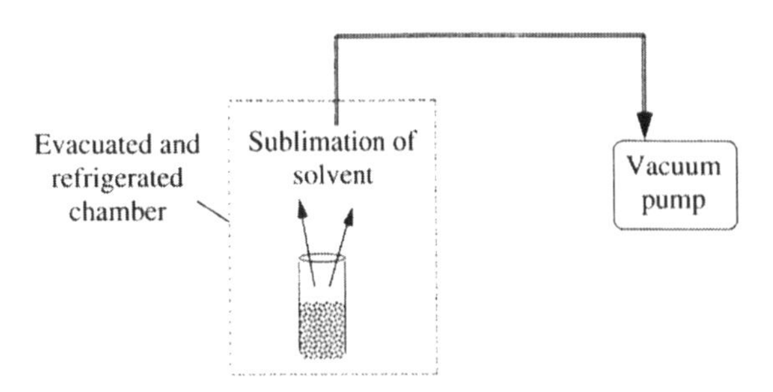

Figure 7.22. Lyophilization. A frozen sample is placed in a refrigerated chamber which is then evacuated. Water, organic solvents and volatile buffer components such as ammonia and carbon dioxide are removed by a process of sublimation. Solute (e.g. proteins or peptides) are collected as a dried and usually denatured residue. Preweighing of sample vial allows estimation of yield by weighing the vial after drying. This process is also referred to as 'freeze-drying'.

A final method for recovery of biomacromolecules from solution is selective removal of the solvent by *lyophilization* or *freeze-drying*. This is facilitated by first transferring the biomacromolecules into a buffer containing volatile components such as ammonium bicarbonate using ultrafiltration or dialysis (Figure 7.22). The sample is frozen to $-198\,^{\circ}C$ by dipping it in liquid nitrogen. It is then placed in an evacuated chamber in which *sublimation* can occur; a molecule passes from the solid phase directly to the gas phase without passage through the liquid phase. Water and ammonium bicarbonate are readily removed from the frozen sample in the form of water vapour, NH_3 and CO_2 leaving the macromolecular component of the sample as a fine solid. The lyophilized sample may lack biological activity as a result of this process involving as it does removal of solvent and exposure to atmospheric oxygen. However, this is a convenient means of preparing large volume samples for subsequent chemical analysis.

7.4 FLOW CYTOMETRY

Flow cytometry is a technique which facilitates analysis of cell populations suspended in a fluid as they pass through a narrow orifice in single file at a rate of many thousands of cells per second. Optical and electronic measurements can be performed on individual cells as they pass through a detector which allows us to fractionate the population into discrete subpopulations and, in this way, to profile the cell population. This can be useful in analysis of complex cell mixtures such as blood or in assessing changes in a population of cells as a result of experimental manipulation. In addition to analytical applications a *preparative* procedure known as *cell sorting* is possible in which a specific subpopulation of cells is purified from the cell population on the basis of measurable biochemical attributes such as expression of a particular receptor or biomacromolecule. In principle, any biochemical association with a cell or particle is amenable to analysis by flow cytometry provided a measurable physical property or fluorescent tracer is available which reacts stoichiometrically and specifically with a target such as a receptor, DNA, RNA or protein. In this section we will look at the experimental design and application of flow cytometry to biochemical analysis.

7.4.1 Flow Cytometer Design

An outline design of a flow cytometer is shown in Figure 7.23. The sample cell population is directed along a narrow quartz tube leading to an *injector* in the *core stream*. This

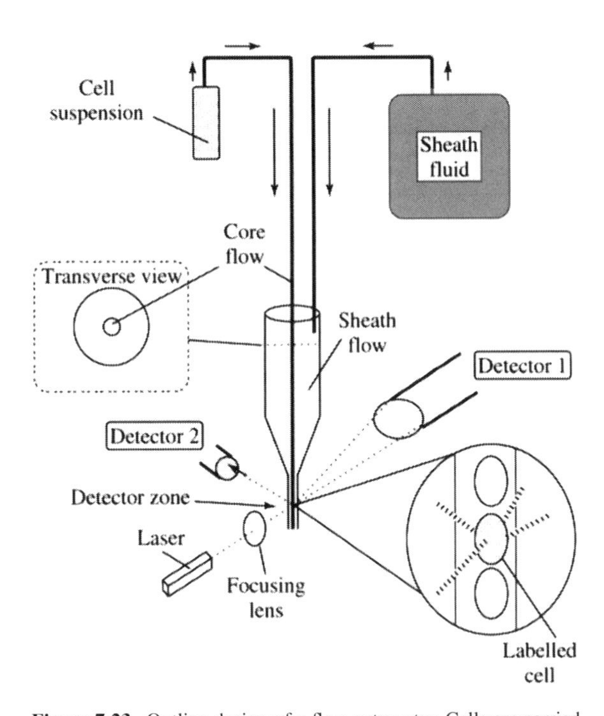

Figure 7.23. Outline design of a flow cytometer. Cells are carried in the core flow which directs them through the detector zone in single file. A transverse view is shown in the dashed box on left. The detector is an area where the beam from a laser is focused. Low angle scattered (forward-scattered) light is detected at detector 1, while right angle scattered and fluorescent light is detected at detector 2. Selective filters allow detection of specific emission wavelengths. The detail on right shows detection of a fluorescently-labelled cell as it passes through the detector.

is surrounded coaxially by a tube containing a stream of buffered medium called the *sheath stream*. The purpose of this arrangement is to concentrate cells in the inner tube (where measurements will be made) while avoiding turbulent or parabolic flow. The injector narrows to a small internal diameter as shown such that cells are compelled to pass in single file through a *flow cell* which places them in the centre of an optical detection system. Typical internal diameters for the tube through which the core stream flows are 3 mm at the top narrowing to between 75 and 200 μm at the injector. This tapering is crucial to the design of a flow cytometer since it avoids the parabolic flow which arises in aqueous solution as a result of viscosity (Section 7.1.1; Figure 7.1).

The *detector* consists of an elliptical laser-beam (e.g. an argon laser; Chapter 6) that interacts with the cells scattering light which may be detected at a low angle (e.g. 0.5–10°) as *forward scattered light* and also at a high angle (e.g. 90°) as *right angle scattered light*. Low angle scattered light is usually taken as a general measure of cell size while light scattered at a right angle is a measure of the refractive index of the cell, in turn reflecting the number of intracellular *granules* it contains. Fluorescence may also be detected at 90° and can be used to quantify particular biochemical events such as receptor binding or binding at specific DNA or RNA sequences (*see* below). The usual arrangement of the optical detection system is *orthogonal* in that the laser is arranged at 90° to the axis of flow while the detector is at 90° to the laser.

This experimental setup allows rapid analysis of over 5000 cells per second. A large number of parameters can be measured for each cell and the data electronically stored. Measurements can be analysed and reanalysed repetitively to identify progressively more highly defined subpopulations of cells within the overall population.

Cells are presented to flow cytometry as *disaggregated suspensions* which may need to be released from the extracellular matrix by enzymatic or physical methods. It is also possible to analyse intact nuclei obtained from cells prepared as paraffin-embedded tissue sections by a combination of mechanical disruption with dewaxing, hydration and exposure to proteases. This latter approach facilitates analysis of nuclear DNA by flow cytometry to identify whether the cells from which the nuclei originated were haploid, diploid or polyploid as well as the stage of the cell cycle reached by the original cells.

7.4.2 Cell Sorting

The flow cytometer can be used to *sort cells* as shown in Figure 7.24. This involves selecting cells which the detector indicates may be of interest (based on fluorescence or granularicity) for collection as droplets on which an electric charge is conferred allowing their later separation in an electric field. This is achieved by vibrating the flow cell or nozzle using a *piezoelectric crystal* (similar to that used to select plane polarized light; Section 3.5.4) at a frequency selected to ensure uniform droplet size and a precise time-interval between detection and droplet formation. Detection

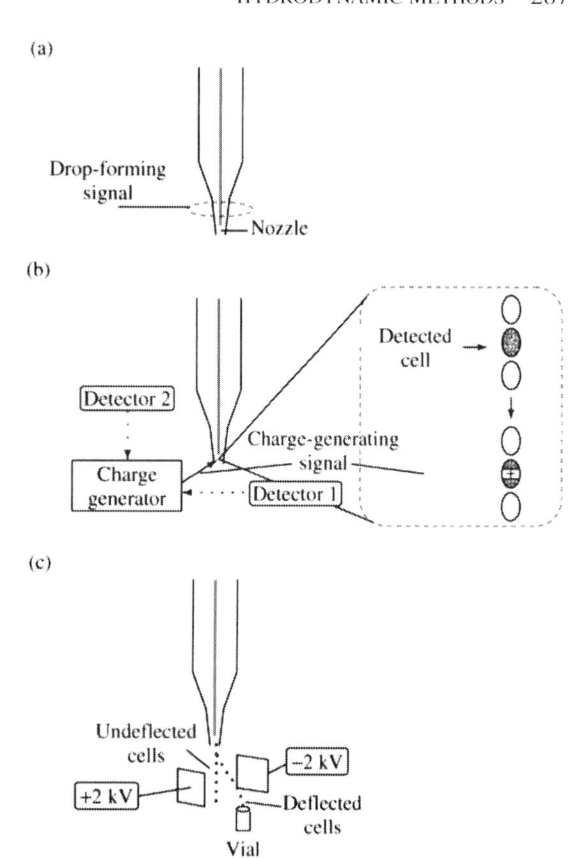

Figure 7.24. Cell sorting in flow cytometry. (a) The nozzle of the flow cytometer is vibrated axially (dashed ellipse) at a particular frequency by the effect of a drop-forming signal on a piezoelectric crystal. 40 000 vibrations per second gives 40 000 drops per second, (b) Labelled cells are detected by detectors 1 and 2 which pass a signal to a charge generator. This sends a drop-charging signal which confers a change (positive or negative) on the drop containing the labelled cell. In practice, drops in front of and behind that containing the detected cell are also usually charged as well, (c) On leaving the nozzle, charged drops are deflected between charged plates and collected. Note that a second population of cells in the same sample could be labelled with a negative charge and collected in a second vial.

of a cell of interest by detectors 1 and 2 triggers a charge-generating signal to the core flow. This applies an electric potential of ~100 V to the fluid containing the cell while still inside the flow cell or nozzle and, when the fluid forms a droplet, it will carry a corresponding electric charge. Two cell populations can be sorted from a single sample by applying a positive charge to one and a negative charge to the other.

The droplets leave the flow cell/nozzle and pass between two electrically-charged plates which deflect them according to their electric charge. Droplets containing cells which are not of interest (which will not be detected by detectors) pass undeflected through the electrically-charged plates. In this way, subpopulations of cells can be purified from the overall population and used for further investigations. In practice, because of possible inaccuracy between detection and sorting, droplets in front of and behind the droplet of interest are also charged and collected.

7.4.3 Detection Strategies in Flow Cytometry

To obtain useful information from flow cytometry, it is essential that cells can be 'seen' by the detector. Detection can take advantage of intrinsic properties of the cell or exploit interactions between cells and extrinsic reagents. Selective information is obtained by controlling the wavelength and other characteristics of the laser light incident on the cells combined with variation in the angle of detection and use of appropriate filters. Modern flow cytometers have the potential for simultaneous detection of scattered light at several wavelengths and angles in individual *channels*.

Intrinsic properties useful in flow cytometry include refractive index and absorption/fluorescence properties of biomolecules within cells. As mentioned above, refractive index measurements quantify granularicity within the cells. Different subpopulations of cells would be expected to have altered granularicity as a consequence of changes in the relative amount and activity of cell organelles. Subtle differences between cell subpopulations can therefore be detected in this way. Absorbance measurements may be used to quantify chromophores such as nucleic acids and haem-containing proteins essentially as described in Chapter 3. Useful intrinsic fluors include pyridine and flavin nucleotides and measurement of these, for example, may be used to detect differences in redox status between different cell subpopulations by measuring NAD: NADH + H^+ ratios. Specific groups of naturally-occurring chromophores such as porphyrins and chlorophylls may also be useful in the study of particular cell types.

Because of its sensitivity, the most versatile detection method in flow cytometry is extrinsic fluorescence (Section 3.4). A number of extrinsic fluors are available with the same excitation wavelength but distinct emission wavelengths. These may therefore be detected independently of each other by use of appropriate filters in the detection system. This means that 3–4 individual fluorescent labels may be detected in a single population of cells, allowing fractionation into a number of subpopulations. Possible strategies for introducing fluors include their accumulation in or on cells, covalent labeling of antigens or antibodies or covalent labeling of oligonucleotides (Example 7.6). This allows simultaneous quantification of expression of several antigens in a single cell population, quantification of changes in such antigen profiles and detection of individual DNA sequences by *in situ* hybridization with labeled oligonucleotides. Studies of nuclei allow quantification of chromosomes such as *karyotyping*, determination of the stage of the cell cycle and measurement of nuclear proteins such as proliferation factors.

7.4.4 Parameters Measurable by Flow Cytometry

Flow cytometry provides a generally-useful strategy for analyzing a wide range of cell parameters. *Physical* attributes of cells such as their volume, diameter, shape and surface area can all be measured and quantified even in complex cell populations. This may involve exposing the cells during flow to light beams arranged in several directions simultaneously and/or analyzing patterns of light scattering at a number of angles. These patterns can be analysed by microcomputer allowing deduction of physical parameters. Moreover, isolation of cells with predefined physical characteristics from complex mixtures is facilitated by flow cytometry using cell sorting. Measurement of *biochemical* attributes is only limited by the specificity and sensitivity of the detection system used as outlined above in Sections 7.4.1 and 7.4.3. Combination of physical and biochemical attributes allows quantification and preparation of cell subpopulations defined by a wide range of parameters.

The literature cited in the bibliography at the end of this chapter (and papers cited therein) describes applications of flow cytometry to measurement of membrane fluidity/permeability, intracellular pH, intracellular ion flux, mitochondrial activity, cell cycle status, expression of specific protein/DNA/RNA species, study of intra/extracellular receptors, detection/sorting of cancer/blood/sperm/protozoan/plant/plankton/fungi/bacterial cells, cell differentiation and chromatin structure making it one of the most versatile and useful tools in the study of cell biology.

Example 7.6 Estimation of apoptosis by flow cytometry

Programmed cell death, apoptosis, is associated with fragmentation of DNA into nucleosome-sized fragments (146 bp). These can be qualitatively detected as 'ladders' in agarose gel electrophoresis (Chapter 5). However, it is sometimes useful to quantify apoptosis in cell populations and flow cytometry is particularly useful for this purpose.

A popular method is the *terminal deoxynucleotidyl mediated dUTP nick end-labelling (TUNEL) assay*. This involves permeabilizing cells with detergent followed by incubation with fluorescently-labeled dUTP (e.g. FITC-dUTP) and an enzyme called terminal transferase. This is a DNA polymerase which does not require a template. It catalyzes the addition of FITC-UMP units to the 3′ end of DNA fragments within the cell. The more nucleosomes in a given cell population, the stronger the fluorescence signal associated with the cells during flow cytometry. The percentage of apoptotic cells in the population can be determined using fluorescence activated cell sorting (FACS) analysis which quantifies DNA-associated FITC.

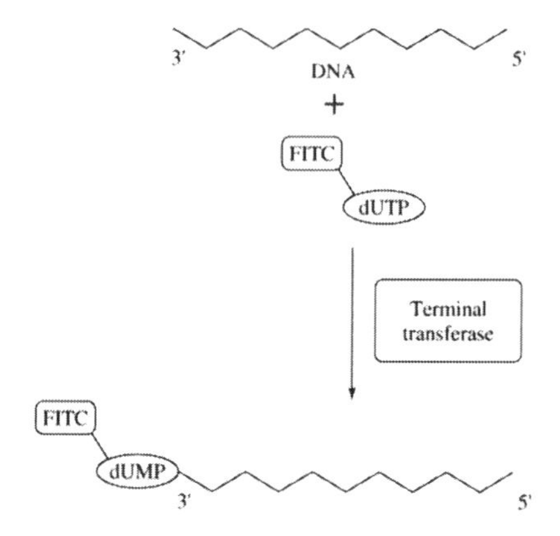

The TUNEL assay

An essential step in the biogenesis of red blood cells is binding of the growth factor *erythropoietin* (EPO) to the erythropoietin receptor. This has effects on intracellular signal transducer/activator of transcription (STAT) proteins such as STAT5 which is rapidly activated. In particular, EPO rescues committed erythroid progenitors from the pathway leading to apoptosis, allowing them to proceed to form mature erythrocytes. Mice lacking both genetic loci for STAT5 (i.e. STAT5a$^{-/-}$ 5b$^{-/-}$) provide a useful model system to study this phenomenon.

Foetuses of STAT5a$^{-/-}$ 5b$^{-/-}$ are found to be far more anaemic than wild-type animals. An explanation for this is provided by TUNEL assay of erythrocyte progenitor liver cells in the presence or absence of EPO as shown below:

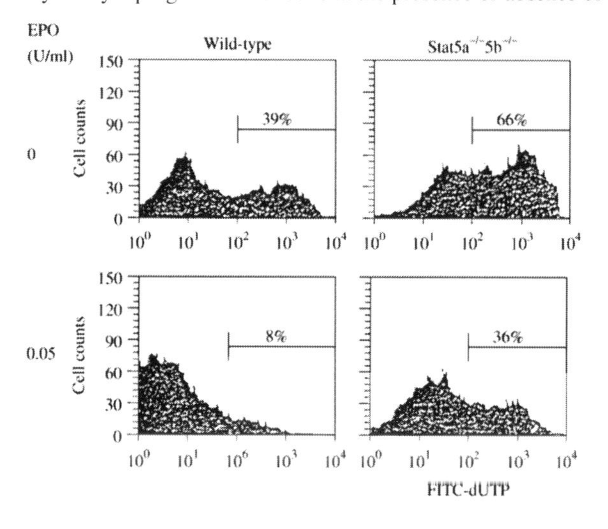

(Continues)

(Continued)

FACS analysis of wild-type and STAT5a$^{-/-}$ 5b$^{-/-}$ liver progenitor cells.

From these data it is clear that STAT5a$^{-/-}$ 5b$^{-/-}$ cells are much less responsive to the antiapoptotic effect of EPO than are wild-type. STAT5a$^{-/-}$ 5b$^{-/-}$ cells are approximately 2.5-fold more apoptotic than wild-type at the time of harvesting. When cultured for 24 h *in vitro* in the absence of EPO they are almost twofold more apoptotic (that is 66% versus 39%). Even in the presence of 0.05 U/ml EPO, 36% of STAT5a$^{-/-}$ 5b$^{-/-}$ cells remain apoptotic compared to 8% of wild-type. These observations account for the fetal anaemia observed in STAT5a$^{-/-}$ 5b$^{-/-}$ animals.

Reproduced from Socolovsky *et al.*, (1999) Fetal anemia and apoptosis of red cell progenitors in Stat5a-/-5b-/-mice: A direct role for Stat5 in Bcl-X-L induction. *Cell* **98**, 181–191, with permission of Cell Press.

REFERENCES

Viscosity

Kuckel, C.L., Lubit, B.W., Lambooy, P.K. and Farnsworth, P.N. (1993) Methylisocyanate and actin polymerisation: the *in vitro* effects of actin carbamylation. *Biochimica et Biophysica Acta*, **1162**, 143–8. An excellent example of the application of viscometry in biochemistry.

Ockendon, H. and Ockendon, J.R. (1995) *Viscous Flow*, Cambridge University Press. A mathematical description of viscous flow.

Veal, J.M. and Rill, R.L. (1991) Noncovalent binding of bis(1,10-phenanthrolene) copper (I) and related compounds. *Biochemistry*, **30**, 1132–40. An elegant illustration of combination of viscometric and spectroscopic methods in the study of ligand binding to DNA.

Sedimentation

Carlson, S. (1998) A kitchen centrifuge. *Scientific American*, **278**, 102–3. A description of how a centrifuge may be constructed cheaply from readily-available materials.

Cole, J.L., Lary, J.W., Moody, T.P. and Laue, T.M. (2008) Analytical ultracentrifugation: Sedimentation velocity and sedimentation equilibrium, in *Methods in Cell Biology*, Vol. **84** (eds J.J. Correia and H.W. Detrich), pp. 143–79. A review of sedimentation methods in analytical untracentrifugation.

Hansen, J.C., Lebowitz, J. and Demeler, B. (1994) Analytical centrifugation of complex macromolecular systems. *Biochemistry*, **33**, 13155–63. A review with examples of advances in data acquisition and analysis in modern analytical centrifugation.

Howlett, G.J., Minton, A.P. and Rivas, G. (2006) Analytical ultracentrifugation for the study of protein association and assembly. *Current Opinion in Chemical Biology*, **10**, 430–6. A review of use of analytical centrifugation in studying protein-protein interactions.

Laue, T.M. and Stafford, W.F. (1999) Modern applications of analytical ultracentrifugation. *Annual Review of Biophysical and Biomolecular Structure*, **28**, 75–100. A review of sedimentation velocity and sedimentation equilibrium analysis.

Laue, T.M. (2001) Biophysical studies by analytical ultracentrifugation. *Current Opinion in Structural Biology*, **11**, 579–83. A review of analytical ultracentrifugation.

Lobert, S., Vulevic, B. and Correia, J.J. (1996) Interaction of *Vinca* alkaloids with tubulin: A comparison of vinblastine, vincristine and vinorelbine. *Biochemistry*, **35**, 6806–14. An excellent example of the application of sedimentation velocity.

Rivas, G. and Minton, A.P. (1993) New developments in the study of biomolecular associations *via* sedimentation equilibrium. *Trends in Biochemical Sciences*, **18**, 284–7. An overview of the application of equilibrium sedimentation analysis to the study of reversible and irreversible association between biomacromolecules.

Rivas, G., Ingham, K.C. and Minton, A.P. (1992) Calcium-linked self-association of human complement C1s*. *Biochemistry*, **31**, 11707–12. An elegant example of the application of seminentation equilibrium to the analysis of calcium binding.

Stafford, W.F. (1997) Sedimentation velocity spins a new weave for an old fabric. *Current Opinion in Structural Biology*, **8**, 14–24. A review of data analysis in sedimentation velocity.

Winzor, D.J., Jacobsen, M.P. and Wills, P.R. (1998) Direct analysis of sedimentation equilibrium distributions reflecting complex formation between dissimilar reactants. *Biochemistry*, **37**, 2226–33. Description of novel procedures for analysis of sedimentation equilibrium data to describe intermolecular interactions.

Dialysis and precipitation methods

Doonan, S. (1996) Bulk purification by fractional precipitation, in *Protein Purification Protocols* (ed. S. Doonan), Humana Press, Totowa, NJ, USA, pp. 135–44. A description of application of precipitation methods to proteins.

Doonan, S. (2004) Concentration of extracts, in *Protein Purification Protocols*, 2nd edn (ed. P. Cutler), Humana Press, Totowa, NJ, USA, pp. 85–90. A description of protein concentration by ammonium sulphate precipitation and forced dialysis.

Seiler, P., Staendker, L., Mark, S. *et al.* (1999) Application of a peptide bank from porcine brain in isolation of regulatory peptides. *J. Chromatog. A*, **852**, 273–283. An example of the application of ultrafiltration and dialysis in large-scale peptide isolation.

Torto, N., Gordon, L., Laurell, T. and Marko-Verga, G. (1999) Technical issues of *in vitro* microdialysis sampling in bioprocess monitoring. *Trends in Analytical Chemistry*, **18**, 252–60. A review of *in vitro* microdialysis.

Flow Cytometry

De Rosa, S.C., Brenchley, J.M. and Roederer, M. (2003) Beyond six colours: A new era in flow cytometry. *Nature Medicine*, **9**,

112–7. A review of the history, applications and future directions of flow cytometry in medical research.

Edwards, B.S., Oprea, T., Prossnitz, E.R. and Sklar, L.A. (2004) Flow cytomtry for high-throughput high-content screening. *Current Opinion in Chemical Biology*, **8**, 392–8. A review of flow cytometry from the perspective of high-throughput screening.

Herzenberg, L.A., Sweet, R.G. and Herzenberg, L.A. (1976) Fluorescence-activated cell sorting. *Scientific American*, **234**, 108–17. An early introduction to the basics of flow cytometry.

Socolovsky, M., Fallon, A.E.J., Wang, S. *et al.* (1999) Fetal anemia and apoptosis of red cell progenitors in Stat5a-/-5b-/-mice: A direct role for Stat5 in Bcl-XL induction. *Cell*, **98**, 181–91. A good example of fluorescence activated cell sorting.

Shapiro, H.M. (1995) *Practical Flow Cytometry*, 3rd edn, Wiley-Liss, NY, USA. A superbly-readable and comprehensive overview of the principles, design and application of flow cytometry.

Chapter 8
Biocalorimetry

Objectives

After completing this chapter you should be familiar with:

- *The* **physical meaning** *of the main thermodynamic parameters.*
- *The physical basis of isothermal titration calorimetry.*
- *The physical basis of differential scanning calorimetry.*
- *Determination of thermodynamic parameters by noncalorimetric means.*

Heat is one the most important physical quantities in biology. Biological systems and processes are highly heat-sensitive and will only function in relatively narrow temperature ranges. These ranges are largely defined by two factors:

1. The exponential effect of temperature on chemical processes described by the *Arrhenius equation*,

$$k = Ae^{-E_a/RT} \tag{8.1}$$

where k is the rate constant, E_a is activation energy (Section 8.1 below), R is the Universal Gas Constant (Appendix 1), T is temperature and A is a constant called the *frequency factor* which reflects the total number of collisions of reactant molecules per unit time in the correct relative orientation for product formation. This means that the rates of most chemical reactions *increase exponentially* with increasing temperature.

2. The sensitivity of biopolymer structure to heat destruction. An example of the latter is heat-mediated denaturation of proteins and DNA. A good example of the consequence of these two factors is that enzyme-mediated reactions tail off at higher temperatures due to loss of structure (Figure 8.1).

Heat is also an important aspect to the full description of chemical reactions because it helps us to understand energy flow within such reactions. *The First Law of Thermodynamics* states that energy cannot be created or destroyed, it can only change form during a chemical process. Heat generated or consumed can therefore, in principle, be directly measured. Such an apparatus is called a *calorimeter*. The usual experimental approach is to maintain the reaction at a single temperature by the addition or removal of heat. The amount of heat requiring to be added or subtracted is automatically measured. Calorimetric measurements of biochemical reactions such as protein–DNA binding, protein–protein complex formation and protein unfolding is called *biocalorimetry*. In this chapter we will review the thermodynamics of biochemical systems and describe some examples of the main biocalorimetric experimental approaches.

8.1 THE MAIN THERMODYNAMIC PARAMETERS

8.1.1 Activation Energy of Reactions

Heat is a form of energy. During a chemical reaction, bonds need to be broken (which releases energy) or to be made (which requires energy). For example, in formation of a peptide bond between two molecules of glycine:

$$\begin{aligned} &NH_2 - CH_2 - COOH + NH_2 - CH_2 - COOH \rightarrow \\ &NH_2 - CH_2 - CO - NH - CH_2 - COOH + H_2O \end{aligned} \tag{8.2}$$

the CO–OH bond of one amino acid and the NH–H bond of the other must be broken while a CO–NH peptide bond must be formed. If the process of bond-breaking requires less energy than that of bond-making, then the reaction is said to be *exothermic*. If the opposite is the case, the reaction is said to be *endothermic*. We can extend this description to other important biochemical processes such as protein folding and complex formation which may not involve formal changes in bonding structure.

The *collision theory* of chemical reaction envisages a population of reactants with a wide range of chemical energy values. Reaction only occurs between reactants when they collide and undergo bond breakage/formation to form

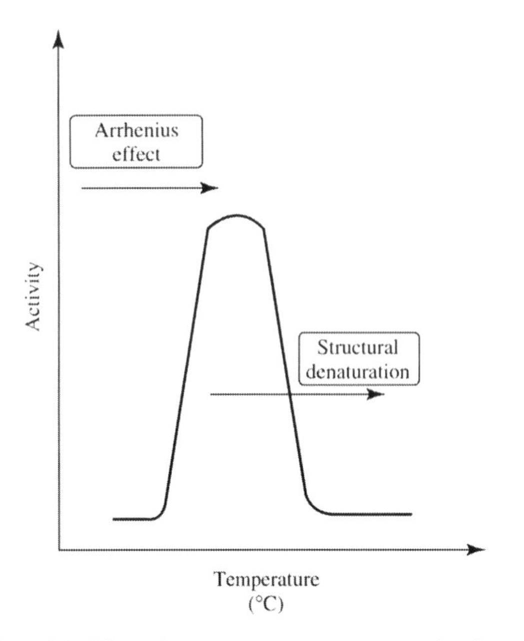

Figure 8.1. Effect of temperature on an enzyme-catalyzed reaction. Temperature increases the rates of chemical reactions including rates of enzyme-catalyzed reactions because it has an exponential effect due to the Arrhenius relationship. Structural denaturation also increases with temperature though which leads to inactivation of enzyme. The temperature profile of enzymes results from these two processes.

the product. Only molecules with energy greater than the activation energy (E_a) can successfully undergo reaction. This is characteristic for the reaction and associated conditions such as temperature and pH. If the molecules in a collision have *less* energy than E_a, they simply bounce off each other without reacting. E_a effectively provides a barrier to reaction called the *potential energy barrier* which can only be overcome by a proportion of reactant molecules. The number of molecules with energy above $E_a(n)$ obeys the Boltzmann distribution,

$$n = n_0 e^{-E_a/RT} \tag{8.3}$$

where n_0 is the total number of molecules in the population. This was used to formulate the Arrhenius equation shown above (Equation (8.1)). If we determine rate constants at a number of temperature values, we can use the Arrhenius equation to determine E_a from a plot of $\log k$ versus $1000/T$ (Figure 8.2) because the slope of this line has a value $E_a/2.303R$. The relationship between the energy of reactants and E_a is shown in Figure 8.3. Exothermic reactions release heat while endothermic reactions absorb heat and both types of reaction occur in Biochemistry. The highest-energy and least-stable species formed in the reaction pathway is called the *transition state*. Some reactions may have several activation energy barriers and there may be several transition states in such pathways.

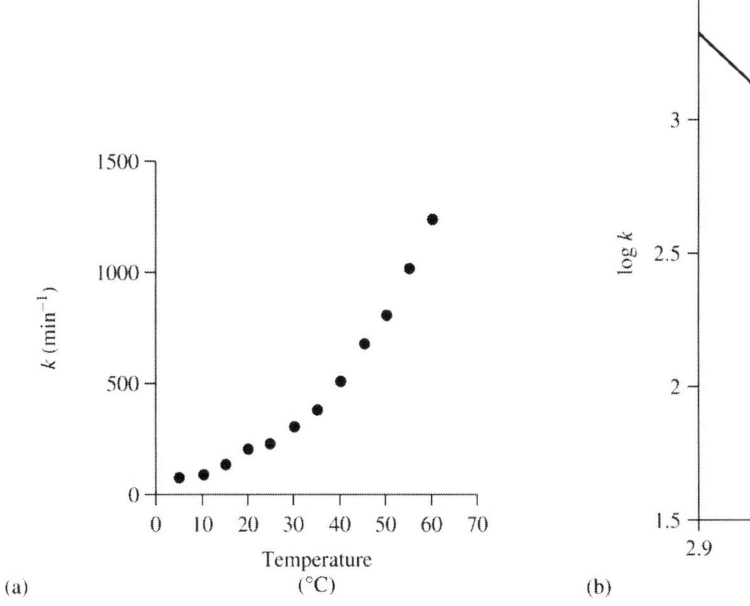

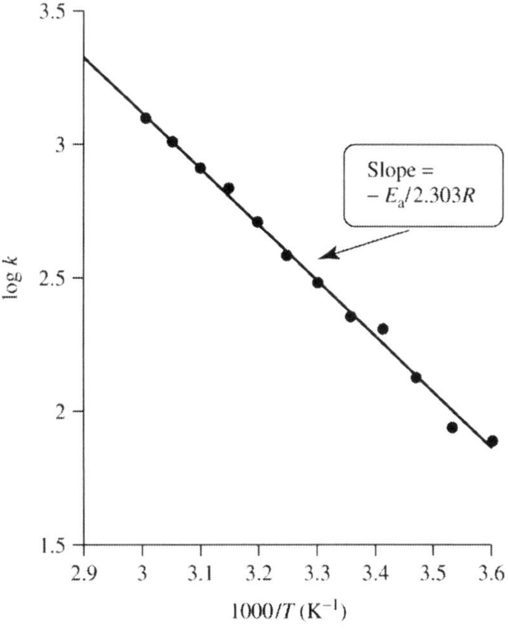

Figure 8.2. Arrhenius plot of a chemical reaction. (a) Chemical reactions depend exponentially on temperature. (b) The activation energy (E_a) can be calculated from the slope of an Arrhenius plot.

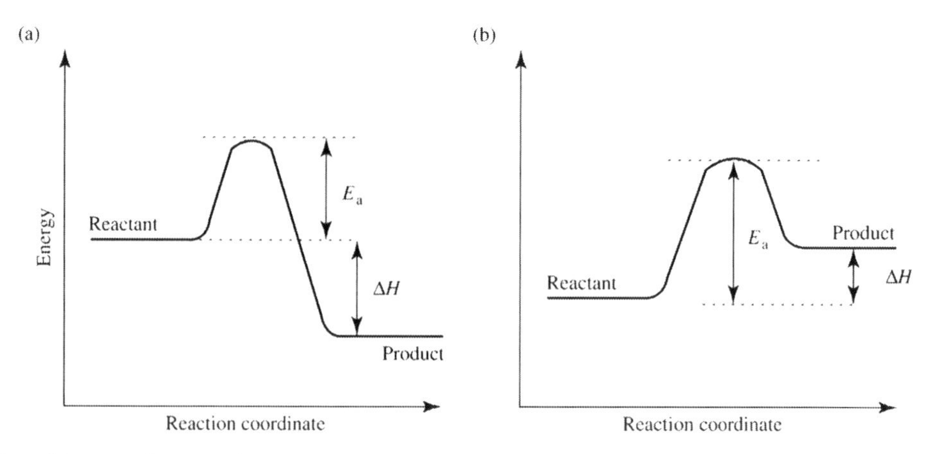

Figure 8.3. Potential energy diagrams of reactions. Potential energy diagrams of reactions are shown. The activation energy barrier is denoted by E_a while the enthalpy of the reaction is shown by ΔH. (a) An exothermic reaction results in positive ΔH. (b) An endothermic reaction results in negative ΔH.

8.1.2 Enthalpy

The energy produced or absorbed by a reaction is called the *change in enthalpy*, ΔH. In endothermic reactions the enthalpy value, H, of the product is *higher* than that of the reactants while in exothermic reactions the opposite is the case. Enthalpy is defined as

$$H = U + pV \tag{8.4}$$

where U is the *internal energy* of the molecule and V its volume at pressure, p.

In a chemical reaction occurring in a thermodynamically *closed* system (i.e. a system to which energy cannot be added or lost during the reaction), a change in both the internal energy, ΔU and the volume, ΔV, can occur as a result of the reaction and this is associated with a quantity of heat, q, either entering or leaving the system:

$$\Delta U = q - (p\Delta V) \tag{8.5}$$

The enthalpy change, ΔH, is defined as

$$\Delta H = \Delta U + p\Delta V + V\Delta p \tag{8.6}$$

where Δp and ΔV are the pressure and volume changes during the reaction, respectively. Combining Equations (8.5) and (8.6),

$$\Delta H = q + (V\Delta p) \tag{8.7}$$

which means that, at constant pressure ($\Delta p = 0$):

$$\Delta H = q \tag{8.8}$$

ΔH values of this type can therefore be directly determined in an *isobaric calorimeter* by direct measurement of q.

Enthalpy is an inherent property of every molecule under specific conditions of temperature, pressure and phase. Change of phase from solid to liquid to gas result in different inherent enthalpy values for a given molecule. Standardization and comparison are achieved by reporting data as standard *enthalpies of reaction* (ΔH°) which are defined as the enthalpy change taking place when molar quantities react at 25 °C in a pressure of one atmosphere (1.013×10^5 Pa). Examples of this include the standard enthalpy of solvation (ΔH°_{solv}, where dipoles in the sample attract counter-ions in the solvent), solution (ΔH°_{sol}, where one mole of sample is completely dissolved) and formation (ΔH°_{form}, when one mole of sample is formed from its constituent elements under standard conditions).

8.1.3 Entropy

Entropy (S) is another inherent thermodynamic value of all molecules including biomolecules. It may be conveniently thought of as a measure of disorder or randomness in the molecule or process. *The Second Law of Thermodynamics* states that a system will always undergo spontaneous change in such a way as to increase entropy. This is especially relevant to biomacromolecules and to biological systems generally, since these are unusually highly ordered and therefore constantly experience a natural tendency to adopt more random forms.

A completely crystalline solid with no vibrational movement would have zero entropy. As temperature is increased, so does the entropy level. Going from the solid to the liquid phase results in a large entropy change as does going from the liquid to the gaseous phase. Similarly, any process resulting in an increase in disorder such as dissolving a solid in a liquid solvent, increasing the number of molecules in a chemical solution or unfolding a protein results in entropy increases.

Absorption of heat by a chemical system results in an increase in disorder and hence an increase in entropy (ΔS). This value is the ratio of the heat absorbed to the absolute temperature:

$$\Delta S = \frac{q}{T} \tag{8.9}$$

If T remains constant, such a process is said to be *isothermal*. A consequence of the Second law of Thermodynamics is that ΔS_{total} for a spontaneous process must be greater than 0 where ΔS_{total} is the sum of the ΔS for the specific system (e.g. the chemical reaction or process) plus the ΔS of its surroundings:

$$\Delta S_{total} > \Delta S_{system} + \Delta S_{surroundings} \tag{8.10}$$

If the reaction is exothermic, the system loses enthalpy to its surroundings. If the surroundings are sufficiently large that this does not result in an increase in temperature (which is usually the case in living cells), then

$$\Delta S_{system} - \frac{\Delta H}{T} > 0 \tag{8.11}$$

or

$$T\Delta S - \Delta H > 0 \tag{8.12}$$

Thus, spontaneous reactions will occur generally when $\Delta H < T\Delta S$. Endothermic reactions will occur provided the ΔS value is sufficiently large to make $T\Delta S$ greater than ΔH which is likely to be the case at higher temperatures.

8.1.4 Free Energy

Just as all molecules have intrinsic enthalpy and entropy values at given conditions of temperature, pressure and phase, so too they all have different amounts of a specific type of energy defined as that *energy available to do work*. This is called the *free energy* and, in honour of the American scientist Willard Gibbs (1839–1903), denoted as G and sometimes referred to as the Gibbs free energy. Spontaneous chemical reactions always result in *decrease* in G, that is a negative ΔG which, combined with Equation (8.12), gives us.

$$\Delta G = \Delta H - T\Delta S < 0 \tag{8.13}$$

This reflects the fact that only a *proportion* of free energy is actually available to do work as some of it is swallowed up in entropy changes.

8.2 ISOTHERMAL TITRATION CALORIMETRY

Biocalorimetry allows direct measurement of enthalpy changes associated with chemical reactions and processes. Two main types of calorimetry are in widespread use nowadays: *Isothermal titration calorimetry (ITC)* and *differential scanning calorimetry (DSC*; Section 8.3). ITC may be used directly to measure ΔH changes associated with processes such as ligand binding and complex formation. DSC is appropriate for the study of heat-initiated phase changes in biopolymers and gives measurements of ΔH values as well as other thermodynamic parameters.

8.2.1 Design of an Isothermal Titration Calorimetry Experiment

An outline design of an ITC calorimeter is shown in Figure 8.4. This system is useful in the protein concentration range of mg/ml. A typical experiment involves measurement of heat change as a function of addition of small quantities of a reagent to the calorimeter cell containing other components of the system under investigation. For example, this reagent could be a protein ligand or substrate/inhibitor of an enzyme. At the beginning of the experiment, there is a large excess of protein compared to ligand. This means that ΔH values associated with each aliquot can be individually measured. Initially, these values are large but, as aliquots are progressively added, eventually decrease to values similar to the ΔH of dilution of ligand into the solution in the calorimeter cell (Figure 8.5). The ΔH measured is the *total* enthalpy change which includes heat associated with processes such as formation of noncovalent bonds

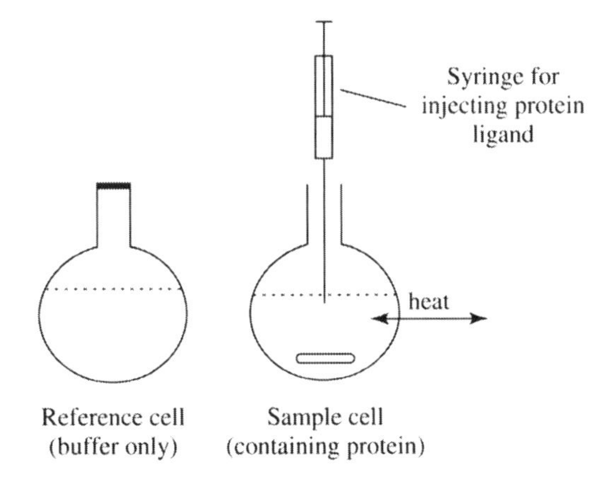

Figure 8.4. Outline design of an isothermal titration calorimeter. Addition of a substance such as a protein ligand into the sample cell causes a release or absorption of a small amount of heat. This is detected and analyzed electronically. Several such injections are made during a typical experiment.

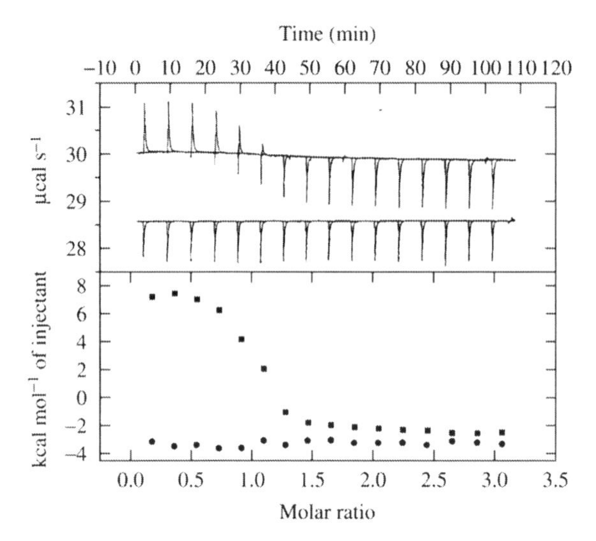

Figure 8.5. A typical ITC experiment. Small aliquots of a drug capable of interacting with DNA are sequentially injected into the calorimeter sample cell which contains DNA. Aliquots of heat resulting from each injection are accurately measured (upper panel). After subtracting the heat of solution for each injection (inverted spikes in upper panel), a binding isotherm may be constructed for this process (square points in lower panel). Reprinted from Ladbury and Chowdhry (1998). Biocalorimetry: Applications of calorimetry in the Biological Sciences. Copyright John Wiley & Sons Limited.

between interacting molecules and with other equilibria in the system such as conformational changes, ionization of polar groups (e.g. deprotonation) and changes due to interactions with solvent. Frequently, the results are reported as 'apparent ΔH' and compared to other experiments performed under identical conditions to allow identification of specific effects. This is an important point because such measurements are readily comparable to DSC data (Section 8.3) which can distinguish binding effects more precisely from related conformational/solvent effects.

8.2.2 ITC in Binding Experiments

ITC provides a useful method for studying binding processes such as those involving a protein (P) and a ligand (L):

$$P + L \leftrightarrow PL \qquad (8.14)$$

It allows estimation of both the binding constant (K_B)

$$K_B = \frac{[PL]}{[P][L]} \qquad (8.15)$$

and of the dissociation constant (K_D);

$$K_D = \frac{[P][L]}{[PL]} \qquad (8.16)$$

K_D is related to ΔG by;

$$\Delta G = -RT \ln K_D \qquad (8.17)$$

Because the ΔH for the binding interaction decreases as the system approaches equilibrium, it is a sensitive probe for the extent of the binding interaction. Subtraction of components such as heats of dilution gives a binding isotherm which may be used to calculate K_D or K_B (Figure 8.5). It is possible for binding of two individual ligands to a single protein to have similar K_D values (and consequently similar ΔGs; Equation (8.17)) but quite different ΔH and ΔS values. ITC allows direct determination of ΔH and therefore helps confirm detailed differences in binding thermodynamics (Example 8.1).

8.2.3 Changes in Heat Capacity Determined by Isothermal Titration Calorimetry

The *heat capacity* (also known as the *specific heat*) of a material is a measure of its ability to absorb heat. It is defined as C_p (the subscript denotes constant pressure) the quantity of heat which will change the temperature of exactly 1 g of a substance by 1 °C. For example, the specific heat of water is 1 cal/g °C while that of ethanol is 0.58 cal/g °C. This means that if identical weights of water and ethanol absorbed the same amount of energy, the temperature of ethanol would rise significantly higher than that of water. For most reactions involving low molecular mass reagents, the values for C_p of reactants and products are not significantly different.

However, for large molecules such as proteins, there may be significant changes in C_p in processes involving large biomacromolecules. Water associated with the surfaces of such biopolymers behaves differently from bulk water. This is even truer of processes resulting in the exposure of hydrophobic regions such as protein unfolding which exposes hydrophobic parts of the protein chain to solvent water (Chapter 6). These molecules are held relatively rigid by *hydrophobic interaction* with the protein. When the temperature is raised, this rigidity is considerably reduced resulting in a large *increase* in C_p for the denatured state versus the native state. A converse effect is important during complex formation as considerable parts of a protein surface may be removed from direct contact with solvent as a result of complex formation resulting in large *decreases* in C_p. In fact, there is a direct and experimentally useful correlation between ΔC_p and surface area removed from solvent during complex formation.

ITC can be used to determine changes in heat capacity as temperature is increased since the instrument can be operated in the temperature range 0–60 °C. The variation in C_p,

Example 8.1 Isothermal Titration Calorimetry in Study of Oligomerization of Vancomycin

Vancomycin is a peptide antibiotic which is thought to function partly by oligomerization into dimers and larger polymers. This was revealed by the construction of ligand binding curves at a range of concentrations in the presence of presumed target molecules such as a small peptide as shown below in the panel on the right.

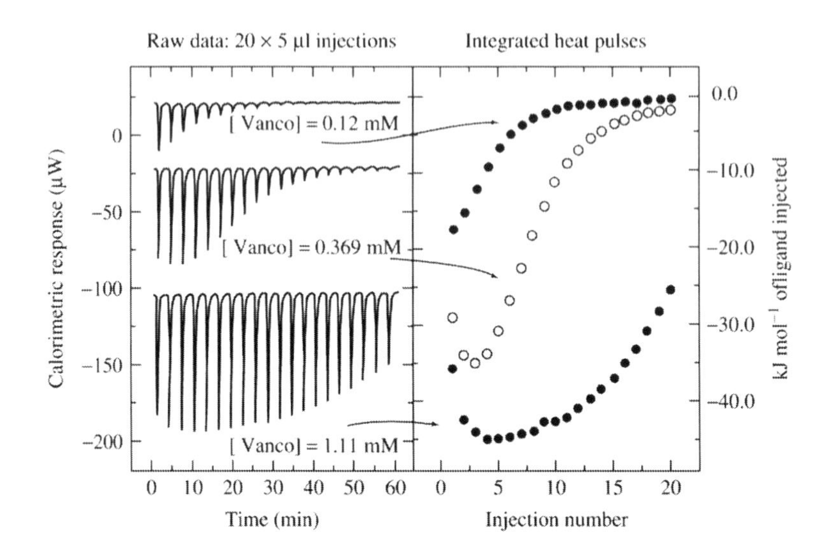

Source: Isothermal titration curves of 0.12 nM, 0.369 mM and 1.11 mM vancomycin in the presence of a target peptide (left) were used to plot binding curves (right). Reprinted from Ladbury and Chowdhry (1998). Biocalorimetry: Applications of calorimetry in the Biological Sciences. Copyright John Wiley & Sons Limited

These ligand binding curves were fitted to a simple dimerization model. From these measurements, the ΔH and K_B for dimerization were calculated and it was shown that this situation could be distinguished from oligomerization by ITC measurements alone.

From Cooper, A. (1998). Microcalorimetry of protein-protein interactions. In, Ladbury, J.E. and Chowdhry, B.Z. (eds.) *Biocalorimetry: Applications of Calorimetry in the Biological Sciences*. John Wiley & Sons Ltd, reproduced with permission.

enthalpy and entropy at two different temperatures, T_1 and T_2 is given by the following equation:

$$\Delta C_p = \frac{\Delta H_{T_2} - \Delta H_{T_1}}{T_2 - T_1} = \frac{\Delta S_{T_2} - \Delta S_{T_1}}{\ln (T_2/T_1)} \quad (8.18)$$

Plotting ΔH versus temperature gives a slope corresponding to ΔC_p (Figure 8.6).

Measurement of ΔC_p by ITC therefore provides a generally useful method for probing structural effects during processes involving biopolymers such as complex formation, protein folding and DNA duplex formation (Example 8.2). The literature cited in the Bibliography at the end of this chapter gives examples of experimental applications of ITC.

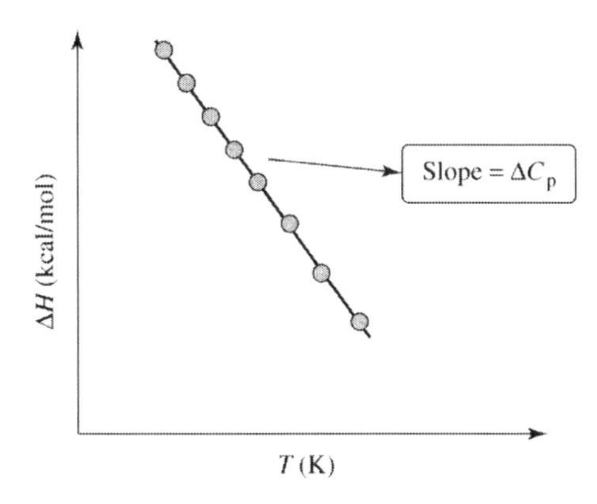

Figure 8.6. Determination of specific heat from enthalpy data. A plot of ΔH versus T is a straight line with slope of ΔC_p.

Example 8.2 Isothermal Titration Calorimetry in Study of DNA Duplex Formation

Since formation of duplexes of DNA from complementary strands depends on a number of factors such as hydrogen bonding and base stacking, it is a highly temperature-sensitive process. Duplex formation was studied in an experimental system involving complementary oligonucleotides which permitted investigation of a range of experimental variables such as salt concentration. The experiment involved injecting a small amount of one oligonucleotide into a bulk solution of the complementary oligonucleotide followed by measurement of the ΔH of duplex formation. The results shown below were obtained at (a) 292.8 K and (b) 312.4 K. Binding curves (bottom panels) were constructed from the ITC data (upper panels).

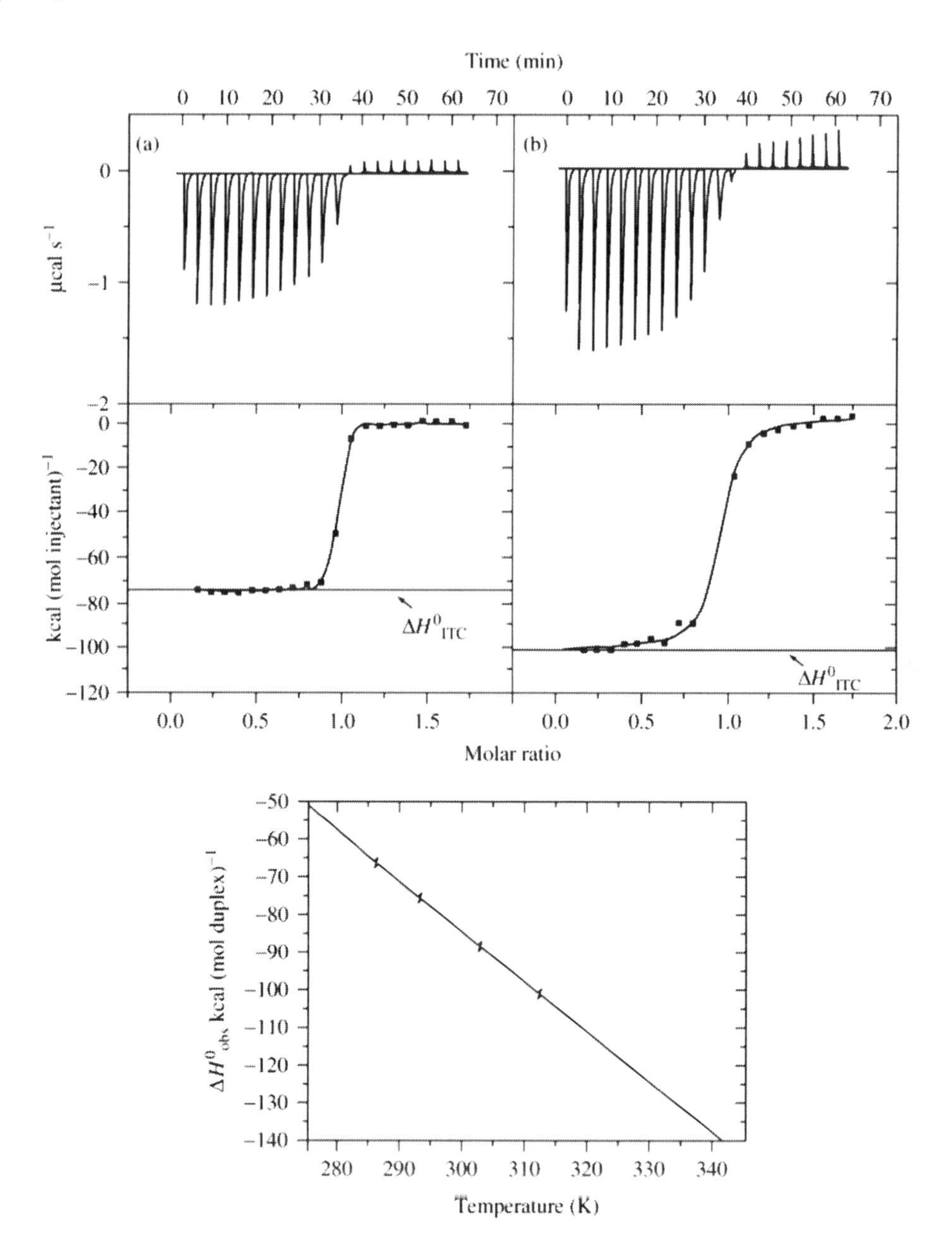

(Continues)

(Continued)

Source: Comparison of ITC profiles for duplex DNA formation at two different temperatures

ΔH values of -74.4 and -102.5 kcal/mol were determined for (a) and (b). ΔH values were plotted against temperature (below) and the slope of this line allowed calculation of ΔC_p of -1.3 kcal K^{-1} mol $duplex^{-1}$.

Source: Plot of ΔH of duplex formation determined by ITC versus T

These data were combined with DSC studies to propose that formation of single-strand DNA helices which then associate to form duplex helices are coupled processes.

Reprinted with permission from Holbrook *et al.*, Enthalpy and heat capacity changes for formation of an oligomeric DNA duplex: Interpretation in terms of coupled processes of information and association of single-stranded helices. *Biochemistry*, **38**, 8409–8422, © (1999) American Chemical Society.

8.3 DIFFERENTIAL SCANNING CALORIMETRY

DSC is the second major type of biocalorimetry experiment and involves measurement of changes in C_p as a function of change in temperature. It is especially useful in determining thermodynamic parameters of phenomena initiated by an increase or a decrease in temperature. The effects of different experimental conditions such as pH, ionic strength, solvent identity, presence/absence of ligands on these parameters can be studied by DSC. The technique is especially sensitive to phase changes in biopolymers arising during phenomena such as protein-protein binding, protein unfolding and ligand binding. Because very small ΔC_p values are usually measured, it is necessary to compare a solution of the sample to a reference solution containing only buffer. Moreover, repeated scans need to be made over a period of time to build up sufficient data for accurate C_p determination.

8.3.1 Outline Design of a Differential Scanning Calorimetry Experiment

An outline design of the experimental apparatus used is shown in Figure 8.7. DSC involves measuring tiny differences in heat (hence *differential* scanning calorimetry) between sample solutions (e.g. 2 ml of a 1–5 mg/ml protein) in the sample cell compared to reference solutions (e.g. buffer) in the reference cell. These two solutions are electrically heated (*up-scans*) or cooled (*down-scans*) and the additional current required to heat/cool the protein solution is recorded. The solutions are then cooled/heated and the experiment repeated several times. The scans (hence differential *scanning* calorimetry) are summed to give an accurate measure of enthalpies of processes such as protein unfolding, ligand binding or complex formation. Most DSC calorimeters operate in the temperature-range 1–98 °C although some operate over a wider range than this. It is usual to perform the experiment over as wide a concentration range as possible, at a variety of scan-rates (i.e. °K/h) and under a variety of experimental conditions such as pH, ion strength, and so on A typical DSC signal is shown in Figure 8.8. This results from subtraction of a significant baseline as shown in Figure 8.9. The area under the curve is the ΔH and the melting temperature (T_m) and ΔC_p are estimated as shown. These thermodynamic parameters ideally should show no dependence on scan rate or sample concentration.

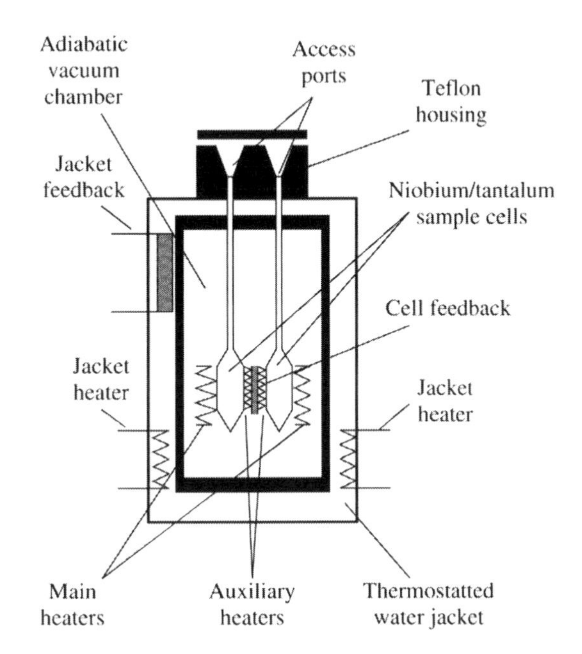

Figure 8.7. Outline design of a differential scanning calorimeter. The outline design of a Microcal MC-2 calorimetric unit is shown. Reprinted from Ladbury and Chowdhry (1998). *Biocalorimetry: Applications of Calorimetry in the Biological Sciences*. Copyright John Wiley & Sons Limited.

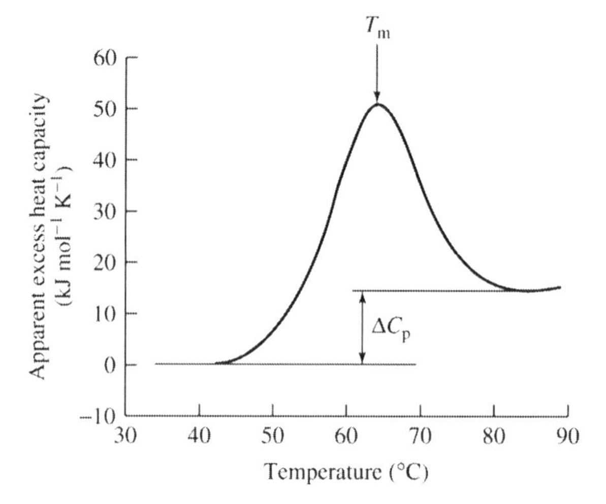

Figure 8.8. DSC of ubiquitin. A typical DSC result for the thermal denaturation of ubiquitin (5 mg/ml) is shown. The melting temperature is denoted by T_m while the change in specific heat (ΔC_p) is calculated as shown Reprinted from Ladbury and Chowdhry (1998). *Biocalorimetry: Applications of Calorimetry in the Biological Sciences*. Copyright John Wiley & Sons Limited.

DSC is really only appropriate to large molecules because these give measurable ΔC_p values and this is an important point of difference with ITC. The technique complements independent spectroscopic methods such as circular dichroism (Figure 8.9).

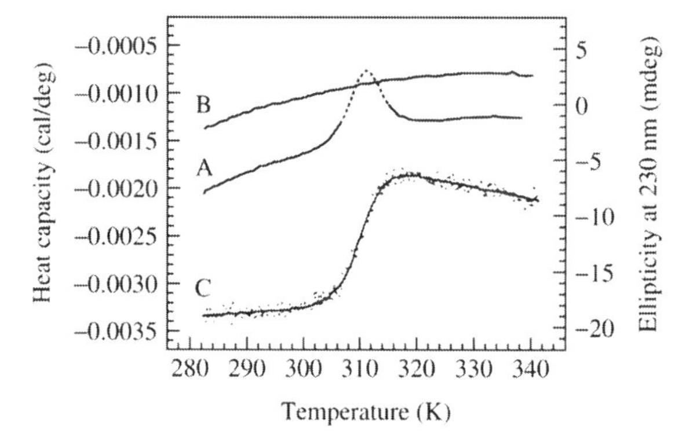

Figure 8.9. DSC of ribonuclease. In arriving at a final DSC profile, the baseline (B) must be subtracted from the thermal denaturation profile of the protein (in this case RNAase from *Bacillus amyloliquifaciens*). Circular dichroism (C) confirms that the melting temperature (T_m) occurs at the peak of the DSC signal. Fersht A. (1999). *Structure and Mechanism in Protein Science: A Guide to Enzyme Catalysis and Protein Folding*. W.H. Freeman & Co, reproduced with permission.

8.3.2 Applications of Differential Scanning Calorimetry

DSC has been used in the study of a wide range of biochemical phenomena. In general, any process involving a temperature-induced change of phase in a biomacromolecule is amenable to study by this technique. These would include double-to single-strand transitions in DNA, unfolding of globular proteins and processes involving multi-domain/multi-subunit proteins. It is possible to vary the experimental conditions used such as pH, ion strength or by including stabilizing molecules. Moreover, comparison of data for mutant and wild-type proteins allows estimation of contributions from single residues to the thermodynamics of specific processes. DSC can also be used to resolve effects involving different protein domains and hence as a tool to detect interdomain cooperativity. The technique is not limited to proteins and has been widely used in studies of polysaccharides and even nonbiological polymers.

8.4 DETERMINATION OF THERMODYNAMIC PARAMETERS BY NON-CALORIMETRIC MEANS

8.4.1 Equilibrium Constants

For any chemical process the equilibrium constant, K_a, is simply the ratio of concentration of reactants to that of products. Provided a means is available to quantify reactant and product concentrations separately, this parameter can be routinely determined. For example, in Chapter 6 we have seen that spectroscopic methods such as fluorescence can be used to measure the fraction of a protein presented in its unfolded form under a variety of chemical conditions. These values are related to the enthalpy change of the reaction by the *van't Hoff equation*:

$$\frac{d \ln K_a}{dT} = \frac{\Delta H}{RT^2} \tag{8.19}$$

This provides a noncalorimetric method for determination of ΔH which is independent of calorimetry. It can be used to *confirm assumptions* underlying analysis of the data such as, for example, showing that unfolding of a particular protein is a 2-state system consisting of various amounts of native and unfolded protein (Example 8.3).

Other means for determining equilibrium and affinity constants include equilibrium dialysis (Chapter 7) and plotting fluorescence quenching data (Chapter 3). Comparison of such data with calorimetric measurements allows independent confirmation of details of binding as well as checking assumptions in modeling these processes.

Example 8.3 Differential Scanning Calorimetry in the Study of Protein Unfolding

Fibroblast growth factors are cytokines which stimulate mitogenic and chemotactic responses in a wide range of cells. These effects arise from binding to receptors located on cells such as myoblasts, endothelial cells, chondrocytes and fibroblasts. Acidic fibroblast growth factor (aFGF) is the only one known to bind to all four characterized receptors. The binding of the protein to heparin renders it more stable to extremes of heat and pH and to degradation by proteolysis or oxidation. In the absence of heparin, the T_m of the protein is near-physiological while the presence of heparin increases the T_m by some 20 °C. The protein contains three cysteines and formation of disulfide bridges between these is responsible for inactivation in the absence of heparin. Treatment with reducing agents such as dithiothreitol regenerates active protein.

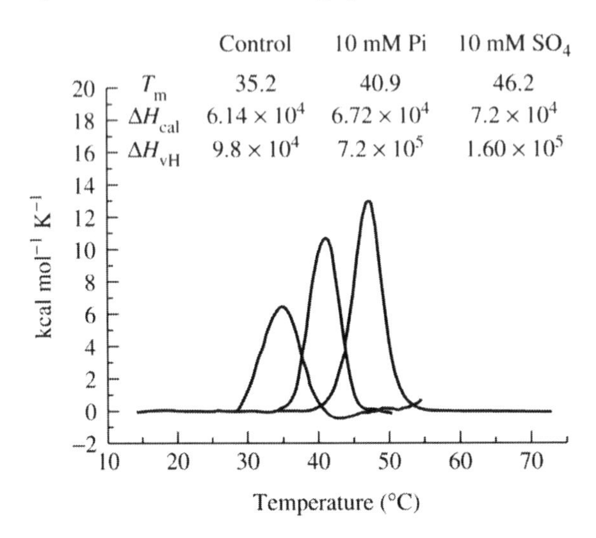

Source: DSC of aFGF under control, 10 mM phosphate and 10 mM sulphate conditions. Corresponding ΔH_{cal} and ΔH_{vH} are shown. Reproduced with permission of John Wiley & Sons, Ltd from Adamek *et al.* (1998). Denaturation Studies of haFGF. In Ladbury, J.E. and Chowdhry, B.Z. (eds.) *Biocalorimetry; Applications of Calorimetry in the Biological Sciences.*

In order to study this unfolding process further, the protein was exposed to low concentrations of the chaotropic agent, guanidinium hydrochloride and calorimetric data were collected by DSC. It was observed (see below) that addition of phosphate and sulfate resulted in a significant increase in the T_m. This is consistent with the structure of this protein determined by X-ray crystallography which shows a concentration of basic groups in a particular location containing a sulfate ion. Note that the ΔH determined by DSC (ΔH_{cal}) is markedly lower than that determined from the van't Hoff relationship (ΔH_{vH}; Section 8.4). This is consistent with the presence of multiple protein states or enthalpy contributions associated with aggregation or precipitation.

REFERENCE

General Reading

Connelly, P.R. (1994) Acquisition and use of calorimetric data for prediction of the thermodynamics of ligand-binding and folding reactions of proteins. *Current Opinion in Biotechnology*, **5**, 381–8. Review of use of calorimetric measurements in study of processes such as protein folding and ligand binding.

Janin, J. (1997) Ångstroms and calories. *Structure*, **5**, 473–9. A thought-provoking exploration of the relationship between calorimetric measurements and structure in biomacromolecules.

Jelesarov, I. and Bosshard, H.R. (1999) Isothermal titration calorimetry and differential scanning calorimetry as complementary tools to investigate the energetics of biomolecular recognition. *Journal of Molecular Recognition*, **12**, 3–18.

Ladbury, J.E. and Chowdhry, B.Z. (eds) (1998) *Biocalorimetry; Applications of Calorimetry in the Biological Sciences*, John Wiley & Sons, Ltd, Chichester. A comprehensive overview of the basic theory of both ITC and DSC with specific examples.

Plum, G.E. and Breslauer, K.J. (1995) Calorimetry of proteins and nucleic acids. *Current Opinion in Structural Biology*, **5**, 682–90. Reviews the contribution of calorimetry to elucidation of structural stability and ligand interactions in proteins and DNA.

Wunderlich, B. (1996) Thermal analysis of macromolecules – a personal review. *Journal of Thermal Analysis*, **46**, 643–79. A historical review of the development of calorimetry in the study of polymers.

Thermodynamics

Holdgate, G.A. and Ward, W.H.J. (2005) Measurements of binding thermodynamics in drug discovery. *Drug Discovery Today*, **10**, 1543–50. A review of the role of thermodynamics in rational drug design.

Orstan, A. (1990) Thermodynamics and life. *Trends in Biochemical Sciences*, **15**, 137–8.

Isothermal titration calorimetry

Buurma, N.J. and Haq, I. (2007) Advances in the analysis of isothermal titration calorimetry data for ligand-DNA interactions. *Methods*, **42**, 162–72. An article considering improvements in ITC data analysis.

Feig, A.L. (2007) Applications of isothermal titration calorimetry in RNA biochemistry and biophysics. *Biopolymers*, **87**, 293–301. A review of ITC as applied to RNA.

Freyer, M.W. and Lewis, E.A. (2008) Isothermal titration calorimetry: Experimental design, data analysis and probing macromolecule/ligand binding and kinetic interactions, in *Methods in Cell Biology*, Vol. **84** (eds J.J. Correia, and H.W. Detrich), Elsevier, pp. 79–113. A review of ITC.

Oda, M., Furukawa, W., Sarai, A. and Nakamura, H. (1999) Construction of an artificial tandem protein of the c-Myb DNA-binding domain and analysis of its DNA binding specificity. *Biochemical Biophysical Research Communication*, **262**, 94–7. An excellent example of use of ITC measurements in the study of protein–DNA binding.

Chung, E., Henriques, D., Renzoni, D. *et al.* (1998) Mass spectrometric and thermodynamic studies reveal the role of water molecules in complexes formed between SH_2 domains and tyrosyl phosphopeptides. *Structure*, **6**, 1141–51. An example of the combination of ITC with MS approaches to understanding the importance of water molecules in interactions between tyrosyl phosphopeptides and the SH_2 domain of tyrosine kinase.

Livingstone, J.R. (1996) Antibody characterization by isothermal titration calorimetry. *Nature*, **384**, 491–2. Application of ITC to characterization of antibodies and related products.

Turnbull, W.B. and Daranas, A.H. (2003) On the value of c: Can low affinity systems be studied by isothermal titration calorimetry? *Journal of the American Chemical Society*, **125**, 14859–66. Study of the potential of ITC for study of low-affinity binding systems.

Differential scanning calorimetry

Bruylants, G., Wouters, J. and Michaux, C. (2005) Differential scanning calorimetry in life science: Thermodynamics, stability, molecular recognition and application in drug design. *Current Medicinal Chemistry*, **12**, 2011–20.

Fisher, H.F. and Singh, N. (1995) Calorimetric methods for interpreting protein-ligand interactions. *Energetics of Biological Macromolecules*, **259**, 194–221. A review of study of ligand binding to proteins with DSC.

Holbrook, J.A., Capp, M.W., Saecker, R.M. and Record, M.T. (1999) Enthalpy and heat capacity changes for formation of an oligomeric DNA Duplex: Interpretation in terms of coupled processes of formation and association of single-stranded helices. *Biochemistry*, **38**, 8409–22. An excellent example of the combination of ITC and DSC with other approaches to resolve DNA duplex formation into intrastrand, and interstrand effects.

Robinson, C.R., Liu, Y.F., O'Brien, R. *et al.* (1998) A differential scanning calorimetric study of the thermal unfolding of apo- and holo-cytochrome b (562). *Protein Science*, **7**, 961–5. An interesting application of DSC.

Spink, C.H. (2008) Differential scanning calorimetry, in *Methods in Cell Biology*, Vol. **84** (eds J.J. Correia and H.W. Detrich), Elsevier, pp. 115–41. A review of DSC.

A useful web site

University of Washington calorimetry centre: http://mozart.bmsc.washington.edu/~hyre/calorimetry.html.

Chapter 9
Bioinformatics

Objectives

After completing this chapter you should be familiar with:

- *The main sequence and three-dimensional structure archives available* via *the www.*
- *How to access and download information from these archives.*
- *How to analyse this information to extract functionally relevant data.*
- *For primary structures, you will be able to align sequences, predict secondary structure, hydropathy and potential post-translational modification sites. For tertiary structures you will be able to view structures in a variety of ways and to carry out homology modelling.*

In Chapter 1 we introduced the worldwide web (www) as a resource for large-scale storage, sharing and analysis of information (Sections 1.5.1 and 1.5.2). In biochemistry, a major category of stored information comprises sequences of nucleotides (DNA) or amino acids (proteins). However, other types of useful information relevant to biochemistry are also available on the web. For example, dedicated databases exist for two-dimensional SDS PAGE and MS, which are used in proteomics (Chapter 10). Specialist databases also exist for peptide and carbohydrate chemistry and for various types of spectroscopy (Tables 9.1 and 9.2). Similar organizing principles apply to the main databases used in molecular life sciences. Historically, individual databases arose independently of each other (which originally posed problems in navigating from one to the other), but they are now very closely interlinked. For example, it is possible to search the *macromolecular structural database* (MSD; also called the protein databank or PDB, Section 9.4.2) for a particular amino acid sequence and to 'translate' a nucleotide sequence into an amino acid sequence. In a real sense these databases now comprise a closely interwoven 'web' of connections within the wider www. The very range and complexity of these interconnections can be bewildering so, in this chapter, we will review the principal databases relevant to molecular life sciences and explain how we can use them to help us solve problems involving physical techniques in biochemistry. Bioinformatics is a dynamic and growing field and it is impossible to give it comprehensive coverage in a short chapter. Instead, it is hoped to describe selected resources (e.g. those listed in Tables 9.1–9.3), which the reader may find especially useful in accomplishing practical tasks such as accessing or analysing sequences or three-dimensional structures. The Bibliography points the reader to some excellent texts, which give a more detailed picture. Because of the highly interconnected nature of resources on the www, there is frequently more than one way to achieve a particular objective (e.g. retrieving a sequence or viewing a structure) so the strategies outlined here are to some extent matters of personal preference and should be regarded by the reader as intended merely to illustrate general possible approaches. All the resources described have extensive literature available on-line to help the novice understand and exploit tools and archives to which they are linked so the reader is encouraged to enter the web sites mentioned and to explore them. To illustrate just how informative and rewarding this can be, the aquaporin family of proteins will be used as an example throughout this chapter. These are a widespread group of transmembrane proteins which form channels for the transport of water across membranes of a wide range of cell types. Roderick MacKinnon and Peter Agre were honoured with the Nobel Prize in Chemistry (2003) for their work on elucidating structure-function relationships in this important protein family.

9.1 OVERVIEW OF BIOINFORMATICS

Until the late 1980s much of the information used in Biochemistry was stored in paper form either in books or research papers in the peer-reviewed literature. For example, an atlas of protein sequences was published annually to contain all the then-known amino acid sequences (*Atlas of Protein Sequence and Structure*; Margaret Dayhoff ed.). This meant that information was hard to find, comparisons were difficult and there was always a significant time-lag

Table 9.1. Selected bioinformatics web sites.

Site	Contents	Url
DNA database of Japan (DDBJ)	Nucleotide sequences	http://www.ddbj.nig.ac.jp/
National Center for Biotechnology Information (NCBI)	Nucleotide sequences (GenBank), Entrez search and retrieval system, sequence/genome analysis tools, EST databases.	http://www.ncbi.nih.gov/
European Bioinformatics Institute (EBI)	Nucleotide sequences (EMBL), protein sequences (TrEMBL and UniProt), sequence/structure analysis tools.	http://www.ebi.ac.uk/
Swiss Institute of Bioinformatics (ExPASy)	Protein sequences (SwissProt and UniProt), Protein families (ProSite) and sequence analysis tools, Proteomics tools, Protein modelling (Swiss PDB viewer).	http://us.expasy.org/
Protein Information Resource (PIR)-international UniProt	Protein sequences, protein structure analysis tools.	http://pir.georgetown.edu
	United protein encompassing SwissProt, PIR and TrEMBL.	http://www.pir.uniprot.org/
Munich Information Centre for Protein Sequences.	Genome resources.	http://mips.gsf.de/services/genomes
Research Collaboratory for Structural Bioinformatics (RCSB)	Files of atomic coordinates for macromolecules, Protein graphics programs	http://www.rcsb.org
Cambridge Crystallographic Data Centre	Atomic coordinates of small molecules and tools for analysis/comparison	http://www.ccdc.cam.ac.uk/

Table 9.2. Selected genome web sites.

Site	Contents	Url
The Sanger Centre	Access to a wide range of microbial and eukaryotic genomes including the Human genome and genome searching tools.	http://www.sanger.ac.uk
National Center for Biotechnology Information (NCBI)	Access to completed and partial genomes and tools for searching and analysis.	http://www.ncbi.nih.gov/Genomes/index.html
European Bioinformatics Institute (EBI)	Access to completed and partial genomes and tools for searching and analysis.	http://www.ebi.ac.uk/genomes/index.html
Protein Information Resource (PIR)	Access to completed genomes.	http://pir.georgetown.edu/pirwww/search/genome.html
University of Bordeaux, Genolevures.	Partial and complete genomes of hemiascomycetous yeasts.	http://cbi.labri.u-bordeaux.fr/Genolevures/
Munich Institute for Protein Sequences.	Microbial, yeast, plant and mitochondrial genomes.	http://mips.gsf.de/
Fungal Genome Resource	Fungal genomes	http://gene.genetics.uga.edu/
The Genome Web	Comprehensive listing of genome links.	http://www.cbi.pku.edu.cn/mirror/GenomeWeb/genome-db.html
Institute of Genome Research (TIGR).	Genome databases and analysis tools	http://www.tigr.org/htdig/

Table 9.3. Selected bioinformatics tools.

Site	Tool	Url
NCBI	BLAST	http://www.ncbi.nlm.nih.gov/BLAST/
PIR	FastA	http://pir.georgetown.edu/pirwww/search/fasta.html
	PIRSF	http://pir.georgetown.edu/pirsf/
	iProClass	http://pir.georgetown.edu/iproclass/
EBI	Clustal W	http://www.ebi.ac.uk/clustalw/index.html
	PANDIT	http://www.ebi.ac.uk/goldman-srv/pandit/
	SRS	http://srs.ebi.ac.uk
Weizmann Inst. Israel	Hydropathy plots	http://www.ebioinfogen.com/cgi-bin/webframe.pl? http://bioinformatics.weizmann.ac.il/
PBIL-IBCP, Lyons, France	GOR IV (Secondary Structure)	http://npsa-pbil.ibcp.fr/cgi-bin/npsa_automat.pl? page=npsagor4.html
	GENO3D	http://pbil.ibcp.fr/htm/index.php?page=pbil_index.html
Columbia University, USA	Predict Protein server (PROF)	http://cubic.bioc.columbia.edu/pp/submit_exp.html#top
Boston University, USA	PSA structure server	http://bmerc-www.bu.edu/psa/
Sanger Institute, UK	Pfam	http://www.sanger.ac.uk/Software/Pfam/
ExPASy	PROSITE	http://us.expasy.org/prosite/
	Swiss-PdbViewer	http://us.expasy.org/spdbv/
	SWISS-MODEL	http://swissmodel.expasy.org/
Berkeley University, USA	SCOP Database	http://scop.berkeley.edu/data/scop.b.html
University College London, UK	CATH Database	http://www.biochem.ucl.ac.uk/bsm/cath/
MDL	Chime	http://www.mdli.com/products/framework/chemscape/
NCBI software page	RasMol	http://www.bernstein-plus-sons.com/software/rasmol/
	RasTop	http://www.geneinfinity.org/rastop/
	PyMOL	http://pymol.sourceforge.net/
	WebMol	http://www.cmpharm.ucsf.edu/˜walther/webmol.html
	UniGene	http://www.ncbi.nih.gov/entrez/query.fcgi?db=unigene
University of California, San Diego, USA	Protein Explorer	http://molvis.sdthat is to say, namelyedu/protexpl/frntdoor.htm
German Cancer Research Centre	Multi-GIF Page	http://www.dkfz-heidelberg.de/spec/pdb2mgif/
University of California, San Francisco, USA	MODELLER	http://salilab.org/license-server/license.cgi
University of Namur, Belgium	ESyPred3D	http://www.fundp.ac.be/urbm/bioinfo/esypred/
Lawrence Livermore Natl. Laboratory, USA	PROCHECK	ftp://www-structure.llnl.gov/Procheck_NT/

between discovery and publication of knowledge. The development of computers made possible rapid recovery, comparison and analysis of very large amounts of information in almost real time. Bioinformatics is the application of computer science to the solution of biological problems. Up to the mid-1990s this was the preserve of specialists trained in fields such as computer science, structural chemistry and advanced mathematics. However, the explosion in computer power (larger memories, faster processors, better and faster programs) in the late 1990s have made bioinformatics increasingly part of the 'toolkit' essential for modern molecular life scientists. Spin-offs of this include large-scale sequencing projects such as the human genome project (HGP), the unprecedented and continuing increase in structures held in the PDB and an activity known as *database mining* where scientists use databases rather than conventional screening programs to discover potential new drugs or new families of proteins. Bioinformatics also supports the new field of *systems biology,* which attempts to link genomics, proteomics and other forms of organized biological information to allow greater understanding of functioning at cell-, tissue- or organ-level arising from complex interactions between multiple elements.

It can be helpful to divide up bioinformatics resources into three distinct components: *web sites*, *databases* and *programs* (Figure 9.1). Web sites represent a single 'location' on the www curated by a scientifically reputable entity such as a university, a national or supranational governmental organization, a professional body or a combination of several of these. Examples include the European Bioinformatics Institute (EBI; an outstation of the European Molecular Biology Laboratory located near Cambridge, England),

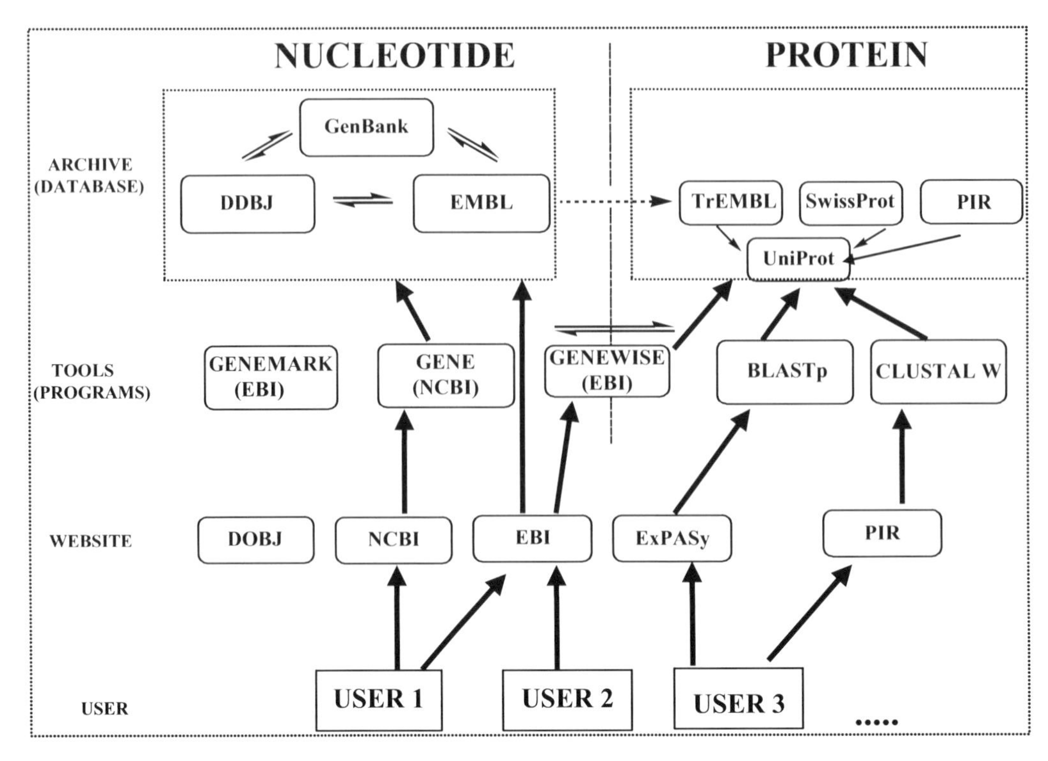

Figure 9.1. Overview of bioinformatics resources on the www. Three types of resources are shown: web sites, programs and databases together with possible routes taken by three separate users (bold arrows). Notice that the same databases can be accessed via different web sites and that user 1 can access the same nucleotide sequence databases through different web sites. Similarly users 2 and 3 can arrive at the same database by different routes. Where programs are unique to a web site this is indicated.

the National Center for Biotechnology Information (NCBI; run by the National Library of Medicine, National Institutes of Health, Bethesda, Maryland, USA) and the ExPASy site (run by the Swiss Institute of Bioinformatics). These sites act as links to the principal databases and software resources for submission, retrieval and analysis of database entries. Information relevant to each site may be submitted (often as a precondition for publication in the peer-reviewed literature) by a means standard for each site. The data are converted into a uniform format and stored in a unique file. Collections of these files are called *archives* which are *searchable* by a file identifier and by a variety of other means. This searchable nature of databases makes them especially useful in molecular life sciences. They can commonly be very rapidly searched by word, author (providing an important link to the peer-reviewed literature) or by similarity to other entries. Some examples of such searches are given below. A list of files is generated which is usually arranged in order of similarity to the search-term(s). Where several similar databases are searched with the same search term(s) we commonly get slightly differing numbers of 'hits' so it is always worthwhile trying several databases and several searches at the beginning of any bioinformatics project. Most web sites also contain access to programs which allow us to make comparisons or carry out further analysis of our target files. Sometimes these programs can be downloaded to our own PC or we can access them in the web site on-line. Some programs are found at more than one web site and these would usually represent the most widely-accepted approach to the particular analysis. Thus the BLAST tool (Section 9.3.1) is the standard program for carrying out rapid alignment of sequences while Clustal W (Section 9.3.1) is currently the standard program for analysing phylogenic relationships between sequences.

At a research level bioinformatics concerns itself with developing better, faster and more informative programs but many of these are not generally accessible or publicly available. In this chapter we will only concern ourselves with publicly available programs, which are free of charge and which work in a Windows operating system. For more powerful programs access to a dedicated workstation may be needed. These often use the UNIX operating system. It is possible to install UNIX on a PC and even to partition a PC hard-drive between Windows and UNIX. Some of the programs described here can run on Macintosh PCs just as well as Windows and the web sites usually provide clear guidance on this.

9.2 SEQUENCE DATABASES

Some of the most useful information in biochemistry comes from the sequences or primary structures of biomacromolecules. Since DNA and proteins are informational molecules, their ultimate shape and hence biological properties arise largely from their primary structures. Originally, individual databases were developed for proteins and nucleic acids in different parts of the world. Nowadays there are close links between all sequence databases, which make it possible to navigate between them. Figure 9.1 also illustrates that it is possible to search a database from more than one web site further emphasizing their high level of interconnectedness. Some of the principal bioinformatics web sites are listed in Table 9.1 and the sequence databases they curate are discussed individually below. The web sites also host a wide variety of powerful programs (often referred to as *tools*), which allow scientists both to submit, retrieve and analyse sequence files on-line. Although web sites may have common tools and generally similar organization there can be differences in emphasis. The NCBI site is mainly concerned with nucleotide and genome sequences while ExPASy is more concerned with protein sequences and their analysis. The EBI site has extensive resources for both nucleotide and protein sequence analysis. The *Entrez* search and retrieval system at NCBI is a particularly powerful tool for finding protein sequence information, notwithstanding the emphasis NCBI otherwise places on nucleotide sequence.

9.2.1 Nucleotide Sequence Databases

The principal nucleotide sequence databases are GenBank (NCBI), EMBL (EBI) and the DNA database of Japan (DDBJ). These currently contain in excess of 60 000 000 individual DNA sequences which have been sequenced in both strands with overlapping sequence runs (Section 5.4). These data come from large-scale genome sequencing projects, patent applications and sequences submitted by individual researchers. All three databases exchange data with each other on a daily basis to give a comprehensive up-to-date collection of nucleotide sequences accessible at EMBL (Figure 9.2). It should be borne in mind that some of the sequences in the nucleotide sequence databases may not have been published in the peer-reviewed literature. Each entry in the database has a unique identifier called an *accession number*. When the database accepts a sequence as meeting its quality requirements, this number, which is unique to the entry, is issued. These are usually quoted in any peer-reviewed scientific publication and, whether published or unpublished, users can access these files directly on-line.

9.2.2 Protein Sequence Databases

One of the oldest and best resources for protein sequences is the *Protein Information Resource* (PIR) located at Georgetown University, Washington DC. The other principal protein sequence databases are *SwissProt* and *TrEMBL*, a database comprising sequences translated from the EMBL nucleotide database. PIR now operates an international collaboration with EBI and the Swiss Institute of Bioinformatics and other sequence databases offering a very high level of annotation of protein sequence files (Figure 9.3). PIR hosts a single database updated every three weeks called *UniProtKB* (Table 9.1; Figure 9.4) which has two sections; UniProtKB/SwissProt (~360 000 entries) and UniProtKB/TrEMBL (~5 500 000 entries). UniProtKB/SwissProt uses human annotation with minimum redundancy whilst UniProtKB/TrEMBL uses computer-based annotation with a high level of redundancy, partly explaining the discrepancy in the number of entries. Some other databases which deal with how primary structure information can be further categorized/analysed in terms of secondary structure, post-translational modification and specific biological properties are discussed below and listed in Table 9.3.

9.2.3 Genome Databases

Availability of greater computer power coupled with developments in robotics and availability of automated methods for sequencing large amounts of DNA allowed completion of large-scale genome sequencing projects over the last 15 years. The most famous of these is the *Human Genome Project,* which sequenced all the genes encoded in the ~3200 million base pairs and approximately 2 m per cell of human DNA. This project was completed ahead of schedule by an eventual collaboration between a public domain academic human genome project consortium and a commercial company *Celera* (which means 'swift') (Figure 9.5). The complete genomes of many other representative organisms are now available. The smallest genomes, those of viruses, can be <200 000 base pairs in length while the genome of *Escherichia coli* consists of 4.6 million base pairs. We now know the sequences of hundreds of other prokaryotes and archaebacteria. Many of these are pathogenic organisms (e.g. causative agents of bubonic plague, diphtheria, pneumonia, etc.) or species of agronomic or biotechnological importance (e.g. lactobacilli). In bacteria, the bulk of DNA encodes genes while most eukaryotic DNA is noncoding. Selected URLs devoted to genomes are listed in Table 9.2.

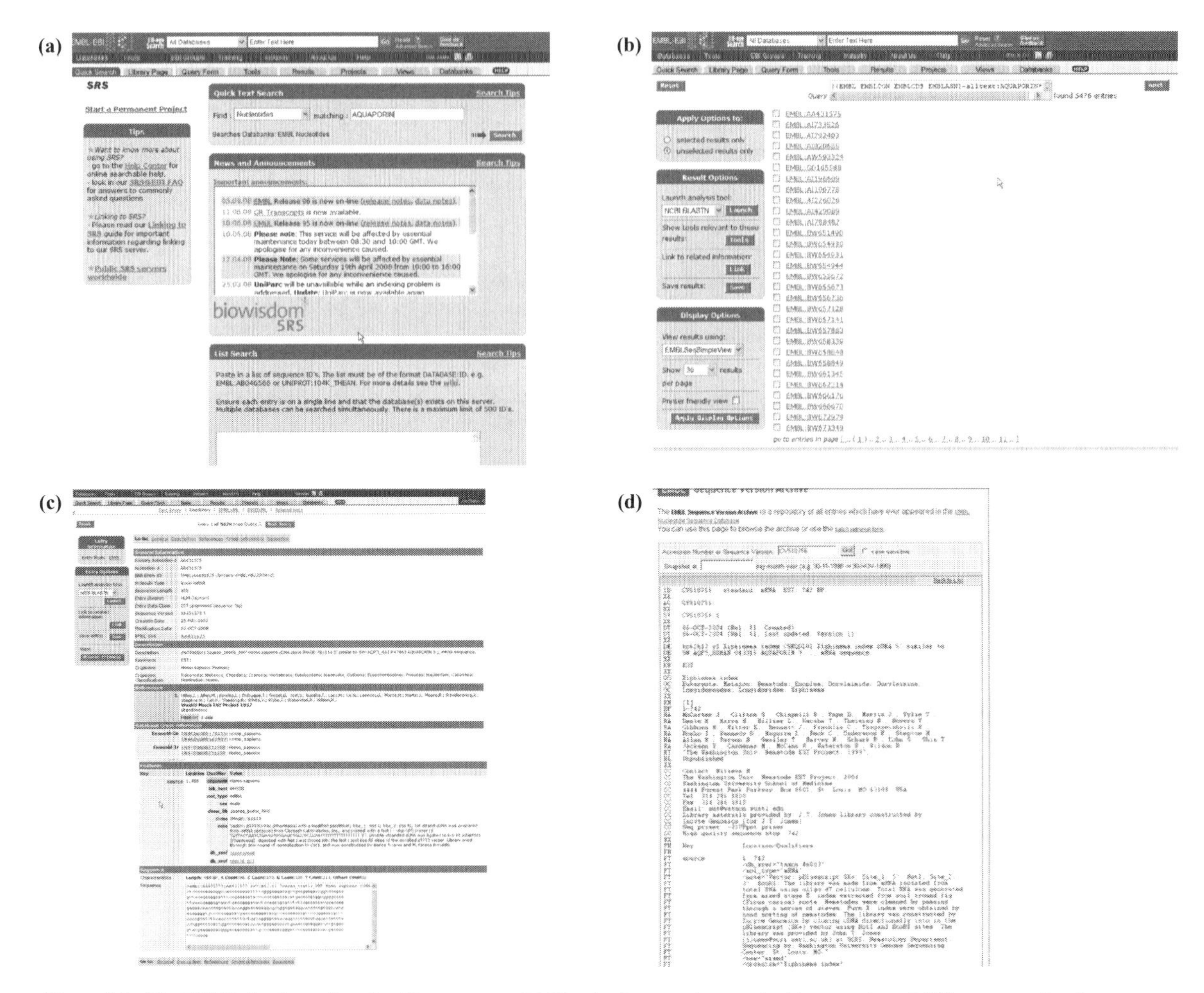

Figure 9.2. The EMBL database of nucleotide sequences. (a) The database can be searched in a number of different ways but the sequence retrieval server (SRS; http://srs.ebi.ac.uk) allows us to search by text. Typing 'aquaporin' into the query box selects all files held for aquaporins. (b) A listing of hits is provided. (c) and (d) We can select just one of these to view in detail.

Computer programs are used to identify *open reading frames* (ORFs) which are candidate genes beginning with a start codon (ATG) and ending with a stop codon. Bacterial genes are easier to identify as they have continuous sequences unlike most eukaryote genes which are interrupted by introns. Eukaryote ORFs are defined as potential genes based on similarity with previously-identified genes from other organisms (*homology*) or based on identification of characteristic sequence features. A TATA box is usually found approximately 30 base pairs upstream of the start exon ending immediately before a GT splice signal. Internal exons begin immediately after an AG and end immediately before a GT splice signal. The 3′ exon starts immediately after an AG splice signal and ends with a termination codon. Species-specific codon usage preference provides a yardstick for distinguishing coding from noncoding regions of genomes.

Two general approaches may be taken to assembling the sequence of a complete genome (Figure 9.5). One approach, taken by the public domain human genome project consortium, was to generate a 'map' of highlights of the genome and to fill in the relationship of other parts of the genome to these highlights by matching overlapping pieces of DNA sequence in a hierarchical manner. Maps of the genome of eukaryotes begin with the physically discrete individual chromosomes. For some genes *genetic maps* could be generated by quantifying frequency of coinheritance or *linkage* of characteristics encoded by pairs of genes. But, for

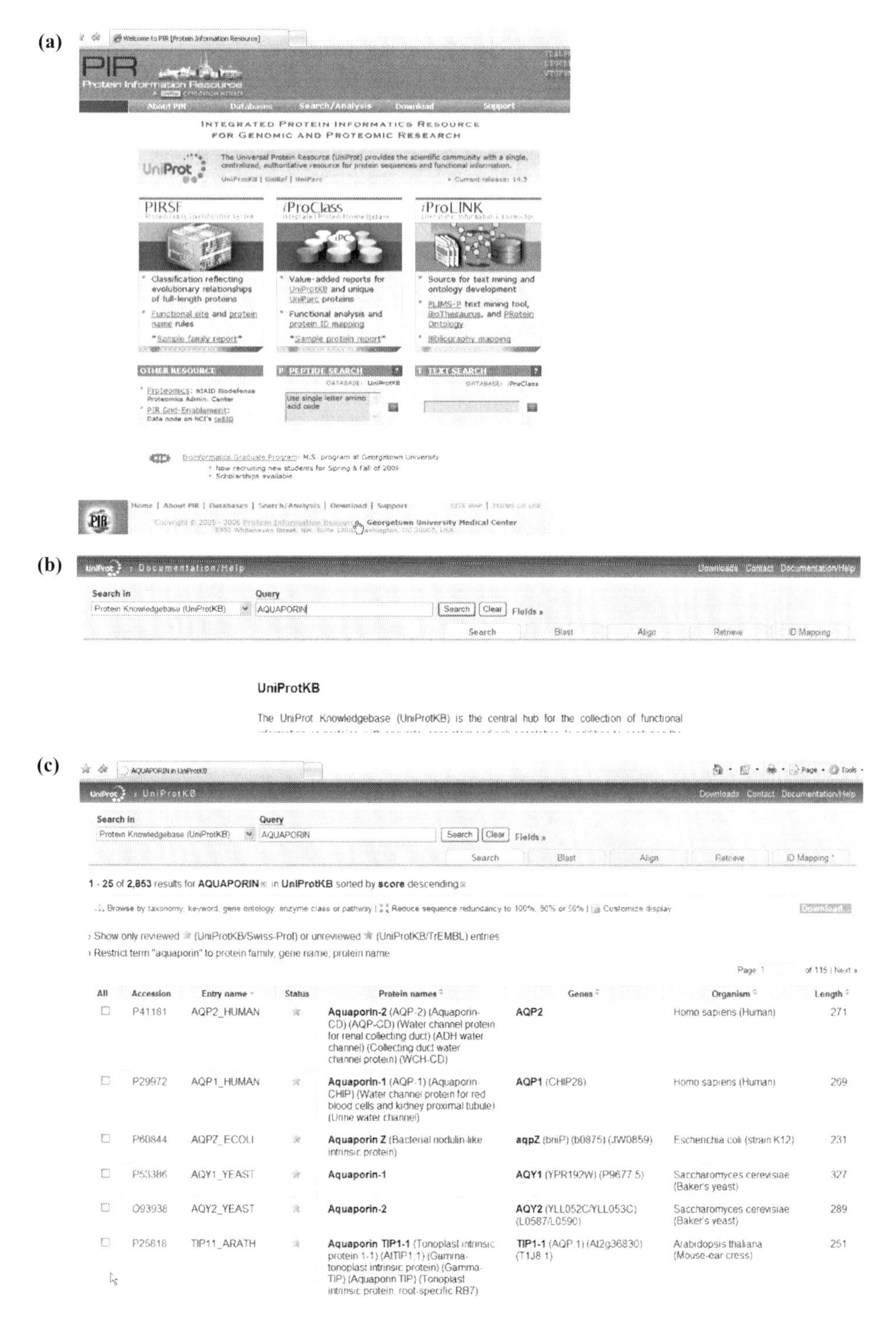

Figure 9.3. The Protein Information Resource (PIR). (a) The PIR homepage, UniProt can be accessed here by selecting "UniProtKB" (i.e. UniProt KnowledgeBase). Note that other resources such as PIRSF and iProClass are accessible here also. (b) This is searchable by text using the pulldown menu; an aquaporin search is shown. (c) A list of hits is generated.

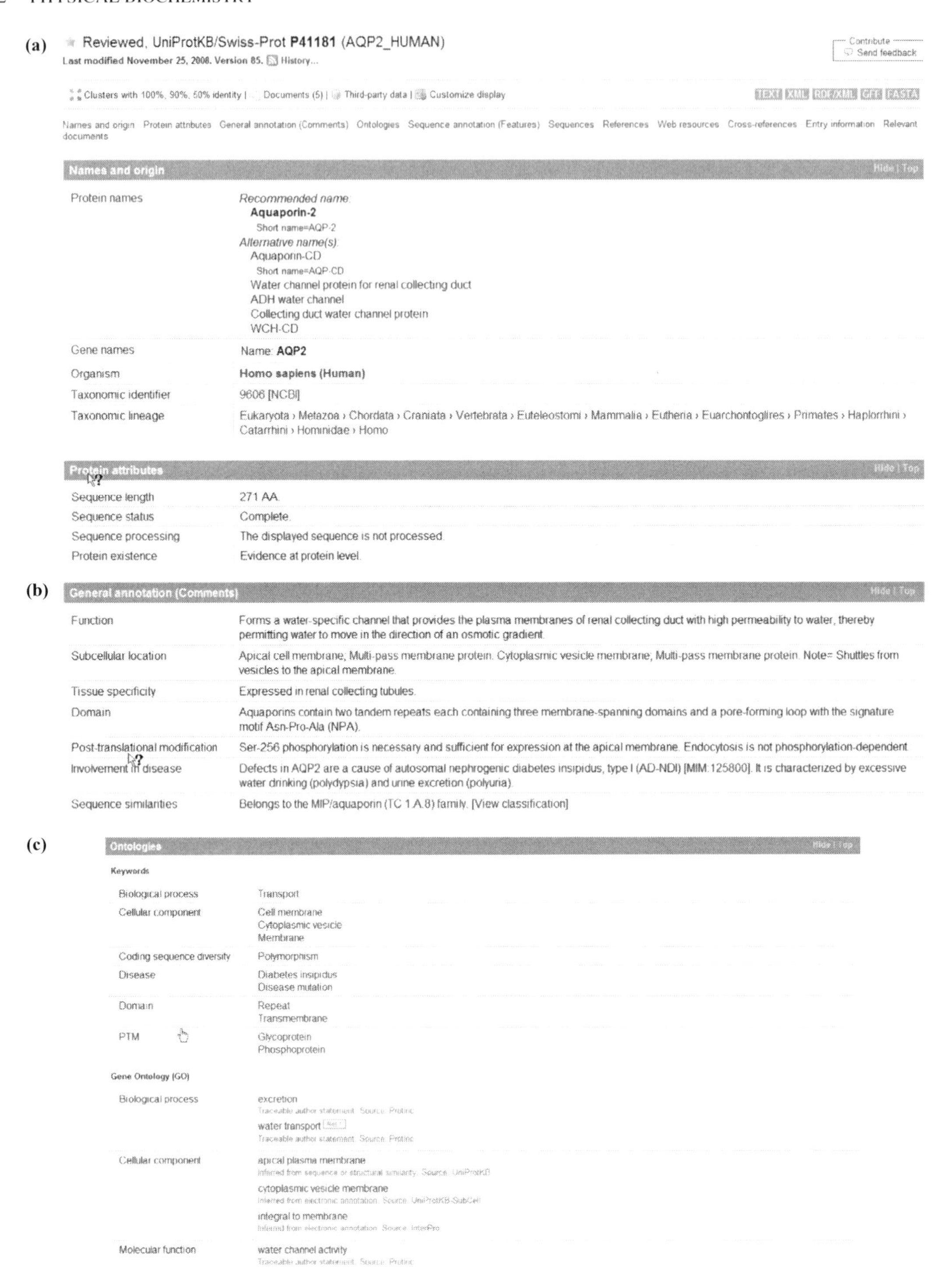

Figure 9.4. The UniProt database. (a) The entry for human aquaporin 2 (P41181). (b) General annotation section of the file describing attributes such as likely functions and subcellular location. (c) Ontologies section of the file giving keywords and gene ontologies. (d) Sequence annotation section of the file describing domain and residue functions/modification. (e) A listing of naturally occurring and artificial mutagenesis with consequences to function.

(d)

Sequence annotation (Features) Hide | Top

Feature key	Position (s)	Length	Description	Graphical view	Feature identifier
Molecule processing					
Chain	1 – 271	271	Aquaporin-2		PRO_0000063934
Regions					
Topological domain	1 – 16	16	Cytoplasmic Potential		
Transmembrane	17 – 34	18	Potential		
Topological domain	35 – 40	6	Extracellular Potential		
Transmembrane	41 – 59	19	Potential		
Topological domain	60 – 85	26	Cytoplasmic Potential		
Transmembrane	86 – 107	22	Potential		
Topological domain	108 – 127	20	Extracellular Potential		
Transmembrane	128 – 148	21	Potential		
Topological domain	149 – 156	8	Cytoplasmic Potential		
Transmembrane	157 – 176	20	Potential		
Topological domain	177 – 202	26	Extracellular Potential		
Transmembrane	203 – 224	22	Potential		
Topological domain	225 – 271	47	Cytoplasmic Potential		
Motif	68 – 70	3	NPA 1		
Motif	184 – 186	3	NPA 2		
Amino acid modifications					
Modified residue	256	1	Phosphoserine, by PKA		
Modified residue	261	1	Phosphoserine By similarity		
Modified residue	264	1	Phosphoserine By similarity		
Glycosylation	123	1	N-linked (GlcNAc...) Potential		

(e)

Feature key	Position (s)	Length	Description	Graphical view	Feature identifier
Natural variations					
Natural variant	22	1	L → V in AR-NDI.		VAR_015239
Natural variant	28	1	L → P in AR-NDI.		VAR_015240
Natural variant	47	1	A → V in AR-NDI.		VAR_015241
Natural variant	57	1	Q → P in AR-NDI. dbSNP rs28931580.		VAR_015256
Natural variant	64	1	G → R in AR-NDI.		VAR_004401
Natural variant	68	1	N → S in AR-NDI.		VAR_015242
Natural variant	71	1	V → M in AR-NDI.		VAR_015243
Natural variant	100	1	G → V in AR-NDI.		VAR_015257
Natural variant	121	1	L → F. dbSNP rs11169226.		VAR_037577
Natural variant	125	1	T → M in AR-NDI.		VAR_015244
Natural variant	126	1	T → M in AR-NDI.		VAR_015245
Natural variant	147	1	A → T in AR-NDI.		VAR_015246
Natural variant	168	1	V → M in AR-NDI.		VAR_015247
Natural variant	175	1	G → R in AR-NDI.		VAR_015248
Natural variant	181	1	C → W in AR-NDI.		VAR_015249
Natural variant	185	1	P → A in AR-NDI.		VAR_015250
Natural variant	187	1	R → C in AR-NDI; mutant protein does not fold properly and is not functional.		VAR_004402
Natural variant	190	1	A → T in AR-NDI; mutant protein does not fold properly and is not functional.		VAR_015251

Figure 9.4. *(Continued)*

large-scale high resolution genome sequencing, *physical maps* based on natural landmarks are required since not all genes have observable phenotypes. These landmarks are often repeating sequences, which form *marker regions* distributed across chromosomes. A good example is provided by *variable number tandem repeats* (VNTRs; also known as *minisatellites*). These contain regions of 10–100 base pair sequences, which repeat a variable number of times in the genomes of different individuals and can be amplified by the polymerase chain reaction. *Short tandem repeat*

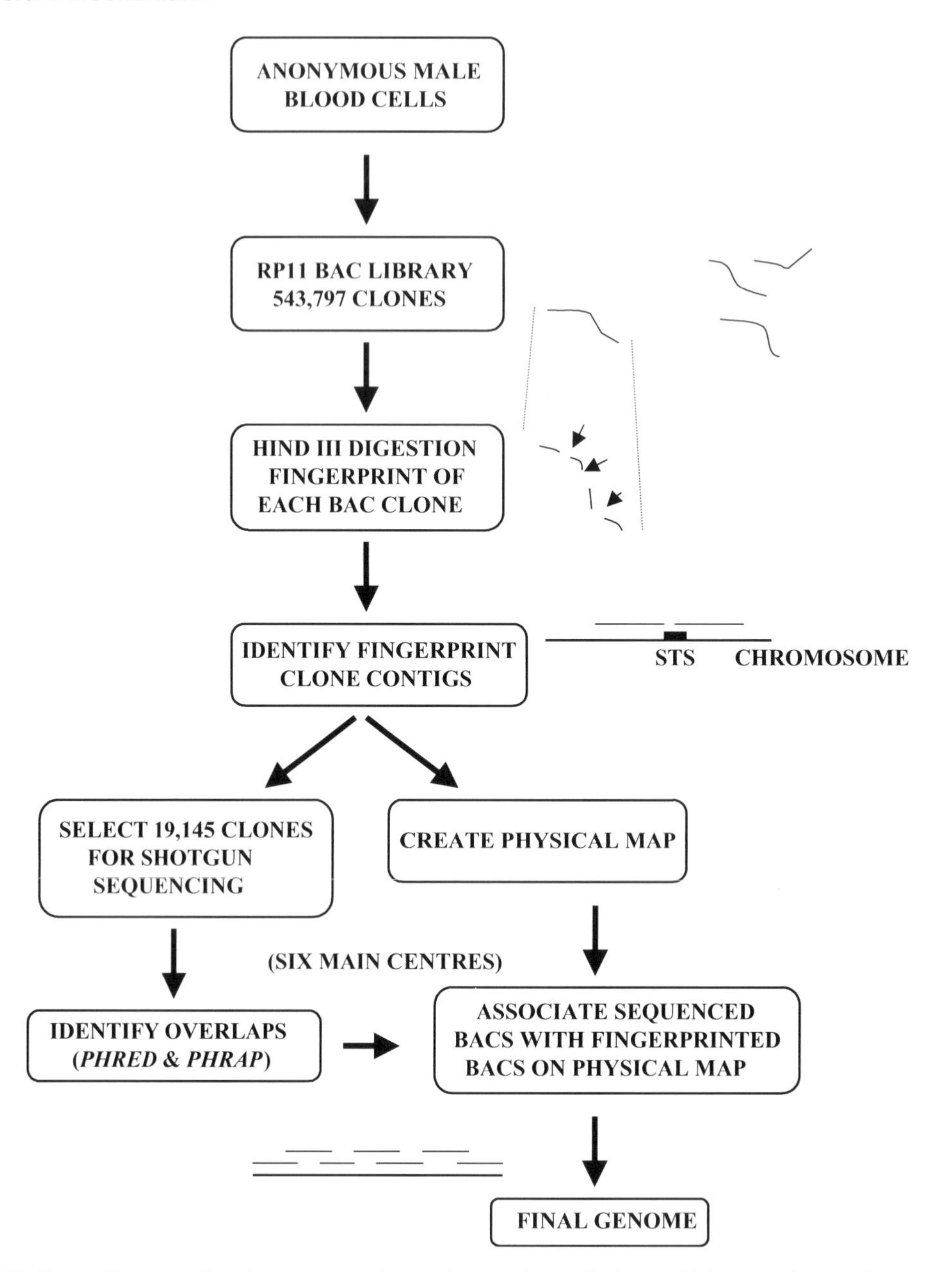

Figure 9.5. The Human Genome Project. An overview of the actual approach taken is shown. This is a combination of the 'map-based' and whole genome shotgun approaches. The bulk of the DNA sequenced came from blood cells of anonymous male donors which generated a RP11 BAC library (although clones from other libraries were also used). Each clone was digested with the restriction enzyme Hind III and used to generate a fingerprint on agarose gels. This allowed unambiguous identification of each clone and of overlaps between clones. These were associated with genetic and physical markers on specific human chromosomes by probe hybridization and contributed to building a physical map. Selected clones (some 19 145) were sequenced by shotgun sequencing and sequence data were assembled with computer programs *PHRED* and *PHRAP*. Sequenced BACs were then associated with fingerprint BACs to generate a sequence contig map and the final genome sequence was assembled with the aid of *GigAssembler*.

polymorphisms (STRPs; also known as *microsatellites*) are sequences of 3–5 base pairs repeated 10–30 times which have been found to be more evenly distributed through the genome than VNTRs. A third type of marker is offered by *sequence tagged sites* (STSs), short sequences of 200–500 base pairs, which have a unique location in the genome. The genome is cut into small pieces with restriction enzymes and each piece individually sequenced in a process called *shotgun sequencing*. Regions of overlap allow the assembly of progressively longer sequences some of which will contain markers. These markers are used to generate the physical map of the genome by gradually building up a hierarchical picture of its physical structure. By contrast, a second approach to assembling the genome (taken by Celera) was to eschew creation of maps and instead to apply shotgun sequencing to the whole genome by randomly shearing it into 2000–10 000 base pair fragments and sequencing these individually. Although this approach had proven remarkably successful in viral and prokaryote genomes, which lack extensive repeat sequences, its application to eukaryote genomes was quite controversial initially.

In practice, the map-based and whole genome shotgun approaches are not mutually exclusive and combining the two approaches assembled the human genome. Human DNA fragments (100 000–200 000 base pairs) were generated by restriction digestion and cloned into bacteria as a library of vectors called *bacterial artificial chromosomes* (BACs). These can accommodate fragments as large as 250 000 base pairs, so a library of 20 000 BAC clones accommodated the entire genome. Restriction mapping of individual fragments allowed identification of overlap and thus the clones could be ordered. Each BAC clone was 'mapped' onto human chromosomes using markers (VNTRs, TSRPs, STSs) as described above to generate a physical map. Each BAC clone was then cut into smaller 2000 base pair pieces and sequenced at each end. Piecing these sequences together assembled the sequence of the BAC clone by generating first a *contiguous clone map* or *contig* and eventually a complete sequence. The physical map allowed assembly of the BACs into the final version of the genome. Surprisingly, the human genome was found to contain significantly less genes than anticipated (25 000–30 000 rather than 50 000–100 000 as previously thought) and contains hundreds of bacterial genes apparently acquired by horizontal transfer during the vertebrate lineage.

In addition to the human, complete and partial genomes are available at NCBI for the following vertebrates: baboon, bat, cat, chicken, chimp, cow, coyote, dog, dolphin, duck-billed platypus, elephant, frog, gibbon, gorilla, hedgehog, horse, lemur, macaque, mouse, orang utang, opossum, pig, rabbit, rat, sheep, shrew, squirrel and wallaby. Genomic data are also available for important insects including the honeybee, wasps, silkworm, fruitfly *Drosophila* and the *Anopheles* mosquito responsible for carrying the causative agent of malaria. The complete genome of this causative agent, an intracellular protozoan parasite called *Plasmodium*, is also now available as are representative plant, worm and fungal species. In addition to this central genome resource many genome projects operate their own web sites independently of NCBI (Table 9.2). This allows combination of species-specific information with genome data.

9.2.4 Expressed Sequence Tag Databases

It is noteworthy that genomes of even the best-studied species contain a large percentage of genes of as-yet unknown function (e.g. 25% of the yeast *Saccharomyces cerevisiae*'s relatively small genome of 5885 genes). These are nucleotide sequences which appear to have in-frame start/stop codons and appropriate up- and downstream regulatory sequences but which show little homology to genes of known function. Moreover, only a proportion of the genes encoded in genomes are expressed during the organism's life. These genes may be no longer necessary for survival but have been retained in the genome during evolution. Genes which are transcribed to mRNA can be detected as *expressed sequence tags* (ESTs) which are short subsequences of cDNAs transcribed from mRNA. Databases of such ESTs provide a powerful resource for identifying new families of proteins and are hosted by NCBI. They are important because they allow us to concentrate research efforts on actively transcribed genes rather than diluting our interest on essentially inactive ones. An illustration of the usefulness of EST databases is given in Example 9.1.

9.2.5 Single Nucleotide Polymorphism (SNP) Database

DNA sequencing has revealed that some locations or *loci* in genes may be mutated in different individuals resulting in an alteration in primary structure at that position. These are called *single nucleotide polymorphisms* (SNPs; pronounced 'snips'). These sites are often of interest for recognizing predisposition to disease in people carrying the polymorphism and the human genome is known to contain 1.4 million SNPs. NCBI hosts a database of human SNPs, *dbSNP*, which is accessible directly through Entrez.

9.3 TOOLS FOR ANALYSIS OF PRIMARY STRUCTURES

The DNA sequence of a gene or even a genome is in itself rather uninformative consisting as it does of a bewilderingly large number of As, Gs, Cs and Ts repeating seemingly at

Example 9.1 Database Mining Reveals Expression of the Neuronal Calcium Sensor Family in Tissues Peripheral to the Nervous System

In study of mammalian biochemistry and physiology we are frequently interested in elucidating the tissue distribution of particular protein families. We often find that specific protein isoforms are expressed in specific tissues and this can give important clues to the role of these isoforms. The neuronal calcium sensor (NCS) protein family has been described as being largely specific to the nervous system. However, some members of this family are known to be expressed in other tissue types: VILIP-3 (haematopoietic system, gut, kidney, spleen, testis) and KChIPs (heart) raising the possibility that they may be involved in functions besides neuronal signalling.

GenBank sequences in NCBI are partitioned into a nonredundant set of gene clusters in a resource called UniGene (Table 9.3). This lists likely expression frequencies for gene clusters by tissue distribution of ESTs. UniGene was probed for VILIP-1, hippocalcin and NCS-1 and the results analysed by subtracting sequences observed in pathological contexts (e.g. cancer).

Breakdown by Tissue		
	Hs. 288654	
Bladder	0	0/21715
Blood	0	0/78292
Bone	0	0/55730
Bone Marrow	0	0/36541
Brain	132	61/462100
Cervix	0	0/41264
Colon	5	1/179987
Eye	89	15/168244
Heart	0	0/58912
Kidney	7	1/135458
Larynx	0	0/27551
Liver	0	0/131463
Lung	17	5/288794
Lymph Node	0	0/128142
Mammary Gland	0	0/128200
Muscle	0	0/109115
Ovary	0	0/95612
Pancreas	11	1/84639
Peripheral	0	0/24996
Placenta	0	0/237797
Prostate	7	1/133636
Skin	0	0/165608

Portion of UniGene data for hippocalcin. Expression profile is suggested by observed EST frequency. Note that the bulk of expression is in brain.

The results (Table Ex1.1) were then viewed with the SAGE AnatomicViewer (http://cgap.nci.nih.gov/SAGE/AnatomicViewer), to derive a 'virtual' Northern blot (Section 5.11.4).

In this paper, it was revealed that VILPI-1 and NCS-1 are also expressed in peripheral tissue while hippocalcin is restricted (98%) to nervous tissues. These in silico results were confirmed by Western blotting. Analysis of EST data by SAGE is no substitute for actual experimentation. However, this example shows how bioinformatic analysis can suggest sensible experiments as well as contributing to understanding results.

Reprinted with permission of Elsevier from Gierke et al., (2004) Expression analysis of members of the neuronal calcium sensor protein family:Combining bioinformatics and Western blot analysis. *Biochem. Biophys. Res. Commun.*, **323,** 38–43.

(Continues)

(*Continued*)

Table 9.4. Tissue-specific expression of NCS proteins as deduced from human UniGene database.

	VILIP-1	Hippocalcin	NCS-1
UniGene cluster	Hs.2288	Hs.288654	Hs.301760
Total no. EST sequences	196	96	299
Pathological	23	26	121
Other (pooled, unknown)	15	3	19
No. EST sequences counted	158 (100%)	67 (100%)	161 (100%)
Tissue			
Brain	117 (74%)	55 (82%)	76 (47%)
Nerves	14 (9%)	1 (1.5%)	10 (6%)
Eye/optic nerve	6 (4%)	9 (13.4%)	18 (11%)
Heart	1 (0.64%)		6 (3.7%)
Liver	2 (1.3%)		
Spleen			
Lung	3 (1.9%)		2 (1.2%)
Kidney	4 (2.5%)		7 (4.3%)
Stomach	1 (0.64%)		1 (0.6%)
Colon	1 (0.64%)		4 (2.5%)
Skin			6 (3.7%)
Testis	2 (1.3%)		
Ovary			1 (0.6%)
Placenta/uterus	2 (1.3%)		5 (3%)
Glands. adrenal gland	1 (1.3%)		9 (5.6%)
Pancreas			5 (3%)
Prostate	4 (2.5%)	1 (1.5%)	4 (2.5%)
Breast			6 (3.7%)

random. In order to extract information of real relevance we need to be able to analyse sequences with fast and efficient computer programs. A simple example of this is the problem of translating a DNA nucleotide sequence into a corresponding protein amino acid sequence. This is a trivial operation for a short sequence but becomes significantly more challenging as the DNA sequence becomes longer since the number of operations required quickly becomes prohibitively large. Fortunately, since sequences are simply a string of building blocks, they lend themselves readily to representation and analysis as a series of numbers and thus may be 'read' by computer programs specifically designed for the task. The following sections give a short introduction to some of the programs currently in use (URLs are listed in Table 9.3). These maximize our possibilities for interpreting DNA and protein sequence data.

9.3.1 BLAST Programs

The *B*asic *L*ocal *A*lignment *S*earch *T*ool (BLAST) was developed for aligning query sequences with sequences in databases to identify locally conserved or similar regions. The program is very fast and overcomes shortcomings of conventional programming methods, which were too slow for complete searches of large databases. The program maximizes local similarity with a *maximal segment pair score* in contiguous regions of the sequence. The output of the search is a reliable measure of sequence identity arrived at rapidly, which is statistically robust. Various forms of the BLAST tool are available at the NCBI, EBI, PIR and ExPASy sites (Figures 9.6 and 9.7). These include: BLASTP (amino acid sequence searches in protein sequence databases), TBLASTN (amino acid sequence searches in translated nucleotide sequence databases), TBLASTX (Translated nucleotide sequences searches in translated nucleotide sequence databases). There is also the possibility of searching with short query sequences, using the BLAST tool to search entire genomes and of using megaBLAST for searches with longer query sequences than BLASTN (nucleotide sequence searches in nucleotide sequence databases). One secret of the speed of BLAST programs is that they do not attempt to match sequences along their full length but rather look for regions of highest similarity. WUBLAST (developed at Washington University and accessible at EBI and

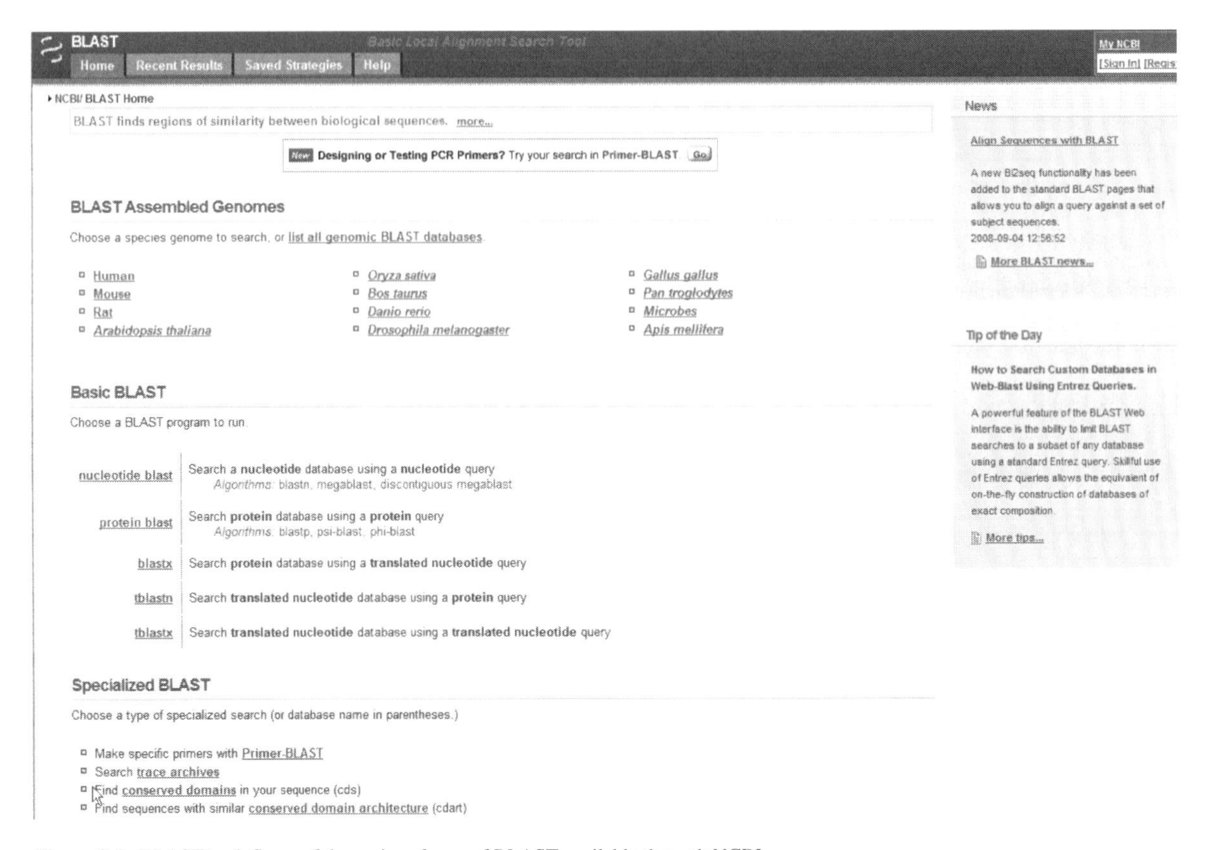

Figure 9.6. BLAST tool. Some of the various forms of BLAST available through NCBI.

ExPASy) was developed to permit *gapped alignments* which speed up searching. PSI-BLAST (available at NCBI) begins with a 'one at a time' search similar to BLAST but then goes on to derive pattern information from multiple alignments on the original 'hits'. The program continuously reprobes the database with the pattern detected in an iterative manner constantly refining the pattern obtained. This allows much more powerful detection of distantly-related sequences than is possible with BLAST. Other programs capable of picking up such distant relationships are *FastA* and *ClustalW*. In using these tools, the user should always be cognizant that default settings may differ depending on the site hosting the tool. It can be worthwhile to familiarize oneself with these settings and, if necessary, to vary them.

9.3.2 FastA

FastA is a family of programs available at EBI and PIR which allows rapid screening of proteomic, genomic, protein and nucleotide sequence databases to reveal sequence similarity. It can be very specific when identifying long regions of low similarity even in highly diverged proteins or genes. Amino acid or nucleotide sequences are entered in the program in a particular text-based convention known as *FastA format* (Figure 9.8). This begins with a single line description commencing with > followed by an arbitrary title (e.g. '>protein 1; Note, there is no gap between '>' and 'protein'). Subsequent lines contain the sequence in single letter convention for amino acids (U is used for selenocysteine) or nucleotides as specified by the International Union of Applied Chemistry. In fact, many bioinformatics programs require query sequences to be entered in FastA format.

9.3.3 Clustal W

Sometimes alignment searches between a query sequence and a database will reveal several genes or proteins with points of sequence similarity. We would then like to know if there is a relationship between the group formed by the statistically significant hits and the query sequence. One approach would be to carry out exhaustive pair-wise alignments with BLAST or FastA but this would be very time-consuming. *Clustal* programs (of which the latest is Clustal W2) allow alignment of multiple sequences in

(a) 

BLASTP 2.2.10 [Oct-19-2004]

```
Reference:
Altschul, Stephen F., Thomas L. Madden, Alejandro A. Schäffer,
Jinghui Zhang, Zheng Zhang, Webb Miller, and David J. Lipman (1997),
"Gapped BLAST and PSI-BLAST: a new generation of protein database search
programs",  Nucleic Acids Res. 25:3389-3402.

RID: 1102933551-3806-200588368848.BLASTQ4

Query= Aquaporin 1
        (269 letters)

Database: Non-redundant SwissProt sequences
          159,656 sequences; 58,786,444 total letters

If you have any problems or questions with the results of this search
please refer to the BLAST FAQs

Taxonomy reports
```

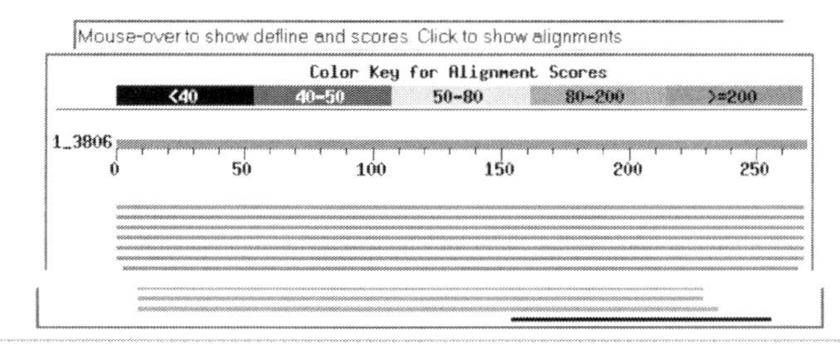

(b) 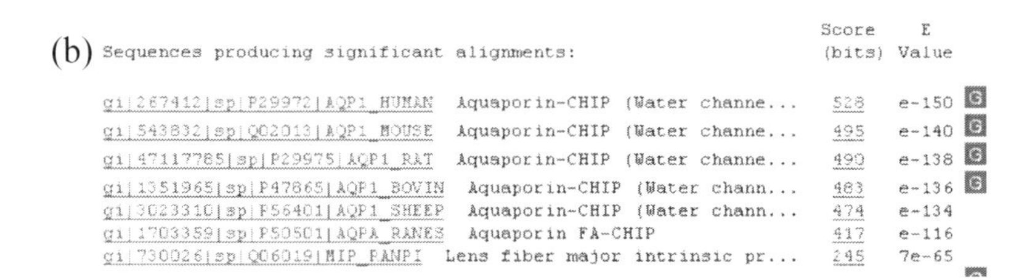

Sequences producing significant alignments:		Score (bits)	E Value
gi|267412|sp|P29972|AQP1_HUMAN	Aquaporin-CHIP (Water channe...	528	e-150
gi|543832|sp|Q02013|AQP1_MOUSE	Aquaporin-CHIP (Water channe...	495	e-140
gi|47117785|sp|P29975|AQP1_RAT	Aquaporin-CHIP (Water channe...	490	e-138
gi|1351965|sp|P47865|AQP1_BOVIN	Aquaporin-CHIP (Water chann...	483	e-136
gi|3023310|sp|P56401|AQP1_SHEEP	Aquaporin-CHIP (Water chann...	474	e-134
gi|1703359|sp|P50501|AQPA_RANES	Aquaporin FA-CHIP	417	e-116
gi|730026|sp|Q06019|MIP_RANPI	Lens fiber major intrinsic pr...	245	7e-65

Figure 9.7. Results of BLAST search. (a) The sequence of human aquaporin 1 is entered in the search box in FastA format. (b) A listing of 100 'hits' is presented with an 'E value'. This is a statistical estimate of the likelihood of a match occurring by chance. The value for the most likely hit in our search is 1×10^{-150} which suggests that we can be very confident indeed that our search sequence is aquaporin 1.

a single step so as to reveal points of similarity within related sequences and to reveal patterns of their divergence (Figure 9.9). This program is available at EBI and can be downloaded to your computer.

9.3.4 Hydropathy Plots

In Chapter 6 we saw that hydrophobic amino acids have a strong preference to be located away from solvent water so these are often found in locations such as the interior of proteins or in membrane-bound protein surfaces. Conversely, polar amino acids have a strong preference for being in contact with water and tend to be found on the surface of proteins. A statistical analysis of the frequency with which each of the 20 common amino acids is found on the surface or interior of proteins for which three-dimensional structures are available was carried out

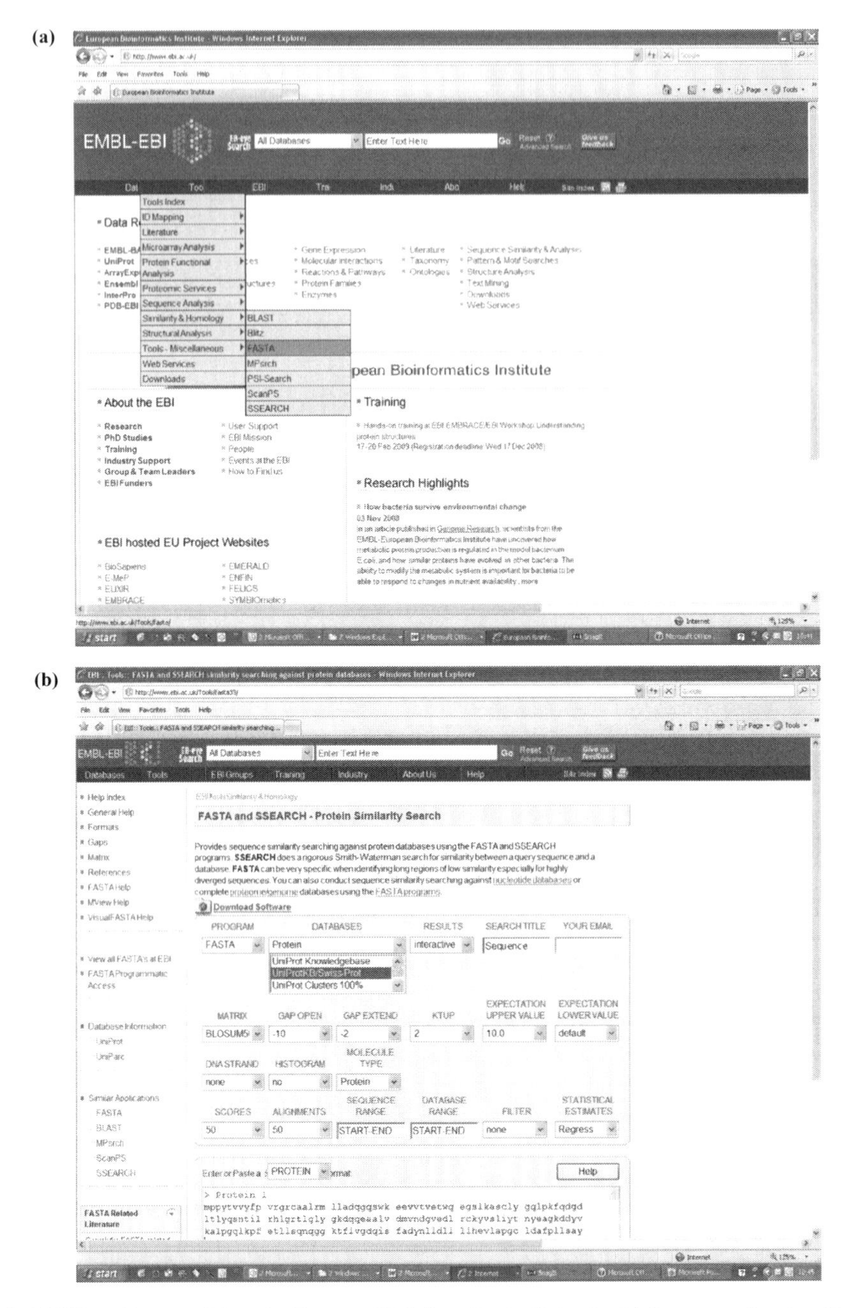

Figure 9.8. Fast A. (a) This tool is accessible at the European Bioinformatics Institute site. (b) A sequence is entered in the search box or else users may browse to a file.

by Kyte and Doolittle in the 1980s and this resulted in a computer program containing an *algorithm* which allowed prediction of the eventual location of stretches of amino acids within a protein sequence (Figure 9.10). Such plots, called *hydropathy plots*, were originally used to predict location of likely immunogenic regions of newly-sequenced genes (these could then be synthesized by Merrifield peptide synthesis and antibodies could be raised to at least some of the putative epitopes). However, hydropathy plots are now more widely used for prediction of transmembrane domains of membrane-bound proteins (Example 9.2).

(a)

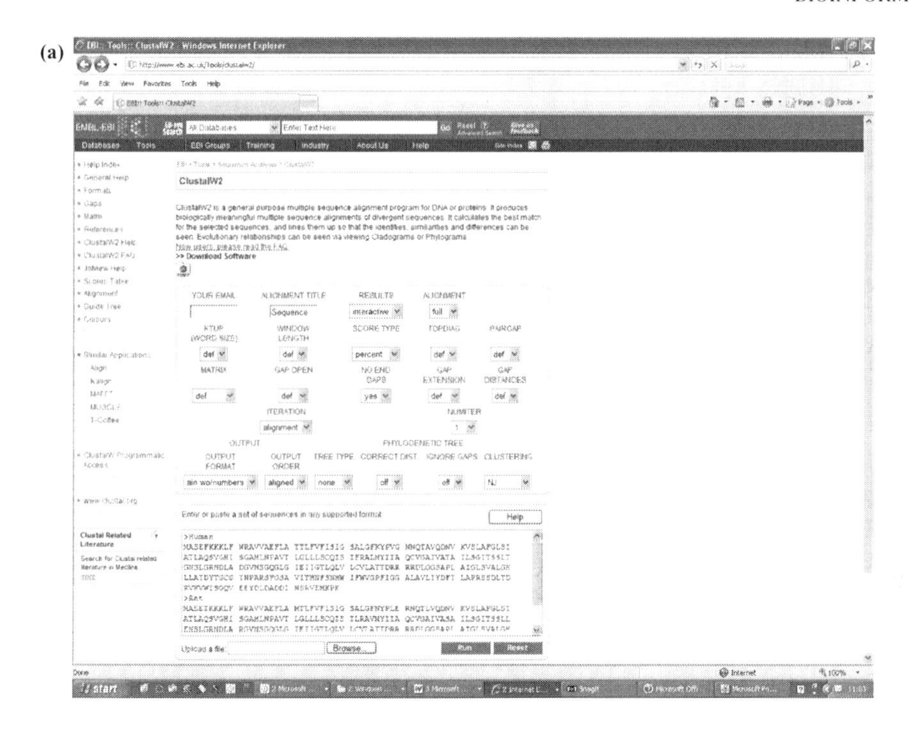

(b)

```
CLUSTAL 2.0.10 multiple sequence alignment


Human           MASEFKKKLFWRAVVAEFLATTLFVFISIGSALGFKYPVGNNQTAVQDNVKVSLAFGLSI
Rat             MASEIKKKLFWRAVVAEFLAMTLFVFISIGSALGFNYPLERNQTLVQDNVKVSLAFGLSI
                ****:*************** **************:**: .*** ***************

Human           ATLAQSVGHISGAHLNPAVTLGLLLSCQISIFRALMYIIAQCVGAIVATAILSGITSSLT
Rat             ATLAQSVGHISGAHLNPAVTLGLLLSCQISILRAVMYIIAQCVGAIVASAILSGITSSLL
                *******************************:**:*************:**********

Human           GNSLGRNDLADGVNSGQGLGIEIIGTLQLVLCVLATTDRRRRDLGGSAPLAIGLSVALGH
Rat             ENSLGRNDLARGVNSGQGLGIEIIGTLQLVLCVLATTDRRRRDLGGSAPLAIGLSVALGH
                 ********* *************************************************

Human           LLAIDYTGCGINPARSFGSAVITHNFSNHWIFWVGPFIGGALAVLIYDFILAPRSSDLTD
Rat             LLAIDYTGCGINPARSFGSAVLTRNFSNHWIFWVGPFIGSALAVLIYDFILAPRSSDFTD
                *********************:*:***************.*****************:**

Human           RVKVWTSGQVEEYDLDADDINSRVEMKPK
Rat             RMKVWTSGQVEEYDLDADDINSRVEMKPK
                *:***************************
```

Figure 9.9. Clustal W. This allows alignment of related sequences. (a) Two sequences (human and rat aquaporin 1) are entered in Fast A format and submitted to Clustal W. (b) The program reports an alignment with identical residues underlined by asterisks and similar residues by colons.

(a)

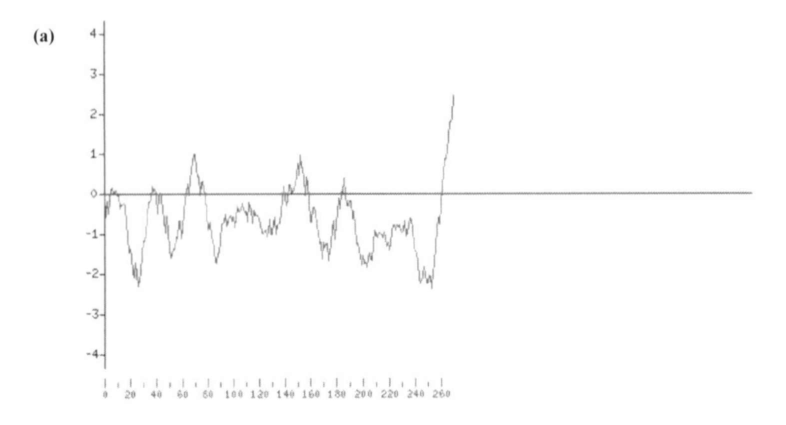

(b)

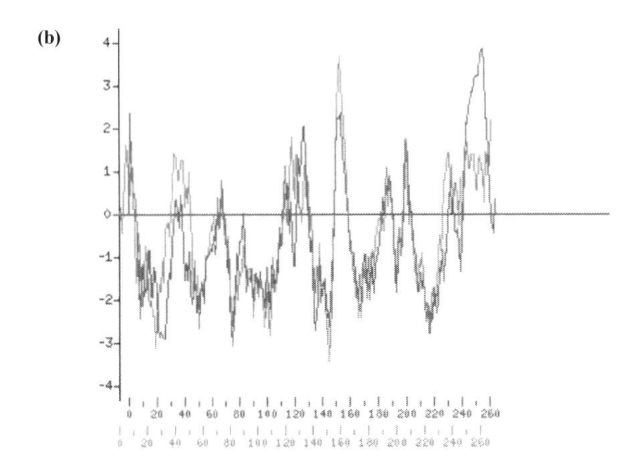

(c)

Query Sequence (protein1)
compared with
P30868 escherichia coli. glucuronide carrier protein (glucuronide permease). (UIDB_ECOLI)

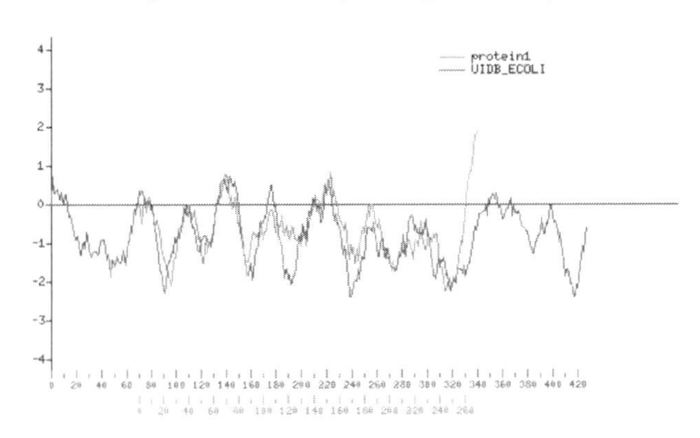

Figure 9.10. Hydropathy searching. Net hydrophobicity varies along a polypeptide chain with hydrophobic regions likely to be found in the interior of proteins or on the surface of membrane-bound proteins. Hydrophobic regions give positive hydropathy while hydrophilic residues are negative. (a) Hydropathy plot from a single sequence. In this analysis a window of 17 residues is selected and hydropathy is averaged across this window. (b) Comparative hydropathy plots for two sequences. The sequences of human aquaporin 1 and rat aquaporin 5 are compared (a window of 7 residues is selected in this case). (c) SwissProt may be searched for sequences with similar hydropathy profiles. This reveals that the glucuronide permease of *Escherichia coli* yields a similar hydropathy profile to human aquaporin 1. (Analyses performed at URL of Ebioinfogen, The Weizmann Institute, Israel; Table 9.3).

Example 9.2 Sequence Alignments Identify a Family of Bacterial Transmembrane Small Multidrug Resistant Proteins

Bacterial resistance to antibiotics is a pressing public health problem. One mechanism for this resistance is expression of transmembrane proteins called small multidrug resistance (SMR) proteins which actively pump antibiotics from cells. This pumping action is either driven by hydrolysis of ATP or else is coupled to inward movement of protons down their transmembrane concentration gradient. Because this family of proteins is transmembrane, it is difficult to solve their structure by X-ray diffraction or NMR (Chapter 6). One of the best-studied bacterial SMRs is EmrE. This was found by CryoEM (Section 6.6.2) to contain four transmembrane α-helices per subunit with a dimeric quaternary structure. In structural modelling, assignment of hydrophobic sequences to these eight helices would result in $>160\,000$ permutations. However, if a symmetry arrangement existed between two parts of the structure, this number reduced to 48. Sequences of similar proteins were used to identify most conserved residues based on the rationale that residues packed against helices are well-conserved through evolution. By contrast, residues facing away from the protein core towards lipid can mutate much more readily. In this way, the helices could be identified and a plausible model generated for their relative orientation in space which revealed a clearly symmetric relationship as shown:

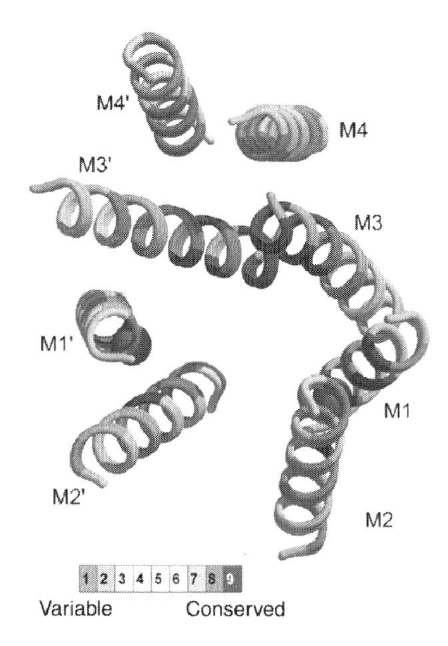

Importantly, this structure accords closely with the known biochemical properties of this protein unlike previous structural studies which seem to have trapped nonphysiological structural variants.

Source: Fleishman *et al.*, (2006) Quasi-symmetry in the cryo-EM structure of EmrE provides the key to modeling it's transmembrane domain *J. Mol. Biol.* **364**, 54–67. Reproduced with permission of Academic Press.

9.3.5 Predicting Secondary Structure

Protein *secondary structure* reflects common strategies for folding up the extended polypeptide chain based principally on the tendency of dihedral angles to occupy limited parts of the *Ramachandran diagram* and formation of hydrogen bonds between carbonyl and amide groups of separate peptide bonds (Figure 9.11, see colour plate section). The requirement to maximize repulsion of 'R' groups and to form straight, undistorted hydrogen bonds leads to formation of two common types of secondary structure: α-helix and β-pleated sheet. Other types of secondary structure exist and are found in proteins with specific structural requirements (e.g. the Poly Pro II structure of the collagen triple helix) but these two types dominate the bulk of secondary structure. Secondary structure components of proteins are linked by regions of *random coil* and by *turns*. Random coil is something of a misnomer as the polypeptide chain in these regions usually has a fixed structure in space but the term simply denotes that the structure in that region does not fall into the other convenient categories. Prolines are disruptors of the main types of secondary structure but allow the polypeptide chain to change direction (i.e. *turn*) because they can adopt

(a)

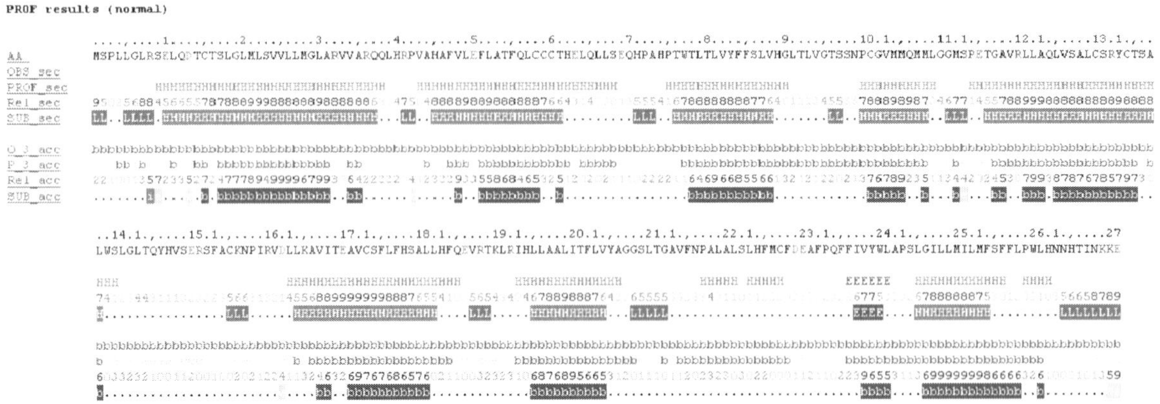

PROF results (normal)

```
         ....,....1....,....2....,....3....,....4....,....5....,....6....,....7....,....8....,....9....,....10.1.,....11.1.,....12.1.,....13.1.,..
AA       MSPLLGLRSELQDTCTSLGLMLSVVLLMGLARVVARQQLHRPVAHAFVLEFLATFQLCCCTHELQLLSEQHPAHPTWTLTLVYFFSLVHGLTLVGTSSNPCGVMMQMMLGGMSPETGAVRLLAQLVSALCSRYCTSA
OBS_sec
PROF_sec             HHHHHHHHHHHHHHHHHHHHHHHHHHHHHH    HHHHHHHHHHHHHHHHHHHHHHHHHHHHHH          HHHHHHHHHHHHHHHHHH         HHHHHHHHHHHH    HHHHHHHHHHHHHHHHHHHHHHHHHHHHHH
Rel_sec  95  5688456655787888999888888988888886  475 4888898898888888764  4      5554 67888888888776       455  788898987  4677 4557888999888888888898888
SUB_sec  LL..LLLL.HHHHHHHHHHHHHHHHHHHHHHHHHHHHH...LL..HHHHHHHHHHHHHHHHHH.........LLL..HHHHHHHHHHHHH........LL..HHHHHHHHHH..LLL..HHHHHHHHHHHHHHHHHHHHHHHHH

O_3_acc  bbbbbbbbbbbbbbbbbbbbbbbbbbbbbbbbbbbbbbbbbbbbbbbbbbbbbbbbbbbbbbbbbbbbbbbbbbbbbbbbbbbbbbbbbbbbbbbbbbbbbbbbbbbbbbbbbbbbbbbbbbbbbbbbbbbbbbbbbbbbb
P_3_acc     bb b    b  bb bbbbbbbbbbbbbbb   bb        b  bbb bbbbbbbbbbbbb bbbbb          bbbbbbbbbbbbbbbbbbbbbbbbbbbbbbbbbbbb    b    bbbbbbbbbbbbbbbbbbbb b
Rel_acc  22     357233527 477789499996799  6422222 4 2322293 558684653 5                    6469668556613    2  3767892 5  3442  2453 7993878767857973
SUB_acc  .......1.....b.bbbbbbbbbbbbbbb..bb.............b..bbbbbbbbb..b.................bbbbbbbbbbbb............bbbbb..b...b....bb..bbb.bbbbbbbbbbbb..

         ..14.1.,....15.1.,....16.1.,....17.1.,....18.1.,....19.1.,....20.1.,....21.1.,....22.1.,....23.1.,....24.1.,....25.1.,....26.1.,....27
AA       LWSLGLTQYHVSERSFACKNPIRVDLLKAVITEAVCSFLFHSALLHFQEVRTKLRIHLLAALITFLVYAGGSLTGAVFNPALALSLHFMCFDEAFPQFFIVYWLAPSLGILLMILMFSFFLPWLHNNHTINKKE
OBS_sec
PROF_sec HHH                         HHHHHHHHHHHHHHHHHHHHHHH        HHHHHHHHHHHHHHH        HHHHH HHHHH             EEEEEE    HHHHHHHHHHHH   HHHH
Rel_sec  74   44            566      45568899999998887655541   5654 4 4678898888764  65555     4                        6775      6788888875          56658789
SUB_sec  H................LLL......HHHHHHHHHHHHHHHHHHH....LLL.....HHHHHHHHHHH...LLLLL........................EEEE....HHHHHHHHHHH.........LLLLLLLL

O_3_acc  bbbbbbbbbbbbbbbbbbbbbbbbbbbbbbbbbbbbbbbbbbbbbbbbbbbbbbbbbbbbbbbbbbbbbbbbbbbbbbbbbbbbbbbbbbbbbbbbbbbbbbbbbbbbbbbbbbbbbbbbbbbbbbbbbbbbb
P_3_acc  b                            b bbbbbbbbbbbbbbbbbbb             bbbbbbbbbbbbbbbbbbb    b bbbbbbbbbbbbbbb           bbbbbbbbbbbbbbbbbbbbbbbbbb
Rel_acc  6 3323                                 4632 6976768657 6          687689566531                          22396553    6999999998666626         59
SUB_acc  b.......................................bb..bbbbbbbbbbbb.............bbbbbbbbbbbb...................................bbbb....bbbbbbbbbbbbbb..b.........
```

(b)

Strand/Turn/Helix Probabilities

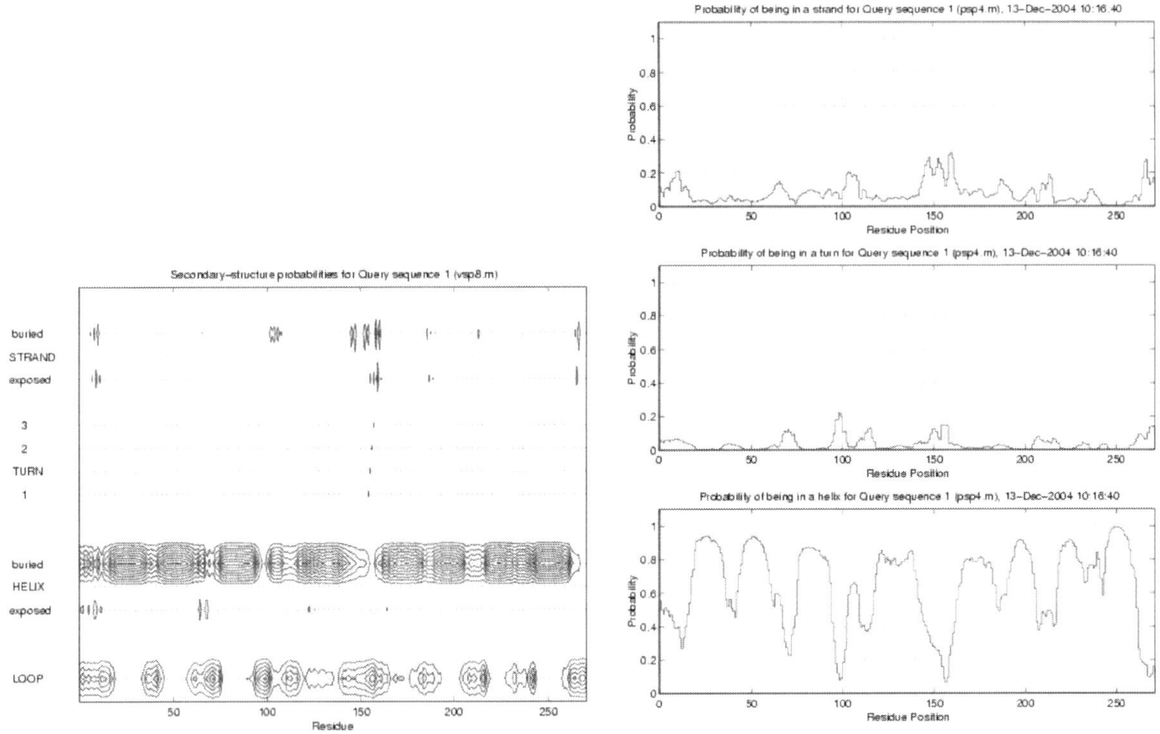

Figure 9.12. Prediction of Secondary structure for human aquaporin 1. (a) Results from the Predict Protein server. This contains an analysis by PROF which predicts 63.47% α-helix (H), 2.2% β-pleated sheet (E) and 34.3% random coil (L). The numbers (minimum 1 and maximum 10) denote the significance of assignment. (b) Graphical representations from the PSA server of probability of formation of α-helix, turns and β-sheet. Notice that the structure is indeed dominated by periodically repeating α-helices which are characteristic of transmembrane proteins.

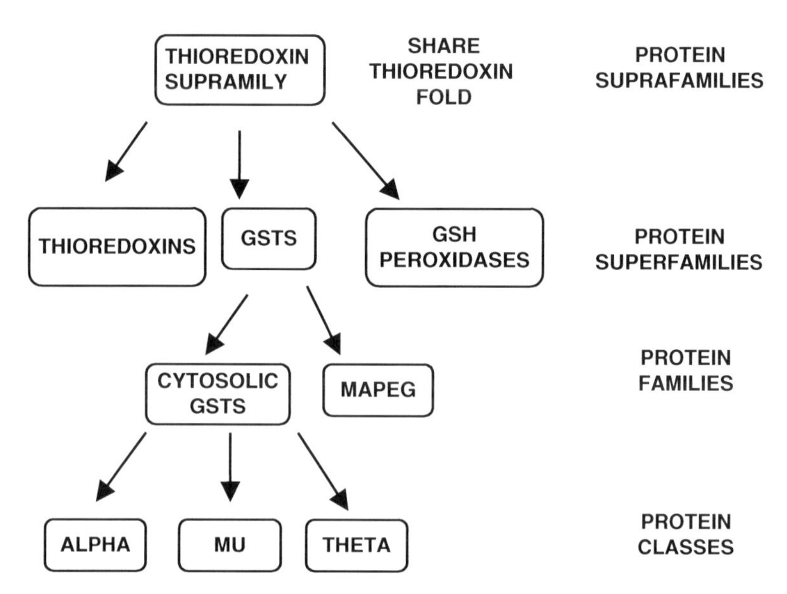

Figure 9.13. Protein families. Based on architecture, fold, domain structure and sequence homology, various levels of similarity can be observed amongst proteins. The thioredoxin suprafamily consist of several protein superfamilies which share the thioredoxin fold presumably because they all bind glutathione. Only three of these are shown. The GSTs are one superfamily which in turn consists of two families: cytosolic GSTs (which are dimeric) and *membrane-associated proteins involved in eicosanoid and glutathione metabolism* (MAPEG) which are membrane-bound and trimeric. The cytosolic GST family consists of at least 13 classes of which three are shown.

a *cis* conformation far more often than other residues. There are many different types of turns in proteins and these also have been subjected to classification. For most purposes α-helix, β-pleated sheet, turns and random coil allow an adequate description of secondary structure. Reference has been made in Chapter 3 to experimental estimation of secondary structure by spectroscopic techniques such as FTIR (Section 3.6.4) and CD (Section 3.5.5). However, using a similar approach to that of Kyte and Doolittle for hydropathy, it is possible to estimate the frequency with which all 20 amino acids are found in α-helix or β-pleated sheet or neither. Some residues are found strongly to promote α-helix or β-pleated sheet (strong formers) while others strongly disrupted their formation (breakers). Still other residues are found to be intermediate or neutral in their effect. This forms the basis for prediction of secondary from primary structure using the *Garnier Osguthorpe Robson (GOR)* algorithm, the *predict protein service* and the *PSA server* (Table 9.3). These algorithms read the amino acid sequence as a series of numbers reflecting the average propensity to form or break a particular secondary structure type at that position. The numbers are usually averaged over a 'window' of residues, which allows us to predict the most likely secondary structure of a query protein for which we may only have the amino acid sequence (Figure 9.12). When using on-line algorithms for prediction of hydropathy plots or secondary structure it should be borne in mind that these are based on the average behaviour of amino acids in the comparatively limited number of proteins for which we have three-dimensional structures. Accordingly, these algorithms sometimes have a relatively low predictive accuracy.

9.3.6 Identifying Protein Families

Just as the secondary structure and hydropathy properties of proteins are encoded in the primary structure, so is the three-dimensional fold, which determines the ultimate biological function of the protein. It is clear now that individual ancestral proteins gave rise to similar but distinct proteins by processes such as *divergence*, which resulted in protein classes, families, superfamilies and even suprafamilies. The GSTs are a good example of this since these enzymes can be classified in such a hierarchical manner within the thioredoxin suprafamily (Figure 9.13). A feature of proteins encoded by the human genome is that structural units such as motifs and domains are used and reused in a particularly versatile way to result in creation of a higher level of biological functionality than in other organisms. Some of the programs described above reveal similarities between related proteins but there is often such a low level of overall sequence identity that special analyses are required to allocate proteins to distinct families. *Pfam* is a searchable database available at the Sanger Institute (and elsewhere), which automatically allocates protein sequences to specific families

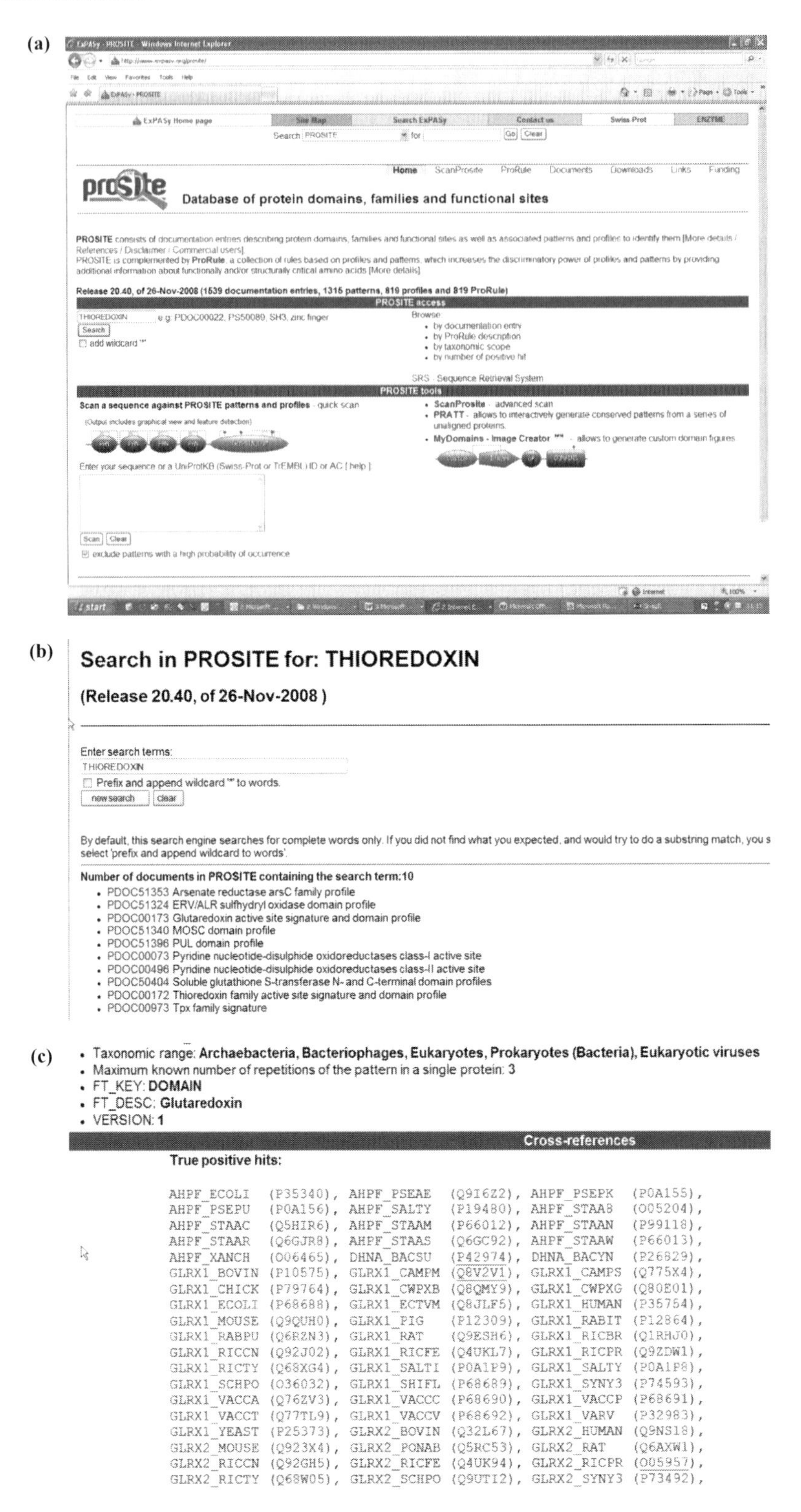

Figure 9.15. The PROSITE database. (a) PROSITE (Table 9.3) can be searched by text for 'thioredoxin'. (b) This reveals nine families. (c) One of listed entries.

(a)

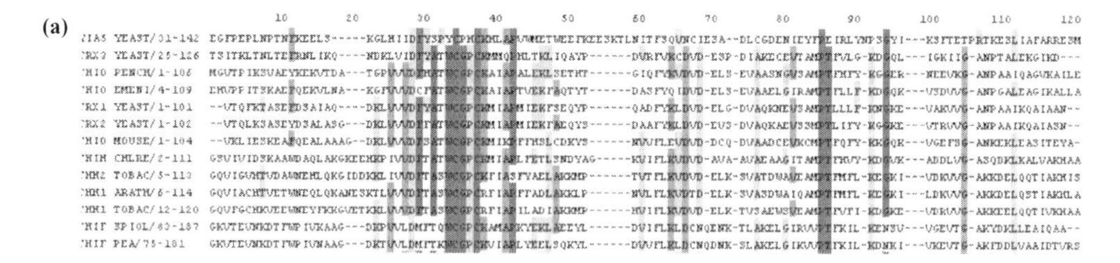

(b)

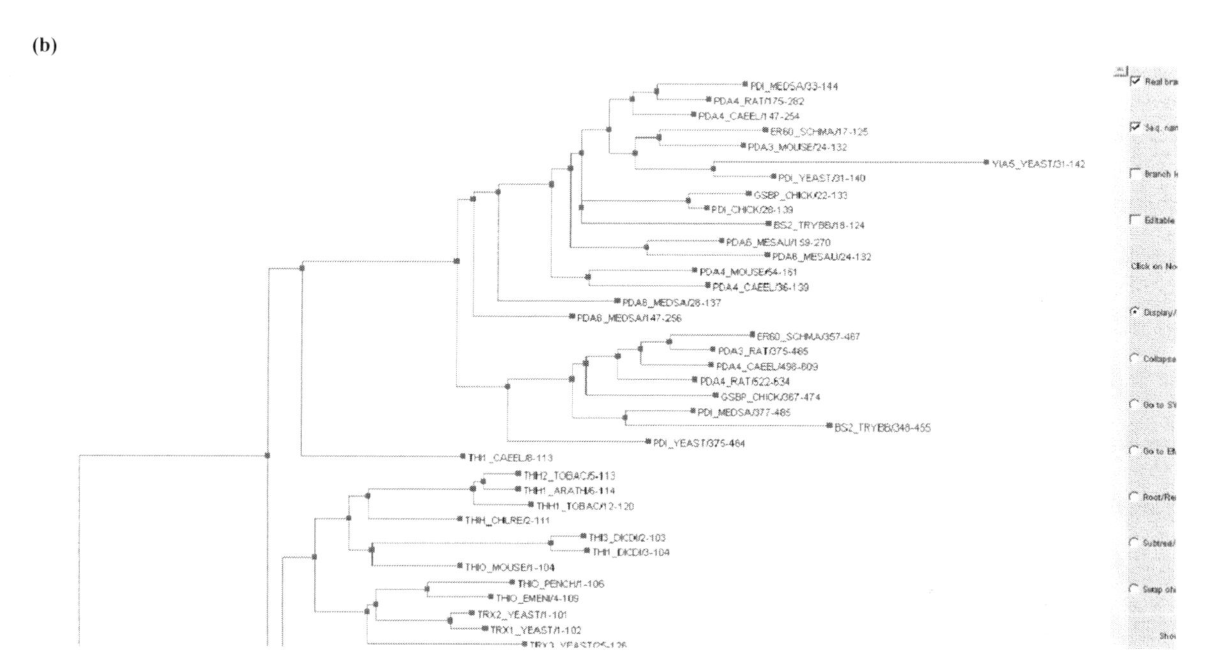

Figure 9.16. The PANDIT database. (a) Portion of the *alignment* option for the thioredoxin family in PANDIT. (b) Portion of the phylogenic tree for the same family.

(Figure 9.14, see colour plate section). This currently (July, 2007) contains more than 9 000 families covering some 75% of known protein sequences (an individual protein may belong to more than one family). An especially useful feature of the resource is that it contains data on possible structural interactions between proteins. Other protein family databases listed in Table 9.3 include *PROSITE* (available at ExPASy; Figure 9.15), *PANDIT* (available at EBI; Figure 9.16) and *iProClass* and *PIRSF* (available at PIR; Figure 9.17).

9.4 TERTIARY STRUCTURE DATABASES

The biological properties of biomacromolecules arise from their three-dimensional structure. In Chapter 6 we saw that, although this structure is 'encoded' in the primary structure, it is not yet possible directly to calculate tertiary from primary structures because we do not fully understand the processes by which biomacromolecules fold up; the protein folding problem (Chapter 6). We also saw that elucidation of three-dimensional structure is a major experimental undertaking involving principally X-ray diffraction through crystals or multi-dimensional NMR. These techniques have severe inbuilt limitations. It is difficult to find precise conditions for crystallizing proteins and not all proteins are fully soluble in unaggregated form at acid pH values. By contrast, high-throughput methodologies are available for sequencing genomes, genes and proteins. Thus, in searching sequence databases we often find a very large number of polynucleotide sequences, somewhat fewer protein sequences and only a handful of three-dimensional structures for any given protein type. More often, we find no

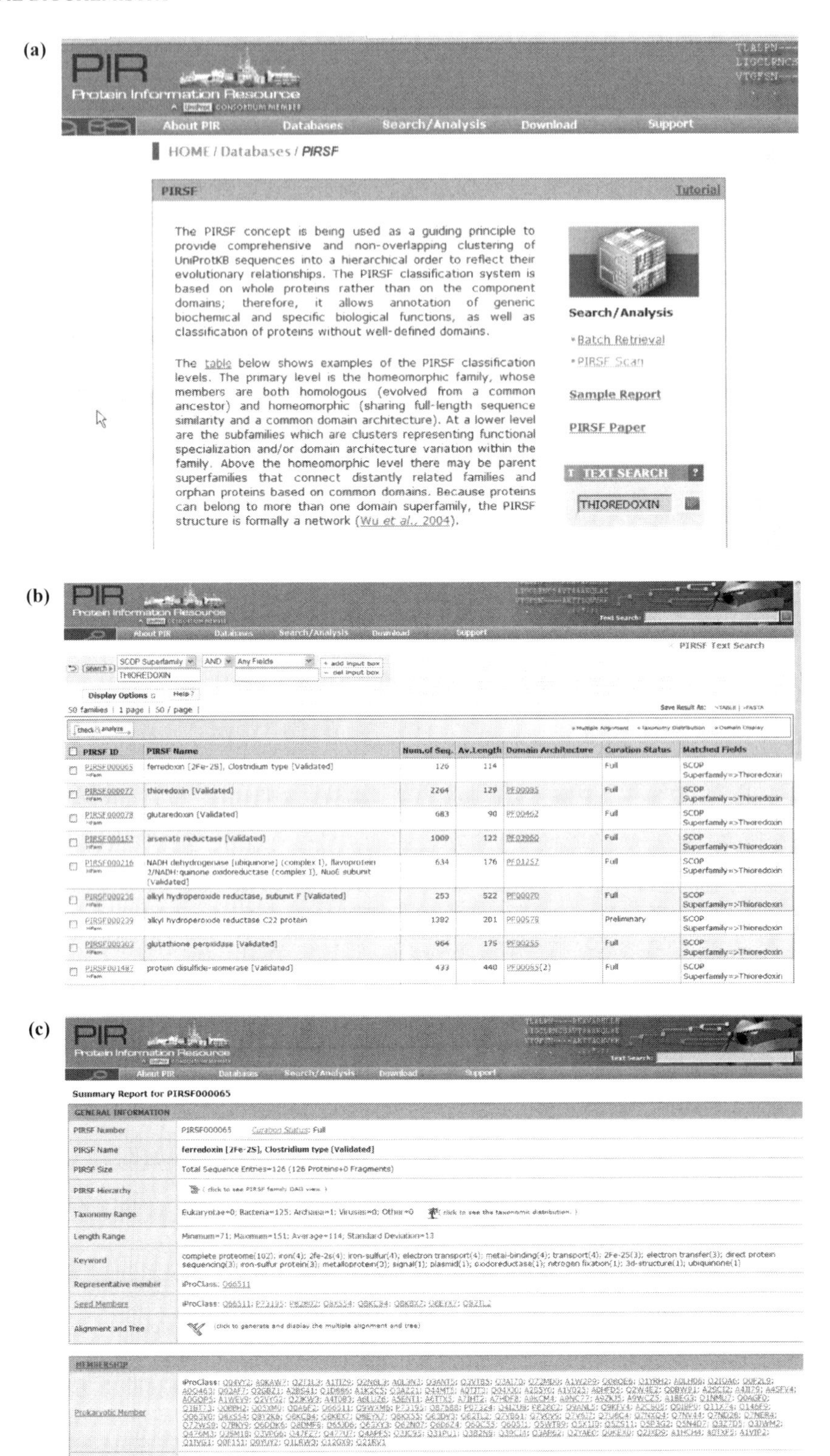

PIRSF ID	PIRSF Name	Num.of Seq.	Av.Length	Domain Architecture	Curation Status	Matched Fields
PIRSF000065	ferredoxin [2Fe-2S], Clostridium type [Validated]	126	114		Full	SCOP Superfamily=>Thioredoxin
PIRSF000077	thioredoxin [Validated]	2264	129	PF00085	Full	SCOP Superfamily=>Thioredoxin
PIRSF000078	glutaredoxin [Validated]	683	90	PF00462	Full	SCOP Superfamily=>Thioredoxin
PIRSF000152	arsenate reductase [Validated]	1009	122	PF03960	Full	SCOP Superfamily=>Thioredoxin
PIRSF000216	NADH dehydrogenase [ubiquinone] (complex 1), flavoprotein 2/NADH:quinone oxidoreductase (complex 1), NuoE subunit [Validated]	634	176	PF01257	Full	SCOP Superfamily=>Thioredoxin
PIRSF000238	alkyl hydroperoxide reductase, subunit F [Validated]	253	522	PF00070	Full	SCOP Superfamily=>Thioredoxin
PIRSF000239	alkyl hydroperoxide reductase C22 protein	1382	201	PF00578	Preliminary	SCOP Superfamily=>Thioredoxin
PIRSF000303	glutathione peroxidase [Validated]	964	175	PF00255	Full	SCOP Superfamily=>Thioredoxin
PIRSF001487	protein disulfide-isomerase [Validated]	433	440	PF00085(2)	Full	SCOP Superfamily=>Thioredoxin

Figure 9.17. The PIRSF database. (a) Text search for superfamilies similar to thioredoxin. This search results in 50 families. (b) Portion of the summary report for one superfamily. (c) Summary report for one entry.

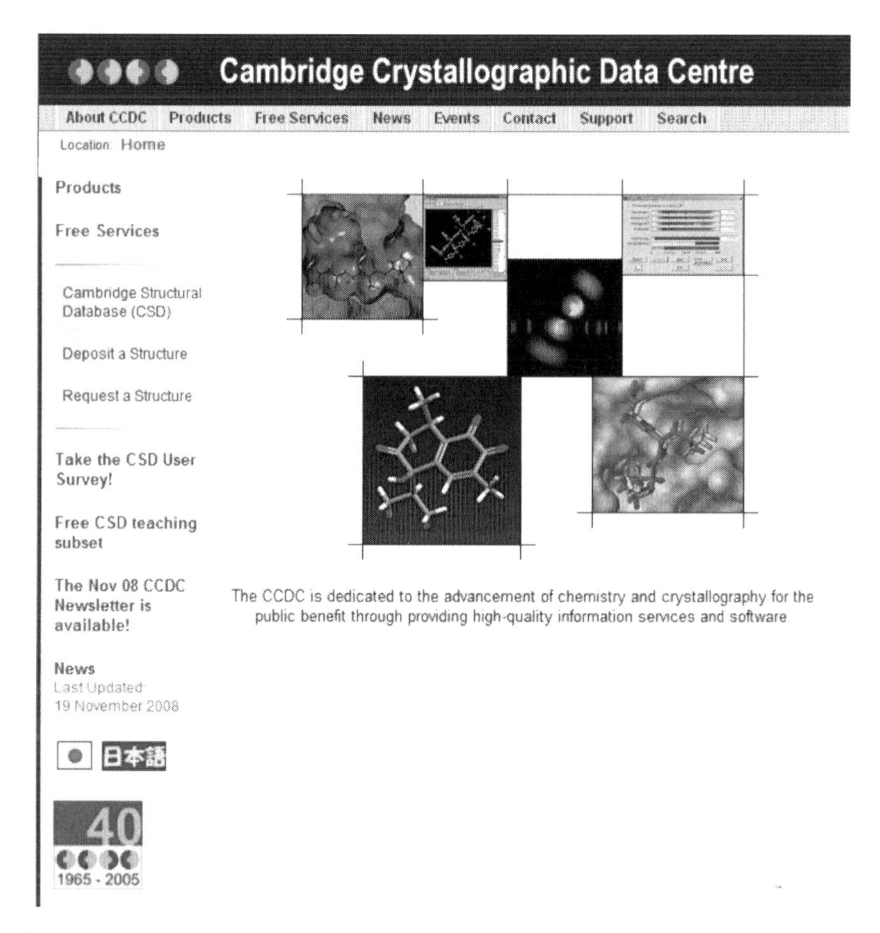

Figure 9.18. The Cambridge Crystallographic Data Centre. Clicking on each icon gives information about the programs used for visualizing and analysing small molecules in this database. Note that CCDC does not contain structures of inorganic or metals/alloys. These are held at the Inorganic Crystal Structure (http://www.icsdwebfiz-karlsruhe.de) and CRYSTME (http://www.tothcanada.com) databases, respectively.

three-dimensional structure for a query protein. This is especially true for proteins associated with membranes, a category representing almost a third of the 70% of genes of known function in the Human genome. Despite great technological advances we still have disproportionately few structures for membrane proteins such as aquaporins. In contrast to this paucity of structural data, more than 400 000 structures are available for small molecules in the Cambridge database and these provide important avenues for new drug discovery. The following sections offer a brief introduction to the main three-dimensional structure databases.

9.4.1 Cambridge Database

Structures of small molecules are much easier to determine by X-ray diffraction than biomacromolecules because they are easier to crystallize, give stronger and simpler diffraction patterns and are more structurally monodisperse. These structures (as well as structures determined by neutron diffraction) are archived in a database at the *Small Molecule Cambridge Crystallographic Data Centre* (CCDC, Cambridge, England). Various tools are available here for comparison and analysis of these structures (Figure 9.18). These structures can be 'docked' onto protein structures and the resulting complex *energy minimized* to reveal effects of the ligand on protein structure. Small molecules capable of binding proteins are of interest as potential drugs and this application of bioinformatics is now an important activity in drug discovery (Section 9.6.3 below).

9.4.2 Protein Databank (PDB)

Three dimensional structures of biomacromolecules are now principally archived by the *Research Collaboratory for Structural Biology* (RCSB) as described in Chapter 6 (Section 6.8.1). While the overwhelming majority of

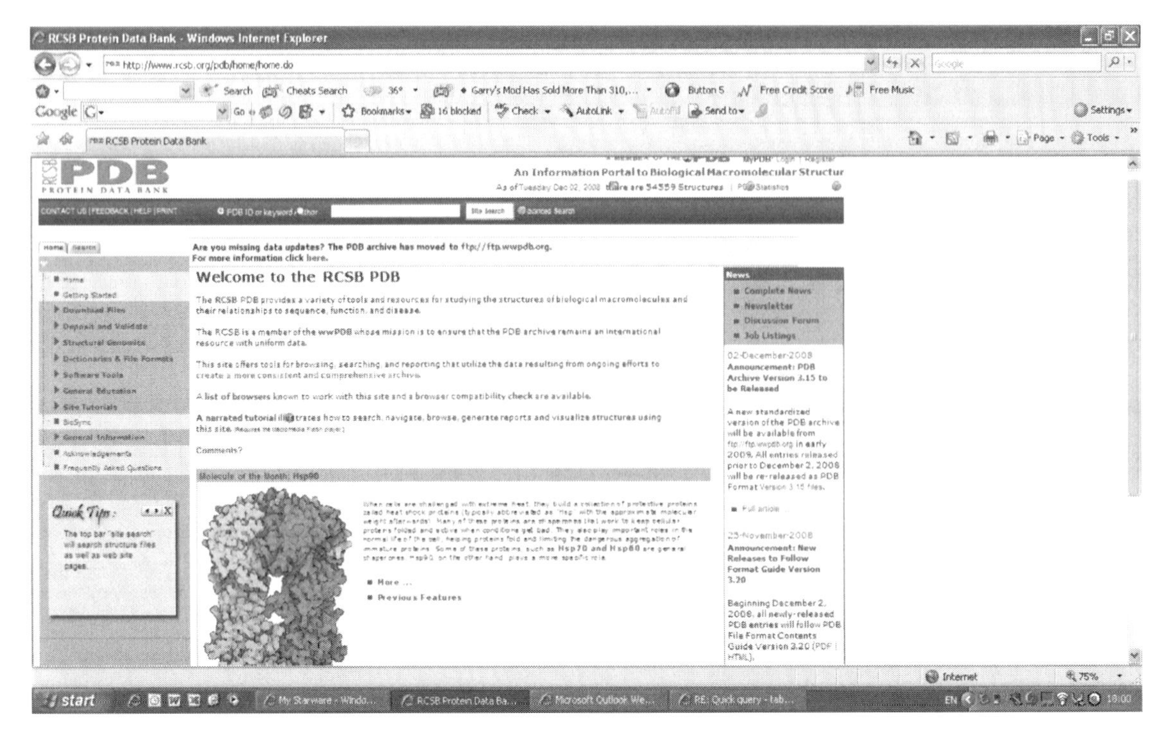

Figure 9.19. The Protein Data Bank (PDB).The PDB can be searched by PDB ID (a 4-letter code unique to each file), by text or by author. Other resources available from the front-page include data on holdings, methodologies for structure determination and downloadable software.

biomacromolecular three-dimensional structures are held in the PDB it should be borne in mind that a small number of atomic coordinate files are not deposited there but are instead held in specialist databases. Also, there is a separate archive of NMR-derived structures *BioMagRes* at the University of Wisconsin. Moreover, some of the older files held by RCSB are bibliographic files only and do not contain atomic coordinates. Access to the PDB is possible from many sequence databases and in these contexts is sometimes referred to as the *Macromolecular Structure Database* (MSD). Structure files are submitted to the PDB ahead of publication in the peer-reviewed literature and receive a unique *PDB ID* or identification code (this code is given in the eventual publication), which is analogous to the accession code of a sequence database. The PDB ID for the file for human aquaporin 1 is 1J4N. If these authors submitted a second structure file for this protein, it would be denoted 2J4N, and so on. The PDB is highly *redundant* with, for example, hundreds of structures for the enzyme lysozyme. It frequently contains variants such as site-directed mutants, proteins crystallized with different ligands or apoproteins (proteins crystallized without ligands). In Chapter 6 we saw that molecular replacement is often used to solve the phase problem in protein crystallography and this has led to a tendency for new structures to 'cluster' around previously solved structures. Conversely, as mentioned above, there is a paucity of structures for proteins associated with membranes. It is instructive to bear in mind that our knowledge of protein structure is limited by the constraints of the experimental techniques available for structure elucidation.

A number of useful resources are available from the PDB front page (Figure 9.19) including a breakdown of current holdings updated weekly, PDB statistics, background information on PDB history and on how PDB curates structure data. The PDB is also a very useful repository and link for software and programs for structure visualization and analysis. Many of these can be downloaded without charge and advice/help on downloading are provided.

The PDB can be searched by keyword (e.g. entering the protein name 'aquaporin'), by PDB ID (e.g. '1J4N' so that one can return to a file previously retrieved without extensive searching) or by author generating a list of files (Figure 9.20). In this listing the experimental method used (e.g. X-ray diffraction) is stated together with the file title and structure resolution (e.g. 2.0 Å in Figure 9.21). Both NMR and X-ray diffraction PDB files are organized in the same way but it should be borne in mind from Chapter 6 that NMR generates a family of structures rather than the single

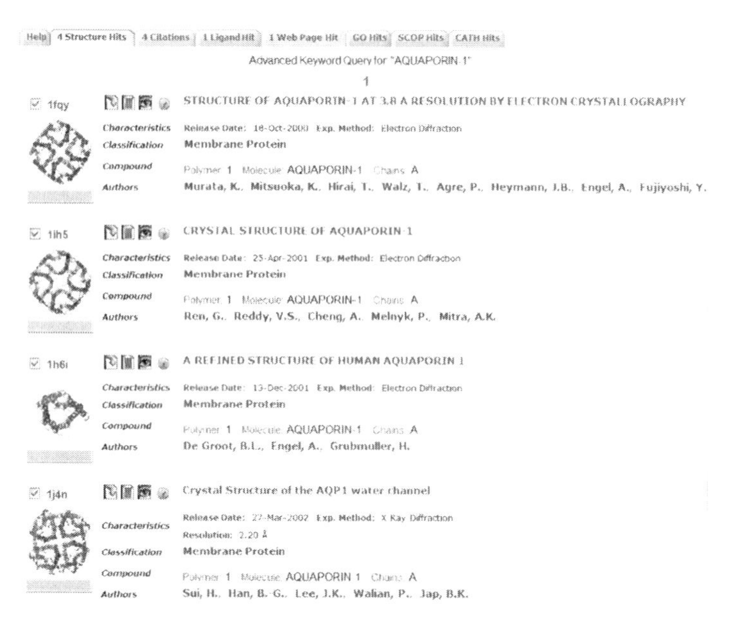

Figure 9.20. Searching the PDB. A list of 4 structure files is found for 'aquaporin-1'. Note that only one of these was determined by X-ray diffraction (2.2 Å resolution) while the others were obtained by lower resolution electron diffraction (3.7–3.8 Å). This is typical of findings with membrane-bound proteins.

structure obtained from crystals. This means that an NMR file will often contain 10–30 structures making these files larger than those for a corresponding single structure. We can select a file by clicking on its title or choose to download or open the file or to visualize the structure with Jmol viewer (see Section 9.5 for discussion of graphics programs) by clicking on appropriate icons. PDB files are all laid out in the same way with bibliographic annotation referring to the peer-reviewed literature first followed by information on the experiment used to generate the file of atomic coordinates (Figure 9.21). In the case of X-ray diffraction this will include crystallographic information such as unit cell architecture, space group and strategies used to generate the electron density map. The bulk of the file consists of a listing of individual atoms beginning with the N-terminal amide nitrogen followed by the α-carbon, carbonyl carbon and 'R'-group atoms denominated sequentially (Figure 9.21). Then the atoms for residue 2 are given, and so on until the carbonyl carbon of the C-terminal residue of that chain. If there is more than one chain this is denoted as A, B, C, and so on. Nonprotein atoms are denoted as *heteroatoms*. Principally this consists of oxygen atoms of water molecules (HOH). A small number of atoms are associated with any ligand cocrystallized with the protein and these are listed after the water molecules. Each atom is associated with three numbers, *the atomic coordinates*, which are essentially Cartesian coordinates denoting the location of the centre of the atom. Comparing coordinates for consecutive atoms will show that these numbers vary slightly in near-neighbour atoms but if, say, atom 1 is compared to atom-20 the numbers alter quite considerably. Protein graphics programs accept PDB files as inputs and interpret their list of coordinates to allow visualization of the structure in three dimensions (Section 9.5; Table 9.3).

9.4.3 Specialist Structural Databases

A key component of hierarchical protein structure classification involves recognition of secondary structure composition and patterns of arrangement within protein structure files in the PDB. Protein folds are quite robust to changes in primary structure with the result that sequence alignments can miss similarities in proteins, which may nonetheless adopt similar folds and hence possibly have similar function. Specialist structural databases allow us to classify proteins into three-dimensional classes based partly on their composition and arrangement of secondary structure elements. The best known of these are CATH and SCOP (Table 9.3).

CATH classifies proteins based on four levels of structural hierarchy (Figure 9.22). *C*lass is automatically assigned as mainly α-helix, mainly β-sheet, mixture of α-helix/β-sheet and proteins with especially low composition of α-helix/β-sheet. *A*rchitecture is manually assigned based on well-known secondary structure arrangements such as β-barrels

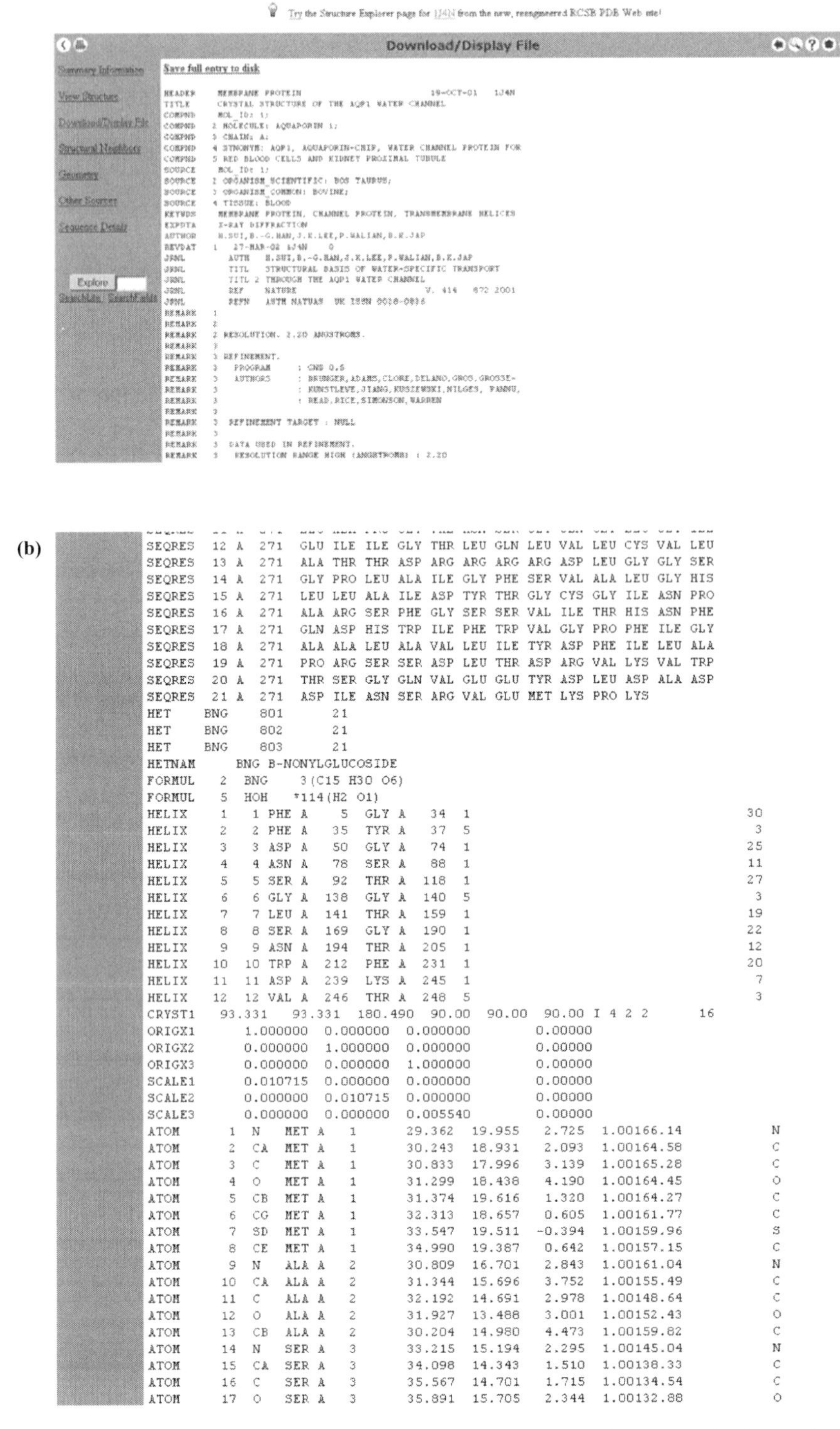

(a)

PDB PROTEIN DATA BANK — Structure Explorer - 1J4N

Title: Crystal Structure Of The Aqp1 Water Channel
Classification: Membrane Protein
Compound: Mol_Id: 1; Molecule: Aquaporin 1; Chain: A; Synonym: Aqp1, Aquaporin-Chip, Water Channel Protein For Red Blood Cells and Kidney Proximal Tubule
Exp. Method: X-ray Diffraction

Try the Structure Explorer page for 1J4N from the new, reengineered RCSB PDB Web site!

Download/Display File

Summary Information
View Structure
Download/Display File
Structural Neighbors
Geometry
Other Sources
Sequence Details
Explore
SearchLite SearchFields

Save full entry to disk

```
HEADER    MEMBRANE PROTEIN                        19-OCT-01   1J4N
TITLE     CRYSTAL STRUCTURE OF THE AQP1 WATER CHANNEL
COMPND    MOL_ID: 1;
COMPND   2 MOLECULE: AQUAPORIN 1;
COMPND   3 CHAIN: A;
COMPND   4 SYNONYM: AQP1, AQUAPORIN-CHIP, WATER CHANNEL PROTEIN FOR
COMPND   5 RED BLOOD CELLS AND KIDNEY PROXIMAL TUBULE
SOURCE    MOL_ID: 1;
SOURCE   2 ORGANISM_SCIENTIFIC: BOS TAURUS;
SOURCE   3 ORGANISM_COMMON: BOVINE;
SOURCE   4 TISSUE: BLOOD
KEYWDS    MEMBRANE PROTEIN, CHANNEL PROTEIN, TRANSMEMBRANE HELICES
EXPDTA    X-RAY DIFFRACTION
AUTHOR    H.SUI,B.-G.HAN,J.K.LEE,P.WALIAN,B.K.JAP
REVDAT   1   27-MAR-02 1J4N    0
JRNL        AUTH   H.SUI,B.-G.HAN,J.K.LEE,P.WALIAN,B.K.JAP
JRNL        TITL   STRUCTURAL BASIS OF WATER-SPECIFIC TRANSPORT
JRNL        TITL 2 THROUGH THE AQP1 WATER CHANNEL
JRNL        REF    NATURE                         V. 414   872 2001
JRNL        REFN   ASTM NATUAS  UK ISSN 0028-0836
REMARK   1
REMARK   2
REMARK   2 RESOLUTION. 2.20 ANGSTROMS.
REMARK   3
REMARK   3 REFINEMENT.
REMARK   3   PROGRAM     : CNS 0.5
REMARK   3   AUTHORS     : BRUNGER,ADAMS,CLORE,DELANO,GROS,GROSSE-
REMARK   3               : KUNSTLEVE,JIANG,KUSZEWSKI,NILGES, PANNU,
REMARK   3               : READ,RICE,SIMONSON,WARREN
REMARK   3
REMARK   3  REFINEMENT TARGET : NULL
REMARK   3
REMARK   3  DATA USED IN REFINEMENT.
REMARK   3   RESOLUTION RANGE HIGH (ANGSTROMS) : 2.20
```

(b)

```
SEQRES  12 A  271  GLU ILE ILE GLY THR LEU GLN LEU VAL LEU CYS VAL LEU
SEQRES  13 A  271  ALA THR THR ASP ARG ARG ARG ARG ASP LEU GLY GLY SER
SEQRES  14 A  271  GLY PRO LEU ALA ILE GLY PHE SER VAL ALA LEU GLY HIS
SEQRES  15 A  271  LEU LEU ALA ILE ASP TYR THR GLY CYS GLY ILE ASN PRO
SEQRES  16 A  271  ALA ARG SER PHE GLY SER SER VAL ILE THR HIS ASN PHE
SEQRES  17 A  271  GLN ASP HIS TRP ILE PHE TRP VAL GLY PRO PHE ILE GLY
SEQRES  18 A  271  ALA ALA LEU ALA VAL LEU ILE TYR ASP PHE ILE LEU ALA
SEQRES  19 A  271  PRO ARG SER SER ASP LEU THR ASP ARG VAL LYS VAL TRP
SEQRES  20 A  271  THR SER GLY GLN VAL GLU GLU TYR ASP LEU ASP ALA ASP
SEQRES  21 A  271  ASP ILE ASN SER ARG VAL GLU MET LYS PRO LYS
HET    BNG    801      21
HET    BNG    802      21
HET    BNG    803      21
HETNAM     BNG B-NONYLGLUCOSIDE
FORMUL   2  BNG    3(C15 H30 O6)
FORMUL   5  HOH   *114(H2 O1)
HELIX    1   1 PHE A    5  GLY A   34  1                                  30
HELIX    2   2 PHE A   35  TYR A   37  5                                   3
HELIX    3   3 ASP A   50  GLY A   74  1                                  25
HELIX    4   4 ASN A   78  SER A   88  1                                  11
HELIX    5   5 SER A   92  THR A  118  1                                  27
HELIX    6   6 GLY A  138  GLY A  140  5                                   3
HELIX    7   7 LEU A  141  THR A  159  1                                  19
HELIX    8   8 SER A  169  GLY A  190  1                                  22
HELIX    9   9 ASN A  194  THR A  205  1                                  12
HELIX   10  10 TRP A  212  PHE A  231  1                                  20
HELIX   11  11 ASP A  239  LYS A  245  1                                   7
HELIX   12  12 VAL A  246  THR A  248  5                                   3
CRYST1   93.331   93.331  180.490  90.00  90.00  90.00 I 4 2 2      16
ORIGX1      1.000000  0.000000  0.000000        0.00000
ORIGX2      0.000000  1.000000  0.000000        0.00000
ORIGX3      0.000000  0.000000  1.000000        0.00000
SCALE1      0.010715  0.000000  0.000000        0.00000
SCALE2      0.000000  0.010715  0.000000        0.00000
SCALE3      0.000000  0.000000  0.005540        0.00000
ATOM      1  N   MET A   1      29.362  19.955   2.725  1.00166.14           N
ATOM      2  CA  MET A   1      30.243  18.931   2.093  1.00164.58           C
ATOM      3  C   MET A   1      30.833  17.996   3.139  1.00165.28           C
ATOM      4  O   MET A   1      31.299  18.438   4.190  1.00164.45           O
ATOM      5  CB  MET A   1      31.374  19.616   1.320  1.00164.27           C
ATOM      6  CG  MET A   1      32.313  18.657   0.605  1.00161.77           C
ATOM      7  SD  MET A   1      33.547  19.511  -0.394  1.00159.96           S
ATOM      8  CE  MET A   1      34.990  19.387   0.642  1.00157.15           C
ATOM      9  N   ALA A   2      30.809  16.701   2.843  1.00161.04           N
ATOM     10  CA  ALA A   2      31.344  15.696   3.752  1.00155.49           C
ATOM     11  C   ALA A   2      32.192  14.691   2.978  1.00148.64           C
ATOM     12  O   ALA A   2      31.927  13.488   3.001  1.00152.43           O
ATOM     13  CB  ALA A   2      30.204  14.980   4.473  1.00159.82           C
ATOM     14  N   SER A   3      33.215  15.194   2.295  1.00145.04           N
ATOM     15  CA  SER A   3      34.098  14.343   1.510  1.00138.33           C
ATOM     16  C   SER A   3      35.567  14.701   1.715  1.00134.54           C
ATOM     17  O   SER A   3      35.891  15.705   2.344  1.00132.88           O
```

Figure 9.21. PDB file for aquaporin 1 (1J4N). (a) The file opens with bibliographic annotation. (b) Portion of PDB file showing part of the amino acid sequence of aquaporin 1 and the beginning of the list of atomic coordinates.

(a)

CATH
PROTEIN STRUCTURE CLASSIFICATION

Home Search

CATH Search

You can search for general text, a CATH identifier (PDB code, domain id, CATH code, etc) or provide a FASTA sequence and perform a BLAST search of CATH domains

Search

Start Browsing at the top of the CATH hierarchy > Browse

Search by ID/Keywords

THIOREDOXIN > Text Search

Search by FASTA Sequence [example]

> Sequence Search

(b)

CATH
PROTEIN STRUCTURE CLASSIFICATION

Home Search 3 30 1720

Topology: 3.30.1720

Thioredoxin fold

Classification Lineage (3.30.1720)

CATH Code	Level Description	Links
3	Alpha Beta	
3.30	2-Layer Sandwich	
3.30.1720	Thioredoxin fold	

Summary of Child Nodes

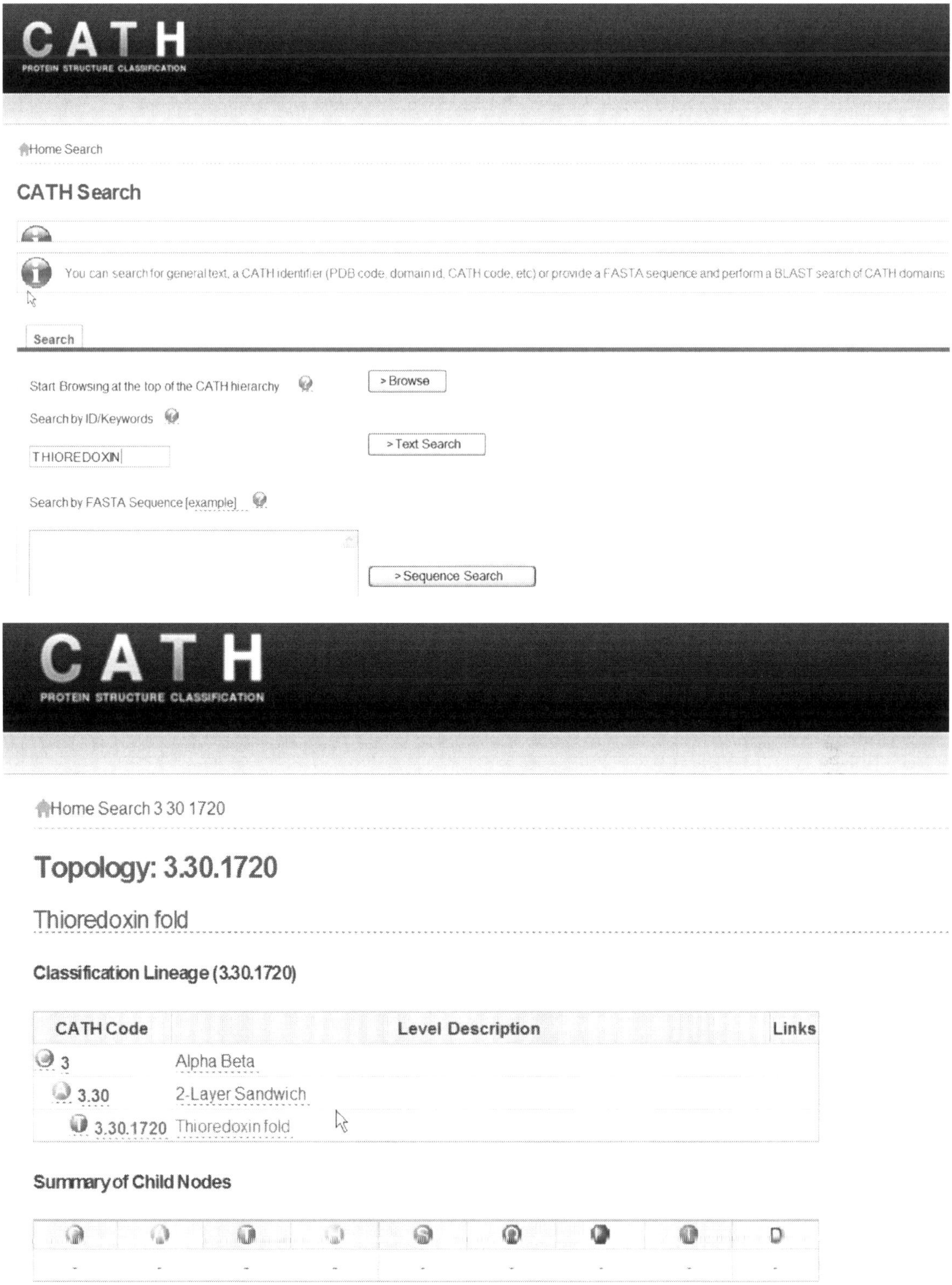

Figure 9.22. The CATH database. (a) The database can be searched by text for thioredoxin. (b) The classification is based on Class, Architecture, Topology and Homology.

and three-layer sandwich but ignoring connectivity between secondary structure elements. *T*opology is based on similar folds recognized by an algorithm involving both secondary structure features and their connectivity. *H*omology recognizes sequence identity suggesting a common ancestor.

SCOP (Figure 9.23) also has a hierarchical structure with currently eleven classes dominated by secondary structure elements similar to CATH (all α-helix, all β-sheet, mainly parallel β-sheets a/β, mainly antiparallel β-sheets α + β, multi-domain proteins, membrane/cell surface proteins and small proteins). Four classes (coiled coil, low resolution protein structures, peptides and designed proteins) are operational rather than true classes. This database is searchable by text and has the facility to represent the protein as a three-dimensional protein graphic with the Chime plug-in (Section 9.5.2).

9.5 PROGRAMS FOR ANALYSIS AND VISUALIZATION OF TERTIARY STRUCTURE DATABASES

Much of the information contained in three dimensional biomacromolecular structures needs to be visualized in three dimensions in the same way that primary structures may be viewed as a unidirectional string of building blocks. It is necessary to interpret structure files with graphics programs, which trace atoms through three-dimensional space. These programs create a sort of optical illusion in that they actually present a two-dimensional view on the PC screen at any given moment in time. However, because the screen can be refreshed more quickly than the eye can detect, the impression of a three-dimensional view is achieved. The viewer can rotate, translate, zoom in/out and choose to represent some or all of the atoms/residues in a range of ways. Some very powerful graphics programs are available but many of these work in specialist environments such as UNIX. The following is a brief description of some graphics programs, which run on windows or Macintosh systems. Unless otherwise stated, they can be downloaded from the *software* section of the PDB (or from links listed there) free of charge provided they are used for academic not-for-profit purposes (Tables 9.1 and 9.3). Once installed on a PC, these programs can be used to view structures downloaded from sources such as the PDB. This sometimes allows us to work *off-line* with both program and file on a local computer.

9.5.1 RasMol/RasTop

RasMol was written in the 1990s by Roger Sayle and is still very useful in that later programs have adopted similar command structures. The first thing to do is to select a PDB file with the *open* command of the *edit* pulldown. By default, this represents the protein as a *wireframe* structure (Figure 9.24, see colour plate section). The view can be altered by one of three methods. 1. Shortcuts with mouse buttons allow zooming, translation or rotation of the image. 2. The pulldown menus allow variation of the structure representation. 3. A *command line* allows typing in of specific commands (a full list is given in the manual available with the *help* pulldown). Examples of useful commands are *select* (which can select a specific residue, run of residues or ligand), *set background* (allows selection of different background colours) and *spacefill*. If a portion of the structure has been selected, any commands subsequently selected (whether by pulldown or command line) are executed only on the selected piece of structure.

RasTop is an updated version of RasMol released in 2000. The graphics are generally superior to RasMol but most of the commands are identical (Figure 9.25, see colour plate section). However, there are more extensive pulldown commands and there is no command line as such. Instead a separate command box is selectable from the *edit* pulldown into which RasMol commands can be entered. From both RasMol and RasTop, views can be exported as GIF files with the *export* command and saved for later use in documents or presentations.

9.5.2 Protein Explorer

Protein Explorer is a more sophisticated derivative of RasMol developed by Eric Martz, which can be accessed through either Internet Explorer (version 5.5SP2 or later) or Netscape (version 4) web browsers (Figure 9.26, see colour plate section). The program operates in conjunction with a separate *plugin* program called *Chime,* downloadable from a company called MDL (see URL at the end of this chapter. NOTE: It is necessary to register online with MDL before downloading Chime). The URL *proteinexplorer.org* brings us to *FrontDoor* which allows us to open up a PDB file by typing a PDB ID into the box and clicking *go*. A structure appears in one of three windows. The image can be modified in two ways. Firstly, by typing commands in the *commands* window we can vary the image with RasMol commands. Selecting *explore more* brings us to a window which allows us to customize the image using pulldown commands *select*, *colour* and *display*.

9.5.3 PyMOL

PyMOL was developed by Warren DeLano and is, in principle, freely available although the authors do request contributions to defray their costs. The programs mentioned so far are not available in *open source* form, which means the user cannot modify the program to improve its performance. On

(a)

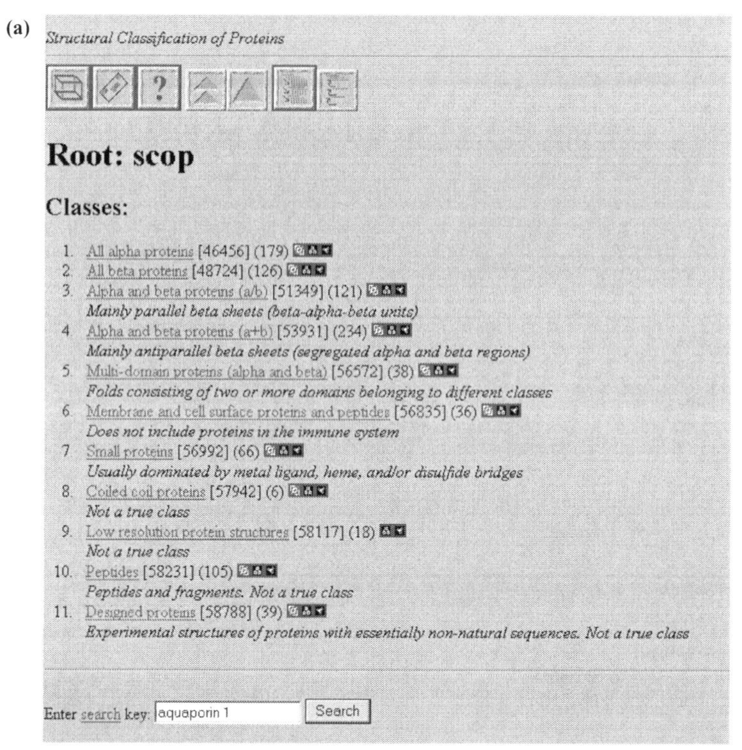

(b)

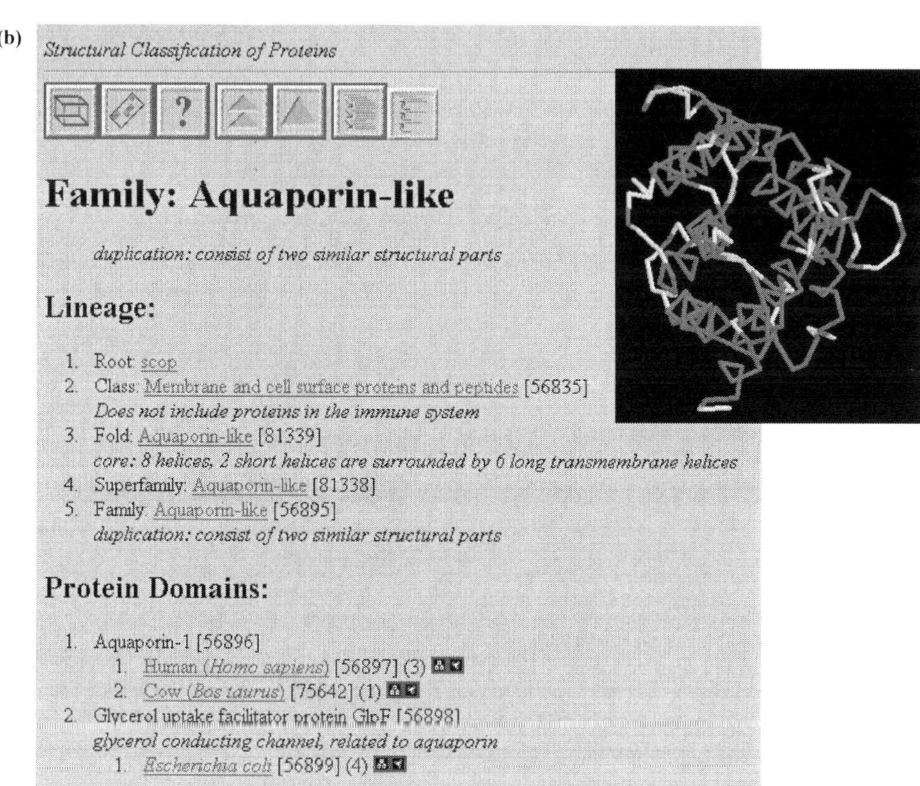

Figure 9.23. The SCOP database. (a) The database can be searched by text (e.g. aquaporin). (b) The result is a classification of the protein from suprafamily down to individual species. Selecting the graphics icon allows us to view the three-dimensional structure by Chime (Section 9.5.2). Clicking on the species listed in (b) reveals a listing of all PDB entries for that species.

the other hand commercially available open source programs are expensive. PyMOL is a generous attempt to provide to the international scientific community the computer power of an open source program in inexpensive form. It allows us to visualize multiple conformations of a single structure, to interface with external programs and to generate high-quality graphics (Figure 9.27, see colour plate section). As with the programs described above, the first step is to use a pulldown to open a PDB file. We can then select *hide*, *show* or *label* so as to display various aspects of the structure. It is possible to open up several PDB files in a single window and to select commands for each one in turn. Clicking on the file name will 'hide' the image for that protein. PyMOL allows saving both of individual images and of movies.

9.5.4 WebMol

WebMol is a viewer program written in the Java computer language by Dirk Walther and designed to act as a browser (although the program may also be downloaded to your local PC). Opening the URL brings you to a window in which the PDB ID is entered (Figure 9.28). A wireframe structure appears by default. Selections can be made on the right hand side of this window which allows us to choose which parts of the structure we wish to show, how we wish to colour it and if we wish to represent surfaces on the structure. A *select* box opens a window which allows us to pick out regions of sequence to represent in the structure window. We can use the *measure* box to measure bond lengths, angles, and so on. A *distance matrix* box allows us to view distances between any two points in the sequence while a *Ramachandran* window allows us to view dihedral angles. A *trace* option draws the three-dimensional structure sequentially. A row of horizontal buttons allows us further options to adapt the image.

9.5.5 Swiss-PdbViewer

Swiss-PdbViewer allows analysis of multiple structures at the same time and in the same window (Figure 9.29). In contrast to PyMOL though, the structures of related proteins can be superimposed to reveal structural alignment. Mutations can be introduced into the sequence at specific points and the effects on structure computed by an energy minimization routine. Measurements can also be obtained of bond angles and distances. The workspace consists of a main window with associated pulldown commands in which the principal manipulations are performed. The *window* pulldown allows selection of supplementary windows; *alignment* shows sequence alignments of files loaded, *control panel* allows selection and manipulation of individual groups and a *Ramachandran* window shows a Ramachandran plot for the whole protein and any parts specifically selected (Example 9.3).

9.6 HOMOLOGY MODELLING

Proteins with similar primary structures adopt generally similar polypeptide folds. In fact, surprisingly often changes in primary structure have little effect on the overall fold of polypeptide chains. The redundancy of the PDB is partly a reflection of the robustness of protein fold to change in primary structure. Although thousands of new three-dimensional structures are added to the PDB each year, very few of these represent entirely new folds. As a general rule of thumb, sequences with a minimum of 50% identity are expected to adopt the same fold over 90% of their structure. Thus the primary structure 'encodes' a three-dimensional fold but the precise rules of this code are not as yet fully understood and there is considerable plasticity built into this relationship. There is major research interest in solving the protein folding problem as this could allow us to create 'designer proteins' with novel properties and it could also help us to understand more about the increasing number of disease states which feature protein *mis-folding* such as Alzheimer's disease and spongiform encephalopathies.

There are two main approaches to predicting three-dimensional fold from primary structure (Figure 9.30). Ab initio *modelling* involves computing a protein structure by estimating the energy of the molecule as it goes through various structural perturbations, a process of *energy minimization*. This requires extensive computer power and is limited by our lack of knowledge of the precise details of energy interactions within folded proteins and between proteins and solvent water. Moreover, for most proteins we know little about the *kinetics* of folding which may mean that the fold with lowest thermodynamic energy may not be kinetically possible. For these reasons, a more empirical approach is often taken called *homology modelling*. This predicts three-dimensional structure for a protein of known sequence (*the target sequence*) by modelling it on the known three-dimensional structure of a protein of related sequence (*the template sequence*). Generally, this approach is not feasible below approximately 30% homology.

A useful parameter in homology modelling is the *root mean square deviation* (rmsd) of the model.

$$\text{rmsd}^2 = \frac{\sum (x_{\text{tr}} - x_{\text{tp}})^2}{n} \qquad (9.1)$$

where x_{tr} is the location of n α-carbon atoms in the target and x_{tp} is the location in the template. This is a numerical reflection of the average deviation of the model from template. The rmsd has a value of 0 for identical structures

(a)

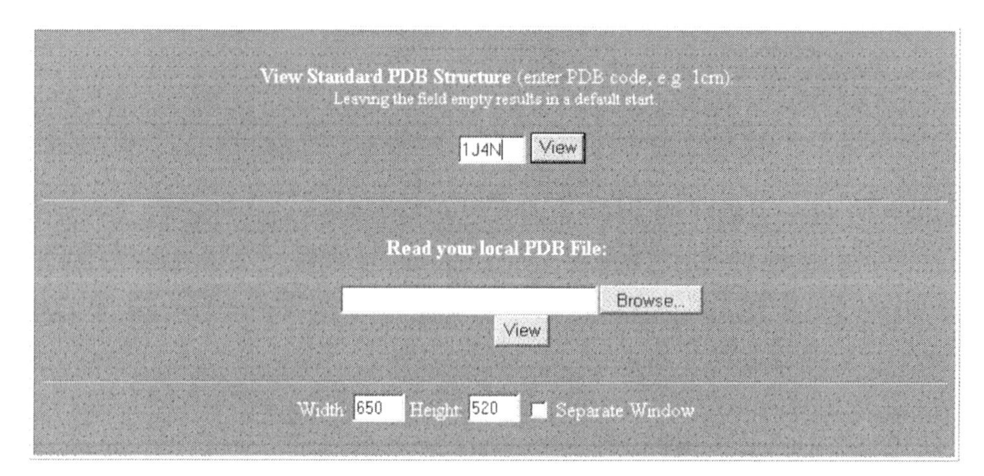

(b)

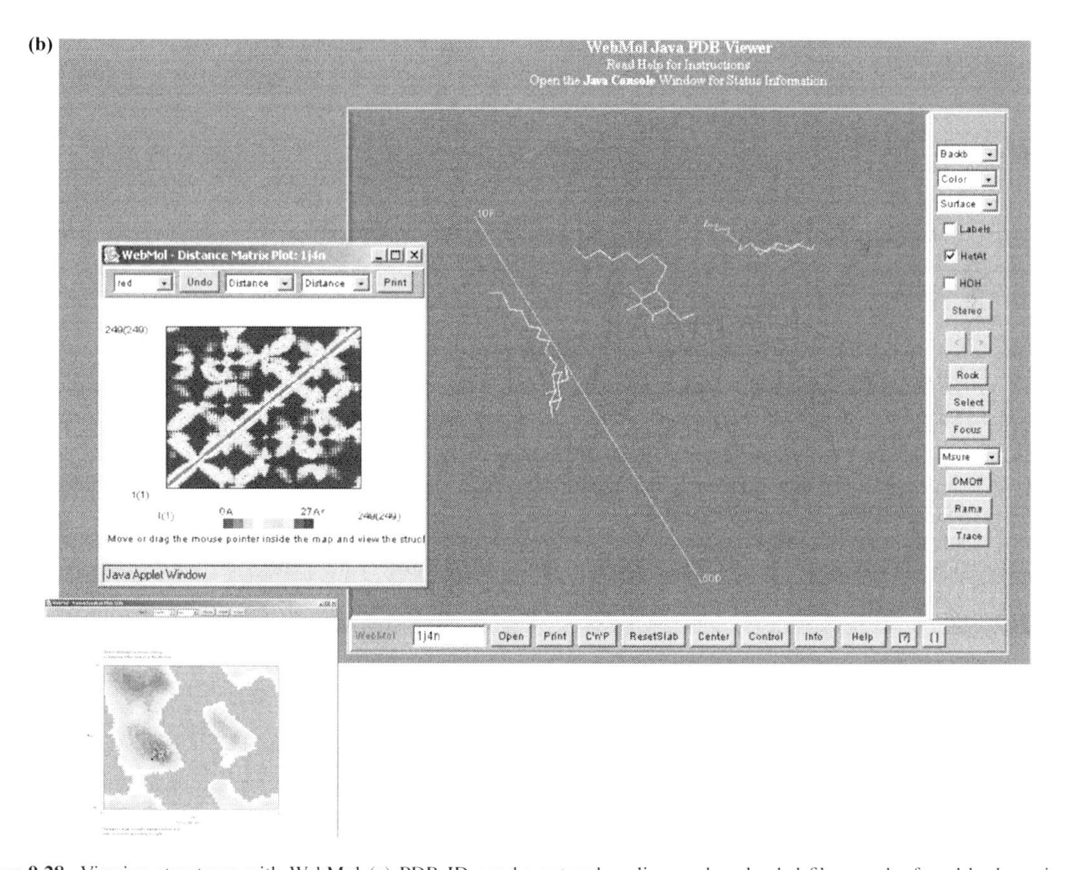

Figure 9.28. Viewing structures with WebMol (a) PDB ID can be entered on-line or downloaded files can be found by browsing. (b) Structure of human aquaporin 1 (1J4N). The distance matrix window shown allows measurement of distances between selected residues. Similarly, Ramachandran diagrams allow measurement of dihedral angles around specific residues.

(a)

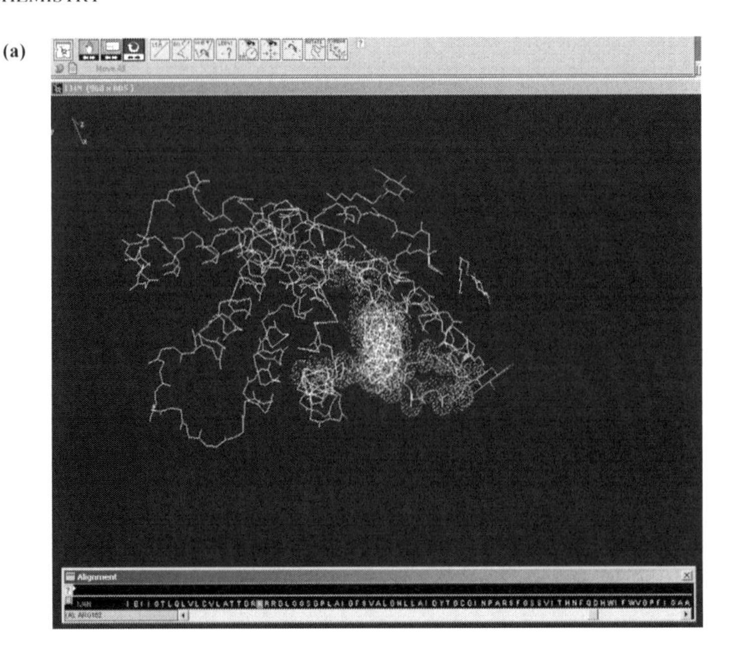

(b)

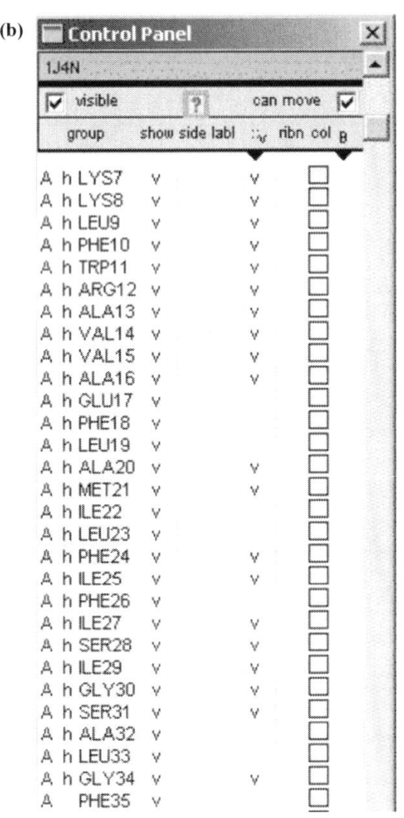

Figure 9.29. Viewing structures with SwissPdbViewer. (a) Opening the file for human aquaporin 1 (1J4N) produces a view of the protein in wireframe. Some selected residues have their van Der Waals radii highlighted in the model. The 'align' window is opened from the *windows* pulldown and in this case Args162 is selected. (b) Residues are selected from the control panel which is also opened from *windows* pulldown. For other possibilities with SwissPdbViewer see Figure 9.32 below.

Example 9.3 Viewing Aquaporin with Swiss-PdbViewer

Entering the accession code 1j4n into the PDB search box allows download of a structure file for bovine aquaporin 1 determined to 2.2 Å resolution by X-ray diffraction (Sui et al., 2001). Selecting 'open' in Swiss-PdbViewer allows navigation to and opening of this file. The structure is viewed with backbone rendered in solid 3D ('Display' pulldown). The view is through the pore and two molecules of the ligand (nonylglucoside) with which the protein was cocrystallized are clearly visible.

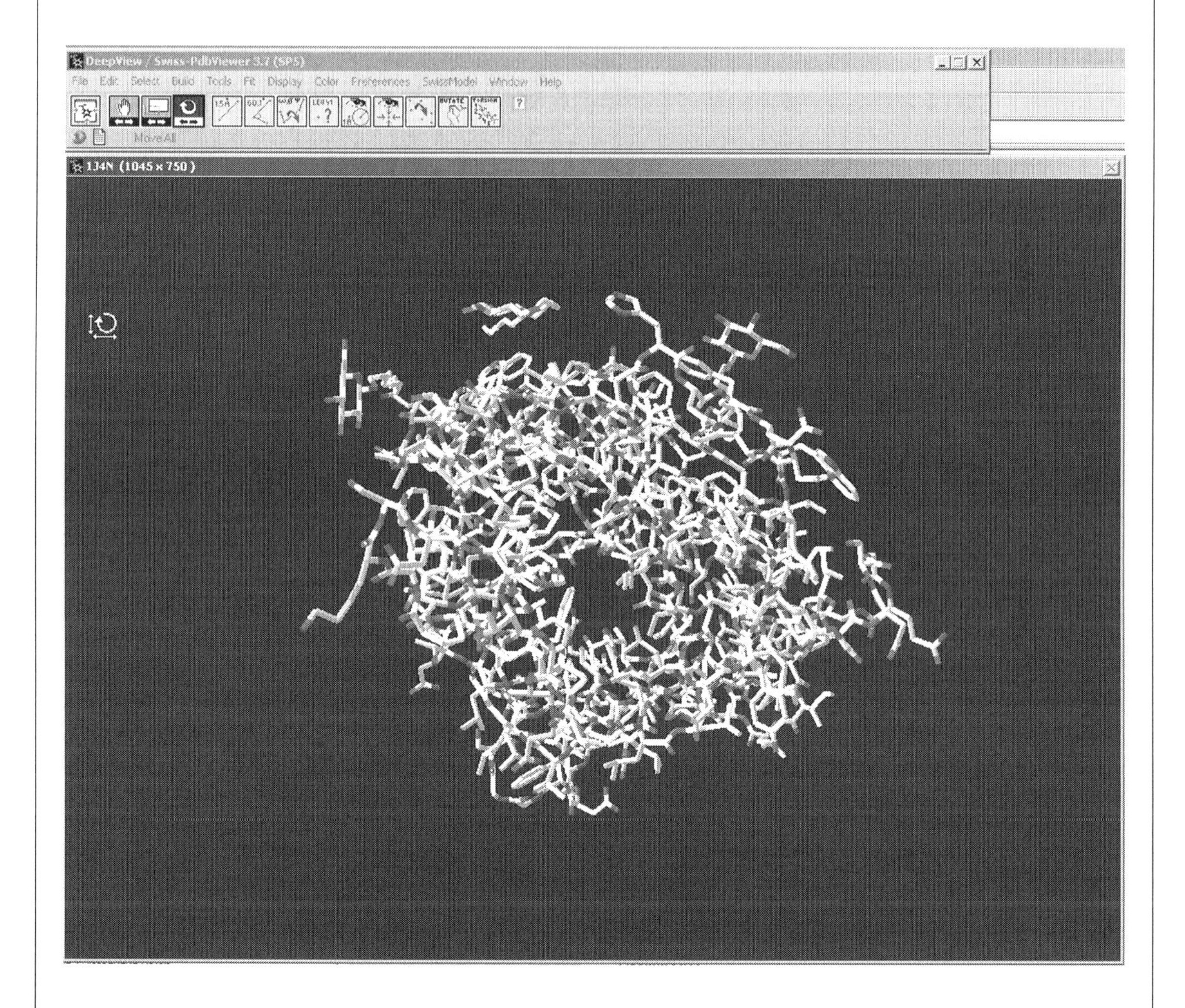

Selecting residues 39–42 on the control panel allows viewing of the dihedral angles for each of these residues in the Ramachandran Plot option of 'window' pulldown:

(Continues)

(Continued)

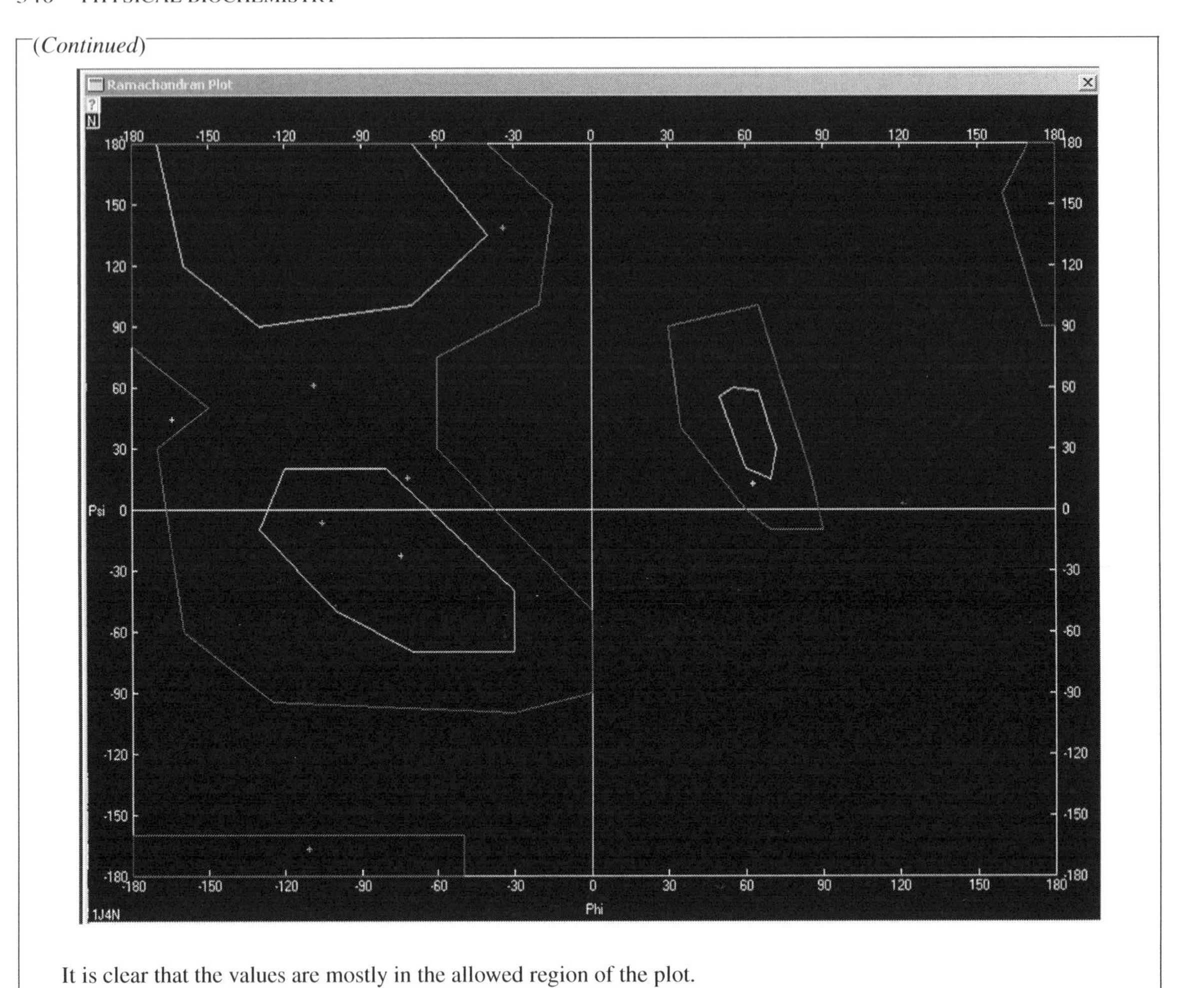

It is clear that the values are mostly in the allowed region of the plot.

(in theory) but the best models can have values of up to ~0.5 Å, approximately the variation found between individual structure determinations of a single protein. A higher number for rmsd reflects increasing deviation between the model and template and suggests that the model should be viewed with increasing caution. Because n varies with protein size, rmsd values are not directly comparable between proteins of differing size but methods for normalizing for this effect are available.

9.6.1 Modelling Proteins from Known Homologous Structures

The first step in homology modelling is to identify *structurally conserved regions* (SCRs) in the two proteins by sequence alignment. The atomic coordinates for SCRs are then copied from the PDB file of the template sequence and used to construct a partial model for the target. In general, in regions of high sequence identity this gives a good overlap between positions of α-carbons but positions of side-chains and nonconserved residues will differ. Amino acids within SCRs differing between the two sequences are now replaced with the correct residue taken from a structure library. Gaps between the SCRs must now be filled in with *loop sequences* in a process known as *loop searching*. These gaps are usually <12 residues long and represent surface regions of the protein. Loop searching uses the geometry at either side of the gap to search the PDB for amino acid sequences of similar sequence and geometry containing approximately the same number of residues as the gap. The atomic coordinates of the loop are taken from the PDB and included in the model.

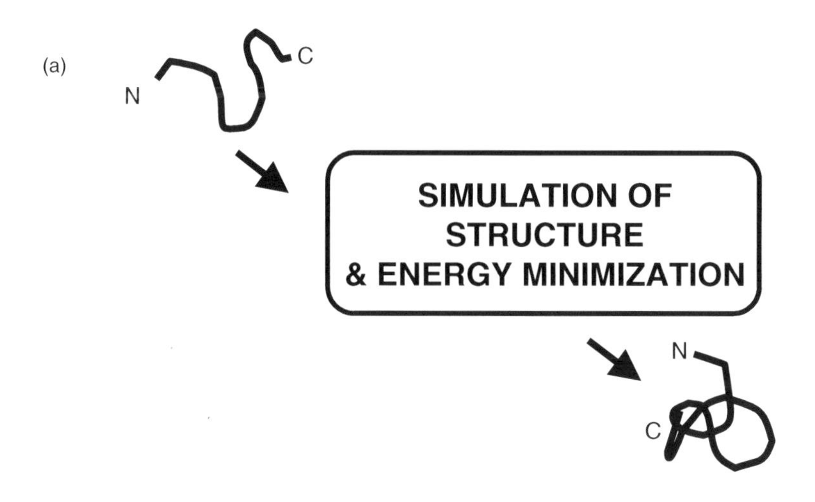

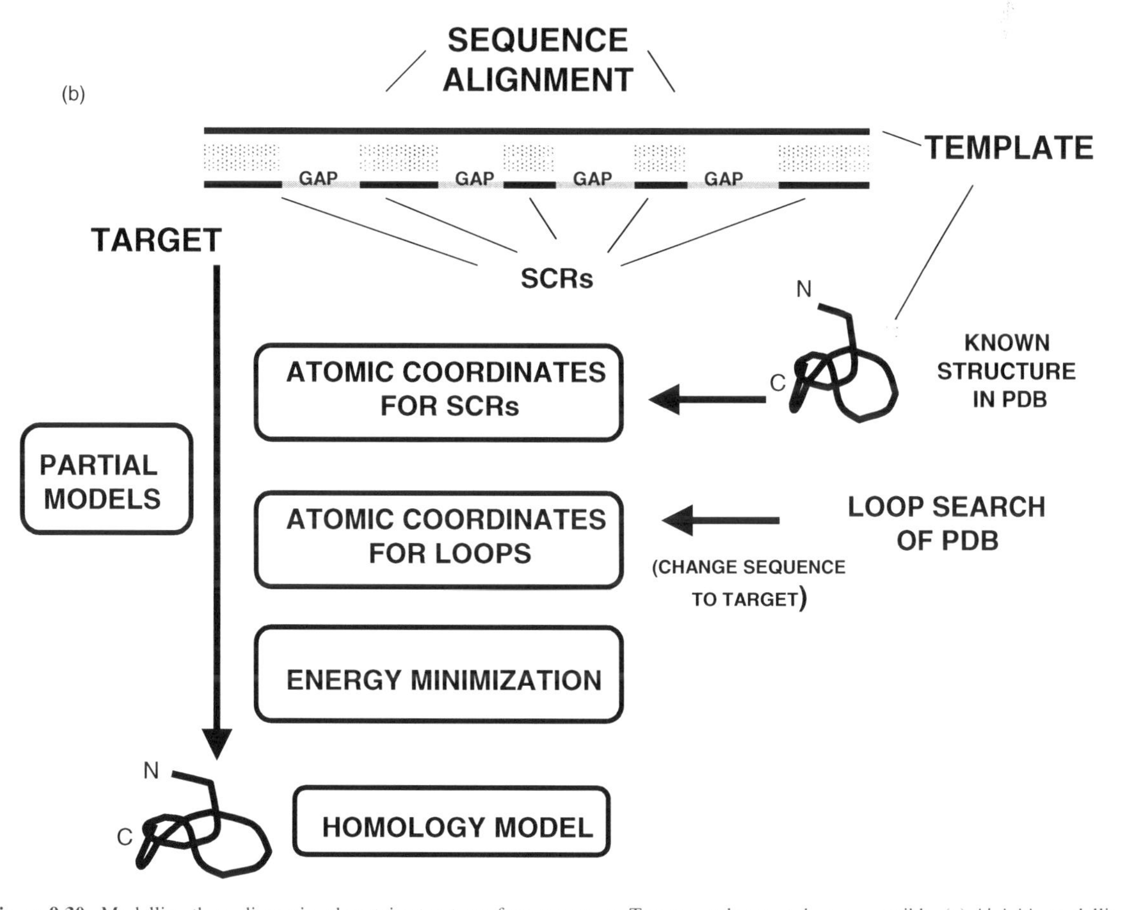

Figure 9.30. Modelling three-dimensional protein structures from sequence. Two general approaches are possible. (a) *Ab initio* modelling is based on calculating a low energy structure based on interactions within the protein and between the protein and solvent. (b) Homology modelling calculates a structure based on a pre-existing structure in the PDB for a protein (template) showing sequence similarity to the target protein. SCRs are found by sequence alignment and coordinates of these are obtained from the PDB. Gap regions are modelled by searching the PDB for loops of similar length, geometry and sequence. The coordinates for these are similarly added to the model. The sequence of loops is converted to that of the target protein and a final structure calculated by energy minimization.

Example 9.4 Homology Modelling of the Motor Componenent of Dynein Heavy Chain

Eukaryotic cells have three distinct types of motors: dyneins, myosins and kinesins which allow movement along the cytoskeleton in a manner driven by ATP hydrolysis. This type of movement underlies change of shape and is responsible for movement of organelles within cells. These proteins are all part of the P-loop ATPase protein family. Although otherwise structurally diverse, there is considerable sequence conservation in the part of the motor ('motor unit') involved in ATP hydrolysis and development of force. Sensitive sequence alignment comparisons suggested similarities between the motor unit part of dynein and members of the AAA class of chaperone-like ATPases. From this it was deduced that the core of the motor unit consisted of 6 AAA-like motifs each of approx. Mr 35-40 kDa.

Three AAA template proteins were selected based on sequence alignment and a homology model created as follows:

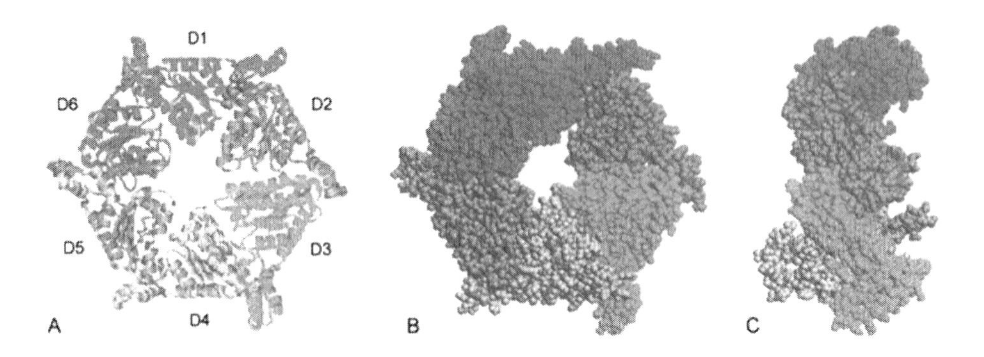

A) View shown in ribbon representation with AAA-like motifs labeled D1-D6. B) Spacefill. C) Side-view of B).

Source: Mocz, G. and Gibbons, I.R. (2001) Model for the motor component of dynein heavy chain based on homology to the AAA family of oligomeric ATPases. *Structure*, **9**, 93–103. Reproduced with permission of Cell Press.

The sequence of the loop now must be corrected to that of the target sequence. Loop searching is performed in this way for all loops. The model resulting from this process is then energy-minimized resulting in a final model.

9.6.2 Automated Modelling

The process of homology modelling involves quite time-consuming decisions on identification of templates, loop searching and varying modelling parameters all of which require the intervention of expert knowledge on the part of the researcher. A quicker approach to homology modelling is to use *automated on-line resources* which will automatically align the target sequence on the template in three dimensions based on the latter sequence's known structure in the PDB. An important *caveat* to this approach is that the resulting model may contain significant errors precisely because all choices are made automatically. Such models must therefore be interpreted with caution. Automated modelling is either performed with a program accessible on-line, for example *MODELLER* or by accessing a *server* which takes the user through the steps required (Table 9.3). In the latter case, the target sequence is entered in a box along with the researcher's e-mail address (Figure 9.31, see colour plate section). If modelling is successful, results are e-mailed in the form of a file viewable with the programs described in Section 9.5 above. The Swiss-PdbViewer is especially useful in that it detects imperfections in the model such as deviant dihedral angles or clashes between side-chains (Figure 9.32). Swiss-PdbViewer can also perform *de novo* modelling of loops, multiple alignment of protein sequences/structures and generation of phylogenic trees. The reader is referred to Gale Rhodes' excellent on-line tutorials for more information on these topics (see References). The structures and sequences of both the template and target are shown in the same windows and options in the *select* pulldown allow identification of clashes, unfavourable bond angles and other indications of the model's quality. On-line homology modelling servers carry out a sequence alignment to find the most similar sequence in the PDB. The target sequence is then modelled on this template using a program such as *SWISS-MODEL* (ExPASy site), *ESyPred3D* (University of Namur, Belgium) and *Geno3D* (CNRS, Lyons, France). The last-named site is especially interactive in that the user can view alignments, select various templates and model on multiple templates (Figure 9.33). The program generates multiple structures and checks their stereochemical appropriateness with *PROCHECK*.

(a)

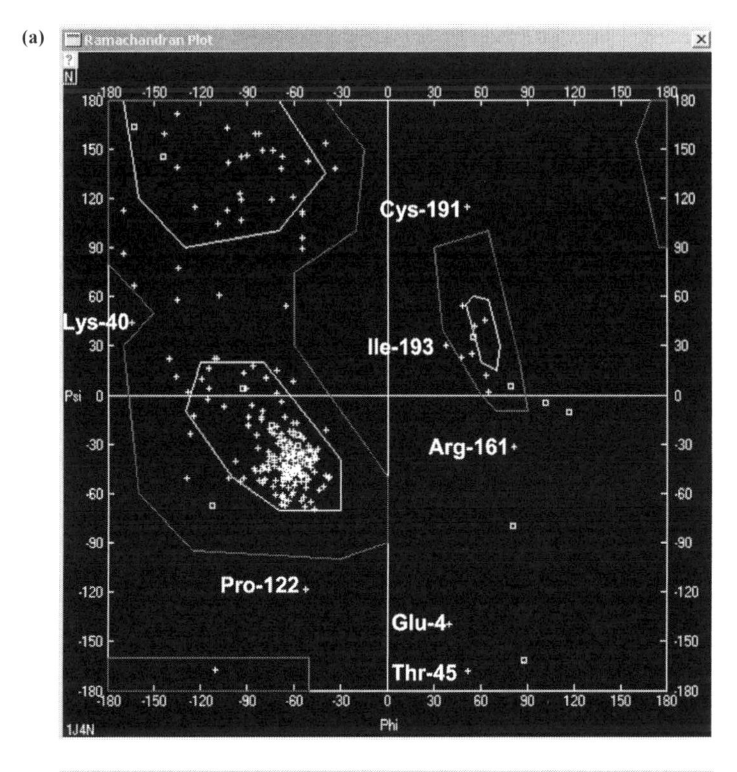

(b)

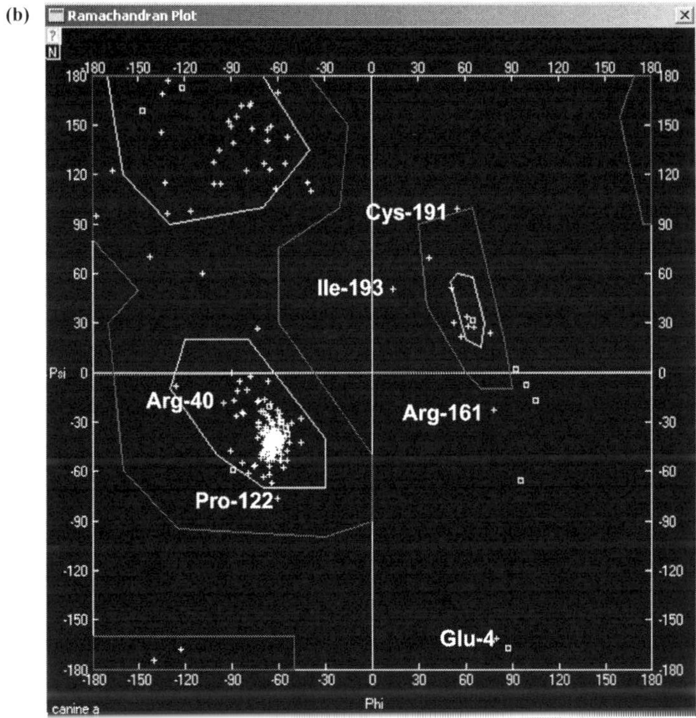

Figure 9.32. Using Swiss-PdbViewer to compare homology model to template. Ramachandran diagrams are shown for (a) Human aquaporin 1 (PDB ID 1J4N) and (b) The model for canine aquaporin generated by ESyPred3D shown in *Figure 9.31*. Note the overall similarities between the range of values occupied and that most of the nonglycine residues outside the three heavily-populated parts of the diagram are common to both structures. These similarities, combined with absence of side-chain clashes, suggest that this model is structurally plausible.

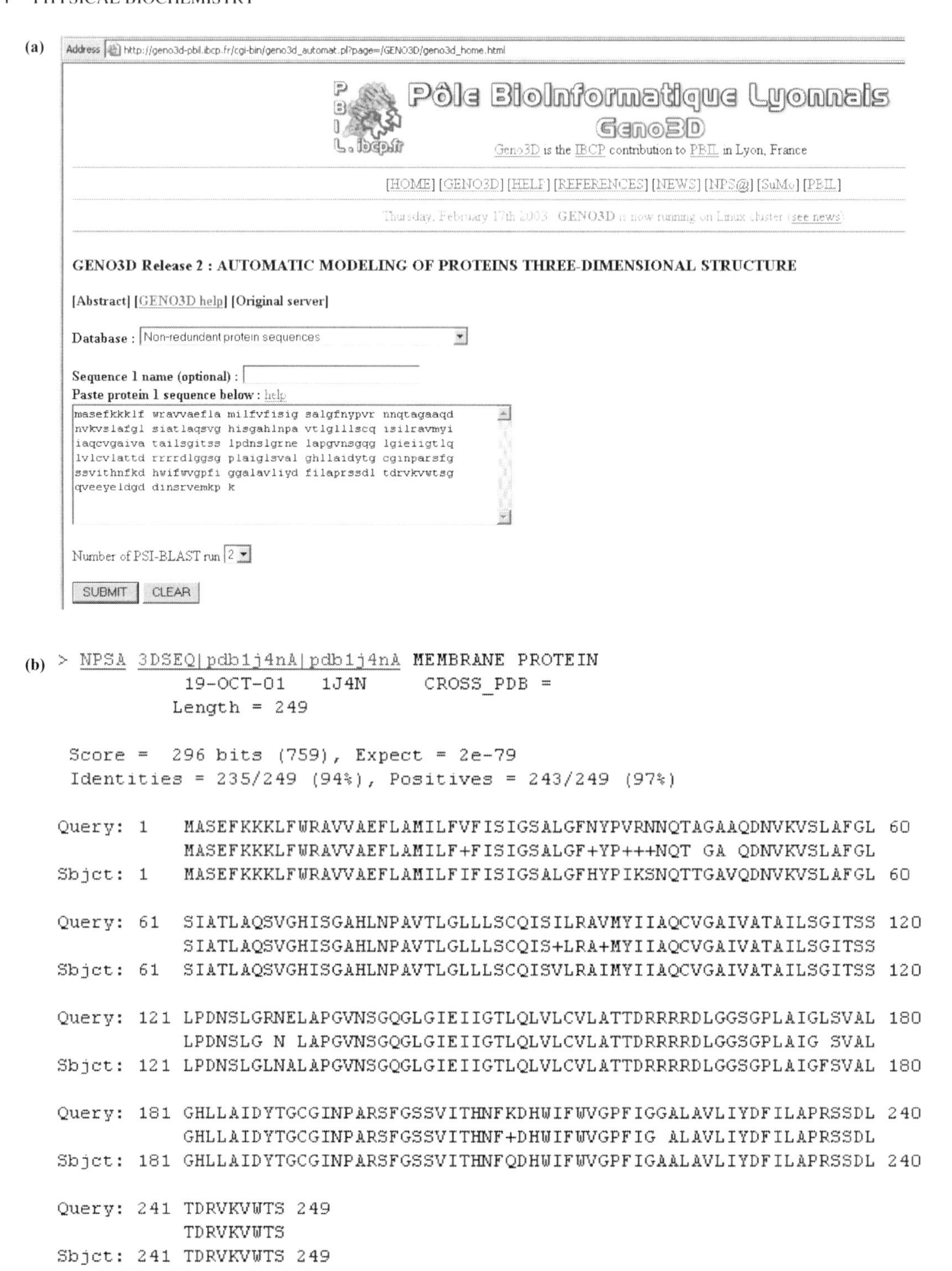

Figure 9.33. On-line homology modelling with Geno3D. (a) The sequence of canine aquaporin is entered in the sequence box and this is submitted for PSI-BLAST. (b) A listing of alignments is returned. The user selects preferred template, in this case human aquaporin 1 which shows 94% identity with the query sequence. (c) Three models were created using a structure human aquaporin 1 (PDB ID 1IH5) determined by electron diffraction as a template. These were returned by e-mail and viewed with PyMOL. Models have rmsd values in the range 0.52–1.05 Å compared to the template. (d) Plot of deviation along the chain confirms that the bulk of the deviation in structure is at the N- and C-termini.

(c)

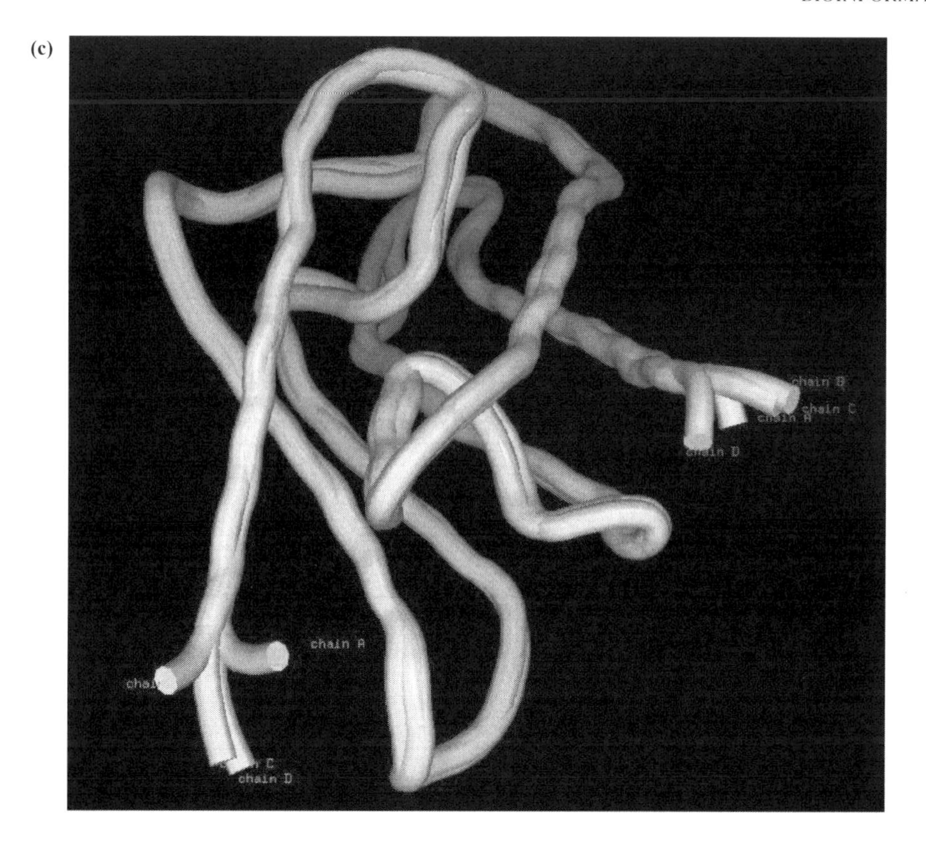

(d)

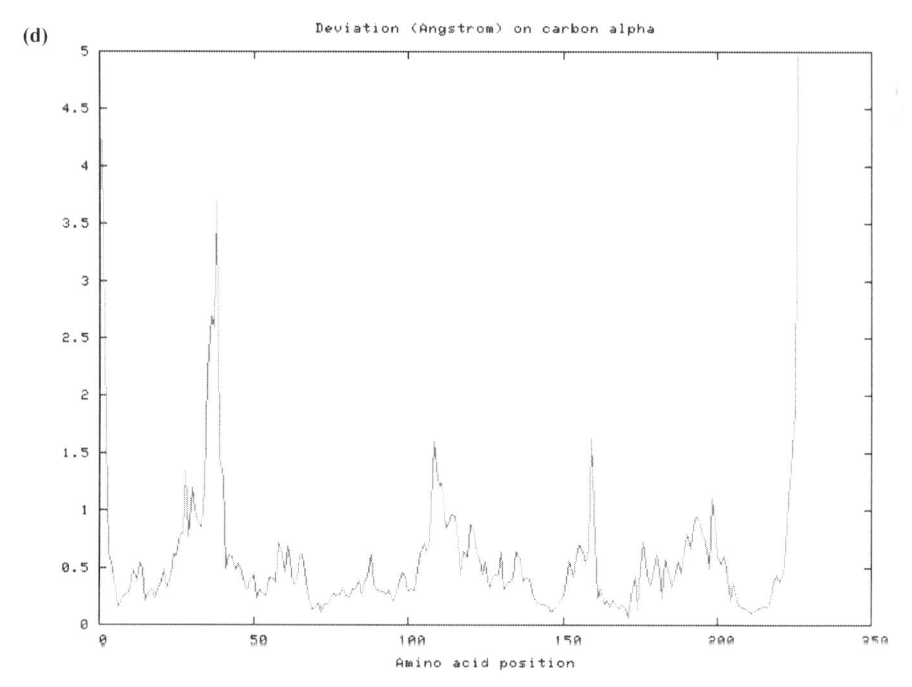

Figure 9.33. (Continued)

9.6.3 Applications of Homology Modelling to Drug Discovery

A major application of homology modelling is in *rational drug design* to create new families of drugs based on structure-function knowledge rather than synthesizing and screening random variants of molecules known to be bioactive. This involves docking variants of drug structures into binding sites of protein models *in silico* with the objective of selecting candidate structures for synthesis. Not alone is a model of the target protein required but the effect of interaction of the ligand on the protein structure must also be modelled. Commercial programs for accomplishing this are available at the CCDC including *GOLD* (protein-ligand docking), *SuperStar* (predicts protein-ligand interactions using experimental data) and *Relibase* (searches for protein-ligand complexes in PDB and proprietary databases). A program called *AutoDock* written originally by David Goodsell and Arthur Olson of the Scripps Institute is downloadable from the *Software* section of the PDB and models docking of flexible ligands to proteins.

REFERENCES

General reading

Lesk, A.M. (2008) *Introduction to Bioinformatics*, 3rd edn, Oxford University Press, Oxford, UK. An excellent general introduction to Bioinformatics focusing on structural bioinformatics in particular.

Barnes, M.R. (2007) *Bioinformatics for Geneticists: A Bioinformatics Primer for the Analysis of Genetic Data*, 2nd edn, John Wiley & Sons, Ltd, Chichester, UK. A beginner's introduction to bioinformatics emphasizing its role in genetics research.

Claverie, J.-M. and Notredame, C. (2003) *Bioinformatics for Dummies*, John Wiley & Sons, Ltd, Chichester, UK. An excellent practical guide to using internet-based resources in bioinformatics.

Aquaporins

Sui, H., Han, B.-G., Lee, J.K. *et al.* (2001) Structural basis of water-specific transport through the AQP1 water channel. *Nature*, **414**, 872–8.

Database mining

Ellegren, H. (2004) Microsatellites: Simple sequences with complex evolution. *Nature Reviews Genetics*, **5**, 435–45. A review of the origins and uses of microsatellite sequences.

Gierke, P., Zhao, C., Brackmann, M. *et al.* (2004) Expression analysis of members of the neuronal calcium sensor protein family:Combining bioinformatics and Western blot analysis. *Biochem. Biophys. Res. Commun.*, **323**, 38–43. A good example of database mining.

Medique, C. and Mozer, I. (2007) Annotation, comparison and databases for hundreds of bacterial genomes. *Research in Microbiology*, **158**, 724–36. A review of bacterial genome databases.

UniProtKB

The UniProt consortium (2007) The universal protein resource (UniProt). *Nucleic Acids Research*, **35**, D193–7.

The UniProt consortium (2008) The universal protein resource (UniProt). *Nucleic Acids Research*, **36**, D190–5. These review the status of UniProt.

CATH

Orengo, C.A., Michie, A.D., Jones, D.T. *et al.* (1997) CATH: A hierarchic classification of protein structures. *Structure*, **5**, 1093–108.

Homology modelling

Mocz, G. and Gibbons, I.R. (2001) Model for the motor component of dynein heavy chain based on homology to the AAA family of oligomeric ATPases. *Structure*, **9**, 93–103. An excellent example of homology modeling.

Sequence alignments

Fleishman, S.J., Harrington, S.E., Enosh, A., Halperin, D., Tate, C.G., Ben-Tal, N. (2006) Quasi-symmetry in the cryo-EM structure of EmrE provides the key to modeling it's transmembrane domain. *J. Mol. Biol.* **364**, 54–67. An good example of use of sequence alignments.

Web sites

Berkeley University, USA: http://scop.berkeley.edu/data/scop.b.html.

Boston University, USA: http://bmerc-www.bu.edu/psa/.

Cambridge Crystallographic Data Centre: http://www.ccdc.cam.ac.uk/.

Columbia University, USA: http://cubic.bioc.columbia.edu/pp/submitexp.html#top.

DNA database of Japan (DDBJ): http://www.ddbj.nig.ac.jp/.

EBI: http://www.ebi.ac.uk/clustalw/index.html; http://www.ebi.ac.uk/goldman-srv/pandit/.

European Bioinformatics Institute (EBI): http://www.ebi.ac.uk/.

European Bioinformatics Institute (EBI): http://www.ebi.ac.uk/genomes/index.html.

ExPASy: http://us.expasy.org/prosite/; http://us.expasy.org/spdbv/; http://swissmodel.expasy.org/.

Fungal Genome Resource: http://gene.genetics.uga.edu/.

German Cancer Research Centre: http://www.dkfz-heidelberg.de/spec/pdb2mgif/.

Institute of Genome Research (TIGR): http://www.tigr.org/htdig/.

Lawrence Livermore Natl. Laboratory, USA: ftp://www-structure.llnl.gov/ProcheckNT/.

MDL: http://www.mdli.com/products/framework/chemscape/.

Munich Information Centre for Protein Sequences: http://mips.gsf.de/services/genomes.

Munich Institute for Protein Sequences: http://mips.gsf.de/.

National Center for Biotechnology Information (NCBI): http://www.ncbi.nih.gov/.

National Center for Biotechnology Information (NCBI): http://www.ncbi.nih.gov/Genomes/index.html.

NCBI software page: http://www.bernstein-plus-sons.com/software/rasmol/; http://www.geneinfinity.org/rastop/; http://pymol.sourceforge.net/; http://www.cmpharm.ucsf.edu/~walther/webmol.html; http://www.ncbi.nih.gov/entrez/query.fcgi?db=unigene.

NCBI: http://www.ncbi.nlm.nih.gov/BLAST/.

PBIL-IBCP, Lyons, France: http://npsa-pbil.ibcp.fr/cgi-bin/npsaautomat.pl?page=npsagor4.html; http://pbil.ibcp.fr/htm/index.php?page=pbilindex.html.

PIR: http://pir.georgetown.edu/pir/www/search/fasta.html; http://pir.georgetown.edu/pirsf/; http://pir.georgetown.edu/iproclass/.

Protein Information Resource (PIR): http://pir.georgetown.edu/pir/www/search/genome.html.

Protein Information Resource (PIR)-international UniProt: http://pir.georgetown.edu; http://www.pir.uniprot.org/.

Research Collaboratory for Structural Bioinformatics (RCSB): http://www.rcsb.org.

Sanger Institute, UK: http://www.sanger.ac.uk/Software/Pfam/.

Swiss Institute of Bioinformatics(ExPASy): http://us.expasy.org/.

The Genome Web: http://www.cbi.pku.edu.cn/mirror/GenomeWeb/genome-db.html.

The Sanger Centre: http://www.sanger.ac.uk.

University College London, UK: http://www.biochem.ucl.ac.uk/bsm/cath/.

University of Bordeaux, Genolevures: http://cbi.labri.u-bordeaux.fr/Genolevures/.

University of California, San Diego, USA: http://molvis.sdsc.edu/protexpl/frntdoor.htm.

University of California, San Francisco, USA: http://salilab.org/license-server/license.cgi.

University of Namur, Belgium: http://www.fundp.ac.be/urbm/bioinfo/esypred/.

Weizmann Inst. Israel: http://www.ebioinfogen.com/cgi-bin/webframe.pl?http://bioinformatics.weizmann.ac.il/.

Chapter 10
Proteomics

Objectives

After completing this chapter you should be familiar with:

- *The principal physical methods used in proteomics.*
- *What type of questions can proteomics answer?*
- *What is the relationship between proteomics and other '-omic' technologies?*
- *Chemical tagging methodologies and their use to detect subproteomes.*
- *Some methods tell us about abundance while others reveal changes in proteins with no net change in abundance.*

Genetic material contains genes many of which encode proteins. This complement of genes is referred to as the *genome* and can be regarded as a relatively static set of instructions (Figure 10.1). When transcribed, these genes direct synthesis of a dynamic population of mRNA molecules that will vary with the requirements of the cell at any given time. This is referred to as the *transcriptome*. When translated, the population of mRNAs directs synthesis of a population of proteins that can be further varied by post-translational processing such as glycosylation or proteolysis. This is the *proteome*, the total complement of proteins expressed in a cell at any given point in time and under a given set of growth conditions. Like the transcriptome, the proteome is a *dynamic* quantity that can change in response to nutritional status, stress and specialization of the cell or with stage of the cell cycle. *Proteomics* is the study of the proteome using methodologies capable of analyzing large numbers of proteins simultaneously. Three main types of techniques have contributed to development of the field of proteomics; electrophoresis, mass spectrometry and chip technologies. These make possible high-throughput analysis capable of revealing alterations in presence or level of individual proteins. These approaches have contributed to development of subspecialties such as *descriptive* proteomics (cataloguing of proteins), *functional* proteomics (focusing on the dynamic state of the proteome) and *interaction* proteomics (exploring protein-protein interactions). This chapter describes these technologies and some of their current applications. The reader should note that some common chemical tools/approaches are used across these methodologies (e.g. use of reducing and alkylating agents; isotope/fluor-labelling; reactions of functional groups with groups on stationary phase beads). Traditional proteomics aimed to catalogue and quantify relative abundance of proteins but increasingly interest is shifting to exploring differences in the properties of proteins associated with modest structural changes (e.g. phosphorylation, glycosylation and oxidation). Often, these changes do not alter the relative abundance of target proteins. It should also be remembered that the availability of large-scale sequence data from genome projects (Chapter 9) is an essential factor for automatic identification of specific proteins within the proteome.

10.1 ELECTROPHORESIS IN PROTEOMICS

The theory and practice of electrophoresis has already been described (Chapter 5). In the 1970s Patrick H. O'Farrell (then at the University of Colorado) combined isoelectric focusing (IEF) in one dimension with SDS PAGE in a second dimension to separate protein mixtures into individual spots (Sections 5.3 and 5.5). This technique is called *two-dimensional SDS PAGE* (2-D SDS PAGE). However, this approach required preliminary formation of a *soluble* pH gradient that tended to migrate during the experiment towards the cathode ('cathodic drift'). It also proved difficult to achieve reproducible pH gradients. For these reasons, 2-D SDS PAGE had limited robustness for general use. As mentioned earlier (Section 5.5.1), in the 1980s a method was developed for *immobilizing* pH gradients on solid supports such as plastic strips. This made 2-D SDS PAGE a reliable and reproducible proposition and it is now the principal electrophoresis format used in modern proteomics (Figure 10.1). Advantages of this method include

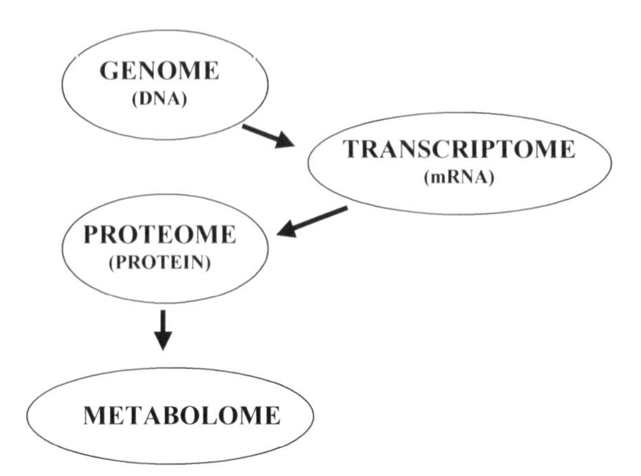

Figure 10.1. The *genome* of most organisms consists of DNA. This is transcribed into the *transcriptome* which varies with cell type and metabolic status of cells. The *proteome* is the complement of proteins expressed at any given moment in time. The *metabolome* is the population of small molecules expressed and is a reflection of the metabolic status of the cell

that it makes up to thousands of proteins amenable to examination in a single experiment and that separated proteins are made available for further analysis by procedures such as N-terminal microsequencing, MS or immunoblotting. However, it should be borne in mind that some categories of proteins are often under-represented in 2-D SDS PAGE separations including membrane, low abundance and extreme pI proteins. Except for the simplest organism, it is therefore not yet possible to separate whole proteomes in a single experiment. For this reason, we frequently concentrate instead on *subproteomes*. These can be selected in various ways including structural (e.g. organelles) and functional (e.g. affinity-purified, glycoproteins, phosphoproteins etc.). Proteomics has harnessed and developed a wide range of selection or detection methodologies.

10.1.1 Two-Dimensional SDS PAGE

In a simple bacterial cell such as *E. coli* there are thought to be more than 2000 individual proteins while in eukaryotes the number may be as high as 100 000. Each of these proteins has a distinct set of physical and structural properties that facilitates their separation in electrophoretic systems. Although electrophoresis gives high-resolution separation, it is usually difficult to distinguish more than 100 or so individual species in a single experiment such as SDS PAGE or nondenaturing electrophoresis. For this reason, most electrophoretic separations are performed on samples that have been previously purified in some way (e.g. by chromatography; Chapter 2). We have previously seen how interfacing of two high-resolution techniques is possible with MS analysis (Chapter 4) allowing greater resolution of complex mixtures than is possible with single techniques alone. Combination of the electrophoretic procedures of IEF and SDS PAGE in 2-D SDS PAGE facilitates virtually complete separation of individual polypeptides even from crude cell extracts.

10.1.2 Basis of 2-D SDS PAGE

In current usage, IEF separation is carried out in commercially-available or laboratory-prepared gel strips containing immobilized pH gradients (IPG; Figure 10.2). IEF strips have relatively large loading capacity (100–200 μg) and result in highly reproducible separation of proteins. It is possible to prepare strips in the laboratory so as to select a specific pH range and commercial strips are also available in a wide range of pH intervals and even covering very narrow increments. The usual approach is to carry out preliminary separations across a broad pH range and then to narrow the range so as to improve resolution of individual spots. After focusing, the IEF strip is placed along the top of an SDS PAGE gel and overlaid to hold it in place (Figure 10.3). SDS PAGE electrophoresis is then performed orthogonally (i.e. at 90°) to the axis of the IEF dimension. When the SDS PAGE gel is stained (e.g. silver staining), individual proteins are visible as spots on the gel. Proteins separate in the first dimension based on differences in their pI values whilst, in the second dimension, they separate based mainly due to differences in their molecular mass. It should be noted that, since both dimensions are usually carried out under denaturing conditions, separation of *individual polypeptides* is achieved in this experiment (Example 10.1).

10.1.3 Equipment Used in 2-D SDS PAGE

Commercially available IEF strips containing an immobilized pH gradient are available in small, intermediate or large formats that match the dimensions of SDS PAGE apparatuses. The larger the format, the more spots are resolvable but the slower the electrophoretic separation and the more protein is required. Thus, small format gels are often used initially to find the optimum pH range and larger format gels may be used to optimize spot resolution. Strips are fitted into a commercial IEF apparatus with a built-in power supply which facilitates handling of multiple samples under fixed temperature conditions (Figure 10.3) and IEF is performed as described earlier (Section 5.5.2). IEF strips are equilibrated in SDS PAGE sample buffer and laid across the top of the stacking gel of the SDS PAGE system. The glass plate used for the SDS PAGE gel is specially shaped to provide a convenient trough into which the IEF strip may be placed. The IEF strip is overlaid with warm agarose that

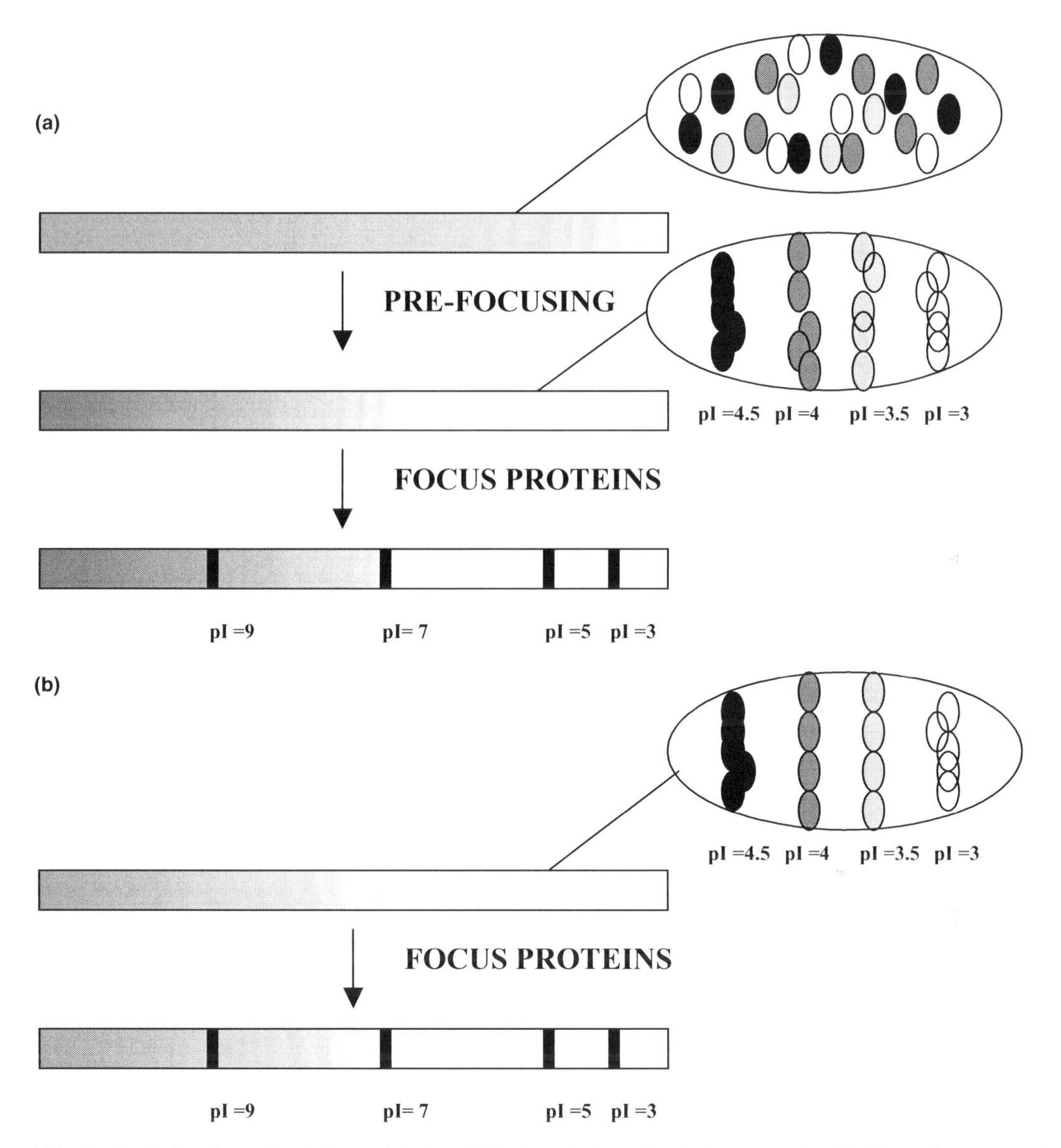

Figure 10.2. Isoelectric focusing with (a) free and (b) immobilized ampholytes. Ampholytes act as local buffers. It is necessary first to prefocus free ampholytes forming a soluble pH gradient. If ampholytes are covalently attached to a gel network on a plastic strip in order of increasing pI, then the gradient is immobilized. In practice a polyacrylamide gel is immobilized on plastic strips and is swollen by addition of sample protein solution in 8 M urea

sets as it cools, fixing the strip in place. While vertical SDS PAGE gels are most commonly used, horizontal gels are also possible as described in Prof A. Görg's on-line manual listed in the bibliography. Flat-bed IEF gels may be treated in the same manner as gel strips except that individual tracks need to be excised with a sharp scalpel and treated as described above.

10.1.4 Analysis of Cell Proteins

2-D SDS PAGE is based on the reasonable supposition that it is unlikely that any two polypeptides will have *exactly the same* pI and molecular mass. Experimentally, proteins that appear as a sharp band on either IEF or SDS PAGE will appear as *spots* in 2-D SDS PAGE analysis. In practice, it is sometimes difficult to distinguish closely migrating

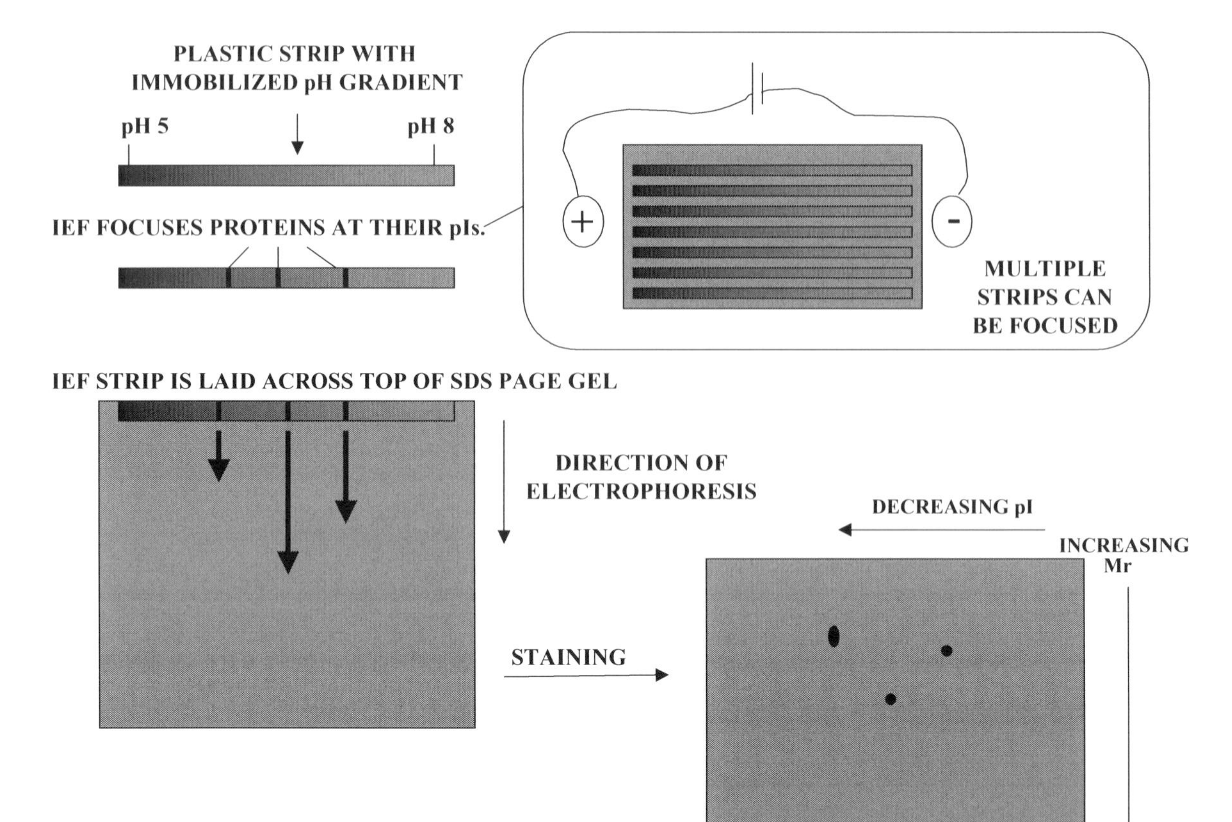

Figure 10.3. 2-D SDS PAGE. Proteins are focused by isoelectric focusing (IEF) in an immobilized pH gradient on a plastic strip. In this case the pH range is 5–8 but other ranges are available. The strip is then soaked in SDS buffer and applied to an SDS PAGE gel. After staining, proteins appear as 'spots'

spots completely but this technique certainly resolves the majority of polypeptides. Because of this, it has been used widely for comparison of complex protein mixtures from crude cell extracts. For example, it is possible to compare extracts of deletion-mutant bacteria to those of corresponding wild-type cells and to identify individual spots absent in the mutant though present in the wild-type. It is also possible to look at a single cell-type at different stages of the cell cycle to identify polypeptides only expressed at particular stages of the cycle.

A feature of 2-D SDS PAGE analysis is that proteins are detectable across three or four orders of magnitude in a single mixture. In order to access this quantitative information *image analysis* systems are used. These convert the image of the stained gel into digital information suitable for storage and comparison of separations (Figure 10.4). In practice, it is found that some proteins are expressed very abundantly in cells giving large, densely-staining spots while other proteins may be barely detectable, even with silver staining. For meaningful results, it is usually necessary to run replicate gels and to use these to generate an 'average gel' for test and control separations. These separations are inspected carefully to identify nonsymmetric, overlapping or saturated spots. In comparing test to controls, either commercially available internal standards or spots common to the two separations can be used to adjust for loading differences. Alternatively, prior to separation, test and control mixtures can be labelled with distinct fluors (Chapter 3). These are then mixed and run on a single gel in a process called *difference gel electrophoresis* (*DIGE*) (Figure 10.5). Excitation at different wavelengths allows us to distinguish between test and control samples and fluorescence intensities are used to quantify specific proteins. Because the data from 2-D SDS PAGE are digitized by image analysis systems, they can be stored electronically and used to create a *database* for specific cell types. These databases annotate spots with their (Mr, pI) and identity (if known). Such databases can be made more widely available to the international scientific community through the Internet (see link to such databases at: http://biohttp://www.net/detail-364.html).

A key activity in proteomics is identification of individual proteins. Usually, spots of interest can be excised, digested

Example 10.1 Use of 2-D SDS PAGE to identify proteins affected by fluoride toxicity in rat kidney

Fluoride is often added in small quantities to drinking water to help prevent tooth decay. However, in larger doses, fluoride is known to be toxic to cells. Kidney is the main organ through which fluoride is detoxified and extensive fluoride intoxication is know to cause lesions in kidney. These workers exposed rats to a reasonably high dose of fluoride and then used 2-D SDS PAGE combined with peptide mapping and MS methods to identify changes in levels of proteins. After 2-D SDS PAGE separations, gels were stained with a dye called Coomassie Brilliant Blue (more sensitive staining is possible with silver). The patterns obtained are shown below.

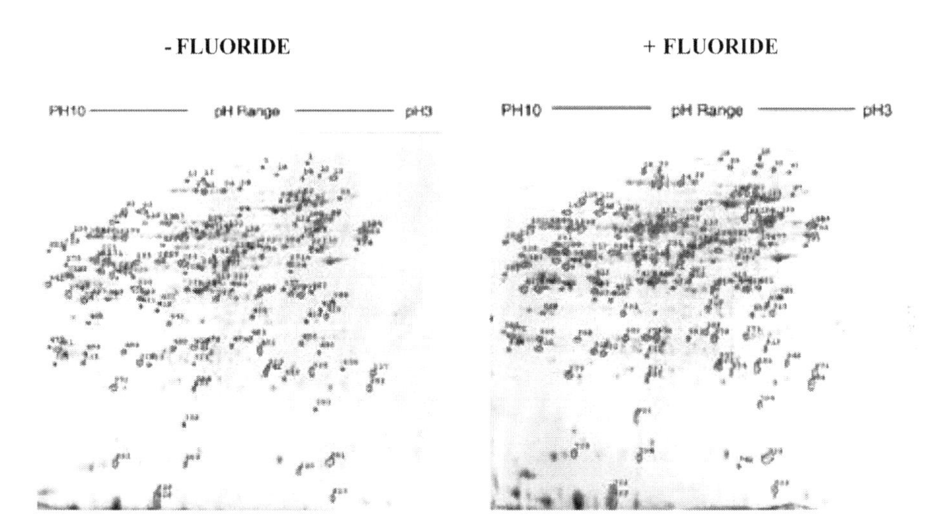

Spots were quantified using image analysis software and it was discovered that 141 spots were at least twofold up-regulated, 8 decreased at least twofold and 280 were unchanged. Of these 12 up-regulated and 2-down-regulated proteins were identified by MALDI-TOF MS. Several up-regulated proteins were associated with intermediary metabolism such as the urea cycle, fatty acid and amino acid metabolism.

Reproduced from Xu, H., Hu, L.-S., Chang, M., Jing, L., Zhang, X.-Y. and Li, G.S. (2005) Proteomic analysis of kidney in fluoride-treated rat. *Toxicology Letters*, **160**, 69–75. Copyright © 2005, Elsevier.

with trypsin and tryptic fragments analysed by tandem MS Robotic workstations are now available which largely automate this activity to allow high-throughput analysis of the proteome. In organisms for which genomes are available (e.g. human, yeast, rat, etc.) tryptic fragments are automatically matched to known proteins. Even in species for which full genomes are as yet unavailable, individual sequences sometimes exist in databases or else there is sufficient sequence conservation for a proportion of tryptic peptides to identify a plausible source for the pattern obtained. Another approach to spot identification is to use western blotting with immunodetection of specific spots (Section 5.10.2).

10.1.5 Free Flow Electrophoresis

In introducing electrophoresis (Chapter 5) it was pointed out that most modern electrophoresis procedures are performed in some type of gel matrix. *Free flow electrophoresis* is an exception to this as the electrophoresis is performed in a gel-free context. The principal of free flow electrophoresis is that proteins are allowed to migrate in an electric field across which a continuous laminar flow of buffer solutions is directed (Figure 10.6). The directions of the electric field and buffer flow are orthogonal (i.e. at right angles) to each other. The buffer is collected in the 96 capillaries of a collection plate at the end of a separation chamber, allowing continuous fractionation of complex mixtures. By varying the composition of the separation buffer, three distinct separation principles can be used: IEF (Section 5.7); zone electrophoresis (Section 5.1.2); isotachophoresis (Section 5.9.1). In contrast to most other electrophoresis formats, in free flow electrophoresis sample is loaded *continuously*. This means small samples (~50 μl) can be loaded in a few minutes while large samples (~100 ml) can be loaded in a day. The technique can be used to separate peptides, proteins organelles and even whole cells and it is possible to vary the

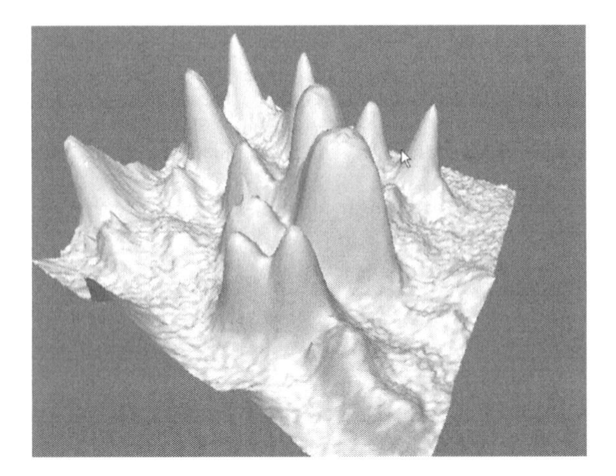

Figure 10.4. Image analysis in 2-D SDS PAGE. Spot intensities can be converted into a digital image from which quantitative information can be obtained. In this representation, individual spots are represented as three dimensional peaks. Care must be taken to ensure spots are not 'saturated' (i.e. with flat tops leading to underestimation of protein), are not overlapping and are symmetric

buffer composition to create denaturing/nondenaturing conditions suitable for the separation principle being used. In the context of proteomics, free flow electrophoresis allows prior fractionation of complex samples which can then be further fractionated or analysed, for example by reversed phase HPLC or tandem MS.

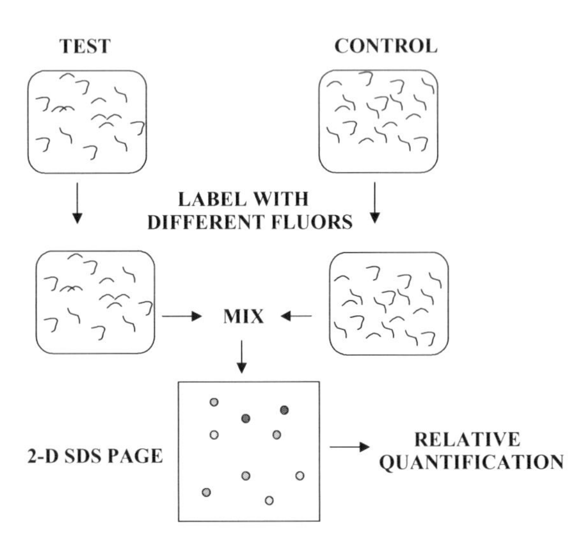

Figure 10.5. Difference gel electrophoresis (DIGE). Two samples (test and control) to be compared are labelled separately with distinct fluorescent dyes (e.g. green and red). They are then mixed and separated on a single 2-D SDS PAGE gel. Intensity of spots shows if they are equal in both samples (yellow), up-regulated (green) or down-regulated (red). A colour version of this figure appears in the plate section of this book

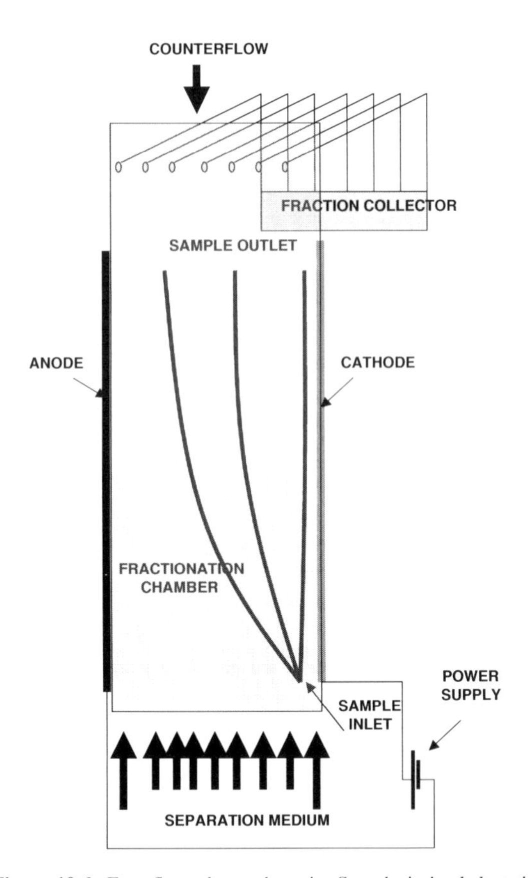

Figure 10.6. Free flow electrophoresis. Sample is loaded at the sample inlet and electrophoreses through the separation buffer. Individual components separate as shown because they are differentially affected by the electric field. They are collected in 96 capillaries at the end of the separation chamber before passing to a fraction collector. By varying the composition of the buffer, sample loading and electrophoresis variables, conditions can be chosen to separate most sample components from each other

10.1.6 Blue Native Gel Electrophoresis

Some 25% of eukaryote proteins are membrane-bound and some of these precipitate under the conditions used in 2-D SDS PAGE leading to their under-representation. Moreover, many of these proteins are associated with other proteins in multi-protein complexes such as, for example, the mitochondrial electron transport chain. Revealing such protein-protein interactions is a key target of proteomics called *interaction proteomics*. Blue native gel electrophoresis separates individual protein complexes from each other in a nondenaturing first dimension followed by separation of individual protein components of complexes under denaturing conditions in the second dimension (Figure 10.7). The first electrophoresis dimension consists of a gradient polyacrylamide gel (e.g. 5–12% acrylamide). Before loading samples in the

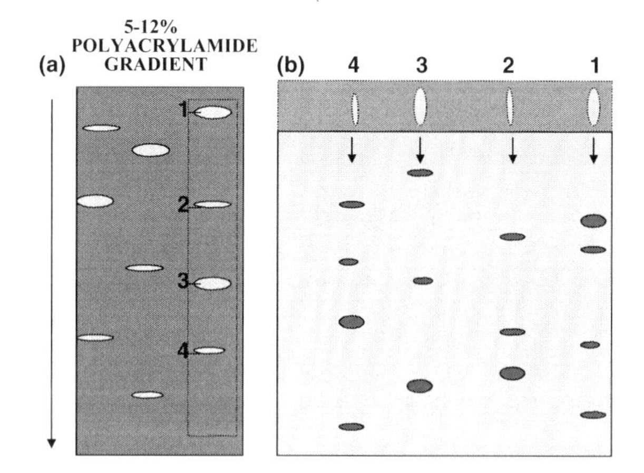

Figure 10.7. Blue native gel electrophoresis. (a) Mixtures of protein complexes are first separated under nondenaturing conditions. (b) A track excised from the first dimension is overlaid with agarose under denaturing conditions and electrophoresed through a second gel, separating protein components of each complex. Four complexes (1–4) are shown

wells of the gel, they are mixed with Coomassie Brilliant Blue G-250 (hence 'blue native gel electrophoresis'). Somewhat like SDS in conventional SDS PAGE, this dye confers negative charges on all proteins but, unlike SDS, the dye is not denaturing. After separation, gel sections containing visible protein complexes may be excised for second dimension analysis. Alternatively, intact electrophoresis lanes may be excised from the gradient gel. These are laid on top of an SDS PAGE gel and overlaid with hot agarose solution containing SDS and 2-mercaptoethanol. When the agarose sets, denaturing electrophoresis is performed which separates the individual proteins in complexes. Separated proteins can be quantified and identified with Coomassie Brilliant Blue or by immunoblotting.

10.1.7 Other Electrophoresis Methods Used in Proteomics

While 2-D SDS PAGE is the principal electrophoresis format used in proteomics, capillary electrophoresis (CE) has also found application. This technique has been described in detail in Section 5.9. Because of the loading limitations of CE, it is largely an analytical technique and lacks the preparative dimension of 2-D SDS PAGE.

10.2 MASS SPECTROMETRY IN PROTEOMICS

The physical basis and some applications of MS have been described (Chapter 4). Availability of ionization methods such as ESI and MALDI has made MS of proteins and peptides an increasingly routine activity. The importance of this was recognized by the award of the 2002 Nobel Prize in Chemistry to John Fenn and Koichi Tanaka for their development of 'soft' ionization methodologies. Use of MS to *identify* electrophoretically-separated proteins has already been described (Section 10.1.4). However, *quantitative* data can be obtained from MS separations using tagging methodologies to compare tryptic peptides from paired samples (e.g. test and control; Example 10.2). Because MS methodologies do not yet allow discrimination of every protein in the proteome, selection methodologies are necessary to select subproteomes for analysis.

10.2.1 Tagging Methodologies Used in MS Proteomics

Digestion with trypsin usually generates dozens of peptides per protein. Thus, the complete human genome could potentially generate ~900 000 individual peptides (Table 10.1). In principle, it ought to be possible to compare this population of peptides from paired samples so as to provide a complete proteomic comparison of relative abundance. In practice, this approach results in the dataset being dominated by a large number of redundant peptides with consequent loss of sensitivity especially for low-abundance proteins. One way of avoiding this is to select from the proteome using *affinity methods*. Thus, *IMAC* (Section 2.4.6) can be used to select for histidine-containing peptides, *thiol exchange chromatography* selects for cysteine-containing peptides and *lectin affinity chromatography* selects glycosylated peptides. Alternatively, proteins can be labelled with a chemically identical *tag*. The samples could be test/control and the labels can be designed so as to be specific for individual amino acid side-chains or other structural features (Figure 10.8). The most common experiment of this type is *stable isotope tagging*. Proteins are labelled with a stable isotope affinity tag that is of two types. One tag is 'light' while the other is 'heavy' (Figure 10.8). Since MS detects differences in mass, identical peptides will appear as pairs separated exactly by the mass difference of the two types of tag. Proteins from both samples are enzymatically digested, labelled and mixed. Samples are automatically separated by C-18 reversed phase HPLC (Section 2.6) coupled to tandem MS (Section 4.3.1). This allows comparison of test to control in a single separation and comparison of corresponding light and heavy tagged peptides allows accurate quantification of relative abundance in the starting proteome. As with 2-D SDS PAGE, some proteins have identical abundance in the test and control samples whilst others are up- or down-regulated.

Example 10.2 Isotope-coded affinity labels in quantification of lysosomal proteins during *in vitro* differentiation

A human cell line HL-60 is widely used as an *in vitro* model for cell differentiation as it can be induced to convert into a morphologically-distinct monocyte-like state on exposure to 12-phorbol 13-myristate acetate (PMA). The differentiated cell has increased cell adherence so it was of interest to study the population of proteins expressed in membrane compartments including microsomes. Control cells and those treated with PMA were used as a source of microsomal fraction. Proteins were tagged either with the isotopically light (d0) or heavy (d8) variant of the ICAT label shown in Figure 10.8, mixed and digested with trypsin. After cation exchange chromatography, labeled peptides were selected by avidin affinity chromatography and separated by reversed phase microcapillary HPLC followed by online electrospray tandem MS. The results are illustrated;

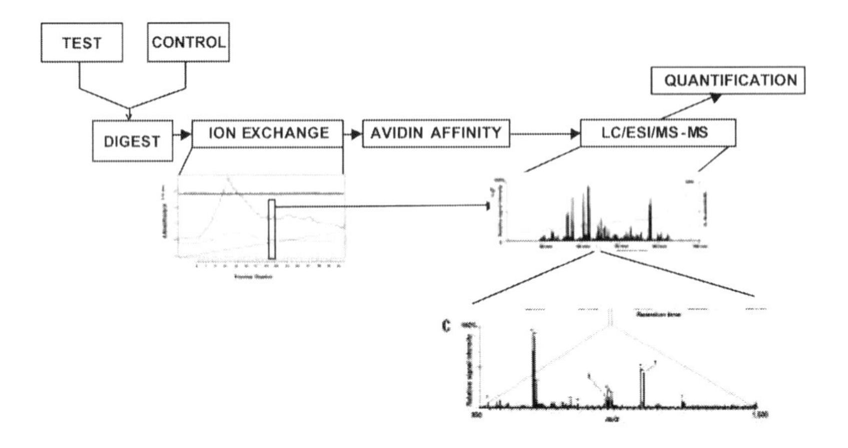

Proteins labeled 1 and 2 in C were identified in the sequence databases using the program SEQUEST and quantified by determination of ratio of light:heavy isotopes with the program XPRESS as follows:

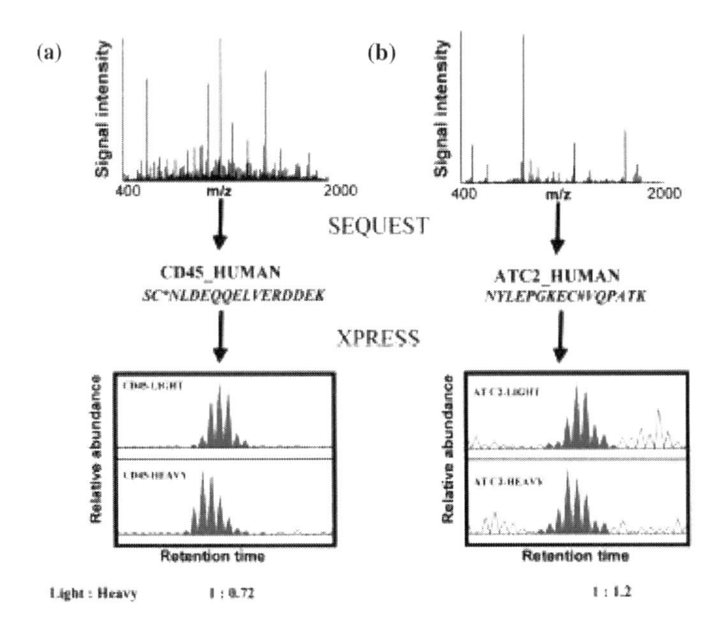

In this way 491 proteins could be identified and their abundance ratios determined.

Reproduced from Han, D.K., Eng, J., Zhou, H. and Aebersold, R. (2001) Quantitative profiling of differentiation-induced microsomal proteins using isotope-coded affinity tags and mass spectrometry. *Nature Biotechnology*, **19**, 946–951. Copyright © 2001, Elsevier.

Table 10.1. Calculated number of tryptic peptides with amino acids targeted by tagging and percentage coverage of the human proteome

Peptides	Total number	% total peptides	% genome Coverage
Total tryptic Peptides[a]	892 584	100	100
N-/C-terminal	52 816	5.9	100
Cys-containing	237 111	26.6	96.1
Met-containing	227 773	25.5	98.9
Trp-containing	157 538	17.6	91.5
His-containing	274 576	30.8	97.3
His/Cys-containing	95 238	10.7	73.2

[a]Peptides end in Arg or Lys but are not followed by Pro. (Adapted from *Zhang et al.*, 2004).

10.2.2 Isotope-Coded Affinity Tagging (ICAT) for Cysteine-Containing Proteins

Stable isotope tagging has most commonly been used to label cysteine side-chains of proteins. This selects approximately 26% of tryptic peptides from the human proteome while covering 96% of total proteins (Table 10.1). As only cysteine-containing peptides are analysed, more proteins can be identified in a given analysis time than if all tryptic peptides were analysed. The traditional tag used in this experiment has three components (Figure 10.8). One part contains a *thiol-specific* chemical group such as iodoacetic acid/iodoacetamide. This is connected to a *linker* that can either contain a 'heavy' isotope (e.g. deuterium) or a 'light' isotope (e.g. hydrogen). The third component is *biotin* which can bind to avidin. Such tags are called *isotope-coded affinity tags* (ICAT) as the isotope identity allows *coding* of the two samples (e.g. as test and control) while biotin allows *affinity selection* for labeled proteins. Samples are exposed to ICATs for long enough to label all –SH groups. The two extracts (one now labelled with the light tag and the other with the heavy) are then mixed and digested with trypsin. Cysteine-containing peptides are selected with avidin and analysed by LC-MS-MS Some limitations associated with the traditional tag shown in Figure 10.8 are that heavy/light peptides may not have identical retention on HPLC, some nontagged peptides can bind to avidin nonspecifically and the biotin moiety makes labelled peptides quite

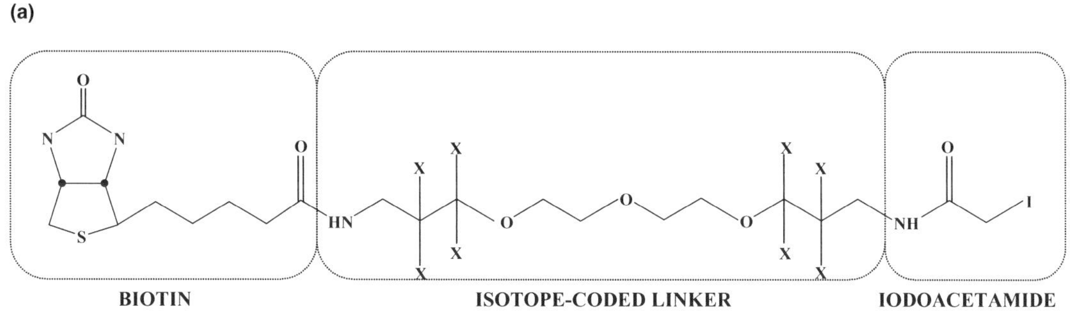

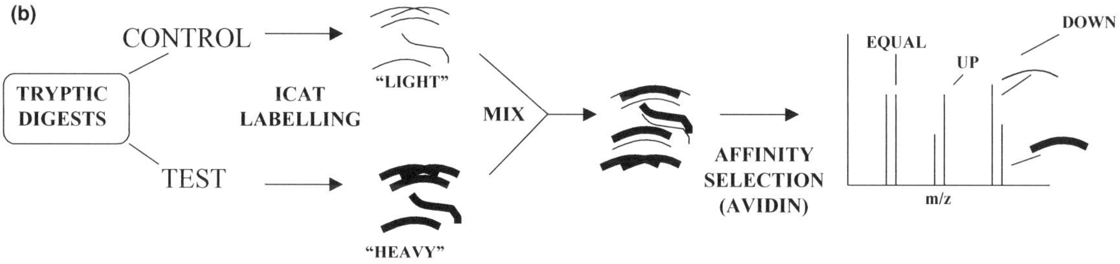

Figure 10.8. Isotope-coded affinity tag (ICAT). (a) ICATs contain moieties for affinity selection (e.g. biotin specifically binds avidin), isotope-coding (e.g. a linker where isotope X can be added) and targeting of protein groups (e.g. a thiol-reactive iodoacetamide). The isotope-coded region can contain light (e.g. hydrogen) or heavy (e.g. deuterium) isotopes at X. (b) Test and control tryptic digests are separately ICAT labelled and mixed. Samples are affinity selected and run on MS. Incremental differences of 8AMU allow matching of corresponding test/control peptides. Simplified examples of a single peptide present in equal amounts, up-regulated in test and down-regulated in test are illustrated. The m/z difference between pairs of peptides is given by the difference in mass of 8X as shown

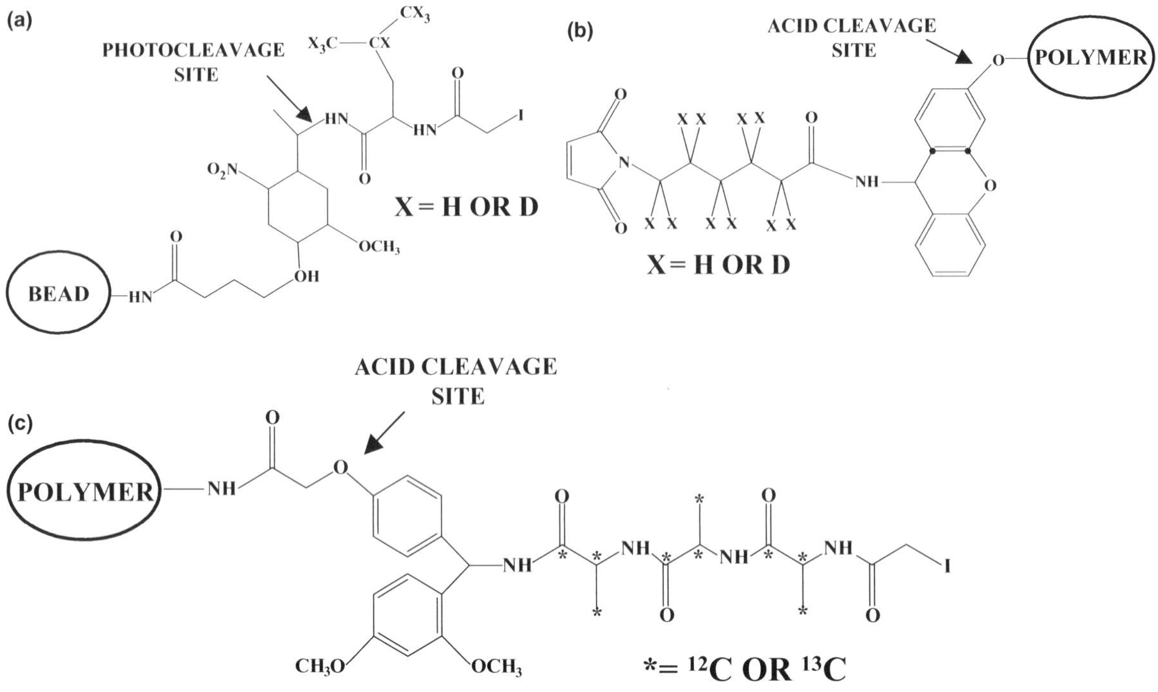

Figure 10.9. Solid phase cysteine tags. These contain groups which react with cysteine thiols as well as a point where the tag can be cleaved from a solid phase for release of captured peptides. (a) Iodoacetamide-based thiol tag with photocleavage site. (b) N-Ethyl maleimide based tag with acid-labile site. (c) Iodoacetamide-based tag with acid-labile site

hydrophobic with the result that they elute in a narrow part of the LC chromatogram. More recent variants of this approach have employed a solid phase capture and release process for trapping cysteine-containing peptides (Figure 10.9). Some of these tags have cleavable groups in the linker which facilitate release of target peptides after other peptides have been efficiently washed away. Acid treatment shortens tags before LC-MS-MS resulting in reduced collision energy loss during ionization with associated improvement in sensitivity. However, acid treatment (50% trifluoroacetic acid) may be so severe as to affect acid-labile post-translational modifications. A selection of smaller cysteine-specific tags is shown in Figure 10.10.

10.2.3 Tagging of N- and C-Termini

Other features of protein structure have been selected for tagging (Table 10.1). The N-terminus can be labelled with N-acetoxysuccinimide or succinic anhydride that are available in 'light' or deuterated forms (Figure 10.11). Thus a population of peptides arising from tryptic digestion provides many sites for tagging. Tags can also react with other primary amines present in the protein (e.g. lysine and arginine side-chains) which may complicate the peptide pattern somewhat. However, by selecting appropriate reaction conditions, it may be possible to exclude these side-reactions. For example, N-terminal amino groups react 100-fold more strongly with nondeuterated/pentadeuterated phenyl isocyanate (Figure 10.11) than do lysine ε-amino groups at pH 8.0.

Labelling of C-termini can be achieved by converting –COOH groups into methyl esters as shown in Figure 10.12. Methanol containing three hydrogen or deuterium atoms provides the methyl group. As with N-terminal labeling, this reaction will also tag aspartate and glutamate residues. Another widely used C-terminal tagging method that does not label these residues is to introduce ^{18}O into the C-terminal group formed during tryptic digestion (Figure 10.12). One protein mixture is digested in $H_2{}^{18}O$ while the other is digested in H_2O. Either one or two oxygen atoms become incorporated into the C-terminus. It is even possible to achieve this effect by isotope exchange after tryptic digestion. This is done by incubating previously-digested proteins in either H_2O or $H_2{}^{18}O$ in the presence of trypsin. ^{18}O exchanges with ^{16}O under these conditions. Drawbacks to these C-terminal labeling approaches are the high cost of isotope-labeled water and the possible overlap of isotopic distributions of light and heavy forms of peptides due to incomplete incorporation of ^{18}O. N- and C-terminal tagging are not mutually

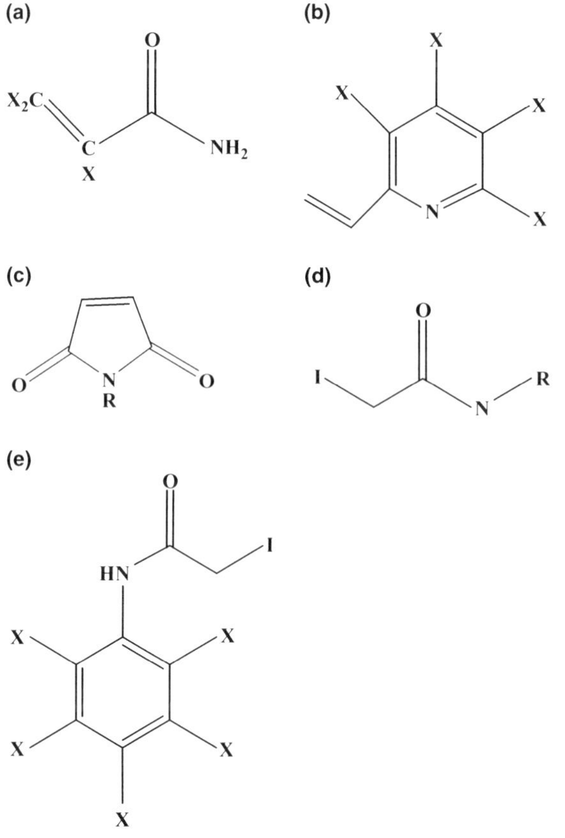

Figure 10.10. Small thiol-specific tags for differential isotope coding (X = hydrogen or deuterium). (a) Acrylamide (b) 2-Vinylpyridine (c) N-substituted maleimides ($R = CX_3$ or CX_3CX_2) (d) N-Substituted iodoacetamides ($R = CX_3CX_2$ or $C(CX_3)_3$) (e) Iodoacetanilide

exclusive so it is possible to label peptides at both N- and C-terminal ends. This ensures that all peptides in a sample can be labeled even if the N- or C-termini of some of them are modified.

10.2.4 Tagging for Tandem MS

The tagging approaches described above (Sections 10.2.1–10.2.3) are mainly aimed at affinity labelling and comparative quantification. It may also be desirable to label peptides so as to improve MS sequencing especially where only small amounts of protein are available (e.g. spots excised from 2-D SDS PAGE). This may be done by introduction of permanent negative or positive charges in such a way as to alter the distribution of a, b or y-ions (Figure 4.19). A negative charge may be introduced to N-terminal ions by *sulphonation* (Figure 10.13). This suppresses the N-terminal fragment ions so that only y-ions are observed, thus simplifying MALDI spectra. It is necessary to modify lysine residues (e.g. by guanidination) prior to sulphonation. A modification of this approach is to guanidinate N-termini in 2-D SDS PAGE gels that can later be tryptic digested and sulphonated. Conversion of lysine to homoarginine means trypsin does not cleave at these sites but this approach nonetheless allows sequencing by detection of y-ions by MALDI MS Phosphorous-containing reagents also introduce positive charge as *phosphonium* groups at the N-terminus using reagents such as phosphines and phosphonium salts (Figure 10.14). This causes preferential formation of N-terminal a- and b-ions which makes sequencing from MALDI fragmentation patterns possible (Figure 4.19).

10.3 CHIP TECHNOLOGIES IN PROTEOMICS

A third innovation feeding into the growing field of proteomics is *chip technology*. High density chips make it possible to immobilize thousands of genes or proteins on a solid surface which can be read by high-throughput instrumentation. This forms the basis of *microarrays* which can be used to screen for gene expression in functional genomics. This technology is now well-advanced and representative arrays are available for certain cell types or groups of proteins. These arrays may represent thousands of individual genes and provide a tool with which it is possible to interrogate the population of mRNAs in a cell or tissue of interest in a single experiment. The readout from the experiment is identification of unchanged, undetected, up- or down-regulated genes in the sample. Based on success with microarrays, there is increasing interest in applying similar technology to proteins. *Protein chips* are increasingly important tools in interaction proteomics.

10.3.1 Microarrays

Microarrays are sometimes also called *DNA arrays*, *DNA chips* or *gene chips*. Typically, a microarray is a glass slide such as a wafer of quartz on which DNA molecules are attached at fixed locations (spots; Figure 10.15). This is achieved by a number of methodologies including printing by a robot, photolithography or inkjet printing. Each spot contains a number of identical molecules of a single DNA varying in length from 20 to hundreds of base pairs. They can originate from genome-wide oligonucleotide libraries of uniform length (25 bp) or else from cDNA libraries. Slightly different results can be obtained from each since oligonucleotide-based arrays provide great specificity as a result of uniform hybridization conditions conferred by

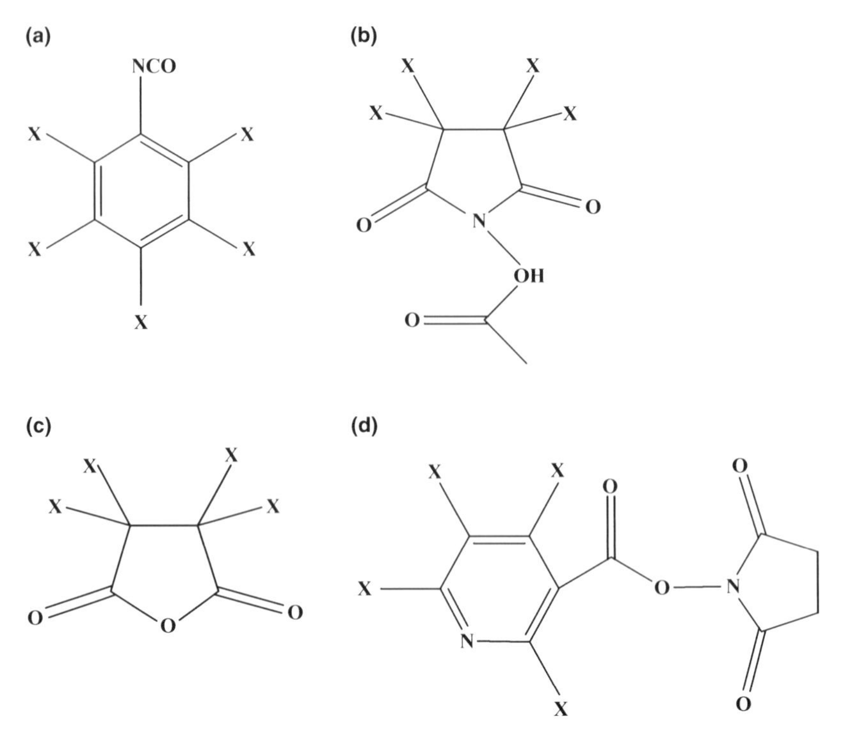

Figure 10.11. Isotope-coded tags for labelling peptide N-termini and lysine ε-amino groups. (X = hydrogen or deuterium. (a) Phenyl isocyanate (b) N-acetoxysuccinimide (c) Succinic anhydride (d) N-nicotinoxyloxysuccinimide

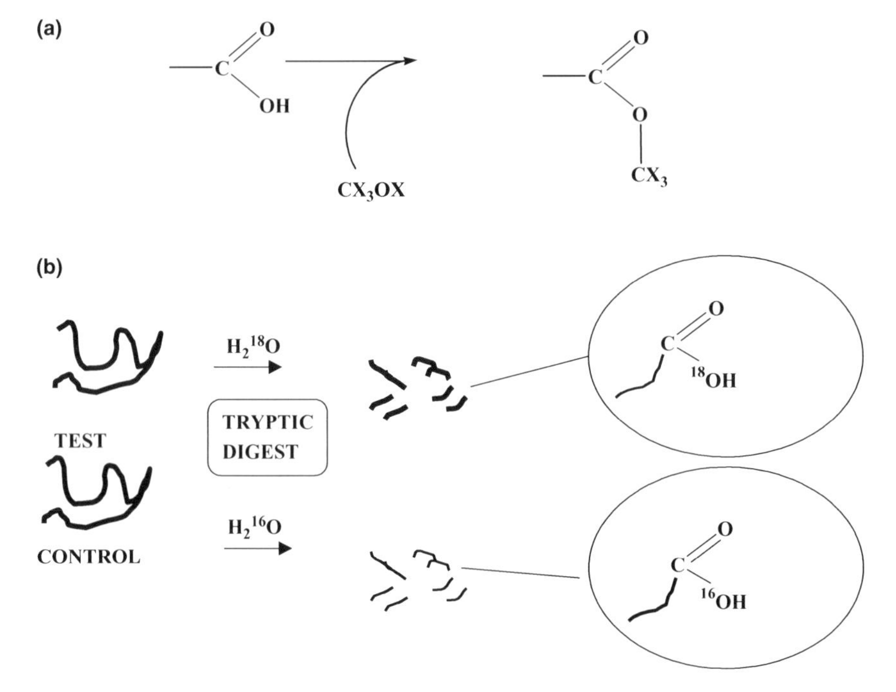

Figure 10.12. Isotope-coded tagging of peptide C-termini. (a) Conversion of –COOH into a methyl ester in methanol (X = hydrogen or tritium). (b) Digestion with trypsin in water with differing oxygen isotopes introduces one or two isotope atoms at the C-termini of peptides

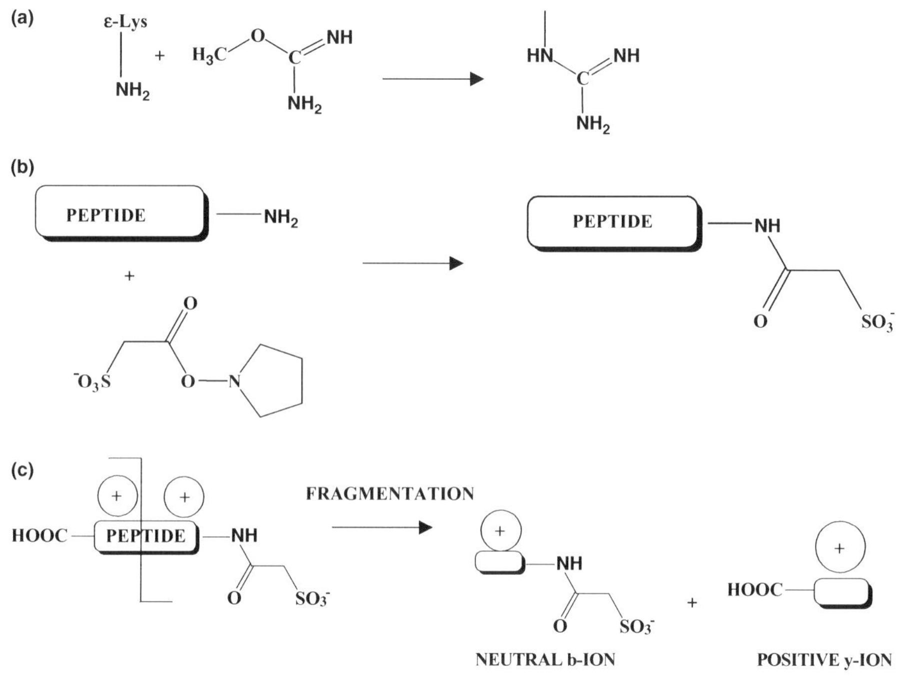

(a) ε-Lys NH2 + H3C–O–C(=NH)NH2 → HN–C(=NH)NH2

(b) PEPTIDE–NH2 + -O3S–CH2–C(=O)–O–N → PEPTIDE–NH–C(=O)–CH2–SO3-

(c) FRAGMENTATION
HOOC–PEPTIDE–NH–C(=O)–SO3- → NEUTRAL b-ION + POSITIVE y-ION

Figure 10.13. N-Terminal sulphonation. (a) Tryptic digests produce C-terminal lysines and arginines. Lysine can be converted to homoarginine by guanidinylation. (b) N-terminal amino groups react with an N-hydroxysuccinimide ester of 3-sulphopropionic acid. (c) C-terminus has positive charge due to guanidinylation. Fragmentation by postsource decay MALDI produces a y-ion from each peptide which is detected and an uncharged b-ion which is not. Mass differences between adjacent y-ions allow calculation of sequence

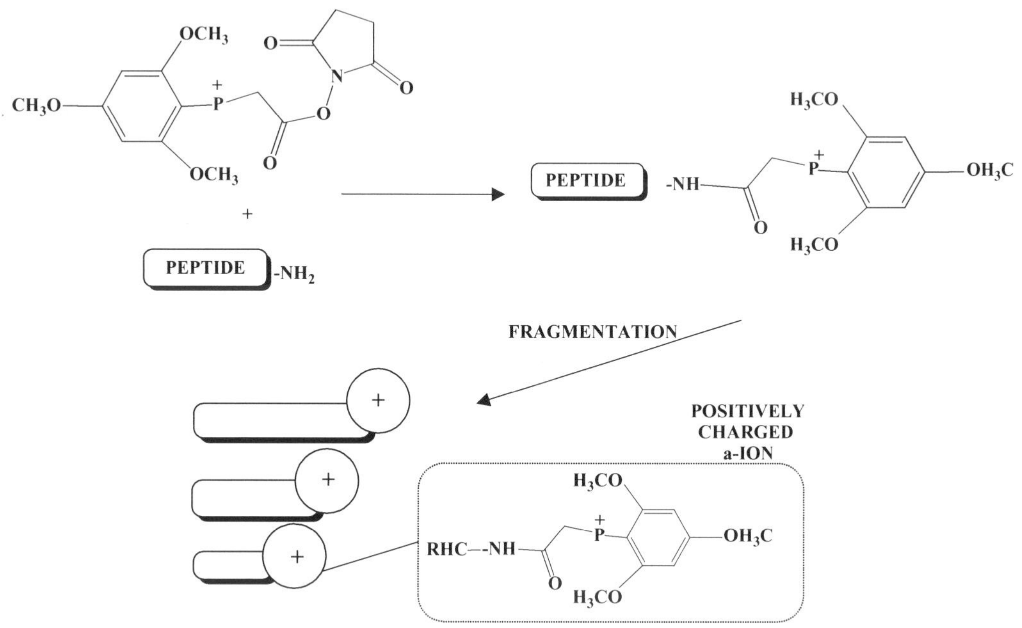

OCH3, CH3O, P+, OCH3, O + PEPTIDE-NH2 → PEPTIDE -NH– C(=O) P+ H3CO OH3C H3CO

FRAGMENTATION

POSITIVELY CHARGED a-ION

RHC–-NH–C(=O)–P+ H3CO OH3C H3CO

Figure 10.14. N-terminal phosphination. Tryptic peptides can be acetylated at the N-terminus with the N-hydroxysuccinimide ester of N-tris(2,4,6-trimethoxyphenyl)phosphine to confer a permanent positive charge on the N-terminus. On fragmentation by MALDI, mainly a-type ions are generated. Incremental mass differences between adjacent a-ions give the amino acid sequence

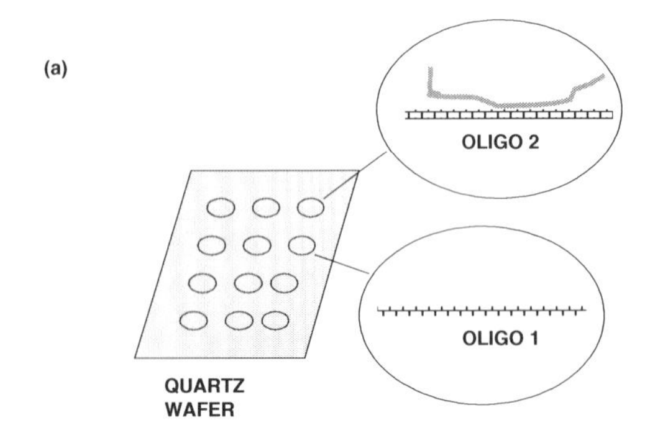

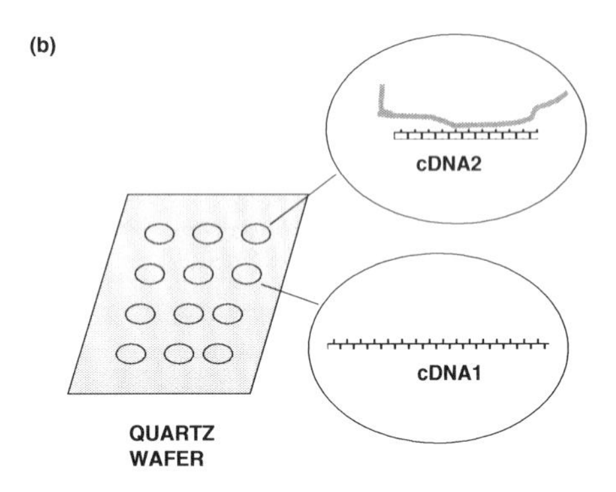

Figure 10.15. DNA Microarrays. DNA molecules are covalently attached at specific spots on a quartz wafer. (a) Oligonucleotide-based arrays consist of DNA of equal length. (b) cDNA arrays contain DNA of variable length. A real microarray contains thousands of spots

identical length. In cases where alternatively spliced transcripts are common (e.g. human mRNA), each transcript is recognized by a unique oligonucleotide. By contrast, such multiple transcripts will hybridize to a single cDNA.

Ideally, each spot represents one gene or one exon of a gene but this is not always possible in practice due to the presence of gene families consisting of closely related sequences. Microarrays of the entire set of ~6000 yeast genes have been available since the late 1990s but most microarrays of human genes contain a subset of the human genome. A common use of microarrays is to measure relative levels of gene expression in two sets of closely related cells (e.g. test and control). This is done by extracting mRNA from cells to give a reflection of its transcriptome. Since mRNA transcription precedes appearance of a new protein, there is a loose correlation between levels of mRNA and its encoded protein. This correlation is not absolute, however, as mRNAs differ greatly in their stability and proteins are turned over at differing individual rates.

The first step is to extract mRNA and to use this directly or else to generate synthesis of a cDNA with reverse transcriptase (Figure 10.16). The mRNA/cDNA population from each cell type is then labelled with a different fluorescent dye (e.g. green for control and red for test). The two populations are washed over the microarray and allowed to hybridize to the DNA printed on the spots. If a gene is preferentially expressed in the test sample, the spot will glow red when illuminated with a laser beam. If a gene is preferentially expressed in the control sample (due to down-regulation), the spot will glow green. If a gene is equally expressed in both test and control, the spot will glow yellow. Image analysis software provides readout of this experiment in terms of spot identification and quantification. Microarrays usually come with built-in blank spots, control spots and at least duplicate spots for each gene. By comparing known levels of genes on a number of microarrays with observed spot intensities for genes under investigation it is possible to generate a *gene expression matrix* for the cells under investigation which can be related to results from other experiments.

10.3.2 Protein Biochips

Although not yet as well-established, it is possible to immobilize proteins in an array format conceptually similar to that of microarrays in such a way that they can recognize and bind protein components of complex biological mixtures (Figure 10.17). One way of doing this is to express individual recombinant proteins as cDNA clones in each spot. An array containing a set of structurally diverse antigens can be used to screen serum of patients with autoimmune disease (e.g. *lupus erythematosus*) allowing identification of autoantibody patterns that may vary from patient to patient and with identity of the autoimmune disease. Other applications that take advantage of the biospecific recognition properties of proteins include protein-ligand and enzyme-substrate arrays. Examples of protein-ligand recognition include an array of ~6000 yeast recombinant proteins in identification of novel calmodulin-binding proteins and an array of 37 000 human recombinant proteins expressed in *Escherichia coli* in which 13 proteins were identified as binding a biotin-labelled peptide (Example 10.3). Enzyme-substrate recognition is exemplified by arrays of protein kinases, phosphatases, peroxidases and restriction enzymes. This complements use of activity-based chemical probes for enzyme classes in MS (Section 10.4.7).

10.3.3 SELDI-TOF MS on Protein Chips

Protein biochips comprise an array of proteins with specific binding properties. An alternative use of protein chips is as a derivatized surface with distinct chromatographic features including hydrophobic, hydrophilic, affinity, reversed phase, ion exchange and IMAC groups (Figure 10.18). This allows fractionation of the proteome so as to overcome limitations of traditional MALDI-TOF MS which can only handle a relatively limited number of proteins in a single analysis.

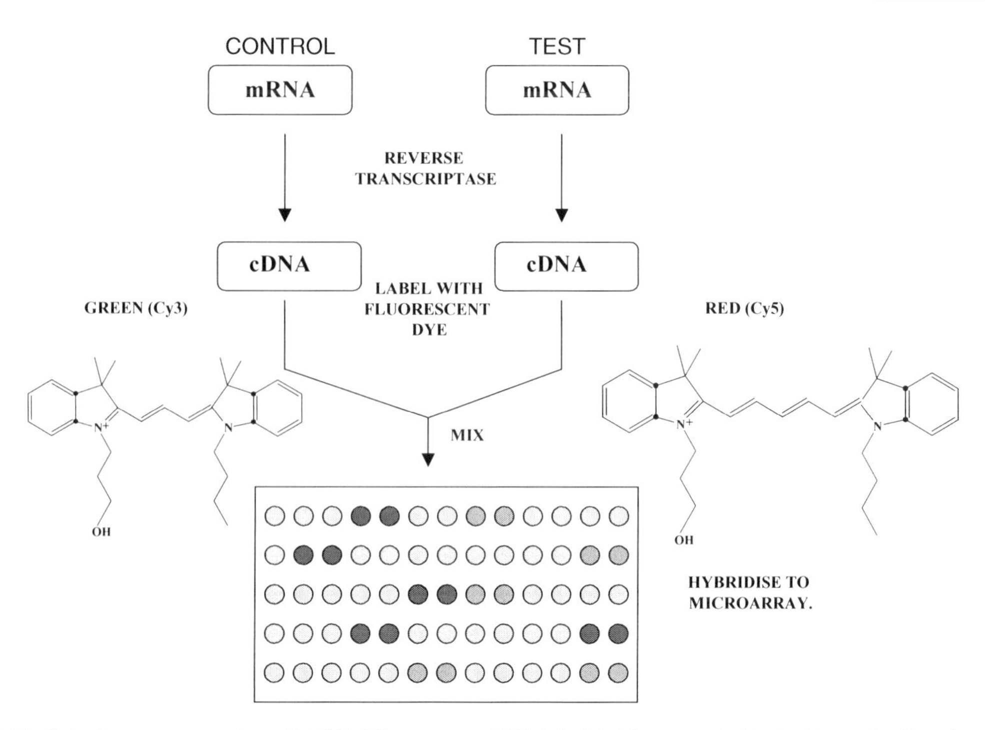

Figure 10.16. Detecting gene expression with DNA Microarrays. mRNA is isolated from matched test and control cells and converted to cDNA by reverse transcriptase. Each cDNA population is tagged with a fluorescent dye (red for test, green for control) and washed over the microarray. cDNA hybridises to its complementary gene. Un-hybridized cDNA is removed by washing. Scanning the microarray with a laser reveals up-regulated (green), down-regulated (red) and equal gene expression (yellow) at each spot of the microarray. Spots are in duplicate and the array contains both positive and negative controls. A colour version of this figure appears in the plate section of this book

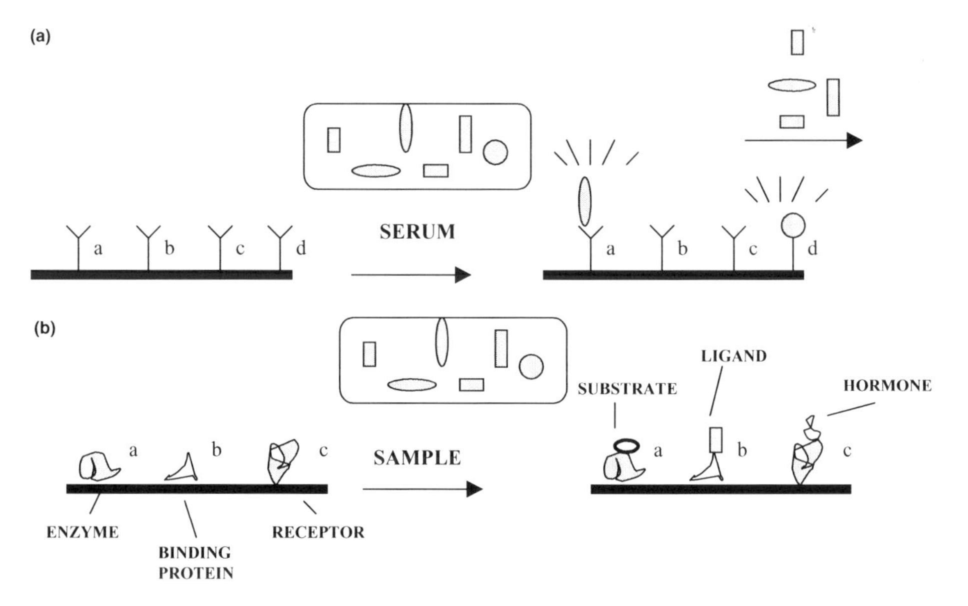

Figure 10.17. Protein arrays. There are two general types of protein arrays. (a) Analytical array for detection of antigens in serum. Multiple antibodies are attached at spots on array. These can bind specific fluorescently labelled antigens in biological fluid (e.g. serum). In this example antigens a and d are detected. (b) Functional protein arrays are based on bioactivity of proteins such as receptor ligand, enzyme-substrate recognition

Example 10.3 Use of activity based probes in profiling rat testis proteomes

Activity-based probes (ABPs) are designed based on known recognition of reactive groups within the probe by the active sites of particularly well-understood enzyme classes (e.g. serine hydrolases). *Nondirected* probes may be able to recognize less well-understood enzyme families and these feature an additional *binding* group. A family or *library* of such probes for serine hydrolases was synthesized based on a sulphonate reactive group with a range of binding groups as shown below:

LABEL/TAG
LINKER
BIOTIN
= R
RG BG
1 2 3 4
RG BG
5 6 7 8
RG BG
9 10 11

These probes were used to analyse testis proteomes. The common biotin moiety allows capture of targeted proteins followed by SDS PAGE. (a) Probes 1–7, (b) Probes 8–11. Pre-boiling of extracts (denoted by Δ) abolishes much of the recognition. Notice that many of the bands recognized are common to several probes and that each probe has its individual specificity.

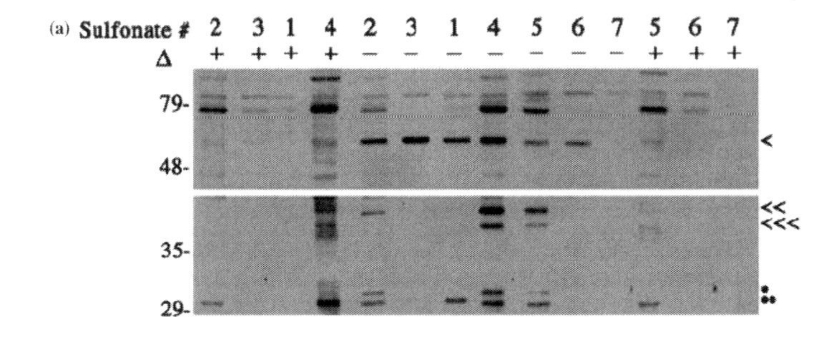

(*Continues*)

(*Continued*)

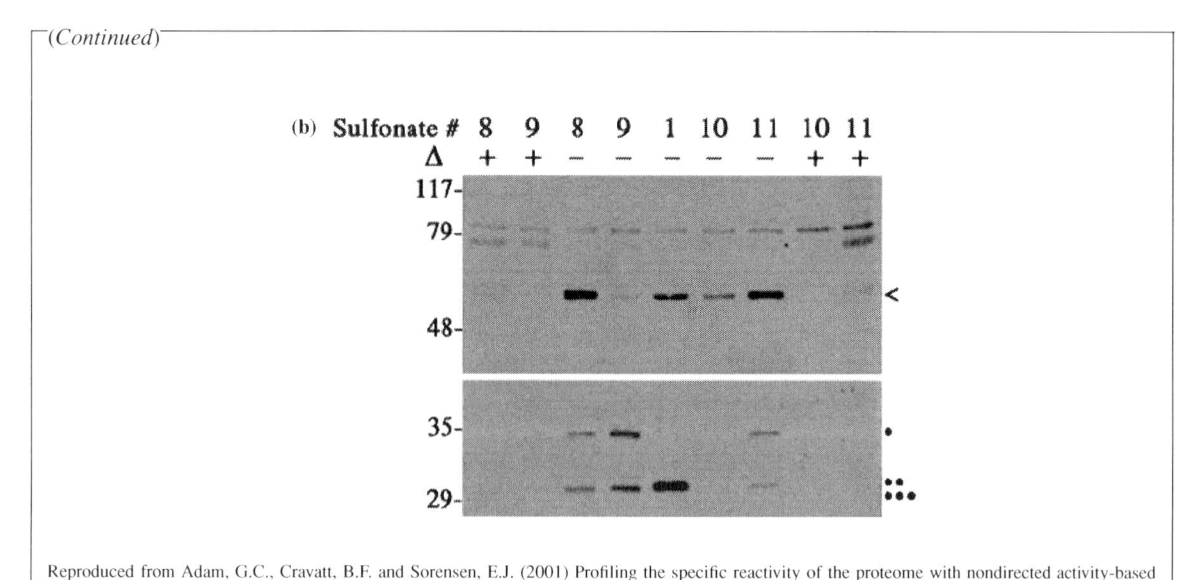

Reproduced from Adam, G.C., Cravatt, B.F. and Sorensen, E.J. (2001) Profiling the specific reactivity of the proteome with nondirected activity-based probes. *Chemistry and Biology*, **8**, 81–95. Copyright © 2001, Elsevier.

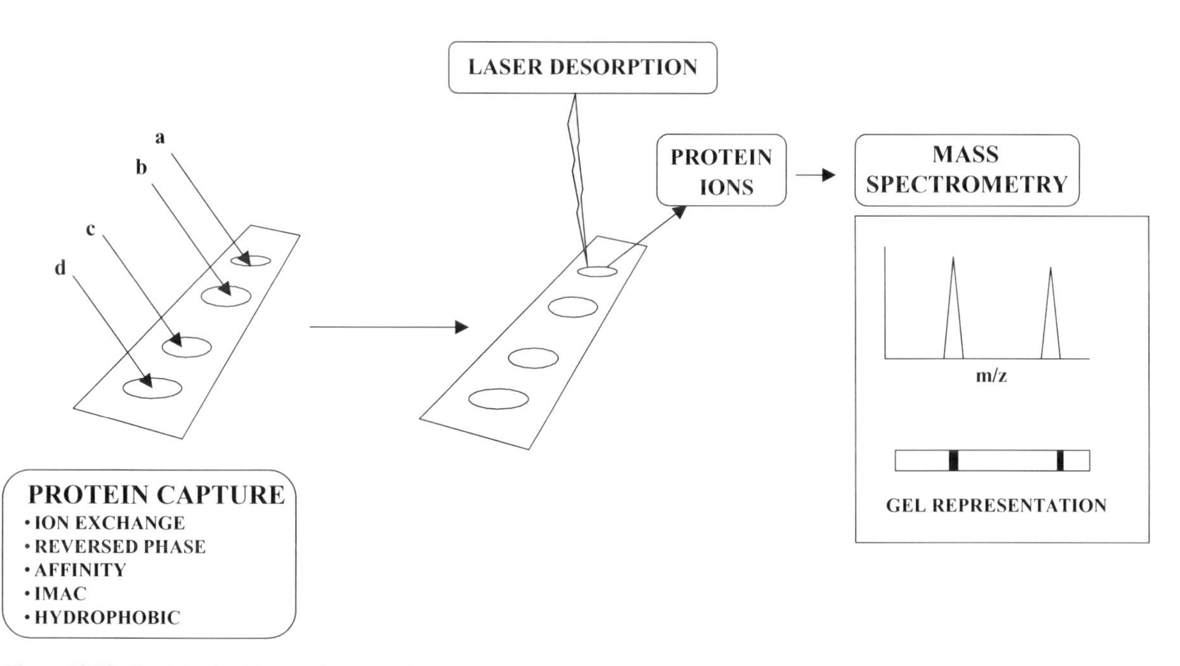

Figure 10.18. SELDI MS. This is distinguished from MALDI MS by the use of protein chips in a range of formats based on chromatograpic interaction. These select subproteomes which can be ionized by laser desorption and analysed by MS. Output of the experiment is presented as an MS spectrum or as if the separation was performed in an electrophoresis gel

This approach has been commercialized (Ciphergen Biosystems Inc., Fremont, California, USA). An energy-absorbing matrix is added to the ProteinChip® so that, when laser energy is applied to the sample, the proteins present become ionized and they may be analysed by MS. The ionization step is called *surface enhanced laser desorption ionization* (SELDI) and detection is in a TOF detector (Section 4.1.4). This enables high-throughput, rapid and sensitive analysis of complex biological mixtures.

10.4 POST-TRANSLATIONAL MODIFICATION PROTEOMICS

In simple bacteria post-translational modifications occur in approximately 25% of gene products. In vertebrates, as many as 3–4 times the number of unique proteins is possible from a given number of genes. This arises from processes such as subunit hybridization and alternative splicing. However, a relatively large number of eukaryotic proteins are also processed by post-translational modification (for examples of modifications see Table 4.3). This provides extensive options for controlling biological properties of proteins allowing greater complexity to be achieved with a given genome. They also provide a means for cells to adapt to changed circumstances such as altered nutrition status or exposure to oxidative stress. Phosphorylation in signal transduction cascades and the conferring of recognition on cell-surface proteins by specific glycosylation provide good examples of the biological importance of post-translational modifications. Elucidation of these structural changes has the potential to give insights into fundamental cellular processes and to events in the proteome.

10.4.1 Proteolysis

Because proteolysis produces smaller products of usually distinct mass and with altered physicochemical properties compared to the substrate protein, these products are readily detectable in 2-D SDS PAGE or MS analyses. *Zymograms* (Section 5.2.4 and Example 5.2) provide a means of identifying the presence of proteases in electrophoresis gels which can have diagnostic value. Affinity probes for specific classes of enzymes have been developed for proteomic analysis (Figure 10.19). Examples

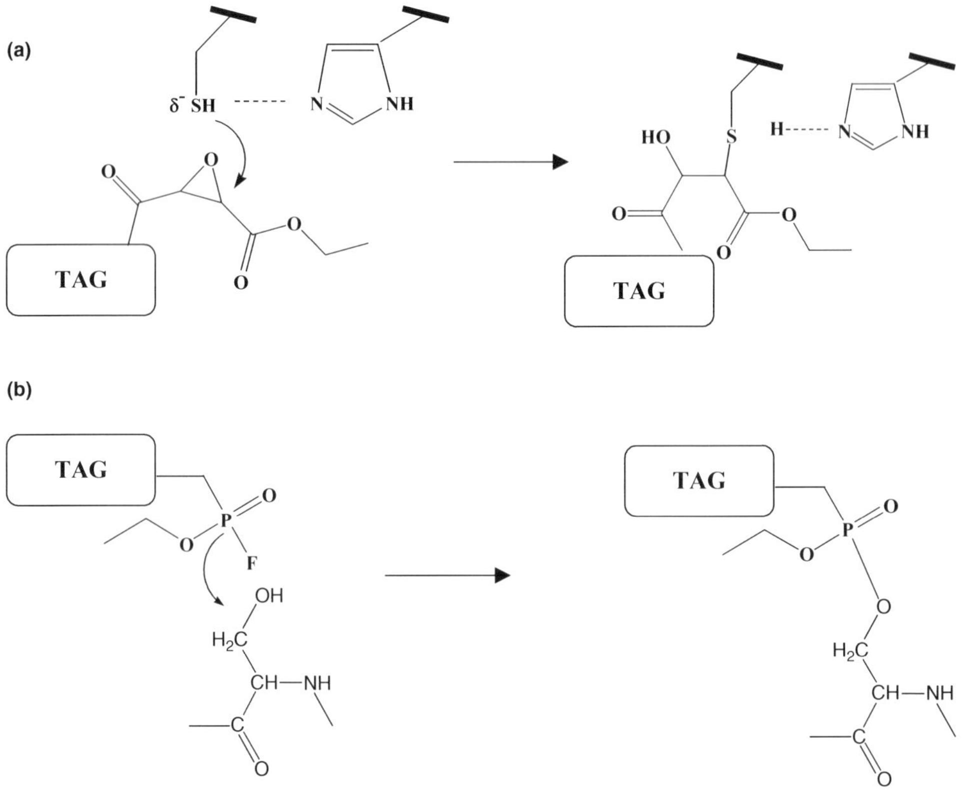

(a)

δ⁻ SH ------ N NH
O O
TAG O
HO S H------ N NH
O O O
TAG

(b)

TAG
O P O
F
OH
H_2C
CH—NH
C
O

TAG
O P O
O
H_2C
CH—NH
C
O

Figure 10.19. Activity-based protein profiling. (a) Cysteine proteases. (b) Serine hydrolases (including serine proteases)

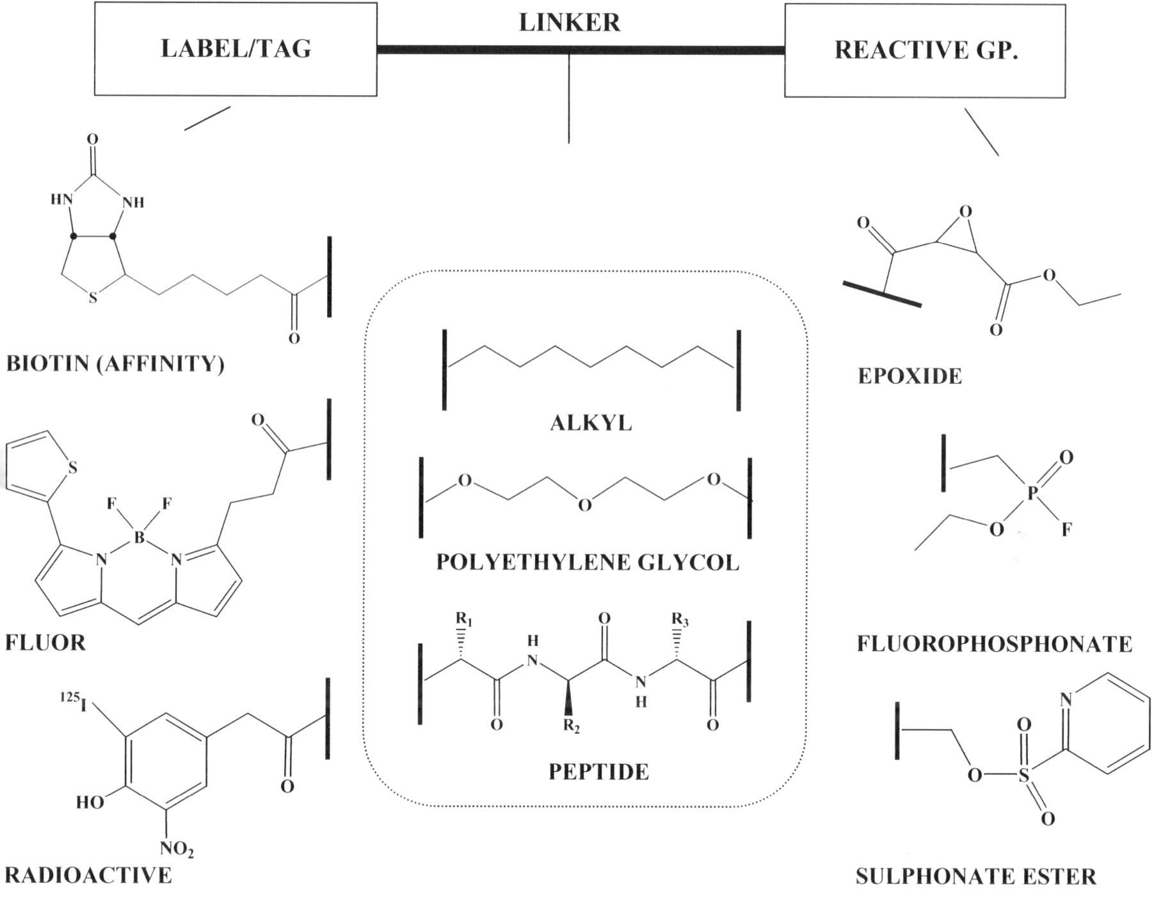

Figure 10.20. Activity-based probes. Activity based probes consist of a reactive group which reacts covalently with the enzyme active site, a linker (which contributes to selectivity) and a label or affinity tag which facilitates detection or capture of affinity-labelled proteins. (Adapted from Jeffery and Boygo, 2003)

include inhibitors of cysteine and serine proteases and components of the 20S proteasome. These agents allow comparative profiling of proteomes in which the target family of proteases is active/inactive. This is an example of *activity based profiling* in which specific enzyme classes in the proteome are selectively targeted by small molecules that bind to their active site. Activity probes have three components: a reactive group (which reacts with the enzyme active site), a linker and fluorescent dye (which facilitates detection) (Figure 10.20). *Directed activity probes* have a well-characterized reactive group which targets a specific enzyme mechanism. In addition to proteases, such probes have been developed for tyrosine phosphatases and glycosidases. *Non-directed probes* have an additional binding group attached to the reactive group. Various binding groups can be used so a family of probes is generated. In the family of sulphonate esters described in Example 10.3, a nucleophile in the enzyme active site can displace the ester and accept the linker which remains covalently attached to biotin (Figure 10.21). In this way, proteases with similar functions can be identified in the proteome.

10.4.2 Glycosylation

Glycosylation is a key modification of proteins which is clinically very relevant in processes such as cell transformation and tumorigenesis. It frequently confers biospecific recognition on proteins especially when expressed on cell surfaces. Two main points of attachment to proteins commonly occur: N-linkage at asparagines in –Asn-X-Ser/Thr- (X is any residue except proline); O-linkage at serine or threonine (Figure 10.22). N-linked glycosylation sites are found in proteins expressed on the extracellular side of the cell membrane, in secreted proteins and in proteins of

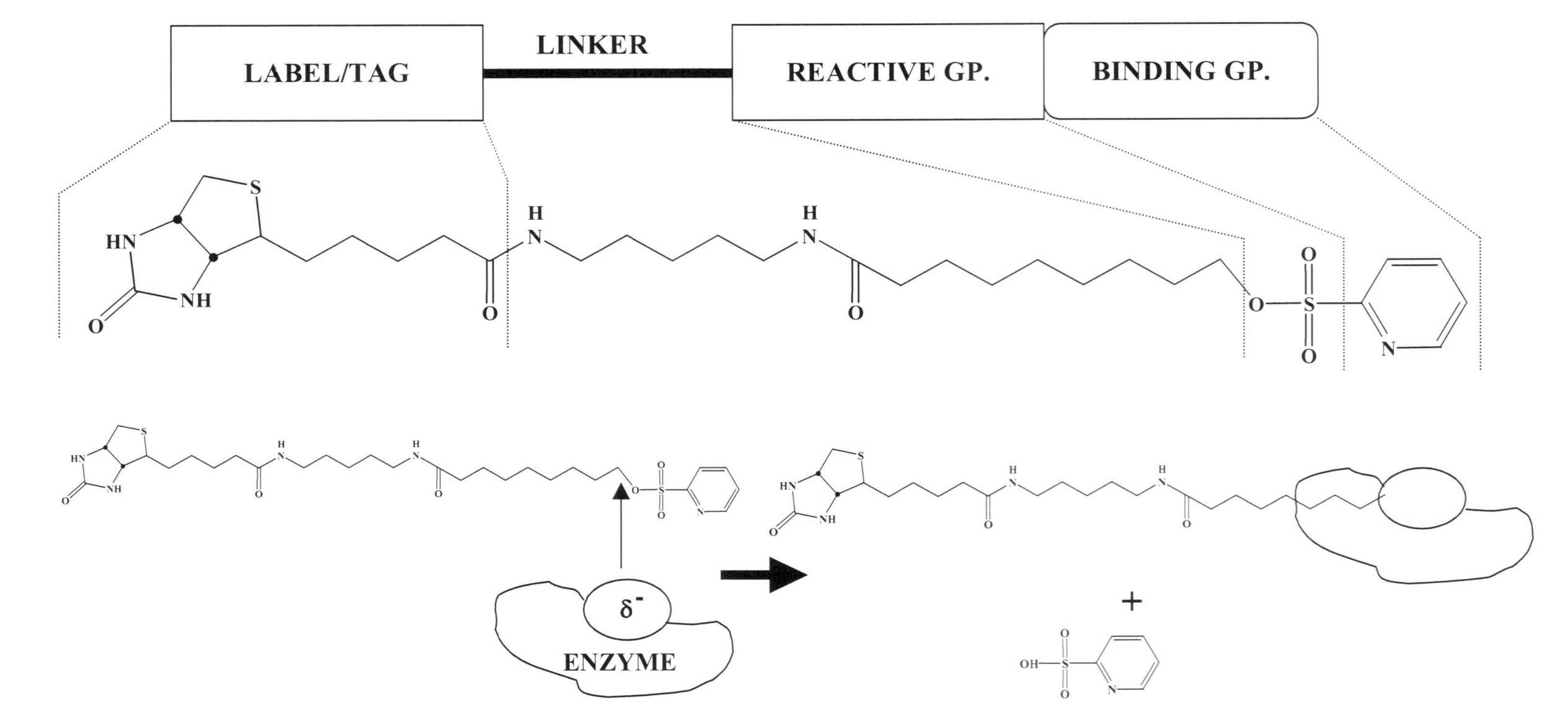

Figure 10.21. Nondirected activity-based probes. Non-directed probes have an additional binding group which directs the probe to target enzymes. The example is a pyridyl sulphonate ester with a biotin label. A nucleophilic group at the enzyme catalytic site attacks at the sulphonate and displaces the ester, linking the probe covalently to the enzyme

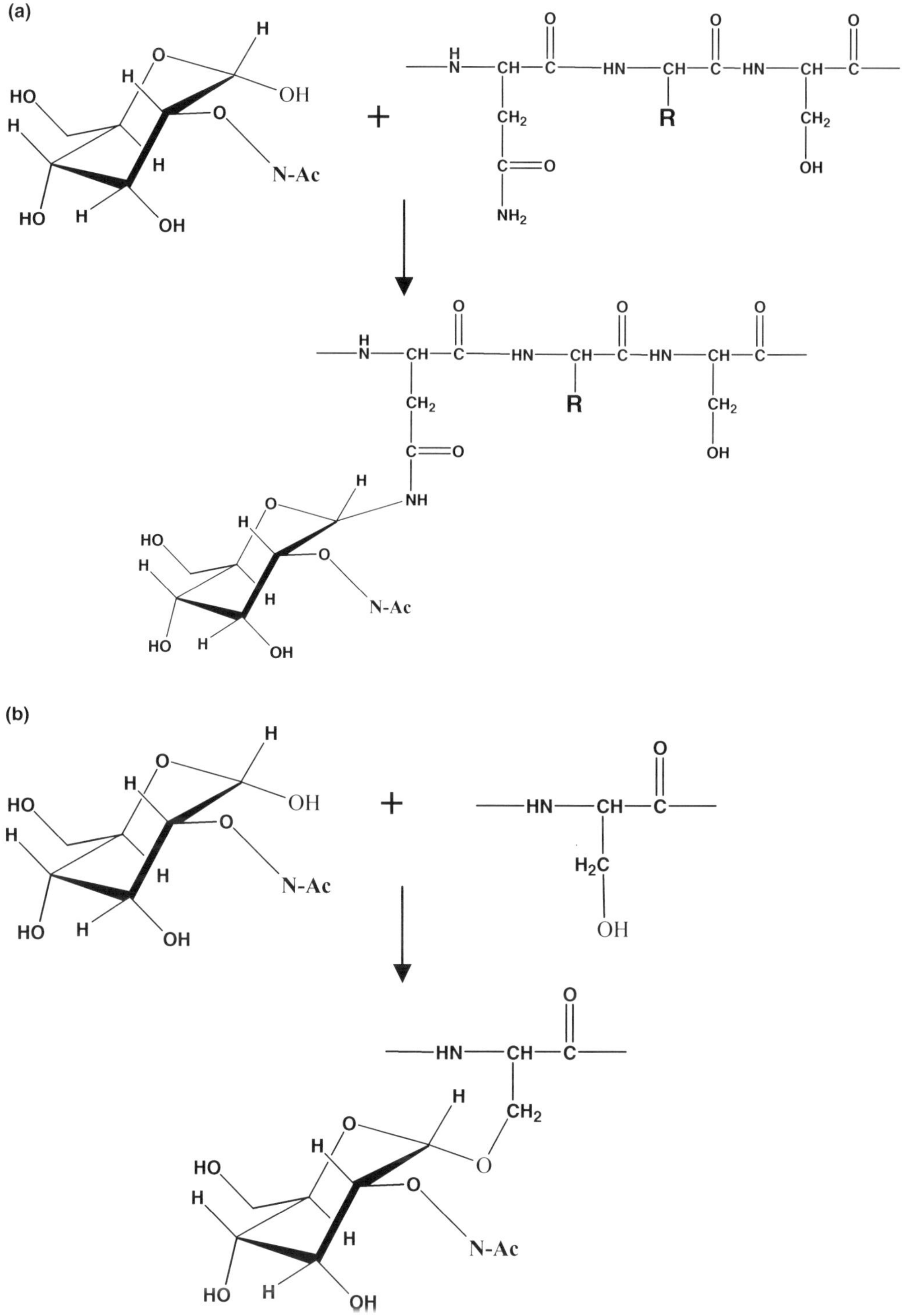

(a)

(b)

Figure 10.22. Glycosylation sites in glycoproteins. (a) N-Linkage at Asn-X-Ser/Thr. X is any amino acid except proline. (b) O-Linkage at serine or threonine

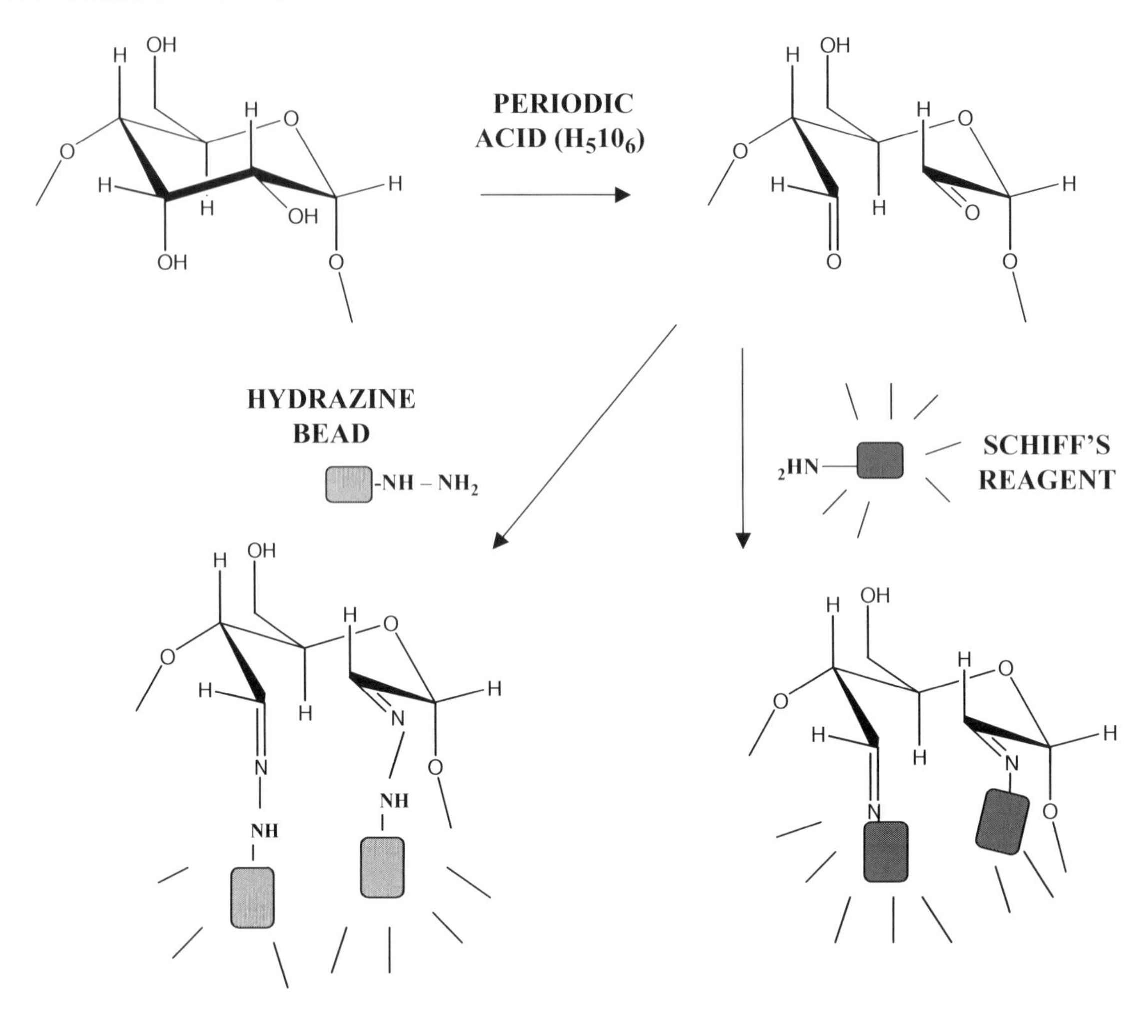

Figure 10.23. Oxidation of glycoproteins. The *cis*-diol groups of monosaccharide components of glycoproteins are conveniently oxidized by periodic acid to form aldehyde groups. These react with either Schiff's reagent or hydrazines to attach a dye or bead as a Schiff's base or hydrazone derivative, respectively. Schiff's reagent is used to stain glycoproteins in gels or after transfer to blotting membranes. Hydrazine capture is used to isolate glycoproteins for MS analysis (see Figure 10.24)

bodily fluids. There is scope not just for modification *per se* but also for complex patterns of glycan branching which presents a major analytical challenge. In 2-D SDS PAGE separations, the principal methods for selectively and sensitively labelling glycoproteins depend on periodic acid Schiff base chemistry (Figure 10.23). After 2-D SDS PAGE, proteins transferred to PVDF membranes (Section 5.11.5) can be visualized with fluorescent dyes. Unlike other categories of biopolymers (e.g. proteins, lipids), the identity, sequence and arrangement of monosaccharide building blocks in glycoproteins are difficult to analyse. Yet the biological properties of glycoproteins depend strongly on these chemical details.

MS methods combined with LC separation have been developed for proteomic comparisons of glycoproteins. In the case of N-glycosylation, these depend on two general types of affinity preselection (Figure 10.24). The first exploits the affinity of glycoproteins for plant lectins which can be immobilized on a stationary phase (Section 2.4.5). Once captured by the lectins, glycoproteins can be eluted, digested with trypsin and reapplied to the lectin column. This allows specific removal of nonglycosylated peptides from the digest. Glycopeptides can be enzymatically deglycosylated by the action of peptide N-glycosidase F (PNGase F) and collected. If performed in $H_2{}^{16}O/H_22^{18}O$, this allows isotope-coded labeling at the point of glycosylation

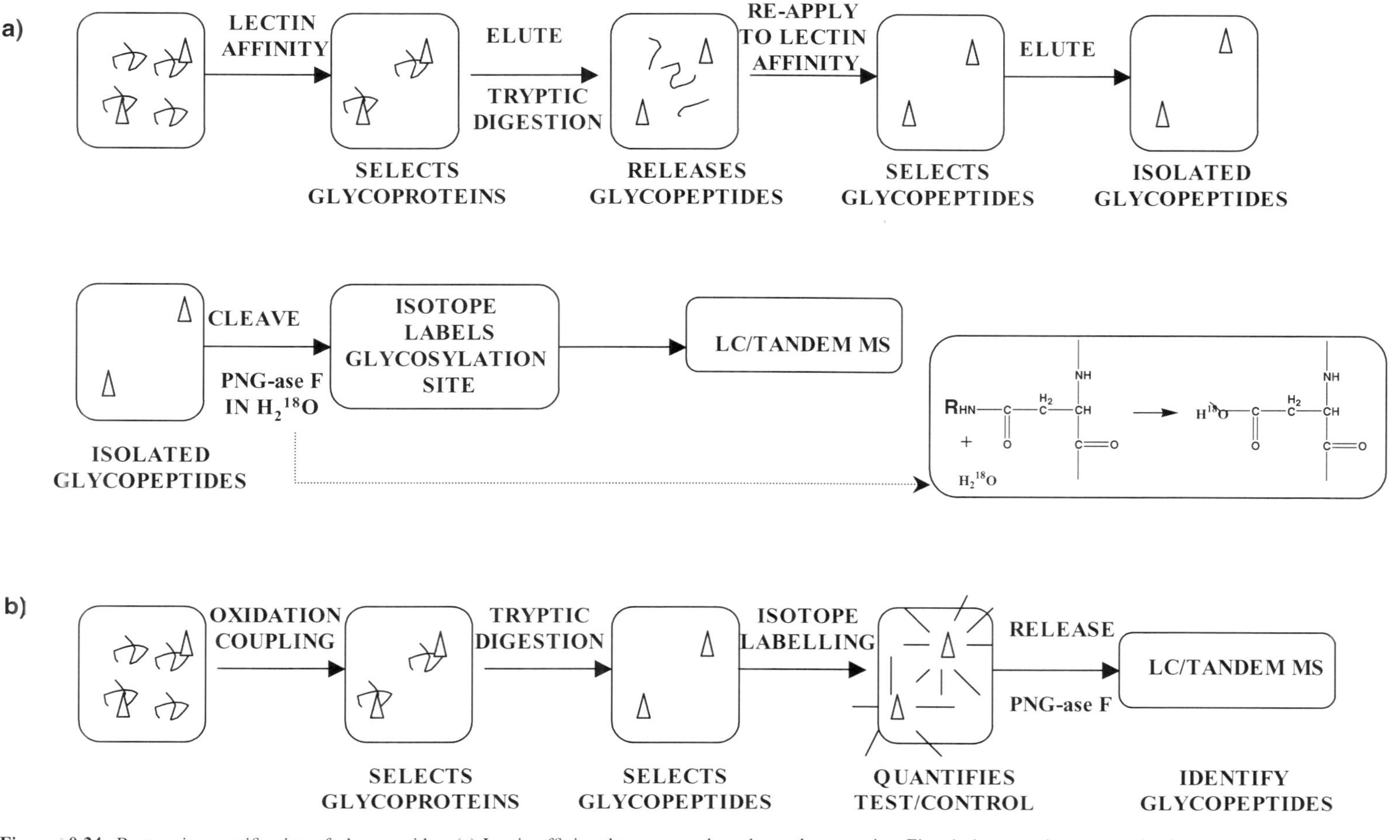

Figure 10.24. Proteomic quantification of glycopeptides. (a) Lectin affinity chromatography selects glycoproteins. Eluted glycoproteins are tryptic digested and reapplied to the lectin column. N-Linkage sites are released and isotope labelled by peptide-N-glycosidase F (PNG-ase F) in $H_2{}^{18}O/H_2{}^{16}O$. Comparison of test/controls on tandem MS will show peptides differing by +2. (b) Proteins are oxidized with periodic acid and N- and O-linked glycoproteins are captured on hydrazine beads. Tryptic digestion releases nonglycopeptides. Isotope labelled N-linked glycopeptides are released by digestion with PNG-ase F and identified by LC/tandem MS

(Figure 10.24). Released glycans can then be separated by normal or reversed phase HPLC (Section 2.4.3) followed by tandem MS (Figure 10.24). This generates a population of fragments with incremental mass differences corresponding to recognizable units (e.g. glucose, mannose) ranging from the intact glycopeptide moiety down to the peptide core of the glycopeptide. Analysis of this population results in a plausible pattern of identity, sequence and branching.

Hydroxyl groups of monosaccharide units can be conveniently oxidized at *cis*-diols to form aldehydes (Figure 10.23). Glycoproteins can be specifically captured with hydrazine immobilized on beads resulting in a hydrazone linkage between the bead and protein. Tryptic digestion can now be performed and nonglycosylated peptides selectively washed away (Figure 10.24). Remaining glycosylated peptides can be isotope coded with succinic anhydride (Figure 10.11) and enzymatically released from the beads by digestion with PNG-ase F. In principle, hydrazide capture will select both N- and O-glycosylated peptides. To remove the latter, it will be necessary to remove oligosaccharide units with a panel of exoglycosidases until a fuschin core is reached. This can then be removed by O-glycosidase. As not all glycoproteins contain this core, a chemical method such as β-elimination would be better for general release of O-glycopeptides.

10.4.3 Oxidation

Oxidation is a naturally occurring process in most cells. It can arise from leakage of free radicals from the electron transport chain, as a consequence of the partially oxidizing microenvironment of the endoplasmic reticulum or as a result of exposure to *reactive oxygen species* resulting in *oxidative stress*. Increased oxidation is a key factor in the ageing of cells and organisms as a result of decreasing efficiency of antioxidant processes. It is also a feature of much pathology including cancer, Alzheimer's and cardiovascular disease. Exogenous factors leading to production of reactive oxygen species include chemical pollutants such as metals. Oxidation has the potential to alter the structure of lipids (lipid peroxidation), DNA (e.g. formation of 8-oxo and hydroxylamine derivatives at bases causing genetic lesions) and proteins. Some amino acid residues are especially susceptible to oxidation and thus provide a 'buffer' to cells experiencing oxidative stress. We have previously seen that methionine can be reversibly oxidized to methionine sulphoxide (a marker of ageing in biology) or irreversibly modified to methionine sulphone (Table 4.3). Similarly, cysteines (–SH) can be irreversibly oxidized to sulfonic (cysteic) acid ($-SO_3H$) and reversibly oxidized to nitrosothiol (–SNO), sulphenic acid (–SOH) sulphinic acid ($-SO_2H$) and thiyl radicals ($-S^{\cdot}$). Other redox-based modifications at cysteines include cysteinylation and S-glutathionylation (Section 10.4.4 below). Some amino acid side-chains can be oxidized to aldehyde or ketone groups (collectively called *carbonyls*) and these modifications can target proteins for turnover. As well as being an indicator of cell ageing or pathology, oxidative modification of proteins is important in *signal transduction* in cells. Nitration of tyrosine to form 3-nitrotyrosine can interfere with the activities of *protein kinases*. There is major interest in being able to identify components of the proteome which are targets for oxidative stress.

Protein thiols can be quantitatively labelled with N-ethyl maleimide, iodoacetamide or iodoacetic acid. Advantage is taken of this in the design of ICAT probes by incorporating an iodoacetoxy group as part of the label as described in Figure 10.8. Thiols involved in mixed disulfides with cysteine or glutathione are highly redox-sensitive and often form part of the cell's response to oxidative stress (Section 10.4.4 below). An approach to their study is described in Figure 10.25. Free thiols are first labelled with iodoacetamide/iodoacetic acid. This is followed by reduction with a reducing agent (e.g. DTT) to reveal thiols involved in disulfides followed by labelling with a tagged reagent. Examples of appropriate tags include the iodoacetoxy and maleimide ICAT probes shown in Figure 10.10 or iodoacetamide-fluorescein. These can be detected respectively in tandem MS systems or 2-D SDS PAGE (by western blotting with antifluorescein).

Carbonylation of proteins has the effect of introducing chemically reactive aldehyde or ketone groups. These can be conveniently labeled with hydrazine attached to dinitrophenol (DNP). Thus, carbonylated proteins can be identified in 1-D or 2-D SDS PAGE separations by western blotting (Section 5.11.2) with anti-DNP (Figure 10.26). This antibody can also be used to 'capture' carbonylated proteins by immunoprecipitation followed by tryptic digestion and protein identification by tandem MS.

Similar approaches can be taken to detection of nitrosylated proteins. Tyrosines are irreversibly nitrosylated to 3-nitrosotyrosines by peroxynitrite, a potent oxidant. An antibody to 3-nitrosotyrosine can be used to find proteins with this modification in 2-D SDS PAGE separations. This antibody can also be used to immunoprecipitate labeled proteins followed by identification by tandem MS. S-nitrosylation at cysteines (SH is converted to SNO) is a result of production of NO and provides a reversible means by which this important molecule can regulate activities of various proteins. Nitosothiols can be detected by a similar method to that described in Figure 10.25 except that nitrosothiols can be reduced by ascorbate (which does not reduce disulfides). An immunoblotting/immunoprecipitation approach is made possible by availability of specific antibodies to –SNO thiols.

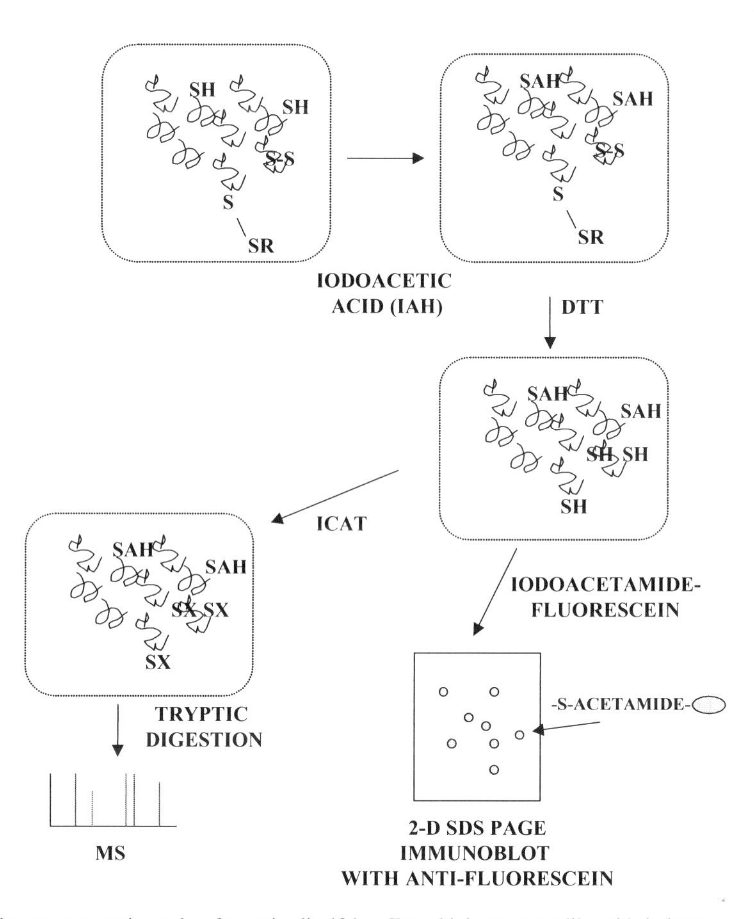

Figure 10.25. Approaches to proteomic study of protein disulfides. Free thiols react readily with iodoacetoxy reagents to form SAH. Treatment with a reducing agent (e.g. DTT) reveals –SH groups previously involved in disulfides. These can be labelled for 2-D SDS PAGE (iodoacetamide-fluorescein) or for MS (ICAT)

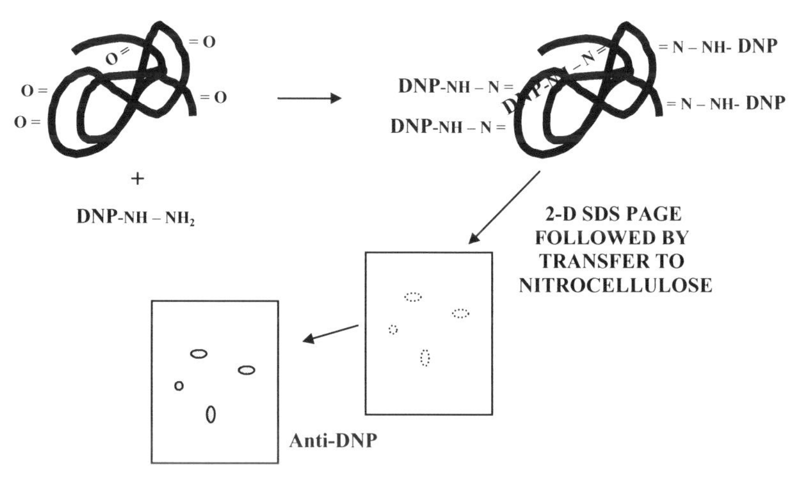

Figure 10.26. Detection of protein carbonyls. Proteins are treated with DNP-hydrazine. This reacts with carbonyls to form hydrazones. After electrophoretic separation in 1-D or 2-D SDS PAGE, carbonylated proteins can be detected with anti-DNP.

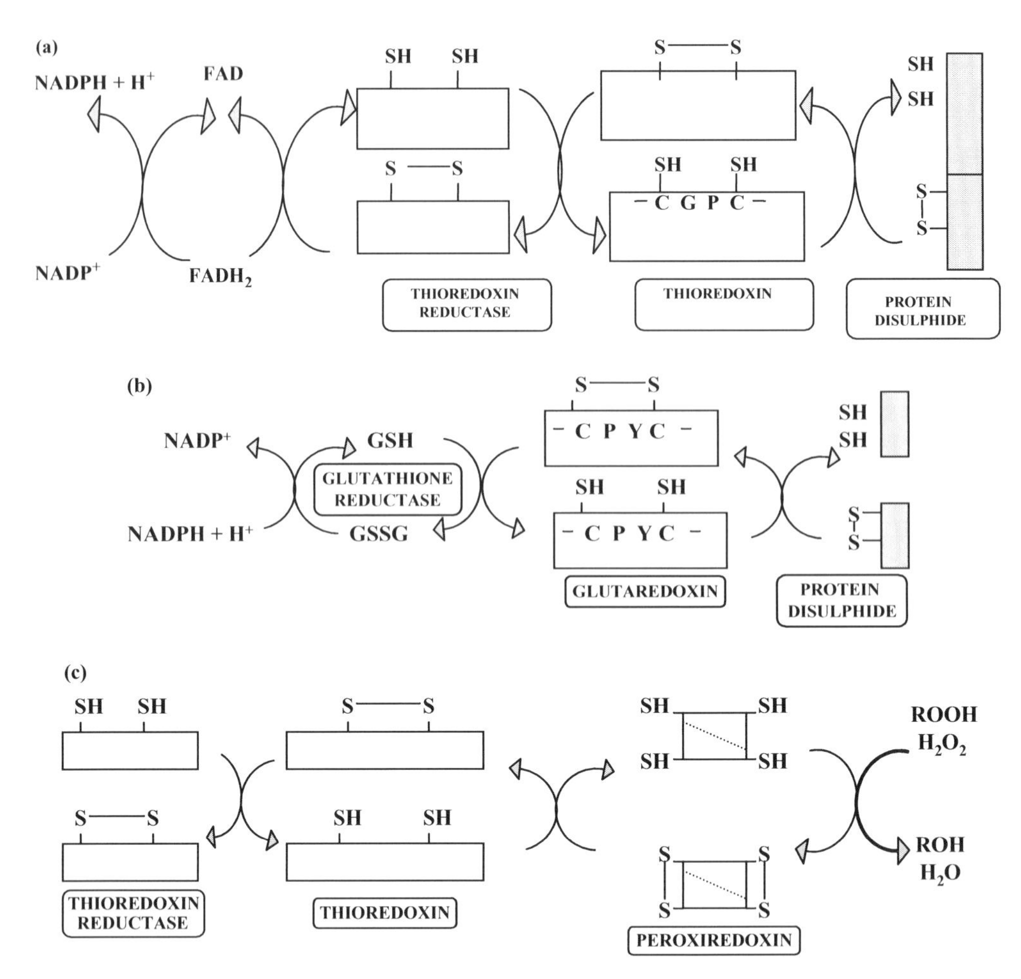

Figure 10.27. Enzymatic reducing systems. Protein disulfides can be reduced by thioredoxin and glutaredoxin while hydroperoxides are reduced by peroxiredoxins. (a) Thioredoxin has a conserved active site motif, reduced by thioredoxin reductase which is in turn reduced by reducing equivalents originating in NADPH $^+$ H^+ from the pentose phosphate pathway. (b) Glutaredoxin is reduced by glutathione which is in turn reduced by glutathione reductase. (c) Peroxiredoxins are of three types. *1-Cys* contain a single cysteine which is oxidized to sulphenic acid by the hydroperoxide and reduced by an unknown reductant. *Dimeric 2-Cys* peroxiredoxins have two pairs of cysteines. The *peroxidatic* cysteine forms sulphenic acid which is reduced by the second (*resolving*) cysteine on the other subunit. In *monomeric 2-Cys* peroxireductases, the cysteine pair is on the same subunit. Peroxiredoxins are regenerated by the thioredoxin system as shown

10.4.4 Protein Disulfides

The –SH group of cysteine can be present in proteins as a free thiol (–SH) or associated in disulfide bridges (–S–S). This can occur naturally (e.g. ribonucleotide reductase is *inactive* in a disulfide bridged form but is activated when –S–S– is reduced to the dithiol form SH HS) or can result from oxidative stress. When –SH is oxidized to sulfenic acid (–SOH), it can react with a nearby –SH or with the –SH of glutathione to form a disulfide. Cells possess several enzyme families, collectively grouped as the *thiol-disulfide oxidoreductase superfamily*, which reduce protein disulfides on specific protein targets (Figure 10.27). This may protect these –SHs against irreversible oxidation just long enough for the cell to recover from an oxidative event or it may be part of sampling possible disulfide bridge patterns involved in protein folding. These enzymes contain one or more disulfide bridge(s) in their active site which is reduced to activate the protein. The pKa of the catalytic –SH groups are often perturbed either far below or far above those of normal cysteines which confers extra reactivity upon them. Each protein family recognizes different substrates, is reduced in different ways and contains specific consensus motifs around the active site disulfide which makes them distinct. The *thioredoxins* reduce disulfides to dithiols using reducing

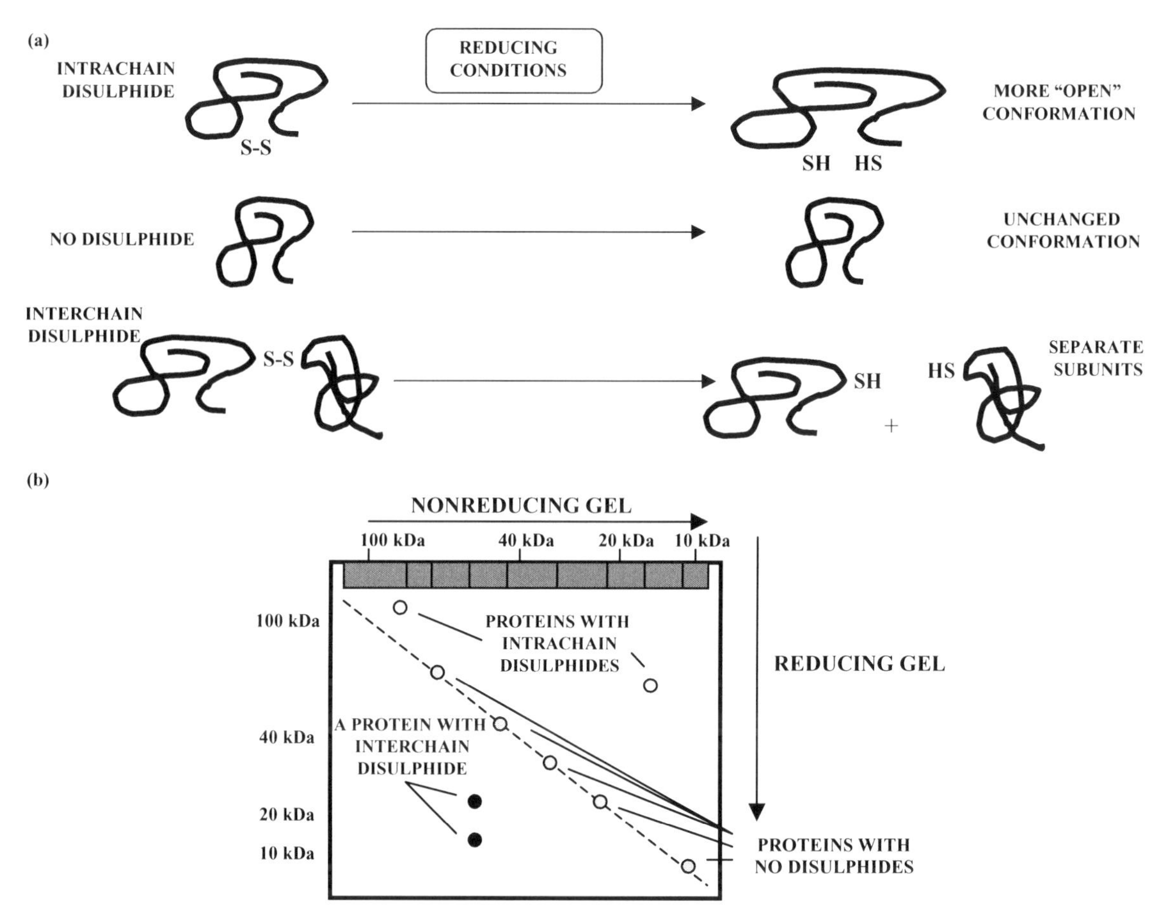

Figure 10.28. Diagonal gel electrophoresis. (a) Proteins with intrachain disulfides have more open conformation and hence electrophorese more *slowly* in SDS PAGE. Proteins lacking disulfides migrate identically in reducing/nonreducing conditions while proteins with interchain disulfides migrate as distinct subunits in reducing conditions. (b) Proteins are electrophoresed in nonreducing conditions. The track is excised and exposed to reducing conditions. Most proteins cluster on the diagonal while those with interchain disulfides migrate below it and those with intrachain disulfides migrate above it

equivalents provided by $FADH_2$; *glutaredoxins* are reduced by glutathione while *peroxiredoxins* (thioredoxin-dependent peroxide reductases) are themselves substrates for thioredoxins and directly reduce hydroperoxides in cells. The peroxiredoxins are subdivided into two classes, *1-Cys* and *2-Cys*, with the latter class further subdivided into monomers and dimers (Figure 10.27). Members of the thioredoxin superfamily *protein disulfide isomerases* catalyze formation of disulfide bonds in the endoplasmic reticulum of eukaryotes while another thioredoxin, *DsbA*, forms disulfide bridges on proteins translocated into the *E. coli* periplasm. In both cases, these enzymes help proteins to fold into their correct disulfide bridge pattern before folding is complete. For this reason, the ER is a partially oxidizing compartment of eukaryote cells. Reducing equivalents for these processes are supplied ultimately by $NADPH + H^+$ from the pentose phosphate pathway. If the requirement for reducing equivalents exceeds the supply capacity of this pathway, then oxidative stress can result.

The importance of disulfide bridges in protein folding has led to development of methods for identifying differences in disulfide patterns within the proteome. One of the most popular is *diagonal gel electrophoresis* whereby samples are exposed sequentially to nonreducing and reducing conditions during SDS PAGE (Figure 10.28). This is a convenient method for distinguishing proteins which have interchain or intrachain disulfides in a single separation where nondisulfide bridged proteins cluster along a diagonal. Variations in disulfide patterns resulting from oxidative stress can be revealed by comparing patterns. Proteins which are

Example 10.4 Use of diagonal gels to study protein disulfides

Proteins may have *intrachain* disulfide bridges which are often an important means of stabilizing tertiary structure. Interchain disulfide bridges may attach separate polypeptides to each other. In cells exposed to oxidative stress the enzymes responsible for maintaining protein thiols in their reduced state can be overcome resulting in changes in the pattern of disulfides within the proteome. Diagonal gel electrophoresis provides a means of distinguishing these changes. The method involves performing one dimensional electrophoresis under nonreducing conditions. A slice of gel containing the separated proteins is then excised and exposed to reducing conditions (e.g. DTT) and electrophoresed 90° to the original electrophoresis direction. The majority of proteins in the proteome lack disulfide bridges and therefore migrate along the diagonal. Proteins containing intrachain disulfides have a more open structure and migrate more slowly in reducing conditions. These therefore appear as spots *above* the diagonal. Proteins with interchain disulfides migrate as separate polypeptides under reducing conditions and thus appear as spots *below* the diagonal.

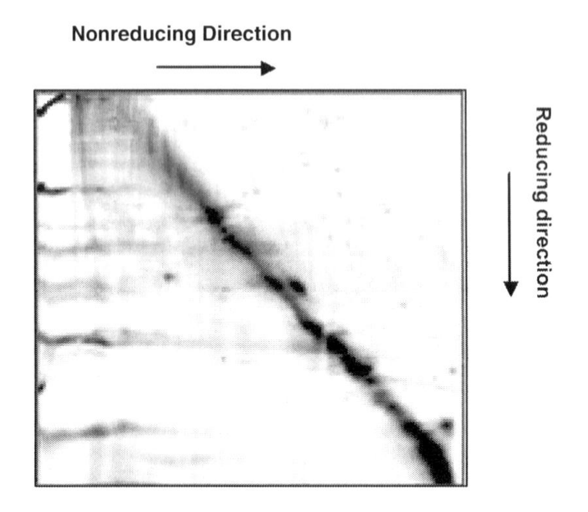

Reproduced from McDonagh, B., Tyther, R. and Sheehan, D. (2006) Redox proteomics in the mussel Mytilus edulis. *Marine Environmental Research*, **62**, S101–S104.

substrates of members of the thioredoxin superfamily can also be routinely identified in this analysis (Example 10.4).

10.4.5 The Phosphoproteome

Phosphorylation is one of the most important post-translational modifications of proteins. The reaction is catalyzed by *protein kinases* which make up almost 2% of the genes encoded in the human genome. *Protein phosphatases* dephosphorylate proteins and many proteins are only active in their phosphorylated form making phosphate a convenient chemical 'on-off' switch. Phosphorylated protein variants are a feature of *signal transduction cascades* in the cell. Phosphate groups are conjugated to serine, threonine or tyrosine side-chains and some 30% of proteins are capable of being phosphorylated. The *phosphoproteome* is the complement of proteins existing in phosphorylated form under a given set of conditions. Because of the central importance of phosphorylation in signal transduction, it represents a key target for study.

A tiny proportion of proteins are phosphorylated at any given time. For 2-D SDS PAGE, phosphoproteins can be directly labeled with ^{32}P and identified using autoradiography of separations. However, this approach does not lend itself to high-throughput analyses. It is often necessary to enrich for phosphoproteins in order to arrive at quantitative data. Antibodies to phosphate groups and metal affinity columns can be used for this purpose but frequently preparations are contaminated with nonphosphorylated proteins. It is also possible to replace the phosphate group with biotin affinity tags allowing for enrichment of phosphopeptides/ proteins on avidin (Figure 10.29). This takes advantage of the fact that phosphates are labile at alkaline pH and can undergo β-elimination forming a reactive *dehydroanalyl* residue. This can act as an acceptor of a sulphydryl

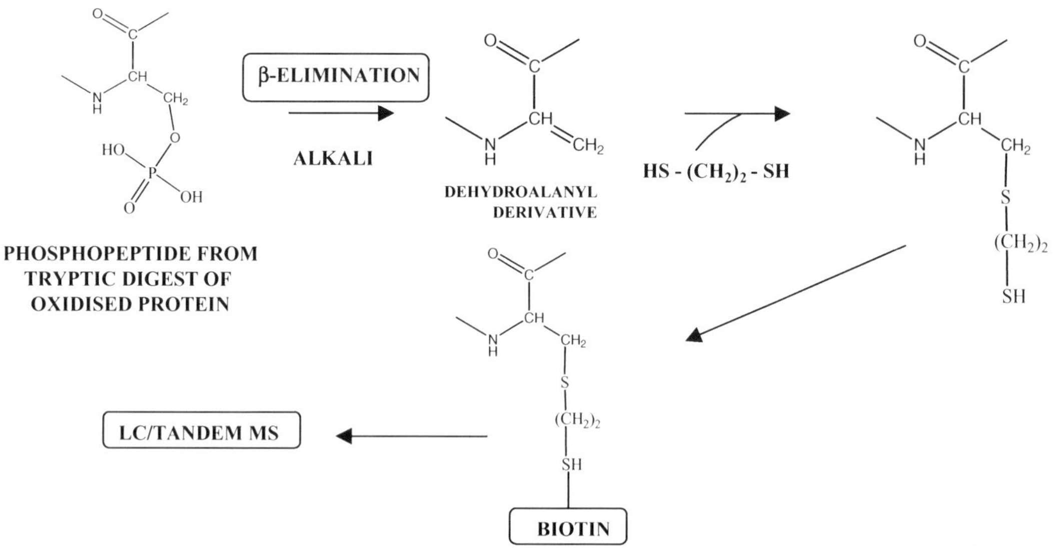

Figure 10.29. Enrichment of phosphopeptides by Michael addition. Protein thiols are first oxidized to cysteic acid with performic acid and the oxidized protein digested with trypsin. Phosphate is eliminated from peptides containing phosphoserine under alkaline conditions. Ethanedithiol adds to the α–β-unsaturated bond in a Michael addition labelling the site of phosphoserine with –SH. This reacts with biotin-NEM. On tandem MS, a fragment characteristic of the biotin label is generated only in phosphopeptides. The Michael addition also occurs with phosphothreonyl peptides albeit at a slower rate

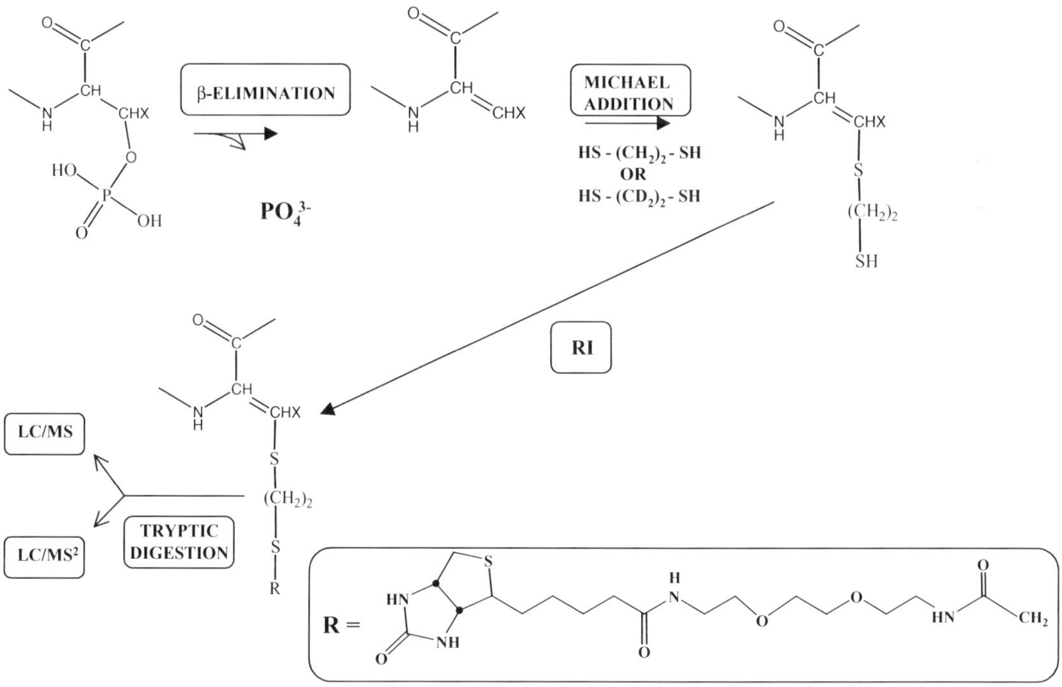

Figure 10.30. ICAT identification of phosphopeptides by Michael addition. A peptide containing phosphoserine (X = H) or phosphothreonine (X = CH_3) is pretreated to block –SH groups (by alkylation). Phosphate is then removed by β-elimination (see Figure 10.27). The α–β unsaturated bond is susceptible to a Michael-type addition of 1,2-ethanedithiol which is available in protonated or deuterated forms resulting in a 4 Da mass difference between matched peptides on LC/MS. Tandem MS identifies site of modification in specific peptides

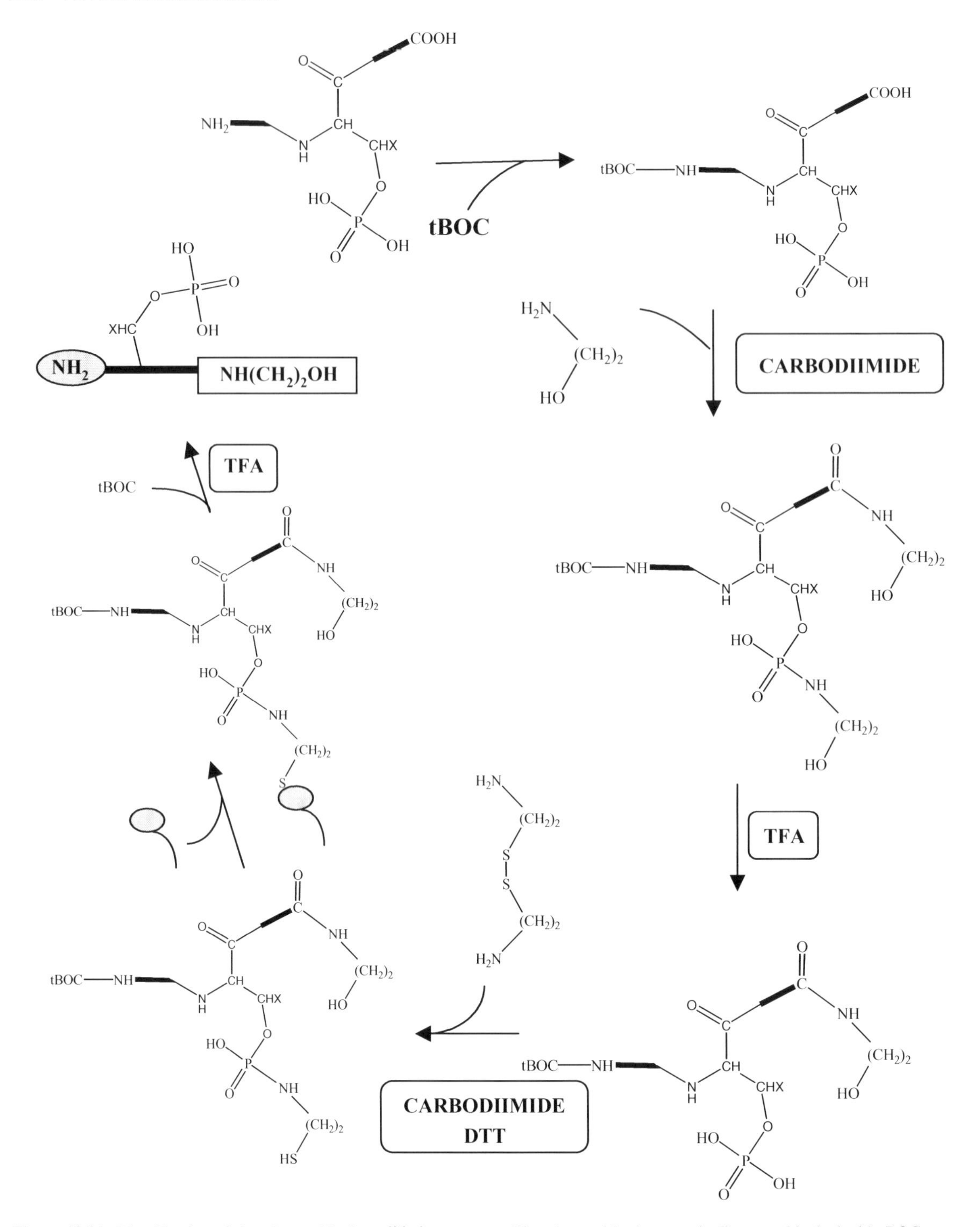

Figure 10.31. Identification of phosphopeptides by solid phase capture. Phosphopeptides in a tryptic digest are blocked with tBOC groups followed by condensation of phosphates/carboxyls to ethanolamine. Trifluoroacetic acid (TFA) regenerates phosphates selectively and these are reacted with cystamine introducing –SH on phosphates. Phosphopeptides are captured by iodoacetic acid coated glass beads. TFA releases and deblocks peptides which are analysed by LC-Tandem MS

reagent in a Michael-type addition. Provided the phosphopeptides originated from oxidized proteins, only phosphopeptides will now carry an –SH and these can be labeled with NEM-biotin and collected on avidin. Tandem MS generates a characteristic fragment from this label.

Methodologies resembling the ICAT approach (Section 10.2.2; Figures 10.10 and 10.11) have been developed for enrichment and MS investigation of phosphopeptides. The most popular methods preoxidize –SH groups with later introduction of new –SH groups at sites of phosphorylation. These are amenable to modification with –SH ICAT probes. Solid phase capture ensures efficient washing steps and enrichment of phosphopeptides. β-Elimination, followed by Michael addition, adds 1, 2-ethanedithiol at the site of phosphorylation (Figure 10.30). This dithiol is available in both protonated and deuterated forms thus facilitating isotope coding. The –SH attached at the phosphorylation site can react with iodoacetetic acid attached to biotin and labeled peptides can be captured on avidin for LC-tandem MS analysis. The biotin label shows up in the tandem MS as a characteristic fragment denoting phosphorylation. A slightly different approach condenses ethanolamine to both phosphate and carboxyl groups with selective regeneration of the phosphates (Figure 10.31). These then react with cystamine which introduces –SH at the phosphorylation site. Labeled peptides can be captured on glass beads coated with iodoacetic acid and peptides eluted on treatment with TFA.

REFERENCES

General reading

Campbell, D.A. and Szardenings, A.K. (2003) Functional profiling of the proteome with affinity labels. *Current Opinion in Chemical Biology* 7, 296–303. Overview of affinity-based methods for detection of proteins in proteomic separations.

Patton W.F. (2002) Detection technologies in proteome analysis. *Journal of Chromatography B* **771**, 3–31. A general review of detection technologies in electrophoresis, MS and arrays.

Reinders, J., Lewandrowski, U., Moebius, J., Wagner, Y. and Sickmann, A. (2004) Challenges in mass spectrometry-based proteomics. *Proteomics* **4**, 3686–3703. A good general review of recent innovations in MS as applied to proteomics.

Isoelectric focusing and two-dimensional SDS-PAGE

Cottrell, J.S. & Sutton, C.W. (1996) The identification of electrophoretically separated proteins by peptide mass fingerprinting. In: J.R. Chapman (ed.): *Protein and Peptide Mass Analysis by Mass Spectrometry* pp. 67–82. Humana Press, Totowa. MS analysis of electroblotted two-dimensional gels.

Garrels, J. (1983) Quantitative two-dimensional gel electrophoresis of proteins. *Methods in Enzymology* **100B** 411–423. Review of 2-D SDS PAGE.

Görg, A., Postel, W. & Guenter, S. (1988) The current state of two-dimensional electrophoresis with immobilized pH gradients. *Electrophoresis* **9**, 531–546. Review of the use of immobilized pH gradients in 2-D SDS PAGE.

Hellman J. (2007) Polyacrylamide lamination enables mass spectrometry compatible staining and in-gel digestion of proteins separated by agarose IEF. *Proteomics* **7**, 3441–3444. An elegant illustration of how agarose gel IEF can by combined with polyacrylamide gels for proteomic analysis of large proteins and complexes.

http://www.weihenstephan.de/blm/deg/manual/manualwork2html 02test.htm. An on-line manual for 2-D SDS PAGE by Prof A. Görg and colleagues, Technical University of Munich.

O'Farrell, P. (1975) High resolution two-dimensional electrophoresis of proteins. *J. Biol. Chem.* **250**, 4007–4021. Classic description of 2-D SDS-PAGE technique.

Righetti, P.G. & Bossi, A. (1997) Isoelectric focusing in immobilized pH gradients: Recent analytical and preparative developments. *Analytical Biochemistry* **247**, 1–10. Review of the use of immobilised pH gradients.

Sun, Q. et al.(1998) AP30, a differential protein marker for perilymph and cerebrospinal fluid in middle ear fluid has been purified and identified as human apolipoprotein D. *Biochimica et Biophysica Acta* **1384**, 405–413. A good example of the power of 2-D SDS PAGE analysis.

Vercauteren, F.G.G., Arckens, L. And Quirion, R. (2007) Applications and current challenges of proteomic approaches, focusing on two-dimensional electrophoresis. *Amino Acids* **33**, 405–414. A critical review of current issues in 2-D SDS PAGE.

Xu, H., Hu, L.-S., Chang, M., Jing, L., Zhang, X.-Y. and Li, G.S. (2005) Proteomic analysis of kidney in fluoride-treated rat. *Toxicology Letters* **160**, 69–75. An excellent example of the use of 2-D SDS PAGE.

Other electrophoresis methods in proteomics

McDonagh, B., Tyther, R and Sheehan, D. (2006) *Marine Environmental Research*, **62**, S101–S104. An example of diagonal gel electrophoresis

Righetti, P.G., Castagna, A., Herbert, B., Candiano, G. (2005) How to bring the "unseen" proteome to the limelight via electrophoretic pre-fractionation techniques. *Bioscience Reports*, **25**, 3–17. A review of electrophoresis for pre-fractionation in proteomics.

Yang, W., Steen, H. and Freeman, M.R. (2008) Proteomic approaches to the analysis of multiprotein signaling complexes. *Proteomics*, **8**, 832–851.

Mass spectrometry

Aebersold, R. and Mann, M. (2003) Mass spectrometry-based proteomics. *Nature* **422**, 198–207. An overview of MS in proteomics.

Han, D.K., Eng, J., Zhou, H. and Aebersold, R. (2001) Quantitative profiling of differentiation-induced microsomal proteins using isotope-coded affinity tags and mass spectrometry. *Nature Biotechnology* **19**, 946–951. A good example of use of ICAT.

Kocher, T., Superti-Furga, G. (2007) Mass spectrometry-based functional proteomics: From molecular machines to protein networks. *Nature Methods* **4**, 807–815. A review of MS methods in identification of protein-protein interactions.

Leitner, A. and Lindner, W. (2004) Current chemical tagging strategies for proteome analysis by mass spectrometry. *Journal of Chromatography B*, **813**, 1–26. A detailed review of chemical tagging approaches in mass spectrometry.

Standing, K.G. (2003) Peptide and protein de novo sequencing by mass spectrometry. *Current Opinion in Structural Biology* **13**, 595–601. An excellent overview of recent developments in use of MS for sequencing peptides and proteins.

Zhang, H., Yan, W. and Aebersold, R. (2004) Chemical probes and tandem mass spectrometry: A strategy for the quantitative analysis of proteomes and subproteomes. *Current Opinion in Chemical Biology* **8**, 66–75.

Microarrays

Auburn, R.P., Kreil, D.P., Meadows, L.A. Fisher, B., Matilla, S.S., and Russell, S. (2005) Robotic spotting of cDNA and oligonucleotide microarrays. *Trends in Biotechnology,* **23,** 373–379. An overview of spotting techniques in creation of microarrays.

Brown, P.O. and Botstein, D. (1999) Exploring the new world of the genome with DNA microarrays. *Nature Genetics,* **21,** 33–37. An overview of gene microarrays.

Curtis, R.K., Orešič, M. and Puig, V. (2005) Pathways to the analysis of microarray data. *Trends in Biotechnology,* **23,** 429–435. An excellent review of data analysis aspects of microarrays.

Protein Chips

Lueking, A., Cahill, D.J. and Müllner, S. (2005) Protein biochips: a new and versatile platform technology for molecular medicine. *Drug Discovery Today* **10**, 789–794.

MacBeath, G, and Schreiber, S.L. (2000) Printing proteins as microarrays for high-throughput function determination. *Science* **289**, 1760–1763.

Predki, P.F. (2004) Functional protein microarrays: Ripe for discovery. *Current Opinion in Chemical Biology* **8**, 8–13. A review of functional protein arrays.

Taussig, M.J. and Landegren, U. (2003) Progress in antibody arrays. *Targets* **2**, 169–176. A review of antibody arrays.

Zhu, H., Bilgin, M., Bangham, R., Hall, D., Casamayor, A., Bertone, P., Lan, N., Jansen, R., Bidlingmaier, S., Houfek T. *et al.*, (2001) Global analysis of protein activities using proteome chips. *Science* **293**, 2101–2105. Description of protein chips in activity-based detection of proteins.

Zhu, H. and Snyder, M. (2003) Protein chip technology. *Current Opinion in Chemical Biology* **7**, 55–63. Review of protein chip technology.

Activity-based profiling

Adam, G.C., Cravatt, B.F. and Sorensen, E.J. (2001) Profiling the specific reactivity of the proteome with non-directed activity-based probes. *Chemistry and Biology*, **8**, 81–95. A good example of activity based profiling.

Barglow, K.T. and Cravatt, B.F. (2004) Discovering disease-associated enzymes by proteome reactivity profiling. *Chemistry and Biology* **11**, 1523–1531. A good description of application of activity-based profiling to clinical scenarios.

Jeffery, D.A. and Boygo, M. (2003) Chemical proteomics and its application to drug discovery. *Current Opinion in Biotechnology* **14**, 87–95. A review of application of activity based profiling to drug discovery.

Jessani, N. and Cravatt, B.F. (2004) The development and application of methods for activity-based profiling. *Current Opinion in Chemical Biology* **8**, 54–59. An excellent review of activity-based profiling.

Post-Translational Modification Proteomics

Ahn, N.G., and Resing, K.A. (2001) Towards the phosphoproteome. *Nature Biotechnology* **19**, 317–318. An assessment of methods for proteome analysis.

Conrads, T.P., Issaq, H.J. and Veenstra, T.D. (2002) New tools for quantitative phosphoproteome analysis. *Biochemical and Biophysical Research Communication* **25**, 885–890.

Dalle-Donne, I., Scaloni, A. and Butterfield, D.A. (Eds.) (2006) Redox Proteomics: From Protein Modifications to Cellular Dysfunction and Diseases. John Wileyand Sons, Hoboken, NJ, USA. A comprehensive volume on proteomics approaches to redox modifications.

Guan, J.-Q. and Chance, M.R. (2005) Structural proteomics of macromolecular assemblies using oxidative footprinting and mass spectrometry. *Trends in Biochemical Sciences* **30**, 583–592. Review of MS for detection of protein lesions arising from oxidation.

Sheehan, D. (2006) Detection of redox-based modification in two dimensional electrophoresis proteomic separations. *Biochemical and Biophysical Research Communications* **349**, 455–462. A review of methods for finding oxidatively modified proteins in two dimensional SDS PAGE separations.

Wuhrer, M., Deelder, A.M. and Hokke, C.H. (2005) Protein glycosylation analysis by liquid chromatography-mass spectrometry. *Journal of Chromatography B*. **825,** 124–133. Application of LC-MS to analysis of glycosylation of proteins.

Organelle proteomics

Gabaldón, T. and Huynen, M.A. (2004) Shaping the mitochondrial proteome. *Biochimica et Biophysica Acta – Bioenergetics* **1659**, 212–220. A review of that status of mitochondrial proteomics.

Pederson, T. (2002) Proteomics of the nucleolus: More proteins, more functions. *Trends in Biochemical Sciences* **27**, 111–112. Review of nucleolus proteomics.

Schirmer, E.C. and Gerace, L. (2005) The nuclear membrane proteome: Extending the envelope. *Trends in Biochemical Sciences* **30**, 551–558. Review of recent developments in organelle proteomics.

Taylor, S.W., Fahy, E. and Ghosh, S.S. (2003) Global organellar proteomics. *Trends in Biotechnology* **21**, 82–88. Excellent overview of issues in organelle proteomics.

Appendix 1
SI Units

Table 1.1. Base and derived units of the SI system

Quantity	Unit name	Symbol
	BASE UNITS	
Mass	Gram	g
Length	Metre	m
Time	Second	s
Current	Ampere	A
Temperature	Kelvin	K
Amount	Mole	mol
Luminous intensity	Candela	cd
	DERIVED UNITS	
Volume	Litre	L
Energy	Joule	J
Voltage	Volts	V
Resistance	Ohms	Ω
Force	Newton	N
Pressure	Torr	t
Frequency	Hertz	Hz
Radioactivity	Curie	Ci

Table 1.2. Prefixes commonly used with the SI system

10^9	giga	G
10^6	mega	M
10^3	kilo	k
10^{-1}	deci	d
10^{-2}	centi	c
10^{-3}	milli	m
10^{-6}	micro	μ
10^{-9}	nano	n
10^{-12}	pico	P
10^{-15}	femto	f
10^{-18}	atto	a

Table 1.3. Some important physical constants

Atomic mass unit (AMU)	1.661×10^{-24} g
Avogadro's number (N)	6.022×10^{23}/mol
Boltzmann constant (k)	1.381×10^{23} J/K
Universal Gas Constant (R)	8.315 J/mol.K
Planck's constant (h)	6.626×10^{-34} J.s
Velocity of light (c)	2.998×10^{10} cm/s
Angstrom (Å)	10^{-10} m
Electron volt (eV)	1.602×10^{-19} J

Table 1.4. Some useful conversion factors

Temperature	k = °C + 273
Energy	1 J = 0.239 calories = 10^7 erg
Pressure	1 t = 1.333×10^2 Pascal = 1.32×10^{-3} atmosphere
Radioactivity	1 Ci = 3.7×10^{10} dps = 37 Bequerel

Appendix 2
The Fourier Transform

Sometimes in science we need to relate two sets of numbers. A simple example is the relationship between the side of a square and the area of a square as follows:

Side of square (m)	Area of square (m^2)
1	1
2	4
5	25
9	81

Note that there is a unique relationship between the side and area numbers and that the value for area can be calculated from that of the side and vice versa. Note also that the *units change* when the side of square data is transformed into area data and that the two data-sets could be related by a mathematical equation which also relates their units (e.g. side^2 = area). This is a general example of a *transform*. Such mathematical operations can be useful in relating some quantity which we can experimentally measure to some other property of the system under investigation which we wish to quantify. For example, using the simple equation above we can calculate the area of any square from its length without needing to measure the area directly.

Joseph Fourier (1768–1830) was a French mathematician and physicist who, because of his interest in heat conduction, developed a mathematical method for the representation of any *discontinuous function* in space or time in terms of a much simpler trigonometric series of *continuous* cosine or sine functions. Such a series is referred to as a *Fourier series* and the process of dissection into cosine and/or sine components is called *Fourier analysis*. The opposite or *inverse* operation, recombining cosine and/or sine functions, is called *Fourier synthesis*.

Figure A2.1 shows an example of a problem studied by Fourier. If part of a metal bar is heated, the temperature of the bar is highest at the point where it is heated and this heat gradually spreads through the whole bar (Figure A2.1(a)). If we plot the temperature as a function of distance along the bar we see that this is an example of a discontinuous function (b). Fourier realized that, if the temperature was decomposed into continuous sinusoidal curves having 1, 2, 3 or more cycles along the bar (c), then summation of these would yield a good approximation of the original discontinuous function (dashed line in (d)). The more curves added into the Fourier series, the better the agreement of the Fourier synthesis with the experimental measurement.

In Chapter 3 (Figure 3.1) we saw that any wave function can be defined in terms of amplitude, A and wavelength, λ, or frequency, ν. In Chapter 6 we further saw that a full description of such a wave also requires definition of a phase, ϕ (Figure 6.46). A simple one-dimensional wave function, $f(x)$, is shown in Figure A2.2. $f(x)$ specifies the height of the wave at any horizontal point x where x is defined in terms of wavelength such that $x = 1 = \lambda$. This wave function could be represented as either a sine or a cosine function as follows:

$$f(x) = A\cos 2\pi(\nu x + \phi)$$

or

$$f(x) = A\sin 2\pi(\nu x + \phi)$$

The term 2π appears because there are 2π radians per wavelength. Fourier analysis of complex wave functions dissociates them into a Fourier series of simple wave functions such as $f(x)$. We could write a Fourier series containing n terms as

$$\begin{aligned} F(x) &= A_0\cos 2\pi(0x + \phi_0) + A_1\cos 2\pi(x + \phi_1) \\ &\quad + A_2\cos 2\pi(2x + \phi_2)\cdots + A_n\cos 2\pi(x + \phi_n) \\ &= \sum_{\nu=0}^{n} A_\nu \cos 2\pi(\nu x + \phi_\nu) \end{aligned}$$

This shows that any complex wave can be expressed as a composite of simpler cosine or sine waves such as those in Figure A2.1.

If a complex discontinuous function is periodic (although this is not essential) with a period T such that $f(t + T) = f(t)$,

Physical Biochemistry: Principles and Applications, Second Edition David Sheehan

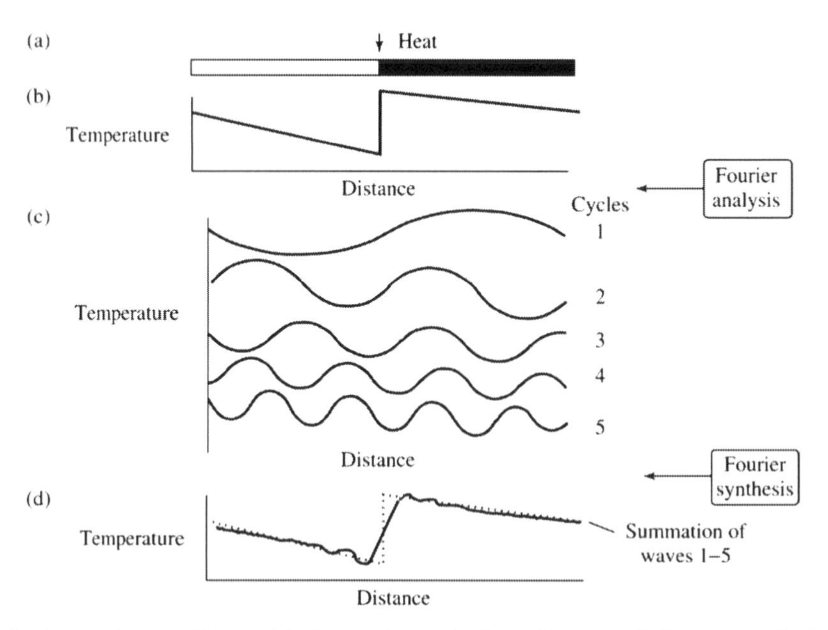

Figure A2.1. Heat Conductance in a metal bar. (a) A bar is heated at a point denoted by arrow, (b) Temperature distribution along the bar. (c) Temperature may be represented as a Fourier series of continuous wave functions along the bar. This dissociation into individual waves is called Fourier analysis, (d) Summation of these continuous functions (Fourier synthesis) gives a good representation of the discontinuous function. The more waves added together (i.e. the more terms in the Fourier series), the closer the fit to the data shown in (b).

it may be represented as a Fourier series of the form

$$
\begin{aligned}
f(t) &= a_0/2 + a_1 \cos 2\pi(t/T) + a_2 \cos 2\pi(2t/T) \\
&\quad + \cdots + b_1 \sin 2\pi(t/T) \\
&= \frac{a_0}{2} + \sum_{n=1}^{\infty} a_n \cos 2\pi(nt/T) + \sum_{n=1}^{\infty} \sin 2\pi(nt/T) \\
&= \sum_{n=-\infty}^{\infty} cne^{2\pi i(nt/T)}
\end{aligned}
$$

where t is time, n is the number of components in the series, i is the *imaginary number* $(i^2 = -1)^{1/2}$ and a, b, c and so on are coefficients describing the individual waves in the series.

This mathematically expresses the fact that a discontinuous function can be dissected into individual sine/cosine wave functions which may in turn be summed to arrive at an expression for it as a function (in this case) of time. The relationship between the two sets of functions can be generalized with a pair of equations which are *Fourier transforms of each other*:

$$f(x) = \int_{-\infty}^{\infty} F(y)e^{2\pi ixy}dy$$

$$F(y) = \int_{-\infty}^{\infty} f(x)e^{-2\pi ixy}dx$$

where $i^2 = -1$.

A good practical analogy for Fourier analysis is the diffraction of white light from the sun after passage through a prism into waves of individual frequency and amplitude which we described in Chapter 3 (see Figure 3.2). The *strength* of sunlight incident on the prism may be expected to vary with time which, in turn leads to variation in the *amplitude* of each diffracted frequency. The prism therefore acts to transform a strength-time domain into an amplitude-frequency domain.

Many of the techniques described in this book result in complicated discontinuous functions which, like white light or the temperature of a heated metal bar, may be regarded as composites of simpler continuous wave functions. The Fourier transform is special because the equation relating sets of numbers which can be interconverted by the Fourier transform contains sine and cosine terms allowing us to

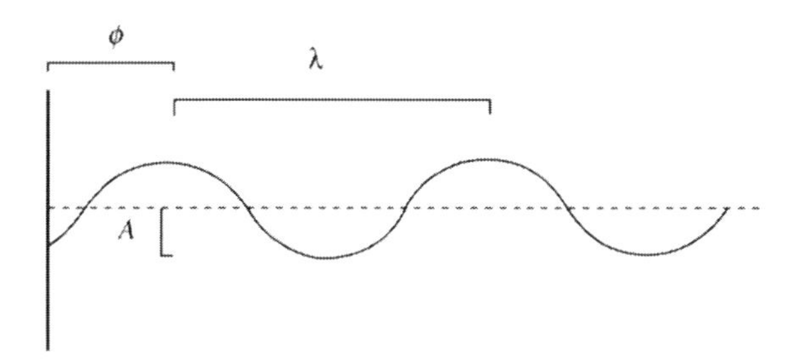

Figure A2.2. A one-dimensional wave. The wave can be uniquely defined in terms of phase (ϕ) amplitude (A), wavelength (λ) or frequency (ν).

relate the complex wave to a series of simpler waves in a fixed and meaningful way. This allows us to determine unknown discontinuous experimental functions such as the electron density of an atom (a function in space) or the electromagnetic spectrum of a molecule (which can be represented as the time-inverse function, frequency).

In X-ray diffraction experiments the electron density in one dimension may be represented as a complex, discontinuous wave function in space (Figure A2.3; see Chapter 6; Figure 6.55). This is composed of many *simpler* electron density waves which could combine together in an infinite variety of ways depending on their relative phases. However, if the relative phases are known, these individual electron density waves represent a Fourier series and can be combined by Fourier synthesis to generate a unique overall one-dimensional electron density wave.

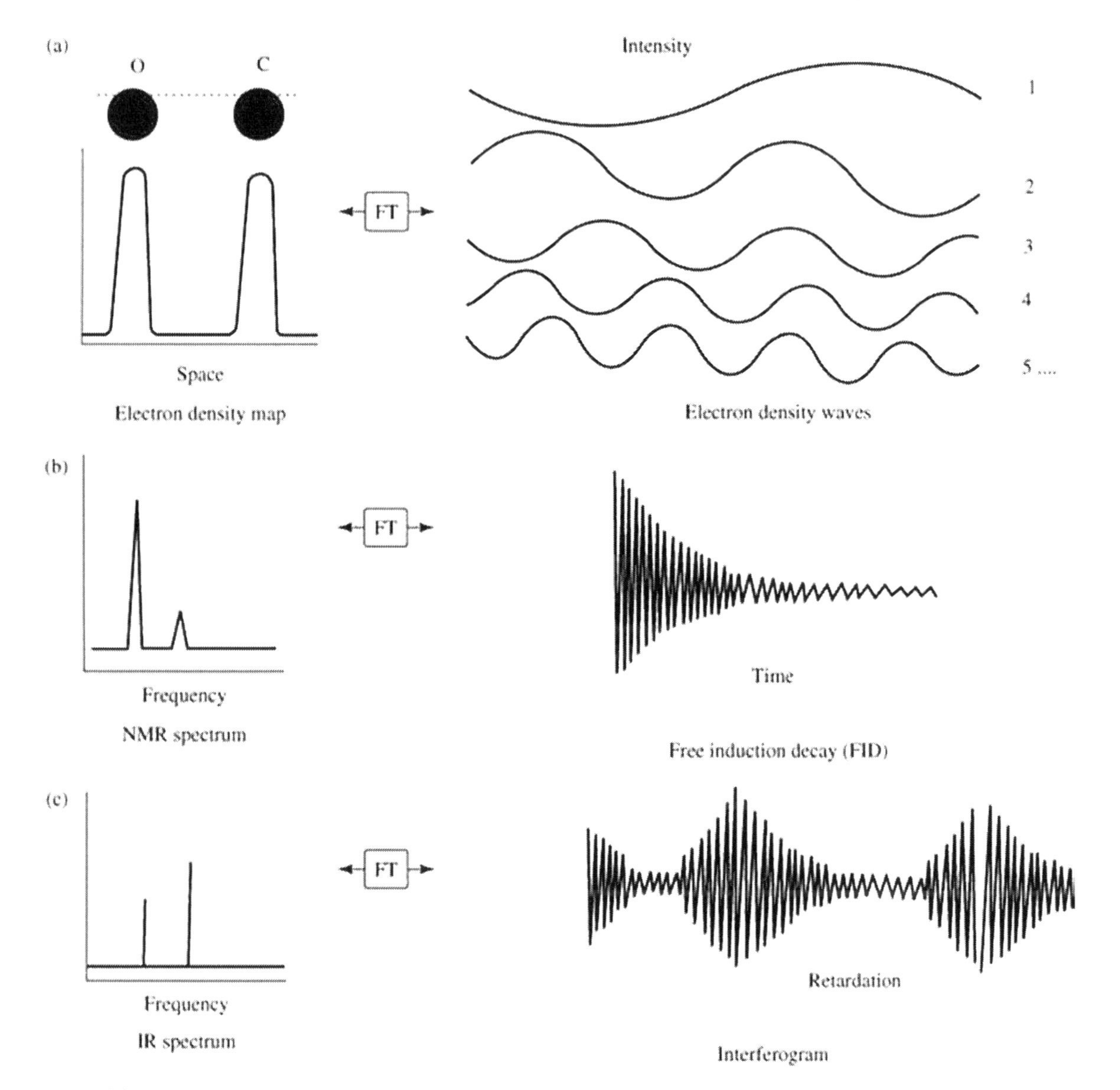

Figure A2.3. Practical applications of the Fourier transform. The Fourier transform allows analysis of discontinuous functions in space or time in terms of continuous wave functions. Examples include the following: (a) Calculation of electron density maps. A one-dimensional section (dashed line) through a C=O group is shown. Summation of electron density waves of known phase (right) allows calculation of a one-dimensional electron density map (left), (b) Multi-dimensional NMR. The free induction decay (right) is the Fourier transform of the NMR spectrum (left), (c) Fourier transform infrared spectroscopy. The interferogram (right) is the Fourier transform of the infrared spectrum (left). Note the change in units commonly found in Fourier transform.

If the situation shown in Figure 6.55 is extended to three dimensions (*xyz*), we can use summation of structure factors and relative phase angles to calculate electron density in three dimensions. The Fourier transform therefore provides a means to move back and forth between these two sets of numbers, one representing reciprocal space where we make our experimental measurements (i.e. intensities of Bragg reflections $= |F_{hkl}|^2$ and relative phase angles, α_{hkl}, determined as described in Sections 6.5.1 and 6.5.3–6.5.6, respectively) while the other represents real space (i.e. the electron density map). Thus the electron density, $\rho(xyz)$ is the Fourier transform of the array of structure factors and the array of structure factors is the inverse Fourier transform of the electron density with both the structure factors and the electron density representing Fourier series of simple waves:

Structure factor:

$$F_{hkl} = \int_V \rho(xyz)\mathrm{e}^{i\phi}\,\mathrm{d}V$$

Electron density:

$$\rho(xyz) = \frac{1}{V}\sum_{hkl} |F_{hkl}|\mathrm{e}^{i\phi}\,\mathrm{d}V$$

where $(\phi = 2\pi)\,(hx + ky + lz)$.

In NMR (Figure A2.3(b); see Chapter 6; Section 6.2.1) the result of the experiment is also a complex wave called a *free induction decay* which is represented as a plot of intensity as a function of time. Fourier analysis of this complex wave allows transformation into a conventional NMR spectrum of intensity as a function of frequency. That they are related by the Fourier transform is clear from the following:

Time domain:

$$f(t) = \int_{-\infty}^{\infty} F(s)\,\mathrm{e}^{2\pi ist}\,\mathrm{d}s$$

Frequency domain:

$$F(s) = \int_{-\infty}^{\infty} f(t)\,\mathrm{e}^{2\pi ist}\,\mathrm{d}t$$

Fourier transform infrared spectroscopy (Figure A2.3(c); see Chapter 3; Figure 3.43) produces a resultant wave called an *interferogram* which represents an interference pattern between simpler waves partially out of phase with each other. Many of these are generated during a single experiment and their pattern, which is plotted as a function of *retardation*, is characteristic for the infrared absorbance spectrum of each sample molecule. Fourier analysis simplifies the calculations necessary to analyse these interferograms by dissecting out the simpler waves from each other and by transforming the retardation domain into a frequency domain which can be plotted as a conventional infrared spectrum. In both these spectroscopic techniques there is a fixed relationship between the complex wave function on the one hand and the simplified spectrum on the other which facilitates their interconversion by the Fourier transform.

The reason the Fourier transform is so widely used is that it offers specific *computational* advantages over other mathematical approaches. However complex the Fourier series, it is related to the original function by a comparatively simple equation making it possible to move from the real-world domain of space or time to the Fourier domain of frequency, phase and amplitude. Moreover, considerable error can accumulate due to shortcomings in real-world experimental systems such as those used in diffraction or spectroscopy. Development of the *fast Fourier transform algorithm* has facilitated very *rapid* calculation by computer in the Fourier domain which can smooth out and diminish this error, resulting in a more accurate overall measurement. The Fourier transform therefore provides a computationally versatile tool to analyse complex functions arising from experimental measurements by dissecting them into simpler wave functions which can be used to determine experimental unknowns.

REFERENCES

Bracewell, R.N. (1989) The Fourier transform. *Scientific American*, **260**, 86–95.

Glasser, L. (1987) Fourier transforms for chemists; Part I. Introduction to the Fourier transform. *J. Chem. Educ.*, **66**, A228–33.

Glasser, L. (1987) Fourier transforms for chemists; Part II. Fourier transforms in chemistry and spectroscopy. *J. Chem. Educ.*, **66**, A260–6.

Index

Printed and bound by CPI Group (UK) Ltd, Croydon, CR0 4YY